# MAIZE CROP

## Improvement, Production, Protection and Post Harvest Technology



Maize is one of the versatile emerging crops with wider a[illegible]y under varied agro-climatic conditions. Globally, maize is known as queen of cereals because it has the highest genetic yield potential among the cereals. It is cultivated on nearly 150 m/ha in about 160 countries having wider diversity of soil, climate, biodiversity and management practices that contributes 36 % (782 m/t) in the global grain production. The United States of America (USA) is the largest producer of maize contributes nearly 35 % of the total production in the world. It is the driver of the US economy.

**Dr. A. Solaimalai** is presently working as Associate Professor (Agronomy). He availed TNAU Merit Scholarship for his Ph.D program. He received Prof. S.Subramanian Award for the Best Thesis in Irrigation Management for his Ph.D during 2000. He authored 10 books, 20 book chapters, 12 review papers, 30 research papers, 11 manuals and 35 popular articles to his credit. He has specialization in Irrigation Agronomy, Agrometeorology and Dryland Agriculture.

**Dr. P. Anantharaju** is working as Assistant Professor in Plant Breeding and Genetics. He has more than 10 years of research, teaching and extension experience in various capacities. He has contributed to release cotton variety SURAJ. He has taught in Plant breeding in field and horticultural crops. He published 15 research papers in scientific journals and co-authored two text books. His Specialization is biotic and abiotic stress breeding in Millets, Pulses and Cotton.

**Dr. S. Irulandi** is presently working as Assistant Professor (Agricultural Entomology). He received ICAR - Junior Research Fellowship award for the M. Sc. (Agri.) programme. He served as Scientist in Central Coffee Research Institute, Coffee Board Research Department, Ministry of Commerce, Chikmagaluru, Karnataka from 1998 to 2009. He published 89 research papers in National and International levels and more than 125 popular articles to his credit. He has specialization in biological control of insect pests' management.

# MAIZE CROP

## Improvement, Production, Protection and Post Harvest Technology

**A. Solaimalai, M.Sc. (Ag.), Ph.D**

Associate Professor (Agronomy)
Agricultural Research Station, TNAU
Kovilpatti, Tamil Nadu – 628 501

**P. Anantharaju, M.Sc. (Ag.), Ph.D**

Assistant Professor (PB&G)
Department of Pulses, TNAU
Coimbatore, Tamil Nadu – 641 003

**S. Irulandi, M.Sc. (Ag.), Ph.D**

Assistant Professor (Agrl. Entomology)
Horticultural College and Research Institute
Periyakulam, TNAU, Tamil Nadu – 625 604

**M. Theradimani, M.Sc. (Ag.), Ph.D**

Dean
College of Agricultural Technology
Theni, Tamil Nadu – 625 562

Routledge
Taylor & Francis Group
LONDON AND NEW YORK

NPH

**NARENDRA PUBLISHING HOUSE DELHI (INDIA)**

First published 2021
by CRC Press
2 Park Square, Milton Park, Abingdon, Oxon, OX14 4RN
and by CRC Press
6000 Broken Sound Parkway NW, Suite 300, Boca Raton, FL 33487-2742



*CRC Press is an imprint of Informa UK Limited*



*British Library Cataloguing-in-Publication Data*
A catalogue record for this book is available from the British Library

*Library of Congress Cataloging-in-Publication Data*
A catalog record has been requested

ISBN: 978-0-367-54694-6 (hbk)
ISBN: 978-1-003-09018-2 (ebk)

NPH

**NARENDRA PUBLISHING HOUSE**
**DELHI (INDIA)**

# Content

# Preface

Maize is one of the versatile emerging crops with wider adaptability under varied agro-climatic conditions. Globally, maize is known as queen of cereals because it has the highest genetic yield potential among the cereals. It is cultivated on nearly 150 m ha in about 160 countries having wider diversity of soil, climate, biodiversity and management practices that contributes 36 % (782 m t) in the global grain production. The United States of America (USA) is the largest producer of maize contributes nearly 35 % of the total production in the world. It is the driver of the US economy.

The USA has the highest productivity (> 9.6 t /ha) which is double than the global average (4.92 t /ha). The average productivity in India is 2.43 t/ ha. In India, maize is the third most important food crops after rice and wheat. It is cultivated in 8.7 m ha (2010-11) mainly during Kharif season which covers 80 % area. Maize in India, contributes nearly 9 % in the national food basket and more than Rs. 100 billion to the agricultural GDP at current prices apart from the generating employment to over 100 million man-days at the farm and downstream agricultural and industrial sectors. The introduction of single cross hybrid which now covers 20 % area under maize cultivation making good dent on Indian scenario.

In addition to staple food for human being and quality feed for animals, it serves as a basic raw material as an ingredient to thousands of industrial products that includes starch, oil, protein, alcoholic beverages, food sweeteners, pharmaceutical, cosmetic, film, textile, gum, package and paper industries etc. In India, the maize is used as human food (23%), poultry feed (51 %), animal feed (12 %), industrial (starch) products (12%), beverages and seed (1 % each).

The Indian hybrid seed sector is now pegged at around Rs. 12,000 crore with hybrid maize accounting for little over Rs 1,500 crore. At present, the two major maize growing states of AP and Karnataka account for a third of the country's maize production.

India's hybrid maize seed sector could more than double to two lakh tonne a year from 90,000 tonne now in the next two years thanks to expanding demand from poultry and industrial sectors. High yields, coupled with remunerative selling prices, are influencing farmers to shift to hybrid maize cultivation from crops like hybrid cotton that have seen near saturation. High hybrid maize cultivation now being witnessed in the country is similar to the success story of hybrid cotton seed sector in the last decade. The Indian seed industry comprises some 14 state seed corporations and two national level corporations, 20 large players including multinationals and around 500 small regional players. Hybrid maize seed adoption is around 50 per cent of maize areas, providing scope for further adoption.

Value addition in maize has a great potential and there are several value added products of maize particularly QPM and baby corn that not only increase the farm income but also provides employment to rural youth and farm women. Value added products developed using normal maize and QPM are baked products (bread and cake), extruded products (Vermicelli and pasta), convenience foods (instant idli and dhokla mix, porridge mix and sprouted chaat), popped products (Popcorn and popped laddoo) and traditional products (laddoo, halwa, kheer, chapatti, sev, mathi, pakora and cheela).

A wide range of products from baby corn have been developed such as sweet products (Halwa, kheer, barfi), preserved Products (Jam, chutney, pickle, candy, murrabba), chinese products (soup, manchurian, babycorn chilly, chowmein sweet and sour vegetables) and traditional products (pakoda, cutlet, chaat, salad,dry vegetables, kofta, mixed vegetable, and raita).

This book is intended for the agricultural courses of graduate, post graduate and doctoral students in the field of Agriculture and extension functionaries. This book is written in simple language dealing with various subject matters of maize crop in 36 chapters. This book has been prepared with a purpose of importing complete information about maize crop. We hope that the students and readers will find this book with much utility. We thank all the authors / publishers from which references were collected on various aspects of maize crop. We are sure that this book will serve as valuable text cum reference book to the B.Sc, M.Sc and Ph.D students of agricultural universities.

We thank Mrs. R. Jeyapandiyammal and N. Kanagalakshmi of ARS, Kovilpatti for their sincere efforts in typing the manuscript. In spite of the best efforts, it is possible that some errors may have crept into the compilation. The readers are requested to send their constructive feedbacks and comments to us so that these could be taken care of in the further edition. Finally we thank our publisher M/s. **Narendra Publishing House, Delhi** for bringing out this book in an attractive format.

**Dr. A. Solaimalai**
**Dr. P. Anantharaju**
**Dr. S. Irulandi**
**Dr. M. Theradimani**

# 1

# Introduction

Maize (*Zea mays* L.) is the most widely distributed crop of the world being grown in tropical, sub-tropical and temperate regions. Being a versatile crop, it adapts easily to a wide range of production environments. In India, maize is the third most important cereal after rice and wheat that provides food, feed, fodder, and serves as a source of basic raw material for the number of industrial products, viz. starch, protein ,oil, alcoholic beverages, food sweeteners, cosmetics, bio-fuel, etc. The grain can be consumed as human food, fermented to produce a wide range of foods and beverages and fed to livestock. The leaves, stalks, and tassels can be fed to livestock either green (in the form of fodder or silage) or dried (in the form of Stover).The roots can be used for mulching, incorporated into the soil to improve the physical structure and burned as fuel. Maize occupies 146.7 million hectares globally with a production of 699.3 million tonnes. World demand in 2020 is predicted to rise to about 699 million tonnes.

In India, maize is cultivated over 8.26 million ha with a production of 19.31 million tonnes having an average productivity of more than 2.4 t/ ha, contributing 8.5 % to the Indian food basket. Maize in India contributes more than Rs.100 billion to the agricultural GDP at current prices apart from the providing employment to over 100 million man days at the farm and industrial sectors. Maize serves as a basic raw material as an ingredient to more than 3000 industrial products. It occupies an important place as a source of human food (25 %), animal feed (11 %), poultry feed (52 %), starch (11 %), brewery (1 %) and seed (1 %). The growth rate of area (2.83 %), production (30.93 %) and productivity (27.35%) over the past years, has shown a remarkable increase as compared to other principal cereal crops. It has yield potential far higher than any other cereal and that's why it is sometimes referred to as the miracle crop or the 'Queen of Cereals'.

In India with the growth in demand of poultry feed the demand for maize is also going up. Hence, there is a need to increase the production of maize in the country otherwise looking at the demand growth. India may well have to import maize in the coming years. Production of maize has been going up in eastern Uttar Pradesh and Bihar. However, in many other parts of the country, farmers are shifting from maize cultivation in search of more lucrative crops. In near future maize may become a staple food for human consumption if the demand for rice and wheat is not fulfilled through increased production. Some estimates indicate that India may have to produce 20 million tonnes of maize to meet its requirement for human consumption, poultry, piggery, pharma industry and fodder by 2020. According to experts a maize revolution is likely in sub-Sahara Africa, South Asia and East Asia. The maize crop is extremely productive in USA.

Maize is a grain domesticated by indigenous peoples in Mesoamerica in prehistoric times. The Aztecs and Mayans cultivated it in numerous varieties throughout central and southern Mexico, to cook or grind in a process called nixtamalization. Later, the crop spread through much of the Americas. Between 1700 and 1250 BC, the crop spread to all corners of the region. Any significant or dense populations in the region developed a great trade network based on surplus and varieties of maize crops. After European contact with the Americas in the late 15th and early 16th centuries, explorers and traders carried maize back to Europe and introduced it to other countries through trade. Maize spread to the rest of the world due to its popularity and ability to grow in diverse climates.

Maize is the most widely grown crop in the Americas with 332 million metric tons grown annually in the USA alone (although 40 % of the crop - 130 million tonnes - is used for corn ethanol). Transgenic maize made up 85 % of the maize planted in the USA in 2009. While some maize varieties grow to 12 metres tall, most commercially grown maize has been bred for a standardized height of 2.5 metres. Sweet corn is usually shorter than field-corn varieties. The term 'maize' derives from the Spanish form of the indigenous Taino word *maiz* for the plant. This was the term used in the UK and Ireland where it is now usually called "sweet corn", the most common form of the plant known to people there. Sweet corn is harvested earlier and eaten as a vegetable rather than a grain.

Maize is a versatile crop grown over a range of agro climatic zones. In fact the suitability of maize to diverse environments is unmatched by any other crop. It is grown from 58°N to 40°S, from below sea level to altitudes higher than 3000 m, and in areas with 250 mm to more than 5000 mm of rainfall per year (Shaw, 1988; Dowswell *et. al.,* 1996) and with a growing cycle ranging from 3 to 13 months (CIMMYT 2000). However the major maize production areas are located in temperate regions of the globe. The United States, China, Brazil and Mexico account for 70 % of global production. India has 5 % of corn acreage and contributes 2 % of world production. The use of maize varies in different countries. In USA, EU, Canada and other developed countries, maize is used mainly to feed animals directly or sold to feed industry and as raw material for fermentation industries (Morris, 1998; Galinat, 1988; Shaw, 1988, Mexico, 1994). In developing countries use of maize is variable. In Latin America and Africa the main use of maize is for food while in Asia it is used for food and animal feed. In many countries it is the basic staple food and an important ingredient in the diets of people. Globally, it has been estimated that approximately 21 % of the total grain produced is consumed as food.

Maize is the third most important food grain in India. In India, about 28 % of maize produced is used for food purpose, about 11 % as livestock feed, 48 % as poultry feed, 12 % in wet milling industry (for starch and oil production) and 1 % as seed (AICRP on Maize, 2007). In the last one decade, it has registered higher growth rate among all food grains including wheat and rice because of newly emerging food habits and enhanced industrial requirements.

## Geographic Distribution

Maize is the world's leading cereal crop. It is cultivated globally over an area of about 147.26 million ha with a production of 724.6 million tonnes of grain. The major maize producing countries are given in Table 1.

**Table 1. Major countries of world producing maize during**

| Country | Area (m ha) | Production (m. tonnes) | Productivity (q/ha) |
|---|---|---|---|
| USA | 29.79 | 299.92 | 100.65 |
| China | 25.47 | 130.43 | 51.22 |
| Brazil | 12.41 | 41.81 | 33.68 |
| Mexico | 7.69 | 21.67 | 28.19 |
| Argentina | 2.33 | 15.00 | 64.35 |

| Country | Area (m ha) | Production (m. tonnes) | Productivity (q/ha) |
|---|---|---|---|
| India | 7.50 | 14.10 | 18.80 |
| Canada | 1.07 | 8.84 | 82.40 |
| Indonesia | 3.36 | 11.23 | 33.44 |
| World | 147.26 | 724.59 | 49.20 |

(FAO Production Year Book, 2004)

In India, the crop is cultivated on 7.4 m ha. Among the states, Karnataka, Andhra Pradesh, Uttar Pradesh and Bihar are leading producers of maize (Table 2).

**Table 2. Area, production and productivity of maize in important states of India in 2004-05**

| State | Area (000 ha) | Production (000 t) | Productivity (q/ha) |
|---|---|---|---|
| Andhra Pradesh | 657.0 | 2064.0 | 31.42 |
| Assam | 19.2 | 13.9 | 7.24 |
| Bihar | 614.4 | 1465.7 | 14.86 |
| Chattisgarh | 96.5 | 131.7 | 13.65 |
| Gujarat | 459.5 | 412.5 | 8.98 |
| Haryana | 16.0 | 40.0 | 25.00 |
| Himachal Pradesh | 324.0 | 736.0 | 22.72 |
| Jammu & Kashmir | 322.7 | 492.3 | 15.26 |
| Jharkhand | 191.0 | 286.0 | 14.97 |
| Karnataka | 850.0 | 2512.0 | 29.55 |
| Madhya Pradesh | 896.2 | 1252.6 | 13.98 |
| Maharashtra | 428.0 | 753.0 | 17.59 |
| Orissa | 65.0 | 106.0 | 16.31 |
| Punjab | 154.0 | 422.0 | 27.40 |
| Rajasthan | 1042.4 | 1262.0 | 12.11 |
| Tamil Nadu | 189.9 | 294.7 | 15.52 |
| Uttarakhand | 30.0 | 44.0 | 14.67 |
| Uttar Pradesh | 876.0 | 1494.0 | 17.05 |
| West Bengal | 46.9 | 139.6 | 29.77 |
| India | 7430.4 | 14172.0 | 19.07 |

(Fertilizer Association of India, 2006)

## References

Aldrich, S.R., W.O. Scott and E.R. Leng (1982). Modern Corn Production, Station A. Champaign, Illinois, 378, USA.

Egrinya Eneji, A. (2009). Agronomy of tropical crops. Studium press, LLC, Houston, Texas, P: 28 - 41.

González Ceniceros, F. and N.N. Singh (2001). Maize an optional crop for nutrition and profitability. Indian Farming 51: 29-32.

Hoeft, R. G., E.D. Nafziger, R.R. Johnson and S.R. Aldrich (2000). Modern corn and soybean production. MSCP Publ., Champaign, IL.

Javad Hussain, H.S. and V. Sethuraman (1980). Maize. Directorate of Agriculture, Chenpauk, Chennai, P: 92.

Lal, S. and R.S. Saini (1985). Technology for increasing maize production. Indian Farm. p.35.

Martin, J.H., W.H. Leonard and D.L. Stamp (1976). Principles of Field Crop Production,, Macmillan Publishing Co., Inc., New York.

Pathic, D.S. (2001). Sustainable maize production system for Nepal. Maize Symposium, Dec-3-5, 2001. NARC and CIMMYT, Kathmandu, Nepal. 11 p.

Patil, S.J., M.C. Wali, S.I. Harlapur, and M.Prashanth (2000). Maize research in North Karnataka. Technical Bulletin-I, University of Agricultural Sciences, Dharwad, April, 2000, India.

Pingali, P.L. and S. Pandey (2001). Meeting world maize needs: technological opportunities and priorties for the public sector, 1999/ 2000 World Maize Facts and Trends.

Thakur, C. (1980). Scientific crop production, Volume: I. Metropolitan Book Co. Pvt. Ltd., New Delhi. P: 145 – 185.

Toshi, P.K., N.P. Singh, N.N. Singh, R.V. Gerpacio and P.C. Pingali (2005). Maize in India : Production. systems, constraints and research Priorities. CIMMTY, Mexico, P: 42.

https://apps.fas.usda.gov/psdonline/circulars/production.pdf

# 2

# Origin of Maize Crop

Several theories have been proposed about the specific origin of maize in Mesoamerica. It may be a direct domestication of a Mexican annual teosinte, *Zea mays* ssp. *parviglumis*, native to the Balsas River valley in south-eastern Mexico, with up to 12 % of its genetic material obtained from *Zea mays* ssp. *mexicana* through introgression. It may have been derived from hybridization between small domesticated maize (a slightly changed form of wild maize) and a teosinte of section *Luxuriantes*, either *Z. luxurians* or *Z. diploperennis*. It may have undergone two or more domestications either of wild maize or of a teosinte. It may have evolved from a hybridization of *Z. diploperennis* by *Tripsacum dactyloides*. The term "teosinte" describes all species and subspecies in the genus *Zea*, excluding *Zea mays* ssp. *mays*.

In the late 1930s, Paul Mangelsdorf suggested that domesticated maize was the result of a hybridization event between unknown wild maize and a species of *Tripsacum*, a related genus. However, the proposed role of *Tripsacum* (gama grass) in the origins of maize has been refuted by modern genetic testing, refuting Mangelsdorf's model and the fourth listed above. The teosinte origin theory was proposed by the Russian botanist Nikolai Ivanovich Vavilov in 1931 and the later by George Beadle in 1932. Though it had experimental support, in particular the ability of teosinte and maize to cross-breed and produce fertile offspring. I It did not explain how the immense diversity of the species of sect. *Zea* originated? how the tiny archaeological specimens of 3500–2700 BC could have been selected from a teosinte? and how domestication could have proceeded without leaving remains of teosinte or maize with teosintoid traits earlier than the earliest known until recently, dating from ca. 1100 BC?.

The domestication of maize is of particular interest to researchers archaeologists, geneticist, ethnobotanists, geographers, etc. The process is thought by some to have started 7,500 to 12,000 years ago. Research from the 1950s to 1970s originally focused on the hypothesis that maize domestication occurred in the highlands between Oaxaca and Jalisco, because the oldest archaeological remains of maize known at the time were found there. John Doebley identified *Zea mays* ssp. *parviglumis*, native to the Balsas River valley and also known as Balsas teosinte, as being the crop wild relative teosinte genetically most similar to modern maize. However, archaeobotanical studies published in 2009 now point to the lowlands of the Balsas River valley, where stone milling tools with maize residue have been found in an 8,700 years old layer of deposits. Some of the earliest pollen remains from Latin America have been found in lake sediments from tropics of southern Mexico and upper Central America, up to Laguna Martinez and have been radiocarbon dated to around 4,700 years ago. Archaeological remains of early maize ears, found at Guila Naquitz Cave in the Oaxaca Valley, date back roughly 6,250 years. The oldest ears from caves near Tehuacan, Puebla, date ca. 2750 BC. Little change occurred in ear form until ca. 1100 BC when great changes appeared in ears from Mexican caves. Maize diversity rapidly increased and archaeological teosinte was first deposited.

Perhaps as early as 1500 BC, maize began to spread widely and rapidly. As it was introduced to new cultures, new uses were developed and new varieties selected to better serve in those preparations. Maize was a major staple of most pre-Columbian North American, Mesoamerican, South American and Caribbean cultures. The Mesoamerican civilization was strengthened upon the field crop of maize through harvesting it. Maize formed the Mesoamerican people's identity. During the 100 AD, maize cultivation spread from Mexico into the Southwest USA and a millennium later into Northeast USA and southeastern Canada, transforming the landscape as Native Americans cleared large forest and grassland areas for the new crop. It is unknown what precipitated its domestication, because the edible portion of the wild variety is too small and hard to obtain to be eaten directly, as each kernel is enclosed in a very hard bi-valve shell. However, George Beadle demonstrated that the kernels of teosinte are readily "popped" for human consumption, like modern popcorn. Some have argued that it would have taken too many generations of selective breeding in order to produce large compressed ears for efficient cultivation. However, studies of the hybrids readily made by intercrossing teosinte and modern maize suggest that this objection is not well founded.

## Geographic Origin and Distribution

The center of origin for *Zea mays* has been established as the Mesoamerican region, now Mexico and Central America (Watson and Dallwitz, 1992). Archaeological records suggest that domestication of maize began at least 6000 years ago, occurring independently in regions of the southwestern United States, Mexico, and Central America (Mangelsdorf, 1974). The Portuguese introduced maize to Southeast-Asia from the America in the $16^{th}$ century. The maize was introduced into Spain after the return of Columbus from America and from Spain it went to France, Italy and Turkey. In India, Portuguese introduced maize during the seventeenth century. From India it went to China and later it was introduced in Philippines and the East Indies. Corn now is being grown in USA, China, Brazil, Argentina, Mexico, South Africa, Rumania, Yugoslavia and India. Various hypotheses have been proposed on the origin/domestication of maize (OECD, 2006). Teosintes (*Z. diploperennis* and *Z. mays* sp. *mexicana*) and Tripsacum species are often described as having roles in the domestication process of maize (Mangelsdorf, 1974; Galinat, 1988). An early hypothesis proposed that *Z. mays* sp. mexicana was the product of a natural hybridization of Tripsacum and Zea (Mangelsdorf, 1974). Further crossings of teosinte with wild maize are thought to have produced the modern races of maize. The possibility of intergeneric hybridization of either *Z. diploperennis* or Tripsacum with extinct wild maize has also been proposed as the ancestral origin of *Z. mays* (Radu *et al.* 1997; Purseglove, 1972). Eubanks (1993, 1997a) suggests that domesticated maize may have arisen via human selection of natural hybrids between *Tripsacum* and *perennial teosinte*.

## References

Arnon, I. (1972). Crop production in dry regions. Leonard Hill Books, London. P: 147 - 187.

Egrinya Eneji, A. (2009). Agronomy of tropical crops. Studium press, LLC, Houston, Texas, P: 28 - 41.

Hoeft, R.G., E.D. Nafziger, R.R. Johnson, and S.R. Aldrich (2000). Modern corn and soybean production. MSCP Publ., Champaign, IL.

Javad Hussain, H.S. and V. Sethuraman (1980). Maize. Directorate of Agriculture, Chepauk, Chennai, P: 92.

Satyanarayana, R.S.Kumar and M.Y.Sarma (1994). Technologies for maize cultivation in Andhra Pradesh. Indian Farming, 41 (7): 9 -11.

Thakur, C. (1980). Scientific crop production, Volume: I. Metropolitan Book Co. Pvt. Ltd., New Delhi. P: 145 – 185.

Yadav, R.P., P. Singh, R.K. Aggarwal and M.K. Yadav (2006). Improved technology for sustainable production of maize *(Zea mays)* in Shivalik foothills of Himachal Pradesh. Indian Journal of Dryland Agricultural Research and Development, 21 (1): 1 -6.

# 3

# Breeding Methodology for Maize

Mass selection is the principle method of breeding open pollinated maize. Several varieties of open pollinated by maize have been developed largely by mass selection. In the mass selection of breeding of maize cobs are chosen on the basis of plants and cob characteristics. Cobs from morphologically identical plants are thus collected and the seeds bulked and planted *enmasse*. The cob is the unit of selection. Selection is exercised again for the specific characteristic and the plants following in that specific objective with the grouped and carried over for further study as a separate culture.

By mass selection it was possible to modify plant type, maturity, grain characteristics etc., Selection could be made for long ears, more number of ears per plants earliness and seed characters. By rigorous selection the appearance of the maize plant and ear could be changed within the limits of the genetic variability of the variety which in most varieties appear to be quite wide selection was useful in adapting varieties to new production area or in developing varieties for special purposes. Selection of plants should be based on the following:

1. Vigorous, stout and healthy plants.
2. Large, sound and well developed cobs.
3. Disease free plants.
4. Proper maturity and good seed setting.

| | | |
|---|---|---|
| **First Season** | **:** | A large population say ranging from 2000 (minimum to 10000 plants are raised from a local bulk variety). |
| Plot size | : | Non-replicated rows of 4 - 5 m length. |
| Spacing | : | 60 x′ 30 cm (15 plants/ row). |
| Selfing | : | All the selected plants. |
| Harvests | : | Single plant basis. |
| Selection | : | Seeds of morphologically identical plants and seed colour are bulked and carried forward. |
| **Second season** | : | Raise again a minimum of 2000 plants in each such groups. |
| Plot size | : | Non-replicated rows of 4.5 m length |
| Spacing | : | 60 x′ 30 cm. |
| Selfing | : | Selection 5 - 10 plants based on the following in each group<br>Vigorous plants<br>Long ear heads<br>Maturity earliness<br>Disease free and self them |
| Harvest | : | Harvest them undividedly. After ascertaining the seed characters bulk them and carry it to Row Yield trial. |
| **Third season** | : | Replicated Row Yield Trial Test 30 - 40 cultures derived from the above. |
| Plot size | : | 4.5 x′ 0.60 m |
| Spacing | : | 60 x′ 20 cm (20 plants) |
| Selfing | : | Selfing is done in 3 - 5 plants |
| Harvest | : | The entire row is harvested and analysed |
| Selection | | The best 5 - 8 are carried formed compared to the local check variety. |
| **Fourth Season** | : | Preliminary yield trial<br>Test 5 - 8 cultures selected in each group. Local varieties are included for comparison. |
| Plot size | : | Gross - 2.4 x 4.5 m; Net - 1.2 x 3.52 m. |
| Spacing | : | 60 x 20 cm or plants / row 20. Net 16 plants. No. of rows 4 / plot. Net 2 rows. |
| Selfing | : | 3 - 5 plants on the border rows. |
| Observations | : | Observations on the plant characteristics have to be recorded in 5 plants. |
| Harvest | : | Plot wise harvest yield analysed and the superior one forwarded to comparative yield trial. |

| | | |
|---|---|---|
| **Fifth season** | : | Comparative yield trial. Replicated only very few will be advanced to comparative yield trial. They are tested against the local strains. |
| Plot size | : | Gross : 4 - 8 x 4 - 5 m<br>Net : 3.60 x 3.52 m |
| No. of rows | : | Gross : 8 Net : 6 |
| No. of plants / row | : | Gross : 20; Net : 16 |
| Selfing | : | 5 - 10 plants in each replication. |
| Observations | : | All morphological and economical attributes are to be recorded. |
| Harvest | : | Plot war and analysed. The best one is recommended for A.R Trials. |

## Head to Row Method of Breeding

This method provides more efficient way of selection than the mass selection. The following procedure is followed.

### First Season

From an open pollinated variety 50 - 100 (50 minimum and 100 maximum) cobs of good size, setting and grain colour are selected and thrashed individually. These are sown in each row.

Row length: 4.5 metre length
Spacing: 60 x 30 cm
Layout: Non-replicated.
Each line is examined for purity of desirable characters and rows showing uniformity are selected. Others are rejected. About 5 to 10 lines are selected.

### Second season

There 5 - 10 lines are again sown in lines but a replicated plot is adopted to minimize the block error. The selected lines are study in replicated row yield trial and the yield assesses. This is compared with local standard strains. The best lines are selected.

Plot size: 60 x 20 cm
Length of the row: 4.5 m
Design: S. R. B. D. The best few are selected.

**Third season**

The best lines are tested in comparative yield trial with local standard strains.

| | | |
|---|---|---|
| Design : | S. R. B. D. | |
| Layout : | Replicated 4 - 6 times. | |
| Plot size: | Gross: 3.6 x 4.5 m | Net: 2.4 x 4.5 m |
| No. of rows: | Gross : 6; Net: 4 | |
| Spacing: | 60 x 20 cm. | |

Plant characters are observed and recorded. The statistical analysis is done and the superior one is selected and released.

## Production of Hybrid Maize

Hybrid maize is the first generation progeny from cross involving inbred lines. The breeding of hybrid maize involves.

a. Development of inbred lines by controlled self pollination,

b. Determination of inbred lines may be combined into high productive crosses and

c. Commercial utilization the crosses for seed production.

### a) Development of Inbred line

An inbred line is a pure line developed by self pollination and selection until apparently homozygous plants are obtained. This usually requires 5 - 6 generations of selfing. Since maize is a naturally cross pollinated crop, open pollinated plants obtained from a variety are always heterozygous. By self pollination (controlled) and followed by selection for vigorous and healthy plants, desirable characters can be fixed and inbred lines are produced.

### Controlled self pollination

Collecting pollen from the tassel and dusting on the stigma of the same plant. Self pollination (sibmating) allows pollination among the plants of the same population. Self fertilization in maize results in inbreeding depression. The inbreeding depression which accompanies self fertilization is greatest in the first generation and become less and less in subsequent generation. A condition is finally reached when there is no loss of vigour and plants breed true. At this stage the plant becomes an inbred.

**Procedure**

Self pollination and vigorous selection for vigorous and healthy plants for 5 - 6 generations until uniform lines are obtained.

**b) Testing inbred lines for general combining ability**

The value of the inbred line ultimately depends upon its ability to produce a superior hybrid in combination with another inbred or hybrid. This ability of the inbred to combine in serious of hybrids is known as general combining ability. The general combining ability of the inbred is generally assessed by top cross tests.

**Top cross test**

The Top cross test measures the general combining ability of the inbred lines tested. In this method, a series of the inbreds are crossed with a common tester. The tester may be an open pollinated variety or a composite or a multiple or single cross hybrid. The progenies are tested for yield. Progenies which are outstanding would presumably involve combinations of the standard variety with inbred line that have the ability to combine well with the variety.

| Inbred | Variety | Performance of Progeny |
|---|---|---|
| 101 × | tester | High yielding |
| 102 × | tester | Poor yielding |
| 103 × | tester | High yielding |
| 104 × | tester | High yielding |
| 105 × | tester | Poor yielding |
| 106 × | tester | High yielding |

Inbred 101, 103, 104 and 106 possess high general combining ability.

**Testing for specific combining ability**

The inbreds which have good general combining ability are then crosses in all possible combinations and the specific ability of the line to combine well with other inbred is determined, from the yield of the progenies as illustrated, below.

| Inbred line crosses | Yield of progeny |
|---|---|
| 101 × 103 | High |
| 101 × 104 | Average |
| 101 × 106 | Average |
| 103 × 104 | High |
| 103 × 106 | Average |
| 104 × 106 | High |

From the above, it is found that inbred 101, 103, 104 and 106 combine well especially inbred 103 combines well with 101 and 104. This is indicated that inbred 103 has high specific combining ability. Such inbreds are selected and then they made use to produce hybrids in different methods.

## Method of Producing Hybrids

After obtaining as many desirable inbred lines as possible, the problem remains as to how best to use these inbreds lines in hybrid seed production. The following types of hybrids are produced.

1. Single cross hybrid
2. Double cross hybrid
3. Three way cross hybrid
4. Multiple cross hybrid

### 1. Single cross hybrids

These involve the promising two inbred lines which combine well. The seeds obtained by crossing 101 x 103 is a single cross hybrid seed. Single crosses can be used only when the inbred lines yield sufficiently well to make seed production economical. The higher yielding inbred is used as female parent.

In the commercial production of single cross hybrid, the two inbred lines to be crossed are planted in an isolated field. Two rows of female inbred lines are planted with one row of male inbred line. The inbred line used as female parent is detasselled and allowed to the open pollinated by the pollen from male parent.

### Detasselling

Removal of tassel from the female plant is known as detasselling. When the tassel has just emerged, the peduncle is held firmly with left hand and by the right hand the tassell is pulled out with a force or jerk. The tassel will come off easily. This process is known as detasselling.

### 2. Double cross

The double cross is a cross of two single cross hybrids. The two single cross F1s are used as parents. It can be represented as below:

(A x B) x (C x D)
Double cross hybrid

The commercial production double cross hybrid consists of first production of single cross hybrids as explained already in isolation. In the next season, the F1 of the single cross which is used as female is sown in 4 lines with one row of male parent, the other F1 of the second single cross.

The female parent is detasselled and allowed for open pollination with male parent. The seeds obtained from female parent from the double cross hybrid seed. Example: Deccan Maize hybrid (CM 104 x CM 105) x (CM 202 x CM 201)

### 3. Triple cross hybrid seed

In three way cross hybrid, a good pollen producing inbred is used as male parent and a single cross hybrid is used as female parent.

(A x B) x C (Pollen parent)

Both the inbreds used in the single cross should combine well with the pollen parent. Commercially the triple cross hybrid can be produced by planting four rows of the single cross hybrid (female parent) a one row of the pollen parent. The female plants are detasselled and allowed for out crossing with the pollen parent. Example: Ganga 5 (CM 202 x CM 111) x CM 500.

## Production of conventional hybrids

Exploiting hybrid vigour for the production of hybrid maize was first reported by Shull (1908). However the exploitation of hybrid vigour in commercial scale of hybrid maize was put forth by Jones (1918). Since then hybrid seed production is in vogue. It was Jones (1918) recommended double cross method of the hybrid seed production for high yield. For the production of hybrid maize the following three steps are to be followed:

1. Developing inbreds.
2. Evaluating inbred lines.
3. Combining the inbred lines into hybrids.

### 1. Developing inbreds

Artificial control of pollination is necessary for effecting inbreeding. Inbred lines can be isolated from open pollinated varieties, composites or synthetics, broad based germplasm pool, single crosses, double crosses or any other heterogenous and heterozygous populations. In order to exercise the minimum scope of selection, the base population should be genetically broad based one with high standard of performance. Any one of the following methods for selection of inbreds may be employed.

a. Standard method of self fertilization and selection.
b. Single hill method.
c. Homozyous diploid.

Or any other methods such as genetic, selection, convergent improvement, recurrent selection etc. It is essential that selection is exercised by visual scoring for vigour, productivity and resistance to pest and diseases. It is observed that more productive inbreds always give more productive hybrids with few exceptions. But it is to be taken note of that visual selection for general combining ability often becomes ineffective. By selfing for 5 - 6 generations, stability is achieved and most of the inbreds become homozygous.

### 2. Evaluating inbreds

The hybrids, thus evolved have to be evaluated for their combining ability. Combining ability (in a hybrid) is the most important criterion for selection of inbreds and this character was found to be more complex. It is always better to test the inbreds early so that a large number of lines may be rejected in the earlier phase itself. Therefore, the inbreds are evaluated in first or second generations of inbreeding by visual scoring and then top crosses and tested for combining ability subsequently. The better promising ones are carried over.

### 3. Testing inbreds for general combining ability

General combining ability represents the average performance of a line (inbred) in a number of hybrid combinations. It gives an idea as to which line or inbred will produce the best hybrids when crossed with the other line or lines. It is assessed by top crossing the inbred line with any one of the standard and open pollinated variety or synthetic or composite or a multiple hybrid that is largely used in the programme. Top cross tests reveal the average general combining ability which expresses the additive gene action.

### 4. Testing for specific combining ability

The individual performance of an inbred in a single specific hybrid combination is known as specific combining ability. The inbred lines which are selected to have high general combining ability are used as tester. Specific single crosses are also used and evaluated. Single cross tests are a measure of non-additive components like dominance and epistatis.

### 5. Production of hybrids

Top cross method helps in identifying the promising inbreds with good gest and therefore limited single crosses will be required to make to assess the specific combining ability. After this it will be necessary to identify the particular single, Three way or double cross hybrid that will produce the highest yield. The number of combinations of inbreds taken 4 at a time to produce a double cross increase rapidly with the increase in the number of inbreds. It will be therefore difficult and time consuming if all the double crosses are made and tested. Jenkins (1934) had proposed the following methods to predict the performance of a double cross on the basis of the yield potential of single crosses.

i. The mean performance of the 6 possible single crosses among any set of 4 inbreds e.g. when four inbreds A, B, C and D are involved, gives the predicted performance of the double cross of (A × B) x (C × D) as follows :

$$\frac{(-(A\times B)+(A\times C)+(A\times D)+(B\times C)+(B\times D)+(C\times D))}{6}$$

ii. The mean performance of the four non parental single crosses gives the double cross performance. Predicted performance of (A × B) × (C × D)

$$\frac{(-(A\times C)+(A\times D)+(B\times C)+(B\times D))}{4}$$

iii. The mean performance of a set of 4 lines over a series of single crosses e.g. mean of the performance of inbreds ABCD separately in a number of single crosses combination with other inbreds EFGH etc. In case of double cross (A × B) x (C × D). Predicted yield:

$$\frac{(AE+AF+AG+AH+BE+BF+BG+BH+CE\text{ etc.})}{N}$$

iv. The average top cross performance of a group of 4 lines in a double cross (A × B) x (C × D) is worked out as follows. Predicted performance:

$$\frac{((A\times V)+(B\times V)+(C\times V)+(D\times V))}{4}$$

Where 'V' is the tester used for top cross.

In these four methods, somewhat different assumptions are required, with regard to the type of gene action involved. Methods (i), (iii) and (iv) assume additive gene action i.e. gene will produce its character effect regardless of the order of pairing. However method (ii) permits the recognition of non additive effects such as dominance and epitasis. In general, the correlation is rather very high between predicted and observed yields with the method (ii). Therefore, this method is widely used for the prediction of the yield of double cross hybrids.

Hybrids involving inbreds of different genetic sources have shown phenomenal vigour. When inbreds of dent and inbreds of flint are crossed then the order of pairing should be as follows:

Dent inbreds: A, B
Flint inbreds: Y, Z

High yielding single cross: (A × Y), (A × Z), (B × Y), (B × Z)
High yielding double cross: (A × B) x (Y × Z)

## Composites

According to Dhawan (1965) and Singh (1965) the main objective of developing a composite is to utilize them as commercial varieties, for developing countries like India which has adequate man power, commercial cultivation of a composite is a economical viable proposition. This has been clearly indicated in India wherein six high yielding composites were developed and released by the coordinated Maize Improvement Project in 1967 and subsequently three opaque-2 composites known as Sakthi, Rattan and Protina during 1970 and another composite variety K 1 in Tamil Nadu during 1978. The successful evolution of a composite depends on the choice of material and methodology for synthesising composites.

### Choice of material

It would be desirable to include varieties, synthetics, advanced generation hybrids, etc. which have wide genetic diversity and high general combining ability. In other words, those varieties that shall give high F1 performance with relatively low depression in F2, the divergence in the varieties is indicated by the geographic origin of the materials, varied grain types and material belonging to the different racial groups. In addition it would be desirable to include suitable sources of resistance to major diseases and pests.

### Methods

A composite may be developed in several methods and the choice of the procedure shall depend upon the number of materials to be incorporated, floral biology, the system of pollination and the use to which the developed composite shall put to.

#### a) Simple method

The simplest method is to take 4 - 6 materials, make all possible crosses and bulk equal number of seeds and the resultant population is grown in isolation. With the increase in number of varieties selected in a cross pollinated species, with the convenience it will be desirable to develop chain crosses whose balanced composite is grown in isolation.

#### b) Roman Square design

When a large number of varieties say 400 - 500 collections are involved, Roman Square Design has been suggested for the development of composites. The collections are planted in isolation in a nursery of the modified cob to row system. The collections are planted in 1-2 row plots as females while the male rows, formed as a balanced composite, are sown on 3-4 dates of sowing to ensure 'nicking' with materials of different maturity group. The female rows are detasseled. At harvest 5 - 6 best ears from each collection are saved for next cycles. None of the female collections are discarded. About 3.5 cycles of such intermating the composite will be homoginised. This procedure has been used in CIMMYT. Mexico and the composite Tuxpeno was synthesized.

It would be more desirable, from a practical point of view, to keep the number of varieties to a lower number preferably 6 - 10, but not more than 20. It is always better to develop all possible crosses excluding reciprocals. Both F1 and F2 are evaluated and the crosses which do not show any marked reduction in F2 are bulked and grown in isolation, preferably on 2 - 3 sowing dates to ensure forced crossing under isolation. Two or three such cycles are practiced. This would help in developing new genetic recombination by breaking the old linkages. In each cycle only mild selection is practiced. Nursery in each cycle of selection is grown from the balance composite made at harvest.

### Full sib. family design

Use of full sib family design has also been suggested to develop composites. In this method, the material is maintained in the family structure. This procedure helps to shorter the period required in random mating. In addition to developing new combinations fairly vigorous selection is also practiced.

### Multiple unit selection

This is another method followed in the development of composites. In this procedure, units are developed from the promising parental varieties which may be grown under desired varied environments. This method also helps to build new recombination with additional advantage of picking out the more desired unit for commercial use, while the other desired units can be recycled. The scheme particularly has the provision of continuously intercrossing raw materials and cutting out the undesirable ones in each cycle of selection. Adequate seed quantity developed for each unit would help testing the varied traits in multilocation trials. This system helps to combine high yield, resistance to diseases and pests as well as a wide adaptability.

## Synthetics

The term synthetic is restricted to cases wherein the constitutional parental materials selected in the development of a synthetic are inbred lines, previously evaluated for their combining ability. The synthetic is derived from the advanced generation material involving multiple crosses. Generally multiple hybrids with another double cross hybrid (8 parents) and sometimes this hybrid is again crossed with such another hybrid (of eight parents), resulting a multiple cross involving 16 parents.

The performance of a synthetic is dependent on the number of inbred lines and their combining ability. Its performance is predicted by the following formula (Wright, 1922).

| | | |
|---|---|---|
| F2 - (F1 - P) / N | : | Number of inbred lines. |
| F1 | : | Mean performance of all possible crosses among N lines. |
| F2 | : | Performance of synthetic. |
| P | : | Mean per se performance of the lines. |

It is clear that the performance of the synthetic can be improved by the selection of better combining inbreds and by increasing the number of inbreds lines. It was reported that highest calculated synthetic yield was obtained when only 4 - 6 best combining lines were involved. The All India Coordinated Maize Improvement Project, had produced number of synthetics among which Syn A. 22, Syn. A. 23 and Syn. E. 13 showed promise. Though they were not released for general cultivation, they were utilized in elite Breeding stocks in subsequent improvement programmes.

**Advantages to synthetics and composites**

The synthetics and composites in comparison to hybrids have a special utility in developing countries since they obviate the necessity of recurrent problem of hybrid seed production. In contrast to the three tier seed production programme of hybrids, the seed multiplication of composites and synthetics is relatively much simpler and relatively inexpensive, and is similar to those for the open pollinated varieties. Farmers can themselves multiply their seed and even practice simple mass selection which would help in upgrading the performance of the variety in every cycle of multiplication, provided suitable assigned isolation facilities are available. Simplicity and inexpensive nature of the seed production of composites and synthetics help to cover up the problems of excessive seed production.

Besides direct commercial utilization composites and synthetics may also serve as a base material for isolating the elite inbred lines to be used in the hybrid programme. Moreover the composites and synthetics can also be further improved for yield and other desirable traits through appropriate intra population improvement programme. The likely genetic advancement to a considerable extent shall depend upon the genetic variability and the nature of gene action for the character under selection. The extent of variability in a composite or synthetic shall depend upon the genetic variability and the nature of gene action for the character under selection. The composites are likely to have greater genetic variability than the synthetics since later involves elite inbred lines (The inbred lines have 50 % or more of the loci homozygous). The composites are thus likely to better respond to the selection than the synthetics. Both additive, additive x additive and like higher other expistatic interaction are exploited in the development and further improvement of synthetics and composites.

## References

Bland, B.F. (1971). Corn production. In: Cereals and legumes. Academic Press, London, P: 63 - 215.

Damborsky, M., O. Chloupek and J. Ehrenbergerova (1994). Variability of maize lines and diallel cross hybrids. Genetika a Slechteni, 30: 297 - 303.

Ebelhar, S.A., E.J. Kanprath and R.H. Moll (1997). Effect of nitrogen and phosphorus on growth and cation composition of corn genotypes differing in average ear number. *Agron, J.* 78 (2): 1986 - 1989.

Edmeades, G.O., J. Bolanos and H.R. Lafitte (1990). Selecting for drought tolerance in maize adapted to the lowland tropics. In: Proceedings of the 4th Asian Regional Maize Workshop, 23–27 September, Islamabad, Pakistan.

Edmeades, G.O., J. Bolanos, S.C. Chapman, H.R. Lafitte and M. Banziger (1999). Selection improves drought tolerance in tropical maize populations. I. Gains in biomass, grain yield and harvest index. Crop Sci., 39: 1306 - 1315.

Emam, Y. (2001). Sensitivity of grain yield components to plant population density in non-prolific maize (*Zea mays*) hybrids. Indian J. Agric. Sci., 71(6): 367 - 370.

Geczki, I. (1976). Effect of testers on yield, straw strength and yield components in maize in 1974. Agrartudomanyi Egyetem Kozlemenyei, P: 197 - 204.

Granados, G. (2000). Maize insects. In R.L. Paliwal, G. Granados, R. Lafitte, A.D. Violic and J.P. Marathee (eds.), Tropical Maize Improvement and Production. FAO Plant Production and Protection Series 28. Food and Agriculture Organization (FAO), Rome.

Gregory, J.R.K. (1997). Hybrid Selection for reducing herbicide rates in corn. Ph.D. Dissert. Department of Agronomy, Univ. of Nebraska, Lincoln, Nebraska, USA.

Ito, G.M. and J.L. Brewbaker (1991). Genetic analysis of pericarp thickness in progenies of eight corn hybrids. J. Am. Soc. Hort. Sci., 116(6): 1072-1077.

Ivanovic, M. (1979). Investigation of factors of lodging resistance in inbred maize and their hybrids of the F1 generation. Arhiv za Poljoprivredne Nauke, 32: 97 - 119.

Ivanovic, M. and K.Rosic (1985). Path co-efficient analysis for three stalk traits and grain yield in maize (Zea *mays* L.). Maydica, 30: 233 - 239.

Ivanovich, M., B. Kerechki, D. Peich, M.Ivanovic, B.Kerecki and D. Peic (1985). Influence of some biochemical factors on the resistance of maize (*Zea mays* L.) stems to lodging, Informatsionnyi Bulletin po Kukuruze No., 4: 177- 183.

Kim, S.M, M.H. Lee and V. Parkinson (1986). Estimates of yield reduction by *Striga hermonthica* on susceptible and tolerant hybrid maize varieties. In: NCMRP conference on maize. Ife, Nigeria, March 17 - 21, 1986. P: 128 – 133.

Liu, G. T., B. G.Li, S. S.Zhao and J. Y. Zhao (1993). Preliminary report on introduction trial for dwarf and compact maize. Bejing Agricultural Sciences, 11: 15 - 18.

Otegui, M.E. and G.A. Slafer (2000). Physiological basis for maize improvement. Food Products Press, New York.

Perry, C.O. (1983). Diallel analysis of rind puncture and grain yield and their interactions with plant densities for twelve elite inbred lines of maize, *Zea mays* L. Dissertation Abstracts, 44: 975.

Poehlman, J.M. (1987). Breeding field crops. AVI publishing company Inc., West Port,, Connecticut. P: 451 - 507.

Redinbaugh, M.G., R.E. Gingery and M.W. Jones (2004). The genetics of virus resistance in maize (*Zea mays* L.). *Maydica,* 49:183-90.

Roggen, K. and J. Gregory (1997). Hybrid selection for reducing herbicide rates in corn. *University of Nebraska Absts.* pp. 2189.

Sprague, G.F. and J.W. Dudley (1988). Corn and Corn Improvement. American Society of Agronomy, Madison, WI.

Xiao, Y.N, X.H. Li, S.H. Zhang, X.D. Wang, M.S. Li and Y.L. Zheng (2004). Identification of quantitative trait loci (QTLs) for flowering time using SSR marker in maize under water stress. Korean J. Genetics, 26 (4): 405 - 413.

Zhang, ZemMin, Jia ChangZhu and C. Z. Zhang Jia (1997). Effects of plant type on genetic gain in maize. Hereditas Beijing, 19: 31 - 34.

Zhao, M., Z. Zhang, S. Zhang, W. Li, D.P. Jeffers, T. Rong and G. Pan (2006). Quantitative trait loci for resistance to banded leaf and sheath blight in maize. *Crop Science*, 46:1039 - 1045.

# 4

# Maize Breeding Activities in India

Maize breeding strategy in India has gone through many phases from the start of All India Coordinated Research Project on Maize in 1957. However, the last three to five years have been the land owny mildewark years since the adoption of high yielding hybrids on farmers' fields proved as critical input for achieving high growth rate in maize. The breeding activities have therefore, been reoriented towards the development of high yielding single cross hybrids for different agroecological regions of the country to tackle the problems posed by biotic and abiotic stresses. This has been duly supported by development of vigorous, productive and genetically diverse inbred lines that have good performance per se as well as in cross combinations.

The major mandate is to evolve and disseminate inbred hybrid technology. With the urbanization, specially corn has gained a great acceptability among the masses. Their demand in Indian market has gone up. Sweet corn and baby corn hybrids/ varieties are sought to cater to the demands of peri-urban agriculture. Over the years 131 hybrids have been developed and released since 1961. Four dozen are public-bred single cross hybrids of different maturity (Extra - early variety: <75 days, early variety: 75 - 85days, Medium variety: 85 - 95 days, Late variety: > 95days) and suitable for cultivation in different agro-climatic conditions of the country. Since 2005, as many as 23 normal, seven Quality protein maize (QPM) and one each of baby corn and sweet corn single cross hybrids have been released.

Work done at Kalyanpur, Ludhiana, Hyderabed and Doli from 1958 to 1961 had resulted in releasing 4 high yielding hybrids. In the year 1962, one more hybrid VL 54 was released. During 1963, Ganga safed and Hi-starch were released followed by Ganga 3 and Him 123 during 1964. Ganga 5 was released in 1968 and is very popular throughout India followed by Deccan 101 in the year 1975. The All India Coordinated Maize Improvement Project also produced as many as six composites namely Amber, Javahar, Kisan, Vikram, Sona and Vijay, of which Kisan and Vijay were very popular. All these six composites were released during 1967.

Apart from the above composites, the All India Coordinated Maize Improvement Project also released three high lysine/tryptophan nutritionally superior composite varieties carrying opaque-2 gene. These three composites Sakthi, Rathan and Protina during released during 1971possess high lysine and tryptophan than the normal maize hybrids and composites. All these hybrids and composites were released through the Central Release Committee. Further, the All India Coordinated Maize Improvement Project has also released a number of hybrids and composites through the State Varietal Release Committees of the different states. Ganga 7 and VL 42 were released as hybrids from Uttar Pradesh in 1968 and 1972 respectively. Sixteen composites were released during the period 1964 to 1978 from different states. The detailed information on these varieties, composites and hybrids including pedigree, characteristic features, area of adaptation, etc, has been given below.

## Hybrids

### COH (M) 4

This hybrid was released during 2002. Parentage: (UMI 90 / UMI 285) / UMI 112. Duration: 94 days. Season: Rainfed - September to October and Irrigated - June to July. Grain Yield: Irrigated - 5694 kg/ha (23 % increase over COH 3 and 40 % increase over CO 1 ), Rainfed 4521 kg/ha (12 % increase over COH 3 and 20 % increase over CO 1). Straw yield: 19 t / ha (irrigated) and 15 t / ha (rainfed). This hybrid is suitable to all maize growing tracts of Tamil Nadu. It is early maturing with high yield potential (11,422 kg/ha). Cobs fully covered with husk and higher percentage of seed set. High starch content (70.9 %) and fat content (5.08 %). It has bold and yellow grains with flint texture. It is moderately resistant to downy mildew diseases and stem borer.

### COH (M) 5

This high yielding single cross hybrid was released during 2006. Resistance to downy mildew disease. Moderate resistance to stem borer. Bold and yellow colour grains. High starch

(68.7%) and protein (8.23 %) content. Hybrid seed production is easy. Parentage: Single cross hybrid of UMI 285 (Female) and UMI 61 (Male). Duration: 100 – 105 days. Season: Irrigated - Throughout the year and Rainfed - September – October. Grain yield: Irrigated - 5400 kg/ha and Rainfed - 4280 kg/ha (% increase over COH (M) 4 - 24.0 %). Highest yield obtained: 8500 kg/ha. Area of adoption: All maize growing districts of Tamil Nadu.

**CO (M) 6**

The parentage is UMI 1200 x UMI 1230 with duration of 110 days. It is a high yielding single cross hybrid. The season is kharif and rabi for irrigated (June - July & November – December) and kharif for rainfed (September – October). It has high starch (76.30 %) and protein (11.25%) and high beta-carotein (0.48 mg/100 g) with moderate level of fat (4.65 %) and crude fibre (1.29%). It has bold, orange yellow semi dent karnal. The shelling is 81% with a test weight 400 g /1000 grains. It has multiple disease resistance viz, Sorghum downy mildew, Maydis leaf blight, Turcicum leaf blight, Post flowering stock rot and Banded leaf and sheath blight. The average yield under irrigated and rainfed condition are 7400 and 5000 kg/ha respectively. Highest yield obtained was 13272 kg/ha. This hybrid is recommended for cultivation in maize growing tracts of Tamil Nadu. It was released during 2012 by TNAU, Coimbatore.

**CoH (M) 7**

It was released by TNAU, Coimbatore. It is a late maturing hybrid. The average yield is 7800 kg/ha. It is suitable for cultivation in Uttar Pradesh, Bihar, Jharkhand, Odisha, Andhra Pradesh, Tamil Nadu, Maharashtra, Karnataka during Kharif season.

**CoH (M) 8**

It was released by TNAU, Coimbatore. It is a medium maturing hybrid. The average yield is 7100 kg/ha. It is suitable for cultivation in Uttar Pradesh, Bihar, Jharkhand, Odisha, Andhra Pradesh, Tamil Nadu, Karnataka, Rajasthan, Gujarat, Madhya Pradesh, Chhattisgarh, Punjab, Haryana, Delhi and Maharashtra during Kharif season.

**CoH (M) 9**

It is a single cross hybrid. It was released by TNAU, Coimbatore. It has a duration of 85-90 days. The average yield is 6279 kg/ha. It is suitable for cultivation in Uttar Pradesh, Bihar, Jharkhand, Odisha, Rajasthan, Gujarat, Madhya Pradesh and Chhattisgarh during Kharif season.

**PEEHM 5**

It was developed from a single cross between SC - 7 - 6 - F - 1 - 1 - Ä - # (Female parent) and IPA 510 - f - # (male parent) through heterosis breeding by Division of Genetics, IARI, New Delhi. It has medium height (180 - 200 cm) with ear placement at 90 - 100 cm. The days to 50% silking and 50 % pollen are 55 and 52 days after sowing. The silk colour, anther colour and heart colour are green - light pink, yellow and white respectively. This hybrid possesses orange flint, bold and attractive grains. In addition, it has stay green quality and is tolerant to high temperature prevalent in Southern states. It has tolerance to tursicum leaf blight, maydis leaf blight, post florets stalk root and *chilo partellus*. Number of kernel rows / ear, number of kernel / rows and tassel size are 16, 25 and 25 - 35 cm. It was released for cultivation under irrigated condition of Punjab, Haryana, Delhi and Uttar Pradesh. This hybrid can also be considered for the northern hills and peninsular zones. PEEHM 5 recorded mean yield of 49 q/ha with a potentiality of 52 q/ha. It showed a superiority of 22.8 and 42.14 % in Jammu and Kasmir, H.P and Uttaranchal, 32.08 and 23.26 % in Delhi, Punjab, Haryana and Uttar Pradesh and 31.02 and 30.23 % superiority in Southern peninsular region over the checks Surya and Hima 129 respectively. It has good adaptability and flexibility to varying moisture regimes and different nitrogen levels. These features combined with its early maturity (80 days), the new maize hybrid is expected to fit in well with specific cropping sequences involving Rabi crops like wheat, pea and gram.

**DHM 117**

The pedigree is BML 6 x BML 7. It is released by ANGRAU, Hyderabad during 2010. Area of adoption is Andhra Pradesh. It is medium maturing type, orange-yellow and flint grains. It is nutrient responsive, tolerant for lodging and stay green at harvest. The average yield is 7500 kg/ha.

**DHM 111**

The pedigree is BML 6 x BML 15. Year of release is 2010 from ANGRAU, Hyderabad. Area of adoption is Andhra Pradesh. It is medium maturity type with yellow and semi-dent grains. It is fertilizer responsive hybrid with tolerance to lodging and stay green at harvest. The average yield is 6500 kg/ha.

**DHM 113**

The pedigree is BML 2 x BML 7. It is released by ANGRAU, Hyderabad during 2010. It is suitable for Andhra Pradesh state. It is late maturing, orange, semi-dent grains. It is nutrient responsive and tolerant to lodging. The average yield is 6600 kg/ha.

**HM 11**

The pedigree is HK1-1128 x HK-163. It is released by CCSHAU, Karnal during 2009. It is recommended for across the country except Himalayan belt during Rabi. It is late maturity type, orange and flint grains. It is responsive to higher doses of fertilizers with average yield of 5500 kg/ha.

**HM 10**

The pedigree is HKI 193-2 x HKI 1128. It is released by CCS HAU, Karnal during 2008. It is suitable for Delhi, Punjab, Haryana and Western Uttar Pradesh, Rajasthan, Madhya Pradesh, Gujarat, Andhra Pradesh, Tamil Nadu, Maharashtra and Karnataka. It is medium maturity group, yellow and semi flint grains. It is highly responsive to inputs and resistant to maydis leaf blight with average yield of 7200 kg/ha.

**PMH 3**

The pedigree is LM 17 x LM 14. It is released by PAU, Ludhiana during 2008. It is suitable for Delhi, Punjab, Haryana and Western Uttar Pradesh. It belongs to late maturity type having orange and flint grains. It is highly responsive to inputs and average yield is 7500 kg/ha.

**PMH 19**

It is popularly called as Parbhani Shakti. It is released during 2003. The yield level is 6000 kg/ha. It is early maturing and high yielding.

**Nithya Shree**

The pedigree is SKV-50 x NAI-105. It is released by UAS, Naganahalli during 2008. It is recommended for Karnataka during Kharif and Rabi seasons. It is late maturity having yellow-orange and semi-dent grains. It is resistant to lodging with average yield of 8000 kg/ha.

**Vivek 33**

The pedigree is V 372 x CM 212. It is released by VPKAS, Almora during 2008. It is suitable for Jammu and Kashmir and Uttarakhand. It is early maturity, yellow, dent grain and average yield of 5000 kg/ha.

**PAU 352**

The pedigree is LM 15 x CML 32. It is released by PAU, Ludhiana during 2007. It is suitable for Punjab, Haryana and Delhi. It is early maturity, resistance to maydis leaf blight, bacterial stripe downy mildew and erwinia stalk rot with average yield of 3500 - 4800 kg/ha.

**Vivek 15**

The pedigree is CM 152 x CM 212. It is released by VPKAS, Almora during 2005. It is suitable for Jummu and Kashmir (Himalayan region), Andhra Pradesh, Tamil Nadu, Maharastra and Karnataka. It is extra early maturity with moderate degree of tolerance against turcicum leaf blight and average yield of 4500 - 5000 kg/ha.

**Vivek 17**

The pedigree is CM 153 x CM 212. It is released by VPKAS, Almora during 2005. It is suitable for across the country, except hill states. It is extra early maturity with moderate degree of tolerance against turcicum leaf blight and maydis leaf blight with average yield of 4000 - 5000 kg/ha.

**NAH 2049**

This hybrid maize has yield potential of 9000 - 10000 kg/ha. It posses high adoptability and tolerance to both turcicum leaf blight and downy mildew diseases has been identified from ARS, Naganahalli which is superior in yielding ability compared to almost all public and private.

**HM 8**

This hybrid is cross between HKI 1105 and HKI 16. It was developed by CCSHAU, Karnal during 2007. It is suitable for Maharashtra, Andhra Pradesh, Tamil Nadu and Karnataka states. It has orange flint with medium maturity with an average yield of 5000 - 6800 kg/ha.

**HM 9**

The pedigree is HKI 1105 x HKI1128 and developed by CCSHAU, Karnal during 2007. It is recommended for cultivation in Bihar, Jharkhand, Orissa, West Bengal and eastern Utter Pradesh. It possesses orange flint grains with medium maturity. The average yield is 6000 kg/ha.

**Malviya 2**

It is a cross between HUZM 185 and HKI1105. It is released by BHU, Varanasi during 2007. The area of adoptability is eastern Utter Pradesh, Bihar, Jharkhand, Chattisgarh, West Bengal and Orissa. It has medium duration, semi flint and yellow grains with resistance to maydis leaf blight disease and responsive to higher doses of fertilizers. The average yield is 5400 kg/ha.

**Vivek 21**

The parentage is CM 212 and V 341. It is developed by VPKAS, Almora during 2007. It is suitable for Uttaranchal, Himachal Pradesh, Jammu and Kashmir, North Eastern Hill regions, Delhi, Punjab, Haryana, western Utter Pradesh, Maharashtra, Andhra Pradesh, Tamil Nadu and Karnataka. It is extra early hybrid (85 - 90 days in hills and 80 - 85 days in plains) and tolerance against turcicum leaf blight disease. The average yield is 6609, 4536 and 4195 kg/ha in Zone I, Zone II and Zone IV respectively which was 41.4, 31.7 and 29.4 % higher over double cross Him 129 in the respective zones. The grains are semi flint, yellow, flat, and bold with average 1000 grain weight of 358 g. The plant is characterized by medium ear placement, cylindrical ear shape, 16 - 18 cm ear length, 16 -17 cm ear girth and 16 - 18 cm regular kernel rows per ear. The tassel is large and open with purplish anther and silk colour.

**Vivek 23**

It is a cross between V 351 and V 341. It is released by VPKAS, Almora during 2007. It is recommended for hills of Uttaranchal. It is early- tall flint, yellow, bold, moderate tolerance against turcicum leaf blight and average yield of 4500 - 5000 k/ha.

**Vivek 25**

The parentage is V 341 and V 346 and developed by VPKAS, Almora during 2007. It is suitable for Uttarakhand, Himchal Pradesh, Jammu and Kashmir and North Eastern Hilly region. It is extra-early maturing type with bold yellow semi dent grain. It is tolerant against turcicum leaf blight. The average yield is 5000 - 5500 kg/ha.

**Vivek 27**

It is a cross between V 335 and V 345 by VPKAS, Almora during 2007. It is recommended for Eastern Uttar Pradesh, Bihar, Jharkhand, Orissa, Chattisgarh, West Bengal, Maharashtra, Andhra Pradesh, Karnataka and Tamil Nadu. It is an extra-early maturing hybrid, yellow and semi-dent grains with an average yield of 5000 - 5500 kg/ha.

## PMH 1

The pedigree is LM 13 x LM 14. It is developed by PAU, Ludhiana during 2007. It is recommended for irrigated areas of Punjab state. It is late maturing with zig-zag stem. It is resistant to maydis leaf blight and stalk rots. The average yield is 5200 kg/ha.

### PAU 352

It is a cross between LM 15 and CML 32 and was released by PAU, Ludhiana during 2007. It is suitable for Punjab, Haryana and Delhi states. It is early maturing, resistant to maydis leaf blight, bacterial stripe, downy mildew and erwinia stalk rot. The average yield is 3500- 4800 kg/ha.

### PMH 2

The pedigree is LM 15 (1102) x LM 16 (SE 539) and was released by PAU, Ludhiana during 2006. It is recommended for Delhi, Haryana, Central and Western UP. It is an early maturing hybrid and resistant to maydis leaf blight, bacterial stripe, downy mildew and post flowering stalk rot.

### Vivek 15

It is a cross between CM 152 and CM 212 and was developed by VPKAS, Almora during 2005. It is suitable for Jammu and Kashmir (Himalayan region), Maharashtra, Andhra Pradesh, Karnataka and Tamil Nadu. It is an extra early hybrid with moderate degree of tolerance against turcicum leaf blight. The average yield is 4500 - 5000 kg/ha.

### Vivek 17

The parentage is CM 153 x CM 212. It was released by VPKAS, Almora during 2005. It recommended for across the country except hill states. It is an extra- early with moderate degree of tolerance against turcicum leaf blight and maydis leaf blight. The average yield is 4000 - 5000 kg/ha

### Buland

It is a cross between CM 146 and CM 147 and was developed by PAU, Ludhiana during 2005. It is recommended for cultivation in Punjab, Uttar Pradesh, Haryana, Delhi and Tarai regions of Uttarakhand. It is late maturing hybrid and resistant to turcicum leaf blight and Common rust. The average yield is 8500 kg/ha.

**HM 5**

It is a cross HKI 1344 between HKI 1348. The year of release is 2005 by CCSHAU, Karnal. It is recommended for Haryana in kharif and Rabi seasons. It is medium tall, white dent grains, medium maturing, responsive to high doses of fertilizers and tolerance to frost. The average yield is 6800 - 7200 kg/ha.

**Pratap 1**

The pedigree is El 116 x EI 364 and was developed by MPUAT, Udaipur during 2004. It is suitable for Rajasthan, Gujarat and Madhya Pradesh. It is an extra early maturing and white semi flint grains. It is moderately resistant to C. partellus, with average yield of 3800 kg/ha.

**Sheetal**

It is released by Punjab Agricultural University, Ludhiana during 2004.

**DMH 2**

The pedigree is CI 4 x KDMI 10. It is released by UAS, Dharwad during 2002. It is suitable for Karnataka state. It is late maturing type, having tall plants, yellow and semi flint grains. It is resistant to sorghum downy mildew.

**Vivek 9**

The pedigree is CM 214 x CM 145 and was released by VPKAS, Almora during 2001. It is suitable for Andhra Pradesh, Karnataka and Tamil Nadu. It is an extra early maturity group (80-85 days). It is tolerant against turcicum leaf blight and maydis leaf blight. It possesses orange semi-flint grain type. The average yield is 4500 - 5500 kg/ha.

**Vivek 5**

The pedigree is CM 212 x V25 and was developed by VPKAS, Almora during 2001. It is recommended for UP under rainfed condition. It is an extra-early maturity (80-85 days). It possesses bold, orange semi-flint grains. It is tolerant against leaf blight and stem borer. The average yield is 4500 - 5000 kg/ha.

**JH 3459**

The pedigree is CM 143 x CM 144. It was released by PAU, Ludhiana during 2001. It is suitable for Punjab state. It belongs to an early maturity group (85-90 days). It has orange flint kernel. It has yield potential of about 5000 - 6000 kg/ha.

**PEHM 3 (AH 58)**

It is a cross between CM213 and CM 142. It is developed by IARI, New Delhi during 2001. It is suitable for cultivation in Andhra Pradesh, Karnataka, Maharashtra, Tamil Nadu, Punjab, Haryana and Delhi. It is an early maturing type (78 - 82 days). It has an attractive yellow flint bold seeds. It is tolerant to high temperature and lodging. The average yield is 4000 - 5500 kg/ha.

**HHM 1**

It is a hybrid between HKI 536 and HKI 295. It is released by CCSHA, Karnal during 2000. It is recommended for cultivation in Haryana state. It is medium maturing type. It is resistant to maydis leaf blight and common rust whereas tolerant to major insect-pests, frost and cold during winter season. The average yield is 5000 - 6000 kg/ha.

**HHM 2**

It is cross a between HKI-1352 and HKI-1344 by CCSHAU, Karnal during 2000. It is recommended for Haryana state under irrigated conditions in kharif and rabi seasons. It is late maturing, tall and resistant to maydis leaf blight in kharif and common rust in Rabi. It has tolerance to major insect-pests, responsive to fertilizers and average yield is 5000 - 6000 kg/ha.

**Vivek 4**

It is a hybrid between CM 212 and CM 141. It is released by VPKAS, Almora during 1999. It is recommended for Gujarat, Rajasthan and Madhya Pradesh. It is an extra early hybrid, tolerant against turcicum leaf blight, maydis leaf blight and average yield is 4500 - 5000 kg/ha.

**HIM 129**

The parentage is (CM 128 x CM 129) x CM 502 and was released by VPKAS, Almora during 1997. It is suitable for Eastern Uttar Pradesh, Bihar, Assam, Orissa, Rajasthan, Madhya Pradesh and Gujarat. It belongs to an early maturing group (80-85 days). It is tolerant to maydis leaf blight, turcicum leaf blight and erwinia stalk rot. The average yield is 4000 - 4500 kg/ ha. It has orange yellow flint type of grain.

**COH 3**

It is a cross of (UMI 101 x UMI 130) x (UMI 90 x UMI 285). It is developed by TNAU, Coimbatore during 1997. It can be grown in Maharashtra, Tamil Nadu, Karnataka and Andra Pradesh. It is late maturing, resistant to downy mildew, moderate tolerance to stem borer, high response to fertilizer and average yield is 4000 kg/ha.

**PEHM 1**

It is a hybrid of CM 135 x CM 136 from IARI, New Delhi during 1997. It is suitable for Andhra Pradesh, Maharashtra, Tamil Nadu and Karnataka. It is an early maturing with moderately resistant to maydis leaf blight, *Rhizoctonia solani*, highly resistant to C. *partellus*, resistant lodging and water logging. The average yield is 4500 kg/ha.

**PEHM 2**

It is a cross of CM 137 and CM 138 from IARI, New Delhi during 1997. It is recommended for Andhra Pradesh, Maharashtra, Rajasthan, Gujarat and Madhya Pradesh. It is an early maturing, flint, orange bold grain, tolerant to moisture stress and average yield is 3000 - 5000 kg/ha.

**Matungha**

It is a hybrid developed from (KDM 14 x KDMI 10) x CM 501 by UAS, Dharwad during 1997. It can be grown in Karnataka state. It is late maturing, tolerant to sorghum downy mildew and maydis leaf blight.

**Parkash**

It is a hybrid of CM 139 x CM 140. It is released by PAU, Ludhiana during 1997. It is suitable for across the country. It is an early maturing, flint corn and resistance to logging, shattering, maydis leaf blight and bacterial stripe downy mildew.

**Rajendra 2**

It is a product of cross between EVM 13 and Jogia Local. It is developed by RAU, Dholi during 1996. It is recommended for Bihar state in Rabi season. It is late maturing with white grains.

**Paras**

It is a cross between LM 5 and LM 6 developed by PAU, Ludhiana during 1996. It is suitable for Punjab state.

**Rajendra 1**

It is developed by three way hybrid of (CM400 x CM300) x P 7421 developed by RAU, Dholi during 1994. It is suitable for Bihar during Rabi season. It is late maturing, semi dent corn and resistance turcicum leaf blight, late wilt as well as rust. The average yield is 3000 - 4000 kg/ha.

**DHM 109**

It is developed by three way hybrid of (CM 119 x CM 211) x CM 133 by ANGRAU, Hyderabad during 1994. It is suitable for Jammu and Kashmir, Uttarakhand, North eastern, Himachal Pradesh, Assam, Peninsular India, Andhra Pradesh, Maharashtra, Rajasthan, Gujarat and Madhya Pradesh except Udham Singh Nagar and Haridwar in Uttaranchal. It is an early maturing (80 - 85 days), orange yellow semi dent grain, tolerant to turcicum leaf blight, maydis leaf blight, sorghum downy mildew and BSR. The yield level is 6000 - 6500 kg/ha.

**DHM 107**

The pedigree is (CM 119 x CM 211) x CM 133 and is released by ANGRAU, Hyderabad during 1993. It is suitable for entire country. It is an early maturing (85 - 90 days), yellow semi-flint type of kernel. It is medium tall tolerant against turcicum leaf blight, maydis leaf blight, sorghum downy mildew and erwinia stalk rot. The average yield is 3000 - 4000 kg/ha with the potential yield of 7000 kg/ha.

**COH 2**

It is hybrid of UMI 810 x UMI 90. It is released by TNAU, Coimbatore during 1993. It is recommended for cultivation in Tamil Nadu. It is resistant to downy mildew disease.

**Trishulata**

The pedigree is (CM 131 x CM 211) x CM 120. It is released by ANGRAU, Hyderabad during 1991. Is is recommended for Maharashtra, Andhra Pradesh, Tamil Nadu and Karnataka states. It is late maturing (110 days), tolerant to maydis leaf blight, turcicum leaf blight and wilts. It yields level of 7000 - 7500 kg/ha.

**Deccan 105**

It is a hybrid of (CM 119 x CM 120) x (CM 211 x CM 107). It is developed by ANGRAU, Hyderabad during1990. It can be grown across the country. It is late maturing hybrid (110 days). It has orange yellow semi dent bold grains. It is tolerant to maydis leaf blight, turcicum leaf blight and wilts. It gives yield level of 6600 - 7000 kg/ha.

**VL 42**

It is a double cross with (V 55B x V 57) x (V 20B x V 50). It is released by VPKAS, Almora during 1988. It is recommended for cultivation in Himalayan belt, Maharashtra, Andhra Pradesh, Tamil Nadu and Karnataka. It is an extra early, orange, flint grains and resistance against stem borer. The average yield is 4000 - 4500 kg/ha.

**Ganga 11**

It is a double cross of (CM 202 x CM 111) x Suwan. It is released by DMR, New Delhi during 1988. It is suitable for across the country. It is late maturing and yellow semi flint grains. It is tolerant against downy mildew and turcicum leaf blight. The average yield is 4500 - 5500 kg/ha.

**Sartaj**

It is four way cross of (7617 x H3191) x (H 104-21 x H 104-15). It is developed by PAU, Ludhiana during 1988. It is recommended for across the country. It is late maturing, yellow orange flint grains. It is resistant against banded leaf sheath blight, maydis leaf blight and tolerant to lodging. The average yield is 5000 kg/ha.

**Deccan 103**

It is four way cross of (CM 120 x CM 118) x (CM 208 x CM 119). It is developed by ANGRAU, Hyderabad during 1982. It is suitable for entire country. It is late maturing hybrid (105 - 110 days) and yellow grains. It produces higher grain yield of 5500 - 6000 kg/ha.

**Sangam**

It is a cross of Makka Safed 1 and Tuxpeno and is released by PAU, Ludhiana during 1982. It can be grown across the country. It matures late in 110 - 115 days. Grain is white and semi- flint. The yield potential is 5000 - 6000 kg/ha.

**Deccan 1**

The pedigree is (CM 119 x CM 120) x (CM 121 x CM 108). It is released by ANGRAU, Hyderabad during 1980. It is recommended for entire country. It is an early maturing hybrid (85 - 90 days). It has yellow orange, semi dent and bold grains. It is tolerant to turcicum leaf blight, charival rot and sorghum downy mildew. The average yield levels are 5000 - 7500 kg/ha.

**Ganga 9**

It is cross between (CM 202 x CM 117) and (CM 116 x CM 207). It is developed by DMR, Delhi during 1979. It is suitable for Himalayan belt.

**HIM 128**

It is hybrid between (CM 301 x CM 401) and (CM 302 x CM 303). It is developed by VPKAS, Almora druign 1979. It is suitable across the country.

**Deccan 101**

It is cross product of (CM 202 x CM 206) and (CM 114 x CM 115). It is released by ANGRAU, Hyderabad during 1974. It is recommended across the country.

**Ganga 7**

It is suitable for Uttar Pradesh and other parts of northern plains. Plants are tall and vigorous, tolerant to downy mildew, leaf blight and rust. Ears are tapering and long with tight husk cover. Grains are orange yellow in colour. This hybrid matures within 95 - 100 days and its yield potential is 5000 - 5500 kg/ha.

**Ganga 5**

It is cross product of (CM 202 x CM 111) x CM 500. It is developed by IARI, New Delhi during 1968. This hybrid is suitable for all major maize growing areas and is more suitable than other hybrids and composites where downy mildew and leaf blight are endemic. It is recommended for cultivation in Punjab, Haryana, Uttar Pradesh, Delhi, Bihar and West Bengal. Stem is thick with broad green leaves. Height of the plant is usually 2.5 to 3.0 m. Husk cover is tight. Plants possess high resistance to leaf blight, rust and stem borer. Grains are bold, flat and light orange in colour. Crop matures 95 to 100 days. Yield potential is 5000 to 5500 kg/ha.

**Ganga 4**

It is cross between (CM 400 x CM 300) and CM 602. Like Ganga 2, this hybrid is also suitable for all major maize growing areas of the country where white grain is preferred. Plants have medium thick stem with long narrow leaves, height is 2.5 to 3.0 m. This hybrid also possesses moderate resistance to downy mildew, leaf blight, rust, stem borer and lodging. It possesses purple grains. It matures in about 100 - 105 days. It has yield potential of 5500 - 6000 kg/ha.

**Ganga 3**

It is a double cross hybrid of (CM 109 x CM 110) x (CM 202 x CM 111). It is developed by IARI, New Delhi during 1964. It is suitable for across the country. It is late maturing type.

**HIM 123**

It is a cross of (CM 202 x CM 205) x (CM 113 x CM 112) and developed by IARI, New Delhi during 1964. It is suitable for Himalayan belt of Himachal Pradesh, Jammu and Kashmir. It matures in about 115 days. Plants are resistant to downy mildew and fairly tolerant to leaf blight. Ears are long, thick and conical. Grains are yellow, flint and bold. The yield potential is 7000 kg/ha. Husk is fairly tight.

**Ganga safed 2**

It is a cross product of (CM 400 x CM 300) x CM 600 by IARI, New Delhi during 1963. This hybrid is suitable for all the major maize growing areas where white grain is preferred such as Uttar Pradesh, Bihar, West Bengal, Punjab, Haryana, Delhi and Rajasthan. Stem is medium thick with long narrow light green leaves. Height of the plant is 2.5 to 3.0 m. Husk cover is medium. Plants are moderately resistant to downy mildew, leaf blight, rust, borers and lodging. Grains are medium round, white and flintish. Maturity varies from 95 to 100 days and yield potential is 5000 - 5500 kg/ha.

**Histarch**

It is a three way cross of (CM 400 x CM 300) x CM 601. It is released by IARI, New Delhi during 1963. This hybrid is suitable for all the maize growing areas of the country where starch industry exist. It is a popular hybrid in North Bihar. Plants have thick stem with broad green leaves, tight husk cover and a height of 2.5 to 3.0 m. Plants are resistant to leaf blight, rust, downy mildew, stem borers and lodging. It is medium maturing (105 to 110 days) type. Grain is bold, flat and white dent. The average yield is 5500 kg/ha.

**VL 54**

It is a cross product of (CM 107 x CM 108) x (CM 203 x CM 204) and is developed by VPKAS, Almora during 1963. It is suitable for growing on high elevation in the Himalayan belt especially in Kumaon hills. Plants grow vigorously and are fairly resistant to lodging. It is resistant to downy mildew and tolerant to leaf blight. Grains are semi-flint and of shining orange colour. It is an early hybrid with average yield of 6000 - 6500 kg/ha.

**Ganga 1**

It is hybrid from (CM l01 x CM 102) x (CM 100 x CM 200). It is developed by IARI, New Delhi during 1961. It is recommended for Indo-Gangetic plains, foothills of Himachal Pradesh, parts of Gujarat, Maharashtra and Karnataka. It is medium maturing, yellow flint, lodging resistance, downy mildew resistance, moderately tolerant to top shoot borer and rust resistance.

**Ranjit**

It is cross between (CM 103 x CM 104) and (CM 202 x CM 106). It is released by IARI, New Delhi during 1961. It is recommended for Gujarat, Maharashtra and South Rajasthan. It is late maturing, orange-yellow bold flint grains, lodging resistance, rust resistance and stay green at harvest.

**Deccan**

It is a double cross hybrid of (CM 104 x CM 105) x (CM 202 x CM 201). It is released by ANGRAU, Hyderabad during 1961. It is suitable for peninsular parts of the country including Andhra Pradesh, Karnataka, Tamil Nadu and Kerala. Its stem is thick with broad dark green leaves. Plants are resistant to lodging and remain green at the time of harvest. Plants are usually 3.0 to 3.5 m tall and husk cover is very tight. This hybrid possesses high resistance to downy mildew and is fairly resistant to leaf blight, rust and stalk borers. Grain is of bold, light orange and semi-dent type. Its yield potential is 5500 - 6000 kg/ha. Crop matures within 110 days during Kharif and 130 days during Rabi.

**Ganga 101**

It is double cross hybrid of (CM 103 x CM 104) x (CM 201 x CM 105). It is developed by IARI, New Delhi during 1961. It is recommended for Indo-Gangetic plains. It is late maturing, orange - yellow flint grains, lodging resistance and downy mildew resistance.

## Composites

### NAC 6004

The pedigree is INDIMYT - 345 populations and released by UAS, Nagenahalli during 2001. It has a yield potential of 5600 kg/ha and has been recommended for Andhra Pradesh, Maharashtra, Karnataka, Tamil Nadu and Karnataka. It is late maturing, resistance to turcicum leaf blight, downy mildew, stalk borer, ear worm and flea beetles.

### NAC 6002

The pedigree is NAI - 103, 104, 109, 113, 114, 117, 118, 125, 132, 133, 134, 135, 136, 108 and 140. This is recommended for southern, central, eastern dry and southern transitional zone of Karnataka. It is developed by UAS, Nagenahalli during 2002. It is an early maturing, resistance to turcicum leaf blight, downy mildew, stalk borer, com ear worm and flea beetles. The average yield is 3000 - 3500 kg/ha and has a yield potential of 5700 kg/ha.

### BH 1620

It is released from ANGRAU, Hyderabad. The pedigree is BML 2 / BML 7. Duration is 112-120 days. Season is *Kharif.* Grain yield is 6600 kg/ha. It is tolerant to *maydis* leaf blight, *turcicum* leaf blight and brown stripe. It is non-lodging type. It can be recommended as baby corn hybrid.

### Jawahar pop corn 11

It is developed from Amber Pop by JNKW, Chhindwara during 2007. It is recommended for Madhya Pradesh. It is medium maturing, orange flint and average yield of 3000 - 4000 kg/ha

### Pratap 4

It is a bulk of national Pool - 2. It is released by MPUAT, Udaipur during 2006. It is suitable for Jammu and Kashmir, Uttarakhand, North East, Himachal Pradesh and Assam. Grains are white semi flint. Plants are tall, medium maturing with average yield of 4500 kg/ha. It is moderately resistance to stem borer.

### Pratap 5

It is a bulk of 11 HS progenies selected from C3 cycle of national Pool-2. It is released from MPUAT, Udaipur during 2006. It is recommended for Rajasthan, Gujarat, Andhra Pradesh,

Madhya Pradesh and Chhattisgarh. It is tall white flint with medium maturing, average yield of 4500 kg/ha, moderately resistant to erwinia stalk rot, post flowering stalk rot, maydis leaf blight, turcicum leaf blight, BLBS and bacterial stripe downy mildew and moderately tolerant to stem borer.

**Pusa Composite 3**

It is developed through population improvement using half- sib system of breeding from Yellow Pool-1. It is released by IARI, New Delhi during 2005 for cultivation in Punjab, Haryana, Central and Western Uttar Pradesh and northern plains zone both rainfed and irrigated conditions. It is tall plant (217 cm) and days to 50 % flowering and dry husk stages are 56 and 86 days respectively. Its stalk possesses good forage quality (stay green). It has long ears with yellow flint grains. It is tolerant to major foliar diseases, medium tolerance to stalk borer, resistance to lodging and tolerance to moisture stress. It has average yield potential of 5111 kg/ha which is 9.6 % higher than best check.

**Pusa Composite 4**

It is composite of Yellow Pool-2 developed by DMR, New Delhi during 2005. It is recommended for cultivation in Punjab, Haryana, Central and Western Uttar Pradesh. It is an early maturing and tolerant to major foliar diseases as well as stalk borer. It is lodging resistance and tolerant to moisture stress. It has average seed yield of 4400 kg/ha. It is medium in maturity. It has good forage quality (stay green) with long ears and yellow flint grains.

**Azad Kamal**

It is developed by crossing of (Azad Uttam x Navjot) x Surya. It is released by CSAUAT, Kanpur during 2005. It is recommended for Rajasthan, Gujarat and Madhya Pradesh. It is an early maturing, yellow-orange grain and tolerant against major diseases as well as insect - pests. The average yield is 3000 - 3500 kg/ha.

**Shalimar KG Maize 1**

It composite developed from Pool 39 and Gurez local. It is released by SKUAST, Srinagar during 2005. It is suitable for Jammu and Kashmir. It is an extra early maturing, tolerant to leaf blight and downy mildew. It is resistant to stem rot and moderately tolerant to stem borer. Yield potential is 3500 - 4000 kg/ha.

**Shalimar KG Maize 2**

It is a composite of Pool-42 x Gurez local. It is released by SKUAST, Srinagar during 2005. It is suitable for Jammu and Kashmir. It is extra early maturing, tassel strongly curved, tolerant to leaf blights and downy mildew, resistance to stem rot and moderately tolerant to stem borer. The average yield is 3500 - 4000 kg/ha.

**Pratap 3**

It is developed from bulk of CEW- 8 Pool. It is released by MPUAT, Udaipur during 2005. It is suitable for Rajasthan, Gujarat and Madhya Pradesh. It is an early maturing, tall plants, white semi flint grains with average yield of 4000 kg/ha. It is moderately resistance to erwinia stalk rot, post flowering stalk rot, maydis leaf blight, turcicum leaf blight, banded leaf sheath blight, bacterial stripe downy mildew and stem borer. It is responsive to higher doses of fertilizers.

**Birsa Vikas 2**

This composite has been developed by crosses of six OPVs - Suwan, BM1, Chandan Makka3, GM 6, Shweta and Rudrapur Local. It is developed by Birsa Agricultural University, Ranchi and Gramin Vikas Trust during 2005. It matures in 70 - 80 days and average yield of 4500 -5000 kg/ha. This variety has also qualified for quality protein maize type because it is nutritionally superior being rich in protein content (lysine and trypotophane). Farmers have great preference for this variety for its cooking quality and sweeter taste, drought tolerance, stay green character and high fodder yield. It is suitable variety for Bihar state.

**Jawahar 216**

It is selection of local germplasm from Ambikapur. It is developed by JNKW, Chhindwara during 2004. It is recommended for Madhya Pradesh. It has medium resistance against maydis leaf blight and turcicum leaf blight. The yield potential is 4000 - 5000 kg/ha

**D 994**

It is a composite of population 31 and one line from Suwan-1. It is released by GBPUAT, Pantnagar during 2004. It is suitable for Bihar, Jharkhand and Orissa. It is an extra early maturing, lodging and water logging tolerance. The average yield is 3500 - 4000 kg/ha.

**Vivek Sankul 11**

It is developed from POP 31, C4 and HS Bulk. It is released by VPKAS, Almora during

2004. It is suitable for across the country. It is an early maturing, yellow and flint grains. The average yield is 3000 kg/ha.

**Gujarat 6**

It is selection made from AAU the crosses performed Godhara between white and yellow maize varieties. It is released by AAU, Godhara during 2003. It is recommended for Bihar, Jharkhand, Orissa and West Bengal particularly general and central Gujarat agro- climatic zone. It is an extra early maturing, tolerance to maydis

**Priya sweet corn**

It is composite developed from Malaysian sweet corn, Thai sweet corn, Madhuri and Madhu Makka. It is released by ANGRAU, Hyderabad during 2002. It is recommended for Andhra Pradesh, Karnataka, Maharashtra and Tamil Nadu. It is medium maturing, yellow dent grain, tolerance to turcicum leaf blight, charcoal rot, late wilt and other stalk rots

**Narmada Moti**

It is cross between LGC-40 and EH-2922 (HS). It is released by MMRS, Godhra during 2002. It is suitable across the country. It is an early maturing, white flint grains and moderately tolerant to C. *partellus*. It is resistant to maydis leaf blight and turcicum leaf blight. Average yield is 3500 - 4000 kg/ha. It is an early maturing, high yielding, white flint type variety and proved good for cultivation under low "N" soil fertility and rainfed situation. It is found superior in grain yield (4038 kg/ha) by 13.7 percent over national check Surya. It is found moderately tolerant to C. *partellus* pest and resistant to Maydiys leaf blight and Turcicum leaf blight. This variety is recommended for cultivation in Zone-II (U.P., Punjab, Haryana and Delhi), Zone-IV (A.P., Maharashtra, Karnataka, Kerala and Tamil Nadu) and Zone-V (Rajasthan, M.P. and Gujarat).

**Gujarat 3**

It is developed from advanced final generations of cross composite - 1 x WRF - 9 (HS). It is released by AAU, Godhra during 2002. It is suitable for Gujarat. It is an early maturing and white flint grains.

**Gujarat 4**

It is a composite variety developed from crossing of LGC-40 x WRF- 15(HS). It is released by AAU, Godhara during 2001. It is recommended for Bihar, Jharkhand, Orissa and West Bengal. It has white flint grains and is tolerant to *R. herculea,* maydis leaf blight, bacterial stripe downy mildew and CLS.

**Amar**

It is an intercrosses of Tarim, D 765, Kanchan, Navin, Kiran, Suwan-1, Pool18, pop 26 and 31. It is released by GBPUAT, Pantnagar during 2001. It is suitable for Andhra Pradesh, Maharashtra, Karnataka, Tamil Nadu, Rajasthan, Gujarat and Madhya Pradesh. It is an early maturing, moderately resistant to major foliar and stalk diseases. It is tolerant to stem borer, fertilizer responsive and yellow semi-flint grains.

**Aravali 1**

It is bulk of early and stress tolerant HS families from X2W Pool. It is developed by MPUAT, Udaipur during 2001. It is suitable for Rajasthan, Gujarat and parts of Maharashtra. It is an early maturing, white, semi flint grains.

**Jawahar Composite 12**

It is derived from early to very early national pool (CDW) material. It is released by JNKW, Chhindwara during 1999. It is suitable for Madhya Pradesh. It is an early maturing, white semi-flint grains, tolerance to turcicum leaf blight and maydis leaf blight. Average yield is 3500 - 4500 kg/ha.

**Gaurav**

It is developed from open pollinated ears selected from various trials. It is released by GBPUAT, Pantnagar during 1999. It is recommended for Punjab, Haryana, Central and western Uttar Pradesh. It is an early maturing, resistant to major foliar and stem diseases. It is tolerant to *Chilo partellus* and fertilizer responsive. It is suitable for rainfed conditions and average yield is 3500 - 4800 kg/ha.

**Shakti 1**

It is composite developed from Antigua, Ver 181 HE02, Amarillo crstallino HE 02, Ant Rep Dom, HE02 and temperate HE02. It is released by DMR, New Delhi during 1997. It is recommended for across the country during Rabi season. It is medium maturing, moderately resistance to turcicum leaf blight and tolerance to *Sesamia* inferens. It has orange flint grains, quality protein maize composite with 0.63 % tryptophan in protein.

**Jawahar 8**

It is developed from early to very early natural pool (CDW) material. It is released by JNKW, Chhindwara during 1997. It is recommended for Madhya Pradesh. It is an early maturing, bold white flint grains, grain filling upto the tip end and highly resistant to maydis leaf blight. Average yield is 3500 - 4000 kg/ha.

**Dewaki Composite**

It is composite variety developed from American early dent x Tuxpeno. It is released by RAU, Dholi during 1996. It is suitable for Bihar, Jharkhand, Orissa and West Bengal. It is late maturing type, white, bold, semi dent grains, resistant against turcicum leaf blight as well as fusarium wilt and highly resistant to *Puccinia sorghi*. Average yield is 6000 kg/ha.

**C-8**

It is a composite developed from A3, Jellicrose, B1, Bajpore-143, Medan-269, Thana Mandi - 598, Kalai 144 and Jhallas 44 (inter crossing). It is released by SKUAST, Srinagar during 1996. It is recommended for higher hills of Jammu and Kashmir. It is medium maturing, resistance to lodging and stem borer and tolerant to maydis leaf blight and turcicum leaf blight.

**C-14**

It is developed by compositing of Abasakaja-Zultaja, Maize No: 8, Bhodipur yellow, Mex 3CB; YU 2DSC4.3 CB, YU 2DSC4, YU 2D-SE3, YU 2D-SC1, UL 1 x C6, P-RICO (CMS3) Hunius, C5, Cl, C6, YU 2D-SC 4 and Bhodipur yellow. It is released by SKUAST, Srinagar during 1996. It is suitable for higher hills of Jammu and Kashmir. It is an early maturing, tolerance to lodging, fertilizer responsive and average yield 3800 - 4200 kg/ha.

**Birsa 1**

It is a composite developed from D 741, Genetic Pool of early yellow. It is released by BAU, Ranchi during 1996. It is recommended for Bihar. It is an early maturing, yellow flint grains, resistance against maydis leaf blight and turcicum leaf blight.

**Pearl pop com**

It is released by PAU, Ludhiana during 1996. It is recommended for Punjab. It is a pop corn variety.

**Mahi Dhawal**

It is developed from broad based gene pool of HSN (RFC) - K-87. It is released by MPUAT, Banswara during 1996. It is suitable for Punjab, Haryana, central and western UP. It is late maturing, semi flint, bold, white grains and moderately resistance to stem borer, downy mildew with an average yield of 3500 - 4500 kg/ha.

**Gujarat 2**

It is developed from (4119 x 4123) x DRM - 6. It is released by AAU, Godhara during 1995. It is suitable for Gujarat. It is an early maturing, drought resistance, yellow flint grains and average yield is 3200 - 4800 kg/ha.

**Kesari**

It is a composite variety developed by crossing of (Ludhiana local x J 2014) x Parbhat. It is released by PAU, Ludhiana during 1994. It is suitable for Punjab state.

**Panchganga**

It is a composite variety developed from Pool 15. It is released by Kolhapur during 1993. It is recommended for Maharashtra state.

**Megha**

It is a composite variety developed by crossing of JS2 x Arun. It is released by PAU, Ludhiana during 1993. It is suitable for across the country. It is an early maturing type.

**Mahi Kanchan**

It is developed from broad based gene pool of CDC (yellow)-C 8-K-86. It is released by MPUAT, Banswara during 1992. It is recommended for Punjab, Haryana, Central and western parts of the country. It is an early maturing, semi-flint and yellow grains, moderately resistance to maydis leaf blight, downy mildew and tolerance against turcicum leaf blight with an average yield of 3000 - 4000 kg/ha.

**J 1006**

It is a composite variety developed by crossing of MS 1 and Tuxpeno PB. It is released by PAU, Ludhiana during 1992. It is suitable for Punjab state. It is fodder variety

**Azad Uttam**

It is compositing 250 open pollinated cases of various trials of KH 1977. It is released by CSAUAT, Kanpur during 1991. It is suitable for Uttar Pradesh state. It is an early maturing, yellow semi flint grains with average yield of 3000 - 4000 kg/ha.

**Varun**

It is compositing of 10 inbreds chain crossed. It is developed by ANGRAU, Hyderabad during 1990. It is suitable foothills of Himalayas and Peninsular India. It is medium duration (90 - 100 days). It has semi flint, orange yellow grains. It is tolerant to water stress. The yield potential is 4500 - 5000 kg/ha.

**Madhuri**

It is developed from Florida sweet corn. It is released by ANGRAU, Hyderabad during 1990. It can be cultivated in Andhra Pradesh throughout the year. It is early maturing (75 - 80 days), yellow dent grains and nitrogen responsive. It is moderately tolerant to water stress. It is sweet corn variety. It contains 30-36 % sugars and 15 % proteins. It is suitable for table purpose after boiling. The average yield is 1600 - 2000 kg/ha.

**Parvati**

It is a composite developed from Syn Pool 200. It is released by CSKHPKVV, Bajaura during 1990. It is recommended for Himachal Pradesh state.

**Parbhat**

It is composite variety developed by crossing (J54 x Suwan) x J 54. It is developed by PAU, Ludhiana during 1988. It is recommended for Punjab state.

**VL 88**

The pedigree is national pool CD yellow. It is released by VPKAS, Almora during 1988. It is suitable for foothills of Himalaya's up to 2200 m altitude. It matures in 85-90 days. The yield potential is 3500 - 4000 kg/ha. It has yellow flint grain and stem borer tolerance.

**Navjot**

The parentage is Pratap and Tarun. It is developed by PAU, Ludhiana during 1988. It is recommended for Punjab state. It is early maturing variety (105-110 days). It has yellow-orange flint grains. It is resistant to frost and tolerant to lodging. The average yield is 3600 - 4200 kg/ha. It can be cultivated up to 1200 m altitude.

**Mansar**

It is F2 seed of intercrosses of Vijay, Kisan, Ambar, Ganga 5, Jawahar, H1M123, VL54 and Deccan. It is released by SKUAST, Srinagar during 1988. It is suitable for Himalayan Belt. It is medium maturing, resistance to stem barer and lodging and nitrogen responsive. The average yield is 3000 - 4000 kg/ha.

**Harsha**

It is an intermix of X 102 x EV Syn, B21 x X102; X 102 x CM 111; X102; (X 102 x SUWAN- 2 x X 102); X102 x CM 202; SYN B33 x X 102. It is released during 1988. It is recommended for Maharashtra, Tamil Nadu, Andhra Pradesh and Karnataka states both for rainfed and irrigated conditions. It is an early to medium maturity type (95 - 100 days). It has semi-flint, orange yellow grains. It is tolerant to turcicum leaf blight and fertilizer responsive. Average yield is 4500 - 5000 kg/ha.

**Surya**

The pedigree is compositing of (Gulat x Ant. GR.II) x D 743, (DI x J 603) x D743, (J 603 x J 602) x D 743; D 765 (D 745 x D743, D743. It is released by GBPUAT, Pantnagar during 1988. It is suitable for Uttar Pradesh up to 2200 m altitude. It is an early maturing group (85- 90 days). It is tolerant to stalk rot and nitrogen responsive variety. The average yield is 2800 - 3200 kg/ha.

**Pusa 1**

The pedigree is compositing of Pop A600, A602, A603 and A604. It was released by IARI, New Delhi during 1988. It is recommended for Delhi, Uttar Pradesh, Madhya Pradesh and Andhra Pradesh. It is medium maturing, yellow flint grain with average yield of 3000 - 4000 kg/ha.

**Pusa 2**

The pedigree is A61. It was released by IARI, New Delhi during1988. It is suitable for Delhi, Uttar Pradesh, Madhya Pradesh and Andhra Pradesh. It is medium maturing, yellow flint grain with average yield of 3000 - 4000 kg/ha.

**Dhawal**

The pedigree is VC80 x Eto Tuxpeno Br 2. It is released by DMR, New Delhi during 1988. It can be grown across the country during Rabi season. It is late maturing type (125 -130 days). It has white, semi flint grain and nitrogen responsive variety. The average yield is 4500 - 6500 kg/ha. It has taller plants (191 cm).

**Amber pop**

It is released by ANGRAU, Hyderabad during 1988. It is suitable for Andhra Pradesh for Kharif and Rabi seasons under irrigated condition. It is pop corn variety. It has yield potential of 3000 kg/ha.

**Arun**

It is a selection from Indigenous and exotic materials developed by IARI, New Delhi during 1987. It is recommended for Delhi and adjoining regions. It is an early maturing type having yellow flint grains with average yield of 3000 - 4000 kg/ha.

**MCU 508**

The pedigree is Indian pool 5 using unit selection method developed by DMR, New Delhi during 1987. It is recommended across the country. It is an extra early variety with yellow flint grains.

**Hemant**

It is cross product of Dholi 7744 (AE) and Tuxpeno. It is developed by RAU, Dholi during 1986. It is suitable for Bihar. It is late maturing type.

**Renuka**

It is variety derived from G15 and G1Y. It is released by UAS, Dharwad during 1986. It is recommended for Karnataka. It is an early maturing, yellow semi flint grains. It is tolerant to downy mildew and rust and yield level is 3000 - 4000 kg/ha

**Kiran**

It is composite of 21 elite Punjab Locals x J 650. It is released by PAU, Ludhiana during 1986. It is recommended across the country. It is an early maturing, resistant to borer and lodging. The average yield is 3000 kg/ha.

**NLD white**

It is diallel of 10 promising genotypes. It is released by DMR, New Delhi during 1985. It is suitable for Sikkim and North Eastern Region. It has white dent grains. The yield potential is 7000 kg/ha. It has tall plants (238 cm) and matures in 125 - 130 days.

**Suwan**

It is developed from exotic germplasm by RAU, Dholi during 1985. It is suitable for Bihar and Jharkhand states. It is late maturing type.

**Deira**

The pedigree is selection from Diara. It was released during 1984. It is an extra early variety.

**D 768**

It is cross product of D747 and local OPVs. It developed by GBPUAT, Pantnagar during 1984.

**Partap 1**

It is developed from J 54. It is released by PAU, Ludhiana during 1983. It is recommended for Punjab state during Rabi season.

**Lakshmi**

It is composite developed by crossing of American Dent and Tuxpeno Yellow. It is released by RAU, Dholi during 1983. It is recommended for Bihar. It is late maturing, white, semi flint grains.

**Ageti 76**

It is derived from Chain crossing among 26 inbreds. It is released by PAU, Ludhiana during 1982. It is suitable for Punjab. It matures in 85 - 90 days. Grains are attractive bold, medium round and orange-yellow flint. The average yield is 3800 - 4000 kg/ha. It fits well with the popular maize - wheat rotation. It gives high yield not only under high fertility conditions but also performs well on medium and low fertility soils.

**Tarun**

It is developed from Syn P200 and Kisan by CSAUAT, Kanpur during 1982. It is a yellow grain, early maturing (85 -90 days) composite released for general cultivation for Uttar Pradesh. It matures in 85 -90 days. It has short plant type (2.0 m) with sparsely arranged narrow leaves and wide space between tassel and flag leaf. It has orange yellow, semi-flint grain type. It has a high degree of resistance to downy mildew, tolerant to bacterial as well as pythium stalk rot. Yield potential is about 4000 - 4500 kg/ha.

**RCM 1-2**

It is released by ICAR Research Complex on NEH region during 1997. It is developed from the population of amber pop corn and Sikkim pop corn following modified mass selection methods. It gives 3714 kg/ha as average yield with 96 % popping percentage. It performs well in Sikkim, Nagaland and Mizoram with average yield of 6.37, 2.29 and 2.50 t/ha respectively. It is suitable for intercropping with French bean. It matures in 110 days in mid latitudes of Meghalaya and attains the plant height of 175 cm at silk stage. It is characterized by yellow flint small seeds in long cylindrical ears. The spacing recommended is 50 x 25 cm. It is a pop corn variety.

**Ashwini**

It matures in 100 - 105 days and yield potential of 4000 - 4500 kg/ha. It can be grown under rainfed and irrigated conditions during Kharif season. It is tolerant to stalk borer.

**PH 3426**

It has yield potential of 4500 kg/ha. It has taller plants (182 cm) and matures in 130 - 135 days.

**VL Amber**

The pedigree is compositing of 6 popcorn inbreds. It is released by VPKAS, Almora during 1982. It is suitable for Uttar Pradesh and Himalayan Belt up to 1600 m altitude. It is pop corn variety. The average yield is 1600 - 3000 kg/ha. It matures in 105 - 110 days.

**Chandan Safed 2**

It is developed from exotic germplasm by JNKVV, Chhindwara during 1982. It is recommended for Madhya Pradesh state.

**Vivek 11**

It is developed by VPKAS, Almora during 1972. It is recommended for foothills of Himalaya's up to 2200 m altitude. The yield potential is 3200 - 4600 kg/ha. It matures in 95 -100 days.

**Chandana 1**

It is developed from exotic germplasm by JNKVV, Chhindwara during 1982. It is recommended for Madhya Pradesh. The grain is white in colour.

**African Tall**

It is selection from 7 varieties. It is released from Kolhapur during 1982. It can be cultivated across the country. It is a fodder variety.

**Partap**

It is compositing of Vijay x Pioneer 102; Ganga5 x Pioneer 104; Pioneer 104 x Vijay. It is released by PAU, Ludhiana during 1980. It is recommended for Punjab. It matures 90 - 95 days.Plant has relatively low ear placement. Its stem is strong, medium thick and resistant to lodging. The ears are long, thick with a tight husk cover. The grains are medium, flint to semi flint and yellow- orange colour. Its yield potential is 4500 - 5000 kg/ha.

**Nishat**

It is composite developed from C3 x C2. It is released by SKUAST, Srinagar during 1979. It is suitable for Jammu and Kashmir.

**C1**

It is composite developed from 18 yellow flint varieties. It is developed by SKUAST, Srinagar during 1978. It is suitable for Jammu and Kashmir.

**C2**

It is composite from 13 yellow dent varieties developed by SKUAST, Srinagar 1978. It is suitable for Jammu and Kashmir.

**C3**

It is composite of 8 inbred lines developed from SKUAST, Srinagar during 1978. It is suitable for Jammu and Kashmir. It is medium maturing, orange-yellow and semi flint grain.

**C6**

It is composite developed by bulking of F2 seeds of crosses among flint USA/Arg 3, 5 and Anantnagar Local. It is released by SKUAST, Srinagar during 1978. It is suitable for Jammu and Kashmir.

**C15**

It is composite developed from open pollinated ears of Yugoslavian material. It is released by SKUAST, Srinagar during 1978. It is suitable for Jammu and Kashmir.

**Makka Safed 1**

It is developed from Tus. It is released by PAU, Ludhiana during 1978. It is recommended for Punjab state. It is white grain type.

**Moti**

It is released by MPUAT, Udaipur during 1978. It is recommended for cultivation in Rajasthan state.

**Trikuta**

It is developed by bulking of F2 seeds of 10 genotypes chain crossed by SKUAST, Srinagar during 1975. It is suitable for Jammu and Kashmir state.

**Sweta**

It is an early maturing composite. It is released for cultivation in Uttar Pradesh by GBPUAT, Pantnagar during 1974. It matures in 85 - 90 days. Grains are white and semi-flint type. It has high degree of resistance to brown stripe disease and also tolerant to stalk rots. It is suitable for growing in areas where white maize varieties are preferred over yellow. Its yield potential is 4500 - 5000 kg/ha.

**D741**

It is released by GBPUAT, Pantnagar during 1974. It is recommended for Uttar Pradesh state.

**D 765**

It was developed by GBPUAT, Pantnagar during 1974. It is suitable for Uttar Pradesh state.

**Kanchan**

It is very early maturing composite (75 - 80 days) released by GBPUAT, Pantnagar during 1974. It is suitable for Uttar Pradesh. This has yellow coloured seed. This variety has resistance against brown and stalk rot. It responds to low level of management. It has yield potential of 3000 kg/ha.

**Type 41**

It was released by CSAUAT, Kanpur during 1973. It is recommended for cultivation in Uttar Pradesh state.

**VL 16**

It is developed by Vivekanand Parvatiya Krishi Anusandhansala, Almora during 1972. It is recommended for foothills of Himalaya's up to 2400 m altitude. The yield potential is 3200 - 3600 kg/ha. It matures in 95 - 100 days. Grains are orange yellow in colour. These composite yields better than VL 42. It is a yellow grain variety

**VL 41**

It is released by VPKAS, Almora during 1972. It is suitable for foothills of Himalaya's upto 2400 m altitude. It matures in 80-85 days. The yield potential is 2800 - 3200 kg/ha.

**K 1**

It is composite variety released during 1978. A yellow seeded variety derived from the high yielding inbreds of Pattukotaai Maize. It is an early in maturity (80-85 days) with high lysine and tryptophan content. It possesses good taste, cooking and keeping quality.

**Ambar**

This composite is suitable for peninsular and hilly regions of the country up to an altitude of 1800 m. Plants ares tall (3-4 m), vigorously growing, bearing broad dark green leaves. It is developed by BICY, Delhi during 1967. It has partial resistance to downy mildew, leaf blight and rust. It matures 110 - 115 days and yield potential is 6000 - 6500 kg/ha. Grains are flintish and yellow orange.

**Jawahar**

The pedigree is (Al x Antigua Gr. 1) and developed by IARI, New Delhi during 1967. It is suitable for across the country, except hilly areas. It matures in 100 - 105 days. Plant possesses dark green leaves and thick stem. It is fairly resistant to lodging. Plants have tendency to bear two cobs per plant, tightly covered with husk. The grain is yellow-orange, semi flint and bold. It is resistant against turcicum leaf blight and maydis leaf blight. Plants remain green at the time of harvest. Yield potential is 4500 - 5000 kg/ha.

### Sona

The pedigree is (J 1 x Cuba 115) and is developed during 1967. It is suitable for Indo-Gangetic areas (Punjab, Haryana, Delhi, U.P and Bihar). Plant possesses broad dark green leaves with thick stem. It is resistant to turcicum leaf blight, maydis leaf blight and lodging. Plants remain green at harvest and thus make good fodder. It matures in 100 - 105 days. Yield potential is about 5000 kg/ha.

### Kisan

The pedigree is (J 1 x Coastal tropical flint) and developed during 1967. It is suitable for Indo-Gangetic areas. Plants are compact with thick stem and long, broad, dark green leaves with lodging resistance. Ears are cylindrical and short with tight husk cover. Grains are medium, yellow, hard and flint. Grain yield potential is 5000 - 6000 kg/ha. It is resistant to top shoot borer, downy mildew, leaf blight and rust. It has a tendency to yield two cobs per plant.

### Vijay

It is developed from Jl (JMLI) by PAU, Ludhiana during 1967. This composite is suitable for northern plains (Punjab, Haryana, Delhi, U.P and Bihar). Plants are tall growing with long narrow comparatively lighter green leaves. It matures from 100 to 105 days. Ears are long and tapering, grain yellow, bold, flint to semi-flint. The average yield is 5500 - 6000 kg/ha. It has a fairly high resistance to leaf blight and rust. It has tendency to bear two cobs per plant.

### Vikram

It is developed from A6 (Bassi Local x Eto Amarillo) by IARI, New Delhi during 1967. It is recommended for Punjab, Haryana, Uttar Pradesh, Delhi, Bihar and West Bengal. It matures in about 95 days. Plant possesses dark green leaves and thick stem. It is fairly resistant to lodging. Plants have a tendency to bear two ears per plant, tightly covered with husk. Grains are yellow orange, flint and bold in size. It is also drought tolerant. Yield potential is about 5000 - 5500 kg/ha.

**Masoor**

It matures in 105 - 110 days. It can be cultivated in hilly areas upto 1200 m altitude. The yield potential is 3600 - 4000 kg/ha.

**Murali**

It has yield potential of 6000 kg/ha. It has taller plants (217 cm) and matures in 120 -125 days.

**PH 410**

It has yield potential of 7500 kg/ha. It has taller plants (210 cm) and matures in 125 - 130 days.

**Laxmi**

It is released during 1980 and suitable for Rabi cultivation in Bihar. It has short plant type (2.0 m). Grains are white and semi-dent. It matures in 150-175 days. Yield potential is very high (6500 - 7000 kg/ha). It is not recommended for kharif season because of its late maturity.

**Navin**

It is a composite released for general cultivation in Uttar Pradesh in 1979. It matures in 85 - 90 days. Grains are yellow and semi-flint type. It has a high degree of resistance to brown stripe and pythium stalk rot. Yield potential is about 4500 - 5000 kg/ha.

**CO 1**

The parentage is unit selection from Indonesian population Suwan 1. Duration is 105 - 110 days. Area of adoption is Coimbatore, Erode, Tirunelveli, Tanjore and Pudukottai. It is suitable for both rainfed and irrigated conditions. Grain yield is 5200 kg/ha for irrigated and 3300 kg/ha for rainfed condition. It is resistant to downy mildew. Sheath colour is green. Node colour is green. Mid rib is white colour. Cob size is big. Cob has full husk coverage. Grain colour is deep orange. Nature of kernels is flint.

**Gujarat Makai 6 (GM – 6)**

It was released for cultivation in the year 2002 for Gujarat State for white maize growing areas.It is developed and thoroughly evaluated through Farmers Participatory Plant Breeding (PPB) approach. It is an extra early (75-80 days), drought escaping and white flint grained composite variety for marginal environment of tribal belt of Gujarat, Rajasthan and Madhya

Pradesh states of India. It has also better resistance to maydis leaf blight, brown stripe downy mildew and Curvularia leaf spot.

## Baby Corn

### HM 4

The pedigree is HKI 1105 x HKI 323. It is released by CCS HAU Karnal during 2005. It is recommended for Haryana, Delhi, Punjab and Eastern Uttar Pradesh. Baby corn, medium maturity, orange, flint, medium tall, tolerance to frost and cold. The average yield is 6000-6800 kg/ha.

### VL Baby Corn 1

The parentage is [{(VL 16 x Murulia) x VL 16} x VL 16] and is released by VPKAS, Almora during 2005. It can be grown across the country. It is extra early maturing, yellow, prolific with average yield of 3800 - 4000 kg/ha with baby corn yield as 1200 - 1300 kg/ha.

### VL 42

It is a double cross hybrid and produced baby corn yield of 1428 kg/ha with an average more than 1 - 2 ears per plant. It is recommended for cultivation at the national level. It is an extra early, orange, flint and resistance against stem borer. It is earliest in producing baby corn (45 days) than other varieties. Fresh weight per green cob at harvest is 35 - 40 g. Fresh weight per baby corn is 4.5 - 5.5 g. Diameter of fresh baby corn is 3.5 - 4.0 cm and length of fresh baby corn is 6.5 - 7.0 cm. Green fodder yield is 20.0 - 25.0 t/ha. It is resistant to leaf blight disease.

### COBC 1

The parentage is composite involving UMI 836 and UMI 836-1-2. Duration is 55 - 65 days (fresh cobs harvest). It is recommended for all maize growing areas. It is suitable for both rainfed and irrigated conditions. Grain yield is 6.7 tonnes of tender cobs and 32.3 tonnes of green fodder / ha for irrigated and 3.0 tonnes of tender cobs per ha for rainfed condition. Special features are 2 to 3 cobs / plant, possessing delightful sweet taste and flavour. Suitable for inland and export market. Green fodder has good forage value. Sheath colour is green. Node colour is green. Mid rib is white colour. Cob size is small. Grain colour is yellow. Nature of kernels is dent.

## References

Albrecht, B. and J.W.Dudley (1987). Divergent selection for stalk quality and grain yield in an adapted x exotic maize population cross. Crop Science, 27: 487 - 494.

Anderson, B. and D.G. White (1994). Evaluation of methods for identification of corn genotype with stalk rot and lodging resistance. Plant Disease, 78: 590 - 593.

Anonymous (2003). New crop varieties developed by BAU, Ranchi, ICAR News, (July - September): 14.

Babic, M. and Z. Pajic (1992) Effect of Genotype x environment Interaction on Expansion Volume in Popcorn Hybrids (*Zea mays* L.). Genetica, 24 (1): 27-32.

Betran, F.J., J.M. Ribaut, D. Beck and D.G. Leon (2003). Genetic diversity, specific combining ability, and heterosis in tropical maize under stress and nonstress environments. Crop Sci., 43 (3): 797-806

Bolanos, J. and G.O. Edmeades (1996). The importance of the anthesis - silking interval in breeding for drought tolerance in tropical maize. Field Crops Res., 48: 65-80.

Buren, L.L. (1970). Plant characteristics associated with barrenness in maize. Ph. D. Thesis. Iowa State University, Ames.

Byrne, P.F., J. Bolanos, G.O. Edmeades and D.L. Eaton (1995). Gains from selection under drought versus multilocation testing in related tropical maize populations. Crop Science, 35: 63-69.

Chapman, S.C. and G.O. Edmeades (1999). Selection improves drought tolerance in tropical maize population. II. Direct and correlated responses among secondary traits. Crop Sci., 39: 1315 - 1324.

Gupta, H.S., V.Mahajan, P.K. Agrawal, G.S. Bisht and M.C.Pant (2010). Vivek QPM 9 maize: a hybrid for Himalayan hills and Peninsular region. Indian Farming, 59 (11): 10 -12.

Mahajan, V., H.S.Gupta, S.R.K.Singh and G.S. Bisht (2010). Vivek Maize Hybrid 21: a hybrid of Himalayan hills, northern plains and Peninsular region. Indian Farming, 60 (8): 17 -18.

Nallathambi, G., N. Jayaraman and M.Rangasamy (1998). COBc 1 – A new baby corn composite for Tamil Nadu. Madras Agric. J., 85 (10-12): 644 - 645.

Palavershich, B., M. Roits, D. Parlov, M. Peinovich, B. Palaversic, M. Rojc and M. Pejnovic (1990). Possibilities of utilizing induced stress in maize breeding. *Botanica Acta*, 103: 144 - 147.

Shah, S.H., M.F. Sallem, S.I. Zamir and Tanveer Hassain (2003). Growth and yield performances of three maize cultivars sown in pure and blent furrows. J. Food and Agric. Envir., 1(3-4): 203-205.

Singh, G., M. Singh and K.R. Dhiman (1994). Some promising varieties of maize for Sikkim. Indian Farming, 41 (5): 13.

Singh, H. and B. Rai (1979). Differences in morphological character and yield of important maize genotype and their recovered brachytic versions. Indian J. Agric. Sci., 49: 758 - 763.

Srivastava, A., J.S. Singh, M.Z.K. Warsi and S.S. Verma (2002). Maize hybrids for Uttaranchal. Indian Farmers' Digest, 35 (5): 17 -18.

5

# Genetics of Maize Crop

A genetic variant that accumulates more sugar and less starch in the ear is consumed as a vegetable and is called sweet corn. Immature maize shoots accumulate a powerful antibiotic substance, DIMBOA (2, 4-dihydroxy-7-methoxy-1, 4-benzoxazin-3-one). DIMBOA is a member of a group of hydroxamic acids (also known as benzoxazinoids) that serve as a natural defense against a wide range of pests including insects, pathogenic fungi and bacteria. DIMBOA is also found in related grasses, particularly wheat. A maize mutant (bx) lacking DIMBOA is highly susceptible to be attacked by aphids and fungi. DIMBOA is also responsible for the relative resistance of immature maize to the European corn borer. As maize matures, DIMBOA levels and resistance to the corn borer decline. Because of its shallow roots, maize is susceptible to droughts, intolerant of nutrient-deficient soils, and prone to be uprooted by severe winds.

Many forms of maize are used for food, sometimes classified as various subspecies related to the amount of starch each had. They are flour corn - Zea *mays var. amylacea,* Popcorn - *Zea mays var. everta*, Dent corn - Zea *mays var. indentata,* Flint corn - *Zea mays var. indurata,* Sweet corn - *Zea mays var. saccharata* and *Zea mays var. rugosa*, Waxy corn - *Zea mays var. ceratina,* Amylomaize - *Zea mays,* Pod corn - *Zea mays var. tunicata* Larrañaga ex A. St. Hil and Striped maize - *Zea mays var. japonica*.

Maize has 10 chromosomes (n=10). The combined length of the chromosomes is 1500 cm. Some of the maize chromosomes have chromosomal knobs highly repetitive heterochromatic domains that stain darkly. Individual knobs are polymorphic among strains of both maize and teosinte. Barbara McClintock used these knob markers to validate transposon theory of "jumping genes". Maize is still an important model organism for genetics and developmental biology today.

The Maize Genetics Cooperation Stock Center, Department of Crop Sciences, University of Illinois, Urbana-champaign is a stock center of maize mutants. The total collection has nearly 80,000 samples. The bulk of the collection consists of several hundred named genes, plus additional gene combinations and other heritable variants. There are about 1000 chromosomal aberrations (e.g., translocations and inversions) and stocks with abnormal chromosome numbers (e.g., tetraploids). In 2005, the U.S. National Science Foundation, Department of Agriculture and Department of Energy formed a consortium to sequence the B73 maize genome. The resulting DNA sequence data was deposited immediately into GenBank, a public repository for genome-sequence data. Sequences and genome annotations have also been made available throughout the project's lifetime. Primary sequencing of the maize genome was completed in 2008. The genome, 85 % of which is composed of transposons was found to contain 32,540 genes. Much of the maize genome has been duplicated and reshuffled by helitrons a group of rolling circle transposons.

## Taxonomy

Maize belongs to the tribe Maydeae of the grass family *Poaceae*. "Zea" (zela) was derived from an old Greek name for a food grass. The genus Zea consists of four species of which *Zea mays* L. is economically important. The other Zea sp., referred to as teosintes, are largely wild grasses native to Mexico and Central America (Doeblay, 1990). The number of chromosomes in *Zea mays* is 2 n = 20. Tribe Maydeae comprises seven genera which are recognized, namely Old and New World groups. Old World comprises Coix (2n = 10/20), Chionachne (2n = 20), Sclerachne (2n = 20), Trilobachne (2n = 20) and Polytoca (2n = 20), and New World group has Zea and Tripsacum. It is generally agreed that maize phylogeny was largely determined by the American genera Zea and Tripsacum. However it is accepted that the genus Coix contributed to the phylogenetic development of the species *Zea mays* (Radu et al., 1997). Kingdom: Plantae. Division: Magnoliophyta. Class: Liliopsida. Order: Poales. Family: Poaceae. Genus: *Zea.* Species: *Z. Mays.*

## Relatives of Maize

The closest wild relatives of maize are the teosintes which all belong to the genus Zea. Outside the Zea genus, the closest wild relatives are from the genus Tripsacum. Information about wild relatives (CFIA, 1994; OECD, 2006) teosintes, tripsacum, coix and other Asiatic genera is presented below:

## Teosintes

The teosintes are wild grasses native to Mexico and Central America and have limited distribution (Mangelsdorf *et al.*, 1981). Teosinte species show little tendency to spread beyond their natural range and distribution is restricted to North, Central and South America. They are not reported to occur in Southeast Asia (Watson and Dallwitz, 1992). The nearest teosinte relative to *Z. mays* is *Z. mays* sp. *mexicana* (Schrader). It is previously classified as *Euchlaena mexicana, Zea mexicana* (2n = 20). This Central Mexican annual teosinte is a large flowered, mostly weedy grass with a broad distribution across the central highlands of Mexico. It does not spread readily. It has limited use as a forage and green fodder crop, but can be problematic due to weedy tendencies (Doebley, 1990; Watson and Dallwitz, 1992). *Z. diploperennis* (Iltis *et al, 1979)* and *Z. perennis* (Hitchcock, 1951) (Reeves and Mangelsdorf, 1939) are diploid (2n = 20) and tetraploid (2n = 40) perennial teosintes, respectively, with narrow distributions in Jalisco, Mexico. *Zea diploperennis* was on the threshold of extinction when found in the late 1970s and has since been used extensively for investigating maize ancestry (Eubanks, 1995). *Zea perennis* is reported to be established in South Carolina, USA (Doebley, 1990; Hitchcock and Chase, 1951).

*Zea luxurians* (Durieu *et* Asch.) is an annual teosinte from southeastern Guatemala (Doebley 1990; Watson and Dallwitz 1992).

*Zea mays* ssp. *parviglumis* (Iltis and Doebley, 1980) is a small-flowered, mostly wild teosinte of southern and western Mexico (Doebley 1990).

*Zea mays* ssp. *huehuetenangensis* (Iltis and Doebley, 1980) is a narrowly distributed teosinte of the western highlands of Guatemala (Doebley, 1990).

## Tripsacum

The genus Tripsacum is comprised of about 12 species that are mostly native to Mexico and Guatemala but are widely distributed throughout warm regions in the USA and South America, with some species present in Asia and Southeast Asia (Watson and Dallwitz 1992). All species are perennial and season grasses.

Species of economic importance to agriculture in Southeast Asia include *Tripsacum dactyloides* (L.) and *T. laxum* Scrib and Merr. *T. dactyloides* (L.) L. (Eastern gama grass) is a warm season grass native to Mexico and Guatemala but now distributed throughout the Western Hemisphere to Malaysia (FAO, 2000d). *T. dactyloides* has a haploid chromosome number of n =18, with ploidy levels ranging from 2n = 2x =36 up to 2n = 6x = 108 (deWet *et al.,* 1972 and 1983). Different accessions exhibit both sexual and apomictic (asexual) reproductive capabilities. *T. dactyloides* has been the focus of extensive breeding work to transfer the apomictic reproductive nature from Tripsacum to *Zea mays* (Kindinger *et al.,* 1995; Savidan *et al.,* 1995).

*T. laxum* Scrib and Merr. (Guatemala grass) is a warm season grass native to Central America but now distributed throughout Mexico, South America, Sri Lanka and Southeast Asia to Fiji. It is used as a fodder grass in Southeast Asia and is reportedly used as a soil binder and organic-matter builder in upland tea estates (FAO, 2000b). It does not flower readily and seed production is unusual except in its native habitat. *T. laxum* has been reported as a significant weed species (Watson and Dallwitz, 1992).

*T. andersonii* Gray (Guatemala) is native to Central America and has 64 chromosomes. It has been suggested to be a hybrid between *Zea luxurians* (n=10) and *Tripsacum latifolium* (n=18, 3x=54) (Talbert *et al.* 1990). It has minor importance as a fodder crop (Watson and Dallwitz, 1992). *T. lanceolatum* occurs in the Southwest USA and is found along the Sierra Madre Occidental north up to Arizona. *T. floridanum* is native to south Florida and Cuba. *T. manisuroides* is known only from Tuxtla Gutierrez, Chiapas, Mexico (deWet *et al.,*1982, 1983).

## Coix and other Asiatic Genera

The Asiatic genera of the Maydeae tribe are native to an area extending from India to Southeast Asia and the Polynesian islands to Australia (Watson and Dallwitz, 1992). They include Coix L. (2n=10, 20 and 40), Sclerachne R. Br. (2n=20), Polytoca R. Br. (2n=20 and 40), Chionachne R. Br. (2n=20), and Trilobachne Schenk and Henrard (2n=20). Species from these genera are annuals or perennials and are commonly found in forest margins. Species of Chionachne and Coix also inhabit streamsides and open habitats and swamp areas, respectively (Watson and Dallwitz, 1992). Coix sp. is the most familiar genera and includes several species. The species *Coix lacryma-jobi* Linn. (Job's Tears) (2n=20) is native to Southeast Asia and exists in the wild and as cultivated races. It is also found wild in Africa and Asia and warmer parts of the Mediterranean. Coix species have been cited as having weediness potential (Watson and Dallwitz, 1992). Chionachne includes several species native to Southeast Asia. The species *C. semiteres* is cultivated as a fodder crop. Polytoca includes a few species, none of which are commonly cultivated. One species has been described for Trilobachne and is not known to be cultivated. Both genera are native to Southeast Asia (Watson and Dallwitz, 1992). Sclerachne has previously been cited to include one species (Soreng *et al.,* 2000).

## Germplasm Diversity

Maize is a cultivated crop throughout the world and accordingly germplasm resources are preserved *ex situ* in many parts of the world. However only in the Meso-American region there still exists, *in situ,* the original ancient maize that gave rise to improved varieties that are grown in all regions of the world. Most of the maize variation can be found in the Meso-American region and the northern part of South America. The great diversity of environments and conditions have created the basis for the development of maize varieties well adapted to harsh conditions of soil and climate as well as to biotic stresses. There is a close correlation among community culture, production system and the type of consumption of maize, with the diversification and variation of maize (Aguirre et al., 1998; Louette and Smale, 1998). There is a growing trend in developing countries to adopt improved maize varieties, primarily to meet market demand. In Mexico, only 20 % of the corn varieties grown 50 years ago remain in cultivation (World Watch Institute, 2000). The narrowing of genetic diversity in modern maize varieties emphasizes the importance of conserving genetic traits for future plant breeding. CIMMYT (International Maize and Wheat Improvement Centre) has taken the lead in preserving maize germplasm. It has the world's largest collection of maize accessions, with over 17,000 lines (CIMMYT, 2000). India also harbours diverse maize germplasm (Singh, 1977; Wilkes, 1981). Landraces with primitive characteristics exist in the northeaster Himalayan region and are called "Sikkim Primitives" (Dhawan, 1964).

An extensive collection of germplasm from the entire NEH region has been made by researchers at the Indian Agricultural Research Institute, New Delhi. It has been shown that the two primitive Sikkim maize strains (Sikkim Primitive 1 and Sikkim Primitive 2) were different from the primitive Mexican races (Mukherjee et al., 1971). Indian maize races have been classified under four categories i.e. primitive group, advanced or derived group, recent introduction and hybrid races (Singh, 1977). The National Gene Bank at New Delhi houses about 6,000 indigenous accessions primarily from the NEH region. Systematic and comprehensive evaluation of this germplasm is being attempted. For agronomically useful traits (Prasanna *et al.,* 2009). In addition to the races, there are several local varieties in India. The genetic variability has resulted by crossing of Indian germplasm with strains imported from other countries particularly USA (Mukherjee, 1989). It has been reported that crosses of Indian x Indian germplasm gave yield superiority of 24-43 %, whereas Indian x US dent germplasm out yielded local varieties by 58 % (Dhawan and Singh, 1961). Highest yielding single cross hybrids were obtained from crosses between Indian x USA germplasm followed by USA x USA and Indian x Indian germplasm, thus highlighting the significance of genetic divergence for obtaining higher yields (Ahloowalia and Dhawan, 1963). Dent x flint crosses involving Indian and Caribbean, and Indian and US germplasm showed highest expression of heterosis over better parent (47-54 %) (Mukherjee and Dhawan, 1970).

## Cytogenetical Aspects of Maize

Maize is the only crop that has been subjected to intensive genetic and Cytogenetical studies. According to Weifer (1952), as many 500 genes have been identified and mapped in all the chromosomes. Linkage maps have been constructed to show the relative position of numerous genes on each of the 10 chromosomes. A number of genetic studies were made possible in maize because the plant is adapted to wide range of environmental conditions and shows many differential characters. Whether cross pollination or self pollination, can be done with case and simplicity. In maize crop a larger quantity of seeds can be obtained from each cob. Many easily observed hereditary characters are available in maize for study. Maize possesses many characters in recessive nature and that these can be well exposed through in breeding as result of its being a cross pollinated crop. Studies on the effects of self and cross-pollination with maize which started early in the present century, have furnished the basis for Mendalian explanation of heterosis. Later studies with particular reference to economic characters, including the combining ability of inbred lines have helped to give a genetic understanding of such complex characters as vigour of growth and yielding ability and have made possible the development of efficient breeding techniques.

## Inheritance Studies

### Endosperm characters

A summary of the mode, of inheritance of the principle normal endosperm character is given in the table below:

| Parental type | F1 | Segregation in F2 |
|---|---|---|
| Yellow vs Colourless | Yellow or intermediate | 3 yellow : 1 colourless |
| Dominant white vs yellow | Ivory somewhat variable in shade | 3 ivory : 1 yellow |
| Brown aleurened vs colourless | Pale yellow partially dominant | 3 coloured: 1 colourless when a single factor pair is involved. |

| | | |
|---|---|---|
| Coloured alcurone (Purple or red) vs colourless. | Purple or red. | 3: 1 or 9: 7 or 27: 37 etc. depending on whether 1-5 factor pairs involved. |
| Purple vs red aleurone | Purple | 3 purple : 1 red |
| Coloured (purple or red) vs Colourless | Colourless became dominant inhibitory factor | Segregation depends on number of factor pairs involved. |
| Starch vs sweet | Starch | 3 : 1 (Starch : Sweet) |
| Starch vs Waxy | Starch | 3 : 1 (Starch : Waxy) |
| Waxy vs sweet | Starch | 9 starch : 3 Waxy : 4 sweet |
| Floury vs cormeous | No immediate effect | 1 floury : corneous |
| Normal vs Shrunken or shrivelled | Normal | 3: 1 (Normal shrunken when one factor pair is involved). |

**Chlorophyll variation**

There are many recessive heritable chlorophyll abnormalities in maize. Many factor have been located in the genetic linkage map. There are several factors in each chromosome that their interaction is responsible for the development of chlorophyll. These recessive are of two types (a) those that appear in seedling progenies and those that appear in mature plants. In a few cases the same factor modifies chlorophyll development in both seedlings and mature plants. The seedling types are frequently lethal and these results due to single gene in homozygous recessive condition.

## Plant colour

There is several different plant colours occur in maize. The plant colour and anther colour result due to the interaction of the genes for aleurone colour with genes B and PI for plant colour. A series of alleles of pericarb and cob colours are of much interest. The following factors have been assigned to each.

Red pericarp and red cob Prr
Red pericarp and white cob Prw
White pericarp and red cob Pwr
White pericarp and white cob Pww
Solid red through various shades of variation Pvv

**Glossy seedling**

Glossy seedlings are recessive to normal seedlings. The leaves have glossy appearance in the early seedling stage. The vigour of glossy seedlings is not greatly different from normals and the character may be used to deduct out cross in an inbred line.

**Linkage studies**

The genes determining the characters in maize that have been studied fell into 10 linkage groups, corresponding to the ten different chromosomes. Cytologically it is demonstrated that all the 10 chromosomes are not identical. They differ morphologically and cytologically. These are of different size, different length and also in the ratio of short arm to long arm. Ten linkage groups have been established and each group is independent of the other. The location of genes on the linkage map of the 10 chromosomes with the description of the characters produced has been well brought out.

**Xenia**

Xenia is the result of double fertilization and is depend as the immediate effect of pollen on the developing kernel, when a yellow maize pollen fertilizers an ovule of white maize, a light yellow kernel develops on the white kernel. When white maize pollen fertilizers an ovule of yellow maize, medium yellow kernels develop. This phenomenon results because the yellow colour is found only in horny starch of the endosperm. The endosperm develops after the fusion of the second sperm with the diploid polar nuclei and thus has triploid chromosome number. The yellow endosperm is conditioned by dominant gene (YY) the recessive ables (yy) produces white endosperm. Since the endosperm derives two sets of chromosomes from the polar nuclei, it will receive two genes for Y or y dependent on the character of mother plant and one gene for Y or y from the pollen. The effect of pollination of the polar nuclei with different kinds of pollen has on the developing kernels may be outlined as follows:

| Colour genes in Polar nuclei | Colour genes in Sperms | Colour genes in endosperm |
|---|---|---|
| YY + | Y | YYY deep yellow |
| YY + | y | YYy medium yellow |
| yy + | Y | yyY light yellow |
| yy + | y | yyy white |

**Inheritance of quantitative characters**

Studies on the inheritance of quantitative characters in maize were started as early as 1906. For quantitative characters, dominance is often incomplete or lacking. When dominance is complete, the expected ratios may be obtained by the expansion binomial (3 + 1) n where n is the number of allebic pairs of factors.

**Inheritance of combining ability**

Hays and Johnson (1939) after studying the inbred lines made single crosses in three groups. The crosses comprised low x low, low x high and high x high yielders. From the crosses between low x low, most of the inbreds selected proved low in combining ability. Crosses of low x high, gave both low and high combining lines whereas crosses with high x high lines gave only high combining lines indicating that the combining ability is always with high yielders.

**Protein content**

Crosses between low ′high protein lines have low protein in F1, It seems probable that many genes are responsible for inheritance of protein.

## References

Lorens, G. C., J. M. Benett and L. B. Loggale (1987). Differences in drought resistance between two corn hybrids II. Component analysis and growth rate. Agron. J., 79: 8008 - 8013.

Maiti, R. K., V.P. Singh, P. Wesche - Ebeling, E. Sanchez- Arreola, T. Hernandez and E, Aguilar- Najera (2004). Research advances on cold, drought and salinity tolerance and its mechanisms of resistance in maize – A review. Crop Res., 27(1): 1 - 29.

Menz, M. A. and A.R. Hallauer (1997). Reciprocal recurrent selection of two tropical corn populations adapted to Iowa. Maydica, 42: 239 - 246.

Mohammad, T. and T.M. Ahmad (1993). Field evaluation of maize genotypes for stalk rot resistance and yield characters. Sarhad Journal of Agriculture, 9: 337 - 339.

Montomery, K.T. (1987). Effects of six cycles of reciprocal full sib selection on several grain, stalk and root traits in BSI 1. Dissertation Abstracts International, B. Sciences and Engineering, 48: 926 - 927.

Muruli, B.I. and G.M. Paulsen (1981). Improvement of nitrogen use efficiency and relationship to other traits in maize. Maydica, 26: 63-73.

Ribaut, J.M., C. Jiang, D. Gonzalez-de-Leon. G.O. Edmeades and D.A. Hoisington (1997). Identification of quantitative trait loci under drought conditions in tropical maize, II. Yield components and marker-assisted selection strategies. Theoretical and Applied Genetics, 94: 887-896.

# 6

# Growth and Development Stages

Maize is a determinate annual C 4 plant varying in height from 1.2 to 4.0 m tall. Maize has a single stem and rarely tillers. Maize has a distinct growth form. The apex of the stem ends in the tassel, an inflorescence of male flowers. The silks are elongated stigmas that look like tufts of hair, at first green and later red or yellow. When the tassel is mature and conditions are suitably warm and dry, anthers on the tassel dehisce and release pollen. Maize pollen is anemophilous (dispersed by wind) and because of its large settling velocity most pollen falls within a few meters of the tassel. Each silk may become pollinated to produce one kernel of maize. Maize usually forms a single ear or cob. The silk is a collection of tubes that run from each potential grain on the ear. The pollen lands on each silk and fertilizies the grain. The ears are female inflorescences, tightly covered over by several layers of leaves, and so closed-in by them to the stem that they do not show themselves easily until the emergence of the pale yellow silks from the leaf whorl at the end of the ear. Plantings for silage are even denser, and achieve a lower percentage of ears and more plant matter. Maize is a long night plant and flowers in a certain number of growing degree days >10 °C in the environment to which it is adapted. The magnitude of the influence that long nights have on the number of days that must pass before maize flowers is genetically prescribed and regulated by the phytochrome system. Photoperiodicity can be eccentric in tropical cultivars, while the long days characteristic of higher latitudes allow the plants to grow so tall that they do not have enough time to produce seed before being killed by frost. These attributes, however, may prove useful in using tropical maize for biofuels. When the seed is sown, it germinates under favourable circumstances within 3 - 4 days. The radical elongates first which is followed by plumules.

Maize is a tall, determinate, monoecious, annual plant. It produced large, narrow, opposite leaves, borne alternatively along the length of stem. All maize varieties follow same general pattern of development, although specific time and interval between stages and total number of leaves developed may vary between different hybrids, seasons, time of planting and location. The various stages of maize growth are broadly divided into the vegetative and reproductive stages as follows:

## Vegetative stages

» Seedling/Sprouting stage comes about one week after sowing, and the plants have about 2-4 leaves at this stage.

» Grand growth stage also called knee height stage which plants arrives about 35-45 days after sowing.

» Tasseling/Flower initiation stage is the stage at which the tassels or male flowers appear. Generally the maize plant would have attained its full height by this stage.

### Reproductive stages

» Silking stage involving the formation of the female flowers or cobs is the first reproductive stage and occurs 2-3 days after tasseling stage. This stage begins when any silks are visible outside the husk. These are auxillary flowers unlike tassels that are terminal ones. Pollination occurs when these new moist silks catch the falling pollen grains.

» Soft-dough/Milky stage commences after pollination and fertilization is over. Grains start developing but they do not become hard. This soft dough stage is noticed by the silks on the top of the cob which remain partially green at this stage. The covering of the cobs also remains green.

» Hard- dough/Maturity stage shows that the leaves get dried; silks get dried completely and become very brittle. Harvesting is done at this stage.

## Root

Normally maize plants have three types of roots, i) seminal roots - which develop from radicle and persist for long period, ii) adventitious roots, fibrous roots developing from the lower nodes of stem below ground level which are the effective and active roots of plant and iii) brace or prop roots, produced by lower two nodes. The seminal roots consist of a primary root or radicle and several lateral roots. Since they arise at the base of the first internode of the stem immediately above the scutellar node they are called as secondary roots. These roots are important in reducing lodging. If the plant is droughty, these roots may not form adequately. The roots grow very rapidly and almost equally outwards and downwards. Favourable soils may allow corn root growth up to 60 cm laterally and in depth.

## Stem

Maize stems superficially resemble bamboo canes. The stem generally attains a thickness of 3 to 4 cm. The internodes can reach 44.5 cm. The internodes are short and fairly thick at the base of the plant. It becomes longer and thicker higher up the stem, and then taper again. Stem is thick and heavy and is filled with solid pith. The ear bearing internode is longitudinally grooved, to allow proper positioning of the ear head (cob). The stem is divided into two segments by the nodes each of which bears a bud, leaf, branch or shoot. The internodes below the ear are straight with a groove on one side, which contains the bud. The arrangement of grooves is alternate. The addition of new nodes terminates with the transformation of tip meristem into the primordium of tassel. The stems are erect, conventionally 2.0 – 3.0 m in height, with many nodes, casting off flag-leaves at every node

## Leaves

Maize forms 16 to 22 leaves per plant. Leaves form at each node and alternate (i.e., they appear on opposite sides of the plant). The lower leaves being like broad flags, generally 50 -100 cm long and 5 - 10 cm wide. The total number of leaves on a mature plant varies but it is always in odd number as the leaves do not grow in pairs. Leaves are arranged alternately on the stem, the sheath of one being wrapped around the stem so that the right half overlaps the left half. Thus the leaf sheaths and leaves are arranged in alternate right and left hand order from base to apex of the stem. At the point where the sheath joins the blade, a small collar (ligule) is formed. The fully developed leaf consist of the sheath, the blade (or lamina) and the ligule. Venation is typically parallel. The leaf is supported throughout its length by a mid rib which is nearly a thickened portion of the blade. Under these leaves and close to the stem grow the ears. They grow about 3 mm a day. The upper leaves in corn are more responsible for light interception and are major contributors of photosynthate to grain.

## Floral Biology

Maize is a monoecious plant, that is, the sexes are partitioned into separate pistillate (ear), the female flower and staminate (tassel), the male flower. It has determinate growth habit and the shoot terminates into the inflorescences bearing staminate or pistillate flowers (Dhillon and Prasanna, 2001). Maize is generally protandrous, that is, the male flower matures earlier than the female flower. Within each male flower spikelet, there are usually two functional florets, although development of the lower floret may be delayed slightly in comparison to the upper floret. Each floret contains a pair of thin scales i.e. lemma and palea, three anthers, two lodicules and rudimentary pistil. Pollen grains per anther have been reported to range from 2000 to 7500 (Kiesselbach, 1949). Within an average of 7000 anthers per tassel and 2000 grains per anther, each tassel could produce 14 x 10-6 pollen grains. Kiesselbach (1949) estimated that 42,500 pollen grains are produced per square inch of cornfield. In terms of the ratio of pollen grains produced per ovules fertilized, it appears that since each ear requires about 1000 pollen grains for fertilization, there are about 20,000 pollen grains per kernel in excess of what is actually needed if pollination were 100 percent efficient. The pollen grains are very small, barely visible to the naked eye, light in weight, and easily carried by wind. The wind borne nature of the pollen and protandry lead to cross-pollination, but there may be about 5 % self-pollination.

The female flower initially is smooth but protuberances soon form in rows. The basal protuberances are formed first and development advances towards the tip of the ears. The part above the attachment of the carpel develops a single sessile ovule, which consists of a nucellus with two integuments or rudimentary seed coats. The united carpel's, which will form the ovary wall or pericarp of the mature kernel, grow upward until they completely enclose the ovule. The two anterior carpel, which face the ear tip, form outgrowths, which develop into the style i.e into long thread, know as silks. Silks are covered with numerous hairs, trichomes which form an angle with the silk where pollen grains are harboured. The base of the silk is unique, as it elongates continuously until fertilization occurs. The cobs bear many rows of ovules that are always even in number. The female inflorescence or ear develops from one or more lateral branches (shanks) usually borne about half-way up the main stalk from auxillary shoot buds. As the internodes of the shanks are condensed, the ear remains permanently enclosed in a mantle of many husk leaves. Thus the plant is unable to disperse its seeds in the manner of a wild plant and instead it depends upon human intervention for seed shelling and propagation.

## Pollination and Fertilization

In maize, the pollen shed is not a continuous process and usually begins two to three days prior to silk emergence and continues for five to eight days. The silks are covered with fine, sticky hairs which serve to catch and anchor the pollen grains. Pollen shed stops when the tassel is too wet or too dry and begins again when temperature conditions are favourable. Under favourable conditions, pollen grain remains viable for only 18 to 24 hours. Cool temperatures and high humidity favor pollen longevity. Under optimal conditions the interval between anthesis and silking is one to two days. Under any stress situation this interval increases. Fertilization occurs after the pollen grain is caught by the silk and germinates by growth of the pollen tube down the silk channel within minutes of coming in contact with a silk and the pollen tube grows the length of the silk and enters the embryo sac in 12 to 28 hours. Pollen is light and is often carried considerable distances by the wind. However, most of it settles within 20 to 50 feet. Pollen of a given plant rarely fertilizes the silks of the same plant. Under field conditions 97 % or more of the kernels produced by each plant are pollinated by other plants in the field. Fertilization of ovules begins about one third of the way up from the base of the ear.

## Spikelets and Flowers

Like all other cereals, maize plant bears its flower in spikelets. The male spikelets are the separate units of which the tassel is mode. They can mostly easily be recognized before flowering just as tassel emerges from the leaves at the top of the plant. The female spikelets are seldom seen as such because they are located by the fact that each one of them ordinarily produces one grain. The female (pistillate) inflorescence, a spike, produces pairs of spikelets on the surface of a highly condensed rachis (central axis, or "cob").

### Male spikelet

The male (staminate) inflorescence, a loose panicle, produces pairs of free spikelets each enclosing a fertile and a sterile floret. The typical or normal staminate spikelet consist of a short rachilla bearing four bracts, the upper most two of which have shorter bearing four bracts, the upper most two of which have shorter bearing branches in their axils known as lemma. The two lowest bracts having empty axils as in most grasses are known as glumes. The short branch in the axil of each lemma has on the side opposite the lemma a single bract, the palea and it is terminated by the flower. The immature flower is thus enclosed between lemma and palea whose edges overlap.

### Flower

The flower consists of three stamans, two lodicules and the rudiment of the pistil. The position of the lodicules indicates that they represent a whorl of flower part outside that formed by the three stamens. They are regarded as remnants of perianth which was present in some ancient ancestors. Their function is to swell within few minutes at the time of flowering and push back the lemma allowing the anthers to emerge on their slender elongating filaments.

### Female spikelets

The pistillate spikelet has the same vegetative parts and the same as the staminate, but its rachilla is so short and its bracts so modified in shape that the homologies between the two spikelets are evident only when section of the two are examined. The upper flower of the spikelet consists of a pistil, the rudementary stamen and two conspicuous and apparently functioning lodicules. The entire lower flower of the spikelet is aborted in most kinds of corns but in a few varieties, it, quite regularly has a functional pistil and the spikelet thus produces two grains instead of one. The female flower is tightly covered over by several layers of leaves, and so closed in by them to the stem that they don't show themselves easily until emergence of the pale yellow silks from the leaf whorl at the end of the ear. The silks are the elongated stigmas that look like tufts of hair initially and later turn green or purple in colour. Each of the female spikelets encloses two fertile florets, one of whose ovaries will mature into a maize kernel once sexually fertilized by wind-blown pollen.

### Inflorescence

Maize is a monoecious plant producing male and female inflorescence separately. The male inflorescence is known as tassel and the female inflorescence is called cob or ear. The male inflorescence or tassel is produced terminally on the main axis. The male inflorescence is much compact and much branched panicle varying in size with variety. It may have stiff branches producing an erect inflorescence or the axis and branches may be more flexous producing a dropping panicle. The male inflorescence generally flowers 1 - 3 days before the female. The cob is surrounded by involucres of leaves known as husk. On the thick pithy spindle which later lignifies are formed 4 to 11 longitudinal ridges on each of which two ears are seated in adjoining rows. Each ear contains two blossoms, one of which is rudimentary and has no ovary. The ovaries bear thread shaped pistils known as "Silks", which hang out of the husks at the time of blossom. The silks become visible and receptive only 3 - 4 days later than the time of blossoming of tassel. The silk is bifurcated at the top and is receptive along most of its length. In about a month's time tassel initiation takes place from the growing point. From the beginning of the tassel initiation, the maize plant normally requires 5 or 6 weeks to reach stage of pollen shedding and silking. After about a week of tassel initiation, ear initiation starts about the $6^{th}$ node below the tassel.

By the end of the growing season, the kernels dry out and become difficult to chew without cooking them tender first in boiling water. The kernel of maize has a pericarp of the fruit fused with the seed coat, typical of the grasses, and the entire kernel is often referred to as the seed. The cob is close to a multiple fruit in structure, except that the individual fruits (the kernels) never fuse into a single mass. The grains are about the size of peas and adhere in regular rows round a white pithy substance, which forms the ear. An ear contains from 200 to 400 kernels, and is from 10–25 cm in length. They are of various colours viz., blackish, bluish gray, purple, green, red, white and yellow. When ground into flour, maize yields more flour, with much less bran, than wheat does. However, it lacks the protein gluten of wheat and therefore makes baked goods with poor rising capability.

**Pollination and fertilization**

Pollination by wind and gravity is the general rule. All the pollen is shed within few hours. Abundant pollen is produced from the opened flowers. An individual plant produces as much as 50,000 pollen grains and each one of them is effective in producing a grain. The silks are receptive as soon as they emerge and remain receptive for some time. The pollen grains fall on the stigma hairs and adhere and with the available moisture, they germinate and the pollen tube may penetrate the body of the style directly and enters the tissue of the style and reaches the ovary by following one of the stands of vascular tissue. The number of pollen grains by a tassel is probably somewhere between 2 and 5 millions. The pollen grains are borne in anthers which split open and shed their pollen into the wind. It is light and is carried to a considerable distance by the wind. One full shedding plant can provide enough pollen for several ears. Pollen shed continues for several days (5 to 8 days) with the peak production coming about the $3^{rd}$ day. Pollen shed begins from the middle of the central spike of the tassel, spreads out over the whole tassel in succeeding days and ends with the shed from the tips and bases of the lower branches.

**Fruit**

The maize cob is a syncarpous fruit. After fertilization, the styles wither away and the grains develop as a broad, obviate wedge shaped caryopsis. The grains are borne in an even number of rows along the length of the cob. There is wide variation in the length, diameter and number of grains per cob. The length varies from 2.5 to 50 cm the diameter from 3 to 7.5 cm.

Cobs are usually 15 to 39 cm long. The average grains per cob ranges from 300 - 1000 and 4 to 48 grains in each row. Maize grains usually weigh around 25 - 40 g per 100 kernels. Colour of the grain generally depends on the nature of husk and the aleuron layer. The white, yellow and red colours are located in the thick cell walls of the husk and the blue and violet colour in the aleuron layers. Young ears can be consumed raw, with the cob and silk, but as the plant matures (usually during the summer months) the cob becomes tougher and the silk dries to inedibility. Certain varieties of maize have been bred to produce many additional developed ears. These are the source of the baby corn used as a vegetable in Asian cuisine.

**Seed**

On an average there are 750 to 1000 potential kennels arranged in an even number of rows around the cob. There are 16 or 20 rows of seeds per cob and about 50 seeds per row. The grain or seed is principally consisting of a thin covering of several layers known as pericarp. The nucellus consists of a single layer of cells bounding the endosperm, which makes up the main bulk of the seed. The outer layer of endosperm is aleuron layer, a layer of cells in which most of the protein is stored, but the bulk of the endosperm consist of large cells packed with starch grains. The endosperm of maize is of two types namely hard flinty and soft starchy. The hard flinty endosperm is opalescent in appearance and contain higher proportion of protein while starchy endosperm is more floury and white in appearance and much softer in texture. However, the proportion and disposition of these two kinds of endosperm in the seed varies with the variety. The embryo is situated in a small portion, at the base of the seed and in close contact with endosperm. The scutellum encloses, the root - apex protected by its sheath the coleorhiza and the stem apex which is enclosed by coleoptile. The embryo itself is rich in fats, minerals, proteins and sugars.

The individual maize grain is botanically a caryopsis, a dry fruit containing a single seed fused to the inner tissues of the fruit case. The seed contains two sister structures, a germ which includes the plumule and radical from which a new plant will develop, and an endosperm which will provide nutrients for that germinating seedling until the seedling establishes sufficient leaf area to become autotrophy.

The germ is the source of maize vegetable oil. Total oil content of maize grain is 4 % by weight. The endosperm occupies about two thirds of a maize kernel's volume and accounts for approximately 86 % of its dry weight. The endosperm of maize kernels can be yellow or white. The primary component of endosperm is starch, together with 10 % bound protein (gluten) and this stored starch is the basis of the maize kernel's nutritional uses.

### Seed dispersal

Seed dispersal of individual kernels naturally does not occur because of the structure of the ears of maize. It, as a thoroughly domesticated plant, has lost all ability to disseminate its seeds and relies entirely on the aid of man for its distribution (Stoskopf, 1985). The kernels are tightly held on the cobs. In case ears fall to the ground, so many competing seedlings emerge that the likelihood that any will grow to maturity is extremely low.

### Mating systems

Under natural conditions, maize reproduces only by seed production. Pollination occurs with the transfer of pollen from the tassels to the silks of the ear. About 95 % of the ovules are cross-pollinated and about 5 % are self-pollinated (Poehlman, 1959), although plants are completely self-compatible. There is no asexually reproductive maize. Cell/tissue culture techniques can be used to propagate calli and reproduce tissues or plants asexually. However, with maize cells and tissues these techniques are difficult (Hoisington et al. 1998).

## Methods of Reproductive Isolation

Various reproductive isolation methods are used by plant breeders and by seed producer to produce genetically pure seed. The isolation of crops using separation distances and physical barriers are common techniques for restricting gene flow and ensuring seed purity for maize seed production. Various experimental practices used to maintain reproductive isolation maize are:

i. **Maintaining isolation distance:** Cross-pollination is controlled in seed lots by separating different lines. The Minimum Seed Certification Standards require a minimum 400m isolation distance for maize seed production (Tunwar and Singh, 1988).

ii. **Detasseling:** Mechanical removal of tassels is another effective method in corn. By remaining the tassel containing the pollen produce male flowers, it is possible to eliminate entirely the source of genetic material from the male flower that can be transferred via pollen.

iii. **Use of barrier crops:** The use of barrier crops have been recommended to decrease the distance, but at the same time achieve the required separation. Barrier crop provides physical barriers for pollinators. The African tall maize has been used as a barrier crop, as it is taller than maize crop, and is dense and thus provides an effective barrier in preventing cross pollination outside the experimental plot. In line with the above, requirements of

400 meters as the isolation distance along with 13 rows of African tall maize has been adopted for conducting confined field trial of genetically engineered maize.

*Zea mays,* Maize, Sweet corn

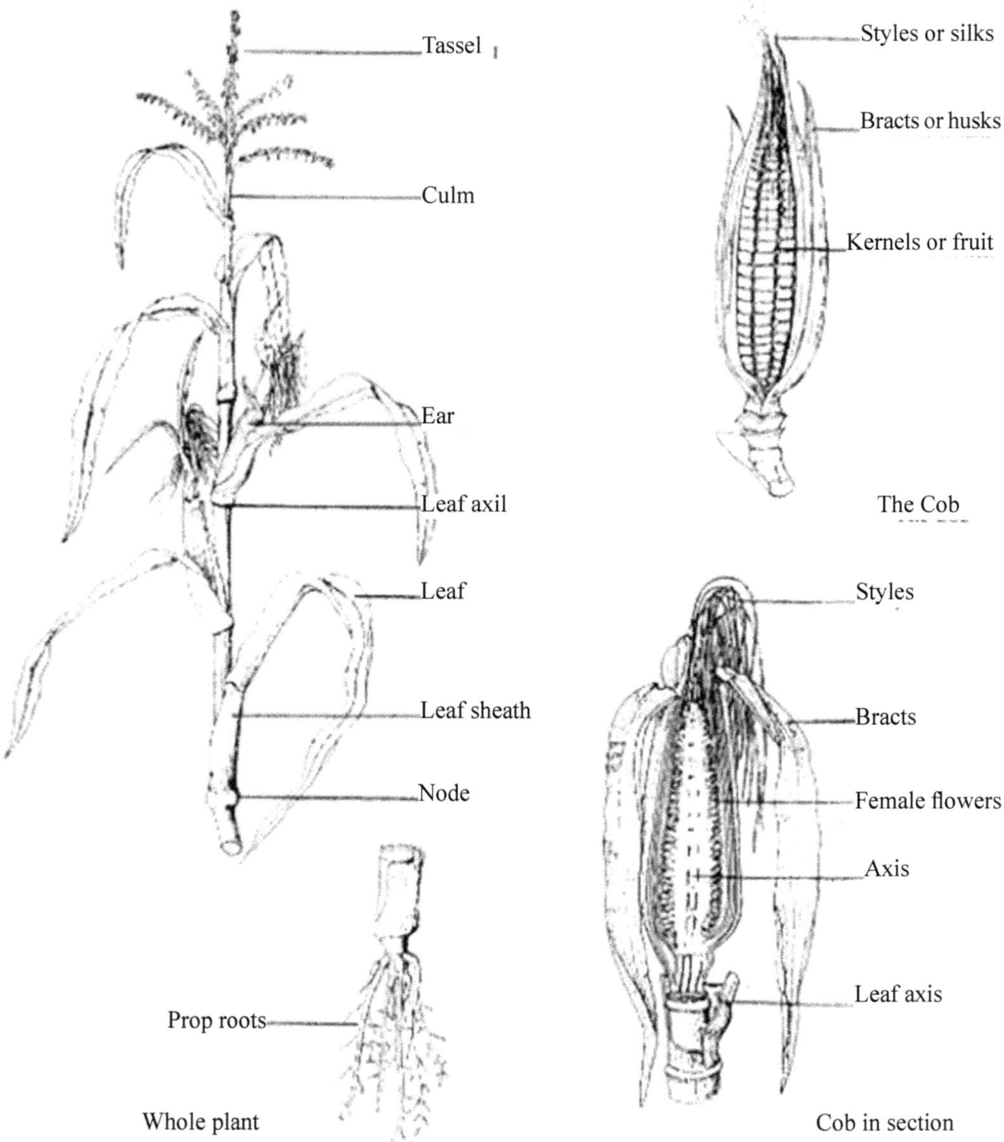


Knowledge of growth and development may help in trouble shooting problems related to abnormal growth caused by pest problems or inappropriate cultural practices. There is two of the most widely used staging systems for corn development. It shows a timeline relating corn growth and development to normal heat unit accumulation and calendar dates during the growing season (Table 1).

**Table 1. Growth stages of maize**

| Vegetative Stage | Reproductive Stage | |
|---|---|---|
| VE: Emergence | R1 | Silking |
| V1: First leaf collar | R2 | Blister |
| V2: Second leaf collar | R3 | Milk |
| V3: Third leaf collar | R4 | Drought |
| V(n): nth-node collar | R5 | Dent |
| VT: Tasselling | R6 | Physiological maturity |

### Five leaf stage (V5)

The young maize plant was cut off 1 cm above ground. The plants produced almost a full yield. The only difference between this and the control was that it flowered one week later. A comparison between hybrids found that the ultra quick hybrid types showed a yield loss of 20 % whereas the long season hybrids had almost none. This results from more leaves being produced by the long season hybrids. Ultra quick hybrids produce 19-20 leaves compared to the 20-23 leaves of the longer growing season hybrids. Environmental stress such as hail damage, drought stress, nutrient or moisture stress may lengthen the vegetative stage but will shorten the reproductive stage.

### Twelve leaf stage (V12)

At this stage the plants showed considerably higher yield loss. Again the longer season hybrids exhibited a smaller yield loss than the short season hybrids. This is also due to the increased number of leaves of the longer season hybrids. The yield loss at this stage was about 40 % for the long season and 55 % for the short season hybrids. With 50 % of the leaves removed the yield loss was 19 % (long season) and 27 % (short season) respectively.

### Pollination period (R1-R3)

At this stage 50 % of the silks of the maize plants are visible and the maize plant is at its greatest risk due to hail damage. Most of the nutrients (N, P and K) have been taken up by the plant. None of the photosynthates (sugars) have been translocated to the cobs. The "factory" of the plant is fully developed. A 100 % leaf loss due to hail will result in about a 100 % yield loss. The leaves are the starch factory of the plant and without leaves the factory cannot function. The most critical time for hail damage, as with drought stress, are the two weeks before and after pollination.

### Grain filling

During the first stages of grain filling i.e. the blister and soft dough stages, yield loss is at its highest. Yield loss progressively lessens until the plant is physiologically mature.

### Hail damage

Early killing frost in the winter season may damage immature corn and reduce yield. The effect of frost damage to corn depends on the severity of the defoliation, stalk damage and stage of growth. The most common damage from hail is loss of leaf area, although stalk breakage and bruising of the stalk and ear may also be severe. Note that the largest yield losses result from defoliation damage that occurs during the late vegetative stages and the reproductive stages (silking and tasselling). Defoliation at early growth stages does not affect yield the same way as it does at later growth stages because much of the plant's total leaf area is not yet exposed. Extensive defoliation of plants in the 10-leaf growth stage (or V8, eight leaf collar stage) does not result in a large yield loss because only 25 per cent of the leaf area is exposed, and the plant can easily recover from early damage. On the other hand, severe damage to plants during tasselling results in a large yield loss because, by that time, 100 per cent of the leaf area has been exposed and cannot be replaced.

| **Growth stage** | **Yield loss (Per cent)** | | | | | |
|---|---|---|---|---|---|---|
| | **10 % area destroyed** | **20 % area destroyed** | **40 % area destroyed** | **60 % area destroyed** | **80 % area destroyed** | **100 % area destroyed** |
| 7 leaf | 0 | 0 | 0 | 4 | 6 | 9 |
| 10 Leaf | 0 | 0 | 4 | 8 | 11 | 16 |
| 13 leaf | 0 | 1 | 6 | 13 | 22 | 34 |
| 16 leaf | 1 | 3 | 11 | 23 | 40 | 61 |
| Tasselled | 3 | 7 | 21 | 42 | 68 | 100 |
| Late Milk | 1 | 3 | 10 | 21 | 35 | 50 |
| Dent | 0 | 0 | 3 | 10 | 17 | 24 |

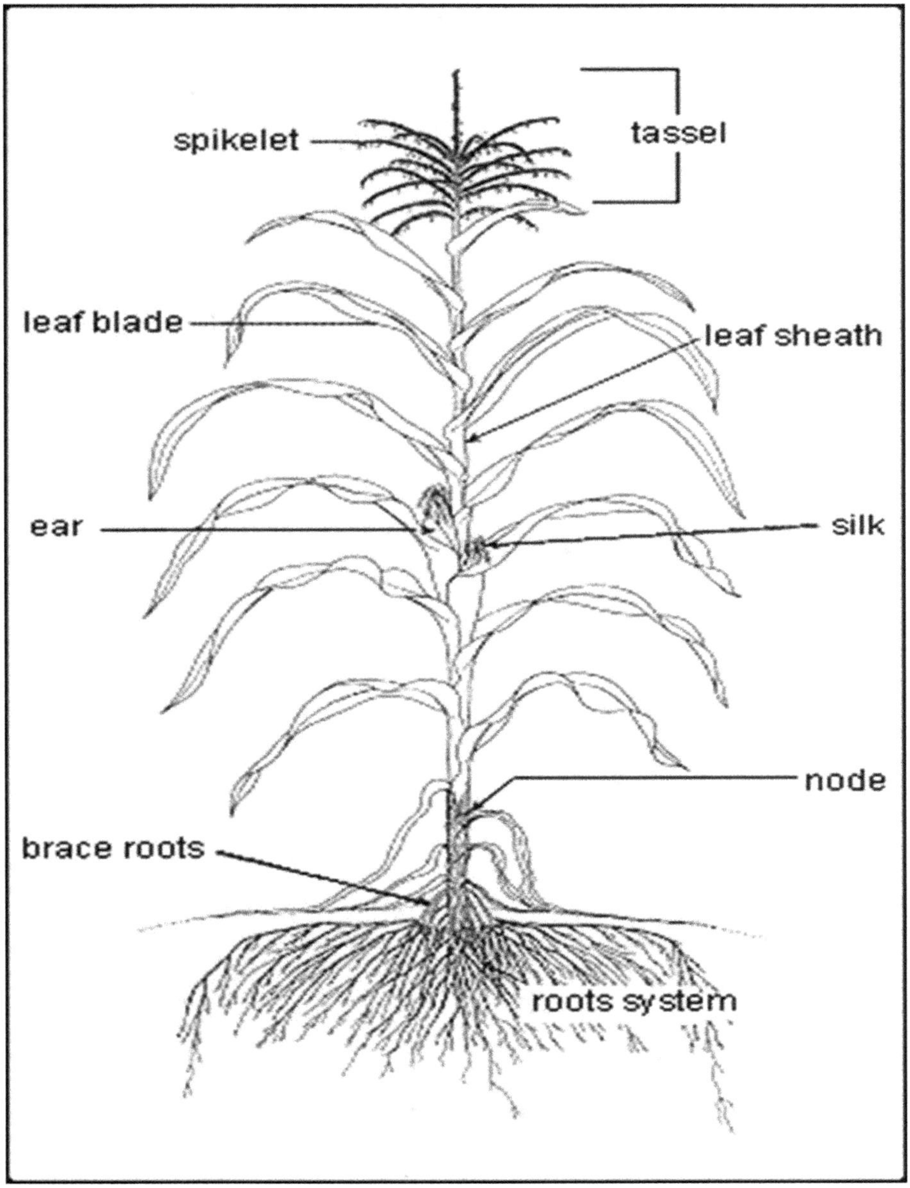
spikelet
tassel
leaf blade
leaf sheath
ear
silk
node
brace roots
roots system

## References

Kiesselbach, T.A. (1999). The structure and reproduction of corn. In: 50th Anniversary Ed. Cold Spring Harbor Lab. Press, Cold Spring. p. 18.

Prine, G.M. (1971).A critical period for ear development in maize. Crop Sci., 11: 782 - 786.

Ritchie, S.W., J.J. Hanway and G.O. Benson (1992). How a corn plant develops? Iowa State Univ. Coop. Ext. Serv., Ames. Spec. Rep. 48.

# 7

# Nutritive Value of Maize Grain

The water free portion of maize kernel contains starch 77 %, sugar 2 %, protein 9 %, fat 5%, pentosans 5 % and ash 2 %. There is greater variation among different varieties of maize in the content of protein and fat. The proportion of protein may be as high as 15 % and low as 6 %. In the maize kernel, about 80 % of the protein is in the endosperm. The germ though constituting only 1/10th of grain contains about 1/5th of the total protein. Maize kernels contain three types of proteins viz., prolamines soluble in alcohol chiefly 'zein', blobulins soluble in neutral salt solutions and glutalin.

The entire protein in maize is known to be low in quality because zein is deficient in tryptophan and lysine which are essential amino acids. Maize is the richest of the cereals for fat with exception of oats. More than 80 % of the fat is concentrated in germ. Yellow maize shows more of vitamin A while white grain possesses lesser quantity, which is due to 'cryptoxanthein'. Maize is fairly rich in Thiamine and more riboflavin is found in maize than rice. The normal maize contains the following principles and energy compared to other cereals like rice, sorghum, cumbu and wheat.

**Table 1. Composition per 100 g of edible portion of maize (dry)**

| Constituents | Content | Constituents | Content |
|---|---|---|---|
| Moisture | 14.9 g | Minerals | 1.5 g |
| Protein | 11.1g | Carbohydrates | 66.2 g |
| Fat | 3.6 g | Calcium | 10 mg |
| Fibre | 2.7 g | Iron | 2.3 mg |
| Phosphorus | 348 mg | Potassium | 286 mg |
| Thiamine | 0.42 mg | Sodium | 15.9 mg |
| Magnesium | 139 mg | Carotene | 90 ug |
| Amino acids | 1.78 mg | Sulphur | 114 mg |
| Copper | 0.14 mg | Vitamin C | 0.12 mg |
| Calories | 342 | Riboflavin | 0.10 mg |

**Table 2. Principle constituents**

| Crops | Moisture (%) | Protein (%) | Fat (%) | Minerals (%) | Fibres (%) | Carbohydrates (%) | Calories per 100 g |
|---|---|---|---|---|---|---|---|
| Rice | 13.3 | 6.4 | 0.4 | 0.7 | 0.2 | 79.0 | 348 |
| Sorghum | 11.9 | 10.4 | 1.9 | 1.6 | 1.6 | 72.6 | 349 |
| Cumbu | 12.4 | 11.6 | 5.0 | 2.3 | 1.2 | 67.5 | 356 |
| Wheat | 12.8 | 11.8 | 1.5 | 1.5 | 1.2 | 71.2 | 346 |
| Maize | 14.9 | 11.1 | 3.6 | 1.5 | 2.7 | 66.2 | 342 |

In general the normal maize does possess protein on par with wheat and definitely higher than rice. Next to cumbu, maize has higher percentage of fat than other cereals. It also possesses higher minerals than rice. Not much of difference has been noticed in calories per 100 g. of those cereals except cumbu, where the calorific value is slightly higher. Compared to rice, maize grains, definitely possess higher percentage of these principles. The mineral content in Maize and other cereals are given in table 2 (Aykroyd *et al.* 1963). It is seen that maize grains possesses higher magnesium (144 mg), phytin (phosphorous - 306 mg) and potassium (286 mg) than other cereals and it is superior to rice in all other elements except ionisable iron and copper.

## Vitamins

The analysis of vitamin content in these crops reveal that maize contains the largest quantity of Vitamin A compared to all the important cereals. Further it also contains high thiamine than rice, sorghum and cumbu.

**Table 3. Vitamins in maize and other cereals (Aykroyd *et al.*, 1963)**

| Crop | Vit. A (IU) | Thamine (mg) | Riboflavin (mg) | Nicotinic acid (mg) |
|---|---|---|---|---|
| Rice | 0 | 0.21 | 0.9 | 3.8 |
| Sorghum | 79 | 0.37 | 0.28 | 1.8 |
| Cumbu | 220 | 0.33 | 0.16 | 3.2 |
| Wheat | 108 | 0.45 | 0.12 | 5.0 |
| Maize | 1502 | 0.42 | 0.10 | 1.4 |

**Table 4. Minerals in maize and other cereals**

| Mineral content | Rice | Sorghum | Cumbu | Wheat | Maize |
|---|---|---|---|---|---|
| Calcium (mg) | 9 | 25 | 42 | 41 | 10 |
| Magnesium (mg) | 38 | 140 | 125 | 139 | 144 |
| Oxalic acid (mg) | 1 | 10 | 21 | 8 | 6 |
| Phytin (mg) | 83 | 172 | 141 | 238 | 306 |
| Ionisable iron (mg) | 1.4 | 1.4 | 4.4 | 2.5 | 0.9 |
| Sodium (mg) | 10.0 | 7.3 | 10.9 | 17.1 | 15.9 |
| Potassium (mg) | 117 | 131 | 30 | 284 | 286 |
| Copper (mg) | 0.33 | 1.78 | 0.55 | 0.49 | 0.19 |
| Sulphar (mg) | 79 | 54 | 147 | 128 | 114 |
| Chlorine (mg) | 13 | 44 | 39 | 47 | 33 |
| Acid balance N/10 | 72 | 27 | 56 | 77 | 101 |

**Table 5. Essential amino acids in different cereals (g per 100 g of protein)**

| Contens | Rice | Sorghum | Cumbu | Wheat | Maize |
|---|---|---|---|---|---|
| Protein | 6.13 | 11.75 | 12.31 | 10.75 | 9.00 |
| Arginine | 8.00 | 6.24 | 7.52 | 5.28 | 4.90 |
| Histidine | 1.28 | 2.40 | 2.56 | 1.44 | 2.40 |
| Lysine | 3.36 | 3.20 | 3.68 | 2.24 | 3.00 |
| Tryptophan | 0.00 | 0.48 | 1.44 | 1.12 | 0.70 |
| Phenylalamine | 5.60 | 4.16 | 3.68 | 5.28 | 5.80 |
| Cystine | 1.28 | 2.40 | 2.08 | 2.24 | 1.70 |
| Methionine | 1.28 | 2.24 | 2.08 | 1.92 | 1.30 |
| Threonine | 3.36 | 3.04 | 3.36 | 2.56 | 4.1 |
| Leucine | 0.0 | 10.40 | 9.28 | 6.56 | 14.60 |
| Isoleucine | 0.0 | 4.80 | 4.96 | 3.52 | 4.20 |
| Valine | 5.44 | 4.80 | 5.44 | 4.00 | 5.70 |

(Aykroyd *et al.* 1963; Joginder singh, 1978)

The essential amino acid content in maize as compared to rice, sorghum, cumbu and wheat are given in table 5. In the normal maize, the general level of most of the amino acids is low, compared to other cereals. Therefore, there is an urgent need to improve the amino acid content especially of the lysine and tryptophan, which are the most important amino acids required for the growth. Fortunately Drs. Mertz, Bates, and Nelson of Purdue University, USA, have discovered the Opaque-2 and Floury 2 genes which are responsible for high lysine and tryptophan content. Opaque-2 genes has a tendency to completely alter the amino acid profile in other ways to an extent that it is comparable to milk. The amino acid profile of the opaque-2, Floury 2 compared with normal maize is given in Table 6. It is quite evident that opague-2 and floury 2 possess high lysine and tryptophan content than the normal maize. Attempts have been made in Tamil Nadu to improve the local maize strains with high lysine and tryptophan content through hybridization with Mexican varieties carrying opaque-2 gene. The progenies with high lysine and tryptophan are in advanced stages of trial.

**Table 6. Amino acid composition for defatted whole kernels of Opaque 2, floury 2 and normal maize (g / 100 gm of protein)**

| Composition | Normal | Opaque | Floury |
|---|---|---|---|
| Lysine | 3.0 | 5.0 | 4.8 |
| Tryptophan | 0.7 | 1.3 | - |
| Histidine | 2.4 | 3.5 | 2.9 |
| Arginine | 4.9 | 7.2 | 6.3 |
| Asparatic | 9.2 | 8.8 | 10.5 |
| Theronine | 4.1 | 3.8 | 4.1 |
| Serine | 5.6 | 4.7 | 5.2 |
| Glutamic acid | 22.6 | 17.2 | 18.5 |
| Proline | 9.6 | 8.4 | 8.9 |
| Glycine | 4.7 | 5.1 | 4.7 |
| Alamine | 9.2 | 6.7 | 8.0 |
| Cystine | 1.7 | 2.0 | 1.6 |
| Valine | 5.7 | 5.2 | 5.7 |
| Methionine | 1.3 | 2.2 | 2.7 |
| Isoleucine | 4.2 | 3.4 | 4.0 |
| Leucine | 14.6 | 9.6 | 12.0 |
| Tryosine | 5.2 | 4.2 | 4.6 |
| Phenylalmine | 5.8 | 4.4 | 5.2 |
| Protein (%) | 9.0 | 10.5 | 17.0 |

Since the introduction of opaque-2 gene, maize is considered as a coarse cereal par excellence. The opaque-2 gene doubled the lysine content in maize. The introduction of opaque-2 gene also altered the amino acid profile in other ways to an extent that maize is comparable to milk. Certain other 'floury' genes also showed this ability. However, the early mutants with opaque-2 genes showed several undesirable characteristics including low yield, increased vulnerability to pests and soft endosperm. Improved varieties incorporating the opaque genes are now in sight with a hard cornerus endosperm that will compare fully with normal commercial hybrids in all other respects while retaining the supreme advantage of the superior amino acid profile. Such maize by and large could fairly be called the cereal par excellence.

## Utilization

Maize is one of the important cereal crop which is cultivated and consumed in nearly all the countries of the world. Maize is put to varied use. The important uses are as follows:

### a. Food

Maize is consumed directly as food and in India 85 % of the maize production is used as food. The commonly known forms are broken grains boiled and cooked as rice meal, chapattis, porridges of various forms, boiled or roasted green ears, break-fast food such as 'Uppuma' etc., and pop-corn. The flour is extensively used in bakery products and also in baby foods. In India commonly grown flint or semi flint varieties are put to varied uses.

### b. Forage

Maize plant is extensively used as a cattle feed both as fresh green fodder, silage and as dry fodder. Maize fodder has much higher digestability than sorghum, bajra and other non-legume forage crops. Maize plant does not possess any of the hodrocyanic acid or prusic acid and therefore the crop can be harvested as green at any stage of the crop and fed to cattle. The most advantageous stage of harvest of green fodder would be, when the crop is in mid-dough stage.

### c. Cattle feed

Maize grains are extensively used as animal feed in most of the developed countries. Maize grains are fed to cattle, chicken, ducks and other domestic animals. The high carotene content of yellow maize is considered very useful in imparting yellow colour to the yolk of the egg and yellow tinge to the cows' milk. No other concentrate is yet known to substitute maize in this regard. Maize hearts can also be used as cattle feed when ground and fortified with mollases.

### d. Industries

Maize is extensively used in the industry for the production of starch and other products. The starch industry using the wet milling process produces starch, maize oil and gluten for feed. The starch in turn is further processed to produce a series of products like dextrose, corn syrup, etc., The brewing industry produce a series of industrial chemicals. Maize hearts are also processed to produce products like furfural. Maize oil when refined is highly valuable as a cooking media. It is a poly-unsaturated oil.

The waxy maize hybrids and high amylose hybrids give recovery of amylopectin and amylose starches. These starches have specialized use in food, textile, laundry, paper and paper board industry.

Corn steep water has been used as culture medium for growing *Pencillium notatum* the organism that produced the antibiotic drug penicillin.

## Manufacture of Alcohol

Brazil has already started extracting alcohol from maize, to be used as a substitute for petrol. In Brazil 'gasohol' now being sold as petrol consist 80 : 20 of petrol and alcohol.

### Production of whisky from maize

Maize is also used in the production of whisky. The Bourbon whisky is almost produced from maize. Scotch whisky is prepared from barley. The dried residue distillers are also used as feed. In addition a large number of other products like high fructose corn syrup, organic cum, synthetic Co-polymers, organic and chemical feed stocks, paper ryon, plastic, textiles, adhesives, dyes, synthetic, rubber, resin, power fuel for torpedoes, artificial leather and boot polish are also produced. The following basic corn products can be made available through a healthy Agricultural Industry to support the corn production.

### a. Food products

The common food products are corn meal, corn chips, confectionary products, corn starch, Corn flakes, baking products, corn oil, corn molasses, salad oil, corn syrup, margarine, lysine (high protein food products), dextrose, canned sweet corn, dextrin pop corn, frozen roasting.

### b. Other products

Paper and paper board, live stock and poultry feed, antibiotic drugs, soaps, paints, varnishes, rubber substitutes and starch for laundry.

## References

Bressani, R., G. Arroyave and N.S. Scrimshaw (1953). The nutritive value of central American Corns. I. Nirtrogen, ether extract, Crude fiber and minerals of twenty -four varieties in Guatemala. Food Res., 18: 261-267.

Krause, V.M. (1988). Rural-urban variations in limed Maize consumption and the mineral content of tortilla in Guatemala centre for studies of sensory Impairment, aging and metabolism, Guatemala, Mc Gill University, Montreal, Canada.

Ranhotra, G.S. (1985). Nutritional profile of corn and flour tortillas. J. Food Sci., 49: 1202-1203.

Saldana, G .and H.E. Brown (1984). Nutritional composition of corn and flour tortillas. J. Food Sci., 49: 1202-1203.

Scott, C.E. and A.L. Eldridge (2005). Comparison of carotenoid content in fresh, frozen and canned corn. J. Food Composition Analysis, 18: 551–559.

Tripathi, S.B. and V.P. Singh (1982). Relationship between dry matter accumulation and nutrient composition at different growth stages of *kisan* composite maize. Indian J. Agric. Res., 16: 149 -152.

Wade, J.H. (1981). Characterization of aroma components of corn. Ph. D. thesis, University of Georgia, Athens.

# 8

# Climatic Effects on Maize Production

Maize in the temperate regions is what sorghum is in tropical areas. This is an interesting situation since maize is believed to have originated in the tropical semi-arid region. The unsuitability of maize to its original habitat may be due to the acquisition of a higher water need for efficient performance during its evolution. Unlike small grain cereals, where post –flowering defoliation has little effect on final yield, foliar photosynthesis is very important for the final yield of maize (Carr and Milbourn, 1976). As the duration of the full leaf of maize extends to the grain – setting stage, dry matter accumulation in maize can virtually proceed at maximum photosynthetic efficiency throughout the life of the crop. Its growth form is such that the optimum leaf-area index is obtained very early in its life-cycle and as explained above maintained till harvest (Bland, 1971). Because of this, maize is the best performer of all cereals under clear weather and under irrigated conditions. For the same reason the production potential of maize is reduced under cloudy conditions. Maize is most susceptible to flooding in active growth stage from 22-45 days after sowing (Pal and Varade, 1984).

Maize is rarely grown in areas where the mean summer temperature is less than that of 19°C and where the mean night summer temperature is less than 13°C. The lowest seasonal precipitation in the maize area is 200 mm. However, available evidence indicates that maize growing, purely as a rainfed crop, may be risky in regions with mean annual precipitation of 400 mm, and that even in areas receiving 600 mm of rain irrigation may have to be resorted to for achieving maximum yields. The greatest intensity of cultivation of maize is in the region bound by the July isotherms of 20° and 26°C. The higher limits of temperature and rainfall which preclude maize cultivation are not definitely known. Shaw (1955) points out that the native natural grasses may give some indication regarding maize culture. Generally, in places where the native grass is short, maize cultivation may be hazardous and that in regions of tall native grasses maize cultivation may be preferred. The climatic requirement of maize has some similarly to that of sorghum and certain distinct features are enumerated below.

## Temperature

Corn can survive brief exposures to adverse temperatures such as low end adverse temperatures being around 0°C and high end temperatures being around 44.5°C. Growth decreases once temperatures dip to 5°C or exceed 35°C. Optimal temperatures range for growth varies between day and night, as well as over the entire growing season. For example, optimal day time temperatures range between 25°C and 32.8°C, and optimal night time temperatures range between 16.7°C and 23.5°C. The optimal average temperatures for the entire crop growing season, however, range between 20°C and 22.8°C. Even though corn seeds germinate and grow slowly at about 10°C, the first spring planting dates usually begin when the average air temperatures reach 12.8°C, and soil temperature at seed depth is more favourable for seedling growth. Poor germination resulting from below-normal temperatures is the greatest hazard of planting too early. The growing point of germinating seedlings remains below or near the soil surface and usually is not vulnerable to freeze damage until plants reach the five-to six-leaf collar stage. By this time, corn is about greatly decreases. The loss of leaves from frost generally does not seriously injure small plants, although such loss may delay plant development.

Temperatures less than 4.4°C reduce photosynthesis, even if the only symptom is a slight loss of leaf colour. Frost injury symptoms may appear on leaves even when night time temperatures do not fall below the mid 30s. Radiational cooling can lower leaf temperatures to several degree below air stalks before physiological maturity (black layer formation) in the fall, sugars usually continue to move from the stalk into moisture may be high because of high grain moisture at the time of frost and slow drying rates following premature death. High temperature stress during ear formation, reproduction, and grain can reduce yield, but temperatures less than 37.8°C usually do not cause much injury if soil moisture is adequate. Under rainfed conditions, corn usually begins to stress when air temperatures exceed 32.2°C during the tasseling –silking (pollination) and grain filling stages. Corn yield may reduce at 35°C temperature or higher during pollination and grain fill. Extended periods of hot, dry winds may cause tassel blasting and loss of pollen.

## Precipitation

A corn crop typically uses 50 – 55 cm of water during the growth season. Water requirements of corn vary according to the stage of development, as shown in table 1. Corn reaches its peak water use during pollination, when plants are silking. Excessive rainfall, resulting in flooding and ponding of soils, may cause serious injury to a corn crop depending on its stage of development. The major stress caused by flooding and ponding is a lack of oxygen needed for the proper function of the root system. When plants are very small (prior to the six-leaf collar stage), they generally are killed after about five or six days of submersion. Death occurs more quickly (within two to four days) if the weather is hot, because warm temperatures speed up the biochemical processes that use oxygen, and warm water has less dissolved oxygen. Cool weather, on the other hand, may allow pants to live for more than a week under flooded conditions.

**Table 1. Water use rates for corn at different growth stage**

| Growth State | Water use rate (cm / day) |
|---|---|
| Prior to 12-leaf stage | < 0.50 |
| 12-leaf | 0.60 |
| Early tassel | 0.70 |
| Silking | 0.75 |
| Blister kernel | 0.65 |
| Milk | 0.60 |
| Dent | 0.50 |
| Full dent | 0.45 |

As soon as plants reach the six-to eight-leaf collar stage and plant's growing point is above the soil surface, plants can tolerate a week or more of standing water-not necessarily without harm. In older plants, total submersion may increase diseases incidence, and plants will suffer from the reduced root growth and function for some days after water recedes. Tolerance of flooding generally increases with plant age, but reduced root function resulting from a lack of oxygen is probably more detrimental to yield before and during pollination than during rapid vegetative growth or grain fill. Nutrient uptake is also reduced in soil saturated by excessive rainfall. Not only does poor aeration inhibit effective root development and function, but the anaerobic conditions associated with saturated soil promote denitrification. Frequent rainfall can also cause nitrate leaching.

For crop moisture to be adequate, available soil moisture must be more than sufficient to meet the atmospheric evaporative demand. On windy, hot sunny days with low humidity, evaporation demand on a crop is high and a high amount of available soil moisture must be present if the crop is high, and a high amount of available soil moisture must be present if the crop is to avoid stress. Under cloudy skies, high humidity, and cooler temperatures, atmospheric evaporative demand is low, and plants can get by lower amounts of available soil moisture.

The soil must provide a corn crop with enough water to offset the amounts lost through transpiration. If these needs are not met, the plant will wilt. The effect of drought on corn grain yield from four consecutive days of visible wilting is presented in Table 2. Corn is fairy tolerant of dry soils during late vegetative stage. Mild drought during June may even be as surface soil dry, and the crop benefits from the greater amount of sunlight that accompanies dry weather. From the two weeks before through the two weeks following pollination, corn is very sensitive to drought. However, dry soils during this period may cause serious yield losses. Most of these losses result from pollination failure, and the most common cause is the failure of silks do not receive pollen. Thus, the kernels are not fertilized and will not develop. Drought later in grain fill has a less serious effect on yield, though root function may decrease and kernels may abort or not fill completely.

**Table 2. Effects of drought on corn yield during several stages of growth**

| State of development | Yield reduction (%) |
|---|---|
| Early vegetative | 5 – 10 |
| Tassel emergence | 10 – 25 |
| Silk emergence | 40 – 50 |
| Blister | 30 – 40 |
| Dough | 20 – 30 |

* After four consecutive days of visible leaf wilting.

Drought stress often leads to plant nutrients stress. The shallow depths where fertilizer is placed are dry under drought situations, which may limit nutrient uptake.

## Seeding practice

The attainment of minimum soil temperature at seeding depth of 15°C is the criterion for sowing of maize in places with Mediterranean type of climates (Arnon, 1972). Sowings can be delayed till June if short-duration varieties are to be used. In the United States, there is a progressive decrease in the sowing of maize from South to North at the rate of about 20 km/day. The sowing starts in late January in the South to North at the rate of about 20 of congenial temperatures, planting dates are less critical in the southern state of the USA. Optimum period is end-February to mid-March. However, if root worms are to be avoided the optimum period is in April. In India, Chele (1960) found that at Jalandhar (Pungab) the maize crop sown in the middle of July gave higher yields of grain and stalk than those sown even 10 days earlier or later. At Karim nagar (Andhra Pradesh), maize should be sown round about 24 May during *kharif* to get maximum yield and to escape stem-borer attack and lodging. During rabi the best day was around 5 November. Maize is likely to suffer if sown in January, March, April and September, while sowings in July and October would lead to average yields (Vittal Rao and Mahboob Ali, 1961). At Karnal, sowings on 1st July consistently gave increased yield over those of sown on 15th June and 15th July (Relwani, 1962). At Delhi, hybrid maize should be planted 10 to 15 days before onset of the monsoon to ensure good emergence and good early stand and that planting after the break of monsoon in the first week of July or later reduces the yields (Gautam et al., 1964).

At Delhi, hybrid maize planted by 23rd June ensures higher yields and better utilization of nitrogenous fertilizers (Shah and Sharma, 1970). At Pantnagar, sowing done on 30th June led to increase in the number of plants with more than one cob per plant and higher yields of grain and stalk than other plantings (Sharma and Mungal, 1966). At Kanpur, sowings of open pollinated maize varieties with the break of monsoon (around 1st July) were better than pre-monsoon (15th June) sowings with irrigation (Pathak et al., 1971). However, high-yielding hybrids and composites of maize should be planted around 22nd May in Kulu valley where the local varieties could be planted by about first June for getting best yields. The crop in later plantings suffered from lack of adequate rainfall in silking stage, while in earlier plantings low minimum temperatures delayed the growth. The number of days to silking in varieties was not significantly affected by dates of planting. The yield of local variety increased with plantings from mid-May to beginning of June. However, sowings of the same after beginning of June reduced the yield (Sandhu et al., 1971).

Jain (1973) has observed that optimum time of sowing of maize in various tracks, viz., 10th April at Kalimpong (West Bengal), 23rd May at Bilaspur (Himachal Pradesh), 25th May at Hyderabad (Andhra Pradesh), 10th June at Jabalpur (Madhya Pradesh), Godhra

(Gujarat) and Pantnagar (Uttaranchal) and around 20 to 25th June at Delhi and Udaipur, is about 10 to 20 days earlier than the normal onset of monsoon. Jain (1973) attributes this to the plants avoiding excess rainfall during the earlier stages of growth when they are very sensitive to soil aeration and also the avoidance of soil moisture stress during flowering stage. Jain (1973) advocates that for the early summer crop planting should be completed as far as possible before mid- January but not later than the end of January to avoid the crop coming into flowering during hot months of March and April. For rabi crops sowings from mid October to mid – November and for rainfed crops in the hilly regions sowings from March to May recommended to get higher yields. Whether maize should be seeded prior to or after the onset of rains would depend on the soil temperature regime (Carr and Milbourn, 1976). If the soil temperature at seeding depth is the optimum range and rainfall would lower if further seeding prior to onset of rains is to be preferred. If the pre-monsoon soil temperatures are supra-optimal for germination, planting after onset of rains may be preferred. However, as yields are reduced when the vegetative growth is too quick, planting after onset of rains under sub-optimal soil temperatures or prior to the rain under supra-optimal soil temperatures may be preferred.

## Germination

The minimum temperature of germination is 10°C (Rush and Neal, 1951). The optimum temperature varies with moisture content and ranges from 30° to 35°C under adequate moisture. Effect of low temperature could be compensated by high soil moisture (Schneider and Gupta, 1985). For uniformity of initial stand, a minimum temperature of 17° to 20°C is required (Arnon, 1972). Jain (1973) considers that the ideal temperature for maize germination is 21°C. In the higher latitudes such as in North Europe, North America, etc., (where maize is sown in the spring season) the low soil temperatures prevent an early seeding. Attempt has therefore been made to evolve varieties capable of germinating at lower temperatures of 6° to 7°C. However, capacity to germinate at lower temperatures is not necessarily related to vigorous growth of the seedlings. On the other hand, earlier germination in these tracts may subject the crop to freezing temperatures associated with late-spring frosts. Also the emerging plants would be subject to a shorter photoperiod which hastens flowering and reduces crop duration. Under these conditions, there is need for the use of day – neutral types (Van Dobben, 1962). Like sorghum, maize growing on cold soils is attacked by pathogens. Too late a sowing to ensure optimum germination temperature may bring in its wake up problems of attack by pests and diseases. The above environmental features affect the seedling time.

## Vegetative growth

The range of temperature for the growth of maize is from 9° to 46°C with the optimum around 34°C. Thus all these values, viz., the maximum, minimum and optimum, are 4° to 6°C higher than those of barley or wheat (Hall, 1945). Rate of emergence of maize leaves is linearly dependent on range of diurnal temperature and day length during leaf emerging period and could be related to mean temperature by a second degree equation (Bishnoi and Sedgley, 1985). The root growth of maize is severely affected by temperature (Hall, 1945). Little root growth occurs at 10°C, while even at 17°C the root growth is less than 1.5 mm per day. Beyond 17°C the root growth is less than 1.5 mm per day. Beyond 40°C root growth is again severely affected. The optimum is around 34°C. Heyne and Laude (1940) considered air temperatures above 35°C as not congenial for maize, especially its root metabolism. Wallace and Breesmann (1937) considered the optimum air temperature to be around 25°C. Jain (1973) considers that the ideal temperature for crop growth is around 32°C. Allmaras *et al.* (1964) have drawn attention to the influence of soil temperature on maize top growth, which increases continuously with soil temperature from 15° to 27°C with an optimum around 20°C. Like sorghum, maize is also a short-day plant and the flowering is accelerated by short photoperiods.

Maize is more resistant to drought. The crop can recover from early season drought. Maize is sensitive to freezing temperatures except in the very early stages and can recover from the effects of frost if the same occurs when the plant is less than 15 cm high (Leonard and Martin, 1963). Very young seedlings are less susceptible to high temperatures (Heyne and Laude, 1940). Maize can also recover from hail damage if the same occurs in early stages. From the above one can state that adverse climatic conditions, if at all it is to be encountered by the crop, are better got through in the early stages. This feature explains the practice of early sowing of maize with the onset of rains, even at the risk of early drought due to late onset of regular rains. Similarly, early sowings to avoid pest and disease attacks may be resorted to even at the risk of freezing temperatures during the early crop period.

Law and Cooper (1974) found that soil temperatures during the vegetative period has a bearing on the time of tasseling (Shaw and Thom, 1951) while soil moisture stress delays tasseling. Shaw (1955) observed that drought in vegetative period than in the very early stages, stunted the crop, delayed silking and led to an increase in sterile tassels. However, both soil moisture stress and temperature have little effect in the time between tasseling and silking (Shaw, 1955). McCalla *et al.* (1939) found that heat unit summations above 10°C can explain 40 to 70 % of the variations in the growth rate of the maize crop.

## Reproductive Stage

In this stage, temperatures above 32°C as well as frosts can reduce yield. Hail is most harmful between jointing and silking state (Shaw, 1955). Soil moisture stress prior to silking, during silking and after silking reduces grain yields, respectively, by 25, 21 and 50 % (Denmead and Shaw, 1960). Strong winds during this period are also harmful. In maize, pollen begins to lose viability at temperatures > 36 °C (Decker *et al.,* 1986). Development of maize is influenced by photoperiod. With decreasing photoperiods hastening of silking takes place (Allison and Daynard, 1979). The photoperiod is affecting maize in the field at the time of tassel initiation (Kinery *et al.,* 1983). Although photoperiod and temperature interact to influence maize development, the thermal models are adequate and equally effective in predicting corn development upto silking (Kinery *et al.,* 1983). However, for predicting physiological maturity the 'Function of Temperature Index' is better (Daughtry *et al.,* 1984). Thus tassel initiation could be assessed non-destructively from data of tassel emergence and temperature.

## Crop-weather relationship

The day degree accumulations above 10°C for maize development are 120 for seedling establishment, 480 for ear formation, 630 for tip of tassel appearance, 775 for anthesis, 1070 for dough stage, 1215 for denting of kernels, 1360 for dented kernels, and 1510 for maturity for long-duration hybrid maize have been indicated by Nield and Seeley (1977). Review of work done estimating maize yields from meteorological data has been made by Leonard and Martin (1963) and Shaw (1955). The main feature that emerges is that as the maize crop is capable of recovering from early seasonal setbacks, the weather prevailing prior to and after tasseling had greater influence on yield and that the effects of temperature and soil moisture stress may either act together or in opposite direction depending on the level of adequacy of the other elements under consideration. From these, Shaw (1955) concludes that the ideal climate for the crop is winter having mean temperature regime of 18° to 23°C with 100 mm of precipitation per month in the crop growth period except the last month. Patil et al. (1971) found that in the Ghataprabha Project Area, Karnataka, the local variety 'Arbhai Local 47' took 20 days less to reach the silking stage. Thus, while the hybrids were superior to the composites, which were in turn superior to the local variety, delayed sowings combined with heavy rains during silking reduced the yields of all varieties and the local yielded as much as the hybrids and composites. These investigations highlight the need for avoiding rainy weather during silking.

Krishnamurthy *et al.* (1972) found higher dry matter production per plant in kharif season than in summer due to the favourable effects of weather on the emergence, development, extension and period of activities of leaves. They found that the differences in yields between the varieties could be explained in terms of leaf-area index and the leaf-area duration. Singh and Leng (1970) studied the growth of hybrids and varieties sown on 15 September, 15 February and 15 June at Patnagar. Hybrids out yielded other varieties in all the three seasons, *viz., zaid, kharif* and *Rabi*. The yield of hybrids was low during *zaid*, while the yield of varieties was higher during *kharif* season. In the temperate regions, variations in grain yield are related to the varied quantum of photosynthesis during or after flowering, while in the tropics the yield variations are due to differences in grain number (Carr and Milbourn, 1976). At Anand (Gujarat), grain yield (Ganga safed 2) was best related with afternoon vapour pressure during silk emergence to dough stage and total biomass with temperature range during emergence to silk emergence stage (Reddy et al., 1995). Ali and Janabi (1980) observed that the temperature explained most of the variation within plant characters for different season and sowing dates. Gardner et al. (1981) found the summation of canopy temperatures and determined the growth stages in maize. Kiniry et al. (1983) reported a linear relationship between tassel initiation and degree days. Bonhomme et al. (1982) found that amount of the photosynthetically active radiation (PAR) trapped by maize was dependant on leaf area index and was higher at silking than maturity. Lee and Jung (1988) reported PAR use efficiency of 4.87 g/MJ for dry matter production. Baktash (1985) observed significant association of grain yield with air temperature and relative humidity at flowering and pollination stages.

**Table 3. Photo thermal index and heat use efficiency of maize crop**

| Treatments | Photothermal index | | | Heat use efficiency ($g^oc^{-1}day^{-1}$) |
|---|---|---|---|---|
| | Emergence to 75 % tasseling | 75 % tasseling to 75 % Silking | 75 % Silking to maturity | |
| 16th May | 28.70 | 30.50 | 27.81 | 35.03 |
| 31st May | 29.76 | 29.58 | 27.65 | 38.05 |
| 15th June | 30.15 | 28.72 | 31.81 | 46.76 |
| 30th June | 27.96 | 28.35 | 32.44 | 42.07 |
| SEM ± | 0.06 | 0.11 | 0.05 | 0.96 |
| (CD 5 %) | 0.19 | 0.35 | 0.17 | 3.14 |

(Kushwaha *et al.*, 2010)

At Pantnagar, minimum significant GDD accumulation at various growth stages were observed at 15$^{th}$ June sowing date. A maximum significant HUE of 49.76 was observed at 15$^{th}$ sowing date (Kushwaha *et al.,* 2010).

**Low temperature stress on maize**

Maize can survive when exposed to adverse temperatures as low as 0°C and as high as 44°C. Growth starts slowing down when the temperature dips below 4.9°C. However, the growth of maize crop is restricted when exposed for the prolonged period to extreme low and high temperatures. The extreme cool temperature affects the maize growth in a number of ways right from emergence till flowering and seed setting. Low temperature at planting greatly affects germination. Prolonged exposure to low temperature during the vegetative phase results in reduction in plant height, yellowing of the leaf and leaf tip firing due to the death of leaf tissue. Cold stress occurring at reproductive phase severely affects flowering and results in reduced tassel size, delayed anthesis, pollen grain death, reduced silk size and seed setting, thereby, considerably affecting the yield and in extreme conditions no yield.

**Injuries due to low temperature**

Exposure to low temperatures can result in two types of injuries. The first is chilling injury that occurs in the temperature range of 0-15°C. This may results in a variety of physiological disruptions in germination, flower, seed development, yield and storage life. Minor chilling at non-lethal temperatures is normally reversible. The second type of injury is called freezing injury. This type of injury occurs when the external temperature drops below the freezing point of water. This may result in interacellular and/or extracellular freezing in plants. In intracellular freezing damage the protoplasmic structure and the ice crystals kill the cell once they grow large enough to be detected microscopically. In extracellular freezing, the protoplasm of the plant becomes dehydrated because a water vapour deficit is created as ice crystals are formed in the intercellular spaces. However, the major limiting factor for winter maize growing regions do not fall below freezing point for a prolonged period for a cause freezing injury.

**Stress during various growth stages**

Cold injury is a physiological damage caused by temperature between 0°C and 16°C (Lylons, 1973). The degree of injury depends on the temperature and the duration of exposure. Wilting and discoloration of the leaves and reduced growth are the most common symptoms. Severe chilling results in killing of the plants or plant parts. Chilling at germination can cause damage to cells as they rupture at imbibitation due to crystal formation. At low temperatures during growth period, the chlorophyll in the juvenile leaves becomes damaged thereby rendering a lighter colour to the plants. If there is chilling at latter stages the leaves will become dark brown and brittle. Chilling may leads to yield losses and lower product quality because of delayed ripening of the maize plant.

**Planting to emergence**

Stress in the period from planting to seedling emergence is usually related to soil temperature, soil moisture, soil aeration conditions, or the interaction among them. Low temperature alone is not likely to enforce a stress on the seedling, but only delays its emergence. However, wet weather along with cold temperature favours development and activity of some soil pathogens that can produce disease in young seedlings and may lead to poor stand due to seed rot and seedling blight. At a temperature between 5 and 15°C and particularly around 10°C, imbibed and germinating seeds are killed by soil fungi (Tatum and Zuber, 1943).

Average seedling mortality was 36 and 21 % at 4 and 6°C, respectively. No damage was found at 8 and 10°C (Segeta, 1964). Young seedlings were killed by a 6 day exposure to 1°C and 8-day exposure to 2.5°C (Wheaton, 1963). Chilling caused ultra structural changes in the meristematic cells of primary roots (Klasova, 1980). Large genetic variations for chilling injury is found when seedlings exposed to 2°C for 6 days (Miedema, 1982). The cold injury caused a slight discolouration of the roots, primary root killing and even killing of the base of mesocotyl. The injury increased with the developmental stage of the seedlings. The imbibed seeds were observed to be the least sensitive. Young maize seedlings were injured at a temperature below 6°C. The severity of damage depends on the temperature and its duration, developmental stage and genotype. Further, seeds of the flint type are in general less sensitive than dent.

**Vegetative growth phase**

During the vegetative growth phase, leaves are most vulnerable to chilling as compared to other organs because of the chilling sensitivity of the chloroplast. There was variation in inbreeds and hybrids for chilling stress during this phase. Hybrids showed more tolerance and were least affected as compared to inbreeds during vegetative growth phase.

**Low temperature on plant processes**

Exposure of maize seedlings after emergence to prolonged low temperature results in poor growth and affects physiological/biochemical processes at 0.3°C and low light intensity in 7 day-old maize seedlings. After chilling, the seedlings were transferred to 21°C to study the effect on physiological and biochemical processes. Leaf injury began to develop after 36 hours of chilling and leaf expansion was markedly reduced. Leaf segments of chilled plants showed increased ion leakage and increased oxygen uptake presumably by uncoupling of oxidative phosphorylation. However, leaf injury in plants chilled for 36 hours disappeared and leaf expansion, oxygen uptake, and ion leakage returned to control levels, 3 to 5 days after transfer to 21°C. This did not occur in plants chilled for 72 hours and the injury was irreversible.

Minimum temperature for shoot development in maize is 32°C. However, the shoot can emerge from the pericarp at 8°C, but its elongation is ceased after some time (Lehenbauer, 1914). The shoot growth increase in temperature above 18° C, due to higher extension rate of mesocotyl (Miedema, 1982). A similar reduction in mesocotyl growth was observed at a diurnal fluctuation of 35/20, 30/15 and 25/10°C (Buckle and Grant, 1974). The shoot growth at constant temperature from sowing to emergence was found to be 23 days at 10°C, 8 days at 15°C, 4 days at 21°C, and 2 days at 32°C (Miedema, 1982). The effect of temperature on the rate of primary root growth was found to be in a nearly linear relationship between 10 and 25°C (minimum temperature 9 °C and optimum 30°C). Elongation rates range from 0.1 mm/hour at 10°C and 3mm/hour at 30°C (Blacklow, 1972). The elongation of primary root was slow and ceased after 2 days when seedlings were reared at 20°C and exposed to a temperature ranging from 2 to 14°C (Crowford and Huxter, 1977).

The optimum temperature required for leaf expansion in maize plants ranged from 7 to 30°C (Tollenar et al., 1979). After the transfer of green seedlings from 20 to 10°C, only leaves and predominantly basal part of the leaf formed at 10°C were choloritic. The cholorofic content in fully emerged green leaves was not lowered by exposure to 10°C (Albera, 1969). Yellowing in seedlings at a day night temperature of 10/16 °C was green. Chlorotic seedlings become completely green within 4 days at 20/10°C (Miedema, 1982). At 16°C, chlorophyll accumulation and carotenoid content strongly decreased when the light intensity was increased from about 12 - 175 $Wm^{-2}$. But at 26°C, chlorophyll was hardly affected by the light intensity. Chlorotic seedlings rapidly greened when temperature was raised above 26°C, temperature was raised above 26°C, indicating that chlorophyll synthesizing mechanism was not damaged (William and Nayler, 1967). Seedlings chilled in a conditioned greenhouse at 2 to 4°C for 60 hours developed transverse chlorotic bands on the leaf blades, 5-10 days after the chilling treatment (Sellshop and Salmon, 1928). Similar necrotic cross bands and other leaf damage in maize seedlings was observed when it was subjected to 4°C temperature in the dark for 3 days. Most of the injury disappeared when transferred to normal temperature. Irreversible leaf damage occurred on exposure of maize seedlings at 4°C for 6 days (Miedema, 1982).

Reproductive stage is the most critical stage in maize development for any type of stress to occur. A combination of moisture and low temperature stress during this period can substantially reduce final grain yield. Any kind of severe crop stress in the early grain filling and maturation stage can reduce final yield as much as 3 - 4 % per day, with the reduction becoming less as the plant approaches physiological maturity. During this period both tassel and silk were affected. The common effects on tassel are reduced size of the tassel, pollen grain death, anther detached from tassel branches and no pollen dispersal/anther dehiscence due to non- bursting of pollen. The common effects on silk are reduced silk size, proliferation due to non-fertilization of the silk, no grain setting in silk and overgrowth of silk due to non-availability of pollen. Low temperature reduces water uptake and transpiration in two ways. Increase in viscosity of water due to low temperature decreases the water flow. Increase in viscosity of water due to low temperature from a temperature decreases the permeability of roots. The water flow though the epidermis and cortex of excised primary roots at 10°C was about one-fourth of that at 20°C (Brouwer, 1965). Low temperature at different plant growth stages from germination onwards restrict nutrient absorption from soil and cause slow growth.

## Growing degree day (GDD) maturity rating system

The growing-degree-day (GDD) maturity rating system is based on heat units. It is more accurate in determining hybrid maturity than the days-to-maturity system because growth of the corn plant is directly related to the accumulation of heat over time rather than the number of calendar days from planting. The GDD system has several advantages over the days-to-maturity system. The GDD system provides information for choosing hybrids that will mature reliably, given a location and planting date; allows the grower to follow the progress of the crop through the growing season; and aids in planning harvest schedules. The GDD calculation method most commonly used for corn in the United States is the 86 / 50 cutoff method. Growing degree days are calculated as the average daily temperature minus 50.

$$\text{GDD} = (\text{T. Max} + \text{T. Min} + 2) - 10$$

If the maximum daily temperature (T. Max) is greater than 30°C, 30 is used to determine the daily average. Similarly, if the minimum daily temperature (T. Min) is less than 10˚C, 10 is used to determine the daily average. The high cutoff temperature (30˚C) is used because growth rates of corn do not increase above 30˚C. Growth at the low temperature cutoff (10˚C) is already near zero, so it does not continue to slow as temperatures drop further. Growing degree days are calculated daily and sum over time to define thermal time for a given period of time.

As with any system, the GDD system has several shortcomings. Growing-degree-days ratings of hybrids with similar days-to maturity ratings don't always agree, especially if the hybrids are from different companies. Some seed companies start counting GDDs from the day of planting, while others begin from the day of emergence. When this occurs, similar maturity hybrids may vary by 100 to 150 GDDs-the average GDDs required for emergence. Some companies use entirely different mathematical methods to calculate GDDs. Although most companies use the "30/10 cutoff method" described earlier, others use different methods to calculate GDDs. Also, under certain delayed planting situations and stress conditions, GDD requirements for maturity may be reduced significantly. Heat unit requirement of rabi maize crops was less than kharif maize crops for all physiological phases (Varughees and Iruthayaraj, 1995).

In maize, variation in temperature is the primary cause of annual variation in its development time. Hence, maturity of maize genotype has been described by the total thermal units for silking or maturity (Medereski et al., 1973). The requirement of thermal unit or growing degree days for various phonological events of winter maize sown at different dates in India vary greatly (100 – 440 growing degree days) (Narwal et al., 1987). The phonological phase, growing degree days where influenced neither by the maize cultivars used nor by the spacings because of the maize hybrids were of similar duration (Narwal et al., 1986). Low temperature increases the number of days between planting to tassel initiation and silking to maturity (Warrington and Kanemasu, 1983) whereas high temperature shortens the intervals of various phonological phases of maize (Coligado and Brown, 1975).

At Rajendra Nagar (Andhra Pradesh), the grain yield can be estimated using satellite image derived NDVI in large areas. The relationship was found to be logarithmic during pre $NDVI_{max}$ and was linear during post $NDVI_{max}$ period (Jayasree et al., 2013).

According to FAO, maize yields currently average 1.5 t/ha in Africa, slightly more than 3 t/ha in Latin America and 1.7 t/ha in India. FAO indicates that grain yields have been recorded as follows: 5–6 t/ha (dryland) and 8–10 t/ha (irrigated). For silage, at 68–70 per cent moisture content, yields of 20 t/ha (dryland) and 42 t/ha (irrigated) have been recorded.

## Yield Gap and Yield Potential

In the developing world, most farmers have to accept low yields, as they are unable to consider the use of improved production methods because they operate at small-scale subsistence levels. Yield gap analyses will draw farmers' attention to lost production potential under the prevailing climatic conditions in their respective environments and which production practices (soil fertility, agronomic measures, cultivar selection, and the like) need to be improved. Yield differences among regions as shown in Table 5 should provide the incentive to manoeuvre toward yield improvement (Yield potential refers to the highest yield achievable on farmers' fields – with the use of improved seed (high yield, tolerance to diseases and pests), appropriate levels of nutrients, water and weed control).

According to Ofori et al. (2004), the difference between the actual and potential yield of a typical maize variety grown during the major cropping season (April through July) on a farm in Ghana over a nine-year period was just over 4 t/ha (that is, the actual yield varied from 0.9 to 1.4 t/ha and the potential for that season should have been 5.5 t/ha). The April–July rainfall varied from 570 to 790 mm over this nine-year period.

Corn hybrids grown in Illinois have planting-to-harvest GDD requirements ranging from 2,200 to 2,400 for early hybrids grown in the northern part of the state to 2,800 to 2,900 for late hybrids grown in the southernmost part of the state. A full-season hybrid for a particular area generally matures in several hundred fewer GDD. Thus, a full-season hybrid for northern Illinois would be one that matures in about 2,600 GDD, while for southern Illinois a hybrid that matures in 2,900 GDD or more would be considered full-season.

Medium-maturity hybrids require 100 to 200 fewer GDD than full-season hybrids. This GDD "cushion" reduces the risk of frost damage and also allows some flexibility n planting time. It is usually not necessary to replace a medium-maturity hybrid with one maturing in fewer GDD unless planting is delayed into June.

**Table 4. Approximate GDD needed to reach different growth stages of a corn crop (planted at the normal time, using a hybrid that requires 2,700 GDD to reach maturity)**

| Stage | GDD from planting | Stage | GDD from planting |
|---|---|---|---|
| VE | 115 | V13 | 995 |
| V1 | 155 | V14 | 1,045 |
| V2 | 235 | V15 | 1,095 |
| V3 | 315 | V16 | 1,140 |
| V4 | 395 | V17 | 1,180 |
| V5 | 475 | V18 | 1,220 |
| V6 | 555 | VT (tassel) | 1,350 |
| V7 | 635 | R1 (silk) | 1,400 |
| V8 | 715 | R2 (blister) | 1,660 |
| V9 | 795 | R3 (milk) | 1,925 |
| V10 | 845 | R4 (dough) | 2,190 |
| V11 | 895 | R5 (dent) | 2,450 |
| V12 | 945 | R6 (mature) | 2,700 |

Research has shown that the number of GDD required for the corn crop to reach particular stages of development tends to be fairly consistent. Table 4 shows the predicted GDD required to reach each vegetative (V) and reproductive (R) stage for a hybrid that requires a total of 2,700 GDD from planting to physiological maturity. These numbers are approximate, especially for R stages, which are not particularly exact. But they should work reasonably well to help predict when, under average temperatures, a crop will reach certain stages.

In some recent work in Indiana and Ohio, researchers found that the GDD requirement for corn hybrids decreased when planting was later than May 1$^{st}$. For each day that planting was delayed after May 1$^{st}$, the reduction in GDD requirement was about 6.5 GDD. Thus, a 2,700 GDD corn hybrid planted on May 20$^{th}$ would require only 2,700 – (20 x 6.5) = 2,570 GDD. While the actual decrease in GDD varied somewhat among years, the fact that there is an expected decrease indicates that changing to a shorter season hybrid when planting is delayed should rarely be done. This decrease in GDD requirement, however, usually comes at the cost of decreased yield; planting on time is still an important goal.

## Agroclimatology of the crop

Climate is interrelated with other production factors and should be understood either as a resource to be managed or a factor that needs to be manipulated. Sustainable use of soil, capital and labour should be balanced with use of climate and weather information.

The response of the maize crop to climate depends on the physiological makeup of the hybrid/variety being grown. Yield differences are the result of the genetic composition of the hybrid, the environmental conditions under which the crop is grown, and the infestation by crop pests (diseases, insects and weeds). The final yield will depend on hybrid selection, soil fertility, soil water and control of crop pests.

Each farmer has to select the hybrids that are most suitable for the climatic region in which his or her farm is located. In temperate regions of the world, the length of the frost-free season dictates the hybrids that are suitable, because the maize plant cannot withstand temperatures below about –2°C. Growing degree-day and heat-unit indexing systems have been developed for most temperate regions of the world, so that the maturing time required by each maize hybrid can be matched with growing degree-day or heat-unit ratings for the frost-free growing season in each climatic region. In tropical regions of the world it is the rainy season onset that dictates the selection and optimum time to plant maize hybrids. They need to be selected to match the anthesis period to the time that the soil moisture is likely to be most adequate, unless there is water available to irrigate the crop when necessary.

The soil fertility needs to have an optimum balance of the three major nutrients, which are nitrogen, phosphorus and potassium, and the necessary micro-nutrients, such as boron, calcium, magnesium, manganese and molybdenum. The farmer should have the soils on his or her farm assessed for the three major nutrients every year or two to determine how much fertilizer should be added to maximize maize production. This is an important management practice in both temperate and tropical regions. The farmer should be aware of fertilization requirements and procedures with respect to the soil nutrient levels, the growth stage, the crop variety, the targeted yield and the agronomic practices.

In order to maximize maize yields, soil moisture should be maintained above 50 % of the available water capacity in the rooting depth of the soil profile throughout the growing season. This is not always possible in either temperate or tropical climatic regions as rainfall can be very scarce and sporadic in some years. It is essential, however, to have at least adequate soil moisture at the time of anthesis in order to have a full set of kernels on the ear at harvest time. This is the time when supplemental water through irrigation would be most beneficial.

**Table 5. Yield potential (t/ha) relative to (current yield) in developing world (Prabhu et al., 2000)**

| Ecological environment | Highland/ transitional | Mid-altitude/ subtropical | Tropical/ lowland |
|---|---|---|---|
| East and South-East Asia | 5.0 (3.5) | 8.0 (3.0) | 5.5 (2.2) |
| South Asia | 5.0 (0.7) | 7.0 (2.6) | 4.5 (1.4) |
| West Asia/North Africa | 4.5 (3.2) | | |
| Sub-Saharan Africa | 5.0 (0.6) | 7.0 (2.5) | 4.5 (0.7) |
| Latin America and Caribbean | 6.0 (1.1) | 10.0 (4.0) | 5.0 (1.5) |

**Table 6. Dominant constraints to bridging the yield gap between potential and actual yields in the developing world (Bellon, 2001)**

| Ecological environment | Highland/ transitional | Mid-altitude/ subtropical | Tropical/ lowland |
|---|---|---|---|
| East and South-East Asia | (a) Limited technological options<br>(b) Banded leaf and sheath blight<br>(c) Borers (*Chilo* spp.) | (a) Drought/moisture stress<br>(b) Soil acidity<br>(c) Downy mildew<br>(d) Borers (*Chilo, Sesamia* spp.) | (a) Limited superior early germplasm |
| South Asia | (a) Low and declining soil fertility<br>(b) Limited technological options<br>(c) Turcicum blight | (a) High temperature<br>(b) Drought/moisture stress<br>(c) Turcicum blight<br>(d) Borers (*Chilo, Sesamia* spp.) | (a) Limited superior early germplasm<br>(b) High temperature<br>(c) Drought/moisture stress<br>(d) Downy mildew<br>(e) Borers (*Chilo, Sesamia* spp.) |
| West Asia/North Africa | | (a) High temperature<br>(b)Drought/moisture stress | |
| Sub-Saharan Africa | (a) Low and declining soil fertility<br>(b) Limited technological options<br>(c) Turcicum blight<br>(d) Rust | (a) Low and declining soil fertility<br>(b) Grey leaf spot<br>(c) Streak virus<br>(d) Weevils<br>(e)Borers (*Chilo, Sesamia* spp.)<br>(f) Drought | (a) Low and declining soil fertility<br>(b) Drought/moisture stress<br>(c) *Striga*<br>(d) Streak virus<br>(e) Borers |
| Latin America and Caribbean | (a) Limited technological options<br>(b) Drought/moisture stress<br>(c) Ear rot<br>(d) Rust | (a) Soil erosion<br>(b)Drought/moisture stress<br>(c) Turcicum blight<br>(d) Borer (S.W. corn borer) | (a) Low soil fertility<br>(b) Soil acidity<br>(c) Drought/moisture stress<br>(d) Fall armyworm<br>(e) Stunt |

**Information to cope with climatic risk for maize production**

According to a study reported by Unganai (2000), any long-range forecast that will potentially benefit farmers has to contain the following information:

a. Onset date of the rainy season;

b. Ending date of the rainy season;

c. Quality of the cropping season rainfall, indicated by the rainfall amount using rainfall percentile studies. Rainfall percentile studies allow for good comparison of rainfall among farming centres or within agroecological zones as well. This also aids in production planning and in the delivery of advisory services;

d. Temporal and spatial distribution of important climate factors;

e. Probability, frequency and timing of adverse weather events (such as dry spells or mid-season drought, floods, windstorms, frost) within the season. This will include climatic risk zoning to determine adaptation or avoidance mechanisms during the season;

f. The patterns of rainfall, temperature, evapo-transpiration, relative humidity, sunshine hours, vapour pressure deficit and other agriculturally significant climate variables. This will include deviations, anomalies and timing of favourable climatic conditions;

g. Characterization of ecological zones suitable for climate manipulation and maize production purposes using the appropriate climate based crop yield or growth models;

h. Interpretation of the above information in terms of which varieties to plant and when to plant.

Such agrometeorological services may be complemented by comparisons with long-term averages and recent seasons and with additional information that was recently shown to be appreciated by farmers and extension agents as described below.

Dry spells that substantially reduce the soil water reservoirs and affect crop yield cause problems for agriculture and other human water needs. Climatic risk zoning must be used to determine the best planting time to avoid or reduce drought effects on crop development. The following determinations and assessments are thus suggested for coping with climatic risk:

a. Potential suitability of a specific variety for a given region;

b. Probability of drought and frost;

c. Phenological stage most susceptible to drought and frost;

d. Availability of meteorological data;

e. Zoning of production districts based on climatic and edaphic conditions;

f. A water requirement index for each phenological stage.

In addition to the above, water requirement and dryness indices, stored soil moisture and the risk of severe drought must be determined. For example:

,

Suitable drought monitoring and characterization indices therefore are:

a. Water requirement index;

b. Drought index;

c. Available stored water;

d. Probability of dry spell;

e. Rainfall anomaly;

f. Soil moisture-holding characteristics;

g. Growing period climate characteristics;

h. Maize fertilization – a methodology for maize fertilization needs to be developed that will have implications for drought/dry spell management or soil moisture storage and nutrient leaching, especially in humid regions.

User requirements for pest and disease management in a climate context are:

a. Pest and disease identification services;

b. Life cycle and mode of attack/infestation, simplified in chart form;

c. Pest and disease mapping for each locality showing the times of the year when climatic conditions favour their survival and multiplication;

d. Monitoring methods to determine pest and disease presence (visual identification methods must be disseminated);

e. Assessment of the effectiveness of control measures. This involves:

   i. Assessment of why the method worked or did not work;

   ii. Close consultation with pest advisers or extension officers;

   iii. "What if" analysis ("what if I lose?", "what if I gain?");

   iv. Reporting of resistance tolerance;

   v. Determination of environmental influences (rain, humidity, wind, soil, temperature);

f. Timing of the application of control methods to optimize their effectiveness;

g. Research on and documentation of social and economic costs of weeds.

## Maize Growing Zones of India

Though maize can be grown almost all over India, it is more popular in the Northern and South-western part. In the Northern part, it is grown only in the rainy season (kharif) but in the south and western parts it can be grown in winter (Rabi), summer (zaid) and rainy (kharif) season. In peninsular India and Southern Rajasthan, maize can be sown at any time of the year if irrigation is available. As a grain and fodder crop, it is extensively grown in Uttar Pradesh, Bihar, Rajasthan, Madhya Pradesh, Punjab and the entire valley tracts of the Himalayas. The important maize growing states are Uttar Pradesh, Punjab, Gujarat, Bihar, Rajasthan, Madhya Pradesh, Andhra Pradesh, Himachal Pradesh and Karnataka, the average productivity of maize was 1606 kg/ha during 1989 - 90. It is relatively high in Karnataka, Andhra Pradesh and Himachal Pradesh and low in Uttar Pradesh and Rajasthan which account for a large part of the maize area in the country. Among the various approaches investigated for increasing yields was the possibility of raising maize during winter (Rabi). This provided a major breakthrough for rapidly increasing production and productivity. The states where Rabi maize is being grown extensively in Bihar, Uttar Pradesh, Andhra Pradesh, Gujarat, Maharashtra, Tamil Nadu, Madhya Pradesh, Punjab and Rajasthan. The yield obtained in Rabi maize is 35 - 80 % higher than that of kharif maize. Over 84 % of maize area in India is rainfed.

**Table 7. Maize growing zones of India**

| Zone | Area |
|---|---|
| I | Hills of Uttar Pradesh, Jammu and Kashmir, Himachala Pradesh, Sikkim, Assam, Eastern Himalayas and West Bengal. |
| II | Western Indo-Gangetic plains of Delhi, Haryana, Punjab, Rajasthan and Western Uttar Pradesh. |
| III | Eastern Indo-Gangetic plains of Bihar and Terai belt of Uttar Pradesh. |
| IV | Southern and Peninsular India viz., Madhya Pradesh, Andhra Pradesh, Gujarat, Karnataka, Maharshtra, Tamil Nadu. |
| V | Riverine areas of east Uttar Pradesh and Diara areas of Bihar. |

## References

Ali, H. C. and S.M. Janabi (1980). Influence of heat accumulation on several traits of five genotypes of corn. Agron. Abstract American Soc. Agron. (1980). p: 75.

Allison, J.C.S. and T.B. Daynard (1979). Effect of change in time of flowering induced by altering photoperiod or temperature on attributes related to yield in maize. Crop Sci., 19: 1 - 4.

Allmaras, R.R., V.C. Burrows and V.E. Larson (1964). Early growth of corn as affected by soil temperature. Proc. Soil Sci. Soc. America, 28: 271 - 275.

Arnon, I. (1972). Crop production in dry regions. Leonard Hill Books, London. P: 147 - 187.

Aspiazu, C. and R.H. Shaw (1972). Comparison of several methods of growing degree day unit calculation for corn. Iowa State Collages J. Sci., 46: 435-442.

Badu-Apraku, B., R.B. Hunter and M.Tollenaar (1983). Effect of temperature during grain filling on whole plant and grain yield of maize (*Zea mays* L.). Canadian J. Plant Sci., 63: 357 - 363.

Baktash, F.Y. (1985). Regression of corn grain yield on air temperature and relative humidity in Iraq. J. Agric. Water Resources, 4 (3): 1-10.

Bland, B.F. (1971). Corn production. In: Cereals and legumes. Academic Press, London, P: 63 - 215.

Bonhomme, R., F. Ruget, M. Derieux and P. Vincourt (1982). Relationship between aerial dry matter production and intercepted solar radiation for various maize genotypes. Competes Rendus Hebdomadaires de Seaces de Academie des Sciences, 294 (8): 393 – 398.

Carr, M.K.V. and G.M. Milboun (1976). Maize: climatic and physiological factors limiting yield. Span, 19: 65 - 67.

Coligado, M.C. and D.M. Brown (1975). Response of corn in the pretassel initiation period to temperature and photoperiod. Agric. Met., 14 (7): 357 – 367.

Daughtry, C.S.T, J.C. Cochran and S.E. Hollinger (1984). Estimating silking and maturity dates of corn for large areas. Agron. J., 76: 415 - 420.

Denmead, O.T. and R.H. Shaw (1960). The effect of soil moisture stress at different stages of corn growth on the development and yield of corn. Agron. J., 52: 272–274.

Elmore, R. W. and B. Doupnik (1995). Corn recovery from early-season frost. J. Prod. Agric., 8: 199-203.

Fakorede, M.A.B and B.O. Opeke (1985). Weather factors affecting the response of maize to planting dates in a tropical rainforest location. Experimental Agriculture, 21:31- 40.

Fourie, A.P. and P.J. Robbertse (1977). Selection criteria for stalk quality in *Zea mays L.* Agroplantae, 9: 55 - 60.

Gallo, K.P and C.S.T. Daughtry (1986). Techniques for measuring intercepted and absorbed photosynthetically active radiation in corn canopies. Agron. J., 78: 752 - 756.

Gardher, B.R., B.L. Blad, R.F. Maurer and D.G. Watts (1981). Relationship between crop temperature and the physiological phonological development of differentially irrigated corn. Agron. J., 73(4): 743 – 747.

Gardner, B.R., B.L. Blad, R.F. Maurer and D.G. Watts (1981a). Relationship between crop temperature and the physiological development and energy requirement of maize. Can. J. Plant Sci., 60 (4): 1451 – 1453.

Hall, A.D. (1945). The soil: An introduction to the scientific study of growth of crops. John Murray, London.

Heyne, E.G. and H.H. Laude (1940). Resistance of corn seedlings to high temperatures in laboratory tests. J. Am. Soc. Agron., 32: 116 - 126.

Jain (1973). New agronomy boosts in maize yield. Indian Farming, 23 (1): 17 - 20.

Jayasree, G., D. Lingaiah, D. Raji Reddy and S.B.S. Narasimha Rao (2013). Relationship between biophysical parameters and normalized difference vegetation index in maize. J. Agrometeorology, 15 (2): 120 – 125.

Johannes, M.M. (1996). Potential effects of climate change on corn production in Zimbabwe. Climate Res., 6: 147 - 151.

Kiniry, J.R., J.T. Ritchie and R.L. Musser (1983). Dynamic nature of the photoperiod response in maize. Agron. J., 75 (4): 700 – 703.

Krishnamurthy, K., A. Bomme Gowda, A. Venugopal, N. Jagannath, G. Raghunatha and B.G. Rajasekara (1972). Pattern of dry matter accumulation and distribution in maize. Indian J. Agron., 17: 104 - 109.

Kucharik, C.J. (2008). Contribution of planting date trends to increase maize yields in the Central United States. Agronomy Journal, 100: 328 - 336.

Lee, Y.S. and Y.S. Jung (1988). Utilization of PAR in maize Research reports of the rural development administration, soils and fertilizers. Agric. Sci. Inst., Sweon, Korea.

Leonard, W.H. and J.H. Martin (1963). Cereal crops. Macmillan & Co., London, P: 449 - 603.

Liang, B.C., A.F. MacKenzie, P.C. Kirby and M. Remillard (1991). Corn production in relation to water inputs and heat units. Agron. J., 83: 794–799.

Mccalla, A.G., J.R. Weir and K.W. Niatby (1939). Effects of temperature and sunlight on the rate of elongation of stems of maize. Canadian J. Res., 17: 388 - 409.

Medereski, H.J., N.E. Miller and C.R. Weaver (1973). Accumulated heat units for classifying corn hybrids maturity. Agron. J., 65: 743 – 747.

Moore, E. (2005). Key drought management tips. Drought Management Factsheet No. 2. British Columbia Ministry of Agriculture and Lands. http://www.agf.gov.bc.ca/resmgmt/publist/600 series/665000-2.pdf.

Morris, M.L., R. Tripp and A.A. Dankyi (1999). Adoption and impacts of improved maize production technology. In: A Case Study of the Ghana Grains Development Project. Economics Program Paper 99–01. Mexico, D.F., CIMMYT.

Muchow, R.C., T.R. Sinclair and J.M. Bennett (1990). Temperature and solar radiation effects on potential maize yield accord locations. Agron. J., 82: 338–343.

Muchow, R.C., T.R. Sinclair and J.M. Bennett (1990). Temperature and solar radiation effects on potential maize yield accord locations. Agron. J., 82: 338–343.

Nafziger, E.D. (1994). Corn planting date and plant population. J. Prod. Agric., 7: 59–62.

Narcico, G., P. Ragni and A. Venturi (1992). Agrometeorological aspects of crops in Italy, Spain and Greece: A summary review of durum wheat, barley, maize, rice, sugar beet, sunflower, soybean, rapeseed, potato, tobacco, cotton, olive and grape. Publication EUR 14124 EN of the Office for Official Publications of the European Communities Series 'Agriculture' Luxembourg.

Narwood, C.A. (2001). Dryland corn in western Kansas: Effect of hybrid maturity, planting date and plant population. Agron. J., 93 (3): 540-574.

Nield, R.E. and M.W. Seeley (1977). Growing degree day predictions for corn and sorghum development and some applications to crop production in Nebraska. Res. Bull. 280. Agricultural Experiment Station, Nebraska.

Nielsen, R.L, P.R. Tomison, G.A. Brown, A.L. Holter, J. Wells and K.L. Wuethrick (2002). Delayed planting effects on flowering and grain maturation of dent corn. Agron. J., 94: 549-558.

Obeng-Antwi, K., P.Y.K. Sallah and P.P. Frimpong-Manso (2002). Performance of intermediate maturing maize cultivars in drought-stressed and non-stressed environments in Ghana. Ghana J. Agric. Sci., 35: 49–57.

Ofori, E., N. Kyei-Baffour and S.K. Agodzo (2004). Developing effective climate information for managing rainfed crop production in some selected farming centres in Ghana. In: Proceedings of the School of Engineering Research Retreat held at HO. Unpublished.

Otengi, S.B.B., C.J. Stigter, J.K. Nganga and D.N. Mungai (2000). Wind protection in a hedged agroforestry system in semi-arid Kenya. Agrofor. Syst., 50: 137–156.

Pal, D. and S.B. Varade (1984). Effect of soil saturation and submergence on maize crop growth. J. Maharashtra Agric. Univ., 9: 129 - 131.

Patil, S.J., M. Mahadevappa, P. Swamy Rao, A.Ramakrishna and R.S. Arathya (1971). Performance of maize hybrids and composites in the Ghataprabha area of Mysore state. Indian J. Agron., 16: 128 - 129.

Prodhan, H.S. (2001). Influence of appropriate cultivars and sowing dates on rabi maize productivity. Indian Journal of Agricultural Science, 71 (1): 55-56.

Reddy, D.R., A.N. Mehta, A.M. Shekh, M.B. Savani, H.R. Patel and H. Venkatesh (1995). Impact of weather variables on yield and yield attributes of Rabi maize. GAU Res. J., 21 (1): 47 – 53.

Rush, C.E. and N.P. Neal (1951). The effect of maturity and other factors on studies of corn at low temperature. Agron. J., 43: 112 - 116.

Sandhu, S.S., V.H. Shah and B.C. Wright (1971). Response of maize germplasms to dates of planting in Kulu valley. Indian J. Agric. Sci., 41: 176 - 181.

Shah, V.H. and R.N. Sharma (1970). Response of maize germplasms to dates of planting at Delhi. Indian J. Agric. Sci., 40: 782 - 794.

Shaw, E.B. (1955). World economic geography. John Wiley and Sons, New York.

Shaw, R.H. (1955). Climate requirement. In: Spraque ve G.F., Dudley J.W. (Eds.), Corn and Corn Improvement, vol. 18. 3rd ed. ASA, Madison, WI.

Singh, C. and E.R. Leng (1970). A note on plant density and season of planting and genotype interaction in maize. Indian J. Agron., 15: 202 - 203.

Singh, V. and R. Singh (1971). Performance of maize varieties at different planting dates. Indian J. Agron., 16: 520 - 522.

Stewart, D. W., C. Costa, L. M. Dwyer, D. L. Smith, R. I. Hamilton and B. L. Ma (2003). Canopy structure light interception and photosynthesis in maize. Agron. J., 95: 1465 - 1473.

Stigter, C.J., J. Kinama, Y. Zhang, K.O. Oluwasemire, D. Zheng, N.K.N. Al-amin and A.T. Abdalla (2005). Agrometeorological services and information for decision-making: some examples from Africa and China. J. Agric. Meteorol. (Japan), 60: 327–330.

Swan, J.B., M.J. Shaffer, W.H. Paulson and A.E. Peterson (1987). Simulating the effects of soil depth and climatic factors on corn yield. Soil Sci. Am. J., 51, 1025–1032.

Tollenaar, M. and A. Agrilera (1994). Radiation use efficiency of an old and a new maize hybrid. Agron. J., 84: 536 – 541.

Unganai, L. (2000). Application of long-range rainfall forecasts in agricultural management: A review of Africa's experiences. In: Proceedings of the International Forum on Climate Prediction, Agriculture and Development, Palisades, New York, 26–28 April 2000. International Research Institute for Climate and Society (IRI), Palisades.

Van Dobben, W.P. (1962). Influence of temperature and light conditions on dry matter distribution, development rate and yield of arable crops. Neth. J. Agric. Sci., 10: 377 - 389.

Vittal Rao, S. and S. Mehboob Ali (1961). Observations on optimum period of sowing of maize crop in Andhra Pradesh. Indian J. Agron., 6: 30 - 36.

Wallace, H.A. and E.N. Bressman (1937). Corn and Corn growing. John Wiley and sons, New York.

Warrington, I.J. and E.T. Kanemasu (1983). Corn growth response to temperature and photoperiod: I seedling emergence, tassel initiation and anthesis. Agron. J., 73: 749 – 754.

Ye, F.H., Y.S. Liu, Y.G. Yang; L.F.F. and Y.S. Jin (2000). Study on planting dates of vegetable corn Suynuo -1 for all the year around. Acta Agriculture Zhejiangensis, 12 (2): 78 – 83.

Yogesh. (2011). Agrometeorological studies in maize and soybean intercropping system. M. Sc. (Agriculture) Thesis, University of Agricultural Sciences, Dharwad, Karnataka (India).

# 9

# Date of Sowing for Maize

Planting should begin before the optimum date if soil conditions will allow the preparation of a good seedbed. Farmers should have the equipment capability to plant more than half of their corn areas prior to the optimum planting date. This should allow planting all the corn areas prior to the calendar date when corn yields begin to quickly decline. Grain yield and test weight were increased by early plantings, whereas grain moisture was reduced, thereby allowing earlier harvest and reducing drying costs. Early planting generally produces shorter plants with better stand ability. Delayed planting increases the risk of frost damage to maize and may subject the crop to greater injury from various insect and disease pest problems, such as European corn borer and gray leaf spot.

Maize should be planted only when soils are dry enough to support traffic without causing soil compaction. The yield reductions resulting from mudding the seed may be much greater than those resulting from a slight planting delay. No-tillage maize can be planted at the same time as conventional, if soil conditions permit. In reality, however, planting may need to be delayed several days to permit extra soil drying. Planting a full-season hybrid first and then alternately planting early-season and mid-season hybrids, allows the farmer to take full advantage of maturity ranges and gives the late-season hybrids the benefit of maximum heat unit accumulation. Full-season hybrids generally show greater yield reduction when planting is delayed compared with short-to mid-season hybrids. Planting early hybrids first, followed by mid-season, and finally the full-season hybrids spreads the pollination interval for all the maize areas over a longer time period and may be a good strategy for some drought-prone areas with longer growing seasons.

Planting hybrids of different maturities reduces damage from diseases and environmental stress at different growth stages (improving the odds of successful pollination) and spreads out harvest time and workload. Consider spreading hybrid maturity selections between early-, mid-, and full-season hybrids-for example, a 25-50-25 maturity planting, with 25 % in early – to mid-season, 50 % in mid-to full-season and 25 % in full-season. Planting a range of hybrid maturities is probably the simplest and most effective way to diversify and broaden hybrid genetic backgrounds. When maize planting is delayed past the optimum dates or if a crop needs to be replanted, it may be necessary to switch hybrid maturities. In most delayed planting situations, however, full-season hybrids still perform satisfactorily and reach physiological maturity (black layer formation) when planted as late as the last week of May. Hybrids planted in late May or early June mature at a faster thermal rate (require fewer heat units) than the same hybrid planted in late April or early May.

Other factors concerning hybrid maturity need to be considered when planting is delayed. For plantings in late May or later, the dry-down characteristics of hybrids should be considered. Although a full-season hybrid may still have some yield advantage over shorter season hybrids planted in late May, the full-season hybrid could have significantly higher grain moisture at maturity than earlier maturing hybrids. In addition, there will be less calendar time for field drying and drying costs will be higher. Later planting dates generally increase the possibility of damage from European corn borer (ECB) and may warrant selection of ECB Bt hybrids.

At Coimbatore (Tamil Nadu), pre-monsoon sown maize crop recorded 40 % higher grain yield (3795 kg/ha) over monsoon sown crop (2678 kg/ ha) (Sakthivel and Balasubramanian, 2003). At Warangal (Andhra Pradesh), maize crop can be sown from 15$^{th}$ November to 15$^{th}$ December in the clay loam soils of rice fallows of Central Telungana region of Andhra Pradesh under no till condition. Sowing of maize on 15$^{th}$ December gave higher net return significantly superior to the rest of the dates of sowing (Malla Reddy et al., 2010).

**Table 1.Effect of time of sowing on rainfed maize**

| Treatments | Plant height (cm) | LAI | CGR ($g\ m^{-2}\ day^{-1}$) | RGR ($g\ m^{-2}\ day^{-1}$) | NAR ($g\ m^{-2}\ day^{-1}$) | Cob length (cm) | Cob weight (g) | Grains/ cob | Grain yield (kg/ ha) | Stover yield (kg/ ha) | Harvest Index |
|---|---|---|---|---|---|---|---|---|---|---|---|
| Premonsoon sowing | 211 | 5.55 | 10.09 | 0.0112 | 0.0189 | 15.60 | 125 | 395 | 3795 | 6153 | 0.38 |
| monsoon sowing | 254 | 4.83 | 2.88 | 0.0036 | 0.0067 | 14.40 | 80 | 280 | 2678 | 4275 | 0.38 |
| CD (P=0.05) | 11 | 0.35 | 0.47 | 0.0006 | 0.0017 | 0.65 | 1.08 | 11 | 258 | 384 | NS |

**Table 2. Performance of maize at different dates of sowing in rice fallow no till condition**

| Date of sowing | No. of kernels/cob | | Kernel Weight /cob (g) | | Dry weight at harvest (g/plant) | | 100-seed weight (g) | | Crop duration (days) | | Grain yield (kg/ ha) | |
|---|---|---|---|---|---|---|---|---|---|---|---|---|
| | 2007 | 2008 | 2007 | 2008 | 2007 | 2008 | 2007 | 2008 | 2007 | 2008 | 2007 | 2008 |
| $15^{th}$ November | 486 | 500 | 117 | 127 | 284 | 297 | 24.8 | 25.0 | 109 | 110 | 5042 | 5414 |
| $30^{th}$ November | 493 | 495 | 120 | 118 | 291 | 289 | 24.8 | 24.8 | 110 | 112 | 5298 | 5265 |
| $15^{th}$ December | 510 | 481 | 134 | 115 | 308 | 278 | 25.1 | 24.8 | 112 | 111 | 6101 | 4965 |
| $30^{th}$ December | 399 | 434 | 96 | 104 | 219 | 243 | 24.2 | 24.4 | 115 | 114 | 3436 | 4076 |

(Malla Reddy et al., 2010)

At Coimbatore (Tamil Nadu), pre-monsoon sowing of maize crop recorded higher plant height (254 cm), LAI (5.55), DMP (10637 kg/ha), length of cob (15.80 cm), cob weight (123 g), grains per row (14.1), grain yield (3947 kg/ha) and (7783 kg/ha) than normal date of sowing under rainfed condition (Ramakichenin et al., 2002). Negative relationship of helio-thermal unit with yield also reported in baby corn (Thavaprakash et al., 2007). Greater reduction in yield of long season maize varieties under delayed sowing situations was attributed to decrease in season length might have an effect of reduction in their potential yield (Johannes, 1996). Higher heat use efficiency was recorded with the crop sown in first fortnight of June and decreased with each delay in sowing. However, at knee high stage (upto 30 DAS) higher heat units were recorded in June first fortnight sowing (Girijesh et al., 2011) (Table 3).

**Table 3. Dry matter accumulation and heat use efficiency (g/GDD) as influenced by date of sowing in maize**

| Crop growth stage | 1st fortnight of June | | | 2nd fortnight of June | | | 1st fortnight of July | | |
|---|---|---|---|---|---|---|---|---|---|
| | GDD | Total dry matter (g/plant) | HUE | GDD | Total dry matter (g/plant) | HUE | GDD | Total dry matter (g/plant) | HUE |
| Knee high stage (30 DAS) | 433 | 20.2 | 0.047 | 421 | 16.1 | 0.038 | 427 | 9.99 | 0.023 |
| Silking stage (60 DAS) | 861 | 138.1 | 0.160 | 860 | 117.8 | 0.137 | 879 | 95.5 | 0.109 |
| Grain development stage (90 DAS) | 1313 | 215.3 | 0.164 | 1306 | 168.2 | 0.129 | 1331 | 129.2 | 0.097 |
| At physiological maturity stage | 1766 | 251.8 | 0.143 | 1761 | 194.6 | 0.110 | 1768 | 146.3 | 0.083 |

(Girijesh *et al.*, 2011)

Helio thermal units and HTUE were higher for the crop sown during 1st fortnight of July. However, the short duration composite was found relatively stable with respect to yield for dates of sowing as the reduction in yield due to delayed sowing was lesser than in remaining genotypes. Further, availability of helio thermal units for short duration variety did not vary much across dates of sowing due to its low maturity period. Grain yield of maize was found inversely related to the helio-thermal units. Thus early sowing escape adverse situation during its life cycle in maize (Girijesh et al., 2011) (Table 4).

**Table 4. Yield response of maize to helio thermal units (HTU) and helio thermal use efficiency (HTUE) across sowing and genotype groups**

| Duration | Date of sowing | Grain yield (kg/ha) | HTU | HTUE (kg/HTU) |
|---|---|---|---|---|
| Long duration (130 – 140 days) | 1st fornt night of June | 7313 | 7142 | 1.02 |
| | 2nd fortnight of June | 5507 | 7231 | 0.76 |
| | 1st fortnight of July | 4023 | 8371 | 0.48 |
| Relatively less duration (120 – 130 days) | 1st fornt night of June | 6530 | 66449 | 1.01 |
| | 2nd fortnight of June | 5154 | 6918 | 0.75 |
| | 1st fortnight of July | 3508 | 7650 | 0.46 |
| Short duration (90 – 100 days) | 1st fornt night of June | 5000 | 4588 | 1.09 |
| | 2nd fortnight of June | 4562 | 4724 | 0.96 |
| | 1st fortnight of July | 4528 | 5910 | 0.77 |

(Girijesh et al., 2011)

**Table 5. Effect of sowing dates on yield and monetary return of winter maize**

| Sowing dates | Temperature at sowing (°C) | Emergence period (days) | Yield (q/ha) | | Harvest index (%) | Net return (Rs/ha) |
|---|---|---|---|---|---|---|
| | | | Grain | Stover | | |
| 15th October | 26.6 ±1.4 | 4-5 | 40.1 | 100.1 | 28.6 | 17753 |
| 25th October | 26.4 ±1.6 | 4-5 | 38.9 | 97.9 | 28.5 | 17087 |
| 5th November | 23.9 ±0.9 | 4-6 | 48.2 | 110.3 | 30.5 | 22109 |
| 15th November | 23.4 ±1.1 | 5-6 | 46.3 | 108.9 | 29.8 | 21117 |
| 25th November | 21.5 ±1.0 | 6-8 | 40.8 | 98.6 | 29.2 | 18058 |
| 5th December | 19.9 ±1.4 | 8-10 | 28.5 | 77.4 | 27.3 | 11278 |
| 15th December | 18.8 ±0.5 | 10-12 | 18.5 | 66.1 | 22.5 | 6033 |
| 25th December | 18.7 ±2.4 | 10-12 | 16.2 | 61.0 | 22.4 | 4730 |
| CD 5 % | - | - | 3.5 | 12.2 | - | - |

At Banswara (Rajasthan), winter maize sown in first fortnight of November gave maximum grain yield of 48.2 q/ha with higher net return of Rs.22109/ha (Table 5). Early and delayed sowings to mid November reduced the grain yield by 3.9 to 66.4 % and net return by Rs.3059-16387/ha. Lower grain yield (16.2 t/ha) and net return (Rs.4730/ha) were realized when maize was sown on 25th December (Porwal and Jain, 1999).

Khandale and Relwani (1991) reported that winter maize yields declined significantly with each successive delay in sowings because environmental factors like temperature, solar radiation and day light gradually diminished during October to December and favoured the growth of early sown crop than that of late sown.

**Table 6. Sowing time on maize based pulses intercropping systems**

| Sowing time | Yield (q/ha) | | | Land equivalent ratio |
|---|---|---|---|---|
| | Maize | Inter crop | Maize equivalent | |
| Early (10th June) | 17.67 | 3.43 | 22.66 | 1.37 |
| Normal (25th June) | 24.57 | 4.76 | 30.10 | 1.41 |
| Late (10th July) | 16.80 | 2.22 | 18.45 | 1.36 |
| CDS% | - | - | 4.30 | NS |

At Dehra Dun, normal sowing time gave higher maize equivalent yield (30.10 q/ha) than early sowing (22.66 q/ha) and late sowing (18.45q/ha) (Table 6) (Khola et al., 1999).

**Table 7. Sowing dates on requirement of growing degree days of winter maize**

| Phenological stages | October, 10 | October, 20 | October, 30 | November, 9 | November, 19 |
|---|---|---|---|---|---|
| 100 % emergence | 204.8 | 213.5 | 307.9 | 326.8 | 342.8 |
| 50 % tasseling | 1824.8 | 2859.9 | 1918.2 | 1794.7 | 1819 |
| 50 % silk emergence | 1965.9 | 2034.3 | 2055.7 | 1995.1 | 2042.2 |
| Maturity | 2869.4 | 3111.8 | 2977.7 | 2957.2 | 3139.2 |

At Parbhani (Maharashtra), maize crop sown on 9th and 19th November took maximum days of seedling emergence. However, days to tasseling, silking and maturity decreased progressively with the delayed sowing (Table 7). Maximum variability among sowing dates was observed for seedling emergence to tasseling interval with the tasseling to silking and silking to maturity interval being identified in different sowing. The crop sown on 10, 20 and 30th October required less growing degree days than that sown on 9th and 19th November sowing (Anita et al., 2003).

**Table 8. Cumulative GDD as influenced by date of sowing of maize (Kisan)**

| Date of sowing | Tasselling | Silking | Grain filling | Physiological maturity | HUE |
|---|---|---|---|---|---|
| 15th December | 165.4 (62.2) | 206.2 (71.3) | 221.5 (82.5) | 485.1 (98.7) | 0.081 |
| 30th December | 183.8 (56.5) | 248.5 (64.7) | 349.4 (85.3) | 552.8 (95.4) | 0.096 |
| 14th January | 247.3 (53.8) | 281.6 (58.2) | 367.5 (81.4) | 668.9 (92.2) | 0.097 |
| 29th January | 362.5 (49.5) | 397.3 (54.3) | 458.2 (78.6) | 774.3 (89.5) | 0.062 |
| CD (5 %) | 6.4 (0.82) | 8.9 (0.75) | 9.2 (0.79) | 17.6 (1.52) | |

* Figures in parenthesis shows number of days to attain respective physiological phases.

At Sonepur (Orissa), when sowing was delayed all the physiological stages required less number of days (Table 8). Sowing on 29th January accumulated maximum heat units followed by 14th January sowing. Heat units required for attaining each stage increased as sowing were delayed. Maximum heat use efficiency was found when sowing was done on 14th January. Maximum conversion of solar heat energy in to dry matter production when crop was sown on 14th January followed by 30th December (Mohapatra et al., 2010).

At Rajendera nagar (Andhra Pradesh), crop sown from 18th June to 21st June recorded significantly more dry matter (Table 9). Sowing beyond 21st July in 2009 and 17th July in 2010 reduced the dry matter production by 11 % over early sown crop (Rani et al., 2012).

**Table 9. Dry matter (g/plant) partitioning at maturity stage of maize as influenced by dates of sowing**

| Dates of sowing | Dry matter (g/plant) | | | | Grains / Cob | Grain yield (g / plant) |
|---|---|---|---|---|---|---|
| | Leaf | Stem | Cob | Total | | |
| 18th June | 34 | 41 | 128 | 203 | 390 | 115 |
| 2nd July | 33 | 40 | 124 | 197 | 364 | 108 |
| 7th July | 34 | 41 | 128 | 203 | 376 | 112 |
| 17th July | 31 | 38 | 120 | 188 | 338 | 99 |
| 2nd August | 26 | 32 | 109 | 167 | 301 | 89 |
| 6th August | 30 | 37 | 120 | 167 | 348 | 102 |
| 22nd August | 25 | 31 | 111 | 167 | 314 | 94 |
| CD 5 % | 4 | 4 | 7 | 10 | 29 | 10 |

Heat units required for maize crop to progress from one phase to other phase has been reported by Girijesj et al. (2011). At Anakapalle (Andhra Pradesh), rate of seedling emergence increased from 0.08 to 0.13 / day with increase in daily mean temperature from 22.9 to 26.6 °C. Significantly higher grain yield, growing degree days and helio thermal units were registered during first fortnight of November sown crop (Gowri et al., 2013).

**Table 10. Yield response of maize to phenothermal index, growing degree days and helio thermal units in different dates of sowing**

| Dates of sowing | Germination to silking | | Silking to maturity | | Phenothermal index | | GDD | HTU | Grain yield (kg/ ha) |
|---|---|---|---|---|---|---|---|---|---|
| | Duration (day) | Mean Temp. (°C) | Days | Mean Temp. (°C) | Germination to silking | Silking to maturity | | | |
| October 15th | 42 | 25.0 | 57 | 22.1 | 15.0 | 12.1 | 1317 | 9087 | 5722 |
| November 1st | 49 | 23.7 | 59 | 21.9 | 13.7 | 11.9 | 1372 | 9604 | 5945 |
| November 15th | 50 | 22.1 | 52 | 23.5 | 12.1 | 13.5 | 1306 | 9077 | 5801 |
| December 1st | 53 | 21.5 | 49 | 24.4 | 11.5 | 14.4 | 1316 | 9007 | 4674 |
| December 15th | 51 | 21.7 | 46 | 25.3 | 11.7 | 15.3 | 1300 | 9230 | 4978 |
| January 1st | 51 | 22.0 | 42 | 27.0 | 12.0 | 17.1 | 1330 | 9709 | 5081 |
| CD 5% | - | - | - | - | - | - | 64 | 107 | 332 |

At Rajendra nagar (Andhra Pradesh), higher grain yield of maize was obtained during kharif season, if the minimum temperature ranged from 18.9 to 22.5°C. At the same time, high range of both morning and afternoon relative humidity during early stage was unfavourable for maize crop which affected the crop dry matter production (Leela Rani et al., 2013).

**Table 11. Cob yield and growing degree days as influence by different dates of sowing**

| Dates of sowing | Cob yield (kg/ha) | | Degree days (°C) | |
|---|---|---|---|---|
| | 2005 | 2006 | 2005 | 2006 |
| June 1st | 5110 | 4591 | 1964 | 1918 |
| June 15th | 4988 | 4646 | 1849 | 1827 |
| July 1st | 4841 | 4292 | 4811 | 1847 |
| July 15th | 3649 | 3633 | 1650 | 1595 |
| August 1st | 2448 | 2403 | 1589 | 1575 |
| CD 5 % | 110 | 111 | - | - |

At Amberpet (Andhra Pradesh), significantly superior yields were registered at early sowing due to accumulation of higher number of growing degree days or heat units (Hemalatha et al., 2013).

**Table 12. Effect of moisture stress and dates of sowing on maize grain yield (t/ha)**

| Dates of sowing | Irrigation as per water requirement crop | Skipping irrigation at vegetative stage | Skipping irrigation at reproductive stage | Rainfed | Mean |
|---|---|---|---|---|---|
| June 11th | 4.19 | 3.93 | 3.13 | 3.09 | 3.53 |
| July 2nd | 3.92 | 3.77 | 2.94 | 2.91 | 3.39 |
| July 18th | 3.69 | 3.31 | 2.87 | 2.86 | 3.18 |
| August 3rd | 3.44 | 3.25 | 2.66 | 2.46 | 2.95 |
| Mean | 3.81 | 3.57 | 2.90 | 2.83 | 3.27 |

CD 5 %: 1.0

In China, the coefficients of variation for ≥10 °C accumulated temperature and GDD requirements were different during different growth periods, with a descending rank order of sowing to emergence > silking to maturity > emergence to silking > sowing to maturity and greater in the north spring maize region than in the Huanghuaihai maize region. The coefficients of variation were lower for ≥10 °C accumulated temperature than GDD requirements for both cultivars in both planting regions. Significant differences existed between locations and years for the ≥10 °C accumulated temperature and GDD requirements. These have implications for appropriate maize cultivars recommendation and high and stable yield achieving by reasonably using accumulated temperature across different regions of China (Peng et al., 2014).

**Table 14. Growth, yield attributes, yield and quality in winter maize as influenced by dates of sowing**

| Dates of Sowing | Plant height (cm) | Days to silking | Days to maturity | Diameter of cob (cm) | Grain yield (q /ha) | Protein (%) in grain |
|---|---|---|---|---|---|---|
| 15 Oct | 179.25 | 51.88 | 141.51 | 4.31 | 35.49 | 8.09 |
| 25 Oct | 186.02 | 53.55 | 143.25 | 4.73 | 39.97 | 8.26 |
| 5 Nov | 186.02 | 49.48 | 138.51 | 4.25 | 30.47 | 8.03 |
| CD 5% | 2.79 | 0.46 | 0.91 | 0.06 | 0.84 | 0.03 |

(Verma et al., 2012)

At Hamirpur (Uttar Pradesh), crop sown on 25 October significantly enhanced the growth and grain yield than early sowing 15 October and late sowing 5th November (Verma et al., 2012).

**Table 15. Yield and yield components of hybrid maize as influenced by different sowing dates**

| Date of sowing | Plant ht. (cm) | Cob length (cm) | Days to maturity | Grains /cob (No) | 100-grain wt.(g) | Grain yield (t/ha) | Stover yield (t/ha) |
|---|---|---|---|---|---|---|---|
| October 30 | 240.3a | 17.9ab | 138 | 498 | 36.8a | 8.44a | 7.30 |
| November 15 | 238.0ab | 18.7a | 144 | 477 | 35.1b | 7.99b | 7.47 |
| November 30 | 238.4ab | 17.3b | 145 | 470 | 34.6b | 7.32a | 7.23 |
| December 15 | 2.33.5b | 16.9b | 137 | 459 | 34.2b | 6.78d | 7.75 |

(Quayyum, 2002)

In Bangladesh, October sowing resulted in higher number of gains/cob and 100-grain weight, which reflected higher grain yield. There was a trend to decrease yield with the advancement of dates. Stover yield was not influenced by different dates of sowing (Quayyum, 2002).

In Bangkok (Thailand), mid-November to mid- December was found best for planting this crop in post-rice condition. There was probably due to maize obtaining better kernel number and kernel weight. To delay planting of maize thereafter, its yield was generally low. The reduction was 24 and 36 % when grown in January and February, respectively (Boonpradub and Kraokaw, 2002).

**Table 16. Kernel yield and yield components of maize (cv. NSX 9210) grown after rice under different planting dates**

| Planting date | Kernel yield (kg /ha) | Kernel / ear | 100 kernel weight (g) | Ears /ha |
|---|---|---|---|---|
| November 1 | 4,864 ab | 456.5 b | 25.9 a | 53,333 a |
| November 15 | 5,279 a | 479.3 a | 24.2 b | 53,333 a |
| December 1 | 5,176 a | 472.6 a | 23.6 b | 53,333 a |
| December 15 | 5,174 a | 470.8 a | 22.2 bc | 53,106 a |
| January 1 | 4,197 bc | 455.3 b | 23.8 b | 51,112 a |
| Januray 5 | 4,316 bc | 449.0 b | 21.3 c | 53,333 a |
| Februrary 1 | 4,228 bc | 446.6 b | 21.2 c | 53,333 a |
| February 15 | 3,823 c | 428.5 c | 21.6 c | 53,887 a |

For instance, the optimum sowing date for corn hybrid 'Pride 5' was determined as July 15th (Tollenaar et al., 1994).

At Jilin, the grain filling rate of Jiyou 11 corn sowed on April 30th was faster and the best harvest date was about October 5th. The protein, fat accumulating rate arrived to maximum on the 25th day after pollination, showing that fat and protein of high oil corn could accumulate simultaneously in grain filling period which made their accumulating rate up to maximum and then their accumulating rate slowed. The peak of starch accumulation appeared on the 30th day after pollination, the content of starch increased fast 10-30 day after pollination, and then the starch accumulation slowed (Wang Yanling et al., 2005).

At Tirupati (Andhra Pradesh), crop sown during second fortnight of August resulted in the highest yield attributes, seed yield and monetary returns over other dates of sowing (Srinivasulu et al., 2008).

**Table 17. Yield attributes, yield and economics of maize as influenced by time of sowing**

| Time of sowing | Cob girth | Cob length | Seed rows/ cob | Seeds / row | Seeds /cob | Cob wt (g) | Shelling (%) | Grain yield (kg /ha) | Stover yield (kg/ ha) | Net returns (Rs/ ha) | B:C ratio |
|---|---|---|---|---|---|---|---|---|---|---|---|
| IInd fortnight of July | 12.34 | 10.48 | 11.76 | 13.76 | 173 | 58.60 | 63.4 | 1595 | 3711 | 3376 | 1.40 |
| Ist fortnight of August | 11.69 | 10.06 | 11.30 | 12.95 | 159 | 53.64 | 59.7 | 1498 | 3555 | 2738 | 1.32 |
| IInd fortnight of August | 13.17 | 12.44 | 12.45 | 15.19 | 194 | 65.33 | 69.8 | 2145 | 4446 | 6861 | 1.82 |
| Ist fortnight of September | 12.79 | 11.32 | 11.84 | 14.21 | 183 | 60.61 | 66.7 | 1772 | 4020 | 4466 | 1.52 |
| CD (5 %) | 0.12 | 0.49 | 0.21 | 0.35 | 12 | 7.6 | 2.1 | 112 | 129 | 1162 | 0.32 |

(Srinivasulu et al., 2008)

At Coimbatore (Tamil Nadu), the weather prevailed during the cropping period was most favourable during Kharif season, followed by Rabi season. Also, based on agrometrics and econometrics, Kharif season was the most potential season for irrigated maize. The next best was Rabi season. Late Rabi season was not suitable for irrigated maize cultivation. Sowing in the second fortnight of August during Kharif season and sowing in the second fortnight of September during Rabi season were identified as the best sowing windows (Karthikeyan and Balasubramanian, 2006).

**Table 18. Seasonal and sowing time efficiencies**

| Season and times of sowing | Grain yield (kg/ ha) | Sowing time efficiency (%) | B:C ratio | Seasonal efficiency (%) |
|---|---|---|---|---|
| **Rabi** | | | | |
| 1st fortnight of September | 2536 | 99.57 | 0.98 | 120.65 |
| 2nd fortnight of September | 3073 | 120.65 | 1.09 | |
| 1st fortnight of October | 2739 | 107.54 | 1.03 | |
| 2nd fortnight of October | 1841 | 72.28 | 0.69 | |
| **Late Rabi** | | | | |
| 1st fortnight of December | 1850 | 153.91 | 0.69 | 56.94 |
| 2nd fortnight of January | 1285 | 106.91 | 0.54 | |
| 1st fortnight of February | 875 | 72.80 | 0.45 | |
| 2nd fortnight of February | 797 | 66.31 | 0.44 | |
| **Kharif** | | | | |
| 1st fortnight of July | 2020 | 78.14 | 0.85 | 122.45 |
| 2nd fortnight of July | 2369 | 91.64 | 0.90 | |
| 1st fortnight of August | 2796 | 108.16 | 1.07 | |
| 2nd fortnight of August | 3156 | 122.09 | 1.24 | |

(Karthikeyan and Balasubramanian, 2006).

At Dharwad (Karnataka), timely (June first fortnight) sown maize took significantly more number of days (111 days) and growing degree days (1569.1) to complete physiological maturity and recorded higher heat efficiency (0.363 g plant-1 °C day-1) as compared to subsequent delayed sowing, but was *on par* with June second fortnight sowing (Thimme Gowda et al., 2013).

**Table 19. Effect of sowing dates on ATT, PTI and HUE in maize at various phenophases**

| Date of sowing | Seedling stage | | | Vegetative stage | | | Flowering stage | | | Maturity stage | | |
|---|---|---|---|---|---|---|---|---|---|---|---|---|
| | ATT | PTI | HUE | ATT | PTI | HUE | ATT | PTI | HUE | ATT | PTI | HUE |
| June I fortnight | 99 | 14.76a | 0.002a | 802 | 14.11a | 0.121a | 908 | 14.14a | 0.326a | 1569 | 14.07c | 0.363a |
| June II fortnight | 106 | 14.29b | 0.002a | 786 | 13.98b | 0.119a | 857 | 13.89a | 0.322a | 1500 | 14.24b | 0.311b |
| July I fortnight | 110 | 14.24bc | 0.001a | 741 | 13.91c | 0.116a | 843 | 13.79a | 0.259a | 1423 | 14.47a | 0.292b |
| July II fortnight | 117 | 13.77bc | 0.001a | 717 | 13.90c | 0.106b | 820 | 14.11a | 0.194b | 1287 | 14.50a | 0.296b |

(Thimme Gowda et al., 2013). ATT: accumulated thermal time (°C days). PTI: phenothermal index (°C day day-1). HUE: heat use efficiency (g plant$^{-1}$ °C day$^{-1}$)

Varughese and Iruthayaraj (1995) showed that heat unit requirement for *rabi* crops was less than for *kharif* crops for all physiological phases.

At Sistan (Iran), higher grain yield and harvest index were obtained from August 5th sowing. July 5th sowing was associated with a decrease in LAI, LAI duration and crop growth rate (Dahmardeh and Dahmardeh, 2010).

**Table 20. Effect of sowing date on grain yield,, biological yield, 1000 kernel weight and harvest index of maize**

| Sowing date | Grain yield (t/ha) | Biological yield (t/ha) | 1000 kernel weight (g) | Harvest index (%) |
|---|---|---|---|---|
| 5th July | 2.30 | 8.08 | 34.48 | 26.42 |
| 20th July | 6.76 | 16.82 | 34.47 | 38.60 |
| 5th August | 8.84 | 20.37 | 33.77 | 45.26 |
| 20th August | 6.17 | 15.9 | 27.26 | 38.22 |
| LSD 5% | 3.14 | 1.8 | 2.23 | 8.48 |

(Dahmardeh and Dahmardeh, 2010)

The optimum planting time for maize cultivation would be from the last week of September to the end of October. Higher grain yield was obtained (2952 kg/ha) during this period (Imam, 1966). The period from November to February is the best time for the higher dry matter production in Khartoum area (Elkarouri and Mansi, 1980). Maize grain yield was reduced when sowing time was delayed to the end of October (MC Cormick, 1974). Delaying sowing date to mid December reduced the 1000 kernel weight (Cirilo and Andarade, 1996). July 15 was optimal sowing date for maize in Peshawar (Pakistan) (Ahmad et al., 2001).

## References

Abdul Rahuman, A.M., E. Lazim Mogboul and E.N. Abdollatief (2001). Effect of sowing date and cultivar on the yield and yield components of maize in Northern Sudan. In: Proc. Seventh Eastern and Sowthern Africa, Regional Maize conference. February 11 – 15, Nairobi, Kenya. P: 295 – 298.

Ahamad, N., A. Waheed and F.S. Hamid (2001). Different sowing date of maize cultivars and their performance. Pakistan J. Biological Sci., 1: 106 – 108.

Anita, C., M.V. Dhoble, D. Usha and M.R. Deshmukh (2003). Effect of sowing dates on requirement of growing degree days of winter maize. Ann. Plant Physiol., 17 (2): 182 – 183.

Aspiazu, C. and R.H. Shaw (1972). Comparison of several methods of growing degree day unit calculation for corn. Iowa State Collage J. Sci., 46: 435-442.

Badu-Apraku, B., R.B. Hunter and M.Tollenaar (1983). Effect of temperature during grain filling on whole plant and grain yield of maize (*Zea mays* L.). Canadian J. Plant Sci., 63: 357-363.

Bali, A.S., M.H. Shah, K.N. Singh and T.S. Raina (1991). Response of maize (*Zea mays* l.) composite to planting dates and fertility level under irrigated conditions of Kashmir Valley. Indian J. Agric. Res., 16(3): 178-182.

Bhrinder, S.S. and J.S. Sawhney (1985). Studies on the time and method of sowing winter maize. Agric. Sic. Digest, 5: 145 - 147.

Biswas, M. and A.A. Begum (2002). Response of hybrid maize varieties to dates of sowing. Annual Res. Rev. Agronomy Division, BARI, Gazipur, P: 34 - 41.

Boonpradub, S. and S. Kraokaw (2002). Maize: A Newly Sequential Crop with Rice in Thailand. Proceedings of the 8th Asian Regional Maize Workshop, Bangkok, Thailand: August 5-8, 2002. P: 307 – 315.

Chele, K.S. (1960). Effect of sowing time on the development and yield of maize crop. Indian J. Agric. Sci., 30: 135 - 141.

Cirilo, A.G. and F.H. Andrade (1996). Sowing date and maize productivity I. Crop growth and dry matter portioning. Crop Sci., 34: 1039 – 1043.

Dahiya, D.S. and S.S. Narwal (1989). Effect of sowing date on phenology and growing degree day requirement of winter maize. Indian J. Agric. Sic., 59 (9): 589 – 592.

Dahmardeh, M. and M. Dahmardeh (2010). The effect of sowing date and some growth physiological index on grain yield in three maize hybrids in Southeastern Iran. Asian J. plant Sci., 9 (7): 432 – 436.

Elkarouri, M.O.H. and M.G. Mansi (1980). Performance of sorghum and maize as forage in irrigated saline soils of the Sudan. Expl. Agric., 16: 431 – 436.

Fakorede, M.A.B and B.O. Opeke (1985). Weather factors affecting the response of maize to planting dates in a tropical rainforest location. Experimental Agriculture, 21:31-40.

Fortin, M.C., F.J. Picrce and M. Edwards (1994). Corn leaf area response to early season soil temperature under crop residue. Agron. J., 85: 355 – 359.

Gautam, O.P., V.H. Shah and Y. Singh (1964). Response of hybrid maize to date of planting and time of nitrogen application. Indian J. Agron., 9: 1 - 10.

Girardin, P. (1980). Leaf development and energy requirements of maize. Can. J. Plant Sci., 60 (4): 1451 – 1453.

Girihesh, G.K., A.S. Kumara Swamy, S. Sridhara, M. Dinesh Kumar, T.S, Vageesh and S.P. Nataraju (2011). Heat use efficiency and helio thermal units for maize genotypes as influenced by dates of sowing under southern transitional zone of Karnataka state. international J. Sci. Nature, 2 (3): 529 -533.

Girijesh, G.K., A.S. Kumaraswamy, S. Sreedhar, M. Dinesh Kumar, T.S. Vageesh and K.S. Rajashekarappa (2011). Heat unit utilization of kharif maize in transitional zone of Karnataka. J. Agrometeorology, 13 (1): 43 - 45.

Gowri, V., B. Bapuji Rao, T. Chitkala devi, M.B.G.S. Kumari and R. Ankaiah (2013). Thermal requirement of rabi maize in North coastal zone of Andhra Pradesh. J. Agrometeorology, 15 (2): 167 – 169.

Hemalatha, S., D. Sreelatha, M. Anuradha and R. Sai kumar (2013). Crop weather relations in maize. J. Agrometeorology, 15 (2): 165 – 166.

Johannes, M.M. (1996). Potential effects of climate change on corn production in Zimbabwe. Climate Res., 6: 147 - 151.

Karthikeyan, R. and T.N. Balasubramanian (2006). Economic analysis of maize (Zea mays L.) production under varied times of sowing in western zone of Tamil Nadu, India. Indian J. Agric. Res., 40 (2): 98 – 103.

Khandale, D.Y. and L.L. Relwani (1991). Effect of sowing date on forage yields of maize, sorghum and oat in Central India. Indian J. Agron., 36 (2): 346-350.

Kolawole, E. and U. Samson (2009). Growth and yield of maize as influenced by sowing date and poultry manure application. Not. Bot. Hort. Agrobot. Cluj, 37 (1): 199-203.

Kucharik, C.J. (2008). Contribution of planting date trends to increased maize yields in the Central United States. Agronomy Journal, 100: 328-336.

Kushwaha, H.S., P.K. Prasoon, H.K. Rai and R.S. Singh (2010). Thermal requirements of maize influenced by planting dates and intra-row spacing. In: Agrometeorological services for farmers (Ed., Vyas Pandey). Anand Agricultural University, Anand, P: 23-28.

Leela Rani, G. Sreenivas and D. Raji Reddy (2013). Evaluating contribution of weather to growth and yield of kharif maize under irrigated conditions. J. Agrometeorology, 15 (2): 156 – 158.

Major, D.J., D.R. Johnson, J.W. Tamer and I.C. Anderson (1975). Maize growth and yield in relation to planting dates in middle Egypt. Bulletin faculty Agriculture, University Cairo, 47 (1): 73 - 85.

Malla Reddy, M., B. Padmaja and D. Vishnu Vardhan Reddy (2010). Performance of maize (*Zea mays* L.) and sunflower (*Helianthus annuus* L.) under different sowing dates in rice fallows. J. Res. ANGRAU, 38 (1&2): 62-69.

MC Cormick, S.J. (1974). The effect of sowing date on maize development and yield of silage and grain. Field Crop Abst., 27 : 33- 34.

Mohammad Moradi, M.R. Naderi and A.R. Golparvar (2009). Effect of planting date and leaf cutting on yield and yield components in maize. Crop Sci., 49**:** 111-115.

Mohapatra, S.C., B.S. Bishoyi and P.K. Panda (2010). Photothermal requirement and photosynthate production with respect to varying sowing date of maize. J. Indian Soc. Costal Agric. Res., 28 (2): 60-61.

Nafziger, E.D. (1994). Corn planting date and plant population. J. Prod. Agric., 7:59–62.

Namakha, A., I.U. Bubakar, I.A. Sadik, A.I. Sharifai and A.H. Hassas (2008). Effect of sowing date and nitrogen level on yield and yield components of two extra early maize varieties (*Zea mays* L.) in Sudan Savanna of Nigeria. ARPN J. Agric. Bio. Sci., 3 (2): 15.

Narwal, S.S., D.S. Dahiya, G. Singh and D.S. Malik (1987). Response of maize cultivars to sowing in winter. J. Agric. Sci. (Camb), 107: 303 -1307.

Narwal, S.S., S. Poonia, G. Singh and D.S. Malik (1986). Response of winter sown maize cultivars to sowing time and row direction in north west India. J. Agric. Sci., (Camb), 106: 169 -173.

Nielsen, R.L, P.R. Tomison, G.A. Brown, A.L. Holter, J. Wells and K.L. Wuethrick (2002). Delayed planting effects on flowering and grain maturation of dent corn. Agron. J., 94: 549-558.

Otegui, M.E., M.G. Nicolini, R.A. Ruiz and P.A. Dodds (1995). Sowing date effects on grain yield components for different maize genotypes. Agron. J., 87: 29 – 33.

Pathak, A.N., V.N. Pathak and K.N.T. Wari (1971). Effect of sowing dates, nitrogen fertilization and varieties on the yield and uptake of nutrients by maize. Indian J. Agron., 16: 159 - 161.

Peng, H., Y. Liu, R. Xie, B. Ming, D.Ma, S. Li and X. Mei (2014). Temporal and spatial variation in accumulated temperature requirements of maize. Field Crops Res., 158: 55 – 64.

Porwal, M.K. and H.K. Jain (1999). Effect of sowing dates on winter maize varieties in command area of southern Rajasthan. Indian J. Agron., 44 (3): 560 -563.

Quayyum, M.A. and S.M. Raquibullah (1987). Effect of seeding dates on the grain yield and yield attributes of maize in Bangladesh. In: Proc. Annual Bangladesh Science Conference, Dhaka, Bangladesh. 10 – 14, January, 1987.

Ramakichenin, B., N. Sakthivel and A. Balasubramanian (2002). Effect of premonsoon sowing and land management practices on growth, yield parameters and yield of rainfed maize. Madras Agric. J., 89 (1-3): 177 - 179.

Rani, P.L, D.R.Reddy, G.Sreenivas, V.P. Rao, K. Surekha and A. Shiva sankar (2012). Dry matter partioning and grain yield potential of maize as influenced by dates of sowing and nilngen application. J. Res. ANGRAU, 40 (1): 30 – 34.

Relwani, L.L. (1962). Studies on the effect of spacing, date of sowing and fertilization on yield of maize. Indian J. Agron., 7: 169 - 170.

Singh, C.P., N.N. Sharma and U.K. Prasad (1991). Response of winter maize (*Zea mays* L.) to seeding date, seed furrow, mulching and fertilization with nitrogen, phosphorus and potassium. *Ind. J. Agric. Sci.*, 61(12): 889 - 892.

Singh, V. and R. Singh (1971). Performance of maize varieties at different planting dates. Indian J. Agron., 16: 520 - 522.

Singh, V.K., R.P. Bajaj and R.J. Sisodia (1992). Effect of sowing dates and plant population on yield of maize. Curr. Res., 21(2): 24-25.

Srinivasulu, A., G. Karuna Sagar, V. Chandrika and D.S. Reddy (2008). Performance of maize genotypes under varied time of Sowing under rainfed conditions in southern agroclimatic Zone of Andhra Pradesh. Agric. Sci. Digest, 28 (3): 233 – 234.

Thavaprakash, N., R. Jagannathan, K. Velayutham and L. Gurusamy (2007). Seasonal influence on phonology and accumulated heat units in relation to yield of baby corn. International J. Agric. Res., 2 (9): 826 - 831.

Thimme Gowda, P., S.I. Halikatti and S.B. Manjunatha (2013). Thermal requirement of maize (*Zea mays* L.) as influenced by planting dates and cropping systems. Res. J. Agric. Sci., 4(2): 207 - 210.

Tollenaar, M., D.E. McCullough and L.M. Dwyer (1994). Physiological basis of the genetic improvement of field crops (Ed. G.A. Slafe). Marcel and Dekker Inc., New York, pp. 183 - 236.

Tripathi, R.S., G.K. Srivastava and S. Malaiya (2004). Effect of variety, sowing time and integrated nutrient management on growth, yield attributes and yield of summer maize (*Zea mays* L.). *Ann. Agric. Res. New Series,* 25(1): 155-158.

Varughese, K. and M.R. Iruthayaraj (1995). Seasonal variation in yield and heat unit requirement of maize. Crop Res., 10(1): 67-73.

Verma, N.K., B.K. Pandey, U.P. Singh and M.D. Lodhi (2012). Effect of sowing dates in relation to integrated nitrogen management on growth, yield and quality of rabi maize (*Zea Mays* L.). Journal of Animal & Plant Sciences, 22 (2): 324-329.

Wang Yanling (2005). Effect of sowing date and harvest time on high oil corn quality. Jilin College of Science and Technology of Agriculture, Jilin.

10

# Physiology of Maize Crop

Maize is one of the most important cereal crops in the world agricultural economy both as food for man and feed animal. It is a miracle crop. It has very high yield potential, there is no cereal on the earth which has so immense potentiality and that is why it is called "queen of cereals". Maize is grown in almost all the states of India.

## Growth and Development

1. Germination: Immediate after maturity - even at plants itself.
2. Radicle emergence 2-3 days.
3. Colespticle and 2, 3 seminal root produced 4-5 days.
4. Mesocotyle elongates 2-3 days.
5. Leaf: Very in length, width, pubescence, liqules, pigmentation, various angle, upper leaf, responsible for high light interception- high PHS – grain development.
6. Inflorescence: Monoceous male and female seen in same plant male tassels female lots.
7. Tassels: Appear 1-3 days prior to silking single anther-shed 2500 pollens, cross pollinated, ear is modified spike produce from shot lateral branch-axil one of large- foliage leaf.
8. Chemical content: Starch - 77 %, Ash - 2 %, Protein - 9 %, Sugar- 4 % and Penotosans - 5 % (Parseglove, 1985).

9. Stem: Thickness 2-4 cm, solid stalk, 2 - 4 m height.
10. Ear bearing – Internodes in longitudinally grooved, to allow proper position of ear head (cob).

## Growth Stages

### i. Larons Hanway (1977) 5 periods

1. Planting to emergence
2. Emergence to tasseling
3. Tasseling to silking
4. Silking to physiological maturity
5. Dry down period.

### ii. Steven et al., (1986) latest 10 stages

1. Visibility of collar of 4$^{th}$ leaf.
2. Visibility of collar of 8$^{th}$ leaf.
3. Visibility of collar of 12$^{th}$ leaf.
4. Visibility of collar of 18$^{th}$ leaf.
5. 75 % tassels produce.
6. Visibility of kernel in blister stage.
7. Very rate.
8. Early dent
9. Full dent
10. Physiological maturity.

14, 10, 23 % in Tulasi and 18, 5.7 % is Rasi of potentially filled grain. Indicating that contribution of panicle reserves stem reserves to grain growth.

## Photosynthesis

Improving the resolution of maize leaf photosynthesis measurements in the field may help to clarify the relationships between environmental stress, carbon assimilation and grain yield (Tollenaar *et al.,* 1994). To date, the cumbersome nature of gas exchange techniques has limited both the time and space resolution of photosynthesis measurement in field studies.

Leaf spectral properties were strongly affected by leaf age, leaf position, and region within the leaf. Greening of young leaves was associated with a decrease in both reflectance and transmittance, especially in the middle wavelengths of the PAR region (Walter-Shea *et al.,*1991). Greening occurred first in leaf tips, followed by the middle and basal leaf sections. Once leaves were fully expanded and their basal regions had achieved their minimum reflectance and transmittance values, reflectance and transmittance of tip, middle and basal thirds were very similar until the onset of leaf senescence, especially in the longer wavelengths of the reduction in mean ABS or a reduction in mean SPAD value over the entire leaf, was not simply a function of leaf age, for instance, leaf 17, the youngest leaf sampled, senesced more rapidly than leaf 13 (one above the leaf subtending the topmost ear).

All sampled leaves reached peak men ABS values between 0.890 (Leaf 17) and 0.904 (leaf 9). Values lower than 0.885 were rarely observed in fully expanded leaves prior to the onset of senescence.

The effect that the spectral photon distribution of the illuminating source can have on leaf absorptance, was calculated from the measured spectral photon distribution of four different light sources and the fractional absorptance of young, mature and senescent leaves.

The mature leaf tissue, the effect of the light source on the calculated leaf absorptance of PPFD was very small (less than 0.01). Thus, the discrepancy between the leaf absorptance values observed in the present work, and those reported by Bjorkman and Demming (1987) and Oberhuber *et al.,*(1993), cannot be attributed to differences in the light sources considered, and must instead be due to differences instrumentation, and /or the plant material used.

**Table 2. Calculated absorptance of PPFD from four different light sources by leaf tissue samples from the middle region of leaf 17. Samples were collected 69 (young), 95 (mature) and 22 (senescent) DAP. Absorptance was calculated by multiplying the leaf sample absorptance in each 5nm waveband between 400and 700nm by the fraction of total incident photons in that waveband from a particular light source, and them summing the products. Each value is the mean of four samples**

| Light source | Leaf tissue sample | | |
|---|---|---|---|
| | Young | Mature | Senescent |
| | Calculated leaf absorptance | | |
| Unweighted | 0.790 | 0.899 | 0.853 |
| Solar irradiance | 0.773 | 0.892 | 0.843 |
| Microwave lamp | 0.773 | 0.889 | 0.844 |
| Tungsten lamp | 0.766 | 0.890 | 0.836 |
| Bjorkman lamp | 0756 | 0.886 | 0.832 |
| 0.016 | | 0.009 | 0.022 |

The accuracy of leaf transmittance and reflectance measurement made with an external integrating sphere depends directly on the accuracy with which the reflectance of the reference is known. In the present work consistent with standard practice, the reluctance of the barium sulphate references was assumed to be equal to 1.00 in the 400 to 700 nm range. However, according to the manufacturer of the integrating sphere, the true mean reluctance of this reference material may be closer to 0.99 in this wavelength range. If a reference reflectance on 0.99 is assumed, then calculated ABS values are increased in the present study by a small amount, ranging from 0.003 to 0.001.

When chlorophyll fluorescence parameters are used to estimate photosyhthetic rates of maize leaves, an accurate estimate of leave absorptance of incident PPFD is required. In the present study, ABS of full expanded, healthy leaves ranged from about 0.88 to 0.91, suggesting that, in most cases, a leaf absorptance value of about 0.895 could be assumed without introducing a great ideal of error into the estimate of leaf photosynthetic rate derived from chlorophyll fluorescence. However, when working with young, chlorite or senescent leaf tissue, leaf absorptance will vary a much broader range, and must therefore be estimated for each individual sample. In these cases, a SPAD 502 chlorophyll meter can be used to provide a rapid and accurate non-destructive estimate of leaf absorptance, one the relationship between SPAD readings and absorptance for that particular chlorophyll meter has been estimated.

Growing of young leaf was associated decreased both reflectance and transmittance with in the middle wavelength of the PAR region, while leaf senescence was associated with decreases in both reflectance and transmittance, primarily in the longer wavelengths of PAR region. Leaf absorptance of incident photons in the PAR region by healthy, fully expanded leaves ranges from approximately 0.88 to 0.91, much lower values were observed for young, chlorotic or senescence leaves.

**Table 3. Maize physiology**

| Photosynthetic | Characteristics | Measured results | References |
|---|---|---|---|
| Light | Compensation | 400-700nm | Murata (1961) |
| Carbon –di-oxide | Compensation point | 10ppm | Baker and Musgrave (1964) |
| Net photosynthetic rate | | 20-40m mol $CO_2$ $m^{-2}$ $s^{-1}$ | Radmer and Larcha (1977) |
| Biochemical pathway | | C-3 pathway and $C_4$ pathway | Akta *et al.* (1969) |
| Bundle sheath | | Present | Akita and Miyasaka (1969) |
| Photo respiration | Absent | | |
| Temperature | | 35-40°C | Harrison (1964) |

Kernel set and flower synchrony within the ear of Maize: 1. Sowing date effects Maria E. Otegi * and Soleded Melon.

## Transpiration

Transpiration water calculated for unit weigh of absorbed N, and N uptake calculated for unit transpired water.

The hybrid Sabrina 3707 transpired the most water (20.19 kg) per 1g N absorbed (Table 4). Its transpiration was at 0.1 per cent level significant above the other four hybrids. The least transpired water per unit of absorbed N (2.09 kg) was recorded for Szentesi 212. Opposite results were obtained when the N uptake was calculated per 1 kg of transpired water. That is, the lowest value was obtained for Sabria 3708 (49.72 mg) and the highest for Szentesi 212 (82.93 mg). The mean value of the five hybrids was 70.78 mg.

**Table 4. Transpiration water calculated for unit weigh of absorbed N and N uptake calculated for unit transpired water**

| Maize hybrids | Transpired water (kg) | Total N uptake (g) | Transpired water calculated per 1g of absorbed N (kg) | N uptake per 1 kg of transpired water (mg) |
|---|---|---|---|---|
| Szentesi 212 | 68.05 | 5.62 | 12.09 | 82.93 |
| Pi 3906 | 62.06 | 4.79 | 12.97 | 77.09 |
| Pi SC 3901 | 80.16 | 6.07 | 13.31 | 76.38 |
| Pi RSC 3732 | 84.11 | 5.67 | 14.94 | 67.80 |
| Sabrina 3707 | 78.35 | 3.88 | 20.19 | 49.72 |
| $LSD_{0.1\%}$ | 25.38 | 1.83 | 4.09 | 20.00 |
| $LSD_{1.0\%}$ | 18.40 | 1.32 | 2.96 | 14.49 |
| $LSD_{0.5\%}$ | 13.28 | 0.96 | 2.14 | 10.46 |

### N/P ratio

There were no significant differences between the N/P ratios of the whole plants and grain. The lowest N/P ratios of the whole plants and grain was at hybrid Pi RSD 3732 (4.81 and 4.18) and the highest at hybrid Szentesi 212 (5.88 and 6.10). Lasztity et al. (1985 a) calculated N/P ratios for the whole plant between 4.4 and 5.4 and for grain 3.8 and 5.0.

## Climatic Relationship

Sowing data had different effects on dry matter accumulation in the vegetative and reproductive growth stages. With late sowing dry matter accumulations was greater during the initial phase of vegetative growth, but terminated sooner due to early flowering, leading to a significantly lower maximum dry mater production per plant than with early sowing dates were favourable for both vegetative and reproductive growth. Sowing date had different effects on changes in and on mean values of the growth indices in the vegetative and reproductive growth. Sowing data had different effects on changes in and on mean values of the growth indices in the vegetative and reproductive growth. In the initial stage of vegetative growth the values of the growth indices were greater after lattice growth stages. However, the values of the indices were greatest in the early and average sowing date treatments and smallest after late sowing. The changes in and mean values of the growth indices calculated with the classic method and using the HP model expressed similar trends for the effect of sowing date. The greatest deviation from the norm was observed in the growth dynamics on CGS and NAR calculated with the classic method. Studies on the correlation between maize grain yields and growth indices revealed the importance of the growth indices examined. On the basis of $R^2$ the multiple determination coefficient, 90-91 % of the variance of the maize grain yield was vegetative growth stage. LAI had a greater effect on CGS than NAR in the reproductive growth stage. However, CGR depended more the NAR than on LAI. Despite the advantages of the Hunt-Parsons growth analysis programme, the combined use of the classic and functional methods is recommended. Light interception reached maximum levels at LAI >3. Leaf area per plant at silking was not affected by different plant populations or nitrogen levels.

Grain yield of maize is mainly dependent on the number of kernels per plant that reach maturity (Cirilo and Andrade, 1994 a; Otegui *et al.,* 1995b). This yield component is very sensitive to environmental conditions around silking (Hall *et al.,* 1981, 1982; Kiniry and Ritchie, 1985; Westgate and Boyer, 1986). Jascobs and Pearson (1991) considered reductions in the potential number of spikelets per ear responsible for stress - induced changes in grain yield. But others have found the number of spikelet primordial at silking not affected by plant density (Ruget, 1989) or mild water deficit (Otegui *et al.,* 1995 a). Derieux *et al.*(1985) observed a significant relationship between kernel set per row and final ovule number per row for certain genotypes sown on different dates. Conversely, Cirilo and Andrade (1994 b) found no effect of sowing date on spikelet primordial counted at silking in the apical ears of two maize hybrids. They suggested that kernel abortion rather than morphogenetic processes was the dominant factor determining final kernel set. This lack of agreement may be partially due to the effects of sowing date on final leaf number and consequently on thermal time between female bud differentiation and silking. While leaf number changed slightly (19.5 – 20.0 leaves per plant) in the experiments of Cirilo and Andrade (1994 a), positive correlations ($P< 0.05$) were established between kernel number per row and total leaf number (Derieux *et al.,* 1985).

**Table 5. Effect of plant growth stimulants on maize**

| Treatments | LAI | Days to silking | Cob length (cm) | Cob diameter (cm) | Rows / cob | Yield (q/ha) | |
|---|---|---|---|---|---|---|---|
| | | | | | | Grain | Stover |
| Tap water | 2.68 | 67.0 | 14.62 | 3.64 | 13.71 | 36.43 | 52.44 |
| Atonik | 2.99 | 65.8 | 15.05 | 3.74 | 13.86 | 39.89 | 51.54 |
| Mixtalol | 3.05 | 64.8 | 16.25 | 4.02 | 13.98 | 43.45 | 56.07 |
| Miraculan | 2.92 | 66.5 | 15.34 | 3.89 | 14.05 | 41.07 | 54.28 |
| CD 5% | 0.18 | NS | 0.82 | 0.17 | 0.22 | 1.81 | NS |

At Udaipur, seed soaked overnight in the solution of mixtalol @ 1 kg seed / litre of solution + foliar spraying of 2 ppm mixtalol at knee high and tasselling stages gave higher grain yield than other growth stimulants (Jain et al., 1987).

Hamid and Nasab (2001) observed that both economic and biological yield are positively correlated with vegetative and reproductive phase duration in maize. Leaf area influences interception and utilization of solar radiation by crop canopies, and consequently dry matter accumulation as well as economic yield.

Photosynthetic efficiency and growth in maize are strongly related to the effect of canopy architecture on the vertical distribution of light within the canopy (Williams et al., 1968). Increasing plant density is one of the ways of increasing the capture of solar radiation within the canopy (Moderras et al., 1998). However, the efficiency of the conversion of intercepted solar radiation into maize yield decreases with a high plant population density because of mutual shading in the plants (Buren, 1970). In addition, a plant population density resulting in interplant competition affects vegetative and reproductive growth (Tetio-Kagho and Gardner, 1988).

## Defoliation

Corn hybrids and inbreds vary in their response to defoliation (Hanway, 1969; Hicks et al., 1977; Ingle et al., 1965). Effects of defoliation on the grain yield of corn hybrids have been well documented in numerous research studies (Egharevba et al., 1976; Hicks et al., 1977; Ingle et al., 1965). Reduction in corn yield has been shown to be directly proportional to the percentage leaf area destroyed (Hanway, 1969). The degree of yield loss caused by defoliation is dependent not only on leaf area destroyed, but also on the growth stage when defoliation occurs. Yield losses from defoliation damage are greatest during the late vegetative and reproductive stages. Although defoliation may cause lower kernel weight and test weight (Hicks et al., 1977; Tollenaar and Daynard, 1978), little information is available on the effects of defoliation on grain quality especially as it relates to the oil, protein, and starch composition of corn.

Singh and Nair (1975) found that defoliation increased grain protein and decreased grain sucrose and reducing sugars. Johnson (1978) reported that complete defoliation of nine commercial corn hybrids at the five-leaf stage reduced grain protein from 11.4 to 10.7 % but had no effect on oil content.

Early-season defoliation can delay anthesis and silking (Dungan and Gausman, 1951;

Cloninger et al., 1974; Singh and Nair, 1975; Vasilas and Seif, 1985a), shorten the duration of pollen shed (Vasilas and Seif, 1985a, 1985b), and reduce the total amount of pollen produced (Dungan and Gausman, 1951). Vasilas and Seif (1985a) found that defoliation at the 14-leaf stage (14L) had differential effect on anthesis and pollen shed by inbred lines and a hybrid. Johnson (1978) defoliated nine hybrids at the five-leaf stage and observed that defoliation delayed pollen shed and silking but did not change the pollen shed-to-silking interval.

**Critical stages of a corn plant**

Nature greatly influences corn growth and yield. For most of hybrids, each plant typically develops 20 - 21 leaves, silks about 65 days after emergence and matures about 120 days after emergence. All normal plants follow this same general pattern of development, but specific time intervals between stages and total leaf numbers developed may vary between different hybrids, seasons, planting dates and locations. The rate of plant development for any hybrid is directly related to temperature, so the length of time between the different stages will vary as the temperature varies. Environmental stress may lengthen or shorten the time between vegetative and reproductive stages. The length of time required for the yield components of ear density, kernel number, kernel weight varies between hybrids and environmental conditions. A producer who understands growth and development of corn will understand the importance of timeliness when using production practices for higher yields and profit.

## Identifying Stages of Development

The staging system most commonly used is the Iowa System. It divides plant development into vegetative (V) and reproductive (R) stages.

| Vegetative Stages | Reproductive Stages |
|---|---|
| VE Emergenece | R1 Silking |
| V1 First leaf | R2 Blister |
| V2 Second leaf | R3 Milk |
| V3 Third leaf | R4 Dough |
| * | R5 Dent |
| * | R6 Physiological maturity |
| * | |
| Vnth | |
| VT tasseling | |

Subdivisions of the V stages are designated numerically as V1, V2, V3, through Vn, where 'n' represents the last stage before VT tasseling. The six subdivisions of the reproductive stages are designated numerically.

Each leaf stage is defined according to the uppermost leaf whose collar is visible. Beginning at about V6, increasing stalk and nodal root growth combine to tear the small lowest leaves from the plant. To determine the leaf stage after lower leaf loss, split the lower stalk lengthwise and inspect for internode elongation. The first node above the first elongated stalk internode generally is the fifth leaf node. The internode usually is about one centimeter in length. This fifth leaf node may be used as a replacement reference point for counting to the top leaf collar. In a corn field all plants will not be in the same stage at the same time. Each specific V or R stage is defined only when 50 % or more of the plants in the field are in or beyond that stage.

Although each stage of development is critical for proper corn production we will focus on VE, V6, V12, V18, R1, and R6. Yield components and the number of growing degree units required at each growth stage are described below.

| Stage | GDU | Potential yield | Actual yield |
|---|---|---|---|
| VE | 125 | ears/area | |
| V6 | 470 | kernels rows/ear | "factory" |
| V12 | 815 | | kernel rows/ear |
| V18 | 1160 | kernels/row | |
| R1 | 1250 | kernel weight | kernel number ears/area |
| R6 | 2350 | | kernel weight |

## Stage VE Determination of potential ear density

Approximately 7-10 days after planting (125 GDU)

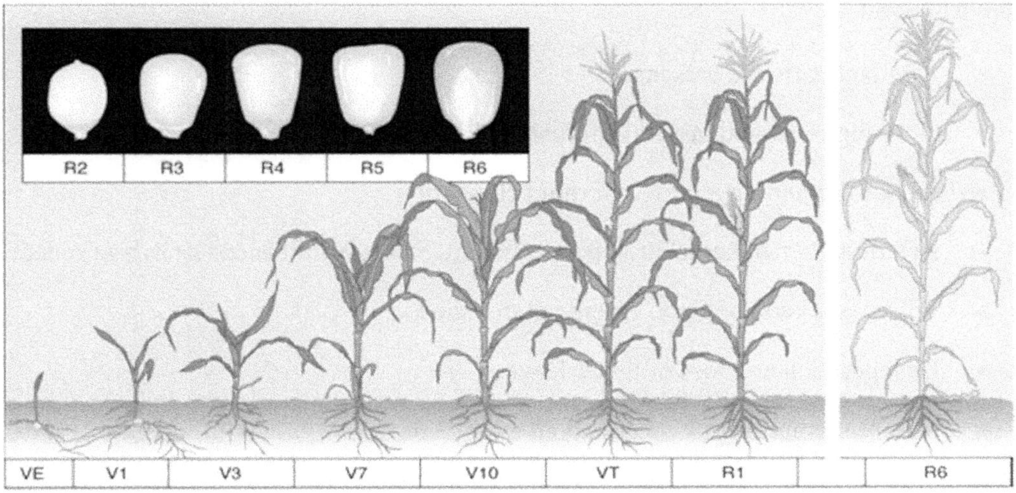


**Corn plant development**

### Above ground

- Coleoptile tip emerges above soil surface
- Elongation of coleoptile ceases
- First true leaves rupture from the coleoptile tip

### Below ground

- Mesocotyl and coleoptile elongation
- Elongation of mesocotyl ceases when coleoptile emerges above soil surface
- Growing point is below the soil surface
- Completed growth of seminal root system (radicle + seminal roots). Seminal root system supplies water and nutrients to developing seedling
- Nodal roots are initiated
- Nodal roots are secondary roots that arise from below ground nodes.

## Stage V6

- Potential plant parts ("factory") developed
- 24-30 days after emergence (475 GDU)

**Above ground**

- All plant parts are present
- Growing point and tassel (differentiated in V5) are above the soil surface
- Stalk is beginning a period of rapid elongation
- Determination of kernel rows per ear begins. Strongly influenced by hybrid genetics
- Tillers (suckers) begin to emerge at this time
- Degeneration and loss of lower leaves
- New leaf emerging (V-stage) about every 3 days

**Below ground**

- Nodal root system is established (approx. 18” deep x 15” wide). This is now the main functional root system of the plant.

## Stage V12 Potential kernel rows determined

- 42-46 days after emergence (815 GDU)

**Above ground**

- Number of kernel rows is set
- Number of ovules (potential kernels) on each ear and size of ear is being determined. Strongly affected by environmental stresses
- New V-stage approximately every 2 days.

**Below ground**

- Brace root formation begins stabilizing the upper part of the plant.

## Stage V18 Potential kernels per row determined

- 56 days after emergence (1160 GDU)

**Above ground**

- Ear development is rapid

- The upper ear shoot is developing faster than other shoots on the stalk

**Below ground**

- Brace roots are now growing from nodes above the soil surface. They will scavenge the upper soil layers for water and nutrients during reproductive stages

## Stage R1 Actual kernel number and potential kernel size determined

- 69-75 days after emergence (1250 GDU)

**Above ground**

- Begins when any silks are visible outside the husks
- Pollen shed begins and lasts 5-8 days per individual plant
- Silk emergence takes 5 days. Silks elongate from base of ear to tip of ear. Silks elongate until pollinated
- Silks outside husks turn brown
- The plant has now reached its maximum height
- First 7-10 days after fertilization cell division occurs within kernel
- Remaining R stages, endosperm cells fill with starch

**Below ground**

- The plant must have a healthy root system because proper uptake of moisture and nutrients are critical at this time.

## Stage R6 Actual kernel weight determined

- 130 days after emergence or 50-60 days after silking (2350 GDU)

**Above ground**

- Physiological maturity is reached when all kernels on the ear have attained their dry matter maximum accumulation
- The hard starch layer has advanced completely to the cob goes from top of kernel to base of cob

- A black abscission layer has formed. This indicates that moisture and nutrient transport from the plant has ceased
- Kernels are at 30-35 % moisture and have attained 100 % of dry weight.

## Crop Lodging

Lodging is recognized to be a severe malady in successful cultivation of many field crops around the world. It may reduce grain yield directly by interfering with dry matter accumulation or indirectly by the difficulties in harvest. Lodging may also adversely affect the grain quality. The severity of lodging and extent of losses resulting from it depend on the crop environment and on the growth stage at which lodging occurs. Generally speaking, favourable growing conditions, promoting crop development and grain yield will evoke lodging and increase its severity. Consequently, lodging should be regarded as an abundance disease that restricts the exploitation of otherwise yield promoting factors. The conjunction of lodging with rather high yields may result in a serious underestimate of its economic importance Losses due to lodging may often amount upto 30 percent of the grain yield. Considering also the other estimates on the frequency and severity of lodging, it may be concluded that in regions where high yields are obtained the lodging damage is as great as that due to diseases and insect pests. Root lodging caused by storms is a serious problem in maize, more so because it makes mechanical harvesting difficult. Mechanical harvesting of lodged plants is possible but slow combine speeds might be necessary to minimize harvesting losses (Carter and Hudelsaon, 1988). It also reduces dry matter accumulation and yield.

The causes of lodging in maize `are stem rots, insect injury, high plant populations, potassium deficiency, 2, 4-D damage and high yield with strong wind and/ or heavy rainstorm (Henry and Bereze, 1977). Lodging varies with population densities, cultivars and stage of the crop (Collaud, 1997; Gamboa et al., 1989). Lodging was greater at harvest than at mid-silk and more severe in hybrids with leaves than composites with lax leaves (Remison and Fajemisin, 1982).

The evaluation of lodging resistance should be based on related morphological or mechanical traits because lodging occurs during occasional severe rainstorms. The level of susceptibility of maize genotypes to root lodging largely depends on architectural parameters of the root system, characters of the above ground parts and environmental conditions. Genotype and environment interaction are generally moderate (Hebert et al., 1997). Ennos et al. (1993) suggested that lodging resistance in maize could be improved by increasing the bending strength and angle of spread of the adventitious roots. It is thought that control may be achieved by matching cultivation practices to the requirements of the particular hybriid grown (Munteanu et al., 1987). Plant growth regulators (PGR) could be an alternate for lodging control on lodging prone cultivars (Harms et al., 1989).

**Table 6. Effect of ethephon on lodging plant height, ear height and grain moisture content of maize**

| Ethephon rate (kg/ha) | Lodged plants (%) | Plant height (m) | Ear height (m) | Grain moisture (%) | Grain yield (Mg/ha) |
|---|---|---|---|---|---|
| 0.00 | 05 | 2.3 | 1.2 | 18.4 | 8.1 |
| 0.14 | 03 | 2.2 | 1.1 | 18.3 | 7.9 |
| 0.28 | 02 | 2.0 | 0.9 | 17.9 | 7.3 |
| 0.56 | 03 | 1.8 | 0.8 | 17.7 | 6.7 |

(Bullock and Raymer, 1989)

## References

Afuakwa, J.J. and N.R. Crookston (1984). Using the kernel line to visually monitor grain maturity in maize. Crop Sci., 24(4): 687 – 691.

Albrecht, B. and J.W. Dudley (1987). Divergent selection for stalk quality and grain yield in an adapted X exotic maize population cross. Crop Science, 27: 487 - 494.

Allan,A.Y. and D. Laycock (1976). Maize yield in western Kenya: effects of plant numbers per hill and stage at thinning. East African Agricultural and Forestry Journal, 42: 141- 148.

Allison, J.C. (1995). The production and defoliation of dry matter in maize after flowering. Ann. Bot., 30: 365 – 381.

Anderson, B. and D.G. White (1994). Evaluation of methods for identification of corn genotype with stalk rot and lodging resistance. Plant Disease, 78: 590 - 593.

Argillier, O., Y. Hebert and Y. Barriere (1995). Relationship between biomass yield, grain production, lodging susceptibility and feeding value in silage maize. Maydica, 40: 125 - 136

Arnold, J. M., L.M. Josephson, W.L. Parks and H.C. Hincer (1974). Influence of nitrogen, phosphorus and potassium applications of stalk quality characteristics and yield of corn. Agronomy Journal, 66: 605 - 608.

Beck, D.L., L.L. Darrah and M.S. Zuber (1988). Relationship of root tensile strength to vertical root pulling resistance in maize. Crop Science, 28: 571- 573.

Below, F. E., R.J. Lambert and R.H. Hageman (1984). Foliar applications of nutrients on maize. I. Yield and N content of grain and stover. Agronomy Journal, 76: 773 - 777.

Bohling, R.W. (1976). The influence of potassium fertilization and plant population upon the performance of several corn hybrid. Dissertation Abstracts International, 36: 3154.

Bratsch, A.D. and H.J. Mack (1990). Ethephon and mechanical topping influence growth, Yield and lodging of sweet corn. Hort Science, 25: 291 - 293.

Bullock, D.G. and P.L. Raymer (1989). Growth, grain yield and tissue mineral concentration of corn treated with ethephon. Agronomy Journal, 81: 480 - 483.

Buren, L.L (1970). Plant characteristics associated with barrenness in maize. Ph. D. Thesis. Iowa State University, Ames.

Burton, L.V. (1982). The measurement of maturity of country gentlemen corn. Canner, 54: 27 – 29.

Caneghem, L.V., X. Ciechanowski, J.F. Ledent and L. Canegham (1984). Effect of application of stem shorteners on the growth and yield of silage maize Mededelingen van de Faculteit Landbouwwenschappen. Rijksuniversiteit Gent, 49: 1121- 1128.

Carter, P. R and KD. Hudelson (1988). Influence of simulated wind lodging on corn growth and grain yield. Journal of Production Agriculture, 1: 295 - 299.

Chen, S.S., T.C. Yang and T.M. Chu (1990). Reduction of maize lodging by irrigation, tillage and nitrogen application. Journal of Agriculture and Forestry, 39: 167 - 174.

Christensen, S.P.L. (1990). Influence of different growth factors on the development of the ears of maize. Tidsskrift for Planteavl., 94: 441 - 447.

Clark, N.A, R.W Hemken, and J.H. Vandersall (1973). Effect of maturity, ripening span and planting rate on the yield, ear stover ratio, and dry matter of corn harvested for silage. Bulletin, University of Maryland Agricultural Experiment Station. 1973. 27 pp.

Cloninger, F.D., M.S. Zuber and R.D. Horrocks (1974). Synchronization of flowering in corn (*Zea mays* L.) by clipping young plants. Agron. J., 66:270–272.

Colbert, T. R. and M. S. Zuber (1974). Effect of sampling dates on estimates of rind puncture values and crushing strength in maize. Agronomy Abstracts, 50 p.

Contreras, M.R. (1978). Soluble solids, pith senescence, and stalk rot development in tropical maize population. Dissertation Abstracts International, 38: 3980.

Crookston, R.K. and D.R. Hicks (1988). Effect of early defoliation on maize growth and yield: An eleven-year perspective. Crop Sci., 28: 371–373.

Csatho, P. (1991). Effect of NPK fertilization and split application of N on lodging due to windstorm and harvestable grain yield of maize. Acta Agronomica Hungarica, 40: 281 - 294.

Csatho, P. (1989). Effect of NPK fertilizers on early development, lodging and yield in maize. Novenytermeles. 38: 335 - 346.

Damborsky, M., O. Chloupek and J. Ehrenbergerova (1994). Variability of maize lines and diallel cross hybrids. Genetika a Slechteni. 30: 297 - 303.

Dorvillez, D. and M. Derieux (1986). Making the most of very early maize grown at high densities by means of a growth regulator. Dissertation Abstracts International, 38: 3822.

Dorvillez, D. and M. Derieux (1983). Effect of ethephon on several morphological characteristics of early maize: effect on lodging and yield. Completes Rendus des Seances de l'Academic d'Agriculture de France, 69: 1393 - 1401.

Dostalek, R. and L. Hruska (1985). Effect of crop density on the production of maize seed, Rostlinna Vyroba, 31: 1103 - 1110.

Dourleijn, C. J. (1989). Selection for resistance to root lodging in forage maize. Prophyta, 43: 92 - 93.

Dow, E.W., T.B. Daynard, J.F. Muldoon, D.J. Major and G.W. Thurtell (1984). Resistance to drought and density stress in Canadian and European maize hybrids. Canadian J. Plant Sci., 64: 575–585.

Dubas, A and Michalski, T. 1983. Effect of sowing rates on yields of maize cultivars grown for grain. Pamietnik Pulawski, 81: 49 - 60.

Dungan, G.H. and H.W. Gausman (1951). Clipping corn plants to delay their development. Agron. J., 43:90–93.

Egharevba, P.N., R.D. Horrocks and M.S. Zuber (1976). Dry matter accumulation in maize in response to defoliation. Agron. J., 68: 40-43.

Eik, K. and J.J. Hanway (1965). Some factor affecting development and longivity of leaves of corn. Agron. J., 57: 7-12.

Ennos, A.R., M.J. Crook and C. Grimshaw (1993). The anchorage mechanics of maize, *Zea mays.* Journal of Experimental Botany, 44: 147 - 153.

Esechie,H.A. (1985a). Relationship between potassium total non-structural carbohydrate accumulation and lodging in maize in the rain forest zone. Samaru Journal of Agricultural Research, 3: 61 - 66.

Esechie,H.A. (1985b). Relationship of stalk morphology and chemical composition to lodging resistance in maize *(Zea mays* L) in a rain forest zone. J. Agric. Sci. (U.K), 104: 429 - 433.

Fourie, A.P. and P.J. Robbertse (1977). Selection criteria for stalk quality in *Zea mays L.* Agroplantae, 9: 55 - 60.

Fritz, C.D. (1989). Reducing sweet corn lodging with ethephon. In: Proceedings of the Plant Growth Regulators Society of America Annual Meeting, 16: 134 - 143.

Gagro, M. (1977). Effect of nitrogen and plant population on stem diameter and proportion of lodged plants in the maize hybrids BcSK 39-41 and OsSK 218. Poljoprivrendna Znanstvena Smotra, 41: 51.

Gagro, M. (1978). The effect of nitrogen application and density of stand on stalk diameter and the percentage of lodged plants in the hybrids BcSK5A. BeSK2l-22 and Pau 280. Poljoprivredna Znanstvena Smotra, 44: 421 - 443.

Gamboa, D.E., F.R. Petez and C.G. Arevalo (1989). Effect of population density on four subtropical maize hybrids. Revista Industrial Y agricola de Tucuman, 6 : 71 - 89.

Gaska, J.M. and E.S. Oplinger (1988). Use of ethephon as a plant growth regulator in corn production. Crop Sci., 28: 981 - 986.

Gaska, J.M. and E.S. Oplinger (1988). Yield, lodging and growth characteristics in sweet corn as influenced by ethephon timing and rate. Agron. J., 80: 722 - 726.

Geczki, I. (1976). Effect of testers on yield, straw strength and yield components in maize in 1974. Agrartudomanyi Egyetem Kozlemenyei. 197 - 204.

Gedge, D.L, W.R. Fehr and A.K. Walker (1977). Inter genotypic competition between rows and within blends of soybean. Crop Science, 17: 787 - 790.

Habitamu, A. *et al.* (1996). In: Conference of Agronomy and Crop Physiology Society of Ethiopia (WoldeyesJJs, S. *et al* ed.), 30-31 May 1995, Addis Ababa, Ethiopia.

Hall, A.J., J.H. Lemoff and N. Trapani (2007). Source limitation in maize at water deficit stress. Agron. J., 47: 560 - 567.

Hames, B.R., S.R. Thomas, A.D. Sluiter, C.J. Roth and D.W. Templeton (2003). Rapid biomass analysis. New tools for compositional analysis of corn stover feed stocks and process intermediates ethanol production. *Applied Biochemistry and Biotechnology,* 105: 5–16.

Hanway, J.J. (1969). Defoliation effects on different corn (*Zea mays* L.) hybrids as influenced by plant population and stage of development. Agron. J., 61:534-538.

Hoggard, A L., J.G. Shannon, and D.R. Johnson (1978). Effect of plant population on yield and height characteristics in determinate soybeans. Agronomy Journal, 70: 1070 - 1072.

Hubner, R. and F. Wagner (1975). Resistance of maize stems to lodging and breaking in relation to cultivar and row width. Wirschaftseigene Futter, 21: 122 - 140.

Ingle, J., D. Beitz and R.H. Hageman (1965). Changes in composition during development and maturation in maize seeds. Plant Physiol., 40: 835 - 839.

Inoue, Y. and T. Okabe (1981). Evaluation of lodging resistance in maize based on density of planting and late planting. Research Bulletin of the Hokkaido Agricultural Experiment Station, 129: 17 - 23.

Ishge, T. (1980). Estimation of the position of the center of gravity of the maize plant by the ear height. Annual report. Division of Genetics, National Institute of Agricultural Sciences, Japan (1981): 10 - 11.

Ivanovic, M. and K. Rosic (1985). Path coefficient analysis for three stalk traits and grain yield in maize (Zea *mays* L.). Maydica, 30: 233 - 239.

Ivanovic, M. (1979). Investigation of factors of lodging resistance in inbred maize and their hybrids of the F1 generation. Arhiv za Poljoprivredne Nauke, 32: 97- 119.

Ivanovich, M., B. Kerechki, D. Peich, M.Ivanovic, B.Kerecki and D. Peic (1985). Influence of some biochemical factors on the resistance of maize (Zea mays L.) stems to lodging, Informatsionnyi Bulletin po Kukuruze No., 4: 177- 183.

Jain, G.L., G.L. Sharma and P.S. Rao (1987). Response of maize to some new plant growth stimulants at varying levels of nitrogen. In: Crop productivity by H.S. Srivastava, S. Bhaskaran and K.K.G. Menon. Oxford and IBH Publishing Co. Pvt. Ltd., New Delhi. P: 153 – 163.

Johnson, R.R. (1978). Growth and yield of maize as affected by early season defoliation. Agron. J., 70: 995-998.

Jones, C.A. and J.A. Kiniry (1986). CERES-Maize, a model of maize growth and development. Texas A & M University Press, Texas.

Kalman, L., A. Korom and I. Maroti (1976). The role of tissue components in the stalk strength of maize. Agrobotanika, 16: 139 - 146.

Karlen, D. L., D.C. Arny and L.M. Walsh (1973). Incidence of chocolate spot (Pseudomonas syringae), northern corn leaf blight (Helminthosporium turcicum) and lodging of corn as influenced by soil fertility. Communications in Soil Science and Plant Analysis, 4: 359 - 368.

Kiesselbach, T.A. (1999). The structure and reproduction of corn. In: 50th Anniversary Ed. Cold Spring Harbor Lab. Press, Cold Spring. p. 18.

Knapp, J.S. (1986). Nitrogen fertilization and plant growth regulator effects on winter wheat yields, grain quality, lodging and chemical composition of culms. Dissertation Abstracts International, B. Sciences and Engineering, 47: 2238.

Knapp, J.S. and C.L. Harms (1988). Nitrogen fertilization and plant growth regulator on yield and quality of four wheat cultivars. Journal of Production Agriculture, 1: 94 - 98.

Koinuma, K., F. Ikegaya and E. Ito (1998). Non destructive and quantitative evaluation method for root lodging resistance by measurement of horizontal pulling resistance in maize. Grassland Science, 43: 424 - 429.

Kumar, N. and C.P. Singh (1999). Yield and yield components of maize (*Zea mays* L.): physiological analysis on seasonal variations. Indian J. Plant Physiology, 4(2): 90-94.

Kushibiki, H. (1979). Studies on the productivity of maize for silage and the cropping combinations of cultivars differing in maturation dates in north eastern areas of Hokkaido 2. Influence of planting density on yield and quality of early and late cultivar groups. Journal of Japanese Society of Grassland Science, 25: 136 - 143.

Leask, W.C. and T.B. Daynard (1976). Changes in pH of corn stover following grain maturity. Crop Science, 16: 505 - 507.

Lizaso, J.H., W.D. Batchelor and M.E. Westgate (2003). A leaf area model to simulate cultivar specific expansion and senescence of maize leaves. *Field Crops Res.,* 80: 1-17.

Mangen, T.F., P.R. Thomison and S.D. Strachan (2005). Early season defoliation effects on topcross high-oil corn production. Agron. J., 97: 823–831.

McGahen, J.H. (1973). The performance of maize hybrid over varying plant populations as affected by cultural tillage treatments of chemically weeded fields. Dissertation Abstracts International, 34: 1341.

Mclchinger, A.E., H.H. Geiger and G.A. Schmidt (1986). Vertical root pull resistance and its relationship to root lodging and forage traits in early maturing European inbred lines and F1 hybrids of maize. Maydica, 31: 335 - 348.

Melchiori, R.J.M. and O.P. Caviglia (2008). Maize kernel growth and kernel water relations as affected by nitrogen supply. *Field Crops Res.,* 108: 198-205.

Melis, M. and M.P.W. Farina (1984). Potassium effects on stalk strength, premature death, and lodging of maize *(Zea mays* L). Crop Production, 13: 26.

Menz, M.A. and A.R. Hallauer (1997). Reciprocal recurrent selection of two tropical corn populations adapted to Iowa. Maydica. 42: 239 - 246.

Milbourn, G.M., G.E.D. Tiley and M.K.V. Carr (1978). Planting density for grain maize in southeast England. Experimental Agriculture, 14: 261 - 268.

Miller, T. L and O.J. Myers (1974). Correlation of pitch cell death with various stalk quality characteristics in two synthetic populations of maize. Crop Science, 14: 215 - 217.

Minami, M. and A. Ujihara (1991). Effects of lodging on dry matter production grain yield and nutritional composition at different growth stages in maize *(Zea mays* L.). Japanese Journal of Crop Science, 60: 107 - 115.

Mohamad, I.H.Z.A. (1978). Interrelationship of plant population, harvest dates and genotypes on lodging and yield of maize (*Zea mays* L). Malaysian Applied Biology, 7: 49 - 54.

Montomery, K.T. (1987). Effects of six cycles of reciprocal full sib selection on several grain, stalk and root traits in BSI 1. Dissertation Abstracts International, B. Sciences and Engineering, 48: 926 - 927.

Moser, S.B., B. Feil, S. Jampatong and S. Peter (2006). Effects of pre-anthesis drought, nitrogen fertilizer rate, and variety on grain yield, yield components, and harvest index of tropical maize. Agric. Water Managt., 81 (1-2): 41-58.

Muchow, R.C. (1989). Comparative productivity of maize, sorghum and pearl millet in a semi-arid tropical environment. II Effects of water deficits. Field Crops Res., 20: 207 -219.

Mulatu, K. (1993). Method and time of intercropping maize *(Zea mays* L) with haricot bean *(Phaseolus vulgaris* L.). IAR Newsletter of Agricultural Research, 8: 2 - 3.

Mulebo, N., T.G. Hort and G.M. Paulsen (1983). Physiological factors affecting maize yields under tropical and temperate conditions. *Tropical Agric. (Trinidad),* 60: 3-10.

Munteanu, I., E. Nagy, I. Timpeanu, I. Bilaus, M. Stefanescu and I. Pasca (1987). Control of breaking and lodging in maize by specific technological measures. Contributii ale cercetarii stiintifice la dezvoltarea agriculturii. P: 263 - 279.

Nesmith, D.S., and J.T. Ritchie (1992). short-and long-term responses of corn to a pre-anthesis soil water deficit. Agron. J., 84:107 - 113.

Norberg, O.S., S.C. Mason and S.R. Lowry (1989). Ethephon alteration of corn plant morphology. Agron. J., 81: 603 - 609.

Norberg, O. S., S. C. Mason and S. R.Lowry (1988). Ethephon influence on harvestable yield, grain quality and lodging of corn. Agron. J., 80: 768 - 772.

Nouna, B.B., N. Katerji and M. Mastrorilli (2000). Using the CERES-Maize Model in a semi-arid Mediterranean environment. Evaluation of model Performance. Euro. J. Agron., 13 (4): 309–322.

Okigbo, B.N. (1973). Maize experiments on the Nsukka plain. V. Effect of pre-planting cultivations, mulching and weeding frequency on the yield and general performance of maize. Agronomie Tropicale, 28: 54 - 74.

Pandey, S.N., B.D. Nautiyal and C.P. Sharma (2008). Pollution level in distillery effluent and its phytotoxic effect on seed germination and early growth of maize and rice. *J. Environmental Biology, 29*: 267-270.

Pellerin, S., R. Trendel and A. Duparque (1990). Relationship between morphological characteristics and lodging susceptibility of maize *(Zea mays* L). Agronomic, 10: 439 - 446.

Pereira, W, J. A. Osuna and S. Miguel (1977). Study of variability in the root system of maize *(Zea mays* L.) cultivars in relation to several phenotypic characters. Cientifica, 5: 40 - 44.

Perry, C.O. (1983). Diallel analysis of rind puncture and grain yield and their interactions with plant densities for twelve elite inbred lines of maize, *Zea mays* L. Dissertation Abstracts, 44: 975.

Pinter, L. and L. Kalman (1981). Effect of leaf area reduction on standability and premature death of leaves in maize (Zea mays L.) hybrid with different genotypes. Acta Agronimica Academiae Scientiarum Hungaricae, 30: 61 - 67.

Pordesimo, L.O., B.R. Hames, S. Sokhhansanj and W.C. Edens (2005). Variation in corn stover composition and energy content with crop maturity. Biomass Bioenergy. 28: 366–374.

Prine, G.M. (1971). A critical period for ear development in maize. Crop Sci., 11: 782-786.

Quanqi, C. Yuhai, L. Mengyu, Z. Xunbo, D. Baodi and Y. Songlie (2008). Water potential characteristics and yield of summer maize in different planting patterns. Plant Soil Environ., 54 (1): 14 - 19.

Rajan, I. and M. Tollenaar (1999). Source: sink ratio and leaf senescence in maize: I Dry matter accumulation and partitioning during grain filling. Field Crops Res., 60 (3): 245 – 253.

Rao, M. (1997). Evaluation of Barbary-Plante on water stress management in maize and sunflower. Progress Report Nagarjuna Agrl. Research and Development Institute. Hyberabad, India.

Remison, S. U. and D. Akinleye (1978). Relationship between lodging, morphological characters and yield of varieties of maize *(Zea mays* L.). Journal of Agricultural Science (U.K), 91: 633 - 638.

Remison, S.U. and J.M. Fajemisin (1982). Comparative growth of maize cultivars with different leaf orientation. J. Agric. Sci. (U.K), 99: 61 - 66.

Riedell, W.E. and G.E. Inglett (1985). Influence of 1, l-dimethyl piperidinium chloride on morphological and physiological characteristics of Zea mays L. In: Proceedings of the Plant Growth Regulator Society of America, Twelfth Annual Meeting, P: 137 - 145.

Ritchie, S.W., J.J. Hanway and G.O. Benson (1992). How a corn plant develops. Iowa State Univ. Coop. Ext. Serv. Spec. Rep. 48. Ames.

Ritchie, S.W. and J.J. Hanway (1997). How a corn plant develops Spec. Rep. 48. Iowa State Univ. Coop. Ext. Serv., Ames.

Rosic, K., M. Ivanovic and D. Fidler (1979). Investigation of stem breakage in maize *(Zea mays* L.) and mathematical modelling of the mechanical stresses. Arhiv za Poljoprovredne Nauke, 32: 13 - 25.

Saxena, M.C. and V. Singh (1965). A note on area estimation intact maize leaves. *Indian J. Agron.,* 10: 437-439.

Schenk, M.K. and S.A. Barber (1979). Root characteristics of corn genotypes as related to phosphorus uptake. *Agron. J.,* 71(6): 921- 924.

Sharma, S.R. (1970). Effect of foliar spray of panchagavya on protein content of maize (*Zea mays* L.). *Madras Agric. J.,* 34(2): 122-125.

Shobha, D., T.A. Sreeramasetty, Puttaramanaik and K.T.P. Gowda (2010). Evaluation of maize genotypes for physical and chemical composition at silky and hard stage. Karnataka J. Agric. Sci., 23 (2): 311-314.

Shun, S. X., W.L. Gu and J.Y. Dai (1989). The effect of plant density on lodging of maize J. Shenyang Agric. Univ., 20: 413 - 416.

Sierra, J., H.O. Lafontaine, L. Dufour, A. Meunier, R. Bonhomme and C. Welcker (2006). Nutrient and assimilate partitioning in two tropical maize cultivars in relation to their tolerance to soil acidity. *Field Crops Res.*, *95* (2-3): 234 - 249.

Singh, H and B. Rai (1979). Differences in morphological character and yield of imprtant maize Genotype and their recovered brachytic versions. Indian Journal of Agricultural Sciences, 49: 758 - 763.

Stamp, P. and C. Kiel (1992). Seedling trails of maize as indicators of root lodging. Agronimie, 12: 157 - 162.

Steiner, J.L., T.A. Howell, J.A. Tolk and A.D. Schneider (1991). Evapotranspiration and growth predictions of CERES-Maize, Sorghum and Wheat in the Southern High Plains. In: Proceedings of the 1991 National Conference, ASCE, Honolulu, Hawaii, USA, 22–26 July 1991. P: 212 – 219.

Stewart, D.W., C. Costa, L.M. Dwyer, D.L. Smith, R.I. Hamilton and B.L. Ma (2003). Canopy structure light interception and photo synthesis in maize. Agron. J., 95: 1465 -1473.

Subedi, K.D. and B.L. Ma (2005). Nitrogen uptake and partitioning in stay-green and leafy-maize hybrids. Crop Sci., 45: 740-747.

Sun, S.X., J.Y. Dai, and W.L. Gu (1989). Effects of N, P and K on lodging and yield in maize. Scientia Agricultura Sinica, 22: 28 - 33.

Teare, I.D. and D.L. Wright (1990). Corn yield starter fertilizer interaction for yield and lodging. Crop Sci., 1298 - 1303.

Thiagarajah, M.R., L.A. Hunt and J.D. Mahon (1981). Effects of position and age on leaf photosynthesis in corn. Can. J. Bot., 59: 28-33.

Thiagarajan, N. (1981). Effect of moisture stress at different growth phases of maize Ganga 5. M.Sc. (Ag.) Thesis, Tamil Nadu Agricultural University, Coimbatore.

Trappeniers, G., M. D. Neuville, D.Mouraux and J. F. Ledent (1987). Effect of ethephone on plant height and yield of maize, Mededelingen van de Faculteit Landbouwwetenschappen. Rijksuniversiteit Gent., 52: 1275 - 1288.

Trooijen, T.P., L.L. Buschman, P. Sloderbeck, K.C. Dhuyvetter and W.E. Spurgeon (1999). Water use efficiency of different maturity corn hybrids and grain sorghum in the central Great Plains. J. Prod. Agric., 12: 377–382.

Twumasi, A.S. and R.B. Hunter (1982). Evaluation of quantitative methods for determining stalk quality in short season corn genotypes. Canadian J. Plant Sci., 62: 55 - 60.

Tyagi, N.K., D.K. Sharma and S.K. Luthra (2003). Determination of evapotranspiration for maize and berseem. Irrig. Sci., 21: 173–181.

Vasilas, B.L. and R.D. Seif (1985). Defoliation effects on two inbreds and their single-cross hybrid. Agron. J., 77: 816-820.

Vasilas, B.L. and R.D. Seif (1985a). Effect of defoliation effects on the timing of anthesis and silking of maize inbred lines. Maydica, 30: 427–435.

Vasilas, B.L. and R.D. Seif (1985b). Pre-anthesis defoliation effects on six corn inbreds. Agron. J., 77: 831–835.

Videnovic, A. and M. Ivanovic (1980). The influence of N, P and K application on stalk rot, lodging and yield in maize (Zea mays L.). Arhiv za Poljoprivredne Nauke, 41: 341 - 346.

Watkins, M.E. and J. Wu (2008). Water use efficiency and sensitive stage to water deficit in maize. Plant Physiol., 74: 526 - 534.

Westgate, M.E. and J.S. Boyer (1986). Reproduction at low silk and pollen water potentials in maize. Crop Sci., 26: 951 - 956.

Whigham, D.K. (1972). Leaf orientation and its effect on *Zea mays* L. Dissertation Abstracts, 32: 1320.

Wilhelm, W.W. and L. Etoran (1995). Yield quality and N use of inbreed corn with very numbers of leaves removed during tasselling. Crop Sci., 35: 209 – 210.

Williams, W.A., R.S. Loomis, W.G. Duncan, A. Dovert and F.Nunez (1968). Canopy architecture at various population densities and the growth and grain of corn. Crop Sci., 8: 303 - 308,

Xianshi, G., T.R. Sinclair and J.D. Ray (1998). Effect of drought history on recovery of transpiration, photosynthesis and leaf area development in maize. Soil Crop Sci. Soc. Fla Proc., 57: 83-87.

Zaidi, P.H. and N.N. Singh (2005). Stresses on maize in tropics. Directorate of Maize Research, PUSA campus, New Delhi, P: 148 - 461.

Zaidi, P.H., G. Srinivasan, H.S. Cordova and C. Sanchez (2004). Gains from improvement for mid-season drought tolerance in tropical maize (*Zea mays* L.). Field Crops Res., 89 (1): 135-152.

Zhang, X.Y. (1997). Field researches on water potential, transpiration and resistance to water flow for wheat and maize. Acta Geogr. Sin., 52: 543–550.

Zuber, M.S. and M.S. Kang (1978). Corn lodging slowed by sturdier stalks. Crops and Soil, 30: 3 - 15.

# 11

# Integrated Nutrient Management

A maize crop producing 9.5 tonnes of grain per hectare under North American conditions can remove the following amounts of nutrients through grain plus stover (IFA, 1992), macronutrients (kg/ha): N 191, $P_2O_5$ 89, $K_2O$ 235, MgO 73, CaO 57 and S 21. micronutrients (g/ha): Fe 2 130, Zn 380, Mn 340, B 240, Cu 110, Mo 9 and also 81 kg Cl.

## Macronutrients

High yield in maize is closely associated with N application, but only where other inputs and management practices are optimal. N interacts positively with plant population, earliness of sowing, variety, weed control and moisture supply.

Neither higher plant population nor high levels of N alone will improve yields where a third factor is limiting. Where moisture supply is inadequate or uncertain, optimal levels of fertilizer as well as plant population will be below those required for top yields. Fertilizer can improve the utilization of soil water by increasing rooting depth. However, the best returns from N fertilizer are only obtained where the water supply, either natural or supplemented by irrigation, is adequate for full crop growth. Under good growing conditions, a yield response of 30 kg grain/kg N can be obtained.

Maize takes up N slowly in the early stages of growth. However, the rate of uptake increases rapidly to a maximum before and after tasseling, when it can exceed 4 kg N/ha/

day. N fertilizer application is best scheduled in accordance with this pattern of uptake in order to avoid serious losses by volatilization or leaching and to ensure that N levels are high in the soil when the crop demand is also high.

An application to the seedbed followed by a side-dressing when the crop is knee high, or for very high application rates two top dressings (the second at tasseling) are usually recommended. The N application rates for rainfed maize are about half of those for the irrigated crop.

Fertilizer requirement in relation to yield level can be calculated directly from crop uptake of N only in specific regions because of the variations in soil N supply and the rather unpredictable efficiency of fertilizer N by the crop. However, N fertilizer requirement may be about 50 kg/ha with unimproved varieties under uncertain rainfall, rising with yield potential to 250–300 kg N/ha where yields of 12 tonnes/ha or more can be expected. Local recommendations on amounts of N should, as always, be based on local experimentation under the prevailing growing conditions. For irrigated HYVs of maize such as hybrids/composites in India, the general recommendation is to apply 60–80 kg N/ha to early-maturing varieties, 80–100 kg N/ha to medium-duration varieties and 90–150 kg N/ha to late-maturing varieties.

Adequate P is very important for maize as the crop cannot readily take up soil P in the large amounts needed for optimal growth and high yield. Best results from N and other inputs will not be obtained without adequate P, which should be applied mainly in water-soluble form. Rates of P application should be varied according to soil test for available P and in relation to yield potential. These can be in the range of 30–100 kg $P_2O_5$/ha. Phosphate application in the highly successful maize production region of Illinois (USA), is based on soil tests and crop removals with the twin objectives of building up the soil P level up to the optimum and replacing the P removed by maize at harvest.

K is taken up in large quantities by maize but only a small proportion of total uptake is removed with the grain. While maize can obtain appreciable amounts of soil K, it is important to ensure that the overall supply is sufficient for high yields. Use of K fertilizer is especially important where high rates of N fertilizer are used and high yields expected. Recommended rates of K application are in the range of 30–100 kg $K_2O$/ha. Where the soils are supplied adequately with K, its application is advocated on the basis of soil analysis and yield potential.

In the intensive maize–wheat annual rotation, fertilizer recommendations in subtropical India suggest that the application of P and K to maize can be omitted where the preceding wheat crop has been regularly fertilized with these nutrients or 12–15 tonnes FYM/ha is applied to maize. Organic manures should be applied 3–4 weeks before planting maize. These can be surface broadcast followed by incorporation in the soil.

Both P and K are most effective where applied as a basal dressing before or at the time of planting through a seed-cum-fertilizer drill. Where suitable equipment is available, sideband application, together with a moderate rate of N will improve effectiveness on many soils. Where mechanical equipment for fertilizer application is not available, the fertilizer can be dropped in open furrows and covered with soil before planting.

## Micronutrients

Maize can suffer from a number of micronutrient deficiencies. However, Zn deficiency is perhaps the most widespread problem. The problem is mostly on alkaline calcareous soils and soils with low organic matter content. Zn deficiencies may be intensified by a high level of P supply from the soil and/or fertilizer. Local experience combined with soil and plant analysis can be used as a basis for Zn application. One example of a recommendation from India is to add 25 kg zinc sulphate (21 percent Zn) mixed with 25 kg soil along the row, followed by hoeing and irrigation. Where Zn deficiency symptoms are seen in the standing crop, foliar spray can be given at a rate of 3 kg zinc sulphate/ha + 1.5 kg of lime/ha in 500 litres of water.

## Nitrogen

For every 100 kg of grain produced 1.8 kg N in the grain and 1.0 kg in the above ground parts of the plant are required and must be supplied by the soil and the fertilizer. The usual rates of application vary from 100 to 200 kg N/ha.

## Phosphorus

Maize accumulates phosphorus throughout the growing season with maximum uptake occurring between 3 and 6 week of growth. Phosphorus uptake is generally paralled to dry matter accumulation, proceeding slightly ahead of dry matter accumulation early in the season and lagging behind somewhat later on. At maturity about 75 % of the total phosphorus of the above ground parts has become translocated to the grain. Phosphorus content of the leaves tends to reach a maximum about the middle of the season and then decreases as the plant matures.

**Table 1. Nutrient uptake by maize crop**

| Grain yield (t/ha) | Part | Nutrient uptake (kg/ha) | | | | | | Micro nutrient uptake (g/ha) | | |
|---|---|---|---|---|---|---|---|---|---|---|
| | | N | $P_2O_5$ | $K_2O$ | Mg | Ca | S | Mn | Cu | Zn |
| 6.3 | Grain | 100 | 40 | 29 | 9.3 | 1.5 | 7.8 | 70 | 40 | 110 |
| | Stover | 63 | 23 | 92 | 28 | 15 | 9 | 940 | 30 | 200 |
| | Total | 163 | 63 | 121 | 37.3 | 16.5 | 16.8 | 1010 | 70 | 310 |

**Table 2. Recommended dose of nutrients for maize crop in various states of India**

| Season | Recommended dose of nutrients (N:P:K kg/ha) | States |
|---|---|---|
| Kharif | 100: 60: 40 | Odisha, Bihar and Madhya Pradesh |
| Kharif | 120:40:0 | Rajasthan and Gujarat |
| Kharif | 120:60:40 | Himachal Pradesh, Maharashtra, Punjab, Uttar Pradesh and Uttaranchal |
| Kharif | 150:80:60 | Chhattishgarh, Haryana, Karnataka, Jammu & Kashmir, West Bengal and Tamil Nadu. |
| Rabi | 120:75:50 | Bihar and Rajasthan |
| Rabi | 175:60:50 | West Bengal |
| Rabi | 225:80:80 | Andhra Pradesh |
| Spring | 80:40:30 | Bihar |
| Spring | 120:75:50 | Punjab and Uttar Pradesh |

* Apply 80 % N in four splits as top dressing in following proportions

**Table 3. Crop stage and nitrogen application**

| Crop stage | Nitrogen rate (%) |
|---|---|
| V4 (4 leaf stage) | 25 |
| V8 (8 leaf stage) | 30 |
| VT (Tasseling stage) | 20 |
| GF (grain filling stage) | 5 |

## Nutrient Deficiencies in Maize

### Nitrogen deficiency

#### Symptoms

Maize is highly sensitive to N deficiency. Deficient plants become stunted and develop small ears. Affected ears have small kernels and low numbers of kernels per ear. The growth of young plants is severely reduced even under mild deficiencies. Nitrogen is mobile in plants and under conditions of low soil supply it is easily mobilised from older to younger leaves. The deficiency symptoms appear first and become more severe, on older leaves. In mature plants, pale green young leaves, pale yellow to pale brown middle leaves and dry older leaves can appear simultaneously. Affected ears have small kernels and low numbers of kernels per ear.

#### Developmental stages

In mild deficiencies, the entire plant appears uniformly light green in colour. If the deficiency becomes severe, a pale yellow chlorosis begins at the tip of older leaves and proceeds towards the base along the midrib in a V-shaped pattern. As the deficiency advances, the whole leaf becomes pale brown and dies. The deficiency is likely to occur in soils having low organic matter, light-textured sandy soils that have been leached by heavy rainfall or excessive irrigation, soils exhausted by intensive cropping and waterlogged conditions.

#### Integrated nutrient management

Analyse the soil before planting to measure the amount of plant available N. Basally apply the analysis-based recommendation for N using organic manures, nitrogenous fertilizer and biofertilizers. Top dress soluble fertilizers such as urea in two split doses. For quick recovery in standing crops, apply urea (2 % solution) as foliar spray. Foliar sprays are required to be repeated every 10 to 15 days. Incorporate legume crops into the rotation.

### Phosphorus deficiency

#### Symptoms

Deficient plants appear stunted, thin and spindly, delayed in maturity and produce low grain yields. Phosphorus is mobile in plants and under conditions of low soil P supply it is easily mobilised from older to younger leaves. The deficiency symptoms appear first on older leaves, while young leave usually remain unaffected. Older leaves develop a characteristic dark to bluish green colouration. Under severely deficient conditions, the bluish green leaves can develop reddish-brown or purple tints.

### Developmental stages

In the early stage of the deficiency, a dark purple colour develops on the margins of older leaves. Symptoms usually begin from the tip and proceed towards the base of the leaves. If the deficiency becomes severe, the entire leaf becomes reddish-brown or purple. As the deficiency advances, the brown necrosis develops on the tip and proceeds along the margins towards the base of the leaves. This deficiency likely to occurs in soils having low organic matter, alkaline and calcareous soils, soils exhausted by intensive cropping, acid soils and highly weathered soils and areas with topsoil removed by erosion.

### Integrated nutrient management

Analyse the soil before planting to measure the amount of plant available N. Basally apply the analysis-based recommendation for P using organic manures, phosphatic fertilizers and phosphorus-solubilising microbial inoculants. In standing crops, apply soluble P fertilizers such as ammonium phosphate with irrigation water.

## Potassium deficiency

### Symptoms

Potassium deficiency causes shortening of the internodes, dwarfing of plants and a general loss of healthy dark green growth. Younger actively growing leaves draw K from older parts of the plant. Deficiency symptoms appear first on older leaves, while younger leaves usually remain green and healthy. Symptoms begin as a pale yellow chlorosis on the tips of old leaves and cover marginal tissue. The chlorosis is followed by pale brown necrosis and both (chlorosis and necrosis) advance down the margins towards the base. Affected plants produce small ears that are often very pointed and underdeveloped at the tip. In severely deficient conditions, prominent red strips develop on lower stem and leaf sheaths.

### Developmental Stages

Mild deficiency causes stunted growth, thin stems and pale green foliage. When deficiency persists and becomes more severe, marginal chlorosis develops on older leaves starting from the leaf tips. Chlorosis is followed by pale brown necrosis. Both chlorosis and necrosis advance down the margins towards the base, leaving the mid-vein and surrounding tissue pale green –

marginal chlorosis and necrosis of older leaves is the specific symptom of K deficiency in maize. In acute deficiency conditions, red strips develop on the lower stem and lower leaf sheaths. Deficiency is likely to occur in soils formed from parent material low in K, light-textured soils where K has been leached by heavy rainfall or excessive irrigation, soils low in organic matter, soils with wide Na:K, or Ca:K ratios, soils with acute sodic/saline conditions, soils receiving irrigation waters with high $HCO_{3-}$ concentration and acid soils having pH below 6.0.

**Potassium deficiency**

**Magnesium deficiency**

### Integrated nutrient management

Analyse the soil before planting to measure the amount of plant available K. Problematic acid/alkaline/saline soils should be reclaimed. Add organic manures well before planting. Apply $KC1$, $K_2SO_4$, or $KNO_3$ to the soil at or before planting as per soil testing recommendations. In standing crops, apply soluble K sources with irrigation water. Foliar sprays are usually not recommended because a number of spray applications are needed to fulfill crop requirements.

## Calcium deficiency

### Symptoms

Deficient plants appear stunted, with short and stout stems. Grain yields are drastically reduced. Calcium is immobile in plants and under conditions of low soil supply, it is not easily mobilised from older to younger leaves. The deficiency symptoms appear first on younger leaves while older leaves usually remain unaffected, dark green and healthy in appearance. Newly formed leaves become torn and malformed, which gives the plant a ragged appearance. Under severely deficient condition, the tips of the youngest leaves are joined together and do not separate from the whorl, which gives the plant a "ladder-like" appearance.

### Developmental stages

Mild deficiency causes stunted growth, thin stems and pale green foliage. Early on, the young leaves become pale green and then develop yellow to white lesions in interveinal (between the veins) tissues. If the deficiency becomes severe, the yellow to white lesions become enlarged and the lamina tears from these areas. In acutely deficient conditions, the newly emerging leaves die. The deficiency symptoms is likely to occur in acid, sandy soils that have been leached by heavy rainfall, sodic soils that have high exchangeable Na, soils having high soluble Al and low exchangeable Ca and strongly acid peat and muck soils that have low exchangeable Ca.

### Integrated nutrient management

Analyse the soil before planting to measure the lime or Ca requirement of the soil. Apply the analysis-based recommendation for Ca-containing fertilizers well before planting using soluble nutrient amendment and fertilizer sources such as gypsum ($CaSO_4 2H_2O$), $Ca(NO_3)_2$ or $CaCl_2$. In low pH soils, lime or $CaCO_3$ should be applied to correct the problem.

## Magnesium deficiency

### Symptoms

Deficient plants appear stunted with thin and long stems, reduced kernel size and small ears. Magnesium is mobile in plants and under conditions of low soil supply, it is easily mobilised from older to younger leaves. The deficiency symptoms appear first on older leaves. A pale yellow interveinal chlorosis develops in the middle portion of the older leaves and then proceeds towards the tip of leaves. The youngest leaves usually remain unaffected and appear healthy.

### Developmental stage

In the early stages of the deficiency, the old leaves become pale green then develop pale yellow interveinal chlorosis. If the deficiency becomes more severe, the chlorotic interveinal tissues turn into brown necrotic lesions. As the symptom advances, rust brown strips develop on older leaves and then leaf tips and margins become dark brown and die. In acutely deficient conditions, the older leaves die and hang down. Deficiency likely occur in acid sandy soils that have been leached by heavy rainfall, peat and muck soils that have low total Mg, soils having high quantities of Ca or K and soils derived from parent material that is inherently low in Mg.

### Integrated nutrient management

Analyse the soil before planting to measure the amount of plant available Mg. Apply an analysis-based recommendation for Mg before planting using soluble fertilizers such as $MgSO_4$ or $MgCl_2$. In standing crops, apply soluble Mg fertilizers along with irrigation water.

## Sulphur deficiency

### Symptoms

Deficient plants appear stunted, spindly, have delayed maturity and produce low grain yields. Sulphur is immobile in plants and under conditions of low soil supply, it is not easily mobilised from older to younger leaves. Deficiency symptoms appear first on younger leaves. The whole plant generally appears pale green or pale yellow, but the youngest leaves appear the most pale. The pattern of yellowing on the entire leaf appears to be uniform, affecting both the veins and the interveinal tissues equally.

### Developmental stages

In mild deficiencies, all leaves on the plant become pale green although old leaves remain darker. If the deficiency becomes severe, the younger leaves turn pale yellow while old leaves remain pale green. In the later stage, the youngest leaves turn pale yellow to white but without necrosis. Deficiency is likely to occur in soils having low in organic matter, light textured sandy soils that have been leached by heavy rainfall or excessive irrigation, soils exhausted by intensive cropping and soils derived from parent material that inherently low in S.

### Integrated nutrient management

Analyse the soil before planting to measure the amount of plant available. Apply an analysis-based recommendation for S well before planting by mixing either elemental S or gypsum ($CaSO_4$) into the soil surface. In standing crops, apply soluble S fertilizers such as $(NH_4)_2SO_4$, or $K_2SO_4$ with irrigation water.

## Iron deficiency

### Symptoms

Deficiency symptoms appear first on younger leaves while older leaves appear normal. A pale yellow chlorosis develops in interveinal tissues, leaving the veins to appear green and prominent.

### Developmental stages

In mild deficiencies, the top-most leaves of plants develop temporary fading of interveinal tissues along with prominent green veins. Plants can recover and regain a normal appearance with time. When deficiency persists and becomes more severe, interveinal tissues of affected leaves become obviously pale yellow with prominent green veins. Interveinal chlorosis uniformly extends the full length of the leaf – a specific symptom of Fe deficiency in maize. In later stages of deficiency, the prominent green veins also fade and become light green to pale yellow. In acutely deficient conditions, the entire leaf bleaches and appears papery white. Deficient is likely to occur in sandy soils having low total Fe, calcareous soils where solubility of Fe is very low, peat and muck soils where organic matter ties up Fe and reduces its availability in soil solution and acid soils with excessively high levels of soluble An, Mn, Cu, or Ni, which can hamper the uptake of Fe by the plant despite high availability of Fe in soil solution.

### Integrated nutrient management

Analyse the soil before sowing to measure the amount of plant available Fe. Problematic alkaline soils should be reclaimed. Add any organic manure well before sowing. Basally apply soluble Fe fertilizer sources such as $FeSO_4$ (commonly at 25 kg/ha) or Fe chelates (10 kg/ha). Use of organic chelates appears most promising as other forms of Fe can be rapidly made insoluble upon their application to affected soils. In standing crops, apply $FeSO_4$ or Fe chelates (0.5 % solution) as foliar sprays. As Fe is immobile in plants, foliar sprays are required to be repeated every 10 to 15 days.

## Manganese (Mn) deficiency

### Symptoms

Deficient plants appear stunted with short, thin stems and pale green to yellow leaves. Harvested ears and kernels are small in size, resulting in low grain yields. It is immobile in plants and under conditions of low soil supply it is not easily mobilised from older to younger leaves. Deficiency symptoms appear first on younger leaves.

### Developmental stages

In mild deficiencies, the leaves become light green with pale yellow interveinal chlorosis. If the deficiency becomes severe, the leaves turn pale yellow and white flecks appear in interveinal chlorotic areas. Deficiency is likely to occur in light-textured sandy soils that have been leached by heavy rainfall or excessive irrigation, calcareous and alkaline soils where solubility of Mn is very low, waterlogged peaty soils where organic matter ties up Mn and reduces its availability in the soil solution and soils derived from parent material that is inherently low in Mn.

### Integrated nutrient management

Analyse the soil before planting to measure the amount of plant available Mn. Problematic alkaline soils should be reclaimed. Add any organic manure well before planting. Apply a basal dressing of soluble forms of Mn fertilizer such as $MnSO_4$. In standing crops, apply $MnSO_4$ (0.2 to 0.3 % solution) as a foliar spray. Repeated sprays may be required if symptoms reappear.

## Zinc deficiency

## Symptoms

Zinc deficiency symptoms in maize commonly occur within two weeks after seedling emergence. Since Zn is not readily transferred from older to younger plant parts, symptoms develop first on young leaves, while older leaves remain green and apparently healthy. White to yellow broad bands of bleached tissue can appear on each side of the midrib beginning at the base of the leaf. The midrib and the leaf margins remain green. Zinc deficiency may cause delayed and irregular silk emergence with distorted tassels devoid of anthers. Deficiency prevents internode elongation resulting in the crowding of the top-most leaves producing a fan-shaped appearance.

### Developmental stages

Mild deficiencies may appear as interveinal striping on young leaves somewhat similar to Fe or Mn deficiency. However, in case of Mn and Fe, the interveinal striping runs the full length of the leaf while Zn deficiency occurs mainly on the basal half of the leaf. If the deficiency persists and become more severe, the youngest leaves turn pale green while broad, white to yellow bands appear between the midrib and margin in the basal half of the leaf – a specific symptom of Zn deficiency in maize. Eventually the affected tissue dies and turns pale gray. Deficiency likely occurs in leached sandy soils where total Zn is low, alkaline soils having pH more than 7.5, recently levelled soils where sub-soil is exposed for cultivation. Available Zn in surface soil is often double that of the sub-soil and soil with excessive P application – a condition that can hamper the uptake and use of Zn by crops.

### Integrated nutrient management

Analyse the soil before planting to estimate the amount of plant available Zn. Problematic alkaline soils should be reclaimed. Add organic manures well before planting. Apply $ZnSO_4$ (commonly at 25 – 30 kg/ha) or zine chelate at 10 kg/ha once every two years in Zn deficient soils. Do not mix Zn fertilizers with P fertilizers. In standing crops, spray 3.0 kg $ZnSO_4$ plus 1.5 kg up-slaked lime/ha in 500 liters water 2-3 weeks after seedling emergence.

**Table 4. Recommended manures and fertilizers for maize based cropping system in India**

| State | Manures / fertilizers | Remarks |
|---|---|---|
| Andhra Pradesh | 60-80kg N + 30-40 kg $P_2O_5$ /ha | N in two splits (50 % as basal and 50 % at knee high stage) |
| Chhattishgarh | 40 kg N + 40 kg $P_2O_5$ /ha (desi)<br>80 kg N + 60 kg $P_2O_5$ /ha (hybrid) | N in two splits (50 % as basal and 50 % after first weeding) |
| Karnataka | 10 t FYM/ha + 75:50:25:15 kg NPKZn/ha | N in two splits (50 % as basal and 50 % at knee high stage) |
| Madhya Pradesh | 80: 60:20 kg NPK/ha (composites) | N in two splits (50 % as basal and 50 % at knee high stage) |
| | 100:60:20 kg NPK/ha (hybrids) | N in two splits (50 % as basal, 25 % at knee high and 25 % at tasselling stage) |
| Maharashtra | 75:30:0 kg NPK/ha | - |
| Rajasthan | 50:30:0 kg NPK/ha | N in two splits (50 % as basal and 50 % at knee high stage) |
| Uttar Pradesh | 80:40:30 kg NPK/ha | - |

(Ramachandrappa and Mudalagiriyappa, 2011)

Over 84 % of Maize area in India is rainfed. The rainfall pattern determines the yield levels with applied fertilizers to a considerable extend. Recommendations given on the basis of zones are summarized in Table 5.

**Table 5. Zone wise general fertilizer recommendations for maize**

| Zone | Crop Maturity | Kharif maize (kg/ha) | | | Rabi maize (kg/ha) | | | Remarks |
|---|---|---|---|---|---|---|---|---|
| | | N | P | K | N | P | K | |
| I | Early | 60 | 40 | 20 | - | - | - | Half N at sowing, half N at knee high stage |
| | Medium | 80 | 40 | 30 | - | - | - | |
| | Late/full | 90 | 40 | 30 | - | - | - | |
| II | Early | 60 | 30 | 20 | - | - | - | As in zone I for kharif. For Rabi, 25 % N at sowing, 50 % at knee high and 25 % N at flowering |
| | Medium | 90 | 40 | 30 | 120 | 60 | 40 | |
| | Late/full | 120 | 60 | 40 | 180 | 60 | 40 | |
| III | Early | 45 | 30 | 20 | - | - | - | As in zone II. |
| | Medium | 60 – 90 | 40 | 30 | 120 | 50 | 40 | |
| | Late/full | 60-120 | 40 | 30 | 180 | 60 | 40 | |
| IV | Early | 60 | 30 | 20 | - | - | - | In kharif in 3 equal splits at sowing, knee high and flowering stages. In Rabi, 25 % N at sowing, 50 % at knee high stage and 25 % at flowering. |
| | Medium | 90 | 40 | 40 | 100 | 50 | 40 | |
| | Full irrigated | 150 | 75 | 40 | 150 | 60 | 40 | |
| | Full Rainfed | 100 | 50 | 25 | - | - | - | |
| V | Early | 80 | 40 | 40 | - | - | - | In kharif of 50 % N in 10 cm deep at sowing, rest at knee high and flowering stages. In rabi, 75 % N placed at 10 – 15 cm deep and 25 % N top dressed at knee high stage. |
| | Medium | - | - | - | 120 | 50 | 40 | |
| | Late/full | 120 | 60 | 40* | 180 | 100 | 100 | |

* Application of 10 – 25 kg $ZnSO_4$ /ha also recommended.

Higher grain yield of 4280 kg/ha was recorded when 12 t FYM/ha was applied along with 120:60: 60 kg N, $P20_5$ and $K_2O$ /ha. However, FYM application increased gain yield to the extent of 532 kg /ha over no FYM, that is at the rate of 44 kg grain/t FYM. Incorporation of poultry manure, castor cake and FYM into soil increased grain yield by 314, 414 and 442 kg/ha respectively over a control yield of 1211 kg/ha. Application of lime equivalent to 100 % of lime requirement gave higher maize yield (5147 kg/ha) and neutralised exchangeable Al. Liming augmented the uptake of P and Ca. Combined application of FYM and biofertilizer was found to be superior to their individual application (Singh et al., 1983).

Increasing levels of potassium upto 60 ppm K increased the dry matter weight and dry weight of roots upto 90 pmm if applied to maize in calcarlous soil (Jadav et al., 1991). Response of maize to K was highly significant upto 60 mg $K_2O$/ kg soil application. The production efficiency to K application decreased with increasing levels of potassium (Grewal and Malik, 2010).

Organic manures have always played pivotal role in revitalizing the soil health. They decompose and lead to formation of humus and substances, which are house of micro flora and fauna in soil. It has crucial role in holding nutrients in liable pool, buffering the soil, source of food and shelter to micro-organisms, improving availability of macro and micro nutrients and so on. Further organic manure plays role in stabilizing physical properties soil like bulk density, water holding capacity, infiltration, etc. Therefore, it has become important component in integrated nutrient supply system. Among the various organic manures, poultry manure has relatively higher proportion of micro and macro nutrients like N (4.5 - 5.0 %), $P_2O_5$ (2.50-2.98 %), $K_2O$ (2.04-2.33 %), Ca (2.4-8.8 %), Mg (0.44-0.67 %), S (0.13-0.15 %), Zn (235-463 ppm) and Mn (225-406 ppm) (Devegowda 1997).

## Nutrient Uptake

Highly significant response of maize of 25 t FYM/ha as composed to control in respect of nitrogen use efficiency (Sahota, 1984). At Dholi (Bihar), in calcareous sandy loam soil, the N, P and K uptakes were significantly superior with FYM 1.5 t/ha over control during the winter seasons (Singh *et al.*, 1991). At Palampur (Himachal Pradesh) on clayey soil containing 378.8, 15.6 and 246.4 kg available N, $P_2O_5$ and $K_2O$/ ha, N, P and K content in grain and stover of maize was significantly superior with 10 t FYM/ha over 5 t FYM/ ha (Verma, 1991). At Palampur (Himachal Pradesh) on clay loam soil with 113, 7 and 100 available N, P and K mg/ kg, respectively, total N and K uptakes were significantly more under 10 t FYM/ha over 5 t FYM/ ha and control (Minhas and Sood, 1994). Application of

FYM to maize increased nitrogen uptake and also reduced the loss of nutrients by leaching (Brar et al., 2001).

Addition of chicken manure produced higher tissue concentration of protein, nitrogen (N) and phosphorus (P) than cattle manure (Materechera and Salagae, 2002). At Rakh Dhiansar (Jammu), maximum K uptake by grain, straw and total K uptake were recorded with FYM @ 10 t/ha over control. At Udaipur, FYM application @ 10 t/ha also had significant and positive effect on nitrogen and phosphorus uptake by the crop over control (Mahala and Shaktawat, 2004). At Kanpur (Uttar Pradesh), on sandy loam soil having 182, 19.50 and 126.31 kg/ha available N, $P_2O_5$ and $K_2O$, respectively application of FYM @ 5 t/ha significantly influenced the N, P and K uptake by grain, stover and whole plant (Prasad et al., 2005). Composed poultry manure recorded higher nutrient uptake, agronomic efficiency and apparent nitrogen recovery and it was found on par with sewage sludge and enriched urban garbage compost (Reddy et al., 2005). In Sweet corn, nitrogen uptake of kernels, leaves and stem and total uptake of nitrogen by the crop increased significantly with subsequent increase in FYM level. However, in case of 20 and 15 t FYM/ha the nitrogen, phosphorus and potassium uptake in plant was significantly superior over 10 t FYM/ha and was at par with each other (Gosavi, 2006).

## Quality Characters

At Bangalore (Karnataka), protein per cent, protein yield and sugar per cent in the grains were significantly higher under 10 t FYM/ha over 5 t FYM/ha and control (Ramamurthy and Shivashanker, 1996). At Anand (Gujarat), the per cent protein content of maize grains recorded significantly superior value of 6.76 % with FYM @ 5 t/ha over control during kharif season (Barevadia and Patel, 1996).

At Anand, during Rabi season 75 % RDN + 25 % N through vermicompost prepared from *Parthenium hysterophorus* L. had prominent effect on content of total soluble sugar of sweet corn (22.0 %) which was statistically at par with (21.7 %) 75 % RDN + 25 % N through vermicompost prepared from *Amaranthus spinosus* in sweet corn (Khadtare et al., 2006). In sweet corn, levels of FYM did not differ significantly in respect of protein content in the kernels. However, FYM @ 20 t/ ha showed numerically higher protein content in the kernels 10 and 15 t FYM/ ha was significantly superior to 10 t FYM/ ha (Gosavi, 2006).

At Dapoli, the percent sugar content in grains of sweet corn was increased with increasing levels of FYM and recorded significantly superior with FYM @ 20 t/ ha. The percent protein content in grains did not reach the level of significance (Zende, 2006). In sweet corn, sugar content in the grains of sweet corn increased significantly with the increase in the levels of poultry manure in the nutrient sources and it was maximum with 100 % N as poultry manure (Pinjari, 2007).

## Economics

At Kanpur (Uttar Pradesh), gross income and net profit were higher with application of FYM over control. The cost of cultivation was maximum with the FYM @ 20 t/ ha as compared to other FYM levels, due to the higher cost of FYM. The gross returns and net returns were maximum with FYM @10 t/ ha. The B: C ratio was recorded maximum with control (Verma et al., 2003). At Shimoga (Karnataka), soil application of zinc sulphate @ 10 kg/ha recorded maximum yield of stover (7.52 t/ ha), grain (6.96 t/ha), total uptake of zinc (551.92 g/ ha) and NPK by maize compared to 0.2 % zinc solution as foliar spray and seed priming with 1 % zinc solution for 8 hours treatments (Asha, 2001). At Bapatala (Andhra Pradesh), application of FYM 5.0 t/ ha and soil inoculation of phosphate solublizing bacteria @ 2.5 kg/ ha showed significantly higher yield attributes, yield and P uptake of maize over other treatment combinations (Gahukar et al., 2011).

**Table 6. Effect of agro techniques on yield attributes and yield of maize**

| Treatments | Cob weight (g) | Shelling (%) | Grain yield (kg/ha) | Stover yield (kg/ha) |
|---|---|---|---|---|
| **Organic/ inorganic** | | | | |
| No organic manure and fertilizers | 157.7 | 74.0 | 5809 | 6469 |
| FYM 5.0 t/ ha | 199.8 | 84.0 | 7112 | 8233 |
| Sun hemp *in situ* green manuring | 184.0 | 79.8 | 6773 | 7690 |
| 50 % RDP (30 kg $P_2O_5$/ ha) | 163.7 | 78.1 | 5778 | 7360 |
| 100 % RDP (60 kg $P_2O_5$/ ha) | 171.2 | 80.3 | 6567 | 7482 |
| CD 5 % | 5.7 | 0.73 | 559 | 491 |
| **Bio fertilizer inoculation** | | | | |
| No- inoculation | 163.7 | 76.4 | 5921 | 6715 |
| Soil inoculation of PSB 2.5 kg/ ha | 185.7 | 81.6 | 6608 | 7674 |
| Arbusicular mycorrhizae 5.0 kg/ ha | 176.4 | 79.7 | 6826 | 7447 |
| CD 5 % | 1.4 | 0.82 | 546 | 497 |

**Table 7. Effect of integrated balanced nutrient management on yield & net profit of maize**

| Treatments | Grain yield (kg/ha) | Net return (Rs/ ha) | B:C ratio |
|---|---|---|---|
| Farmers practices (110:60:0 kg NPK/ha) | 2436 | 11642 | 1.01 |
| RDF (150 : 60: 0 kg NPK /ha) + FYM 5t/ ha + Biofertilizers (Azos + Phospho) | 3157 | 16741 | 1.26 |
| Fertilizer application based on soil test value (150 : 60 :30 kg NPK/ ha) + vermicompost 5t/ha + Biofertilizers (Azos + Phospho) | 3675 | 19912 | 1.32 |

At Banswara (Rajasthan), application of fertilizers based on soil test value + vermicompost + biofertilizers recorded higher yield, net return and B: C ratio (Singh et al., 2012).

## Micronutrient

At Coimbatore (Tamil Nadu), significant response was obtained to B application on dry matter production, yield attributes and yield of maize at the level of 2 kg B /ha followed by 1.5 kg B /ha. Foliar application also significantly increased the yield of maize over control. On comparing the sources of boron, Agribor was found to be effective in influencing the yield of maize (Sujatha, 2005).

**Table 8. Effect of sources, levels and methods of B application on yield characters and yield of maize**

| Levels and methods of B application | Length of cob (cm) | Grains/ cob | 100 seed weight (g) | Grain yield (q /ha) | Straw yield (q /ha) |
|---|---|---|---|---|---|
| 0 kg/ha | 17.1 | 414 | 21.0 | 56.5 | 108 |
| 0.5 | 17.2 | 419 | 21.2 | 56.9 | 110 |
| 1.0 | 17.4 | 426 | 21.3 | 58.0 | 113 |
| 1.5 | 18.4 | 438 | 21.7 | 60.8 | 115 |
| 2.0 | 19.2 | 455 | 22.9 | 62.4 | 120 |
| Foliar spray of 0.2 % | 17.5 | 432 | 21.2 | 59.7 | 115 |
| Foliar sprayof 0.3 % | 18.2 | 437 | 21.7 | 60.0 | 115 |
| CD 5 % | 0.10 | 12.0 | 0.43 | 1.542 | .89 |

| Levels and methods of B application | Length of cob (cm) | Grains/ cob | 100 seed weight (g) | Grain yield (q /ha) | Straw yield (q /ha) |
|---|---|---|---|---|---|
| **Sources of B** | | | | | |
| Borax and boric acid | 17.7 | 428 | 21.4 | 58.8 | 113 |
| Agribor | 17.9 | 434 | 21.7 | 59.5 | 115 |
| CD 5 % | 0.05 | 6.51 | 0.23 | 4.47 | 8.37 |

(Sujatha, 2005)

At Coimbatore (Tamil Nadu), application of 5.0 kg Zn + 40 kg S + 1.5 kg B /ha registered higher grain yield (6410 kg/ ha) which produced comparable to 5.0 kg Zn + 40 kg S + 1.5 kg B + 0.5 kg Mo /ha which produced 6342 kg grain /ha (Table 9) was found superior over. Application of Mo did not influence the yield of maize (Duraisami et al., 2007).

**Table 9. Effect of treatments on grain and stover yield and nutrient uptake of maize**

| Treatments | Yield (kg /ha) | | Nutrient uptake | | |
|---|---|---|---|---|---|
| | Grain | Stover | B (g/ ha) | Zn (g/ ha) | S (kg /ha) |
| Control | 4214 | 5300 | 351.3 | 351.3 | 17.16 |
| 1.25 kg Zn /ha as enriched FYM | 4703 | 6867 | 444.6 | 358.7 | 22.77 |
| 2.5 kg Zn /ha as enriched FYM | 4721 | 5967 | 350 | 381.1 | 22.66 |
| 5.0 kg Zn /ha as enriched FYM | 5104 | 6433 | 566.3 | 585.8 | 23.58 |
| 2.5 kg Zn + 1.5 kg B /ha as enriched FYM | 5632 | 6500 | 518.7 | 665 | 27.6 |
| 2.5 kg Zn /ha as enriched CCP | 5218 | 6433 | 538.2 | 893.6 | 27.63 |
| 5.0 kg Zn + 1.5 kg B + 5.0 kg Fe /ha | 5901 | 6600 | 619.6 | 759.8 | 29.84 |
| 5.0 kg Zn /ha alone | 5097 | 6450 | 458.4 | 803 | 26.36 |
| 5.0 kg Zn+ 20 kg S /ha | 5708 | 7867 | 581.1 | 1075.9 | 33.94 |
| 5.0 kg Zn + 40 kg S /ha | 5979 | 7900 | 502.4 | 1132.5 | 37.39 |
| 5.0 kg Zn + 40 kg S + 1.5 kg B ha | 6410 | 7933 | 477.1 | 1242.4 | 35.08 |
| 5.0 kg Zn + 40 kg S + 1.5 kg B + 0.5 kg Mo /ha | 6342 | 7783 | 551.2 | 1154.7 | 31.93 |
| CD 5 % | 417 | 1079 | 56 | 89.2 | 2.92 |

(Duraisami et al., 2007)

### Sulphur

Sulphur application significantly increased N uptake by maize plant when sulpur was applied from 0 to 30 kg S/ha. Significantly higher N uptake was observed when 30 kg S/ha was applied over 20 kg S/ha but the differences among the rates below 20 kg S/ha were found to be non-significant (Muhammad *et. al.*, 2004).

### Hybrid maize

The application of 60 kg $P_2O_5$ per ha gave higher yield of 6.21 tonnes followed by 5.92 with 30 kg $P_2O_5$ per ha (Thanki *et al.* 1988).

## Green Manuring

A significant increase in maize grain yield (35.09 q/ha) was recorded with the incorporation of 20 t /ha of cowpea over the application of 20 t FYM / ha (30.2 q/ha). However, yield obtained with cowpea incorporation was equal to application of 120 kg N / ha (Singh and Brar, 1985). Green manuring with *Sesbania cannabina* and *Sesbania rostrata* with 90 kg N / ha to maize crop increased the grain yield by 10.69 and 10.49 q per ha over control, respectively. *Sesbania cannabina* and *Sesbania rostrata* alone increased maize grain yield by 2.87 and 2.77 q per ha over control, respectively. Green manuring either of the Sesbania species + 30 kg N per ha gave maize yield on par with 90 kg N per ha. It indicated that green manuring was equivalent to 60 kg nitrogen (Tiwari *et al.*, 2004).

## References

Ahmed, M.A. (1989). Response of maize varieties to nitrogen rates. *Annals of Agricultural Research*, 34 (2): 873-887.

Anonymous (1994). Delineation of micro and secondary nutrients deficiency areas. Twenty five years of micronutrients in soils and crops of Tamil Nadu. Department of Soil Science and Agricultural Chemistry, TNAU, Coimbatore.

Arulmozhiselvan, K., R. Vengatesan and M. Deivanai (2009). Nutriseed holder technique for increasing nutrient use efficiency and yield under wetland and upland situations in rice and maize. Res. Crops, 10 (3): 473 – 480.

Arya, K.C. and S.N. Singh (2000). Effect of different levels of P and Zn on yield and nutrient uptake of maize (*Zea mays* L.) with and without irrigations. Indian J. Agron., 45(4): 717-721.

Ashok Kumar, (2008). Productivity, economics and nitrogen use efficiency of specialty corn (*Zea mays* L.) as influenced by planting density and nitrogen fertilization. *Indian J. Agron*., 53(4): 306-309.

Athar, M. (1979). Influence of nitrogen fertilizer and plant density on some physiological aspects and grain yield of maize (Zea mays L.). Dissertation Abstracts International, 40:2524.

Badiyala, D. and S.P. Verma (1991). Integrated nitrogen management in maize (*Zea mays)* + soyabean (*Glycine max*) - wheat cropping sequence under midhills of Himachal Pradesh. *Indian Journal of Agronomy,* 36 (2) 496-501.

Baenziger, P.S. and D.V. Glover (1980). Effect of reducing plant population on yield and kernel characteristics of sugary-2 and normal maize. Crop Sci., 20: 444-447.

Bagal, P.K. and A.K. Shingte (1986). Effect of foliar and soil application of nitrogen on composition of maize grains. J. Maharashtra Univ., 11(1): 115-116.

Bagheri, N., H. Ahmadi and S.K. Alavipanah (2012). Soil-line vegetation indices for corn nitrogen content prediction. International Agrophys., 26: 103-108.

Balyan, J.S. and L.K. Idnani (2000). Fertilizers management in maize (Zea mays L.) – wheat (Triticum avestium) sequence. Indian J. Agron., 45: 648:652.

Brar, B.S., N.S. Dhillon and H.S. China (2001), Integrated use of farm yard manure and inorganic fertilizers in maize (*Zea mays*). *Indian Journal of Agricultural Sciences*, 71 (9): 605-607.

Channabasavanna, A.S., M. Hegde and D. P. Biradar (2007). Evaluation of poultry manure in combination with inorganic fertilizers for its productivity, sustainability and economic feasibility in maize. *Research on Crops* 8(1): 95-98.

Channappa, S.L. (1998). Integrated nutrient management in hybrid maize (*Zea mays* L.). *M.Sc. (Agri.) Thesis*, University of Agricultural Sciences, Dharwad.

Dahiya, S.S. and B.K. Bhatia (1982). Effect of iron and nitrogen levels on dry matter yieldand nutrients uptake by maize. *Haryana Agricultural University Journal of Research*, 12 (2): 254-258.

Das, M., B.P. Singh, M.Ram and R.N. Prasad (1991). Response of maize to phosphorus-enriched manures grown in P-deficit Alfisols on terraced land in Meghalaya. *Indian Journal of Agricultural Science*, 61 (6): 383-388.

Devarajan, R., P. Savithri and K.R. Kumaresan (1988). Response of maize to FYM and zinc. *Madras Agricultural Journal*, 75 (5-6): 204-205.

Devarajan, R., P. Savithri, T.S. Manickam, G.V. Kothandaraman and M. Guruswamy (1987). Residual effect of manures and zinc on maize. *Madras Agricultural Journal*, 74 (3): 150-153.

Duncan, W.G. (1958). The relationship between corn population and yield. Agron. J. 50: 82–84.

Duraisami, V.P., T. Chitdeshwari, K.S. Subramanian and R. Rajeswari (2007). Effect of micronutrients and sulphur on yield and nutrient uptake by maize in an alfisol. Madras Agric. J., 94 (7-12): 283-288.

Dwivedi, S.K., R.S. Singh and K.N. Dwivedi (2002). Effect of sulphur and zinc nutrition on yield and quality of maize in Typic Ustochrept soil of Kanpur. J. Indian Soc. Soil Sci., 50 (l): 70-74

Gajri, P.R., V.K. Aror and M.R. Chaudhary (1994). Maize growth responses to deep tillage, straw mulching and FYM in coarse textured soils of North-Western India. *Soil Use and Management*, 10 (5): 15-20.

Geh, R.J., J.P. Schnidt, L.D. Maddux and W.B. Gordon (2005). Corn yield response to nitrogen rate and timing in sandy irrigated soil. Agron. J., 97:1230-1238.

Jat, S.L., C.M. Parihar and A.K. Singh (2014). Higher productivity and profitability in maize systems by nutrient management. Indian Farming, 64 (4): 12 – 15.

Jayaprakash, T.C. (2001). Role of organics and inorganics in yield maximization of hybrid maize (*Zea mays* L.). *M.Sc. (Agri.) Thesis*, University of Agricultural Sciences, Dharwad.

Kamalakumari, K. and P. Singaram (1996). Effect of continuous application of FYM and NPK on fertility status of soil yield and nutrient uptake in maize. *Madras Agriculture Journal* 83 (5): 181-184.

Kapur, M.L. and D.S. Rana (1980). Response of maize hybrid to the application of fertilizers on cultivators fields. *Indian Journal of Agronomy*, 25 (2): 299-301.

Klausner, S.D. and R.W. Guest (1981). Influence of Ammonia ($NH_3$) conservation from daily manure on the yield of corn. *Agronomy Journal*, 73: 720-723.

Krishnasamy, V. and K.R. Ramaswamy (1986). Effect of N, P and K on maize hybrid seed crop II. Seed weight, germination and vigour. Madras Agric. J. 73: 42-45.

Kumar, K., J.N. Kaul and H.S. Gill. (1992). Optimization of nitrogen fertilization on wintermaize (*Zea mays*). Indian J. Agron., 37(2): 359-360.

Lamani, C.S., Y.B. Palled and H. Shenoy (2000). Soil organic carbon as influenced byintegrated nutrient management in hybrid maize. *Current Research*, 29 (3-4): 37-38.

Madhumitadas, B.P. Singh, M. Ram and R.N. Prasad (1992). Mineral nutrition of maize and groundnut as influenced by P-enriched manures on acid alfisols. *Journal of Indian Society of Soil Sciences*, 40: 580-583.

Mahamoud, M.H., R. Ekshuder and I.M.A. Aziz (1980). Corn production on calcareous sols as influenced by N application and plant population. *Agricultural Research Review*, 58(4): 199-210.

Mallesham, A., K.G. Reddy and A.T. Reddy (1992). Effect of lime, farm yard manure and zinc-sources on zinc content, uptake and dry matter yield in maize crop. *Andhra Agricultural Journal*, 39 (3-4): 198-201.

Mandoff, F.R. and J.R. Amadon (1980). Yield trends and soil chemical changes resulting from N and manure application to continuous corn. *Agronomy Journal*, 72: 161- 164.

Mehta, S.C., S.R. Poonia and B.S. Pawar (1994). Effect of FYM, Gypsum and zinc on the performance of maize in sodic soil. *Crop Research*, 8 (3): 480-485.

Memon, S.Q, M. Zakria, G.R. Mari, M.H. Nawaz and Z. Khan (2011). Effect of tillage methods and fertilizer levels on maize production. *Pakistan Journal of Agricultural Sciences,* 48 (2): 117–20.

Mundra, S.L., A.K. Vyas and P.L.Maliwal (2002). Effect of weed and nutrient management on nutrient uptake by maize (Zea mays L.) weeds. Indian J. Agron., 47: 378-383

Muthukrishnan, P. and S. Subramanian (1980), Weed control in maize under nitrogen levels. *Madras Agricultural Journal*, 67: 785-789.

Muthukumar, V.B., K. Velayudham and N, Thavaprakash (2007). Plant growth regulators and split application of nitrogen improves quality parameters and green cob yield of baby corn. J. Agron., 6 (1): 208 – 2011.

Nanjappa, H.V., B.K. Ramachandrappa and B.U.Mallikarjuna (2001). Effect of integrated nutrient management on yield and nutrient balance in maize (Zea mays L.). Indian J. Agron., 46: 698-701.

Negi, S.C., K.K. Singh and R.C. Thakur (1992), Economics of phosphorous and farm yard manure application in wheat-maize sequence. *Indian Journal of Agronomy* 37 (5): 30-33.

Parasuraman, P. (2005).Studies on integrated nutrient requirement of hybrid maize (Zea mays L.) under irrigated conditions. Madras Agric. J. 92 (1-3): 89-94.

Patel, H.R., R.S. Joshi, and K.R. Patel (1985), Response of maize to various levels of nitrogen and irrigation during summer on heavy black soils. *Indian Journal of Agronomy*, 30 (3): 381-383.

Pawar, S.D., A.S. Jadhav and B.K. Jagtap (1996). Effects of biogas slurry and nitrogen (urea) on maize yield, nutrient uptake and soil properties. *Journal of Maharashtra Agricultural Universities*, 16 (3): 316-319.

Prasad, U.K., Dadansingh, N.N. Sharma, and T.N. Prasad (1985). Effect of soil moisture and nitrogen levels on the grain yield, water requirement, WUE and growth of winter maize. *Indian Journal of Agricultural Sciences*, 55 (4): 265- 268.

Prasad, U.K., H.G. Thakur, S.S. Pandey and N.N. Sharma (1987). Effect of irrigation and nitrogen on winter maize in calcareous saline-alkaline soils. *Indian Journal of Agronomy*, 32 (3): 217-220.

Rai, M.M and Dighe. (1971). Response of added B to maize and choice of extractant for B. JNKVV Res. J., 5(2): 120-122.

Reddy, B.B., R.N. Reddy, V.M. Reddy, M.R. Reddy and K.B. Swamy (1987). Effect of plant population on the performance of maize hybrids at different fertility levels in a semiarid environment. Indian J. Agric. Sci., 57: 705 – 709.

Reddy, M.R. and S.J. Patil (1982). Response of hybrid maize to different sources, levels and split dose of nitrogen application under irrigated conditions. *Mysore Journal of Agricultural Sciences*, 16 (4): 404-407.

Sakal, R., A.P. Singh and. R.B. Sinha (1989). Differential susceptibility of maize varieties to B deficiency in a calcareous soils. J. Indian Soc. Soil Sci., 37 (3): 582-584.

Sandhu, B.S. and S. S. Hunda (1991). Effects of method and date of sowing on productivity of winter maize (Zea mays L). Ind J. Agric. Sci., 61: 178-181.

Sanjeev, K., A.S. Bangarwa and S. Kumar (1997). Yield and yield components of winter maize (Zea mays L.) as influenced by plant density and nitrogen levels. Agric. Sci. Dig. (Karnal), 17: 181-184.

Sankaran, N., S. Meena And N. Sakthivel (2005). Input management in maize. Madras Agric. J., 92 (7-9): 464-468.

Selvaraju, R. and M.R. Iruthayaraj (1994). Influence of irrigation scheduling, methods of irrigation and nitrogen levels on growth and yield of maize. *Madras Agricultural Journal*, 81 (8): 418-420.

Shanmugasundaram, R. and P. Savithri (2005). Effect of levels and frequency of zinc sulphate application on yield and DTPA - Zn availability in maize-sunflower cropping system. Madras Agric. J., 92 (4-6): 252-257.

Sharma, J.P. (1983). Economy in fertilizer use through organic manures in growing maize. *Indian Journal of Agronomy*, 28 (2): 154-155.

Shilpashree, V.M., H.M. Chidanandappa, R, Jayaprakash and B.C. Punitha ( 2012). Effect of integrated nutrient management practices on distribution of nitrogen fractions by maize crop in soil. Indian J. Fundamental and Applied Life Sci., 2 (1): 38 - 44.

Shivay, Y. S. and R. P. Singh (2000). Growth, yield attributes, yields and nitrogen uptake of maize (Zea mays L.) as influenced by cropping systems and nitrogen levels. Ann. of Agric. Res., 21: 494-498.

Singh, N.B. and P.P. Singh (1984). Effect of intercropping with legumes on grain yield of maize and its residual effect on succeeding wheat. *Indian Journal of Agronomy*, 29: 295-298.

Singh, U.K. and O.P. Dubey (1991). Response of maize to the application of nitrogen and phosphorus. *Current Research*, 20 (8): 153-154.

Singh, V.K., R.P. Bajaj and R.J. Sisodia. (1992). Effect of sowing dates and plant population on yield of maize. *Cur. Res.*, 21(2): 24-25.

Sood, B.R., O.P. Awasthi and S.K. Sharma (1981). Effects of cultural practices on yield and grain quality in maize. *Indian Journal of Agricultural Research*, 13: 119- 121.

Srinivas, G and V. Satyanarayana (1996). Nutrient removed by weeds and maize (Zea mays L.). Indian J. Agron., 41: 160-162.

Srinivasan, K. (1992). Effect of amendment and zinc level on the growth and yield of maize. *Indian Journal of Agronomy*, 37 (2): 246-249.

Subramanian, K.S. and C. Charest (1997). Nutritional, growth and reproductive responses of maize (*Zea mays* L.) to arbuscular mycorrhizal inoculation during and after drought stress at tasselling. *Mycorrhiza*, 7: 25-32.

Sujatha, S. (2005). Effect of sources, levels and methods of boron application on dry matter production, yield attributes and yield of maize (*Zea mays* L.) in Inceptisol. Madras Agric. J., 92 (7-9): 479-483.

Surendra Singh and A.K. Sarkar (2001). Balanced use of major nutrients for sustaining higher productivity of maize (Zea mays) – wheat cropping system in acidic soils of Jharkhand. Indian J. Agron., 46: 605-610.

Thakur, D.R., O. Prakash, P.C. Kharwara, and S.K. Bhalla (1997). Effect of nitrogen and plant spacing on yield, nitrogen uptake and economics in baby corn (Zea mays L.). Indian J. Agron., 43(4): 668 - 671.

Venkatesh, R, (1999). Response of hybrid maize (*Zea mays* L.) to row application of organics and nitrogen in transitional tract of northern Karnataka. *M.Sc. (Agri.) Thesis*, University of Agricultural Sciences, Dharwad.

Vos, J., P.E.L Vanderputten and C.J. Birch (2005). Effect of nitrogen supply on leaf appearance, leaf growth, leaf nitrogen economy and photosynthetic capacity in maize (*Zea mays* L.). *Field Crops Research*, 93: 64-73.

Yadav, D. N. (1990). Growth and productivity of maize under different crop sequences and nitrogen rates. Ph.D. Thesis, G. B. Pant Univ. of Agric. and Tech., Pantnagar, India.

# 12

# Water Management

Maize is grown throughout the country under varied soil and climatic conditions. It is grown from sea level to an attitude of 2000 m or more in colder zones as well as in tropical areas. Maize is mostly grown in kharif in northern region both as rainfed or irrigated and during summer season under irrigated conditions on a smaller scale. In southern India, because of favourable weather conditions, three crops of maize can be taken in a year. The recent advances have made it possible to grow maize under irrigated conditions during winter season also in north India.

The optimum depletion level of soil moisture for maize has been found to be 50 % of availability from the surface 30 – 60 cm depth of soil in north India on sandy loam to clay loam soil on lighter soils only 20 – 25 % depletion is adequate. In order to get higher yield on sandy loam soils in kharif season, it is desirable to apply irrigation at a soil moisture tension of 0.6 bar at 15 cm soil depth (Table 1).

**Table 1. Irrigation requirement of maize**

| States | Soil type | Season | Irrigation requirement | |
|---|---|---|---|---|
| | | | Numbers | Amount (cm) |
| Delhi | Sandy loam | kharif | 2 – 3 | 10 – 15 |
| Haryana | Sandy loam | -do- | 5 – 6 | 30 – 36 |
| Uttar Pradesh | Sandy loam | -do- | 4 | 22 – 27 |
| Rajasthan | Sandy loam | -do- | 4 | 30 |
| Karnataka | Clay loam | Kharif | 3 | 15 |
| | | Rabi | 5 | 33 |
| | | Summer | 10 | 51 |
| Orissa | Loam | Kharif | 2 | 10 – 15 |
| | | Rabi | 11 | 50 – 60 |
| | | Summer | 18 | 900 |
| Tamil Nadu | Loamy sand | Summer | 25 | 125 |

(Anonymous, 1977)

The critical physiological stages for applying irrigation in maize in north and central India have been found to be early vegetative stage (20 – 40 DAS) and tasselling and silking stage (45 to 60 DAS). In south India, on clay soils higher yields were obtained by irrigating at 25 % depletion of available soil moisture during the vegetative stage upto about 60 DAS and 50 % depletion after wards. The irrigation requirement of maize varies with the type of soil and season in which grown depending upon the rainfall received. In north India, the crop is normally sown after a pre-sowing irrigation as the monsoon arrives late and hence crop requires 2 – 3 irrigations during pre-monsoon period. During the monsoon, irrigation may be required if soil moisture falls below the desired levels.

In case of winter maize, due to erratic and uncertain rains, the dependence on irrigation is more pronounced. In eastern region of the country for winter maize, the optimum irrigation schedule is to apply irrigation at IW/CPE ratio of 0.9 to 1.0 with 6 cm depth requiring 4 – 6 irrigations. Likewise in Central and South India on black soils, good yield of rabi maize (about 6 t / ha) has been obtained with 5 – 6 irrigations of 6 cm depth scheduled at IW/CPE ratio of 1.0. However, in red soils of south India as well as sandy loam soils of North India, 8 – 10 irrigations are needed for good production of rabi maize with irrigation applied at IW/CPE ratio of 1.2 to 1.5 having 6 to 7.5 cm depth of water (Table 2).

**Table 2. Yield and water requirement of maize**

| States | Grain yield (q/ha) | Water requirement (cm) | Basis of irrigation |
|---|---|---|---|
| Bihar (Rabi) Pusa | 38 | 33 | IW/CPE = 1.00 |
| Rajasthan (Kota) | 27 | 55 | IW/CPE = 0.80 |
| Punjab (Ludhiana) | 39 | 20 | One bar soil moisture tension at 20 cm depth |
| Delhi | 50 | 30 | 0.25 bar soil moisture tension |
| Haryana (Hissar) | 31 | 45 | 50% available soil moisture |
| Majarashtra (Rahuri) | 34 | 20 | IW/CPE = 0.60 |
| Andhra Pradesh (Hyderabad) (Rabi) | 33 | 40 | IW/CPE = 0.90 |
| West Bengal (Kharagpur) | 47 | 36 | IW/CPE = 0.90 |
| Gujarat (Navsari) | 27 | 64 | 25% depletion of available soil moisture |

Rajput (1987)

**Table 3. Effect of irrigation on spring maize**

| Irrigation (ID / CPE ratio) | Cobs/ plant | Grain yield (kg/ha) | Stover (kg/ha) | Harvest index (%) |
|---|---|---|---|---|
| 0.2 | 1.14 | 2412 | 8176 | 22.78 |
| 0.4 | 1.38 | 4065 | 10224 | 28.45 |
| 0.6 | 1.40 | 4087 | 11236 | 28.08 |
| CD 5 % | 0.11 | 328 | 765 | 1.35 |

(Tyagi et al., 1998)

At Hisar, frequent irrigation (0.6 ID/CPE ratio) resulted in higher yield and yield attributes of maize (Tyagi et al., 1998).

**Table 4. Performance of irrigation methods on upland maize**

| Irrigation methods | Grain yield (k/ha) | Stover yield (kg/ha) |
|---|---|---|
| Surface irrigation | 2924 | 6014 |
| Micro sprinkler | 2622 | 5726 |
| Drip irrigation | 2234 | 5289 |
| CD 5 % | 65 | 194 |

(Arulmozhiselvan et al., 2009)

At Coimbatore (Tamil Nadu), surface irrigation recorded higher grain yield and stover yield under upland condition (Arulmozhiselvan et al., 2009).

At Coimbatore (Tamil Nadu), drip fertigated maize with 100 % RDF with 50 % P and K as water soluble fertilizer recorded significantly higher grain yield (7309 kg/ha) followed by 150 % RDF (7212 kg/ha). Higher net return (Rs. 56858 /ha) and B:C ratio (3.24) were obtained under drip fertigation with 150 % RDF + radish intercrop (Anitta Fansh et al., 2011).

**Table 5. Irrigation types on maize crop**

| Irrigation method | Grains/ grain row | Shelling (%) | Grain yield (t/ha) | Harvest index (%) | Soil Moisture extra (%) | | | | WUE (kg/ ha/mm) |
|---|---|---|---|---|---|---|---|---|---|
| | | | | | 0-15 | 15-30 | 30-60 | 60-90 | |
| Life saving irrigation | 22.06 | 70.2 | 3.71 | 45.11 | 96.8 | 144.5 | 74.2 | 55.5 | 10.01 |
| Assured irrigation | 26.19 | 80.3 | 4.38 | 46.64 | 132.3 | 179.8 | 82.7 | 63.2 | 9.57 |
| CD 5 % | 2.35 | 9.1 | 2.9 | NS | - | - | - | - | - |

(Sepat and Ashok Kumar, 2007)

At New Delhi, assured irrigation recorded higher growth, yield attributes and yield of crop than life saving irrigation while WUE decreased with assured irrigation (Sepat and Ashok Kumar, 2007).

**Table 6. Effect of irrigation on winter maize**

| IW/CPE ratio | Soil moisture depletion (%) (0-30 cm) | Grains /cob | Grain yield (kg/ha) | WR (cm) | WUE (kg/ ha-cm) | Net return (Rs./ha) |
|---|---|---|---|---|---|---|
| 0.6 | 39.45 | 365 | 3651 | 386 | 416 | 16926 |
| 0.8 | 40.45 | 379 | 4264 | 444 | 348 | 20969 |
| 1.0 | 42.06 | 388 | 4881 | 438 | 373 | 24163 |
| 1.2 | 41.97 | 390 | 5285 | 499 | 309 | 25685 |
| CD 5 % | - | 21 | 397 | 4.4 | 31 | 3106 |

At Pusa (Bihar), IW/CPE ratio 1.2 of 6 cm depth gave higher grains/cob, grain yield and net return. However, water use efficiency was decreased with increase in IW/CPE ratio and was maximum at IW/CPE ratio 0.6 (Bharati et al., 2007).

## Maize Water Use Characteristics

Evapotranspiration ($ET_c$), or crop water use, is the water removed from the soil by evaporation from the soil surface and transpiration by the plant. For maize, evaporation can account for 20 percent to 30 percent of growing season $ET_c$. Transpiration is the last step in a continuous water pathway from the soil, into the plant roots, through the plant stems and out through leaf surfaces and into the atmosphere. Approximately 70 percent to 80 percent of crop water use results from plant transpiration. The amount of daily corn water use varies with atmospheric conditions: air temperature, humidity, solar radiation and wind speed. High air temperatures, low humidity, clear skies and high wind speed will result in high $ET_c$ demand. High humidity, clear skies and low wind speed will result in lower $ET_c$ demand.

Atmospheric demand must be adjusted for the stage of crop development to estimate crop water use on a daily basis. A high atmospheric demand day in early May will result in little $ET_c$ because the corn plant is small with a limited root zone and little leaf area to transpire water. The same atmospheric demand day in mid-July will result in near peak crop water use because the corn roots are fully developed and the plant leaf area is sufficient to transpire water at rates equal to the atmospheric demand. This adjustment is referred to as applying a crop coefficient to the atmospheric demand or potential evapotranspiration.

Seasonal water use is affected by climatic conditions, relative maturity range, soil fertility, water availability and the interaction of these factors. Although the total amount of water used by maize will vary from season to season and location to location, it will generally follow the pattern dictated by seasonal trends in weather variables and corn development. The smooth curve in *Figure 1 (Curve A)* illustrates the long-term average water use pattern for maize. This average water use pattern shows typical daily $ET_c$ levels throughout the growing season based upon the average daily $ET_c$ over a 10 plus year period. The jagged curve in *Figure 1 (Curve B)* illustrates the fluctuation possible in daily $ET_c$ values for an individual year. Thus, irrigation managers must be familiar with the long-term trend but more importantly is able to determine what the daily $ET_c$ was over the last few days. Knowledge of the long-term trend and actual daily crop water use rates are critical to determining when to irrigate and how much water to apply.

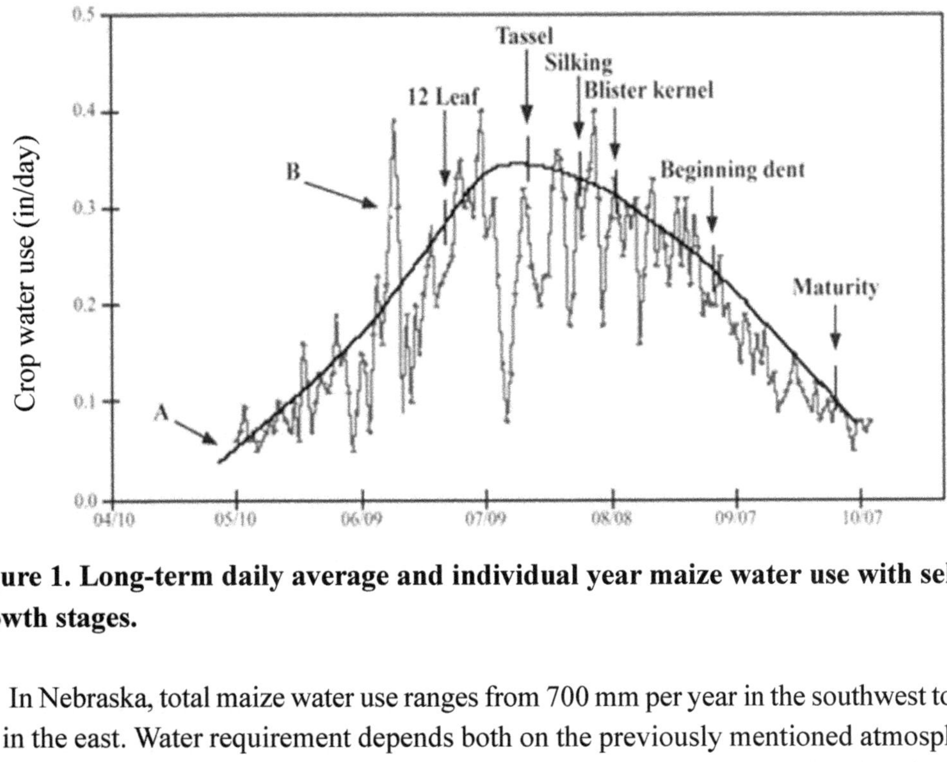


**Figure 1. Long-term daily average and individual year maize water use with select growth stages.**

In Nebraska, total maize water use ranges from 700 mm per year in the southwest to 600 mm in the east. Water requirement depends both on the previously mentioned atmospheric conditions and corn variety. The relative maturity range of a particular variety has the most impact on seasonal $ET_c$. For example, at the same location and in the same year, a corn hybrid with a 113-day maturity will use more water than a 100-day hybrid. Longer-season corn hybrids use more water, but they also have the potential to produce more grain if the heat units and water supply are available. If both varieties are able to mature fully, the grain produced for each mm of $ET_c$ is approximately equal. The difference in seasonal water use is due to total days of water use, and in some cases a difference in daily water use.

Due to variation in weather conditions in a given year and from year to year, the long-term average crop water use per day (Curve A) could vary by plus or minus 0.75 mm per day. Day-to-day variation within a given year (Curve B) can be more than 5.00 mm per day. Thus, for reasons described above, use of soil water sensors and accessing daily crop water use estimates based on the weather conditions during the current growing season are highly recommended. In addition, the duration of each growth stage could vary by plus or minus two days when compared to the long-term average values provided in Table 7.

**Table 7. Average crop water use ($ET_c$) by growth stage for 113-day maturity corn grown in South Central Nebraska**

| Growth stage | Average water use rate /day | Duration[1] (days) | Water needed to reach stage (cm) | Water needed cumulative (cm) |
|---|---|---|---|---|
| Emergence (VE) | 0.08 | 0-10 | 2.0 | 2.0 |
| 4-leaf (V4) | 0.10 | 11-29 | 4.5 | 6.5 |
| 8-leaf (V8) | 0.18 | 30-46 | 7.25 | 13.75 |
| 12-leaf (V12) | 0.26 | 47-55 | 11.25 | 25.0 |
| Early tassel (R1) | 0.32 | 56-68 | 9.5 | 34.5 |
| Silking (R2) | 0.32 | 69-81 | 9.5 | 44.0 |
| Blister Kernel (R3) | 0.32 | 82-88 | 4.75 | 48.75 |
| Beginning Dent (R4.7) | 0.24 | 89-104 | 9.5 | 58.25 |
| Full Dent (R5.5) | 0.20 | 105-125 | 9.5 | 67.75 |
| Maturity (R6) | 0.10 | 126-140 | 3.5 | 71.25 |

[1]Long-term average number of days since planting required to progress from the previous growth stage to the next. For example, to go from the blister kernel stage to the beginning dent stage requires approximately 15 days (day 89 to day104). Days to each growth stage were determined using the Hybrid-Maize Corn Growth Model for the period 1982-2005 at Clay Center, Neb.

Maize does not extract water uniformly throughout its rooting depth. Generally, more water is extracted from shallow depths and less from deeper depths. If water is applied to the soil surface, the typical extraction pattern follows the 4-3-2-1 rule: 40 percent of the water comes from the top 1/4 of the root zone, 30 % comes from the second 1/4 and so on (Figure 2). Water applied using subsurface drip irrigation systems will result in more water removal from the depth where the drip lines are placed. In addition, though corn roots can reach depths of 150 to 180 cm, until late in the season conservative irrigation management assumes a 90 cm effective root zone. Later, when predicting the timing and amount of the last irrigation, the effective root zone is expanded to 120 cm.

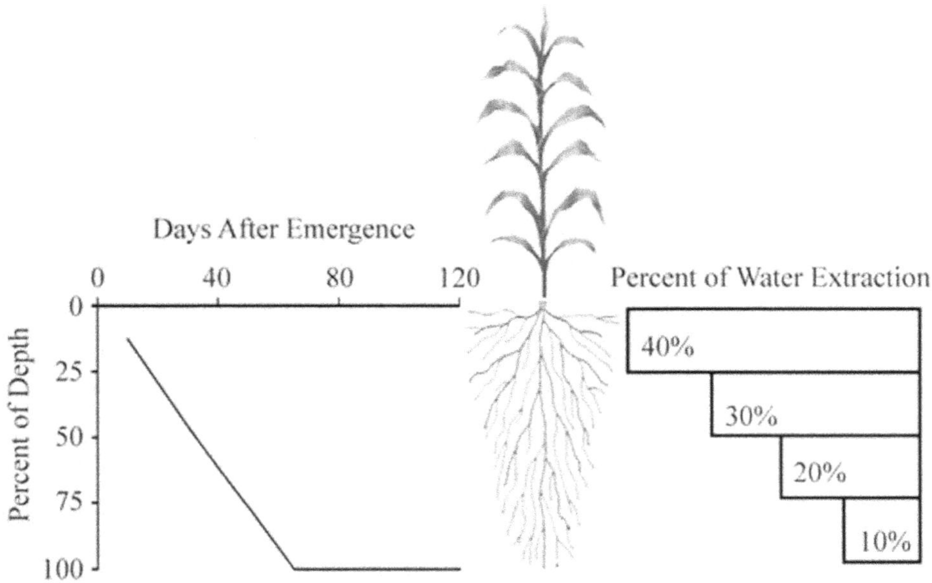


**Figure 2. Root zone soil water extraction and plant root development patterns**

## Matching Crop Demands with Water Application

In general, irrigation water is meant to supplement water stored in the soil profile and any effective rainfall recorded during the growing season. Thus, a water balance approach for estimating the irrigation requirement would be: Effective rainfall + Soil water removed - Seasonal $ET_c$ = Irrigation requirement. On a long-term average, corn grown on a deep silt loam soil in southeast Nebraska requires about 150 mm of net irrigation, central Nebraska requires about 225 mm, and the Panhandle of Nebraska requires about 350mm of net irrigation per year (Department of Natural Resources, 2006). These values will vary from year to year as a function of climatic conditions (rainfall, solar radiation, air temperature, wind speed and relative humidity) that impact $ET_c$ as described earlier.

When planning an irrigation system, sufficient system capacity should be available to meet the crop water use rates in column 2 of *Table 7*. However, simply converting the crop water use rate to a system capacity may result in unnecessarily oversized system components. NebGuide G1851 *Minimum Center Pivot Design Capacities in Nebraska* provides a method for estimating system flow rates needed to meet crop water demand while accounting for system down time, soil water holding capacity and the potential for rainfall during the growing season. Using this approach, system capacities can be determined to meet the management scheme of the operator and minimize the installation and operating cost of the system.

The goal of irrigation management should be to provide supplemental water while considering the economic and environmental consequences. One reason is that maize yield response to irrigation water application follows the law of diminishing returns (i.e., as the crop approaches maximum yield, the last mm or two of water applied will net less additional grain yield than the first mm applied). In addition, because irrigation systems are not 100 percent efficient at delivering water, it is impossible to convert all of the applied water to $ET_c$ and ultimately to grain yield. Thus, managers need to consider the potential for increased grain yield and cost of water application when deciding whether to apply the last mm of irrigation.

Applying several mms of excess water will lower the net return for the irrigated field potentially due to depressed grain yield resulting from leaching nutrients below the active root zone and inhibiting soil aeration. In addition, the cost of pumping the extra water would add to maize production costs. Pumping costs can be estimated by multiplying the fuel use rate per hectare - mm by the cost of fuel per unit. For example, the estimated fuel use to pump water from a depth of 100 feet at a pump outlet pressure of 50 psi is approximately 7.45 litres of diesel fuel per hectare-mm. Thus, the pumping cost for each mm of water applied would be approximately 1.97 times the cost of diesel fuel.

In situations where irrigation water supplies and natural precipitation are not adequate to meet crop $ET_c$ demands, limited or deficit irrigation can be effective in increasing yields above dryland levels. However, applying less water than is needed by the crop will typically result in yield reductions when compared to fully irrigated corn. Under extreme water limited conditions, water applications should be targeted to critical growth stages such as between pollination and early dough, thereby making the most efficient use of the water available.

**Growth Stages**

Immediately after planting, crop water use consists almost entirely of evaporation from the soil surface. Estimated $ET_c$ rates will typically be less than 2.5 mm per day unless the soil surface is wetted by a water application event. Following a rainfall event soil evaporation rate could be more than 5.00 mm per day depending on the soil texture and residue cover and the recorded amount of rainfall. Irrigation during this period is discouraged because water application via sprinkler systems could cause a crust to form on the soil surface that will reduce water infiltration and in some cases impede plant emergence. Maintaining good residue cover will limit crust development by absorbing the water droplet energy and also minimize soil evaporation by reflecting some of the incoming solar radiation.

About two weeks after emergence, the maize plant grows to a height of about 15 cm (4-leaf stage). The permanent root system begins to develop from the nodes and growing point during this time. At a plant height of about 25 to 35 cm (6-8 leaf stage), the tassel and ear are beginning to form inside the stalk. The number of kernel rows and kernels in a row are being determined at this time. Leaf area index (LAI) increases to 2.0 for fully irrigated maize. Daily maize water use averages between 3.75 and 10.00 mm per day (Table 7). The plant's roots are concentrated in the top 45 cm of soil if there are no soil-limiting factors such as compaction, impermeable layer, or gravel. Assuming an effective root depth of 45 cm, plant available water equals 7.5 cm in a silt loam soil and 3.75 cm in sandy soils. First irrigation depths should account for rainfall and be light to limit deep percolation losses.

Between the 8-leaf stage and tassel emergence maize root depth, leaf area and water use grow rapidly reaching peak daily water use rates during pollination. Root depth increases from 45 cm to 120 cm during this period doubling the amount of soil water available for plant growth. LAI increases to over 5.0 under fully irrigated maize with plant populations above 60,000 plants per hectare. Water use rates grow to 8.0 mm per day when averaged over a three- to five-day period (Table 7). Individual day water use rates can reach over 10.00 mm per day when high air temperature, low humidity and windy conditions prevail. Due to the high water use rates, it is important to avoid water stress during the reproductive stage. Severe water stress during silking tends to desiccate the silks and pollen grains causing poor pollination. Water stress during silking will result in the greatest yield reduction. Irrigation depths should be increased to match maize $ET_c$ minus rainfall while leaving some room in the soil for future potential rainfall events (12.05 cm for a silt loam soil).

Water requirements remain high during the early reproductive stages, often remaining in the 7.50 to 8.75 cm per day range until the dough stage (Table 7). During this time, kernels are growing as the plant transfers dry matter to the grain. Root development begins to slow down during the blister kernel stage and remains nearly constant at 135 to 150 cm for the remainder of the season and plant available soil water reaches a maximum value. Bottom leaves begin dying back during this period of time but have little impact of crop water use rates or yield. Irrigation depths should match maize water use minus rainfall while continuing to leave room for future rainfall events.

Maize water use rates decline beginning with the dough stage in response to lower atmospheric demand (shorter days and cooler temperatures and lower solar radiation), loss of transpiring leaf area, and changes in plant physiology as the grain approaches maturity. Maize water use drops from 7.50 to 5.00 cm per day by full dent (Table 7). Lower leaf loss continues during this period as well. It is assumed that the crop root zone remains constant at approximately 135 cm under fully irrigated conditions. Due to reduced crop water requirements, soil water levels can be drawn down below 50 percent depleted toward the end of the dent stage without affecting grain yield. However, maize does require some water right up to physiological maturity so one should continue to monitor maize water use rates and soil water levels. As crop approaches physiological maturity more water can be removed from the soil profile without impacting final grain yields. Determining when to stop irrigating is an important economic decision. Saving one irrigation event with a furrow irrigation system reduces production costs associated with pumping the irrigation water.

**Irrigation Management**

Irrigation management is basically deciding when to irrigate and how much to apply. The decision must be based on the available irrigation water supply, the soil water holding capacity and water intake rate, and the maize water needs. Well-timed irrigations provide enough water to prevent crop stress while fully using water from rainfall and available in the soil. Thus, irrigation scheduling accounts for all plant sources of water to produce economic yields.

Irrigation management also includes considering the economic and environmental consequences of each irrigation event. However, irrigation management must also evaluate the environmental aspects of irrigating. For example, if the plant uses 5.00 mm of the applied water, what are the potential environmental effects of the remaining 20.00 mm of water? If the extra water makes its way to the aquifer and carries 3.63 kg of nitrogen with it, will the increase in net income be worth it?

**Soil Textures**

Soils classified as coarse-textured include: fine sands, loamy sands and fine sandy loams. These soils generally have plant available water holding capacities less than 37.50 mm per 30 cm the plant available soil water at field capacity can be between 37.50 and 67.5 mm. Some sandy soils in Nebraska also have root-restricting layers at shallow depths that can restrict root development. The combination of low plant available water capacity and shallow rooting depth results in a relatively small soil-water reservoir. Small soil-water reservoirs create challenging water management scenarios. Application depths greater than 18.75 mm could result in plant stress in the event of an unexpected system failure. Frequent irrigations of less than 12.50 mm reduce the water application efficiency due to the amount of water lost during each application event. A compromise is to apply relatively frequent light (12.50 to 18.75 mm) water applications.

Medium- and fine-textured soils generally have plant available water capacities of more than 45 mm per 30 cm. In the top 90 cm, the plant available soil water at field capacity can be between 13.5 and 19.00 cm. Because these soil textures can store more water, irrigators have more scheduling flexibility. If no leaching or surface runoff occurs, water applications of 1.88 to 3.25 cm are appropriate for these soils.

**Irrigation systems**

In Nebraska, approximately 70 percent of the irrigation water is supplied by center pivot sprinkler systems and about 30 percent using furrow irrigation. An estimate of the water application efficiency is needed to determine the amount of water that might need to be pumped per season. Conventional gated pipe irrigation with no reuse pit typically has an efficiency of about 50 percent. This means if 25 cm of net irrigation is required, an irrigator using conventional gated pipe and no reuse would have to pump 50 cm of water. With a reuse system the maximum efficiency increases to near 70 percent, and the depth pumped would decrease to 35 cm. When furrow irrigating, it is important to irrigate the entire field quickly. Every-other-furrow irrigation supplies water to more area in a given amount of time and produces yield comparable to those achieved when every-furrow is irrigated. In addition to saving time, the water application depth may be reduced 20 to 30 percent by implementing every-other-furrow irrigation. With no reuse system, every-other-furrow irrigation is about 60 percent efficient. With a reuse system, maximum efficiency increases to near 75 percent.

Properly designed and well-maintained center pivot and lateral move sprinkler irrigation systems can be 85 percent to 90 percent efficient. A sprinkler irrigation system that is 90 percent efficient would require a gross application of 11 inches to deliver a net application of 10 inches. If farming practices are changed to accommodate increased water application rates, and sprinkler spacing is less than 2.25 m, system efficiencies can be improved by dropping the sprinkler nozzles nearer the soil surface. Center pivots also allow more precise application depths and timing than are possible with furrow irrigation systems. The ability to consistently apply only the depth of water needed allow center pivots to take full advantage of rainfall and minimize the potential for plant stress.

**Table 8. Influence of irrigation schedules on yield, WUE, net return and nitrogen uptake of maize**

| Treatment | Grain yield (kg / ha) | Stover yield (kg / ha) | WUE (kg / ha mm) | Net return (Rs. /ha) | Nitrogen uptake (kg / ha) | | |
|---|---|---|---|---|---|---|---|
| | | | | | Grain | Stover | Total |
| 0.6 1W/CPE | 5640 | 7214 | 22.55 | 28698 | 37.8 | 21.5 | 59.3 |
| 0.8 1W/CPE | 6141 | 7744 | 18.37 | 32204 | 42.7 | 24.8 | 67.5 |
| 1.0 1W/CPE | 6629 | 8282 | 15.88 | 35605 | 47.8 | 27.4 | 75.2 |
| CD 5 % | 639 | 769 | - | - | 3.5 | 1.5 | 3.7 |

(Reddy et al., 2012)

At Warangal (A.P), irrigating maize crop at an IW/CPE ratio of 1.0 was significantly superior to 0.6 but at par with 0.8 regarding grain yield and stover yield. Higher net return and N uptake were observed with IW/CPE ratio of 1.0 (Reddy et al., 2012).

At Karnal (Haryana), average daily $ET_c$ of maize varied from <2.8 mm $day^{-1}$ in the early growing period to >4 mm $day^{-1}$ at development and reproductive stages. The peak daily $ET_c$ of maize was 7.7 mm $day^{-1}$ and this occurred 9 weeks after sowing (WAS) at the silking stage of maize when leaf area index (LAI) was 5.5. The measured seasonal $ET_c$ of maize was 354 mm. The estimated values of $K_c$ for maize by the Penman–Monteith method at crop growth stages namely, initial, crop development, mid-season and maturity, were 0.55, 1.00, 1.23 and 0.64, respectively (Narendra et al., 2003).

In Damodar Valley, capillary contribution from ground water increased from $0.38 \times 10^{-2}$ to $1.4 \times 10^{-2}$ m under IW/CPE at 1.2 over 0.6. Relatively drier soil water regimes induced maize plant to develop more extensive root systems at 0.60 to 0.90 m depth with 0.08 to $0.11 \times 10^{-5}$ m $m^{-3}$ more root length density values than wetter soil regime. Significantly higher yields were obtained with five post-sowing irrigations applied at IW/CPE = 1.2. Total actual evapotranspiration was maximum under IW/CPE of 1.2 followed by 0.9 and then 0.6. The yields under three water regimes were 2.22, 1.85 and 1.77 Mg $ha^{-1}$ and the values of water use efficiency were 7.23, 6.41 and 6.32 Mg $ha^{-1}$ $m^{-1}$, respectively (Bandyopadhyay and Mallick, 1996).

Braun worth (1987) reported 311 - 604 mm of water for sweet corn to 21 t/ha green cob. Costa et al. (1988) obtained higher grain yield under irrigation at 0.8 Epan. Higher green cob yield was obtained with higher moisture level (Braun worth and Mack, 1989).

At Sanliurfa (Turkey), a 2-day irrigation frequency, with 100 % ET water application by a drip system was optimum for sweet corn grown in semi-arid regions (Abdullah *et al.*, 2003).

Kaliappa *et al,* (1974) reported that the total water requirement of maize was 382 mm when irrigation scheduled at 75 per cent available soil moisture depletion during vegetative period and 50 per cent during maturity period. The water requirement ranged from 1.94 to 5.96 mm $day^{-1}$ during the crop growth period in dark red latosol soil (Andrade *et al.,* 1988). Eneva and V'Lchanov (1987) registered the maximum yield with 100 % evaporation with the total water use of 272 mm $ha^{-1}$ over 60 and 80 per cent evaporation. Panchanathan *et al,* (1987) recorded the average grain yield of 36.6 q $ha^{-1}$ with the application of 553 mm of water over other regimes (449, 429 and 351 mm). Costa *et al,* (1988) observed that increased application of water from 180 to 520.6 mm increased the seed yield from 10.0 to 63.0 q /ha.

Bowman *et al,* (1991) stated that in mid west region of USA the water requirement of maize was generally vary according to evaporative demand and rainfall. Similarly, the water used by maize was 30.4 cm during summer season in silty clay loam soil of Kashmir valley at 0.75 IW/CPE ratio with maximum yield (Khan *et al.,* 1996). Stegman (1982) observed that WUE of maize generally declined when compared to that of well watered crop because maize yield was shown to be linearly related to ET. Stewart *et al.* (1975) observed maximum WUE of 0.17 q $ha^{-1}$ $mm^{-1}$ at Davis (California). Although maximum maize yield differed only slightly from semiarid to more humid environment, seasonal ET rates were considerably different. As a result of the lower seasonal ET in the humid regions, WUE's as high as high 0.30 q $ha^{-1}$ $mm^{-1}$ have been reported for maize grown in the south Eastern USA.

Hussaini *et al.* (2002) stated that the water use of maize crop in sandy soil of Nigeria during dry period was 355.8 to 444.4 mm and recorded higher yield with scheduling of irrigation at 1.0 IW/CPE (21.19 q /ha) as compared to 0.6 and 0.8 ratio (18.04 and 19.02 q/ ha, respectively). Mahdi *et al.* (2003) noticed that water requirement of maize was 38 to 64 cm with 0.6 to 1.0 ET. Application of water at 1.0 ET recorded higher yield (108.2 q /ha) compared to 0.6 and 0.8 ET (71.5 and 94.6 q /ha, respectively) at Yuma on sandy loam soil.

At Samaru, Nigeria recorded higher WUE with irrigations scheduling at 0.6 IW/CPE ratio (6.96 kg /ha mm) over 0.8 and 1.0 ratio (6.37 and 6.33 kg/ ha mm, respectively) in sandy soil during *rabi* season (Hussaini et al., 2002).

Jiotode *et al.* (2002) observed that water use efficiency was more in irrigations scheduled as per critical growth stages (5.86 kg /ha mm) over 40 mm, 60 mm and 80 mm CPE (3.29, 3.86 and 4.19 kg /ha mm, respectively) in clay loam soil of Akola in Maharastra. Mahdi *et al.* (2003) observed that in sandy loam soil of Yuma, scheduling of irrigation and its interaction with plant population WUE was higher at 1.00 ET (20.02 kg /ha/ mm) with 81,000 plant population over 0.6 and 0.8 ET (17.62 and 18.17 kg /ha/mm, respectively). Balaswamy *et al.* (1986) observed that net return was maximum in scheduling of irrigation at 40% (Rs. 56000 /ha) over 60 and 80 % (Rs. 38000 and Rs. 22000 /ha, respectively) ASMD in sandy loam soil of Hyderabad. Bharthi *et al*, (1997) recorded higher EC, sugar and amino acid with scheduling of irrigation at 0.50 IW/CPE ratio in maize (242.9 dS $m^{-1}$, 436.8 mg 100 $g^{-1}$ and 75.5 mg 100 $g^{-1}$) over 0.75 ratio (185.3 dS $m^{-1}$, 284.5 mg 100 $g^{-1}$ and 55.5 mg 100 $g^{-1}$) where as protein content and dehydrogenase activity was higher in 0.75 ratio (9.69 and 0.835, respectively) over 0.50 ratio (9.00 and 0.636) in sandy soil at Coimbatore (Tamil Nadu).

In Nigeria, Hussaini *et al.* (2001) recorded maximum total dry matter, plant height, CGR and RGR (96.3 g $plant^{-1}$, 171.0 cm, 1.12, 85.1 g /$m^2$ /week and 610 mg /g/ week, respectively) with scheduling of irrigation at 1.0 IW/CPE ratio over 0.6 and 0.8 ratios in sandy soil during *rabi* season. Manish Kumar *et al.* (2001) reported that scheduling of water at 1.0 IW/CPE ratio recorded higher grain yield (34.38 q ha-1) of winter maize over 0.7 ratio (28.78 q $ha^{-1}$) on sandy loam soil of Raipur (Madhya Pradesh). Mahdi *et al.* (2003) observed that grain yield was maximum due to scheduling of irrigation at 1.00 ET over 0.6 and 0.8 ET at Yuma (USA) in sandy loam soil. Further the yield at 0.6 ET was 45 % lower than 1.0 ET.

In Turkey, Okten and Simsek (2004) revealed that maximum grain yield of dent corn was observed at application of 10 per cent deficit irrigation and four day irrigation frequency, while minimum yield was at application of 30 per cent deficit irrigation and eight day irrigation frequency. Yield characteristics were affected negatively by 30 % deficit irrigation and eight day irrigation frequency in clay soil under arid condition. Abu-Awwad (1994) recorded higher dry matter production of sweet corn 144.0 q /ha due to irrigation at 1.0 Epan over 0.25 and 0.5 Epan (70.0 and 104.0 q /ha, respectively) in clay soils. McPherson and Boyer (1977) imposed water deficit that reduced photosynthesis to near zero during entire kernel development period. Grain yields were reduced by only 47 to 49 % of the well watered control as a result to remobilization of assimilates, accumulated before initiation of stress period in sandy loam soil at London. Lehrsch *et al.* (2000) noticed that the application of water in alternate furrow with 0.76 m row spacing recorded higher yield (207.0 q /ha) over same furrow (196.0 q $ha^{-1}$) whereas alternate furrow with 0.56 m recorded (193.0 q /ha) over same furrow (182.0 q/ ha), which was due to increased available water as well as solar radiation interception in 0.76 m. Lorens *et al,* (1987) reported that pioneer brand 3192 recorded higher crop growth rate (31.8 g $m^{-2}$ $day^{-1}$) in irrigation over vegetative stress (28.0 g $m^{-2}$ $day^{-1}$) in fine sandy soil of Florida (USA). Harvest index was also more in irrigated (0.61) over vegetative stress (0.55).

Maize crop responds very well to water and nutrient application. Maize is one of the amenable crops for drip irrigation system, which is an efficient method of irrigation (Zhu *et al.*, 2007). Many scientists reported enhanced growth and development of maize under drip irrigation (Lamm *et al.,* 2001 and O'Brien *et al.,* 2001). A single irrigation at tassel initiation stage increased maize yield by 29 % over no irrigation and an additional irrigation during the vegetative and grain-filling stages increased maize yield an additional 11 and 13 %, respectively (Norwood, 2000). Hergert et al. (1993) reported maize yields of 5.6, 10.1 and 11.8 t /ha with dry land, limited irrigation and full irrigation, respectively. Trooijen et al. (1999) found water use efficiency (WUE) of maize to be greater with limited irrigation, but full irrigation of maize was more profitable than limited irrigation. Musick and Dusek (1980) found that stress during tasseling and silking of maize was most harmful and stress during grain filling was more harmful than that of during vegetative stage.

## Water Use Efficiency by Maize

Maize originated from Mexico, and is mainly grown in the warmer temperate regions and humid subtropics. It is a C4 plant, which confers potentially more efficient use of CO2, solar radiation, water and N in photosynthesis than C3 crops. Water use efficiency (WUE) of maize is approximately double that of C3 crops grown at the same sites. Its transpiration ratio (molecules of water lost per molecule of $CO_2$ fixed) is 388, corresponding to 0.0026 in WUE (Jensen, 1973), while that of wheat is 613, soybean 704. Even though maize makes efficient use of water, it is considered more susceptible to water stress than other crops because of its unusual floral structure with separate male and female floral organs and the near-synchronous development of florets on a (usually) single ear borne on each stem.

Maize has different responses to water deficit according to development stages (Cakir 2004). Drought stress is particularly damaging to grain yield if it occurs early in the growing season (when plant stands are establishing), at flowering, and during mid to late grain filling (Heisey and Edmeades 1999). At the seedling stage, water stress is likely to damage secondary root development. During stem elongation (after floral initiation) leaves and stems grow rapidly, requiring adequate supplies of water to sustain rapid organ development, water stressed plants being shorter and with reduced individual and cumulative leaf area

(Muchow, 1989). The most critical period for water stress in maize is 10 to 15 days before and after flowering, with grain yield reduced two to three times more when water deficit coincides with flowering compared with other growing stages (Grant et al., 1989). During this period, ear growth is susceptible to competition from other organs that are still growing as its growth is source limited, often leading to low grain number per ear and occasionally barren ears. Grain yield of maize suffering water stress at flowering and during grain fill is highly correlated with kernel number per plant (Bolanos and Edmeades, 1996) indicating the importance of adequate water supplies during flowering.

### Improvement of water use efficiency in maize

WUE of maize is a function of multiple factors, including physiological characteristics of maize, genotype, soil water holding capacity, meteorological conditions and agronomic practices. To improve WUE, integrative measures should aim to optimise cultivar selection and agronomic practices.

## References

Agarwal, L.G., N.K.Shekon, A.S. Sidhu and K. Mahant (1995). Plant water status of maize and succeeding wheat grown on organically amended soils. J. Indian Soc. Soil Sci., 43(2): 152-155.

Aggarwal, P., R.N. Garg, A.K. Singh, S. Singh, A.L. Yadav and A.M. Sharma (2000). Assessment of bed planting system in maize for enhancing water use efficiency. In: Proceedings of International Conference on Managing Natural Resources for Sustainable Agricultural Production in 21st century. February 14–18 at IARI Indian Society Soil Sciences New Delhi, India, 175–176.

Ahmad, R., M.Asif, I. Muhammad and H.Bilal (2002). Influence of different irrigation methods and band placement of nitrogen on maize productivity. International J. Agric. and Biolo., 4: 540-543.

Alevi Fazak, M., A.Naderi, Y.Emam, A.Aineband and S.H. Lack (2010). Response of corn to water deficit at different stages of development. Res. on Crops, 11(2): 298 – 301.

Alizadeh, H., A.Liyaghat and F. Abbasi (2009). The study of the effect of the Irrigation fertilizer -Furrow on the efficiency of water and fertilizer consumption and the yield component of corn. Journal of Soil and Water (Agricultural Science and Technology), 23 (4): 132-138.

Al-Kaisi, M.M. and X. Yin (2003). Effects of nitrogen rate, irrigation rate, and plant population on corn yield and water use efficiency. Agron. J., 95: 1475-1482.

Andrade, F.H., L. Echarte, R. Rizzalli, A.M. Della and M. Casanovas (2002). Kernel number prediction in maize under nitrogen or water stress. Crop Sci., 42: 1173 - 1179.

Anitta Fanish, S. and P. Muthukrishnan (2011). Effect of drip fertigation and intercropping on growth, yield and water use efficiency of maize (*Zea mays* L.). Madras Agric. J., 98 (7-9): 238-242.

Anitta Fanish, S., P. Muthukrishnan and P. Ayyadurai (2013). Root Characters of Maize as Influenced by Drip Fertigation Levels American J. Plant Sci., 4: 340-348.

Anitta Fanish, S., P.Muthukrishnan and S. Manoharan (2011). Drip fertigation in maize based intercropping system. Indian J. Agric. Res., 45 (3): 233 – 237.

Anonymous (2004). Annual Progress Report, Water Management Research Centre, Belvatagi, Univ. Agric. Sci., Dharwad.

Anonymous (2006). Annual Progress Report - 2006, Water Management Research Centre, Belavatagi, Univ. Agric. Sci., Dharwad.

Balaswamy, K., B.B. Reddy, K.A. Reddy, B.B. Reddy and M.B. Reddy (1986). Effect of planting patterns mulches and soil moisture regimes on growth and yield attributes of maize. Madras Agric. J., 73 (10): 553-557.

Banziger, M., G.O. Edmeades, D.Beek and M.Bellon (2007). Effect of water deficit and row planting in maize genotypes. Agron. J., 52: 398 - 405.

Basetti, P. and M.E. Westgate (1993). Water deficit affects receptivity of maize silks. Crop Sci., 33: 279- 282.

Bharathi, A., K.K. Vadivelu and R. Umarani (1997). Effect of irrigation management and nitrogen fertilization on shelf life of maize. Madras Agric. J., 84: 356-360.

Burrows, W.C. and W.E. Larson (1992). Effect of amount of mulch on soil temperature and early growth of corn. *Agronomy J.,* 54: 19–23.

Cakir, R. (2004). Effect of water stress at different development stages 1 on vegetative and reproductive growth of corn. Field Crops Res., 89: 1-6.

Camp. C.R., D.R. Karlen and J.R. Lambert (2006). Irrigation scheduling and row configuration for corn in thesoutheaster coastal plain. Trans. ASAE. 28:1159-1165.

Craciun, I. and M. Craciun (1999). Water and nitrogen use efficiency under limited water supply for maize to increase land productivity. In: Kirda, C., Moutonnet, P., Hera, C., Nielsen, D.R. (Eds.), Crop Yield Response to Deficit Irrigation. Kluwer Academic Publishers, The Netherlands, pp. 87–94.

Cyrus, M., S. Ali, and F.S. Seyed (2010). Maize yield response to deficit irrigation during low sensitive growth stages and nitrogen rate under semi-arid climatic conditions. Agr. Water Manage., 97 (1): 12-22.

Daniel, W.S. and C.W. Marr (2005). Supplemental irrigation at reproductive growth stages to improve popcorn grown at different populations. Agron. J., 97: 741–745

Dean, D.S., C.S. Earl and E.K. Raymond (2000). Irrigation management for corn in the northern Great Plains, USA. Irrig. Sci., 19: 107–114.

Devender Singh, R.C. Tyagi, I.S. Hooda and O.P.S. Verma (1997). Influence of plant population, irrigation and N levels of growth of spring maize (*Zea mays* L.). Haryana J. Agron., 13(1): 54-58.

Dhanpal, A.H. (1992). Simulation of soil moisture regime: application of the SWATRE model to a maize crop on the reddish brown earths in the dry zone of Sri Lanka. Agric. Sys., 38: 61–73.

Dhillon, B. S., H.S. Thind, N.S. Malhi and R.K. Sharma (1998). Effect of excess water stress on grain yield and other traits of maize hybrids. *Crop Improvement,* 25: 209–214.

Doerge, T. (2008). Safely delaying the first irrigation of corn. Crop Insights, Vol. 18, No. 7. Pioneer Hi-Bred, Johnston, IA.

Duvick, D.N. (2008). The evaluation of kernel yield with tassel growth at water deficit condition. Soil Sci., 56: 175 - 184.

Elzubeir, A.O and A.M.E. Mohamed (2011). Irrigation scheduling of maize under desert area conditions – North of sydan. Agric. Bio. J. North America, 2 (4): 645 -651.

Farre, I. and J.M. Faci (2009). Deficit irrigation in maize for reducing agricultural water use in a Mediterranean environment. Agricultural Water Management, 96: 383-394.

Fisher, P.D. (1995). An alternative plastic mulching system for improved water management in dryland maize production. Agric. Water Management, 27 (2): 155-166

Foroughi, F. (2006). The deficit irrigation of maize on the basis of water production function - yield. National Conference on Management of the Irrigation and Drainage Networks. Chamran University of Ahvaz.

Gajri, P. R., V.K.Arora and M.R. Chaudhary (1994). Maize growth response to deep tillage, straw mulching and farm yard manure in coarse textured soil of NW India. *Soil Use and Management,* 10: 15–20.

Girijesh, G. K., A.S. Kumara swamy, M. Dinesh dumar, S.P. Nataraju, T.S. Vageesh and K.S. Rajashekarappa (2011). Response of maize (Zea mays L.) to constraint irrigation under late sown conditions. Karnataka J. Agric. Sci., 24 (5): 633-635.

Hebsur, N.S., T.D. Bharamgoudar, B.B. Channappagoudar and S.D. Tamgale (2007). Effect of Sub-Surface Drainage and Amendments on Uptake of Nutrients by Maize (Zea mays L.) in Ghataprabha Command Area of Karnataka. Karnataka J. Agric. Sci., 20 (3): 514 - 517

Hefner, S.G. and P.W. Tracy (1995). Corn production using alternate furrow nitrogen fertilization and irrigation. J. Prod. Agric., 8: 15-16.

Howe, O.W. and H.F, Rhoades (1995). Irrigation practice for corn production in relation to stage of plant development. Soil Sci. Soc. Am. Proc., 19: 94–98.

Howell, T.A., J.A. Tolk, D.S. Arland and R. Evertt (1998). Evapotranspiration, yield and water use efficiency of corn hybrids differing in maturity. Agron. J., 90: 3–9

Humphreys, E., D.J. Smith, R.J.G. White, E. Xevi, and D.C. Godwin (2003). Maize Man – a decision support system for water and nitrogen management, pp.127-134. In: Birch, C. J. and Wilson, S. R. (Eds.) Versatile Maize, Golden Opportunities, Proceedings, 5th Australian Maize Conference, 18-20 February 2003, Toowoomba, Australia, Maize Association of Australia, Darlington Point, NSW.

Hussain, M.A., V.B. Ogunlela, A.A. Ramalan and A.M. Fa-laki (2001). Growth and development of maize (*Zea mays* L.) in response to different levels of nitrogen, phosphorus and irrigation. Crop Res., 22: 141-149.

Jadhav, B.S., A.S. Bhosale and B. R. Patil (1994). Consumptive use water use efficiency and moisture extraction pattern maize as influenced by irrigation scheduling mulching and layouts. J. Soils and Crops, 4(1): 25-30.

Jiotode, D.J., D.L. Lambe and C.S. Dhawad (2002). Growth parameters and water use studies of maize on influenced by irrigation levels and row spacings. Crop Res., 24(2) : 292-295.

Kang, S., W. Shi and J. Zhang (2000). An improved water use efficiency for maize grown under regulated deficit irrigation. Field Crops Res., 67 (3): 207-214.

Kang, S., Z. Liang, W. Hu and J. Zhang (1998). Water use efficiency of controlled alternate irrigation on root divided maize plants. Agric. Water Manag., 38: 69–76.

Kang, S.Z., Z.S. Liang, Y.H. Pan, P.Z. Shi and J.H. Zhang (2000b). Alternate furrow irrigation for maize production in an arid area. Agricultural Water Management, 45 (3): 267-274

Kapusta, G., R.F. Krausz and J.L. Matthews (1996). Corn yield is equal in conventional, reduced and no tillage after 20 years. *Agronomy J.,* 88: 812–817.

Kaur, T. and R.K. Mahey (2005). Effect of planting method and irrigation levels on water use of maize (*Zea mays* L.). *Indian J. Environment and Ecoplanning,* 10: 373–376.

Kipkorir, E.C. and D. Raes and Massawe (2002). Seasonal water production functions and yield response factors for maize and onion in Perkerra, Kenya. Agric. Water Managt., 56: 229–240.

Ko, J. and G. Piccinni (2009). Corn yield responses under crop ET based irrigation management Agric. Water Management, 96: 799 – 808.

Lyle, W.M. and J.P. Bordovsky (1995). LEPA corn with limited water supplies. Trans. ASAE, 38: 2455–2462.

Mahal, S.S., D.G. Dejene and M.S. Gill (2000). Growth and yield of maize (*Zea mays* L.) as influenced by flood under different planting methods and nitrogen levels. Environ. Ecol., 18: 789-92.

Mahdi, M. Al-Kaisi and Xinhua Yin (2003). Effect of nitrogen rate, irrigation rate and plant population on corn yield and water use efficiency. Agron. J., 95: 1475- 1482.

Mahmood, T. (1994). Impact of Water and Nutrient Management on Growth Yield and Quality of Maize (Zea mays L.). *PhD Thesis*, Department of Agronomy, University of Agriculture, Faisalabad.

Martin, E.C., T.L. Loudon, J.T. Ritchie and A. Werner (1994). Use of drainage lysimeters to evaluate nitrogen and irrigation management strategies to minimize nitrate leaching in maize production. Trans. ASAE., 37: 79-83.

Mousa, M., A.Ali and P.P. Abbas (2013). The yield functions of maize toward the deficit irrigation in the tropical climate of Dezfoul. Bull. Environment, Pharmacology and Life Sci., 1 2 (2): 1-4.

Mubeen, M., A.Ahmad, A. Wajid, T.Khaliq and A. Bakhsh (2013). Evaluating CSM-CERES-maize model for irrigation scheduling in semi-arid conditions of Punjab, Pakistan. International J. Agric. Biol., 15(1): 1-3.

Nagy, J. (1997). The effect of fertilization on the yield of maize (Zea mays L.) with and without irrigation. Cereal Research Communications, 25: 69-76.

Narendra, K.T, K.S. Dinesh and K.L Surendra (2003). Determination of evapotranspiration for maize and berseem clover. Irrigation Sci., 21 (4): 173-181.

Norwood, C.A. (2000). Water use and yield of limited-irrigated and dry land corn. Soil Sci. Soc. Am. J., 64: 365–370.

O'Brien, D.M., F.R. Lamm, L.R. Stone and D.H. Rogers (2001). Corn yields and profitability for low-capacity irrigation systems. Applied Engineering Agric., 17: 315- 321.

Ogindo, H.O. and S. Walker (2005). Comparison of measured changes in seasonal soil water content by rained maize-bean intercrop and component cropping in semi arid region in South Phys. Chem. Earth, 30: 799-808.

Ogola, J.B.O., T.R. Wheeler and P.M. Harris (2002). Effects of nitrogen and irrigation on water use of maize crops. Field Crops Res., 78: 105-117.

Okten and Simsek (2004). Effect of irrigation frequencies on yield characteristics of dent corn (*Zea mays* indentata) and water-yield relationships under semi-arid region. Indian J. Agron., 49(3): 174-178.

Otegui, M.E., F.H. Andrade and E.E. Suero (1995). Growth, water use, and kernel abortion of maize subjected to drought at silking. Field Crops Res., 40 (2): 87–94.

Pandey, R.K., J.W. Maranvilla and M.M. Chetima (2000). Deficit irrigation and nitrogen effects on maize in a Sahelian environment. Part II. Shoot-growth, nitrogen uptake and water extraction. Agric. Water Manag., 46: 15 – 27.

Pandey, R.K., J.W. Maranville and A. Admou (2000). Deficit irrigation and nitrogen effects on maize in a Sahelian environment. I. Grain yield and yield components. Agric. Water Manage., 46 (1): 1–13.

Patel, J.B., V.J. Patel and J. R. Patel (2006). Influence of different methods of irrigation and nitrogen levels on crop growth rate and yield of maize (*Zea mays* L.). Indian J. Crop Sci., 1(1-2): 175-177.

Patil, P.L., B.M. Radder, S.G. Patil, Y.R.A ladakatti, C.B. Meti and A.B. Khot (2005). Effect of moisture regimes and micronutrients on yield, water use efficiency and nutrient uptake by maize in Vertisol of Malaprabha command. Karnataka J. Indian Soc. Soil Sci., 54(3): 261-264.

Payero J.O., D.D. Tarkalson, S. Irmak, D. Davison and J.L Petersen (2008). Effect of irrigation amounts applied with subsurface drip irrigation on corn evapotranspiration, yield, water use efficiency, and dry matter production in a semiarid climate. Agric water Manages., 95: 895 – 908.

Payero, J.O., S.R. Melvin, S. Irmak and D. Tarkalson (2006). Yield response of corn to deficit irrigation in an semiarid climate. Agricultural Water Management, 84: 101-112.

Reddy, M. M., B.Padmaja and D. V. V.Reddy (2012). Response of maize (*Zea mays* L.) to irrigation scheduling and nitrogen doses under no till condition in rice fallows. J. Research ANGRAU, 40 (1): 6-12.

Rhoads, F.M. and J.M., Bennett (1990). Corn in irrigation of agricultural. Crops ASAE Agron. Monogr., 30: 569–596.

Sabata, R.J. and S.C. Masen (1992). Corn hybrid interaction with soil nitrogen levels and irrigation regime. J. Produc. Agric., 5: 137-142.

Sampathkumar, T. and B.J. Pandian (2011). Nutrient uptake rate and relationship with grain yields of hybrid maize under drip fertigation. Madras Agric. J., 98 (4-6): 151-153.

Santolla, R., N.C. Tuner and A. Leen (2008). Water management and its relationship to yield improvement in maize. Crop Sci., 84: 654 - 661.

Seema Sepat and Ashok Kumar (2007). Nitrogen management in maize (*Zea mays* L.) under life saving and assured irrigations. Indian J. Agric. Sci., 77(7): 451-454.

Selvaraju, R. and M.R. Iruthayaraj (1994). Influence of irrigation scheduling, methods of irrigation and nitrogen levels on growth and yield of maize. Madras Agricultural J., 81: 418 - 420.

Selvaraju, R. and M.R. Iruthayaraj (1993). Effect of irrigation scheduling, methods of irrigation and nitrogen levels on growth analysis parameters of maize. Madras Agric. J., 80: 562-565.

Selvaraju, R. and M.R. Iruthayaraj (1995). Effect of irrigation, methods of irrigation and nitrogen levels on nutrient uptakes by maize. Madras Agric. J., 82 (3): 215-216.

Shaozhong, K., S. Wenjuan and Z. Jianhua (2000). An improved water-use efficiency for maize grown under regulated deficit irrigation. Field Crops Res., 67(3): 207–214.

Singh, B.R. and D.P. Singh (1995). Agronomic and physiological responses of sorghum, maize and pearl millet to irrigation. Field Crops Res., 42 (2-3): 57–67.

Singh, Y. and R. Bahadur (1997). Effect of distillery effluent irrigation on maize grain crop and soil fertility. Indian J. Eco., 24: 53-59.

Steele, D.D., B.L. Gregor and J.B. Shae (1997). Irrigation scheduling methods for popcorn in the northern Great Plains. Trans. ASAE, 40:149–155.

Sukanya, T.S. and S.S. Meli,(2004). Effect of distillery effluent irrigation on growth, yield and quality of maize grown on sandy loam in Northern Transitional Zone of Karnataka. Karnataka J. Agric. Sci., 17(3): 405-409.

Tandon, H.L.S. and D.K. Kundu (1990). Management of fertility, variety, planting density and irrigation for maximum yield. In: Proceedings for Symposium on Phosphorus requirement for sustainable agriculture in Asia and Oceania,. IRRI, pp: 461-472.

Tara Devi, B. (2007). Response of maize (*Zea mays* L.) to ferti-irrigation of spent wash. M.Sc (Ag) Thesis. Department of Soil Science and Agricultural Chemistry, University of Agricultural Sciences, Dharwad.

Tara Devi, B. and S.G. Patil (2007). Response of Maize (*Zea mays* L.) to Ferti-Irrigation of Spent wash. Karnataka J. Agric. Sci., 20(4): 921.

Tolk, J. A., T.A. Howell and S. R. Evett (1999). Effect of mulch, irrigation and soil type on water use and yield of maize. *Soil and Tillage Research,* 50: 137–147.

Tyagi, R.C., Devender Singh and I.S. Hooda (1998). Effect of plant population, irrigation and N on yield and its attributes of spring maize (*Zea mays* L.). Indian J. Agron., 43(4): 672-676.

Vijay Shankar, M., C.S.P. Ojha and K.S. Hari Prasad (2012). Irrigation Scheduling for Maize and Indian-mustard based on Daily Crop Water Requirement in a Semi-Arid Region. World Academy of Science, Engineering and Technology, 63: 563 – 572.

Yang, X.G., H.L. Liu and L.N. Yu (2003). The changing of water transfer potential in soil-plant-atmosphere continuum system of maize yield. Chin. J. Eco-Agr., 11: 27–29.

Zhu, H., Y. Lan, M. Lamb and C. Butts (2007). Corn nutritional properties and yields with surface drip irrigation in topographically variable fields. Agricultural Engineering International: The CIGR Ejournal. Manuscript L.W. 07 005. Vol. IX. September, 2007.

13

# Maize based Intercropping Systems

To meet ICMR recommendation, pulse production has to be increased from present 16 to 28 million tonnes by 2020 (Ali, 2002). One of the options to increase production is to increase the area under pulses. Since pulses due to economic considerations on their own cannot replace other profitable cereals, oilseeds or commercial crops, the area under the crop could be increased only by including pulses under various cropping systems along with the regular crops by exploiting their intrinsic potential of growing along with other crops without competitive interferences. In this, pulses like French bean hold promise.

French bean (*Phaseolus vuglaris*) being fertilizer responsive has a yield potential of 2.5 to 3.0 t/ ha (Ali and Kushawaha, 1987). Moreover, it is less susceptible to storage pests, quite nutritious with some medicinal values. It acts as anti aging factor, helps to prevent obesity, lowers the cholesterol content of blood and therefore recommended for curing heart and kidney related ailments. Since it is a short duration (85-90 days) crop, it can very well form a component of intercropping and sequence cropping in many of the agro-climatic zones. Being wider spaced, maize crop provides an opportunity for introducing a short duration pulse crop like French bean as an intercrop in additive series. Experiments carried out elsewhere in India clearly indicated that urdbean, soybean, French bean (Bhopal Singh and Awasthi, 1987) and cowpea (Shivay et al., 1999, Padhi, 2001. Rana et al., 2001) as intercrops in maize had no detrimental effect on main crop. Weeds which otherwise would become major production constraint could also be brought under control due to intercropping. Moreover, such a system also helps in efficient utilization of natural resources (space, moisture and light) to harness maximum productivity. Further, to avoid adverse effect on main crop by addition of intercrop, suitable adjustment in plant population and crop geometry has to be worked out.

## Performance of Component Crops

Legumes in maize based cropping systems are considered to be better alternatives for securing nitrogen economy. They are either increasing or not reducing the yield of maize besides bonus legume yield. This leads to greater productivity of land per unit time and space and higher net return of intercropping system over monoculture. Maize yield in pure stand and when intercropped with urdbean, soybean and French bean was of 3.29, 3.47, 3.8 and 3.11 t /ha, respectively and indicated that there was no reduction in intercropped maize yield as compared to sole crop in Himachal Pradesh. In addition, the intercrops gave seed yield of 260, 310 and 340 kg/ ha respectively (Bhopal Singh and Awasthi, 1987). Chui (1988) observed that dry matter yields at maturity in intercrops was reduced by 24.4 and 11.6 % when maize and French bean were grown in same row and french bean between two maize rows, respectively, over sole maize. Maize grain yield was not significantly affected while french bean yield was significantly reduced in intercropping systems. However, intercropping systems increased total grain yield significantly over pure stand. French bean yields were higher with lower maize densities (sowing rates, 20,000 - 60,000 plants/ ha) and reductions in maize yields were compensated by the french bean yields (Rezende et al., 1990).

### Light use efficiency of intercropping of maize with french bean and other pulses

Light is one of the important growth factors influencing the yield of crops. Unlike other resources, light is instantaneously available and has to be instantaneously intercepted, if it is to be used for photosynthesis (Donald, 1961). Low availability of light for any component crop in the mixtures reduced the photosynthetic rate and crop growth rate, finally leading to drastic reduction in grain and straw yields of component crops. Donald (1963) observed that competition for light occurred when one plant casts shadow on another or within a plant when one leaf shades another leaf. Willey and Roberts(1976) emphasized that, light was the most important factor when better temporal use of resources was achieved due to the better distribution of leaf area over time. Yield advantage of more than 20 per cent was achieved in sunflower (tall) and radish (dwarf) intercropping (resembles maize and french bean) but peak light interception values were not higher than those of the sole crops (Lakhani, 1976).

At ICRISAT, measurement of light interception showed that intercropping did not intercept more light energy than sole cropping, but this energy was more efficiently converted into dry matter (Reddy et al., 1980). In maize based intercropping system, decrease in interception of solar radiation from top of maize crop to top of legume intercrop was observed due to shading by maize canopy. But legumes used that radiation efficiently and produced some yield (Scarle et al., 1981). Gardner and Cracker (1983) found increased light interception in intercropping system of maize + french bean. Intercropping of maize and soybean increased interception of incident photosynthetic photon flux density (PPFD) compared with pure stands of soybean at 35 DAS (Clement et al., 1992). In an intercropping system of soybean and maize at Pantnagar, the light transmission was decreased by 0.5 to 2.3 and 0.1 to 0.5 per cent respectively, in single row (60 cm) and paired rows (30/90 cm) (Chandel et al., 1993). Sharma et al., (1994) observed that alternate paired row spatial arrangement of maize + soybean intercropping system provided more light to soybean without declining the desired light utilization by the maize. In South Africa, Tsubo and Walker (2002) measured photosynthetically active radiation above and below the maize + french bean intercrop canopy in both north – south and east – west rows. They observed that the intercropping was equivalent in growth efficiency of maize to the sole cropping. Whereas, french bean had greater radiation use efficiency in intercropping than in sole cropping. Mohan (2003) also found increased light interception in intercropping system of maize + french bean.

## System efficiency of intercropping of maize with french bean

It has been widely recognized in recent years that intercropping can often produce higher yields with greater efficiency of resources than sole crops. There are number of ways to assess the intercropping advantages accurately. Most common methods are land equivalent ratio (LER), area time equivalent ratio (ATER) and crop equivalent yield (CEY).

Trenbath (1979) opinioned that intercropping systems have greater potentialities for interception of more light and judicious use of limited resources compared to the sole stands. Land equivalent ratio is defined as the relative land area under sole crops that is required to produce the yields achieved in intercropping (Willey, 1979). The greater values of land equivalent ratio indicate greater biological efficiency of crops grown in association (Maltra et al., 2001). Land equivalent ratios of the intercropping systems involving maize, potato and sugarcane with french bean were greater than unity indicating greater biological efficiency in utilization of land, space and time by intercrops and there by yield advantages over the respective sole crops (Pilbeam et al., 1994, Dua et al., 2005 and Sharma et al., 2006). Faris et al. (1983) from Brazil concluded that, sorghum or maize with cowpea or french bean produced a land equivalent ratio (LER) greater than 1.0 and indicated the merits of ntercropping over monoculture. In intercropping system, potato and french bean yielded

263 and 15 q/ha respectively with land equivalent ratio of 1.47 as compared to sole potato yield of 326 q $ha^{-1}$ (Ali and Kushwah, 1987). Among the intercrops viz., cowpea, french bean and soybean in 4:1, 2:1 and 1:1 row proportion maize + french bean (1.81) was superior to others followed by maize + cowpea (1.74) and maize + soybean (1.59) all sown in the ratio of 2:1. Maize equivalent yields were also higher in the intercropping stands of maize and legumes in a 2:1 row ratio than in 1:1 or 4:1 row ratio or in their pure stands (Shah et al., 1991). Naik and Singh (1992) at Chotanagpur under rainfed condition observed that maize + french bean intercropping was profitable over monoculture of either of the crops. This system recorded LER of 1.31 and 54 per cent higher french bean equivalent yield over sole crop of french bean. Reposotade et al. (1995) reported that higher yields were obtained with 2 to 3 rows of french bean with a maize population of 30,000 to 40,000 plants/ ha.

**Economics of intercropping of maize with French bean**

Singh and Singh (1993) obtained higher net return (Rs. 10032 /ha) and gross return (Rs. 11941/ ha) with maize + french bean (1:2) intercropping system. In Bangladesh, Mandal et al. (1998) observed that the higher gross return was recorded when 3 rows of soybean were grown between paired rows of maize. This system increased the total production in terms of maize equivalent yield by 114.5 per cent over normally sown sole maize. Sinha et al. (1999) observed that normal planting of maize (1:2) intercropped with rajmash recorded higher net return (Rs.19,260 /ha) and was significantly superior over paired row planting of maize + potato in 2:2 ratio (Rs. 17,550/ ha). Padhi (2001) observed that an intercropping of maize with french bean in 2:2 row ratio realized higher net return (Rs.12,677/ ha) and monetary advantage index (6,604.4) and B.C. ratio compared to other intercropping systems. This system increased the net profit by Rs.10,711 /ha and relative net return by 1.91 over normally sown sole maize. Growing rajmash, french bean and pea as intercrops in potato under recommended NPK /ha (100 kg N, 100 kg $P_2O_5$ and 50 kg $K_2O$ / ha) recorded higher net return to the tune of 14 to 30 % ha as compared to sole potato (Bindra and Thakur, 2005). Vedprakash et al. (2005) reported that the net return increased by 633 and 972 % due to relay intercropping of hybrid tomato and french bean in maize during 2002 and 2003, respectively. Birbal (2006) concluded that maize + soybean grown either in 1:1 or 2:2 row ratio recorded significantly higher net return (Rs.10,934 and 10,711/ ha, respectively) over other intercropping systems and sole crops (Rs. 5,993 – 9,745/ ha). Bharati et al. (2007) reported that among intercropping systems, maize + potato generated the highest net return (Rs.28,781 and 35,661/ ha) followed by maize + french bean (Rs.22,270 and 21,250/ ha) in both the years of experimentation.

## Main aspects to be considered in intercropping system

Successful intercropping needs several considerations before and during cultivation. Silwana and Lucas (2002) found intercropping affects vegetative growth of component crops, therefore have to consider the spatial (Willey and Rao, 1981a), temporal and physical resources. Economically viable intercropping largely depends on adaptation of planting pattern and selection of compatible crops (Seran and Brintha, 2009a). Cereal-legume intercropping, potential to provide nitrogen depends in densities of crop, light interception, crop species and nutrients (Francis, 1989). Compatible crop selection is vital in intercropping. The choice of compatible crops for an intercropping system depends on plant growth habit, land, light, water and fertilizer utilization (Brintha and Seran, 2009). Hardarson Atkins (2003) found legume-cereal intercropping increase the fixation of nitrogen by legumes. Silwana and Lucas (2002) reported different crop species in mixtures increased capture of growth limiting resources and stated that different planting time of component crops improve the resource utilization and reduce the competition (Andrews, 1972).

### Maturity of crop

When two or more crops are grown together the peak period of growth of components do not coincide. The biggest complementary effects and thus biggest yield advantages seen to occur when the component crops have different growing periods so make their major demands on resources at different times. Crops of varying maturity duration should be chosen therefore a rapidly maturing crop completes its life cycle before the major growth period of other crop commence. Selecting crops or varieties with different maturity time can also assist staggered harvesting and separation of grain commodities. By this, the time of peak nutrient demands of component crops should be differed. Crops which mature at different times thereby separating their periods of maximum demand to nutrient and moisture aerial space and light could be suitably intercropped (Enyi, 1977). In maize-greengram, peak light demand for maize is around 60 days after planting, while greengram is ready to harvest (Reddy and Reddi, 2007).

### Compatible crops

Choosing of the crop combination plays vital role in intercropping. Plant density, shading and nutrition competition between plants reduce the yield of monocrop. Plant competition could be minimized not only by spatial arrangement, but also by choosing those crops best able to exploit soil nutrients (Fisher, 1977b). Groundnut is usually intercropped with maize in South East Asia and Africa (Kassam, 1976). Popondo (*Phaseolus lunatus*) and mucuna

(*Mucuna utilis*) lowered maize yield, while calopo (*Calopogonium tnucunoides*), cowpea and greengram had much less effect on maize and were themselves tolerant to maize shade (Agboola and Fayemi, 1971). In tropics, maize-cowpea intercropping is often practiced (Van Kessel and Roskoski, 1988: Mpangane et al., 2004). Maize is easy to manage in maize-pegionpea intercropping (Krantz, 1981). In Central and South America and parts of East Africa, maize is intercropped with bean (Singh et al., 1998).

### Plant density

The planting pattern of the maize and legumes (intercropping or growing maize after the legume harvest) did not affect the yield of maize (Agboola and Fayemi, 1971). Increasing maize plant density had significant effect on LAI in maize soybean intercropping (Prasad and Brook, 2005). Maize dry matter was reduced with increasing lablab population (6-10 plants per $m^2$) (Maluleke et al., 2005). In maize-okra intercropping, high plant density reduced number of leaves due to competition for light and other resources (Muoneke and Asiegbu, 1997).

## Time of planting

Planting cowpea simultaneously with maize gave better yield (Mongi et al., 1976). Simultaneously planting maize and sweet potato not influenced maize grain yields, whereas late planting of sweet potato negatively affects maize yield (Amede and Nigatu, 2001). Several researches have been focused on bush bean and maize planted simultaneously in alternate rows (Frankis et al., 1978: Pilbeam, 1996; Santalla et al., 1999).

## Benefits of Intercropping

### Resource utilization

Maize and cowpea intercropping is beneficial on nitrogen poor soils (Vesterager et al., 2008). Maize-cowpea intercropping increases the amount of nitrogen, phosphorous and potassium contents compared to monocrop of maize (Dahmardeh et al., 2010). Nutrient uptake and utilization more efficient in maize-rice and maize-soybean intercrops than in those crops as monocrop (Suryanta and Harwood, 1976).

Maize-groundnut intercropping enhanced the efficient utilization of strong light by maize and weak light by groundnut lead to provide yield advantages (Jiao et al., 2008). Soybean and maize intercropping has been attributed to better use of solar radiation (Keating and Carberry, 1993), nutrients (Willey, 1990) and water (Morris and Garrity, 1993) over the monocrop. Water use efficiency was the highest under soybean-maize intercropping compared with monocropping maize and monocropping soybean (Barhom, 2001). Soybean-maize intercropping was the best combination system during water scarcity periods (Tsubo et al., 2005).

## Weed control

Intercropping might better control of weeds. Evidence of better weed control is reasonably clear where intercropping provides a more competitive effect against weeds either in time or space than does monocropping. Leafy greens can be intercropped with maize to control weeds in the tropics and increase productivity (Makindea et al., 2009). Weed suppression in maize-groundnut intercropping was reported by Steiner (1984). Intercropping maize and legumes considerably reduced the weed density compared with the monocropping maize by decrease in available light for weeds compared to monocrops (Dimitrios et al., 2010). Maize-cowpea intercropping suppresses weeds and insures against total crop failure when one crop fails (Mongi et al., 1976). Maize-pumpkin and maize-bean intercropping reduced weed biomass by 50 - 66 % when established at a density of 12,300 and 222,000 plants/ ha for beans (Mashingaidze, 2004). Maize-rye intercropping reduces weed biomass by 50 % (Samson, 1991).

## Pest and disease

Maize is susceptible to many insects (Drinwater et al., 2002) and diseases (Flett et al., 1996). Intercropping appears to be a very promising cultural practice for this purpose. It is generally believed that one component crop of an intercropping system may act as a barrier or buffer against the spread of pests and pathogen. Intercropping maize-cowpea reduces the stem borer (Henrik and Peeter, 1997). Maize leafhopper (*Dalbus maindis* L.) was reduced under intercropping reported by Power (1990). In Louisiana, bud worm infestation in pure maize was greater than in maize associated with soybean (Brown, 1935). Maize-groundnut and maize-soybean mixed crop reduced the number of corn borer in maize (Sastrawinata, 1976). Insect problems are less on crops grown in mixture, especially with cowpea, pigeon pea, maize and some legumes (Caswell and Raheja, 1972; Hayward, 1975). In chilli + maize intercropping, the incidence of *Anthonomus eugenii* was lower and yield was greater compared to chilli alone (Gutierrez, 1999). Pest and disease were less in tomato + maize intercropping (Pino et al., 1994).

## Erosion control

Intercropping controls soil erosion by preventing rain drops from hitting the bare soil where they tend to seal surface pores, prevent water from entering the soil and increase surface erosion. Maize-cowpea intercropping, cowpea acts as best cover crop and reduced soil erosion (Kariaga, 2004). Rows of maize in a field with a shorter crop will reduce the wind speed above the shorter crops and thus reduce desiccation (Beets, 1990).

**Yield advantages**

Yield is taken as primary consideration in the assessment of the potential of intercropping practices (Anil et al., 1998). Amede and Nigatu (2001) consistently received LER of 1.5 or greater when using the early maturing variety of maize. Maize-Kenaf-African Yam bean gave LER of 1.12 (Adeniyan et al., 2007). LER increased to a maximum of about 48 % by intercropping maize with soyabean compared with the cereal sole crops (Mohta and De, 1980). Maize yield was not affected by intercropping with soyabean (Mehta and Dey, 1980). Maize-soybean intercropping gave LER of 1.18 reported by Putnam et al. (1985).

**Economic benefits**

Intercropping often provides higher cash return than growing one crop alone (Kurata, 1986). Intercropping occupies greater land use and thereby provides higher net returns (Seran and Brintha, 2009a).

**Maize legume intercropping**

Intercropping of legumes and cereals is an old practice in tropical agriculture that time back to ancient civilization. Systems that intercrop maize with a legume are able to reduce the amount of nutrients taken from the soil as compared to a maize monocrop. During absence of nitrogen fertilizer, intercropped legumes will fix nitrogen from the atmosphere and not compete with maize for nitrogen resources (Adu-Gyamfi et al., 2007). The mixture of nitrogen fixing crop and non fixing crop give greater productivity than monocropping (Seran and Brintha, 2009b). Maize-french bean gave high maize equivalent yield over sole maize yield (Hugar and Palled, 2008b) and kernel yield of maize was unaffected in maize-french bean intercropping (Pandita, 2001). Without nitrogen fertilizer application, gliricidia-maize intercropping system gave high maize yield (Akinnifesi et al., 2006). Maize yield was increased by 26 % in maize-soybean strip intercropping (West and Griffith, 1992; Ghaffarzadeh et al., 1994). In maize-bean intercropping, maize yield was not affected (Tsubo et al., 2005).

**Maize vegetable intercropping**

Tomato intercropped with maize increased the number and weight of fruit per plant, total yield (Sharma and Tiwari, 1996). In maize-okra intercropping, yield and yield components of okra was increased (Muoneka and Asiegbu, 1997). Cauliflower in maize produced 7 t /ha, cauliflower in conjunction with 2.1 t /ha maize (Khatiwada, 2000). Maize-Kenaf-African yam bean intercropping gave higher value of LER compared to monocroppings (Adeniyan et al., 2007). Maize with pumkins gave high LER was reported by Cortes and Los (1997). Maize-cassava intercropping gave LER advantages (Mutsaers et al., 1993). Maize-potato intercropping performed better than sole potato (Begum et al., 1999). Maize yield not affected by maize-potato intercropping (Ifenkwe et al., 1989).

At Bangalore (Karnataka), aggressivity index value was low in 2:1 row ratio (maize: urd bean) (0.27) as compared to 2:2 ratio (1.20) and 1:1 row ratio (1.48) indicated low competition in 2:1 row ratio. However yield realized was more in 2:2 ratio (6845 kg/ha) compared to that of 2:1 ratio (6499 kg/ ha) (Naveena et al., 2012). At Samastipur (Bihar), all intercropping system recorded significantly higher maize equivalent yield and productivity than sole cropping of maize. Among the intercropping systems, maize + turmeric (1:2) showed significantly higher maize equivalent yield and productivity. However, significantly lower values of these were associated with maize + sesame (Pandey et al., 2003).

At Samastipur, maize intercropped with early potato variety was more compatible and advantageous than other potato varieties. Growth and yield attributes of winter maize were superior under sole cropping followed by winter maize + early potato intercropping system. Maize + early potato intercropping system recorded higher maize equivalent yield (178.96 q/ha) and yield advantage (163 %). 100 % RDF (120:75:60 kg NPK/ha) to maize + 75 % RDF (75:30:50 kg NPK/ha) to potato proved economical with 2.27 B:C ratio and maize equivalent yield (190.02 q/ha) but 100 % RDF both to maize and potato recorded higher maize equivalent yield of 194.23 q/ha (Jha et al., 2000).

**Table 1. Effect of genotypes and fertilizer levels on yield and economics of maize + potato intercropping system during winter season**

| Treatments | Yield (q/ha) | | | LER | B:C ratio |
|---|---|---|---|---|---|
| | Maize | Potato | Maize equivalent | | |
| **Intercropping** | | | | | |
| Maize + early potato | 46.80 | 142.56 | 178.96 | 1.63 | 2.21 |
| Maize + medium potato | 43.96 | 158.06 | 173.81 | 1.62 | 2.18 |
| Maize + late potato | 40.42 | 178.83 | 170.32 | 1.62 | 2.09 |
| CD 5 % | 2.32 | 13.19 | NS | NS | NS |
| **Fertilizer levels** | | | | | |
| 100 % RDF to maize + No RDF to potato | 30.95 | 142.25 | 145.96 | 1.30 | 1.91 |
| 100 % RDF to maize + 25 % RDF to potato | 38.48 | 154.02 | 164.21 | 1.51 | 2.08 |
| 100 % RDF to maize + 50 % RDF to potato | 45.06 | 163.02 | 182.23 | 1.67 | 2.21 |
| 100 % RDF to maize + 75 % RDF to potato | 50.83 | 170.13 | 190.02 | 1.80 | 2.27 |
| 100 % RDF to maize + 100 % RDF to potato | 53.30 | 172.14 | 194.23 | 1.85 | 2.25 |
| CD 5 % | 2.80 | 14.12 | 10.24 | 0.11 | 0.12 |

(Jha et al., 2000).

**Table 2. Performance of maize based intercropping systems during kharif season**

| Intercropping | Yield (q/ha) | | | LER | Aggressive index | B:C ratio |
|---|---|---|---|---|---|---|
| | Maize | Intercrop | MEQ | | | |
| Sole maize | 49.5 | - | 49.5 | - | - | 1.72 |
| Sole beans | - | 1.59 | 47.8 | - | - | 1.34 |
| Sole green gram | - | - | 27.3 | - | - | 0.41 |
| Maize + beans 1:1 | 49.3 | 6.1 | 67.8 | 1.38 | -0.16 | 1.75 |
| Maize + beans 1:2 | 52.0 | 8.9 | 78.8 | 1.61 | 0.21 | 1.87 |
| Maize + beans 1:3 | 49.6 | 5.0 | 64.6 | 1.32 | 0.69 | 1.13 |
| Maize + beans 2:1 | 48.8 | 2.8 | 57.3 | 1.16 | -0.36 | 1.71 |
| Maize + beans 3:1 | 48.9 | 1.6 | 53.5 | 1.08 | -0.46 | 1.61 |
| Maize + green gram 1:1 | 49.1 | 3.3 | 62.0 | 1.47 | -0.44 | 1.65 |
| Maize + green gram 1:2 | 50.8 | 3.7 | 65.7 | 1.58 | 0.20 | 1.74 |
| Maize + green gram 1:3 | 49.8 | 2.3 | 58.1 | 1.33 | 0.65 | 1.09 |
| Maize + green gram 2:1 | 48.3 | 1.5 | 54.4 | 1.20 | -0.70 | 1.66 |
| Maize + green gram 3:1 | 48.5 | 0.7 | 51.1 | 1.07 | -0.44 | 1.56 |
| CD 5 % | - | - | 4.5 | - | - | - |

(Pandita et al., 2000).

At Shalimar (Jammu & Kashmir), intercropping of maize with field bean in a row ratio of 1:2 recorded maximum land equivalent ratio (1.61), area time equivalent ratio (1.48) and aggressive index (0.69) and B:C ratio (1.87) (Pandita et al., 2000).

At Faizabad (Uttar Pradesh), maize growth with lentil was more advantageous than rajma and Indian mustard and gave higher maize yield, intercrop yield, maize equivalent yield, net return and B:C ratio. Paired row planting pattern gave significantly higher yield, monetary return and B:C ratio (Misra et al., 2001).

**Table 3. Intercropping in winter maize**

| Treatments | Yield (kg/ha) | | | Monetary return (Rs/ha) | B:C ratio |
|---|---|---|---|---|---|
| | Maize | Intercrop | Equivalent | | |
| **Intercropping** | | | | | |
| Maize + rajma | 7102 | 823 | 8926 | 27078 | 1.37 |
| Maize + lentil | 7772 | 1143 | 10817 | 32571 | 1.91 |
| Maize + Indian mustard | 5250 | 904 | 7071 | 16811 | 1.00 |
| CD 5 % | 181 | 58 | 923 | 946 | 0.17 |
| **Planting pattern** | | | | | |
| Alternate planting (60 cm furrow) | 6657 | 935 | 8848 | 24512 | 1.39 |
| Paired row planting (70 cm and 50 cm apart in paired row of maize and intercrop) | 6845 | 979 | 9808 | 26463 | 1.48 |
| CD 5 % | 65 | 34 | 113 | 1037 | 0.05 |

(Misra et al., 2001).

**Table 4. Intercropping in kharif maize**

| Intercropping | Yield (t/ha) | | | LER |
|---|---|---|---|---|
| | Maize | Intercrop | Maize equivalent | |
| Sole maize | 21.35 | - | 28.00 | - |
| Sole cowpea | - | 5.77 | 15.62 | - |
| Sole black gram | - | 4.69 | 14.69 | - |
| Sole soybean | - | 3.63 | 8.40 | - |
| Maize + cowpea | 19.55 | 2.59 | 35.28 | 1.36 |
| Maize + black gram | 18.97 | 2.29 | 33.50 | 1.37 |
| Maize + soybean | 18.86 | 1.87 | 30.66 | 1.41 |
| CD 5 % | - | - | 4.30 | NS |

(Khola et at., 1999).

At DehraDun (Uttaranchal), intercropping of cowpea and black gram with maize gave higher maize equivalent yield (35.28 and 33.50 q/ha respectively) than sole maize (28.00 q/ha). Intercropping also resulted in higher maize equivalent yield than respective sole legume stands (Khola et at., 1999).

**Table 5. Relay intercropping of tomato in maize**

| Relay intercropping | Yield (kg/ ha) | | | LER | Net return (Rs/ha) | B : C ratio |
|---|---|---|---|---|---|---|
| | Maize | Tomato | Maize equivalent | | | |
| Sole maize (50 cm) | 7656 | - | 7656 | 1.00 | 16278 | 1.74 |
| Paired new maize (40/80 cm) | 7022 | - | 7022 | 1.00 | 13549 | 1.63 |
| Paired new maize (40/60 cm) | 7619 | - | 7619 | 1.00 | 16051 | 1.73 |
| Maize and Tomato in alternate rows | 7914 | 37477 | 52883 | 1.89 | 198820 | 5.03 |
| Maize paired row (40/80 cm) + one row tomato within | 6579 | 38050 | 52244 | 1.82 | 196055 | 4.01 |
| Maize paired row (40/60 cm) + one row tomato within paired rows of maize | 7794 | 31542 | 45343 | 1.73 | 161117 | 3.46 |
| CD 5 % | 710 | 3518 | 2420 | - | 22383 | - |

(Ved Prakash et al., 2004).

At Almora (Uttaranchal), one row of tomato within paired rows of maize (40/80 cm) recorded higher tomato fruit yield (38052 kg/ha). All the relay intercropping combinations enhanced total production in terms of maize equivalent yield by 4.5 – 4.9 times over sole maize crop. Relay intercropping of tomato (transplanted during mid August in standing maize) in maize in row ratio of 1:1 (sown in first fortnight of June and harvested in first fortnight of September) recorded higher maize equivalent yield (52882 kg/ha), net return (Rs. 198820/ ha) and land equivalent ratio (LER) (1.86) (Ved Prakash et al., 2004).

**Table 6. Maize based intercropping systems**

| Intercropping systems | Suitable area/situation |
|---|---|
| Maize + Pigeon pea | All maize growing areas |
| Maize + Cowpea | |
| Maize + Mungbean | |
| Maize + Urdbean | |
| Maize + Sugarcane | |
| Rice + Maize | |
| Maize + Soybean | |
| Maize + high value vegetables | Peri-urban interface |
| Maize + flowers | |
| Baby corn + vegetables | |
| Sweet corn + vegetables | |

## References

Adhikari, S., T. Chakraborty and D.K. Bagchi (2005). Bio-economic evaluation of maize and groundnut intercropping in drought prone areas of Chotonagpur plateau region of Jharkhand. Indian J. Agron., 50 (2): 113-115.

Agarawal, G.C. and N.K. Sekhon (1991). Changes induced by cowpea green manures and farm yard manure in the timing of phonological events in maize (*Zea mays* L.). Ann. Agric. Sci. (Cambridge), 117: 157-163.

Ali, M. (2002). Annual Report (2002), Directorate of Pulses Research, IIPR, Kanpur.

Ali, M. and B.L. Kushawaha (1987). Intercropping of pulses in maize. Indian Farming, 37(2): 20-23.

Anil Kumar and K. S. Thakur (2005). Influence of intercropping and weed control measures on weeds and productivity of rainfed maize (*Zea mays*). Indian J. Weed Sci., 37 (1&2): 65 -67.

Anil Kumar, K.S. Thakur and S. Manuja (2002), Effect of fertility levels on promising hybrid maize under rainfall conditions of Himachal Pradesh. *Indian J. Agron.,* 47(4): 526-530.

Anil Kumar, K.S. Thakur and S. Sharma (2005), Integrated nutrient management in maize (*Zea mays* L.) – gobhi sarson (*Brassica napus*) cropping system under rainfed condition. *Indian J. Agron.,* 50(4): 274 - 277.

Anonymous (2005). Maize -vegetable - mushroom intercropping under drip fertigation. All India Coordinated Research Project on Water management, Agricultural College and Research Institute, Tamil Nadu Agricultural University, Madurai, P:107 - 108.

Ashok Kumar, R.K. Chillar and R.C. Gautam (2006). Nutrient requirement of winter maize (*Zea mays* L.) - based intercropping systems. Indian J. Agric. Sci., 76 (5): 315-318.

Balayan, J.S. (1997). Performance of maize based intercropping systems and their after effect on wheat. Indian J. Agron., 42 (1): 26-28.

Balyan, J. S., 1997, Performance of maize (*Zea mays*) based intercropping systems and their after effect on wheat (*Triticum aestivum*). *Indian J. Agron.*, 42(1): 26-28.

Behera, B., S.K. Mohanty and A. Mishra (2010). Productivity, profitability and sustainability of integrated nutrient management practices in yam (*Dioscorea esculenta*) + maize (*Zea mays*) intercropping system. Indian Journal of Soil Conservation, 38 (3): 217 – 222.

Bhuvaneshwari, J., A. Muthusankaranarayanan and S. Avudaithai (2002). Weed smothering effect of intercropping in maize. Madras Agric. J., 89 (10-12): 714 - 716.

Bilalis, D.J., N. Sidiras, I. Kakampouki, A. Efthimiadou, Y. Papatheohari and P. Thomopouloas (2005). Effects of organic fertilizer on maize / legume intercrop in a clay loam soil and Mediterranean climate – can the land equivalent ratio index be used for root development? J. Food Agric. Environ., 3(3&4): 117 – 123.

Bilalisa, D., P. Papastylianoua, A. Konstantasa, S. Patsialia, A. Karkanisa and A. Efthimiadoua (2010). Weed-suppressive effects of maize-legume intercropping in organic farming. Int. J. Pest Manage., 56: 173-181.

Brij Nandan, B. C. Sharma and Anil Kumar (2013).Effects of mustard-maize intercropping system on productivity of maize in moisture deficit sub-tropical areas of Jammu and Kashmir. Scholarly J. Agric. Sci., 3(2): 66-72.

Chalka, M.K and V. Nepalia (2005). Production potential and economics of maize (*Zeamays*) intercropped with legume as influenced by weed control. Indian J. Agron., 50 (2):119-122.

Chittapur, B.M., H.B. Bablad, S.M. Hiremath, G.A Kulkarni and D.R. Sardeshpande (1994). Intercropping of fodder legumes in grain maize. *J. Maharashtra Agric. Univ*., 19(3): 454-455.

Dasaraddy, S.V. (1998). Studies on residual effect of in situ green manuring on the safflower crop in maize + legume - safflower sequence. *M. Sc. (Agri.) Thesis,* Uni. Agric. Sci., Dharwad (India).

Dasaraddy, S.V., S.M. Hiremath and R.H. Patil (2001). Studies on residual effect of in situ green manuring on the safflower crop in maize + legume-safflower sequence. *Mysore J. Agric. Sci*., 35: 323-327.

Deshveer C.L. and Amar Singh (2002). Weed management studies in maize based intercropping systems. Indian J. Weed Sci., 34 (3&4): 236 – 240.

Dev, A.K. and S.C. Sexsena (2002). Response of soybean and maize intercropping to phosphorus fertilization in the *tarai* soils of Uttaranchal. Ann. Agric. Res., 23(2): 323-330.

Dhingra, K.K., M.S. Dhillon, D.S. Grewal and K. Sharma (1991). Performance of maize and mung bean intercropping in different planting pattern and row orientations. *Indian J. Agron*., 36(2): 207-212.

Gangwar, K. S. and S.K. Sharma (1994). Fodder legume intercropping in maize (*Zea mays*) and its effect on succeeding wheat (*Triticum aestivum*). *Indian J. Agric. Sci*., 64 (1): 38-40.

Gautam, I.P., K. P. Upadhyay, J.N. Choudhary and B. Khatri (2004). Study on intercropping vegetables in maize based cropping system in *Tar* area of western Nepal. Proceedings of IV National Conference on Science and Technology, RONAST, Kathmandu Nepal. pp. 110-114.

Grewal, H.S., J.S. Kolra and J.S. Kang (1992). Effect of combined use of green manure and nitrogen on the productivity of maize (*Zea mays*)-wheat (*Triticum aestivum*) system. *Indian J. Agron.*, 37(4): 635-638.

Hugar, H.Y. and Y.B. Palled (2008a). Effect of intercropped vegetables on maize and associated weeds in maize-vegetable intercropping systems. Karnataka J. Agric. Sci., 21 (2): 159-161

Hugar, H.Y. and Y.B. Palled (2008b). Studies on maize-vegetable intercropping systems. Karnataka J. Agric. Sci., 21(2): 162-164.

Intodia, S.K., L.R. Yadav and O.P. Tomar (1996). Effect of herbicides on weed control efficiency and yield in maize (*Zea mays*) – soybean (*Glycine max*) intercropping system. *Indian J. Agric. Sci.,* 60(12): 730 -731.

Jamwal, J.S. (2005). Productivity and economics of maize-wheat cropping systems under integrated nutrient supply system in rainfed areas of Jammu. *Indian J. Agron.*, 50 (2) : 110-112.

Jat, R.L. (1996). Studies on weed management and fertilizer levels of intercrop with and without inoculation in maize + soybean intercropping system. Ph.D thesis. Rajasthan Agricultural University, Udaipur.

Jha, D., D.P.Singh and R.B.Thakur (2000). Production potential of maize + potato intercropping as influenced by fertilizer and potato genotypes. Indian J. Agron., 45 (1): 59-63.

Kalyan Singh, U.N., R.S.,Singh, Chandel and K.K. Singh (2004). Effect of intercropping in maize and life saving irrigation in garden pea on yield and nutrient uptake by maize garden pea cropping system. *Crop Res*., 28(1): 28-33.

Khola, O.P.S., S.K. Dube and N.K. Sharma (1997). Biological and economical feasibility of intercropping legumes with maize (Zea mays L.) on sloping valley lands. Indian J. Soil Cons., 25: 141-146.

Kins Varghese and Gautam Ghosh (2006). Comparative study of maize cultivars (*Zea mays* L.) with intercrop (*Vigna mungo* L.) under two levels of fertilizers. Mysore J. Agric. Sci., 40(1): 14-20.

Kumar, A. and S.N. Singh (2002). Production potential and economics of winter maize based intercropping systems. Annals Agric. Res., 23(4): 532-534.

Kumar, P, C.R. Sunil Rawat and N.P. Melkani (2002). Nutrient management in maize (*Zea mays*) + cowpea (*Vigna unguiculata*) intercropping system. Indian J. Agric. Sci., 72: 665-666.

Kumar, R.A., K. Chillar and R.C. Gautam (2006). Nutrient requirement of winter maize *(Zea mays)*-based intercropping systems, Indian J. Agric. Sci., 76 (5): 315-318.

Kuruvilla Varughese and M.R. Iruthayaraj (1996). Biological and economic indices for measuring resource utilization in maize – soybean intercropping system. Madras Agric. J., 83 (9): 556 - 559.

Lamani, C. (1998). Integrated nutrient management in maize. *M. Sc. (Agri.) Thesis*, Uni. Agric. Sci., Dharwad (India).

Laxminarayana, K. and G.C. Munda (2004). Performance of rice and maize based cropping systems under mid hills of Mizoram. *Indian J. Agron.*, 49(4): 230-232.

Lingaraju, B.S., S.B. Marer and S.S. Chandrashekar (2008). Studies on Intercropping of Maize and Pigeon pea Under Rainfed Conditions in Northern Transitional Zone of Karnataka. Karnataka J. Agric. Sci., 21 (1): 1-3.

Madar, K.S. (2001). Studies on maize based intercropping as influenced by maize and pigeonpea row proportions and population levels of cowpea. M. Sc. (Agri.) thesis, University of Agricultural Sciences, Dharwad, India.

Meena, O. P., B.L. Gaur and P. Singh (2006). Effect of row ratio and fertility levels on productivity, economics and nutrient uptake in maize + soybean intercropping system. *Indian J. Agron.*, 51(3): 178-182.

Meyyappan, M. and R.M. Kathiresan (2005). Integrated Weed Management in Maize+ black gram Intercropping System. Indian J. Weed Sci., 37(3 and 4): 209 – 211.

Mishra, A.S. and S. Elamathi (2009). Response of maize (*Zea mays* L.) + legume Intercropping system to different weed control practices. Madras Agric. J., 96 (7-12): 322-324.

Misra, B.N., B.Singh and A.L. Rajput (2001). Yield, quality and economics as influenced by winter maize based intercropping system in eastern Uttar Pradesh. Indian J. Agron., 46 (3): 425-431.

Mohapatra, B.K. and L. Pradhan (1992). Intercropping fodder legumes with maize in different planting patterns. Annals of Agric. Res. 13(4): 366-371.

Mpangane, P.N.Z., K.K Ayisi, M.G. Mishiyi and A. White bread (2004). Grain yield of maize grown in sole and binary cultures with cowpea and lab lab in the Limpopo province of South Africa. In: AICAR proc. No: 115.

Mulatu, K. (1993). Method and time of intercropping maize *(Zea mays* L) with haricot bean *(Phaseolus vulgaris* L.). IAR Newsletter of Agricultural Research, 8: 2 - 3.

Muoneke, C.O. and J.E. Asiegbu (1997). Effect of okra planting density and spatial arrangement in intercrop with maize on the growth and yield of the component species. J. Agron. Crop Sci., 179: 201-207.

Naveen Kumar and Naleeni Ramawat (2006). Production potential and economics of maize + guinea grass cropping in different proportions under sub - tropical conditions. Forage Res., 32 (3): 159-162.

Naveena, K., G.B. Mallikarjuna, H. M. Jayadeva and H. S. Surendra (2012). Statistical evaluation for identification of productive row ratio in intercropping system of maize + urd bean using Aggressivity index. Mysore J. Agricultural Science, 46(4): 733 -737.

Padhi, A.K. and R.K. Panigrahi (2006). Effect of intercrop and crop geometry on productivity, economics, energetics and soil- fertility status of maize based intercropping systems. *Indian J. Agron.*, 51(3): 174-177.

Pande, I.B., V. Bharti and S.S. Mishra (2003). Effect of maize (*Zea mays*) - based intercropping systems on maize yield and associated weeds under rainfed conditions. Indian J. Agron., 48(1):30-33.

Pandey, A.K. and Ved Prakash (2002). Weed Management in Maize and Soybean intercropping System. Indian J. Weed Sci., 34 (1&2): 58 – 62.

Pandey, A.K., V. Prakash, R.D. Singh and V.P. Mani (1999a). Effect of intercropping patterns of maize (Zea mays L.) and soybean (Glycine max (L.) Merrill) on yield and economics under mid hills of NW Himalayas. Ann. Agric. Res., 20: 354-359.

Pandey, A.K., V. Prakash, R.D. Singh and V.P. Mani (1999b). Effect of intercropping patterns of maize and soybean on yield and economics under mid hills of North western Himalayas. Annals Agric. Res., 20 (2): 154 -159.

Pandey, I.B., V. Bharati and S.S. Mishra (2003). Effect of maize – based intercropping systems on maize yield and associated weeds under rainfed condition. Indian J. Agron., 48: 30-33.

Pandita, A.K., M.H. Sha, and A.S., Bali (1998). Row ratio in maize (*Zea mays*) legume intercropping in temperate valley condition. *Indian J. Agric. Sci.*, 68(10): 633-635.

Pandita, A.K. (2001). Effect of vegetable intercropping on productivity, economics and energetics of maize (*Zea mays* L.). Indian J. Agron., 46: 204-210.

Pandita, A.K., M.H.Shah and A.S.Bali (2000). Effect of row ratio in cereal – legume intercropping systems on productivity and competition functions under Kashmir condition. Indian J. Agron., 45 (1): 48-53.

Parmeet Singh, Prshotam Singh and Joy Dawson (2007). Correlation and regression studies of winter maize and weed interactions. Indian J. Weed Sci., 39 (1&2): 21-23.

Pathak, K. and N.P. Singh (2008). Growth and yield of black gram (*Phaseolus mungo*) varieties under intercropping systems with maize (*Zeamays*) during rainy season in North India. Farming System Res. Dev., 14 (1):29-34.

Patra, B., B.K. Mandal and B.B. Mandal (1990). Probability of maize legume intercropping system. Indian Agriculturist, 34 (4): 227 - 233.

Patra, B.C., B. B. Mandal, B.K. Mandal and A. K. Padhi (1999). Suitability of maize based intercropping systems. Indian J. Agric. Sci., 69(11): 759 – 762.

Pattanashetti, V.A., CA. Agasimani and HB. Babalad (2002). Effect of manures and fertilizers on yield of maize and soybean under intercropping system. *J. Maharashtra Agricultural University,* 27(2): 206-207.

Prakash, V., A.K. Pandey, A.K. Srivastava and H.S. Gupta (2004). Relay intercropping of hybrid tomato (*Lycopersicon esculentum*) in maize (*Zea mays*) under mid-hill condition of North-Western Himalayas. Indian J. Agric. Sci., 74: 405-408.

Prakash, V., N. Kumar and A.K. Srivastava (2005). Viability of relay intercropping of tomato (*Lycopersicon esculentum*) and French bean (*Phaseolus vulgaris*) with maize (*Zea mays*) in north-west Himalaya. Indian J. Agric. Sci., 75:794-796.

Prasad K. and A. Rafey (1995). Weed Control in Maize (*Zea mays*) intercropped with Soybean (*Glycine max*). Indian J. Weed Sci., 27 (3&4): 167- 169.

Prasad, A., G.Singh and R.K.Upadhyay (2008). Integrated weed management in maize and maize + black gram intercropping. Indian J. Weed Sci., 40 (3&4): 191 - 192.

Prasad, R.B. and R.M. Brook (2005). Effect of varying maize densities on intercropped maize and soybean in Nepal. Exp. Agric., 41: 365 – 382.

Purushotham, S., R. Siddaraju, and G.C.Narayanaswamy (2003). Effect of seed rate and fertility levels in sole and mixed cropping of fodder maize and cowpea. *Mysore J. Agric. Sci.,* 37(1): 51-55.

Quayyum, M.A. and A.F.M. Maniruzzaman (1995). Intercropping of maize and rice with black gram. Indian J. Agron., 40 (1): 20 - 25.

Rajendra, A.N. (2000). Bioefficacy of herbicides in maize + soybean intercropping system in vertisols. *M.Sc.(Agri.) Thesis*, Univ. Agric. Sci., Dharwad.

Rajshekhar, M.G. (2001). Effect of maize plant geometry, row proportion and time of planting on maize - lucerne intercropping system. *M. Sc. (Agri.) Thesis*, Uni. Agric. Sci., Dharwad (India).

Rajshekhar, M.G., Y.B. Palled and S. C. Alagundagi (2004). Performance of maize-lucerne intercropping system. *Karnataka J. Agric. Sci.,* 17(2): 196-202.

Ram prasad (2001). Weed management in fodder maize + cowpea intercropping system. Indian J. Agron., 30 (4): 267 - 268.

Ramaswamy, M., K. Vairvan and K. Srinivasan (1996) Production potential and economics of cereal based cropping system in red lateritic soils of Pudukottai district. *Madras Agric. J.*, 83(4): 236-239.

Rana, K,S. and R.S. Chodhary (2002). Productivity, nitrogen uptake and water use in maize (Zea mays L.) + mung bean intercropping system as affected by nitrogen levels under rainfed conditions. Crop Res., 32(3): 306-308.

Ranbir Singh, R.K. Ashok Kumar Chhillar and K.S. Rana (2002). Production potential and economics of maize (*Zea mays*) based cropping system. *Indian J. Agric. Sci.,* 72(3): 156-157.

Ranbir Singh, R., Bhupinder Singh and S.C. Negi (2001). Studies on intercropping of legumes with maize at varying levels of NPK to legume component. *Indian J. Agric. Res.*, 35(2): 100-103.

Rao, M.R. and M.N. Mathuva (2000). Legumes for improving maize yields and income in semiarid Kenya. Agric. Ecosystem Environ., 78: 123 – 137.

Rathod, P. S., S.I., Halikatti, S.M. Hiremath and S.T. Kajjidoni (2004). Influence of different intercrops and row proportions on yield and yield parameters of pigeonpea in vertisols of Dharwad. Karnataka Journal of Agricultural Sciences, 17**:** 652-657.

Reddy, K. and G. Bheemaiah (1991). Effect of cropping systems and soil types on yield of maize. *Indian J. Agron.*, 36(2) : 139-142.

Sanjay, K.D., G.K. Shrivastara, A.P. Singh and S.S. Kolhe (2011). Weed population, removal by weeds and crop yield under Maize + Black gram inter cropping system in chhatisgarh plains. Indian J. Weed. Sic., 43 (3&4): 203-210.

Sanjay, K.D., G.K. Shrivastava, A.P. Sigh and R. Laspale (2012). Weeds and crop productivity of maize + black gram intercropping system in Chhattisgarh plains. Indian J. Weed Sci., 44(1): 26 – 29.

Saren, B.K. and P.K. Jana (1999). Effect of irrigation and intercropping system on yield, water use, concentration and uptake of nitrogen, phosphorous and potassium in maize (*Zea mays* L.) and groundnuts *(Arachis hypogea*) grown sole and intercrop. Indian J. Agric. Sci., 69: 317-32.

Sarkar, R.M. and D. Shit (1990). Effect of intercropping cereals, pulses and oilseeds with maize on production competition and advantage. Indian Agriculturist, 34**:** 88-89.

Selvakumar, T. and A. Sundari (2006). Effect of intercropping and weed management practices on weeds in maize. Indian J. Weed Sci., 38 (1&2): 133 - 134.

Shapoo, F.A. (2006). Relative efficacy of different herbicides for weed control in maize-pulse intercropping system under sub-tropical conditions of Jammu.M.Sc. Thesis, Division of Agronomy, SKUASTJ, Jammu, India.

Sharma, J. and S.C. Nayital (1993). Crop weed competition in maize (*Zea mays*) + black gram (*Phaseolus mungo*) intercropping system. *Indian J. Agron.,* 36: 64-67.

Sharma, J. and S.C. Nayital (1993). Weed Management in Maize + black gram Intercropping in Mid-Hills Sub-Humid Zone of Himachal Pradesh. Indian J. Weed Sci., 25(1&2): 43 – 46.

Sharma, R.S. and S.D. Choubey (1991). Effect of maize (zea mays) legume intercropping system on nitrogen economy and nutrient status of soil. *Indian Journal of Agronomy,* 36:743-745.

Sharma, V.M., I.S. Chakor and A.K. Manchana (1998). Effect of maize (Zea mays)-based legume intercropping on growth and yield attributes of Succeeding wheat (Triticum aestivum) and economics. *Indian J. Agron.,* 43(2): 231 - 236.

Shekhawat, V.S., M.S. Shaktawat and S.P.S. Tanwar (2002). Effect of weed management on growth and productivity of maize-black gram intercropping system. Agric. Sci. Digest, 22(1): 36 – 38.

Shivay, Y.S., R.P. Singh and S.C. Pandey (1999). Response of nitrogen in maize (Zea mays) based intercropping system. Indian J. Agron., 44: 261- 266.

Shivay, Y.S., R.P. Singh and Madanpa (2001). Productivity and economics of maize as influenced by intercropping with legumes and nitrogen levels. *Ann. Agric. Res.*, 22(4): 576 - 582.

Shivran, R.K. (2003). *Studies on* intercropping and moisture conservation practices on growth,yield and water use efficiency of maize under rainfed conditions. M.Sc. Thesis. Indian Agricultural Research Institute, New Delhi, India.

Singh, M., P. Singh and V. Nepalia (2005). Integrated weed management studies in maize based intercropping system. Indian J. Weed Sci., 37 (3&4): 205 - 208.

Singh, M.K., S.K. Pal, U.N.Verma, R. Thakur and S. Paspalak (1997). Effect of fertilizer and plant density on yield attributes and productivity of maize + black gram intercropping. J. Res (BAU), 9(2): 139 - 143.

Singh, S.N. and Ashok kumar (2001). Performance of maize based vegetable intercropping systems. Indian J. Agron., 30 (4): 263 - 264.

Singh, U., A.A. Saad, and S.R. Singh (2008). Production potential, biological feasibility and economic viability of maize (Zea mays)- based intercropping systems under rainfed conditions of Kashmir valley. Indian J. Agron., 38(12):1023-1027.

Singh, V.P. and R.P. Sharma (1987). Intercropping in *rabi* maize for higher production. *Seeds and Farms*, 13(12): 12-14.

Singh. M.K., S.K. Pal, U.N. Verma, R. Thakur and S. Pasupalak (1997). Effect of fertilizer and plant density on yield attributes and productivity of maize + black gram intercropping. J. Res (BAU), 9(2): 139-143.

Sinha, K.K., S.S. Mishra and S.J. Singh (1999). Yield and economics as influenced by winter maize based intercropping systems in North Bihar. Indian J. Agron., 44 (1): 30-35.

Sivaraman, K. and SP. Palaniappan (1995). Turmeric – maize and onion intercropping systems II. Leaf area index and dry matter accumulation. J. Species Aromatic Crops, 4: 145 – 155.

Sivaraman, K. and Sp. Palaniappan (1996). Turmeric – maize and onion intercropping systems. IV. PAR interception. J. Species Aromatic crops, 5(2): 139 – 142.

Sivaraman. K. and SP. Palaniappan (1994). Turmeric maize and onion intercropping systems. I. yield and land use efficiency. J. Species Aromatic Crops, 3: 19- 27.

Sood, B.R., and V.K. Sharma (1996). Effect of intercropping and planting geometry on the yield and quality of forage maize. *Forage Research*, 24(4): 190-192.

Sunil Kumar, C.R., Rawat and N.P. Melkania (2005). Forage production potential and economics of maize and cowpea intercropping under rainfed conditions. *Indian J. Agron.*, 50(3): 184-186.

Sunilkumar, S.N. (2000). Effect of seed proportion mixtures on yield and quality of forage maize-legume mixed cropping system. *M. Sc. (Agri.) Thesis*, Univ. Agric. Sci., Dharwad (India).

Sur, H.S., A.S. Sidhu, R. Singh, G.C. Aggarwal and K.S. Sandhu (1993). Long-term effect of green manuring on soil physical properties and production potential in green manure-maize-wheat sequence. Ann. Agric. Res., 14(2): 125-131.

Suresha, B.A., T.B. Allolli, M.G. Patil, B..K. Desai and S.A. Hussain (2007). Yield and economics of chilli based intercropping system. Karnataka J. Agric. Sci., 20: 807 - 809.

Thakur, D.R. (1994). Weed management intercropping in intercropping systems based on maize under rainfed mid hill conditions. Indian J. Agron., 39: 203-206.

Tiwari, R. C., P. K. Sharma and S. K. Khandelwal (2004). Effect of green manuring through *Sesbania cannabina* and *Sesbania rostrata* and nitrogen application through urea to maize (*Zea mays*) in maize-wheat cropping system. Indian J. Agron., 49(1): 15-17.

Tripathi, A.K., N.S. Shukla, Kumar Kaushal and S.N. Pandey (2010). Production potential, weed dynamics and energetics of maize + greengram/blackgram intercropping system. Journal of Food Legumes, 23 (1): 30 – 34.

Tripathi, A.K., A. Kumar and S.Nath, (2010). Production potential and monetary advantage of winter maize (*Zea mays*)-based intercropping systems under irrigated conditions in central Uttar Pradesh. Indian J. Agric. Sci., 80(2):125-128.

Varughese, K. and M.R. Iruthayaraj (1996). Biological and economic indices for measuring resource utilization in maize-soybean intercropping system. Madras Agric. J., 83 (9): 556 - 559.

Ved prakash, A.K. Pandey, A.K. Srivastva and H.S. Gupta (2004). Relay intercropping of hybrid tomato in maize under mid-hill condition of North-Western Himalayas. Indian J. Agric. Sci., 74(8): 405 – 408.

Venkatesh, L. (2007). Studies on maize-lucerne intercropping and its residual effect on Bengal gram. Master of Science (Agronomy). Department of Agronomy, University of Agricultural Sciences, Dharwad.

Venkatesh, L. and Y.B. Palled (2007). Studies on Maize Lucerne Intercropping and Its residual effect on Bengalgram. Karnataka J. Agric. Sci., 20(4): 897 - 898.

Venkatesh, L. and Y.B. Palled (2009). Studies on maize + Lucerne intercropping. J. Maharashtra Agric. Univ., 34 (1): 49 - 51.

Venugopal, N. and K. Shivashankar (1991). Influence of maize (Zea mays) crop residue and nitrogen on the productivity and economics maize + soyben (Glycine max) intercropping under paired system of planting. *Indian Journal of Agronomy,* 36: 502-507.

Yadav, M.P., J. Rai, S.P. Kushwaha and G.K. Singh (2005). Production potential and economic analysis of various cropping systems for central plains zone of Uttar Pradesh. Indian J. Agron., 50(2): 83-85.

Zhang, F.S., L. Li and J.H. Sun (2003). Interspecific interactions reduce phosphorus fertilizer rates in the faba bean / maize intercropping. In: proc. 2nd International symposium on phosphorus dynamics in soil – plant continuum. September 21 – 26, Perth, Australia. P: 184 – 185.

Zuofa, K., N.M. Tariah and N.O. Isinimai (1992). Effects of groundnuts, cowpea and melon on weed control and yields of intercropped cassava and maize. Field Crops Res., 28: 309 - 314.

# 14

# Integrated Weed Management in Maize

The problem of weeds is relatively much more in the maize crop grown during Kharif season because frequent rains keep the soil surface moist, which encourages several flushes of weeds and it is often impossible to control the weeds through hand weeding. More over the hot and humid condition during kharif is very much congenial for weed growth and development. Weeds reduce crop yield from 20 - 40 % depending upon weed species and density (Ashique et al., 1997). Yield loss due to weeds in maize varies from 28 to 93 % depending on the type of weed flora and the intensity and duration of crop – weed competition (Sharma and Thakur, 1998). At New Delhi, loss in grain yield due to unchecked weed infestation was 43.5 – 46.4 % (Singh et al., 2007).

In Kashmir, uncontrolled weed growth caused 73.41 % reduction in grain yield of maize (Bahar et al., 2009). At Bajaura (Himachal Pradesh), unchecked weed growth reduced maize yield by 61.3 % as compared to weed free check (Sharma and Nayital, 1993). Yield loss due to weeds in maize varies from 28 to 93 % depending on the type of weed flora and the intensity and duration of crop - weed competition (Sharma and Thakur, 1998). Herbicides use to manage weeds increased 150 % more maize yield than the weedy check (Abid *et al.* 1991; Cavero *et al.,* 2002; Miller and Libby, 1999). Weed interference in maize leads to 25 to 80 % reduction in crop yield (Ford and Pleasant, 1994; Chikoye and Ekeleme, 2003). Herbicide application can increase yield by 77 to 96.17 % than the weedy check (Khan *et al*., 1998).

## Common Weeds

The weed flora differs widely with the environment and soil conditions. Generally, weeds are found in large numbers with greater vigour, because of their wider adaptability even under extremities of climatic, edaphic and biotic stresses. High persistence nature of weeds is attributed to their ability of high seed production and longer seed viability.

At Samastipur (Bihar), *Eleusine indica, Setaria glauca, Echinochloa colonum, Cynodon dactylon, Ageratum conyzoides, Amaranthus viridis, Euphorbia hirta, Phyllanthus niruri, Trianthema monogyna, Leucas aspera* and *Cyperus rotundus* were dominant weed flora in the field (Pandey et al., 2003). At Raipur (Chhattisgarh), *Cyperus rotundus, Altenanthera sessilis, Cynodon dactylon, Brachiaria ramosa, Convolvulus arvensis, Phyllanthus niruri* and *Cynotis axillaries* were dominant weed flora in maize + black gram intercropping system (Dwivedi et al., 2011).

At Annamalai Nagar (Tamil Nadu), *Trianthema portulacastrim, Cyperus rotundus, Cynodon dactylon* and *Phyllanthus niruri* were dominant weed flora in maize field (Selvakumar and Sundari, 2006). At Pantnagar (Uttaranchal), dominant weed species were *Cynodon dactylon, Cyperus rotundus, Echinochloa crusgalli, Echinochloa colonum, Agropynon repens, Parthenium hysterophorus, Digitaria Sanguinalis, Eclipta alba, Euphorbia hirta* and *Commelian benghalensis* (Sarma and Gautam, 2006). At Tirupathi (Andhra Pradesh), major weed flora of the site was *Cynodon dactylon, Digitaria sanguinalis, Dactyloctenium aegyptium, Cyperus rotundus, Cleome viscosa, Commelina benghalensis* and *Trichoderma indicum* (Mahadevaiah *et al.,* 2010).

At Coimbatore (Tamil Nadu), major weed species present in the field were *Echinochloa colonum* (44.92 %), *Digitaria longiflora* (5.31 %), *Dactyloctenium aegyptium* (3.11 %), *Cyperus rotundus* (4.55 %) and the annual broad leaved weeds constituted 39.64 % of the total weed population comprising of *Parthenium hysterophorus* (27.77 %), *Digera arvensis* (4.40 %), *Trianthema portulacastrum* (3.60 %) ((Mynavathi et al., 2008).

At Almora, Pandey *et al.* (2000) at Almora observed that the dominant weeds spp. in maize were, *Echinochloa colonum, Bracharia ramosa, Digitaria sanguinalis, Dactyloctenium aegyptium, Elusine indica, Setaria glauca, Sorghum halepense* and *Panicum* spp. among the grassy weeds, *Ageratum conyzoides, Galinoga parviflora, Commelina benghalensis, Lindernia ciliate, Polygonum hydropiper, Euphorbia geniculata* and *Oxalis latifolia* among the broad leaved weeds and *Cyperus rotundus* among sedges. At Dharwad (Karnataka), Rajendra (2000) observed the important weeds in maize + soybean intercropping system were *Brachiaria eruciformis, Cynodon dactylon, Digitaria marginata* and *Dinebra retroflexa* among grasses, sedge *Cyperus rotundus* and *Ageratum conyzoides, Amaranthus viridis, Acanthospermum hispidum, Alternethra sessilis, Euphorbia* spp., *Commelina benghalensis, Cynotis cyculata, Digera arvensis, Phyllanthus niruri, Portulaca oleracea, Parthenium hysterophorus* and *Physalis minimia.*

The maize field was infested with *Amaranthus spinosis, Commelina benghalensis, Cynodon dactylon, Cyperus rotundus, Digera muricata, Echinochloa colonum, Parthenium hysterophorus* and *Trianthema portulacastrum* (Mundra *et al.*, 2003). Sinha *et al.* (2003) recorded important weed flora *Cyperus rotundus, Cynadon dactylon, Sorghum helepense, Chenopodium album, Convolvulus arvensis, Anagallis arvensis, Cannabis sativa and Melilotus alba* during winter season in North Bihar. Kamble *et al.* (2005) found the major weed flora in the maize field comprised the broad leaf weeds *Digera arvensis, Celosia argentia, Euphorbia hirta, Lagasca mollis* and *Parthenium hysterophorus* and the narrow leaf grassy weeds *Cyperus rotundus, Cynodon dactylon* and *Dinebra arabica.* The major weeds in the experimental field included *Digera arvensis, Trianthema partulacastrum, Amaranthus viridis, Cynodon dactylon* and *Cyperus rotundus* (Ramesh and Nadanassababdy, 2005).

The weeds present were *Cynodon dactylon, Sorghum halepense, Cyperus rotundus, Convolvulus arvensis, Anagallis arvensis, Chenopodium album, Melilotus alba, Lathyrus aphaca, Cichorium intybus* and *Cannabis sativa* (Sinha *et al.*, 2005). Patel *et al.* (2006) observed the major weed flora such as *Elusine indica, Digitaria sanguinalis, Eragrotis major* and *Dactyloctenium aegyptium* as monocot weeds and *Digera arvensis, Phyllanthus niruri, Boerhavia diffusa* and *Euphorbia hirta* as dicot weeds. At Cluj-Napoca, *Echinochloa crus-galli, Setaria glauca, Chenopodium album, Amaranthus retroflexus, Cirsium arvense, Convolvulus arvensis* and *Sonchus arvensis* were important species in maize crop (Ileana et al., 2007).

At Devihosur (Karnataka), important weed species observed were *Cynodon dactylon, Digitaria marginata, Dinebra retroflexa, Brachiaria eruciformis, Cyperus rotundus, Ageratum conyzoides, Alternanthera sessilis, Amaranthus viridis, Euphorbia spp., Commelina benghalensis, Cynotis cyculata, Digera arvensis, Oldenlandia diffusa, Phyllanthus niruri, Portulaca oleracea* and *Parthenium hysterophorus* (Kumar, 2008; Mallikarjun, 2008). At Dharwad (Karnataka), the important grassy weeds observed were *Cynadon dactylon, Dinebra retroflexa, Echinocloa colanum, Elusine indica. Cyperus rotundus* was alone in sedge category. Among the broad leaved weeds *Parthenium hysterophorus, Commelina bengalensis, Portulaca oleracea, Cynotis cuculata, Phylantus niruri and Amaranthus viridis* were the dominant weeds (Ishrat et al., 2012).

At Palampur (Himachal Pradesh), weed flora was composed of *Commlina benghalensis, Ageraturn conyzodies, Echinochloa colona, Panicum dichotomiflorum, Cyperus iria, Digitaris Sanquinalsis, Polygonum alatum* and *Aeschynomene indica* (Suresh Kumar et al., 2012). At Raipur (Chatisgarh), *Alternanthera triandra, Cyperus rotundus, Cynodon dactylon, Cynotis axillaries, Alternanthera triandra* and *Brachiaria ramose* were predominant weeds in maize field (Sanjay et al., 2012). At Navsari (Gujarat), the predominant weed flora was *Echinochloa crusgalli, C. dactylon, C. rotundus, Amaranthus viridis, Digera arvensis, Portulaca oleracea, Alternenthara sessili* and *Trianthema sp* in sweet corn (Aravadiy et al., 2012).

## Competition for moisture

Stainforth (1957) found that yellow foxtail grass competed with maize mainly for moisture. Wiese and Vandiver (1970) stated that *Echinochloa crusgalli* and *Digitaria sanguinalis* depleted soil moisture greatly and resulted in heavy competition with maize crop. Hosmani (1976) studied that influence of weeds on evapotranspiration loss in maize and found that moisture loss due to weed infestation was more than 66 % from 0-45 cm soil layer compared to weed free treatment.

Under condition of limited moisture, weeds thrived better than groundnuts, producing dry-matter yields that were higher than those recorded under adequate rainfall for most of weeding regimes (Sibuga et *al.,* 1989). Royal *et* al. (1997) stated that when moisture levels were above normal, the impact of common cocklebur *(Xanthium strumarium)* on groundnut yield was 9 to 24 % less as compared to normal soil moisture.

## Competition for nutrients

In general, nutrient removal by weeds was more in unweeded check than weeded maize plot. Removal of nutrients by weeds showed a great impact on the availability of nutrients to the crop than affecting its dry matter accumulation and on an average, weeds removed 35.3 kg N, 4.9 kg P and 29.2 kg K / ha in maize (Saikia and Pandey, 1999). Whereas, Sreenivas and Satyanarayana (1996) suggested that uncontrolled weeds removed 52.3, 9.6, 38.9 kg N P K /ha in maize. Channabasappa and Nanjappa (1994) found that weeds absorbed fewer nutrients whereas crops take more nutrients in the intercropping systems than in the monocrops and nutrient uptake by crops decreased with the delay in weeding. Sinha *et al.* (2000) reported that nutrient depletion by weeds was minimum under hand weeding while it was maximum (23.18 kg N, 4.92 kg $P_2O_5$ and 27.83 kg $K_2O$/ ha) under weedy check in maize in calcareous soils of North Bihar.

## Competition for space

Weeds compete for space and reduce the yield of crop plants (Hidayatullah and Sen, 1942). King (1966) stated that fast growth in terms of height and leaf development of weeds suppressed the crop by competing for space. Forcella *et al.* (1992) found that the narrow rows increased crop competition enough to replace intercultivation for controlling weed growth. Harvey *et al.* (1997) stated that planting maize in narrow rows might aid weed control. Teasdale (1998) observed that velvet leaf survival, growth and seed production were reduced as maize density increased. Tollenaar *et al.* (1994) and Murphy *et al.* (1996) reported that higher maize densities reduced the weed biomass compared to their lower densities.

## Critical period of crop-weed competition

At Islamabad (Pakistan), Shad *et al.* (1993) reported that, in maize critical period of crop weed competition was found in between 3 and 5 weeks. Tygai *et al.* (1993) reported that rainy season maize requires a weed free environment for 15-45 DAS to produce yields on par with season long weed free plots. The weedy environment beyond 30 days and upto 45 days after sowing was detrimental to maize growth and causes yield loss (Porwal, 2000). The critical period of crop weed competition was recorded between 3 and 6.5 weeks after corn emergence, where the field was severely infested with *Sorghum halepense* (Ghosheh *et al.,* 1996).Critical stage of maize weed competition in maize was 30 to 45 days from sowing (Kamble *et al.,* 2005). Porwal (2000) observed that in the rainy season, emergence of maize and weeds were simultaneous and found that the first 20-30 days were the most critical period of competition for the crop. Whereas Nayital *et al.* (1989) reported that in maize during rainy season, critical stage for crop weed competition is between 20 and 60 days after sowing (DAS). In irrigated winter maize, beyond 30 days and up to 45 days after sowing was detrimental to maize growth and caused yield loss in command area of Southern Rajasthan (Porwal, 1998).

## Effect of crop-weed competition

Presence of *Amaranthus retroflexus* in maize field reduced grain yield by 7, 20 and 32 per cent in the presence of 20, 10, and 20 weed plants per $m^{-2}$ (Rola and Rola, 1992). Santos *et al.* (1993), reported that in silty clay loam texture soils at Sonning (UK), maize grain yield loss ranged from 40 to 80 per cent if weeds were uncontrolled. According to Vargessel *et al.* (1994) the weed distribution in the maize field was not a critical consideration in determining yield loss. Maize grown in the presence of weeds *viz. Amaranthus palmeri, Erigrostis diffusa, Bidens odorata, Brassica campestris* for the whole cycle of the crop showed 86 to 90 per cent reduction in grain yield (Amador, 1995). He also reported increased grain yield with an increase in the period of weed absence.

Nearly 30 to 80 % reduction of grain yield was reported and it was depending on species, density, and duration of competition. Sen *et al.* (2000) found that, due to weed competition yield losses occur upto 15 to 75 % in maize crop. The loss in grain yield due to unchecked weed growth was to the extent of 33 to 50 % (Sharma *et al.*, 2000). Pandey *et al.* (2002) proved that grassy weeds offered maximum competition to crop reducing grain yield by 77.4 % followed by non-grassy weeds of about 44.2 per cent and sedges 38.4 %. Maize, being a rainy season and widely spaced crop, gets infested with variety of weeds and subjected to heavy weed competition, which often inflicts huge losses ranging from 28 to 100 % in maize (Patel *et al.,* 2006). Efficiency of maize crop to partition the dry matter into its economic yield was higher in herbicides treated plots (Wang, 1994; Porwal, 1995; Dixit and Gautam, 1996).

### Weed control methods in maize

Weeds growing in association with maize crops are of wide range and intensity, varying in their growth habit and life cycles. Therefore, no single weed control method gives continuous and effective weed control. Satisfactory weed control is achieved through direct methods (manual, mechanical, chemical) used within systems of indirect methods (land, input and crop management).

### Tillage

Tillage is to control weeds and check their seed bank. Deep summer ploughing to expose vegetative propagules to solar heat has been found to have a good impact on the control of perennial weeds. Inter row tillage is often used for controlling weeds.

**Table 1. Effect of tillage systems on maize crop**

| Tillage system | Weed count (No/m²) | Weed DMP (g/m²) | Grain yield (t/ha) | Stover yield (t/ha) | Net return (Rs/ha) | B : C ratio |
|---|---|---|---|---|---|---|
| Conventional tillage - flat bed | 91 | 251 | 2.74 | 6.98 | 4840 | 1.30 |
| Conventional tillage - narrow bed | 82 | 270 | 2.71 | 7.44 | 5790 | 1.35 |
| CT - broad bed | 80 | 225 | 2.33 | 7.23 | 5470 | 1.33 |
| Minimum tillage - flat bed | 102 | 318 | 2.90 | 6.51 | 4370 | 1.29 |
| Stale Seed bed-flat bed | 60 | 198 | 0.24 | 7.65 | 6370 | 1.38 |
| CD 5 % | 14.7 | 38 | 1.73 | 0.79 | - | - |

At New Delhi, adoption of stale seed bed practice caused reduction in weed count and dry weight and increased grain yield than conventional tillage with flat bed system. Minimum tillage gave lower yield due to poor growth of maize plant and associated with higher weed infestation (Singh et al., 2007).

**Table 2. Effect of tillage practices on maize**

| Tillage | Plant height (cm) | Dry matter production (kg/ha) | Grain yield (kg/ha) | Weed density on 60 DAS (No/m$^2$) | Weed dry matter (g/m$^2$) | Weed control efficiency (%) |
|---|---|---|---|---|---|---|
| Conventional tillage | 254 | 7560 | 4472 | 4.81(27.5) | 5.39 (42.84) | 80.7 |
| Zero tillage | 235 | 6803 | 4178 | 5.78 (37.81) | 6.09 (52.53) | 80.5 |
| CD | 7.7 | 488 | NS | 0.081 | 0.43 | NS |

At Bapatala (Andhra Pradesh), in conventionally tilled strips the land was ploughed twice with tractor drawn cultivator followed by once with rotavator to obtain a good tilth of soil. In case of zero tilled strips no preparatory tillage operations were taken up but non-selective and contact herbicide paraquat @ 0.6 kg a.i/ha was applied to desiccate the weeds before sowing. Higher weed density and dry matter recorded under zero tillage proves its lower weed competing ability compared to that of conventional tillage (Srividya et at., 2011).

**Table 3. Effect of tillage seed rates and weed control on weeds in maize**

| Tillage methods | Weed density (No/m$^2$) | Weed dry weight (g/m$^2$) | Grain yield (kg/ha) | Net return (Rs/ha) | B:C ratio |
|---|---|---|---|---|---|
| Tilled | 11.4 (157) | 11.6 (167.7) | 4576 | 15901 | 1.51 |
| No tilled | 12.5 (183) | 12.5 (191.1) | 3536 | 10756 | 1.09 |
| LSD (5 %) | 0.46 | 0.37 | 116 | - | - |

Tillage operation resulted in significant reduction in weed dry weight and increased grain yield over no tillage practice (Sarma and Gautam, 2006). Improved crop competitiveness against weeds and higher yield were observed with raised beds planting with application of atrazine 1.5 kg/ha or acetachlor 1.25 kg/ha (Chopra and Angiras, 2008.)

## Cultivars

Tall maize cultivars were not found effective in suppressing weeds as compared to some shorter modern cultivars (Cosser *et al*., 1997). Eisele and Köpke (1997) indicated that maize leaf area was directly correlated to grain yield variability caused by weed competition and interaction of maize with recommended and reduced rates of pre-emergent herbicides (Roggen and Gregory, 1997; Gregory, 1997). Gregory *et al*. (1994) found yield reduction occurred for all cultivars as herbicide rates decreased which conclude that herbicide rate directly affect the grain yield.

## Seed rate

**Table 4. Effect of seed rates on weeds in maize**

| Seed rates (kg/ha) | Weed density (No/m²) | Weed dry weight (g/m²) | Grain yield (kg/ha) | Net return (Rs/ha) | B:C ratio |
|---|---|---|---|---|---|
| 16 (60 × 25 cm) | 12.4 (183) | 12.5 (193.2) | 3868 | 12374 | 1.22 |
| 20 (60 × 20 cm) | 12.1 (173) | 12.1 (180.6) | 4082 | 13477 | 1.31 |
| 24 (60 × 15 cm) | 11.4 (156 | 17.5 (164.4) | 4218 | 14135 | 1.36 |
| LSD (5 %) | 0.57 | 0.46 | 142 | - | - |

(Sarma and Gautam, 2006). Data in parentheses are original values.

Lower weed dry weight and higher grain yield were recorded with higher seed rate (24 kg seed / ha) (Sarma and Gautam, 2006).

## Planting method

**Table 5. Planting methods on weeds, yield attributes and yield of maize**

| Planting methods | Weed density (Nos/m2) | Cobs / plant | Cob length (cm) | Kernels/ cob | Grain yield (t/ha) |
|---|---|---|---|---|---|
| Ridges and furrow | 146 | 0.97 | 19.23 | 511 | 2.89 |
| Broadcast | 180 | 0.97 | 17.21 | 375 | 1.56 |
| Flat bed | 169 | 0.99 | 18.04 | 433 | 2.11 |
| LSD 5 % | 14 | NS | 1.17 | 61 | 0.62 |

(Ashique et al., 1997)

At Peshawar (Pakistan), ridges and furrow planting recorded lower weed population ($146/m^2$), cobs length (19.23 cm), kernels/ cob (511) and grain yield (2.89 t/ha) as compared to other planting methods (Ashique et al., 1997).

## Intercropping system

Growing of intercrops in maize reduced weed dry weight significantly (Prasad and Rafey, 1996).

**Table 6. Influence of intercropping on weeds in maize crop**

| Treatments | Weed density (60DAS) ($Nos/m^2$) | Weed dry Weight ($g/m^2$) (60 DAS) | Weed control efficiency (%) | Maize equivalent yield (kg/ha) | Land equivalent ratio |
|---|---|---|---|---|---|
| **Intercropping system** | | | | | |
| Sole maize | 8.89 (90.23) | 49.19 | 88.5 | 3692 | - |
| Maize + black gram | 7.61 (69.80) | 39.62 | 89.9 | 4956 | 1.34 |
| Maize + cowpea | 6.84 (59.17) | 35.44 | 90.0 | 5284 | 1.44 |
| LSD (5 %) | 0.24 | 0.97 | - | - | - |

(Selvakumar and Sundari, 2006). Figures in parentheses indicate original values.

At Annamalai nagar (Tamil Nadu), higher maize equivalent yield (5284 kg/ha) and income equivalent ratio (1.44) were recorded in maize + cowpea intercropping system (Selvakumar and Sundari, 2006).

**Table 7. Weed smothering efficiency (%) as influenced by maize based intercropping**

| Intercropping | Weed smothering efficiency (%) | | |
|---|---|---|---|
| | 20 DAS | 45 DAS | Harvest |
| Maize (sole) | - | - | - |
| Maize + Green gram (1:1) | 20.62 | 34.21 | 27.55 |
| Maize + Soybean (1:1) | 30.37 | 55.21 | 51.16 |

(Shah et al., 2011).

At Anand (Gujarat), weed smothering efficiency calculated at 20 and 45 DAS and at harvest clearly indicated that intercropping of maize with soybean having higher WSE than the maize with green gram (Shah et al., 2011).

At Raipur, maize + black gram (2:1) was found to be best in terms of maize yield and sole black gram in terms of black gram yield. Application of alachlor 1.5 kg/ha + HW at 40 DAS resulted in significantly higher grain yield of maize and black gram (Sanjay et al., 2011).

## Mulching

**Table 8. Effect of mulching on weeds in maize**

| Weed control measures | Weed density (No/m$^2$) | Weed dry weight (t/ha) | Light in at perception on 90 DAS | | Nutrient uptake (kg/ha) | | | Grain yield (t/ha) | Harvest index (%) |
|---|---|---|---|---|---|---|---|---|---|
| | | | Middle | Bottom | N | P | K | | |
| Weedy check | 10.2 (107) | 5.73 | 54.34 | 71.55 | 33 | 5.48 | 38 | 1.36 | 21.8 |
| Atrazine 1.0 kg/ha | 8.92(81) | 5.12 | 59.48 | 70.15 | 83 | 13.69 | 103 | 3.90 | 27.55 |
| HW 30 and 60 DAS | 7.73 (62) | 4.63 | 60.66 | 69.13 | 96 | 16.53 | 115 | 4.76 | 30.90 |
| Straw mulch 15 t/ha | 6.90 (49) | 3.16 | 61.85 | 67.82 | 102 | 19.09 | 121 | 4.98 | 30.95 |
| Polythene mulch | 5.92 (36) | 2.09 | 63.63 | 65.42 | 108 | 20.28 | 133 | 5.14 | 31.00 |
| CD 5 % | 0.52 | 0.55 | - | - | 7 | 1.78 | 8 | 0.29 | 2.30 |

(Bahar and Singh, 2012)

In Kashmir, minimum total weed density, dry matter accumulation and nitrogen uptake were observed under polyethylene mulch. On the contrary, higher grain yield (5.14 t/ ha) and WCE (67.3 and 66.54 %) were recorded in polyethylene mulch than rest of the weed control measures. Benefit cost ratio was higher atrazine @ 1.0 kg/ ha treatment. Uncontrolled weed growth caused 73.41 % reduction in grain yield of maize (Bahar and Singh, 2012). At Kashmir, polythene mulch being at par with straw mulch @ 15 t/ha resulted in significantly high grain yield (5.12 t/ha) of maize than weedy check (1.36 t/ha). Uncontrolled weed growth caused 73.14 % reduction in yield of maize (Bahar et al., 2009).

### Manual method of weed control

It is the oldest method of controlling weeds. It is still a practical and efficient method of eliminating weeds in cropped and non-cropped lands. It is very effective against annual and biennial weeds. In Hamirpur (Himachal Pradesh), hoeing + manual weed control twice increased maize grain yields by 0.36 t /ha (Bhopal Singh *et al.*, 1991). Hand weeding twice in maize field gave yield similar to that of application of atrazine upto 5 weeks (Sandhu and Bhatia, 1991). At Ranchi, adoptation of hand weeding at 20 and 40 days after sowing resulted minimum weed infestation in rainfed maize (Saha and Srivastava, 1992). During rainy season at Bajaura (H.P), hand weeding at 20 and 40 DAS gave higher weed control efficiency and comparable with metolachlor (1.25 kg/ ha) sprayed treatment (Thakur, 1994).

Sharma *et al.* (2000) reported that hoeing at 15 days and earthing up at 30 days of growth resulted in weed free condition. Kolage *et al.* (2004) opined that lower weed intensity was observed with two hand weedings at 20 and 40 DAS. Sinha *et al.* (2003) showed that weed control efficiency was maximum under hand weeding (91.9 %). The pooled data of 3 years revealed that 2 hoeings + 2 hand weedings at 20 and 40 DAS recorded substantially higher grain (69.29 q /ha) and fodder (103.61 q/ ha) yields over the rest of the treatments, but was found at par with pre-emergence application of atrazine + one hoeing + one hand weeding at 20 DAS (Kamble *et al.*, 2005). Tripathi *et al.* (2005) reported that manual weeding twice at 15 and 30 DAS registered 22.2, 17.6, 22.0 and 20.3 % reduction in *C. rotundus, D. arvensis, P. niruri* and *C.benghalensis* population and also heavy decline in weed dry weight (WCE 87.8 %). Hand weeding twice at 3 and 6 weeks after sowing showed significant effect in reducing weeds and also increased the yield when compared with weedy check (Nagalakshmi *et al.*, 2006).

**Table 9. Effect of mechanical weed control methods on weeds in maize**

| Treatment | Plant height (cm) | Grain yield (kg/ha) | Weed density (No/m$^2$) | Weed dry matter (g/m$^2$) | Weed control efficiency (%) |
|---|---|---|---|---|---|
| Weedy check | 222 | 3550 | 9.4<br>(88.5) | 15.65<br>(245) | - |
| HW on 3&6 WAS | 273 | 5608 | 1.89<br>(3.67) | 2.26<br>(5.54) | 82.9<br>(97.8) |
| Atrazine 1.25 kg a.i/ha | 232 | 3822 | 5.64<br>(31.92) | 5.13<br>(25.84) | 71.7<br>(89.4) |
| Pendimethalin 1.5 kg a.i/ha | 224 | 3791 | 7.32<br>(53.17) | 5.71<br>(32.17) | 68.9<br>(86.8) |
| Paraquat 0.6 kg a.i/ha at 3 WAS | 239 | 4048 | 5.43<br>(29.25) | 4.49<br>(19.74) | 74.0<br>(91.9) |

| Treatment | Plant height (cm) | Grain yield (kg/ha) | Weed density (No/m²) | Weed dry matter (g/m²) | Weed control efficiency (%) |
|---|---|---|---|---|---|
| Pre –emeregence Atrazine 1.25 kg a.i/ha + Paraquat 0.6 kg a.i/ha at 3 WAS | 255 | 4466 | 3.77 (13.84) | 4.19 (17.06) | 75.0 (93.0) |
| Pre –emeregence Pendimethalin 1.5 kg a.i/ha + Paraquat 0.6 kg a.i/ha at 3 WAS | 246 | 4290 | 5.23 (27.75) | 4.43 (19.28) | 73.8 (92.1) |
| Intercultivation with power weeder at 4 WAS | 264 | 5027 | 3.68 (13.17) | 4.13 (16.72) | 75.9 (93.7) |
| CD 5 % | 15 | 213 | 0.46 | 0.43 | 5.2 |

(Srividya et al., 2011).

At Bapatla (Andhra Pradesh), lower weed density and dry matter with higher yield of maize were recorded under hand weeding twice on 3 and 6 WAS which was followed by intercultivation with power weeder at 4 WAS (Srividya et al., 2011).

## Chemical method of weed control

Inadequate weed management is one of the major causes for low yield in rainfed maize. The conventional method of weed control through hand weeding is costly and nonavailable at critical stages. If the cost of hand weeding is high, herbicide treatment becomes a competitive and promising way to control weeds, at least in the first few weeks. Good weed control and increased crop yield were observed with herbicides in maize (Khan *et al.,* 1991; Khan *et al.,* 1993; Khan and Hassan, 2003). Weed control in maize with herbicides has been suggested by many researchers (Devender *et al.*, 1998; Toloraya *et al*., 2001).

**Table 10. Recommended herbicides for maize cultivation**

| Chemical Name | Trade Name | Dose (kg/ha) (a.i) | Formulation | Time of application | Most effective against weed flora |
|---|---|---|---|---|---|
| Atrazine | Atrataf | 0.75 – 1.00 | 50 WP | Pre emergence | Broad spectrum weed control. Some grass weeds are tolerant. Suited to sole maize. |
| Pendimethali | Stomp | 0.75 – 1.50 | 30 EC | Pre emergence | Broad spectrum weeds control. Ideal for intercropped maize. |
| Alachlor | Lasso | 1.50 – 2.00 | 50 EC | Pre emergence | Broad spectrum weeds control. Ideal for intercropped maize. |
| Metalachlor | Dual | 1.00 – 2.00 | 50 EC | Pre emergence | Broad spectrum weed control. |
| 2, 4 – D | 2, 4 – D | 0.50 – 0.75 | 34 EC | Post emergence (4 – 6 weeks) | Effective against dicot weeds only. Good as sequential application to pre-emergence application. |
| Atrazina + Alachlor | - | 0.75 + 0.75 | - | Pre – emergence | Excellent control of all weeds for sole maize only. |
| Atrazine + pendimethalin | - | 0.75 + 1.00 | - | Pre – emergence | Excellent control of all weeds for sole maize only. |
| Atrazine + Metalachlor | - | 0.750 + 0.50 | - | Pre – emergence | Excellent control of all weeds for sole maize only. |

**Pre-emergence application of herbicides**

Pre-emergence herbicides control grassy and broad leaved weeds. Their use would be appropriate not only for minimizing early weed competition, but also for reducing the work load during the peak labour period (Baker and Terry, 1991).

### Pre-emergence application of atrazine

Atrazine belongs to the heterocyclic nitrogen compounds and comes under triazines. It is a soil and leaf herbicide. The triazines move in the transpiration stream mostly in the apoplast of treated plants and inhibit photosynthetic electron transport (Baker and Terry, 1991).

At mid hills of Meghalaya, application of atrazine to maize at 2.0 kg /ha proved quite effective in controlling weeds (WCE 67 %) and enhancing grain yield by 143 per cent (3636 kg /ha) over control (1351 kg /ha) (Varshney, 1990). Gaur *et al.* (1991) also reported that pre-emergence atrazine at 0.5 kg/ ha completely controlled the broad leaved weeds. However all grassy weeds were not controlled. In general, application of pre-emergence atrazine at the rate of 0.25-1.00 kg/ ha followed by manual weed control (40 DAS) under rainfed condition was the most profitable weed control practice and resulted in net return of Rs.5977 / ha compared with net return of Rs.5616 with the farmer's practice (Bhopal Singh *et al.,* 1991).

Sharma *et al.* (2000) showed that atrazine applied as pre-emergence spray reduced the weed population at 15 days of growth significantly (74.8 – 83.61 %) as compared to control (75 – 80 weeds $m^{-2}$). Shukla *et al.* (2001) showed that pre-emergence application of atrazine at 0.75 kg/ ha had higher fodder yield (344.8 q /ha) compared to the control (267.9 q /ha) in forage maize. Similarly the efficiency of atrazine was more in controlling the weeds and thus increasing maize yield (Salarzai, 2001). Patel *et al.* (2006) reported that pre-emergence application of pendimethalin at 0.5 kg/ ha with atrazine at 0.5 kg/ ha gave significantly lower density of monocot and dicot weeds at all the intervals and also recorded higher grain yield of maize as compared to all other treatments. At Nigeria, the new formulation doses of atrazine from 1.0 to 3.5 kg /ha gave 85 to 98 % control of *Commelina benghalensis, Passiflora foetida,* sedges and *Talinum triangulare* at 4 weeks after treatment (Chikoye *et al.*, 2006).

At Palampur (Himachal Pradesh), atrazine at 2.0 and 1.5 kg/ha provided lowest weed dry weights. Atrazine at 1.0 kg/ha, however, did not provide adequate control with weed control efficiency (69.5 %). As the time of atrazine application was delayed, weed control efficiency decreased (from 75.1 % in the pre-emergence treatment to 25.2 % in the 18 DAS treatment) and weed dry weight increased (from 5.4 g/$m^2$ in the pre-emergence treatment to 9.4 g/$m^2$ in the 18 DAS treatment). However, there was no significant difference in dry weight of weeds and weed control efficiency between pre-emergence, 6 and 12 DAS applications. Atrazine at 2.0 and 1.5 kg/ha, recorded higher number of cobs, cob length, grains /cob, and grain yield. There was no significant difference in 1000-grain weight between application rates. Grain yield and yield attributes were statistically similar when atrazine was applied either pre-emergence or post-emergence at 6 or 12 DAS (Saini, 2000).

## Pre-emergence application of butachlor

Dubey (1995) reported that lower weed biomass, higher weed control efficiency (87.76 %), crop growth and yield parameters were on par with hand weeding, fluchloralin and butachlor (2.5 kg /ha) + all cultural methods. However, higher grain yields were achieved with hand weeding alone and Butachlor (2.5 kg /ha) + hand weeding (0.996 and 1.025 t /ha, respectively). Application of Butachlor at 1.0 kg /ha recorded lower weed dry weight and higher seed yield of compared to unweeded control (AICRPWC, 1996). Rout and Satapathy (1996) opined that pre-emergence application of butahclor at 1.5 kg/ ha was found to be the next best alternative to metolachlor and atrazine in maize. According to Rajendra (2000) application of butachlor at 1.5 , 1.0 kg and alachlor at 2.0 kg/ ha recorded significantly lower population and dry weight of weeds and higher weed control efficiency and maize equivalent yields in maize + soybean intercropping system.

## Other herbicides

Simazine and Ametryn showed superiority to the other treatments in controlling grasses, broad leaved and the weeds. These herbicides increased maize growth and yield through their effect in controlling the weeds, more so with hand hoeing (Hassan and Ahmed, 2005). Greater maize yield was obtained by controlling weeds with paraquat plus extavon and ametryn. Post-emergence herbicide application was effective in controlling weeds with no phytotoxic effect (Maglhaes *et al.*, 2000). Pre-emergence herbicides viz., metolachlor plus atrazine, pendimethalin and cyanazine plus atrazine decreased weed population and increased grain yield over weedy control (Khan *et al.*, 1991).

**Table 11. Influence of weed management practices on weeds in maize based intercropping systems**

| Weed management | Weed density (60DAS) (Nos/m$^2$) | Weed dry weight (g/m$^2$) (60 DAS) | Weed control efficiency (%) | Maize equivalent yield (kg/ha) | Land equivalent ratio |
|---|---|---|---|---|---|
| Weedy check | 13.63 (189.07) | 164.56 | - | 2909 | 1.47 |
| Twice hand weeding | 4.45 (20.20) | 4.11 | 97.6 | 5690 | 1.35 |
| Alachlor 2.0kg/ha | 10.55 (111.23) | 39.70 | 75.7 | 3837 | 1.42 |
| Pendimethalin 1.0 Kg/ha | 8.75 (76.67) | 30.19 | 81.0 | 4419 | 1.40 |

| | | | | | |
|---|---|---|---|---|---|
| Alachor + HW on 30 DAS | 5.00 (25.33) | 6.31 | 96.0 | 5153 | 1.36 |
| Pendimethalin + HW on 30 DAS | 4.31 (18.90) | 3.64 | 97.8 | 5856 | 1.36 |
| LSD (5 %) | 0.37 | 1.05 | - | - | - |

(Selvakumar and Sundari, 2006). Figures in parentheses indicate original values.

At Annamalai nagar (Tamil Nadu), among the weed management practices, pre-emergence application of pendimethaline 1.0 kg/ha + hand weeding on 30 DAS recorded higher weed control efficiency (97.9 %) and maize equivalent yield (5856 kg/ha) in maize based intercropping systems (Selvakumar and Sundari, 2006).

**Post emergence application of herbicides**

At Ludhiana (Punjab), post - emergence application of 2, 4 - D dimethyl amine at 0.44, 0.58 and 0.73 kg/ha was found more effective for controlling *Cyperus rotundus* as compared to other formulation viz., sodium and ethyl ester. These treatments increases maize grain yield by 23.8, 31.1 and 26.7 % over unweeded control respectively (Singh *et al.,* 2010).

**Table 12. Effect of different herbicides on Cyperus rotundus in maize**

| Treatments | Cyperus rotundus | | | Grain yield (q/ha) |
|---|---|---|---|---|
| | Before spray (No/m²) (15 DAS) | After spray (No/m²) (30 DAS | Dry matter (at harvest) (q/ha) | |
| 2, 4-D Sodium salt 0.6 kg/ha | 57.4 | 8.5 (42.2) | 8.6 (72.7) | 52.3 |
| 2, 4-D Sodium salt 0.80 kg/ha | 59.8 | 7.0 (48.0) | 5.9 (33.8) | 59.7 |
| 2, 4-D Sodium salt 1.00 kg/ha | 58.5 | 5.8 (41.3) | 5.1 (30.7) | 58.4 |
| 2, 4-D dimethyl amine 0.44 kg/ha | 59.8 | 1.0 (0.0) | 1.0 (0.0) | 66.5 |
| 2, 4-D dimethyl amine 0.58 kg/ha | 56.3 | 1.0 (0.0) | 1.0 (0.0) | 67.6 |
| 2, 4-D dimethyl amine 0.73 kg/ha | 61.1 | 1.0 (0.0) | 1.0 (0.0) | 58.8 |
| 2, 4-D ethyl ester 0.29 kg/ha | 59.4 | 4.30 (22.2) | 4.2 (19.8) | 66.5 |

| Treatments | Cyperus rotundus | | | Grain yield (q/ha) |
|---|---|---|---|---|
| | Before spray (No/m$^2$) (15 DAS) | After spray (No/m$^2$) (30 DAS | Dry matter (at harvest) (q/ha) | |
| 2, 4-D ethyl ester 0.38 kg/ha | 61.8 | 4.12 (19.4) | 4.1 (18.8) | 63.5 |
| Hand weeding 15 and 30 DAS | 61.9 | 4.6 (24.9) | 4.4 (22.2) | 64.3 |
| Unweeded control | 67.6 | 14.7 (216.5) | 8.2 (66.1) | 48.3 |
| CD 5% | NS | 2.1 | 2.1 | 10.8 |

(Singh *et al.,* 2010). Figures in parentheses are original values. After sowing maize, uniform application of atrazine 1.0 kg/ha as pre-emergence given to entire experiment. 2, 4 - D was sprayed on 20 - 25 DAS.

Post emergence application of topramezone at 2-4, 4-6 and 6-8 maize leaf stages produced similar yields as that of hand weeded control (Thoma et al ., 2012). Lesser number of *Cyperus sp* was observed with inclusion of post emergence herbicide ghyposate (Thakur and Singh, 1989) while the intensity of this weed species was greatly suppressed by the inclusion of post emergence herbicide 2, 4-D Na salt (Reddy, 2003).

## Adjuvants

The addition of adjuvants enabled the reduction in herbicide concentration by 30-60 % in maize without affecting its yield (Rola *et al.*, 1999b). Herbicide applied in combination with urea gives better result up to 12 to 13.5 % than the use of herbicide alone (Getmanetz *et al.*, 1991).

## Sequential application of herbicides

Pre-emergence herbicides inhibit seed germination and arrest seedling growth of weeds depending on their mode and mechanism of action. Normally a follow up hand weeding is done after pre-emergence herbicide application to achieve satisfactory weed control at least upto the critical period of crop-weed competition. Alternatively sequential application of post-emergence herbicide after pre-emergence application found to be active on late emerging weeds, especially on broad leaved weeds (Singh and Arya, 1994).

Application of pre-emergence atrazine (0.33 kg /ha) as 20 cm wide bands over the crop row followed by directed spray of paraquat (0.25 kg /ha) 21 days after sowing gave the same level of weed control and grain yield as blanket application of atrazine (1.0 kg/ ha) and the hand weeding treatments (Gill *et al.,* 1985). Sandhu *et al.* (1991) highlighted that the use of atrazine at 0.75 kg /ha and post-emergence at 30 and 50 days after sowing gave a season long control of grasses and broad leaved annual weeds and produced higher grain yield than hand hoeing. The inhibitory effect on growth of *Parthenium histerophorus* was more pronounced when atrazine (3.0 kg/ ha) was followed by 15 days after spraying of 2, 4-D ethyl ester (1.0 kg/ ha) and diquat (0.5 kg /ha) combinations (Dwivedi *et al., 1987).*

Sharma *et al.* (2000) suggested that atrazine can be useful as sand mix post emergence application followed by hoeing at 15 days of growth for effective control of weeds and increase the grain yield of maize by 1.0 t /ha in Khandari area of Punjab. Bogdan *et al.* (2004) reported that pre-emergence application of pendimethalin and atrazine in combination with the post-emergence application of dicamba and 2,4-D, as well as the application of metolachlor, atrazine, dicamba and 2,4-D provided the best control of weeds.

**Table 1. Grain yield, stover yield, harvest index, of maize weed index and monetary return as influenced by sequential application of herbicides**

| Treatments | Grain yield (kg /ha) | Stover yield (kg /ha) | Harvest index (%) | Weed index (%) | Net income (Rs./ha) | B:C ratio |
|---|---|---|---|---|---|---|
| Atrazine (1.25 kg/ ha Pre-em) + one IC at 30 DAS | 5297 | 7289 | 42.08 | 24.75 | 19602 | 2.27 |
| Atrazine (1.25 kg /ha Pre-em) fb Atrazine (0.50 kg/ ha Post-em) | 5876 | 7754 | 43.11 | 16.53 | 22991 | 2.45 |
| Atrazine (1.25 kg /ha Pre-em) fb Atrazine (1.00 kg/ ha Post-em) | 6275 | 8171 | 43.43 | 10.86 | 25130 | 2.55 |
| Butachlor (1.00 kg/ ha Pre-em) + one IC at 30 DAS | 4898 | 6818 | 41.8 | 30.42 | 17526 | 2.17 |
| Butachlor (1.00 kg /ha Pre-em) fb Atrazine (0.50 kg/ ha Post-em) | 5817 | 7685 | 43.07 | 17.37 | 23005 | 2.48 |
| Butachlor (1.00 kg /ha Pre-em) fb Atrazine (0.50 kg /ha Post-em) | 6020 | 7949 | 43.10 | 14.48 | 23962 | 2.52 |
| Atrazine (1.25 kg /ha Pre-em) fb 2,4 - D (0.5 kg /ha Post em) | 6440 | 8200 | 43.98 | 8.52 | 26601 | 2.67 |
| Butachlor (1.00 kg /ha Pre-em) fb 2,4-D ( 0.5 kg /ha Post em) | 6320 | 8146 | 43.68 | 10.22 | 26224 | 2.70 |
| Farmer's practices (IC at 20 and HW at 30 DAS) | 5583 | 7385 | 43.05 | 20.69 | 21082 | 2.33 |
| Weed free check | 7040 | 8785 | 44.48 | 0.00 | 27821 | 2.50 |
| Weedy check | 4195 | 6206 | 40.33 | 40.41 | 14046 | 2.01 |
| CD(0.05) | 841 | 855 | NS | 11.27 | 3903 | 0.26 |

(Mallikarjun, 2008). Pre-em: Pre-emergent herbicides sprayed at one DAS. Post-em: Post-emergent herbicides sprayed at 40 DAS. IC: Intercultivation. HW: Hand weeding. DAS: Days after sowing

At Devihosur (Karnataka), sequential application of either atrazine (1.25 kg /ha) or butachlor (1.00 kg /ha) fb 2, 4-D (0.50 kg/ ha) and atrazine (1.25 kg /ha) fb atrazine (1.00 kg /ha) recorded significantly lower weed population and dry weight of weeds and higher weed control efficiency at all the growth stages with good crop growth. Effective control of weeds, higher grain yield and net income and B: C ratio were obtained with sequential application of either atrazine (1.25 kg /ha) or butachlor (1.00 kg /ha) fb 2, 4-D (0.50 kg /ha) and atrazine (1.25 kg /ha) fb atrazine (1.00 kg /ha) in Northern Transition Zone of Karnataka under rainfed condition (Mallikarjun, 2008).

### Herbicide mixtures

At Bajaura (Himachal Pradesh, combination of metalachlor 1.5 kg/ha + atrazine 0.75 kg/ha gave significantly higher grain yield and stover yield and was at par with metolachlor 1.5 + atrazine 0.5 kg/ha in influencing grains/cob and metalochlor 1.0 + atrazine 0.5 kg/ha, hand weeding twice and hand weeding followed by Halod in influencing stover yield (Shathna et al., 1998).

**Table 14. Effect of various herbicidal treatments on yield attributes and yields of maize**

| Treatment | Cob length (cm) | Cob Girth (cm) | No. of grains/ cob | Grain yield (kg /ha) | Weed density (No./m2) | Weed dry matter (kg /ha) | WCE (%) |
|---|---|---|---|---|---|---|---|
| Atrazine 1.00 kg a.i./ha | 19.66b | 12.64b | 413cde | 3386bcd | 5.36def (28.00) | 633cd | 81 |
| Alachlor 1.00 | 19.50b | 12.53b | 406cdef | 3254de | 7.33bc (53.13) | 1042b | 69 |
| Metolachlor 1.00 kg a.i./ha | 18.95c | 12.25b | 378f | 3065e | 7.87b (61.38) | 1116b | 66 |
| Metribuzin 0.30 kg a.i./ha | 17.34d | 11.68c | 328g | 2439f | 7.45bc (54.88) | 1169b | 65 |
| Pendimethalin 0.50 kg a.i./ha | 19.63b | 12.63b | 415bcde | 3314cd | 6.58cd (42.50) | 822c | 75 |
| Atrazine 0.50 + Alachlor 0.50 kg a.i./ha | 20.49a | 13.43a | 435abc | 3582ab | 3.15hi (9.25) | 65g | 98 |
| Atrazine 0.50 + Pendimethalin 0.25 kg a.i./ha | 20.66a | 13.58a | 456a | 3652a | 2.28i (4.38) | 34g | 99 |

| Treatment | Cob length (cm) | Cob Girth (cm) | No. of grains/ cob | Grain yield (kg /ha) | Weed density (No./m2) | Weed dry matter (kg /ha) | WCE (%) |
|---|---|---|---|---|---|---|---|
| Atrazine 0.50 + Metolachlor 0.50 kg a.i./ha | 20.50a | 13.33a | 428abc | 3548ab | 3.25ghi (9.75) | 198fg | 94 |
| Atrazine + Metribuzin 0.50 + 0.15 kg a.i./ha | 20.25a | 13.19a | 424abcd | 3489abc | 4.55ef (19.88) | 313f | 90 |
| Metolachlor 0.50 + Metribuzin 0.15 kg a.i./ha | 20.46a | 13.29a | 434abc | 3494abc | 4.40efg (18.63) | 224fg | 93 |
| Metolachlor 0.50 + Pendimethalin 0.25 kg a.i./ha | 20.44a | 13.33a | 427abc | 3491abc | 4.23fgh (17.25) | 302f | 91 |
| Alachlor 0.50 + Metolachlor 0.50 kg a.i./ha | 19.53b | 12.42b | 394def | 3271d | 5.94d (34.38) | 607d | 82 |
| Alachlor 0.50 + Pendimethalin 0.25 kg a.i./ha | 20.45a | 13.33a | 439abc | 3509abc | 5.96d (34.88) | 380ef | 88 |
| Alachlor 0.50 + Metribuzin 0.15 kg a.i./ha | 19.53b | 12.50b | 391ef | 3240de | 6.02d (35.63) | 552de | 83 |
| Weed free (HW at 20 & 40 DAS) | 20.68a | 13.66a | 448ab | 3658a | 5.54de (29.88) | 56g | 98 |
| Weedy check | 14.64e | 10.49d | 243h | 1947g | 13.43a (179.88) | 3331a | -- |

(Patel et al., 2006)

**IWM practices**

Covering soil with plant residues, plastic, etc, can affect weed growth by checking sunlight. Covering of soil with transparent polyethylene films for 4-6 weeks during summer months immediately before planting a can be done to control weeds. Growing of suitable intercrop (maize + groundnut, maize + black gram, (1:2), maize + pigeon pea (1:1)) can help in weed control to some extent besides providing additional income. Maximum grain yields of 4.54, 4.44 and 4.41 t / ha were obtained in plots treated with pre-emergence application of acetochlor (Dichlor 50 EC) @ 2.5 litre/ha + post emergence application of carfentrazeo ethyl (Aim 40 DF) @ 0.05 kg/ha + Atrazine (Atratox 38 EC) @ 0.75 litres/ha, post emergence application of Aim + Atiatox compact + one hoeing and pre-emergence application of Dichlor + one hoeing respectively.

Prasad et al. (1990) obtained maximum weed control and economic returns with atrazine 0.5 kg/ha + hand weeding. Eadle et al. (1992) found that one cultivation combined with herbicide applied as band gave maximum weed control and maize yield on no tillage system. Singh et al. (1990) reported that atrazine 1.0 kg/ha + manual weeding 40 DAS and pendimethaline + manual weeding 40 DAS resulted in higher grain yield. Biolly (1996) found that cyanazine @ 4.75 litres /ha and hoeing twice were equally effective in controlling weeds. Post emergence application of topramezone at 2 – 4, 4 – 6 and 6 – 8 maize leaf stages produced similar yields as that of hand weeded control (Toma et al., 2010). Lesser number of *Cyperus sp* was observed with inclusion of post emergence herbicide glyposate (Thakur and Singh, 1989) while the intensity of this weed species was greatly suppressed by the inclusion of post emergence herbicide, 2, 4 – D Na salt (Reddy, 2003).

**Table 15. Agronomic traits of maize as influenced under different weed management practices**

| Treatments | Weed | | Seed yield (kg/ha) | Stalk yield (kg/ha) | Net return (Rs./ha) | B:C ratio |
|---|---|---|---|---|---|---|
| | Density (No/m$^2$) | DM g/m$^2$) (60 DAS) | | | | |
| Atrazine 1.0 kg/ha PE | 76.3 (8.8) | 68.9 | 4132 | 6876 | 30055 | 1.73 |
| Atrazine 1.0 kg/ha PE + IC on 20 DAS | 37.6 (6.2) | 32.6 | 4852 | 7383 | 35493 | 1.82 |
| Atrazine 1.5 kga.i/ ha PE | 63.7 (8.0) | 54.2 | 4315 | 6949 | 31421 | 1.76 |
| Atrazine 1.0 kg/ha PE + Atrazine 1.0 kg/ha at 16 DAS | 39.6 (6.4) | 37.7 | 4488 | 7261 | 33418 | 1.82 |
| Atrazine 1.0 kg/ha PE + Topramezone 40 ml/ha (both at 25 DAS) | 33.3 (5.8) | 28.6 | 4646 | 7304 | 34470 | 1.88 |
| Atrazine 1.0 kg/ha PE + Topramezone 40 ml/ha (both at 12 DAS) | 63 (7.9) | 41.7 | 4317 | 7045 | 35490 | 1.76 |

| Treatments | Weed | | Seed yield (kg/ha) | Stalk yield (kg/ha) | Net return (Rs./ha) | B:C ratio |
|---|---|---|---|---|---|---|
| | Density (No/m²) | DM g/m²) (60 DAS) | | | | |
| Metribuzin 0.75 kg/ ha (16 DAS) | 68.3 8.3 | 46.4 | 1917 | 3893 | 3915 | 0.20 |
| Inter cultivation on 20 and 30 DAS | 29 (5.4) | 26.0 | 5122 | 7934 | 37549 | 1.77 |
| Control | 288 (16.9) | 240.7 | 2585 | 5383 | 14401 | 0.88 |
| CD 5 % | 1.3 | 9.5 | 277 | 839 | 0 | 0 |

(Aleem Ahmed and Susheela, 2012).

At Rajendranagar (Telungana), pre-emergence application of atrazine @ 1.0 kg a. i/ha followed by topramezone @ 40 ml /ha as post emergence spray gave higher productivity and profitability under scarcity and non-availability of labour and slushy soil conditions (Aleem Ahmed and Susheela, 2012).

**Economics of weed control methods**

Application of simazine at 0.25 kg /ha registered a lower weeding cost of Rs. 30/ ha with additional income of Rs. 806 /ha over weedy check (Rajan *et at.,* 1974*)*. Hand weeding twice at 20 and 40 DAS gave maximum net returns of Rs. 6198/ ha and net returns per rupee investment of 1.57 over atrazine applied at 0.5 kg /ha which recorded a net return of Rs. 4651/ ha with net return per rupee investment of 1.26 (Padmavathi and Gopalaswamy, 1995). Pandey *et al.* (2002) reported that all the weed control treatments recorded considerably higher returns over weedy check, which ranged from Rs.4010 to 15,576 /ha. Higher additional returns of Rs.15,576/ ha was recorded under pre-emergence alachlor at 1.00 kg/ ha applied in integration with hand weeding at 30 DAS in baby corn. Kolage *et al.* (2004) reported that pre-emergence application of atrazine at 1.0 kg /ha recorded maximum net returns of Rs.12, 766/ ha and B: C ratio of 2.50 as compared to other herbicides. The most economical and effective method of weed control was the pre-emergence application of atrazine + one hoeing + one hand weeding at 20 DAS which recorded higher cost: benefit ratio of 1:3.68 (Kamble *et al.*, 2005).

**Table 16. Effect of weed control treatments on yield and economics of maize**

| Treatment | Weed count (0.25 m²) | Weed dry weight (g) | Grain yield (t/ha) | HI (%) | Weed Index (%) | Net return (Rs/.ha) | B:C ratio |
|---|---|---|---|---|---|---|---|
| Weedy check (no weeding and interculti vation) | 7.20* (53.68) | 3.71* (12.54) | 5.13 | 41.74 | 33.25 | 42670 | 2.59 |
| Mechanical weeding (at 25 and 45 DAS + IC at 30 DAS) | 5.35 (28.15) | 2.67 (6.61) | 6.94 | 43.82 | 9.96 | 64510 | 3.18 |
| Atrazine (1.00 kg /ha pre-em) + IC at 30 DAS | 6.27 (38.88) | 3.10 (9.14) | 6.31 | 43.70 | 18.09 | 56235 | 2.92 |
| Atrazine (1.25 kg/ hal pre-em) + IC at 30 DAS | 6.13 (37.11) | 3.02 (8.63) | 6.49 | 43.42 | 16.00 | 58371 | 2.97 |
| Atrazine (0.75 kg/ ha pre-em) + 2,4- D (1.00 kg/ ha post- em) at 30 DAS | 5.30 (27.60) | 2.64 (6.49) | 7.57 | 45.19 | 1.78 | 72993 | 3.47 |
| Oxyfluorfen (0.25 kg /ha post – em) at 20 DAS | 5.59 (30.73) | 2.79 (7.31) | 5.45 | 42.28 | 29.37 | 44726 | 2.53 |
| Weed free check (IC at 20 DAS and 40 DAS + HW at 30 DAS and 50 DAS) | 4.69 (21.53) | 2.36 (5.07) | 7.72 | 45.11 | 0.00 | 73785 | 3.39 |
| C.D 5 % | 0.23 | 0.13 | 0.87 | NS | 11.26 | 11841 | 0.42 |

(Shantveerayya Hawaldar and Agasimani, 2012). Pre-em: Pre-emergent herbicide. HW: Hand weeding. Post-em: Post-emergent herbicide. IC: Intercultivation. DAS-: Days after sowing. * V x+0.5 Transformed Values

At Dharwad (Karnataka), higher net returns (Rs. 73784/ ha) was recorded in weed free check followed by sequential application of atrazine @ 0.75 kg /ha followed by 2, 4-D @ 1.00 kg /ha and mechanical weeding weeding alone Rs.72992 and 64509/ ha respectively (Shantveerayya Hawaldar and Agasimani, 2012).

### Herbicides resistanat maize

Herbicide resistant maize plants confer tolerance to glyphosate by production of the glyphosate-tolerant CP4 5-enolpyruvylshikimate-3phosphate synthase (*CP4 EPSPS*) proteins.

Transgenic stack hybrid maize (MON 89034 x NK603) was developed for preventing yield losses of maize crop due to pests and weeds and to improve productivity. The stacked maize crop having both insect protection and herbicide tolerant traits will provide protection to the crop from target pests and also provide tolerance to glyphosate herbicide (Bradshaw et al., 1997). MON 89034 is second generation Bt maize technology effective against lepidopteron insect pests with a unique and innovative dual mode of action. NK 603 is the glyphosate tolerant technology for the effective weed management system (Erickson et al., 2003). Tharp et al. (1999) had earlier reported that maize yields of herbicide resistant hybrids were maximum with glyphosate at 0.84 kg a.i/ha of glyphosate when applied at fifth leaf stage of maize. Additionally, it has also been reported that herbicides proved effective in controlling weeds and produced higher grain yield (Riaz et al., 2007). At Coimbatore (Tamil Nadu), significant improvement in growth and yield attributes of transgenic maize hybrids were obtained with glyphosate at 1800 and 3600 g a.i/ ha and which was comparable with glyphosate at 900 g a.i /ha. Higher grain yield was recorded with POE application of glyphosate at 1800 g a.i /ha in transgenic 900 M Gold and 3600 g a.i/ ha in transgenic Hishell (Ravisankar et al., 2011).

**Table 17. Effect of weed control treatments on maize**

| Treatment | Cob length (cm) | Cob girth (cm) | Rows / cob | Grain (t/ha) | Stover (t/ha) |
|---|---|---|---|---|---|
| Atrazine 1.5 kg/ha as pre-emergence | 14.2 | 13.6 | 13.6 | 4.34 | 16.0 |
| Pendimethalin 1.5 kg/ha as pre-emergence | 14.5 | 13.8 | 14.3 | 4.87 | 16.5 |
| Atrazine 1.5 kg pre - emergence and Atrazine 0.5 kg/ha as post emergence | 15.5 | 14.0 | 14.3 | 5.47 | 17.2 |
| Pendimethalin 1.5 kg/ha pre-emergence and Atrazine 0.75 kg as post emergence | 15.9 | 14.0 | 13.2 | 5.19 | 16.9 |
| Atrazine 0.75 kg/ha + Pendimethalin 0.75 kg/ha as pre - emergence and 2, 4-D 0.75 kg as post emergence | 15.5 | 13.2 | 12.9 | 5.16 | 17.1 |
| Atrazine 0.75 kg/ha + Pendimethalin 0.75 as pre -emergence and metsulfuron methyl 0.004 as post | 14.3 | 14.5 | 14.6 | 3.50 | 13.5 |
| Atrazine 1kg/ha + Pendimethalin 0.5 kg/ha as pre-emergence + 2, 4-D 0.75 kg as post emergence | 16.1 | 14.2 | 15.1 | 5.63 | 16.8 |
| Metsulfuron methyl 0.004 kg/ha as post emergence | 14.6 | 13.6 | 13.7 | 2.93 | 12.1 |
| Hand weeding | 14.3 | 13.9 | 13.8 | 5.46 | 16.2 |
| Untreated check | 11.8 | 12.3 | 11.3 | 2.68 | 12.1 |
| CD 5% | 2.8 | 1.3 | 1.2 | 0.55 | 0.8 |

(Kumar et al., 2012).

At Palampur (Himachal Pradesh), atrazine 1.5 kg /ha as pre-emergence – Atrazine 0.5 kg/ha as post emergence application and atrazine 1 kg/ha + Pendimethalin. 0.5 kg/ha as pre-emergence followed by 2, 4-D 0.75 kg/ha gave significantly higher grain yield and net return. Weeds reduced maize grain yield by 50.3 % (Kumar et al., 2012).

**Chemical weed control in maize based cropping system**

Jat and Gaur (2000) reported that highest NPK uptake in maize + soybean intercropping system observed in pre-emergence pendimethalin at 1.0 kg /ha. Avil Kumar and Reddy (2000) observed that application of pendimethalin at 1.0 kg /ha resulted in maize grain yield which was on par with hand weeding in maize + turmeric intercropping system. Prasad (1995) stated that combination of pigeon pea + maize intercropping with pendimethalin at 1.0 kg / ha gave the higher pigeon pea equivalent yield with better weed control efficiency.

Pendimethalin at 1.0 kg /ha as pre emergence and fluchloralin at 1.0 kg /ha as pre-plant incorporation proved as good as hand weeding in increasing the grain yield of maize and soybeans in maize + soybean intercropping system by controlling the weeds more effectively (Prasad and Rafey, 1996). Saikia and Pandey (2001) reported that in maize- mustard sequential cropping system, weed population and weed dry weight were reduced greatly due to application of pre-emergence atrazine at all doses right from 1.0 kg/ ha to 2.0 kg /ha in maize followed by fluchloralin at 0.75 kg/ ha in mustard. Similarly, in maize-groundnut based sequential cropping system, maize + cowpea intercrop (1: 1) with pre-emergence pendimethalin at 0.5 kg /ha followed by pre-emergence pendimethalin at 1.0 kg /ha in groundnut recorded weed control efficiency and yield of maize as well as groundnut (Audi Reddy, 2003).

**Residual effect of herbicides in soil and succeeding crops**

Excessive use of herbicides leave residue in soil and cause damage to the succeeding crops. Continuous use of single herbicide leads to evolution of herbicide resistant weed species and shift in weed flora (Sinha and Sinha, 1970; Thakur and Sharma, 1996). Sandhu *et al.* (1999) reported that atrazine (0.25 and 0.5 kg/ ha) applied as pre-emergence to winter maize persisted up to 20 and 100 days in plant and soil, respectively. Saikia *et al.* (2000) reported that atrazine content in sandy loam soil at different intervals showed initial slow rate of loss of atrazine in soil up to 20 days and thereafter faster dissipation and at maize harvest (90 days) no detectable residue. They also reported that atrazine application right from 1.0 kg /ha to 2.0 kg/ ha in maize did not leave any detectable residues to succeeding crops of chickpea and Indian mustard. Saikia and Pandey (2001a) stated that atrazine was biologically active in sandy loam soil up to 90 days and its biological activities (impairment of vital growth processes like photosynthesis and chlorophyll synthesis) were well pronounced at 2.0 kg/ ha.

Shukla and Das (2001) reported that pendimethalin at all concentrations right from1ppm to 70 ppm persisted up to 90 days and influenced the microbial activity in the soil. Padmavathi Devi *et al.* (1994) reported that germination was affected by pendimethalin beyond 0.1 ppm in finger millet, 0.5 ppm in Italian millet, pearl millet and wheat and beyond 2-ppm concentration in maize, sunflower and green gram. Aruna Rajagopal *et al.* (1994) stated that the total weed population was drastically reduced by the additive effect of the residual atrazine (0.25 to 1.0 kg /ha) and pendimethalin (0.75 kg /ha) in the succeeding soybean crop in maize-soybean sequential cropping system. Audi Reddy (2003) observed that pre-emergence atrazine (0.25 to 0.5 kg/ ha) and pre-emergence pendimethalin (0.5 kg /ha) applied to maize did not leave any significant amount of residues to adversely affect the germination and yield of succeeding groundnut.

## References

Abdulkadir, I., S. Mohamoud and Sharanappa (2002b). Integrated weed management in maize. *Mysore J. Agric. Sci.*, 36: 203-207.

AICRPWC (1983). Fertilizer use efficency through weed control in maize and associated weeds. Annual Report 1984-85. GBPAUT, Pantnagar, P: 13-14.

AICRPWC (1996). ICAR Project Coordinators Report (Jan 1st 1996 to Dec. 1996). NRCWS, Jabalpur (MP), p.64.

Aleem Ahmed, M.A. and R. Susheela (2012). Weed management studies in Kharif maize. J. Res. ANGRAU, 40 (3): 121 – 123.

Arvadiya, L.K., V.C. Raj, T.U. Patel and M.K. Arvadia (2012). Influence of plant population and weed management on weed flora and productivity of sweet corn. Indian J. Agron., 57 (2): 162 – 167.

Arvadiya, L.K.., V.C.Raj, T.V. Patel, M.K. Arvadia and A.M. Patel (2012a). Effect of plant population and weed management practices on productivity of sweet corn. Indian J. Weed Sci., 44(3): 168 – 171.

Avilkumar, K. and M.D. Reddy (2000). Integrated weed management in maize + turmeric intercropping system. Indian J. Weed Sci., 32 (1&2): 59 – 62.

Bahar, F.A., K.N. Singh and M.A. Malik (2009). Integrated weed management in maize under different nitrogen levels. Indian J. Agric. Sci., 79(8): 641 -644.

Balyan, R.S., R.K. Malik and R.S.Dhankar (1994). Atrazine as post-emergence herbicide for weed control in maize (*Zea mays* L.). Indian J. Weed Sci., 26(3&4): 35-39.

Bhopal singh, C.M. Singh, B.Manoj and R.D.Sood (1991). Integrated weed control in maize. Himachal J. Agric. Res., 17: 8-10.

Bhuvaneshwari, J., A.Muthusankaranarayanan and S.Avudaithai (2002). Weed smothering effect of intercropping in maize. Madras Agric. J., 89 (10-12): 714 - 716.

Bially, M.E. (1995). Efficacy of atrazine with other herbicides used alone in sequence or as tank mix in maize. Annals Agric. Sci. (Cairo), 40: 709 -721.

Chalka, M.K. and V. Nepalia (2005) Production potential and economics of maize (Zea mays) intercropped with legumes as influenced by weed control. Indian J. Agron., 50 (2): 119-122.

Chalka, M.K. and V. Nepalia (2006). Nutrient uptake appraisal of maize intercropped with legumes and associated weeds under the influence of weed control. Indian J. Agric. Res., 40 (2): 86 – 91.

Chopra, P. and N. N. Angiras (2007). Influence of tillage and weed control methods on growth, productivity and quality of maize (*Zea mays* L.). *Journal of Crop and Weed 3* (2): 10–14.

Chopra pankaj and M.N. Angiras (2008), Influence of tillage and weed control methods on weeds, yield and yield attributes of maize (*Zea mays* L.). *Indian J. Weed Sci.,* 40(1&2): 47-50.

Dass, S., M.L. Jat, R.G. Singh and R. Gupta (2009). Resource conserving technologies for maize systems in India. Indian Farming, 58 (11): 47 -52.

Devender, S., R.C. Tyagi, S.K. Agarwal and D. Singh (1998). Weed control methods in spring maize. *Haryana Agric. Uni. J. Res.,* 28(1): 21-25.

Dwivedi, S.K. and G.K. Shrivastava (2011). Planting geometry and weed management for maize + black gram intercropping system under rainfed vertisols. Indian J. Agron., 56 (3): 202 - 208.

El–Bially, M.E. (1995). Efficiency of atrazine with other herbicides used alone in sequence or as tank mix in maize. *Ann. Agric. Sci.*, 40: 709–721.

Gaur, B.L., D.S. Rao and M.K. Kaushik (1991). Comparative efficacy of pre-and post emergence herbicides in controlling weeds in rainy-season maize (*Zea mays*). *Indian J. Agron.,* 36: 261-262.

Geetha Jebarathnam, T. and R.M. Kathiresan (2006). Influence of organic manures on the weed seed bank in maize. Indian J. Weed Sci., 38 (3 & 4): 247 - 249.

Intodia, S.K., L.R. Yadav and O.P. Tomar (1996). Effect of herbicides on weed control efficiency and yield in maize (*Zea mays*) – soybean (*Glycine max*) intercropping system. *Indian J. Agric. Sci.,* 60(12): 730 - 731.

Ishrat D. H., C.S. Hunshal, L.H. Malligwad, B. Basavaraj and V.P. Chimmad (2012). Effect of pre and post emergence herbicides on weed control in maize (*Zea mays.* L.). Karnataka J. Agric. Sci., 25 (3): 392-394.

Jat, R.L. (1996). Studies on weed management and fertilizer levels of intercrop with and without inoculation in maize + soybean intercropping system. Ph.D thesis. Rajasthan Agricultural University, Udaipur. Rajasthan, India.

Kamble, T.C., S.U. Kakade, S.U. Nemade, R.V. Pawar and V. A. Apotikar (2005). Integrated weed management in hybrid maize. *Crop Res.* (Hisar), 29(3): 396-400.

Kannur, S. and Mallikarjun (2008). Effect of sequential application of herbicides in maize (*Zea mays* L.) in northern transition zone of Karnataka. *M.Sc.(Agri.) Thesis*, Univ. Agric. Sci., Dharwad (India).

Keshtkar, Alizadeh and Abbasi (2008). Response of corn to herbigation and conventional metheod of eradicane application. Crop Res., 36 (1-3): 54 - 59.

Kumar, B. and U.S. Walia (2003). Effect of nitrogen and plant population levels on competition of maize with weeds. Indian J. Weed Sci., 35: 53 - 56.

Kumar, D., S. Alagudurai, and K.R. Pushpanathan (2004). Effect of soil solarization on weed dynamics and yield of hybrid maize (Cargil 633). Madras Agric. J. 91 (7-12): 505-509.

Kumar, R.T. (2008). Evaluation of different herbicides and method of application on maize (*Zea mays* L.) under Northern Transition Zone of Karnataka. M.Sc (Agronomy). University of Agricultural Sciences, Dharwad. Karnataka, India.

Kumar, S., N.N Angiras and S.S. Rana (2011). Integrated weed management in maize. Himachal J. Agric., 37(1): 1-9.

Mahadevaiah, K., G.K., Sagar and V. Sumathi (2010). Effect of herbicides on weed dynamics and yield of baby corn in southern agro-climatic zone of Andhra Pradesh. Indian J. Weed Sci., 42 (3&4): 238-240.

Mallikarjun, S.K. (2008). Effect of sequential application of herbicides in maize (*Zea mays* L.) in Northern Transition Zone of Karnataka. M.Sc (Agronomy). University of Agricultural Sciences, Dharwad, Karnataka, India. Karnataka, India.

Mandal, R.C. (1995). Weed, weedicides and weed control, Published by Agrobios (India), Agro House, Jodhpur.

Meyyappan, M. and R. M. Kathiresan (2005). Integrated Weed Management in Maize+ black gram Intercropping System. Indian J. Weed Sci., 37(3&4): 209 – 211.

Mundra, S.L., A.K. Vyas and P.L. Maliwall (2003). Effect of weed and nutrient management on weed growth and productivity of maize. Indian J. Weed Sci., 35 (1&2): 57-61.

Mundra, S.L., A.K. Was and P.L. Maliwal (2002). Effect of weed and nutrient management on nutrient uptake by maize (*Zea mays*) and weeds. Indian J. Agron., 47 (3): 378 – 383.

Mynavathi, V.S., N.K. Prabhakaran and C. Chinnusamy (2008). Evaluation of mechanical weeders in irrigated maize. *Indian J. Weed Sci.,* 40 (3&4): 210-213.

Padmavathi, P. and N. Gopalaswamy (1995). Effect of time of nitrogen application, methods of weed control and sowing on yield and economics of maize. *Madras Agric. J.,* 82: 460-461.

Pandey A.K., Prem Singh, Ved Prakash, R.D. Singh and V.S. Chauhan (1999). Direct and Residual effect of weed control measures in Maize (*Zea mays*) and Wheat (*Triticum aestivum*) cropping system under Mid-Hill conditions of N-W Himalayas. Indian J. Weed Sci., 31 (3&4): 204- 209.

Pandey A.K., V. Prakash, R.D. Singh and V.P. Mani (2001). Integrated weed management in maize (*Zea mays*). Indian J. Agron., 46 (2): 260 – 265.

Pandey, A.C, V. Prakash, R.D. Singh and V.P. Mani (2000). Response of maize varieties to nitrogen levels and sulphydryl compounds. Crop Res., 19(1): 28-33.

Pandey, A.K. and Ved Prakash (2002). Weed management in maize and soybean intercropping system. Indian J. Weed Sci., 34 (1&2): 58 – 62.

Pandey, A.K., V. Prakash, R.D. Singh and V.P. Mani (2002). Studies on crop - weed competition and weed dynamics in maize under mid hill conditions of north-west Himalayas. Indian J. Weed Sci., 34: 63 - 67.

Pandey, I.B., V. Bharati and S.S. Mishra (2003). Effect of maize (*Zea mays)*-based intercropping systems on maize yield and associated weeds under rainfed condition. Indian J. Agron., 48 (1): 30-33.

Pankaj, C. and N.N. Angiras (2008). Influence of tillage and weed control methods on weeds, yield and yield attributes of maize. Indian J. Weed Sci., 40 (1&2): 47 - 50.

Paradkar, V.K. and R.K. Sharma (1993). Integrated weed management in maize. Indian J. Weed Sci., 25 (3 & 4): 80 - 83.

Parmeet, S., P. Singh and D. Joy (2007). Correlation and regression studies of winter maize and weed interactions. Indian J. Weed Sci., 39: 21 - 23.

Patel, G.N., G.J. Patel, S.N.Goyal and B. G. Patel (2000). Integrated weed management in rabi maize. Gujarat Agric. Univ. Res. J., 25 (2): 88-90.

Patel, V. J., P. N. Upadhyay, J. B. Patel and B. D. Patel (2006). Evaluation of herbicide mixtures for weed control In maize (Zea mays L.) under Middle Gujarat conditions. J. Agric. Sci., 2 (1): 81-86.

Patel, V.J., P.N. Upadhyay, J.B. Patel and M.I. Meisuriya (2006). Effect of herbicide mixture on weeds in kharif maize under middle Gujarat condition. Indian J. Weed Sci., 38 (1&2): 298 – 300.

Patel, V.J., P.N. Upadhyay, S.U. Zala and B.D. Patel (2006). Residual effect of herbicide applied as alone and mixture to kharif maize on succeeding rabi oat and mustard. Indian J. Weed Sci., 38 (3 & 4): 258 - 262.

Porwal, M.K. (2000). Economics of weed control measures in winter maize (*Zea mays*). *Indian J. Agron.,* 45(2): 344-347.

Prasad K. and A. Rafey (1995). Weed control in maize (*Zea mays*) intercropped with soybean (*Glycine max*). Indian J. Weed Sci., 27 (3&4): 167- 169.

Prasad, A., G. Singh and R.K.Upadhyay (2008). Integrated weed management in maize and maize + black gram intercropping. Indian J. Weed Sci., 40 (3&4): 191 - 192.

Prasad, K., A. Quayum and A. Rafey (1995). Weed control in cropping sequence based on single and mixed crops. Indian J. Agric. Sci., 65 (8): 142-147.

Prasad, T. V. R., Dwarakanath, N. Narasimah and N. Krishnamurthy (1990). Integrated weed management in maize effect on weed, crop growth and yield. Mysore J. Agric. Sci., 24: 39 - 44.

Rajendra, A.N. (2000). Bioefficacy of herbicides in maize + soybean intercropping system in vertisols. *M.Sc.(Agri.) Thesis*, Univ. Agric. Sci., Dharwad, Karnataka, India.

Ram prasad (2001). Weed management in fodder maize + cowpea intercropping system. Indian J. Agron., 30 (4): 267 - 268.

Ramachandra Prasad, T.V., Dwarakanath, N. Narasimha and Krishnamurthy (1990). Integrated weed management in maize. Effect on weds, crop growth and yield. *Mysore J. Agric. Sci.,* 24: 39-44.

Ramachandran, A., A. Veeramani and P. Prema (2012). Effect of brown manuring on weed growth, yield and economics of irrigated maize. Indian J. Weed Sci., 44(3): 204 – 206.

Ramesh, G. and T. Nadanassababady (2005). Impact of herbicides on weed and soil ecosystem of rainfed maize (*Zea mays* L.). *Indian J. Agric. Res*., 39(1): 31-36.

Rana, S.S., J.J. Sharma and S.S. Manuja (1998). Evaluation of promising herbicide mixtures for weed control in maize. New Agriculturist, 9 (1&2): 1 – 5.

Reddy, A.D., K. Ramamoorthy and O.S. Kandasamy (2004). Integrated weed management in rabi maize and its residual effect on succeeding groundnut. Andhra Agric. J., 51: 517-522.

Saikia, T.P. and J. Pandey (1999). Weed shift in maize – chickpea cropping system. Indian J. Agron., 44 (2): 246-249.

Saikia, T.P., J. Pandey and G. Kulshrestha (2000). Investigation on residue of atrazine and fluchloralin in maize – chickpea and maize – Indian mustard croppig sequence. Indian J. Agron., 45: 653 – 657.

Saini, J. P. and N. N. Angiras (1998). Efficacy of herbicides alone and in mixtures to control weeds in maize under mid-hill conditions of Himachal Pradesh. *Indian J. Weed Sci.,* 30(1/2): 65-68.

Sandhu, K.S. and R.K. Bhatia (1991). Chemical weed control in transplants winter maize. *Indian J. Weed Sci.,* 23(3&4): 53-55.

Sandhu, K.S., J. Singh and S. Singh (1999). Weed composition of maize fields in Punjab. Indian J. Weed Sci., 31: 18 - 24.

Sandhu, K.S., S.P. Mehra and H.S. Gill (1996). Studies on weed control in winter maize through herbicides. *J. Res. Punjab Agric. Univ.,* 23(4): 549-556.

Sanjay, K.D., G.K. Shrivastara, A. P. Singh and S. S. Kolhe (2011). Weed population, removal by weeds and crop yield under maize + black gram intercropping system in Chatisgarh plains. Indian J. Weed. Sic., 43 (3&4): 203-210.

Sanjay, K.D., G.K. Shrivastava, A.P. Sigh and R. Laspale (2012). Weeds and crop productivity of maize + black gram intercropping system in Chhattisgarh plains. Indian J. Weed Sci., 44(1): 26 – 29.

Sarma, C.K. and R.C. Gautam (2010). Weed growth, yield and nutrient uptake in maize (*Zea mays* L.) as influenced by tillage, seed rate and weed control methods. *Indian J. Agron.,* 55(4): 299 - 303.

Selvakumar, T. and A. Sundari (2006). Effect of intercropping and weed management practices on weedsin maize. Indian J. Weed Sci., 38 (1&2): 133 - 134.

Shantveerayya, H. and Agasimani (2012). Effect of herbicides on weed control and productivity of maize (Zea mays L.). Karnataka J. Agric. Sci., 25 (1): 137-139.

Sharara, F.A., T.A. Elshahawy and K.G. Elkokiek (2005). Effect of some novel herbicides on the controlling weeds associated with maize plants. J. Agron., 41(2): 88 - 95.

Sharma, A.R., A.S. Toor and H. Sur (2000). Effect of interculture operations and scheduling of Atrazine application on weed control and productivity of rainfed maize (*Zea mays* L.) in Shiwalik foot hills of Punjab. *Indian J. Agril. Sci.,* 70(1): 757-761.

Sharma, C.K. and R.C. Gautam (2006). Effect of tillage, seed rate and weed control methods on weeds and maize. Indian J. Weed Sci., 38 (1&2): 58 - 61.

Sharma, J. and S.C. Nayital (1993). Crop weed competition in maize (*Zea mays*) + black gram (*Phaseolus mungo*) intercropping system. *Indian J. Agron.,* 36: 64-67.

Sharma, J. and S.C. Nayital (1993a). Weed Management in Maize + black gram Intercropping in Mid-Hills Sub-Humid Zone of Himachal Pradesh. Indian J. Weed Sci., 25(1&2): 43 – 46.

Sharma, S.K., and R.C. Gautam (2003).Effect of dose and method of atrazine application on no-till maize (*Zea mays* L.). Indian J. Weed Sci., 35(1-2): 131-133.

Sharma, V. and D.R. Thakur (1998). Integrated Weed Management in Maize (*Zea mays*) under Mid-Hill Conditions of North-Western Himalayas. Indian J. Weed Sci., 30 (3&4): 158 – 162.

Sharma, V., D.R. Thakur and J.J. Sharma, (1998). Effect of metalachlor and its combination with atrazine on weed control in maize. Indian J. Agron., 43(4): 677 – 680.

Shekhawat, P.S. and R.C. Gautam (2002). Effect of row spacing and weed control methods on the growth attributes and grain yield of maize under tilled and untilled conditions. *Ann. Agric. Res.,* 23 (4): 626-629.

Sher Singh and A.L. Rajput (1995). Chemical weed control in winter maize (*Zea mays*). *Harana J. Agron.,* 11 (2): 103-107.

Shinda, S.H., A.K. Kolage and R.L.Bhilara (2001). Effect of weed control on growth and yield of maize. J.Maharashtra Agric. Uiv., 26 (2): 212–213.

Singh, C.M., A. A. Andrabi, S. Kumar and Suresh Kumar (1994). Integrated weed management in Maize. Indian J. Weed Sci., 20: 201-204.

Singh, M., P. Singh and V. Nepalia (2005). Integrated weed management studies in maize based intercropping system. Indian J. Weed Sci., 37 (3&4): 205-208.

Singh, R., A.R. Sharma and U.K. Behera (2007). Tillage and crop establishment practices for improving productivity of maize under different weed control methods. Indian J. Agric. Sci., 77(11): 731–737.

Singh, S., U.S. Walia, R. Kaur and L.S. Shergill (2010). Chemical control of *Cyperus rotundus* in maize. Indian J. Weed Sci., 42 (3&4): 189-192.

Singh, V.P., S.K. Guru, A. Kumar, A. Banga and N. Tripathi (2012). Bioefficacy of tembotrione against mixed weed complex in maize. Indian J. Weed Sci., 44(1): 1–5.

Sinha S.P., S.M. Prasad and S.J. Singh (2005). Nutrient utilization by winter maize (*Zea mays*) and weeds as influenced by integrated weed management. Indian J. Agron., 50 (4): 303 – 304.

Sinha, S.P., S.K. Prasad and S.J. Singh (2000). Effect of IWM on growth yield attributes and yield of winter maize. J. Application Biol. 10 (2): 158 – 162.

Sinha, S.P., S.M. Prasad, S.J. Singh and K.K. Sinha (2003). Integrated weed management in winter maize (*Zea mays*) in North Bihar. *Indian J. Weed. Sci.,* 35 (3/4): 273-274.

Sreenivas, G. and V. Satyanarayana (1994). Nutrient removal by weeds and maize (*Zea mays*). *Indian J. Agron.,* 41(1): 160-162.

Srividya, S., K. Chandrasekhar and R. Veeraraghavaiah (2011). Effect of tillage and herbicide use on weed management in maize. Andhra Agric. J., 58 (2): 123 - 126.

Subramanyam, D., M. Reddy and R. Reddy (2001). Economics of weed management practices in irrigated maize (*Zea mays*). Andhra Agric. J., 48 (3&4): 318-320.

Suresh Kumar, N.N. Angiras and S.S. Rana (2011). Integrated weed management in maize. Himachal J. Agric. Res., 37 (1): 1 - 9.

Suresh Kumar, S.S. Rana, N.Chander and N.N. Angiras (2012). Management of hardy weeds in maize under mid hill conditions of Himachal Pradesh. Indian J. Weed Sci., 44(1): 11-17.

Tamadaddi, K.R. and S.C. Alagundagi (2008). Evaluation of different herbicides and method of application on maize (*Zea mays* L.) under Northern Transition zone of Karnataka. Karnataka J. Agric. Sci., 21(4) : 658.

Tamadaddi, K. R.and S.C. Alagundagi (2008). Evaluation of Different Herbicides and Method of Application on Maize (*Zea mays* L.) under Northern Transition Zone of Karnataka. Karnataka J. Agric. Sci., 21(4) : 658.

Thakur, D.R. (1994). Weed management in intercropping systems based on maize under mid hill conditions. Indian J. Agron., 39: 203 – 206.

Varshney, J.G. (1990). Efficacy of atrazine with adjuvants on weed suppression in maize in mid hills of Meghalaya. Indian J. Weed Sci., 22(1&2): 27-31.

Varshney, J.G. (1991). Studies on critical stage of crop -weed competition in maize. Indian J. Agron., 36: 153 - 158.

# 15

# Maize based Cropping Systems

As maize has wide adaptability and compatibility under diverse soil and climatic conditions and hence it is cultivated in sequence with different crops under various agro-ecologies of the country. Hence, it is considered as one of the potential driver of crop diversification under different situation. Among different maize based cropping systems, maize - wheat ranks first having 1.8 million ha area mainly concentrated in rainfed ecologies which is the thrid most important cropping systems after rice - wheat and rice - rice and contributes about 3 % in the national food basket. The other major maize systems in India are maize - mustard, maize - chickpea, maize - maize, cotton - maize, etc. Recently, due to changing scenario of natural resource base, rice - maize has emerged a potential maize based cropping system in peninsular and eastern India. In peri-urban interface, maize based high value intercropping systems are also gaining importance due to market driven farming. Further, maize has compatibility with several crops of different growth habit that led to development of various intercropping systems. Compared to existing cropping systems like rice - wheat and rice - rice, maize based cropping systems are better user of available resources and the water use efficiency of maize based cropping systems was about 100 to 200 % higher at different locations.

**Table 1. Maize based sequential cropping systems in different ago-climatic zones of India**

| Agro-climatic region | Cropping system | |
|---|---|---|
| | **Irrigated** | **Rainfed** |
| Western Himalayan Region | Maize-wheat<br>Maize-potato-wheat<br>Maize-wheat-green gram<br>Maize-mustard<br>Maize-sugarcane | Maize-mustard<br>Maize-legumes |
| Eastern Himalayan Region | Summer rice-maize-mustard<br>Maize-maize<br>Maize-maize-legumes | Sesame-Rice+maize |
| Lower Gangetic Plain region | Autumn rice-maize<br>Jute-rice-maize | Rice-maize |
| Middle Gangetic Plain region | Maize-early potato-wheat-mung bean<br>Maize-wheat<br>Maize-wheat-mung bean<br>Maize-wheat-urd bean<br>Maize-sugarcane-mung bean | Maize-wheat |
| Upper Gangetic Plain region | Maize-wheat<br>Maize-wheat-mung bean<br>Maize-potato-wheat<br>Maize-potato-sunflower<br>Maize-potato-onion<br>Maize-potato-sugarcane-ratoon<br>Rice-potato-maize | Maize-wheat<br>Maize-barley<br>Maize-safflower |
| Trans Gangetic Plain region | Maize-wheat<br>Maize-wheat-mungbean<br>Maize-potato-wheat<br>Maize-potato-sunflower<br>Maize-potato-onion<br>Mung bean-maize-toria-wheat<br>Maize-potato-mung bean | Maize-wheat |
| Eastern plateau and hills region | Maize-groundnut-vegetables<br>Maize-wheat-vegetables | Rice-potato-maize<br>Jute-maize-cowpea |
| Central plateau and hills region | Maize-wheat | Maize-groundnut |
| Western plateau and hills region | Sugarcane + Maize | |
| Southern plateau and hills region | Rice-maize<br>Maize-rice | Sorghum-maize<br>Maize-sorghum-Pulses<br>Maize-potato-groundnut |

| Agro-climatic Region | Cropping system | |
|---|---|---|
| | **Irrigated** | **Rainfed** |
| East coast plain and hills region | Rice-maize-pearl millet<br>Maize-rice<br>Rice-maize<br>Rice-rice-maize | Maize – maize - pearl millet<br>Rice-maize + cowpea |
| West coast plain and hills region | Maize-pulses<br>Rice-maize | Rice-maize<br>Groundnut-maize |
| Gujrat plains and hills region | Maize-wheat | Rice-maize |
| Western dry region | Maize-mustard<br>Maize-chickpea | Maize + legumes |
| Island region | Rice-maize | Maize-rice<br>Rice-maize + cowpea<br>Rice-maize-urdbean<br>Rice-rice-maize |

## Area expansion

Cultivation during winter season spreads in entire plain region of the country where temperature during the growth periods does not go below 10$^0$C. The sensitivity of a crop species to low temperature and chilling frequently restricts the environment in which it can be cultivated. Hence, *Rabi* maize cultivation is possible in Eastern Uttar Pradesh, Bihar, Odisha, Tamil Nadu, Karnataka, Andhra Pradesh, Gujarat, Madhya Pradesh, South and east Rajasthan, lower elevations of Sikkim, West Bengal, Punjab and Haryana. Rabi maize can be taken in lower foothills which usually do not receive frost.

## Diversification of agro ecosystems

Under the changing climate scenario maize being a photo-insensitive crop has better options for adaptation and mitigation of these climatic changes. The limitation of rising temperature during grain filling of wheat particularly in eastern India and declining yield of boro rice in West Bengal and Orissa and water scarce areas in Peninsular India (Andhra Pradesh and Tamil Nadu) affecting yield of *Rabi* rice has shown a path to maize as better option. Peninsular India is considered to be a neutral environment for maize wherein maize can be cultivated in either of the seasons. Therefore, it is emerging as a potential driving force for diversification i.e. diversification of rice-rice with rice-maize and other maize based high value cropping systems in water scarcity/lowering of water table is a major concern in rice growing belt of India and making rice cultivation non-remunerative. Hence, maize has emerged as a potential as well as profitable crop in these areas. The *Rabi* rice in Peninsular India and upland rice in Odisha and NEH region has low productivity. Therefore maize is the only suitable alternative crop and more area is likely to shift towards maize cultivation in near future in these non-traditional areas.

Wheat crop adversely affected with terminal heat due to sudden rise in temperature during crop growth and maturity but this favours maize crop positively. Rice-rice is common in tropical climate with distinct dry and wet seasons such as in South India and in subtropical areas with mild cool winter climate such as in Bangladesh, Eastern India, and Eastern Nepal. Rice-wheat systems is extensive in the sub-tropical areas of the Indo Gangetic Plains (IGP) while rice-maize systems exist in all climates ranging from tropical to sub tropical to warm temperate.

Rice - maize systems, however, are less extensive as compared to rice - wheat or rice – rice if total area under these cereal systems is considered. The productivity level of winter maize is higher as compared to *Kharif* sown maize because of comparatively favourable environmental conditions in peninsular India. Rice-maize systems are practiced mostly in the south (Andhra Pradesh, Tamil Nadu and Karnataka) and in the northeast (Bihar and West Bengal) parts of India with acreage of more than 0.5 Million ha.

There is a vast opportunity for intensification of winter maize in flood prone areas as to compensate for the loss during *Kharif* season with proper planning for seeds, inputs and improved management practices and crop diversification. The medium and uplands where subsistence yield of wheat, *Rabi* rice and other winter crops is obtained could be substituted by winter maize in Bihar, West Bengal, Eastern UP, Orissa, parts of Jharkhand, etc. In general, any late maturing single cross-bred variety of *Kharif* season is equally good for winter season. Winter maize (170-180 days duration) has the clear cut comparative advantages of low incidence of diseases and insect pests, is not affected by temperature rise during winter (as the wheat is) and do not suffer on account of heavy rainfall.

**Indo-Gangetic plains**

The IGP countries grew maize in an area of 7.99 million ha in 2002 (FAO, 2003). Maize is traditionally grown as *kharif* crop in these countries, either rainfed or irrigated. In India, the *kharif* maize represents 89 % of the total area under maize. The largest area under the *kharif* maize is in Uttar Pradesh (1.4 million ha), followed by Bihar (0.96 million ha). The spring and *rabi* maize are gaining popularity with farmers and with multinationals showing interest in it because of higher yield potential because of the assured irrigation facilities. The trend is more clear in Pakistan where spring maize, today, occupy 12-15 % of area planted in *kharif* and represents 30-35 % of the annual maize production (Pakistan Agricultural Research Council, 2003a). In India, maize sector is going more towards winter maize. While spring maize has the threat of facing high temperatures during silking resulting possible poor seed setting and grain yield. The winter maize is relatively free from these difficulties and diseases and insects. Winter maize in India is a success in eastern IGP, especially in the state of Bihar. The success of winter maize in Bihar is due to sunny days, long growing season, dry and cool temperatures which are more suitable to the crop and less for the pest. This encouraged farmers to take *rabi* maize to a higher share (Chaudhary and Shahi, 1990). Incidentally, Bihar is the second largest maize growing state in India which grows maize in 0.58 million ha of area with a productivity of 2.6 t/ha. The productivity is higher in the Indian IGP states mainly because of the dominant *rabi* maize in this state.

**Table 2. Key emerging rice - maize agro-ecosystem in South Asia**

| Key features | Current systems | Emerging systems | Key examples |
|---|---|---|---|
| A. Tropical, warm, semiarid. no winter Tropical monsoon with longer dry season; both, rice and maize not limited by low temperatures and can be grown all year round | Rice-rice-pulses | Rice-maize | Cauvery Delta (Tamil Nadu), Kamataka and A.P., India |
| | Rice-wheat | Rice-maize | |
| B. Sub-tropical, subhumid. warm summer mild cool winter Subtropical monsoon with cool winter and summer rainfall; rice but not maize maybe limited by low temperatures | Rice-*Boro* rice | Rice-maize | Central, western, and N\W Bangladesh. Eastern Terai, Nepal. West Bengal, Eastern UP and Bihar, India. |
| | Rice-wheat | Rice-rice - maize<br>Rice-potato-maize | |
| C. Sub-tropical to warm temperate, subhumid, semiarid, warm summer, mild to severe cold winter | | | |
| i) Sub-tropical monsoon with cold winter and summer rainfall; both rice and maize limited by low temperatures and cannot be grown for some time in winter | Rice-wheat | Rice-maize | North and NW India. Central and western Terai and mid-hills, Nepal |
| ii) Sub-tropical to warm temperate, semiarid. with hot summer and cool to cold winter; very low rainfall; both rice and maize limited by low temperatures and can't be grown for some time in winter | Rice-wheat | Rice-maize | Punjab and Sindh, Pakistan |
| | Cotton-wheat<br>Sorghum-wheat | Rice -potato-maize | |

(Timsina et al., 2010)

## Maize as possible candidate of rice-wheat diversification

Late planting is a major constraint in cultivating winter maize after kharif rice crop. The problem is acute when farmers grow long duration scented basmati cultivars in kharif. Sowings are further delayed when farmers resort to thorough field preparation which takes more time than a week time. At Kalyani (West Bengal), higher maize yields could only be obtained when it is sown in first fortnight of November (Prodhan, 2001). The yields were substantially reduced by a ton per week of delay in sowing. Still higher yield reductions are expected if the silking and grain filling stages coincide with hot westerly winds. Late planting can be avoided by adopting zero-till sowing of maize in either *kharif* after harvesting wheat or in *rabi* after harvesting rice crop. In both cases, the maize crop can be planted by using multi-crop zero-till-cum-ferti-drill capable of sowing maize, rice, wheat, pigeon pea, peas, grams, mustard, etc. Paliwal (2003) demonstrated how a successful zero-till maize crop can be grown using various best management practices. Zero-till sowing of maize is a good option worth exploring in areas where it follows the late harvested rice crop. Various on-farm participatory trials revealed little or no difference in zero-till maize when compared to best managed conventional crop. Despite the similar yields, the economic advantage the farmers benefit was with zero-till maize due to savings in tillage and first irrigation as this crop utilizes the residual moisture. Successful establishment of rice on permanent beds would not only bring the benefits of zero tillage but also allows growing of various *rabi* crops, including maize, on raised bed system.

The south Asian farmers are yet to grow maize on raised beds, though it is a common practice in many western countries. Bed planting of maize helps establish proper plant stand, increases input efficiency, increases yields and opens up avenues for further crop intensification through intercropping.

The traditional cropping systems in which maize is grown in Indian IGP are given in Table 3 (Yadav and Rao, 2001). It can be seen that maize-wheat is the major crop rotation followed by maize-mustard or toria-wheat and maize-potato-wheat. This is to be noted that rice-wheat is primary cropping system in all these areas and these cropping systems stand at secondary or tertiary level. *Kharif*, *rabi* and spring maize were promoted in eastern and western Indian IGP on sustaining rice-wheat rotation. Successful farmer participatory experiments in eastern IGP in Bihar, India revealed number of profitable rice based cropping systems where maize could be grown in winter. Some of these rotations are given in Table 4 (Kumar, 2003). Among these rotations, rice-potato-onion-maize (green cobs) was found to be most profitable in south Bihar because of additional higher income from the green cobs. Similarly, rice-winter maize-potato gave higher net returns (USD 1011/ ha) in North Bihar. In comparisons involving traditional sole crop and intercropping of maize and potato on raised beds (RWC, 2003), intercrops performed better than the sole crops mainly due to the reason they were sown on flat fields in improper crop geometry, a common problem in these areas. Pandey et al. (2001) obtained higher net returns of USD 1473 /ha by growing maize (for green cobs) – radish - onion crop rotation under irrigated conditions. The next best system was found to be maize (for green cobs) – rice - cabbage with a net profit of USD 719/ ha.

**Table 3. Area (000' ha) under maize based cropping systems in Indian Indo-Gangetic Plain states (Yadav and Rao, 2001)**

| Cropping system | Area (000' ha) | | | | |
|---|---|---|---|---|---|
| | Uttar Pradesh | Punjab | Haryana | Bihar | Total area |
| Maize-wheat | 149.6 | 453.6 | 20.9 | 106 | 730.1 |
| Maize-mustard | 155.8 | - | - | - | 155.8 |
| Maize-toria-wheat | 130.8 | - | - | - | 130.8 |
| Maize-potato-wheat | 91.8 | 37 | - | - | 128.8 |
| Maize-maize | 54 | - | - | 61 | 115 |
| Maize-barley | 20 | - | - | - | 20 |
| Maize-sugarcane | - | - | - | 15 | 15 |
| Maize-gram | 7.1 | - | - | - | 7.1 |
| Maize-mustard-wheat | 7 | - | - | - | 7 |
| Maize-lentil | 4.5 | - | - | - | 4.5 |
| Maize-potato-sunflower | - | 2.8 | - | - | 2.8 |
| Total area (000' ha) | 620.6 | 493.4 | 20.9 | 182 | 1316.9 |

**Table 4. Net returns (USD) from various cropping systems tested in farmer participatory research in the Eastern Gangetic Plains (Kumar, 2003)**

| System tested | | | | Yields (t/ha) | | | | System profit ($/yr) |
|---|---|---|---|---|---|---|---|---|
| Crop 1 | Crop 2 | Crop 3 | Crop 4 | Crop 1 | Crop 2 | Crop 3 | Crop 4 | |
| Rice | Potato | Maize | - | 3.8 | 23.2 | 7.3 | - | 1161 |
| Rice | Potato | Wheat | - | 3.9 | 21.7 | 3.6 | - | 1281 |
| Rice | Potato | Onion | - | 4.7 | 22.0 | 17.3 | - | 815 |
| Rice | Wheat | Moong | - | 4.6 | 4.4 | 7.0 | - | 570 |

| System tested | | | | Yields (t/ha) | | | | System profit ($/yr) |
|---|---|---|---|---|---|---|---|---|
| Crop 1 | Crop 2 | Crop 3 | Crop 4 | Crop 1 | Crop 2 | Crop 3 | Crop 4 | |
| Rice | Wheat | Urd | - | 4.4 | 4.0 | 5.0 | - | 447 |
| Rice | Potato | Onion | Maize* | 4.0 | 16.0 | 12.8 | 13.5 | 2685 |
| Rice | Wheat | - | - | 4.2 | 4.0 | - | - | 366 |

* harvested for green cobs

Maize is a most versatile crop for growing intercrops, because of the wide row spacing it needs, providing higher income to the farmers. It was found that soybean and groundnut make good intercrops with maize (Prakash et al., 1993). Growing of maize + groundnut in 2:2 row proportion by paired row planting gave higher combined intercrop yield when compared to any other intercropping tested. At Faizabad (Uttar Pradesh), potato + maize intercropping in 2:1 ratio gave higher net returns than when sown in 1:1 proportion. Though the potato-maize sequence gave higher net returns in this experiment, farmers could be benefited by intercropping which gives similar returns in the single season as an added advantage (Singh et al., 1998). Higher yields could be obtained when the maize was sown at earthing-up of potato which will reduce the competition for light and nutrients at the initial stages of potato.

Farmers and researchers quantified the bed planting benefits in terms of savings of resources like irrigation water. In Bangladesh, bed planted maize out yielded the conventional crop by 1.03 t/ha and saved 18 % of water (Khan, 2003). Not only this, replacing one hectare of rice, whose water requirement is 4,240 mm (Sharma et al., 2002) with maize, whose water requirement is 550 - 650 mm (Violic, 2000), allows the saved water to be used to raise six hectares of maize or slightly more hectares of pulses. Bed planting further improves this efficiency, which only means this water can be used in increasing the cropping intensities, area and ultimately the food production. Bed planting of QPM maize + pulse crops will provide complete nutritional security to farming communities and dependent farm animals.

Rice fallows occupy nearly 4 million ha in Indian-IGP alone and are potential areas for increasing the cropping intensity. The available conventional crop establishment techniques can seldom allow these areas to be cropped due to their high moisture content hindering the mechanical manipulation of soil. Improved subsurface drainage system seems to be a possibility to bring these lands into cultivation. A successful establishment of *kharif* rice on raised beds in these areas will open them to cropping in subsequent seasons. Growing winter maize on raised beds in subsequent season with appropriate intercrop will be a good proposition.

## Maize - Wheat Cropping System

It is one among the predominant cereal – cereal cropping systems being followed in India and occupies 1.8 million ha. Among maize-wheat growing areas, maize is the principal crop of Kharif season in northern hills of the country but plains of northern states like Uttar Pradesh, Rajasthan, Madhya Pradesh and Bihar also have sizeable acreage under this crop. It is highly productive cropping system and may yield as high as 14 t/ha/year. However, a productivity of more than 10.5 t/ha /year (6.0 t + 4.5 t) may be realized under sub-montane conditions of Jammu. Likewise, a productivity of 7-8 t/ ha /year at Palampur, 6.28 t/ha per year (2.00 + 4.28 t) at Indore and more than 7 t/ha /year at Ludhiana has been recorded in maize-wheat system. Maximum yield research in maize-wheat sequence at Palampur has shown a potential of 14.21 t/ha per year by optimization of various resources like plant population, manures and fertilizers. Poor maize - wheat yield has been reported from Andhra Pradesh, Assam, Gujarat, Madhya Pradesh, Maharashtra, Rajasthan, Tripura, Eastern Uttar Pradesh and Tamil Nadu. There are numbers of reasons for poor yield but the most significant are sowing time, poor plant population, poor weed management, poor use of organic and inorganic fertilizers and large area under rainfed.

As most of the area in maize-wheat system is in rainfed conditions where uncertainty of rainfall is a major limitation. This also acts as a deterrent to farmers to adopt intensive input use, leading to deterioration of soil health and depletion of major plant nutrients from soil. Farmers in general, tend to grow low yielding traditional varieties. Maize is a widely spaced crop and its cultivation on slopping fields leads to soil erosion, especially when sowing is not done across the slope, maize in very sensitive to water logging and drought. Any delay in weed control in maize can reduce the yield of the system significantly. Since both the crops of maize – wheat system are fertility exhaustive, the declining yield trends under long term fertilizer experiments have indicated that the productivity of the system suffers due to emerging deficiencies of plant nutrients. Intensive input use, continuous over mining of nutrients from soil and imbalanced use of fertilizers lead to deterioration of soil health and stagnation in productivity of the cropping system.

At New Delhi, Sesbania green manure in combination with wheat straw followed by Sesbania green manure alone recorded significant improvement in system productivity (12.5 t/ha), protein content of maize (8.51 %) and wheat (10.24 %) and N, P and K uptake of both the crops and the system. Addition of Sesbania green manure along with wheat straw improved the available N, P, K and organic carbon in soil to the tune of 25.3, 49.3, 5.9 and 11.9 % over control respectively. A significant response of fertilizer N up to 120 kg/ha was recorded on the system productivity, quality and nutrient uptake of maize and the residual effect of that also influenced significantly the N, P and K uptake by wheat and system as a whole. Wheat also responded significantly to direct application of fertilizer N and protein content in grain up to 80 kg N/ha and up to 120 kg N/ha in terms of nutrient uptake (Jet et al., 2012).

**Table 5. Direct and residual effect of organic sources and fertilizer N levels on maize–wheat cropping system**

| Treatments | Maize equivalent yield (t/ha) | Total nutrient uptake (kg/ha) | | | | | |
|---|---|---|---|---|---|---|---|
| | | Maize | | | Wheat | | |
| | | N | P | K | N | P | K |
| **Organic manure** | | | | | | | |
| Control | 10.6 | 73.7 | 18.8 | 55.4 | 78.9 | 9.9 | 107.1 |
| Wheat straw 5 t/ha 15 DBS | 11.4 | 77.5 | 20.5 | 58.2 | 83.4 | 10.3 | 114.0 |
| Sesbania green manure in situ (5.25 – 5.65 t/ha) | 12.1 | 85.4 | 22.5 | 61.6 | 85.9 | 10.9 | 116.0 |
| Wheat straw + Sesbania green manure | 12.5 | 92.1 | 24.0 | 65.9 | 89.0 | 11.5 | 119.0 |
| CD 5 % | 0.5 | 3.1 | 3.4 | 2.1 | 4.9 | 0.4 | 2.8 |
| **Nitrogen level to Maize** | | | | | | | |
| 0 kg/ha | 10.7 | 64.2 | 17.2 | 49.5 | 82.2 | 9.9 | 110.7 |
| 60 kg/ha | 11.7 | 84.4 | 22.0 | 61.9 | 84.2 | 10.7 | 114.2 |
| 120 kg/ha | 12.4 | 98.0 | 25.1 | 69.5 | 87.0 | 11.2 | 116.3 |
| CD 5 % | 0.5 | 2.9 | 2.4 | 1.7 | 3.2 | 0.6 | 4.4 |
| **Nitrogen level to Wheat** | | | | | | | |
| 0 kg/ha | 10.2 | - | - | - | 63.7 | 8.1 | 96.5 |
| 40 kg/ha | 11.5 | - | - | - | 79.0 | 10.1 | 111.4 |
| 80 kg/ha | 12.3 | - | - | - | 94.6 | 11.7 | 121.7 |
| 120 kg/ha | 12.6 | - | - | - | 100.1 | 12.5 | 125.2 |
| CD 5 % | 0.7 | - | - | - | 3.3 | 0.4 | 2.9 |

* DBS: Days before sowing. (Jet et al., 2012).

**Table 6. Major nutrients on production of maize – wheat cropping system in acidic soil**

| Fertility for both crops (kg NPK/ha) | Maize yield (q/ha) | | | Wheat yield (q/ha) | | |
|---|---|---|---|---|---|---|
| | Grain | Straw | B:C ratio | Grain | Straw | B:C ratio |
| 180:90:150 | 34.0 | 67.0 | 1.14 | 40.0 | 66.0 | 2.5 |
| 210:90:150 | 40.0 | 76.0 | 2.51 | 44.0 | 72.0 | 3.5 |
| 240:90:150 | 35.0 | 70.0 | 1.12 | 39.0 | 64.0 | 1.7 |
| 210:0:150 | 27.0 | 49.0 | -2.34 | 26.0 | 37.0 | -6.8 |
| 210:60:150 | 39.0 | 68.0 | 2.92 | 34.0 | 64.0 | 0.4 |
| 210:120:150 | 37.0 | 60.0 | 1.43 | 42.0 | 70.0 | 2.3 |
| 210:90:0 | 23.0 | 36.0 | -3.21 | 30.0 | 56.0 | -1.72 |
| 210:90:100 | 39.0 | 64.0 | 2.66 | 42.0 | 69.0 | 3.32 |
| 210:90:200 | 37.0 | 70.0 | 1.52 | 39.0 | 66.0 | 1.63 |
| 100:60:40* | 30.0 | 55.0 | - | 33.0 | 56.0 | - |
| CD 5 % | 8.0 | 8.5 | - | 7.0 | 9.0 | - |

* State recommended level for both crops

At Pune (Maharastra), application of 120 kg N/ha ((¼ Vermicompost+ ¾ urea + BNF) to kharif maize recorded significantly higher maize grain (6599 kg / ha) and straw (9901 kg/ha) yields whereas application of 120 kg N/ha to rabi wheat has produced significantly higher grain (5016 kg/ha) and straw (6484 kg/ha) yields in maize – wheat cropping system (Hankare et al., 2005).

**Table 7. Effect of bio compost on wheat (Rabi) – Maize (kharif) cropping system**

| Fertilizer management | Plant height (m) | | Test weight (g) | | Yield (t/ha) | |
|---|---|---|---|---|---|---|
| | Wheat | Maize | Wheat | Maize | Wheat | Maize cob |
| Bio compost 10 t/ha | 0.84 | 2.41 | 31.16 | 209 | 3.24 | 17.73 |
| Biocompost 10 t/ha + 50 % N & P | 0.80 | 2.06 | 30.79 | 203 | 3.12 | 15.29 |
| Biocompost 5t/ha | 0.80 | 2.22 | 32.87 | 212 | 4.00 | 19.79 |
| Biocompost 5 t/ha + 50% N&P | 0.83 | 2.24 | 33.83 | 230 | 4.24 | 20.01 |
| Biocompost 2.5t | 0.76 | 1.76 | 29.94 | 197 | 2.94 | 15.05 |
| Biocompost 2.5 +50% N&P | 0.79 | 2.09 | 32.81 | 201 | 3.76 | 19.32 |

| Fertilizer management | Plant height (m) | | Test weight (g) | | Yield (t/ha) | |
|---|---|---|---|---|---|---|
| | Wheat | Maize | Wheat | Maize | Wheat | Maize cob |
| Biocompost 2.5 + 75% N&P | 0.81 | 2.13 | 32.71 | 221 | 4.00 | 17.96 |
| Control (NPK 120: 60:60 kg/ha) | 0.80 | 2.07 | 32.66 | 211 | 3.48 | 18.81 |
| CD 5% | NS | 0.34 | NS | 22 | 0.21 | 2.22 |

At New Delhi, grain yield of wheat recorded higher when bio compost applied @ 5 t/ha with 50 % N and P fertilizers. Cob yield of maize was higher when biocompost applied @ 5 t/ha + 50 % N and P fertilizers which was at par with Biocompost 2.5 t/ha with 75 % N and P. Use of biocompost 5 t/ha with 50 % N and P improved physical and chemical status of the soil particularly nitrogen and potassium content (Tripathi et al., 2007).

**Table 8. Long term effect of fertilizer management practices on maize – wheat rotation**

| Fertilizer | Mena yield (t/ha) | |
|---|---|---|
| | maize | wheat |
| Control | 0.69 | 1.16 |
| 100 % RDF of N | 1.67 | 2.98 |
| 100 % RDF of NP | 2.16 | 4.12 |
| 50 % RDF of NPK | 1.98 | 3.06 |
| 100 % RDF of NPK | 2.56 | 4.78 |
| 150 % RDF of NPK | 2.76 | 4.99 |
| 100 % RDF of NPK + hand weeding | 2.81 | 4.82 |
| 100 % RDF of NPK + Zn | 2.90 | 4.85 |
| 100 % RDF of NPK + FYM 10 t/ha (before sowing for maize only) | 3.52 | 5.11 |
| 100 % RDF of NPK without sulphur source (through DAP) | 2.65 | 4.85 |
| CD 5 % | 0.098 | 0.096 |

RDF: 150: 32.7: 31.2 kg NPK / ha for maize and wheat, $ZnSO_4$ @ 50 kg/ha before sowing of maize after 3 to 6 years.

At Ludhiana (Punjab), application of FYM along with 100 % RDF is suitable for both maize and wheat crop. Without FYM, 100 % RDF yielded significantly less maize and wheat yield and was non sustainable for wheat yield. For maize, 100% RDF + Zinc was the next best and for wheat it was 150 % RDF alone (Dhaliwal et al., 2007).

**Table 9. Effect of tillage and mulching in maize – wheat cropping system**

| Tillage practice | Maize yield (t/ha) | Wheat yield (t/ha) | Wheat equivalent yield (t/ha) | Productivity efficiency (kg/ha/days) | Net return (Rs/ha) | B:C ratio |
|---|---|---|---|---|---|---|
| Conventional method | 2.00 | 0.54 | 2.32 | 11.19 | 9988 | 1.05 |
| Minimum method | 1.95 | 0.50 | 2.23 | 10.8 | 11187 | 1.41 |
| No tillage method | 1.76 | 0.45 | 2.02 | 9.7 | 9680 | 1.23 |
| Raised bed method | 1.84 | 0.49 | 2.12 | 10.3 | 9740 | 1.15 |
| CD 5% | 0.13 | NS | 0.10 | 1.62 | 648 | 0.07 |
| **Mulching** | | | | | | |
| No mulch | 1.36 | 0.39 | 1.60 | 7.74 | 6105 | 0.82 |
| Straw mulch | 2.00 | 0.53 | 2.31 | 11.16 | 11240 | 1.34 |
| Polythene | 2.23 | 0.56 | 2.54 | 12.28 | 12252 | 1.28 |
| Soil mulch | 1.97 | 0.50 | 2.24 | 10.85 | 10999 | 1.41 |
| CD 5% | 0.15 | NS | 0.12 | 1.70 | 740 | 0.09 |

At Rakh Dhiansar (Jammu), grain yield of maize and wheat was statistically at par in conventional and minimum tillage system and significantly higher than no tillage system. Higher grain yield of maize and wheat were recorded from polythene mulch followed by straw mulch. Minimum tillage in conjunction with polythene mulch or straw mulch was economically profitable and improves production and soil quality in maize-wheat crop sequence under rainfed condition (Sharma et al., 2009).

## Maize – Maize Crop Rotation

At Pantnaga (Uttarakhand), tillage systems had significant influence on grain yields in both the seasons. Maximum grain yield was obtained in fresh raised beds formed without crop residues incorporation, followed by conventional tillage without residue incorporation and least with zero tillage without residue incorporation in both the seasons. Raised fresh bed being statistically at par with Conventional tillage without residue incorporation resulted in significant increase in mean grain yield by 20.8 and 19.6 % (*kharif*) and 22.5 and 15.3 % (*rabi*), respectively, over the zero tillage without residue incorporation. The maximum saving of irrigation water in *kharif* (12.3 %) and *rabi* (10.6 %) was noticed with RFB (Veer Singh et al., 2011).

**Table 10. Effect of tillage practices on yield and yield parameters of maize in maize–maize crop rotation during *kharif* and *rabi* seasons**

| Treatment | Kharif | | | | Rabi | | | |
|---|---|---|---|---|---|---|---|---|
| | Plant height (cm) | Dry root mass (kg/ ha) | ASI (days) | Yield (t/ha) | Plant height (cm) | Dry root mass (kg/ ha) | ASI (days) | Yield (t/ha) |
| Conventional tillage without residue incorporation | 210 | 115 | 4.7 | 4.08 | 170 | 607 | 4.9 | 4.21 |
| Zero tillage with one-third residue incorporation | 210 | 524 | 4.6 | 3.64 | 166 | 561 | 4.7 | 3.87 |
| Zero tillage without residue incorporation | 195 | 519 | 4.7 | 3.41 | 154 | 385 | 4.8 | 3.65 |
| Fresh raised beds formed without crop residues incorporation | 212 | 938 | 4.7 | 4.12 | 172 | 997 | 4.9 | 4.47 |
| Zero tillage with continued reuse of existing raise beds with one-third residue incorporation | 211 | 605 | 4.3 | 3.88 | 169 | 602 | 4.8 | 3.93 |
| CD 5 % | 11 | 224 | **NS** | **0.59** | 7.74 | 208 | NS | 0.61 |

(Veer Singh et al., 2011)

At Pantnagar (Uttarakhad), maximum grain yield was obtained in raised fresh bed (4.12 and 4.47 t/ha) followed by conventional tillage (4.08 and 4.21 t/ha) and least with zero tillage without residue (3.41 and 3.65 t/ha) in maize – maize cropping system (Veer singh et al., 2011).

## Finger millet – Maize Cropping Systems

At Kabre (Nepal), higher maize grain yield (5951 kg/ha) was produced by 50 % basal + 50 % at knee high stage followed by 100 % at knee high stage. Lower maize grain yield (4327 kg/ha) was obtained by 100 % basal. Likewise, higher finger millet grain yield (977 kg/ha) was produced by 1/3 knee high + 1/3 tasseling + 1/3 millet transplanting stage followed by the treatment with 50 % basal + 50 % at knee high stage. Lower millet grain yield (715 kg/ha) was obtained in 100 % basal treatment (Shakya et al., 2002).

**Table 11. Yield and other character of maize and finger millet in maize – finger millet relay system**

| Treatments | Maize | | | | Finger millet | | |
|---|---|---|---|---|---|---|---|
| | Silking (days) | Tasseling (days) | Plant height (cm) | Grain yield (kg/ha) | Plant height (cm) | Straw yield (kg/ha) | Grain yield (kg/ha) |
| 100 % basal | 76 | 72 | 231 | 4327 | 51 | 1383 | 715 |
| 50 % basal + 50 % at knee high stage | 77 | 74 | 235 | 5951 | 55 | 1787 | 887 |
| 1/3 basal + 1/3 knee high + 1/3 tasseling stage | 76 | 73 | 227 | 5226 | 57 | 1671 | 817 |
| 50 % knee high stage + 50 % at tasseling | 77 | 74 | 230 | 5329 | 55 | 1917 | 855 |
| 50 % knee high stage + 50 % millet transplanting stage | 77 | 74 | 221 | 5242 | 55 | 1771 | 906 |
| 1/3 knee high + 1/3 tasseling + 1/3 millet transplanting stage | 77 | 74 | 229 | 4656 | 57 | 1921 | 977 |
| 1/3 knee high + 1/3 MTP stage + 1/3 40 days after MTP stage | 78 | 75 | 226 | 4405 | 59 | 1983 | 861 |
| 100 % at knee high stage | 77 | 73 | 238 | 5680 | 54 | 1654 | 827 |
| CD 5 % | 1.98 | 2.49 | 26.21 | 1010.11 | 7.94 | 698.05 | 321.66 |

(Shakya et al., 2002). RDF: 60:30:30 N, $P_2O_5$, $K_2O$ kg/ha; P and K was applied at the time of maize planting. Nitrogen application was split as per treatment. Thirty-five days old finger millet seedlings were transplanted in each plot when maize was at silking stage.

## Pulses – Maize Cropping Systems

**Table 12. Maize yield and economics under different crop sequence**

| Crop sequence | Maize grain yield (kg/ha) | N uptake (kg/ha) | Net return (Rs/ha) | B:C ratio |
|---|---|---|---|---|
| Urd bean – Maize | 3920 | 92.0 | 7261 | 0.83 |
| Mung bean - Maize | 4208 | 100.1 | 9150 | 1.04 |
| Cow pea (Fodder) – Maize | 4404 | 109.0 | 10363 | 1.40 |
| Cow pea (Grain) – Maize | 3594 | 83.2 | 5392 | 0.63 |
| Maize (Fodder) – Maize | 3476 | 77.8 | 6094 | 0.81 |
| Fallow – Maize | 3947 | 94.3 | 5922 | 1.31 |
| CD 5% | 504 | 8.6 | 1020 | 0.08 |

At Pantnagar (U.P.), maize grown after fodder cowpea gave higher grain yield, nitrogen uptake whereas when grown after maize for fodder showed lower value. Similarly higher net return and B:C ratio were also recorded in cowpea (fodder) – maize crop sequence (Ghosh and Singh, 1996).

## Maize - Groundnut in Alley Cropping System

At Chatha (Jammu), higher grain yield (35.6 q/ha) and stover yield (106 q/ha) of maize were recorded under 50 % NPK through inorganics + 50 % through vermicompost + 25 kg sulphur + 5 kg $ZnSO_4$ + 1 kg boron /ha as compared to control (18.7 and 54.6 q/ha). The residual effect of organics and S, Zn and B was also registered on succeeding gobi sarson crop which produced 12.97 and 37.33 q/ha of seed yield and straw yield respectively (Gupta and Sharma, 2010).

**Table 13. Yield of maize and groundnut in alley cropped system with subabul under irrigated red loam soil (6 m apart)**

| Treatments | Dry matter of subabul (kg/ha) | Yield (q/ha) | |
|---|---|---|---|
| | | Maize | Groundnut |
| Green manuring alone for maize (3 times) and groundnut | 1792 | 16.54 | 25.21 |
| Green manuring + 50 % N to maize and 25 % NPK groundnut | 1659 | 20.18 | 27.64 |
| Green manuring + 50 % N to maize and 50 % NPK groundnut | 1609 | 25.97 | 28.90 |
| Green manuring + 75 % N to maize and 75 % NPK groundnut | 1610 | 32.23 | 30.34 |
| Green manuring + 100 % N to maize and 100 % NPK groundnut | 1595 | 34.11 | 31.74 |
| RDF alone to maize and groundnut | 1698 | 31.34 | 29.07 |
| CD 5 % | 30 | 5.32 | NS |

RDF: 150:75:75 kg NPK/ha (Maize); 25:75:37.5 kg NPK/ha (groundnut)

At Raichur (Karnataka), grain yield of maize and groundnut with RDF and green manuring were higher by 8.84 and 9.18 % respectively over the RDF alone. Grain yield of maize at RDF alone was on par with that of 75 % N with green manuring. Pod yield of groundnut at 50 % RDF with green manuring was on par with that of RDF alone (Palled et al., 2000).

At Coimbatore (Tamil Nadu), lower total weed density and dry weight were recorded with twin wheel hoe weeding at 20 DAS + hand weeding at 40 DAS which was on par with twin wheel hoe weeding at 20 and 40 DAS. Twin wheel hoe weeding at 20 DAS + hand weeding at 40 DAS registered higher weed control efficiency in maize (92.4 and 93.5 per cent) and sunflower (91.9 and 94.7 per cent) during 2007-08 and 2008-09, respectively. Higher yield was obtained with twin wheel hoe weeding at 20 DAS + hand weeding at 40 DAS in maize (58.2 and 61.8 per cent) and sunflower (49.4 and 61.1 per cent) over unweeded control during first and second year, respectively (Bhuvaneswari et al., 2010).

**Table 14. Effect of weed management methods on weed control efficiency, grain and seed yield in maize – sunflower cropping system**

| Treatments | Maize | | | Sunflower | | |
|---|---|---|---|---|---|---|
| | Weed dry weight ($g/m^2$) | Weed control efficiency (%) | Grain yield (kg /ha) | Weed control efficiency (%) | Weed dry weight ($g/m^2$) | Seed yield (kg /ha) |
| Stale seedbed technique | 2.02 (101.88) | 34.6 | 4029 | 35.1 | 1.89 (75.00) | 966 |
| *In situ* green manure cowpea incorporation at 45 DAS | 1.65 (43.17) | 70.7 | 4177 | 58.4 | 1.68 (45.54) | 1002 |
| Crop residue mulch at 5 t /ha | 1.64 (41.26) | 71.9 | 5174 | 64.5 | 1.61 (38.84) | 1226 |
| Twin wheel hoe weeding twice at 20 and 40 DAS | 1.11 (10.79) | 92.3 | 5794 | 93.7 | 0.95 (6.86) | 1395 |
| Hand weeding twice at 20 and 40 DAS | 1.19 (13.61) | 90.7 | 5365 | 90.8 | 1.08 (10.08) | 1309 |
| Stale seedbed technique + *in situ* green manure cowpea incorporation at 45 DAS | 1.58 (35.86) | 75.6 | 4266 | 67.9 | 1.57 (35.10) | 1029 |
| Stale seedbed technique + crop residue mulch at 5 t /ha | 1.55 (33.78) | 77.0 | 5281 | 72.4 | 1.51 (30.23) | 1276 |
| Stale seedbed technique + twin wheel hoe weeding at 40 DAS | 1.30 (18.02) | 87.8 | 4691 | 86.2 | 1.23 (15.14) | 1122 |
| Stale seedbed technique + hand weeding at 40 DAS | 1.16 (12.52) | 91.5 | 4747 | 90.5 | 1.09 (10.42) | 1137 |
| Twin wheel hoe weeding at 20 DAS + hand weeding at 40 DAS | 1.07 (9.62) | 93.5 | 5831 | 94.7 | 0.89 (5.83) | 1410 |
| Unweeded control | 2.17 (147.08) | | 3604 | | 2.05 (109.41) | 875 |
| CD 5 % | 0.03 | - | 424 | - | 0.04 | 85 |

(Bhuvaneswari et al., 2010).

## Maize - Sunflower – Beetroot Cropping System

At Coimbatore (Tamil Nadu), gross income (Rs. 3,09,554/ha) was higher with 100 % WRc with 125 % RDF whereas, higher benefit cost ratio of 4.07 was recorded by drip irrigation at 100 % WRc with soil application of RDF. Drip irrigation at 75 % WRc with 125 % RDF recorded higher net profit per mm of water used (Rs. 274), which was followed by same irrigation regime with 100 % RDF (Ramah et al., 2010).

**Table 15. Economics of Maize - Sunflower – Beetroot cropping system under drip fertigation System**

| Treatment | Gross income (Rs./ ha) | Net income (Rs. /ha) | B:C ratio |
|---|---|---|---|
| Drip irrigation at 75 % WRc + 75 % RDF through drip | 2,45,261 | 1,73,542 | 3.42 |
| Drip irrigation at 75 % WRc +100 % RDF through drip | 2,81,619 | 2,03,620 | 3.61 |
| Drip irrigation at 75 % WRc +125 % RDF through drip | 3,08,382 | 2,24,102 | 3.66 |
| Drip irrigation at 75 % WRc +100 % RDF by soil application | 2,40,771 | 1,76,244 | 3.73 |
| Drip irrigation at 100 % WRc + 75 % RDF through drip | 2,73,040 | 2,00,721 | 3.78 |
| Drip irrigation at 100 % WRc +100 % RDF through drip | 2,92,578 | 2,13,979 | 3.72 |
| Drip irrigation at 100 % WRc +125 % RDF through drip | 3,09,554 | 2,24,674 | 3.65 |
| Drip irrigation at 100 % WRc +100 % RDF by soil application. | 2,65,154 | 2,00,027 | 4.07 |
| Drip irrigation at 125 % WRc + 75 % RDF through drip | 2,41,504 | 1,68,585 | 3.31 |
| Drip irrigation at 125 % WRc +100 % RDF through drip | 2,58,446 | 1,79,247 | 3.26 |
| Drip irrigation at 125 % WRc +125 % RDF through drip | 2,67,408 | 1,81,928 | 3.13 |
| Drip irrigation at 125 % WRc + 100 % RDF by soil application | 2,32,965 | 1,67,238 | 3.54 |
| Surface Irrigation | 2,12,623 | 1,55,883 | 3.75 |

WRc: computed water requirement of crop. (Ramah et al., 2010).

## Maize - Soybean Cropping System

Reddy and Reddy (1999a) found that the available macro nutrients (N, P and K) were significantly increased with the integrated use of manures and fertilizers. Nutrient availability was higher in treatments with vermicompost closely followed by poultry manure, and FYM. Reddy and Reddy (1999b) found increased plant height, number of leaves, grain and fodder yield of maize with increasing levels of nitrogen upto 120 kg N /ha.

Grewal *et al.* (1992) studied the effect of combined use of green manure and nitrogen on the productivity of maize-wheat system. The maize yield was considerably higher owing to combined use of 60 kg N per ha and green manuring (37.4 and 37.5 q/ha for cowpea and dhaincha, respectively) compared with the yield obtained owing to 120 kg N per ha in the absence of green manure (29.69 q/ha). Mean response of wheat to residual effect of green manuring was 5 q per ha. Thus, green manuring augmented total productivity of maize – wheat system (21 q/ha).

### Maize - Mustard Cropping Sequence

Laxminarayana and Munda (2004) reported that maximum reduction of soil fertility was observed in maize-mustard cropping sequence because maze was heavy feeder of different essential nutrients as a sole crop than intercropped with groundnut and soybean. However, inclusion of legume crops in both rice and maize based cropping systems showed slight improvement in the soil fertility than the sole cropping of cereals.

At Samba (Jammu and Kashmir), intercropping system where maize in maize + cowpea preceded by mustard + field pea registered statistical higher grain and stover yield as that recorded under sole maize preceded by mustard based intercropping systems and recorded higher nutrient removal by this treatment combination. Maize in maize + cowpea produced more number of gains/cob and 1000-grain weight than other intercropping systems preceded by mustard based intercropping systems. Higher system productivity in terms of mustard equivalent yield, net returns and B:C ratio was also observed with maize in maize + cowpea preceded by mustard + field pea (Brij Nandan et al., 2013).

**Table 16. Effect of mustard-maize system based intercropping sequences on system productivity, net returns and benefit cost ratio**

| Cropping sequences | System productivity (q/ha) | Net returns (Rs/ha) | B:C ratio |
|---|---|---|---|
| Sole mustard-sole maize | 25.9 | 32720 | 1.37 |
| Sole mustard-maize + mung bean | 32.1 | 45334 | 1.78 |
| Sole mustard-maize + urd bean | 29.6 | 40803 | 1.62 |
| Sole mustard-maize+ cowpea | 33.6 | 50822 | 1.97 |
| Mustard + field pea - sole maize | 34.8 | 44745 | 1.71 |
| Mustard + field pea - maize + mung bean | 46.5 | 69051 | 2.48 |
| Mustard + field pea-maize + urd bean | 43.3 | 61843 | 2.25 |
| Mustard + field pea-maize + cowpea | 47.0 | 71608 | 2.54 |
| Mustard + chickpea - sole maize | 34.0 | 44273 | 1.71 |
| Mustard + chickpea - maize + mung bean | 43.6 | 61775 | 2.24 |
| Mustard + chickpea - maize + urd bean | 42.1 | 59188 | 2.17 |
| Mustard chickpea - maize + cowpea | 45.1 | 63196 | 2.27 |
| Mustard lentil - sole maize | 28.8 | 45520 | 1.80 |
| Mustard + lentil - maize + mung bean | 37.0 | 44464 | 1.65 |
| Mustard + lentil - maize + urd bean | 34.6 | 52075 | 1.95 |
| Mustard + lentil - maize + cowpea | 38.1 | 57669 | 2.11 |

(Brij Nandan et al., 2013).

## Maize – Vegetable Cropping Systems

**Table 17. Effect of irrigation management on potato – maize cropping system**

| Treatments | Potato fresh tuber yield (t/ha) | WDE (kg/mm/ day) | Maize grain yield (kg/ha) | Maize stover yield (t/ha) | Consumptive use (cm) |
|---|---|---|---|---|---|
| **Irrigation methods** | | | | | |
| Conventional furrow irrigation | 17.6 | 340.5 | 1420 | 5780 | 54.2 |
| Alternate furrow irrigation | 15.5 | 338.0 | 1480 | 6020 | 46.7 |
| CD | 0.50 | - | - | - | - |
| **Irrigation levels** | | | | | |
| 60 mm CPE | 17.5 | 319.5 | 1480 | 6110 | 55.0 |
| 80 mm CPE | 16.5 | 333.5 | 1450 | 5910 | 49.7 |
| 100 mm CPE | 15.9 | 364.5 | 1420 | 5690 | 46.7 |
| CD 5% | 1.50 | - | 70 | 240 | - |

At New Delhi, significant improvement in crop height, yield attributes, yield, nutrient uptake and consumptive use of water were with conventional furrow irrigation. Irrigation at 60 mm CPE recorded higher yield levels in potato – maize cropping system (Sunilkumar and Idnani, 2012).

**Table 18. Level of sulphur and FYM on onion –maize cropping system**

| FYM levels (t/ ha) | Winter onion (t/ha) | | Kharif Maize (t/ha) | | Economic yield of onion + maize (t/ ha) | Stover of maize + foliage of onion (t/ ha) |
|---|---|---|---|---|---|---|
| | Maize | Forage | Grain | Stove | | |
| 0 | 27.7 | 0.89 | 3.5 | 7.3 | 31.2 | 8.1 |
| 25 | 31.5 | 0.97 | 3.9 | 7.9 | 35.4 | 8.9 |
| CD 5 % | 1.1 | NS | 0.4 | 1.0 | 1.5 | 0.9 |
| **Sulphur (kg/ ha)** | | | | | | |
| 0 | 28.4 | 0.88 | 3.2 | 7.1 | 31.6 | 8.0 |
| 15 | 30.1 | 0.92 | 3.6 | 7.3 | 33.7 | 8.2 |
| 30 | 30.8 | 0.98 | 4.3 | 8.7 | 35.1 | 9.7 |
| 45 | 29.1 | 0.90 | 3.7 | 7.4 | 32.8 | 8.3 |
| CD 5 % | 1.6 | NS | 0.3 | NS | 1.0 | 0.6 |

At Palampur (Himachal Pradesh), application of FYM 25 t/ha significantly increased bulb and foliage yield of onion. Positive effects of 30 kg S/ha was observed on bulb yield over previous levels but addition of sulphur beyond 30 kg S/ha reduced bulb yield of onion. In maize (residual effect) and also in onion – maize sequence as a whole (direct + residual effects) the same trend was visible (Raina and Jaggi, 2008).

## Alfalfa – Corn Cropping Sequence

Morris *et al.* (1994) reported that little yield response to applied N should be expected on first year maize after alfalfa. Other studies indicated that farmers frequently apply substantially more fertilizer nitrogen to maize after alfalfa than required to attain maximum yields. Over application of fertilizer N reduced the profitability of maize production and can cause nitrate contamination of water supplies.

## References

Ahlawat, I.P.S., B. Gangaiah and O.P.Singh (2005). Effect of fertilizer and stover management on productivity and soil fertility in chickpea *(Cicer arietinum)* - maize *(Zea mays)* cropping system. Indian J. Agri. Sci., 75: 400-403.

Anil Kumar Singh (2010). Soil quality parameters as influenced by management practices in rice-wheat and maize-wheat cropping systems. In: Proc. Of 19th World Congress of Soil Science, Soil Solutions for a Changing World, 1 – 6 August 2010, Brisbane, Australia.

Anonymous (1996). Annual Progress Report - 1996, Directorate of Cropping Systems Research, Modipuram, Meerut, Uttar Pradesh.

Arvind Verma, V. Nepalia, and , P.C. Kanthaliya (2006). Effect of integrated nutrient supply on growth, yield and nutrient uptake by maize (*Zea mays* L.)-wheat (*Triticum aestivum*) cropping system. *Indian J. Agron.,* 51(1): 3-6.

Ashok Kumar and Shiva dhar (2010). Evaluation of organic and inorganic sources of nutrients in maize and their residual effect on wheat under different fertility levels. Indian J. Agric. Sci., 80 (5): 364 – 371.

Ashok Kumar, R.C., Guatam, Ranbir Singh and K.S. Rana (2005). Growth, yield and economics of maize (*Zea mays* L.) – wheat (*Triticum aestivum*) cropping sequence as influenced by integrated nutrient management. *Indian J. Agric. Sci.*, 75 (1): 709-711.

Asok kumar and Shiva dhar (2010). Evaluation of organic and inorganic sources of nutrients in maize and their residucal effect on wheat under different fertility levels. Indian J. Agric. Sci., 80 (5): 364 – 371.

Badiyala, D. and S.P. Verma (1991). Integrated nitrogen management in maize (Zea mays L.) + soybean (Glycine max) – wheat cropping sequence under mid hills of Himachal Pradesh. Indian Journal of Agronomy, 36: 496 - 501.

Balayan, J.S. and Ashok kumar (2001). Fertilizer management in maize - wheat cropping system. Indian J. Agron., 30 (4): 257 - 258.

Balyan, J.K. and L.K.Idnani (2000). Fertilizers management in maize (*Zea mays* L.) – wheat (*Triticum aestivum*) sequence. J. Agron., 45: 648 - 652.

Beets, W.C. (1994). Multiple cropping of maize and soybean under a higher level of crop management. Netherlands J. Agric. Sci., 25: 95 – 102.

Begum, N., M.M. Ullah, M.F. Haq, M.A.R. Khatuna and S. Yasmin (1999). Performance of potato intercropped with maize. Bangladash J. Sci. Ind. Res., 34: 183-187.

Behera, B. and P.C. Senapati (2001). Maize + cowpea – toria boosting tribal economy in Eastern Ghat of Orissa. Indian Farming, 51 (9): 7 – 8.

Behera, B., G.S. Singh, P.C. Pradhan and P.C. Senapati (1998 a). Effect of intercropping and residue incorporation on maize + cowpea - niger cropping system in Eastern Ghats. Indian J. Soil Consn., 26(2): 133 - 134.

Bhajan Singh and S.P.S. Brar (1994). Effect of organic manures and nitrogen on grain yield and soil properties in a maize-wheat rotation. J. Punjab Agric. Res., 22: 243-252.

Bhandari, A.L., S.S. Walia and S. Tarlok (2000). Production sustainability of maize – wheat system in a Typic Ustipsamment soil as influenced by integrated nutrient sources. In: Proceeding of International Conference on Managing the Natural Resources for sustainable crop production in the 21$^{st}$ century held at IARI, New Delhi during 14-17 February, Volume 3, P: 789 – 790.

Bharat bhusan, V., B.S. Sidh, S. Singh and N.B. Walkar (2013). Effect of different mulches on soil erosion and carryover of residual soil moisture for sowing of rabi crop in maize – wheat cropping sequence in rainfed Shivaliks of Punjab, Indian J. Soil Conservation, 41(2): 136 -140.

Bhatnagar, G.S., G.L. Sharma and P.G. Chaplot (1994). Long term effect of continuous maize-wheat cropping system and manuring on soil fertility and stability in crop yield. *Indian Journal of Agricultural Science*, 64 (12) : 821-823.

Bhuvaneswari, J., C. Chinnusamy and S.P. Sangeetha (2010). Evaluation of non-chemical methods of weed management in organically grown maize-sunflower cropping system. Madras Agric. J., 97 (7-9): 242-244.

Bodake, P.S., Y.J.Patil, R.B. Hile and H.M. Patil (2006). Response of N, P and K application on productivity and soil fertility in maize – wheat cropping system. J. Farming systems Res. Dev., 12 (1&2): 107 – 108.

Bodale, P.S. and D.S. Rana (2009). Evaluation of Jetropha and castor cake as a source of nutrients and soil amendment in spring sunflower - maize sequence. Indian J. Agron., 54(3): 284 - 290.

Brar, B.S., N.S. Dhillon, D.S. Benipal, Jagadeep Singh and M.S. Mavi (2006). Balanced use of inorganic fertilizers and FYM for higher crop yields and better soil health in maize – wheat cropping system. J. Res. Punjab Agric. Univ., 43(2): 104-107.

Dhaliwal, B.S, M.S. Virk and B.S. Brar (2007). Effect of long term fertilizer use on yield and yield system ability in maize – wheat rotation. Indian J. Agric. Sci., 72(2): 79 – 83.

Dinesh Badiyala and Pankaj Chopra (2011). Effect of Zinc and FYM on productivity and nutrient availability in maize - linseed cropping sequence. Indian J. Agron., 56 (2): 88-91.

Duraiswamy, P. (1992). Efficacy of fertilizer nitrogen with and without coir pith and biofertilizers on role and inter cropped sorghum – maize – soybean cropping sequences. *Ph.D Thesis,* Tamil Nadu Agricultural University, Coimbatore.

Gable, D.B., K.J. Kubde, J.R. Katore, A.V. Fiske and M.R. Deshmukh (2008). Effect of integrated nutrient management on growth and yield of maize-chickpea cropping system. *J. Soils and Crops,* 18(2): 392-397.

Gahukar, T. S., R. Veeraraghavaniah, C. Pulla Rao and G. V. Lakshmi (2011). Effect of agro technique on yield, yield attributes and phosphorus uptake of maize in mining soil reserve phosphorus. Andhra Agric. J., 58(3): 279 - 282.

Gangwar, B., S.K. Duhoon and D.K. Pandy (2006). Diversification of maize based cropping systems. Indian Farming, 56 (6): 20 – 24.

Gassity P.D. and C.G. Hermenegildo (1990). The development of rice-corn rotations in tropical lowland environments: A systems research approach. Food and Fertilizer Technology Center. Taipei City, Rep. of China on Taiwan. Extension Bulletin No.318. 10 pp.

Ghosh, B.N., N.K. Sharma, Dogra and K.S. Dadhwal (2012). Effect of integrated organic input management on resource we efficiency in maize wheat cropping system in stoping lands of the north west Himalayas. Indian J. Soil conservation, 40(1): 84 – 89.

Ghosh, B.N., O.P.S. Khola and K.S. Dashwal (2011). Potassium and crop rotation impact on productivity and resource conservation in maize and cowpea based cropping systems. Indian J. Soil Cons., 39 (2): 136 -141.

Ghosh, P.K. and N.P. Singh (1996). Production potential of summer legumes – maize sequence under varying levels of nitrogen. Indian J. Agron., 41(4): 525 – 528.

Gopal Reddy, B. and M. Suryanarayana Reddy (1998). Effect of organic manures and nitrogen levels on soil available nutrient status in maize – soybean cropping system. Journal of the Indian Society of Soil Science, 46 (3): 474-476.

Gupta, R.K., G.C. Fernando, S. Paliwal, S.V.R.K, Prabhakar and A. Srivastava (2002). Maize in the Rice-Wheat rotation of Indo-Gangetic Plains. Proceedings of the 8th Asian Regional Maize Workshop, Bangkok, Thailand: August 5-8, 2002. P: 207-211.

Hadda, M.S. and S. Arora (2006). Soil and nutrient management practices for sustaining crop yields under maiz - wheat cropping sequence in sub-mountain Punjab, India. Soil and Environment, 25 (1): 1 - 5.

Hanare, R.H., A.S. Jadhav, H.T. Patil and P.P. Pawar (2005). Studies on integrated N management in maize – wheat cropping system. J, Maharashtra Agric. Univ., 30(1): 122 – 124.

Hanegave et al. (2011). Effect of tillage on maize-wheat system. Indian J. Agron., 56 (2): 82-87.

Hari Ram, Y. Singh, K.S. Saini, D.S. Kler, J. Timsina and E.J. Humphreys (2012). Agronomic and economic evaluation of permanent raised beds, no tillage and straw Mulching for an irrigated maize-wheat system in Northwest India. Expl Agric., 48 (1): 21–38.

Hartkamp, A.D., J.W. White, W.A.H. Rossing, M.K. Ittersum, E.J. Bakker and R Rabbinge (2004). Regional application of a cropping systems simulation model: crop residue retention in maize production systems of Jalisco, Mexico. Agric. Systems, 82 (2): 117-138.

Hile, R.B., H.M.Patil, Y.J.Patil and S.S.Bhosale (2007). Effect of N, P and K on productivity and soil fertility in maize (*Zea mays* L.) – wheat (*Triticum aestivum*) cropping system. International J. Agric. Sci., 3(2): 205-207.

Hosamani, M.H., D.P. Biradar and A.K.Singh (2001). Response of nutrients to maize - Bengal gram cropping system in the upper Krishna Project command in Karnatak. Indian J. Agron., 30 (4): 155 - 156.

Hussain Nazim, Imran Haider Shamsi, Sherin Khan, Habib Akbar and Wajid Ali Shah (2003). Effect of legume intercrops and nitrogen levels on the yield performance of maize. *Asian J. Pl. Sci.* 2 (2) : 242-246.

Jamriska, P. (1996). Effect of bio compost on the yield of selected crops. Vedecke prace Vyskumneho Ustavu Ras hinnej vgroby V piest any, 28: 13 – 20.

Jat, M L, A. Srivastava, S.K. Sharma, R.K. Gupta, P.H. Zaidi, H.K. Rai and G. Srinivasan (2005). Evaluation of Maize - wheat cropping system under double-no-till practice in Indo-Gangetic Plains of India. *Japan Association for International Collaboration of Agriculture and Forestry,* 28(1): 25–42.

Jat, M.L., A. Srivastva, S.K. Sharma, R. K. Gupta, P.H. Zaidi and G. Srinivasan (2005). Evaluation of maize-wheat cropping system under double-no-till practice in Indo-gangetic plains of India. *Paper presented in the 9th Asian Regional Maize Workshop, Beijing, China, September 6–9, 2005.*

Jat, M.L, R.K. Gupta, S.K. Sharma, M.S. Gill, S. Dass and R.P. Singh (2008). Evaluating conservation tillage systems under maize-wheat rotation in Indo-Gangatic plains of India (in): Proceedings of the 10th Asian Regional Maize Workshop, 20 - 23 October 2008, Makassar, Indonesia, p: 7.

Jat, N.K., A. Kumar, S.R. Meena, D.S. Rana, B.P. Meena and K.S. Rana (2012). Influence of INM on the productivity, quality and soil health of maize - wheat cropping system. Indian J. Agronomy, 57(4): 327 -332.

Jat, S.L., C.M. Parihar, A.K. Singh, M.L Jat and R.K. Jat (2010). Carbon sustainability and productivity of maize based cropping system under Conservation Agriculture practices in Indo- Gangetic plains. Directorate of Maize Research (DMR) farm, New Delhi. P: 1-5.

Juang, T.C. (1993). Combined use of organic and chemical fertilizers. II Effect of combined application of compost and chemical fertilizers on soil fertility and crop yield under a rice - corn rotation. *Extension Bulletin ASPAC, Food and Fertilizer Technology Center*, July, 1993.

Kedar Prasad, C.P. Verma, R.N. Verma and Ram Pyare (2005). Effect of FYM, gypsum and fertility levels on nutrient uptake by wheat crop in maize – wheat system. Crop Res., 29 (1): 28-33.

Kher, D. and R.S. Minhas (1991), Effect of fertilizer application, manuring and liming on the yield and micronutrients uptake by maize and wheat under mid-hilly conditions of Himachal Pradesh. *Crop Research*, 4 (1): 165-168.

Khola, O.P.S., N.K. Sharma, K.S. Dadhwal and A.K. Khullar (2002). Evaluation of maize based growing systems under varying fertility levels in western Himalayan region. Extended summaries in Second International Agronomy Congress on Balancing Food and Environmental Security- a continuing challenge. Vol. 1, November 26-30, New Delhi, pp. 201-202.

Kler, D.S. and S.S. Walia (2006). Organic, integrated and chemical farming in wheat under maize – wheat cropping system. Indian J. Agron., 51 (1): 6 – 9.

Kumar Ashok, R.C. Gautam, Singh Ranbir and KS. Rana (2005). Growth, yield and economics of maize-wheat cropping sequence as influenced by integrated nutrient management of New Delhi. *Indian. J. Agric. Sci.*, 75(1): 709-711.

Kumar, A. (2008). Direct and residual effect of nutrient management in maize - wheat cropping system. Indian J. Agron., 53 (1): 37 - 41.

Kumar, V. and I.P.S. Ahlawat (2004). Carry over effect of biofertilizers and N applied to wheat and direct applied N in maize in wheat – maize cropping system. Indian J. Agron., 49 (4): 233 – 236.

Kumaresan, M. (2002). Crop residue management in maize - sunflower - cowpea (fodder) cropping system as influenced by phosphorus management practices. Madras Agric. J., 89 (4-6): 340-342.

Kumaresan, M. and M. Rajavel (2002). Soil available nutrients and uptake as influenced by sources and levels of phosphorus in maize – sunflower cropping system. Madras Agric. J., 89 (4- 6): 266 - 268.

Kumaresan, M., V.S. Shanmugasundaram and T.N. Balasubramanian (2003). Influence of phosphorus management on soil available phosphorus and uptake of NPK in a maize-sunflower–cowpea cropping system. *Indian J. Agric. Res.,* 37(1): 34-38.

Kumaresan, M., V.S. Shanmugasundharam, T.N. Balasubramanian and M. Rajavel (2002). Yield and economics of maize – sunflower – fodder cowpea cropping system as influenced by phosphorus management practices. Madras Agric. J., 89 (4- 6): 269 - 271.

Kumpawat, B. S. (2004). Integrated nutrient management for maize (*Zea mays*)–Indian mustard (*Brassica juncea*) cropping system. Indian J. Agron., 49: 18-21.

Kumpawat, B.S. and M.L. Jat (2005). Profitable and energy efficiency integrated nutrient management practices for maize – mustard cropping system in southern Rajasthan. Curr. Agric., 29 (1-2): 97-102.

Kumpawat, B.S. and S.S. Rathore (1995). Response of maize – wheat cropping sequence to fertilizer application. Indian J. Agron., 40: 26 -29.

Lan, F. W. and Z. Quyan (2006). Effect of long-term fertilization on soil fertility of wheat (*Triticum aestivum*) – corn (*Zea mays* L.) double cropping system in China. Indian J. Agric. Sci., 76(9): 521-525.

Lantin, M.M. (1986). Development of corn varieties for rice-based farming systems. *In* Report of the Upland Crops Varietal Improvement Monitoring Tour, May 26-June 4, 1986. IRRI, Los Banos, Laguna, Philippines.

Latha, M.R., P.Savithiri, R.Indirani and S. Kamaraj (2002). Residual effect of Zinc enriched organic manures of micronutrients in maize - sunflower cropping system. Madras Agric. J., 89 (1-3): 120 - 121.

Laxminarayana, K. and G.C. Munda (2004). Performance of rice and maize based cropping systems under mid-hills of Mizoram. Indian J. Agron., 49 (4): 230 -232.

Limon-Ortega, A., Sayre, K. D. and Francis, C. A. (2000).Wheat and maize yields in response to straw management and nitrogen under a bed planting system. *Agronomy Journal* 92: 295–302.

Liu, X.J., X.T. Ju, F.S. Zhang, J.R. Pan and P. Christie (2003). Nitrogen dynamics and budgets in a winter wheat-maize cropping system in the North China Plain. Field Crop. Res., 83: 111-124.

Ma, Y.Q. and Q.H. Han (1995). Effect of wheat straw mulching on the growth, development and yield of maize. *Acta Agricultre Boreali Sinica* 10: 106–110.

Mahua Banerjee, R.K. Rai, Debtanu Maiti and Shiva Dhas (2006). Impact of chemical fertilizers and biofertilizers on yield and fertility build-up of soil under maize

(*Zea mays* L.) – wheat (*Triticum aestivum*) cropping system. Indian J. Agric. Sci., 76 (12): 753-755.

Mallikarjuna, G.B., H. Chandrappa, M. Basavaraja and B.N. Prabhakaran (2004). Effect of nitrogen substitution on rice – maize sequence. Indian J. Agric. Sci., 74 (1): 36 – 37.

Meena, K.N., Ashok Kumar, D.S. Rana and M.C. Meena (2011). Productivity and nutrient uptake of maize - wheat cropping system under different bio-source and nitrogen levels. Indian J. Agron., 56 (3): 182 - 188.

Minhas, R.S. and Gulab Singh (1998). Utilization of weed plants alone or in conjugation with fertilizers in wheat – maize cropping sequence on a Hapludalf in hilly region of Himachal Pradesh. J. Indian Soc. Soil Sci., 46(1): 50-54.

Mishra, V.K. (2000). Water expenses and nutrient use efficiency of wheat and winter maize as influenced by integrated nutrient management. *Agropedology*, 10 (1): 1-5.

Munda, G.C. and M. Islam (2007). Effect of IPNS on productivity, profitability and economic feasibility of maize based cropping system on farmers' field. Indian J. Agric. Res., 41(3): 200 - 204.

Munda, G.C., Mokidul Islam and L.K. Nath (2011). Integrated nutrient management approach for enhancing Productivity and economics of maize (Zea mays L.) – toria (Brassica campestris L.) cropping system. Agric. Sci. Digest., 31 (3): 188 – 192.

Negi, S.C., K.K. Singh and R.C. Thakur (1988). Response of maize-wheat cropping sequence to phosphorus and farm yard manure. *Ind. J. Agron.,* 33(3): 270-273.

Negi, S.C., K.K. Singh and R.C., Thakur (1992). Effect of phosphorus with and without FYM on yield and uptake of N, P and protein content in wheat – maize sequence. Himachal Journal of Agricultural Research, 18 (1-2): 5-8.

Palled, Y.B., B.K. Desai and A.S. Prabhakar (2000). Integrated nutrient management in alley cropped maize – groundnut system with subabul. Indian J. Agron., 45 (3): 520-523.

Pandey A. K., Singh Prem, Ved Prakash, R.D. Singh and V.S. Chauhan (1999). Direct and Residual effect of weed control measures in Maize (*Zea mays*) and Wheat (*Triticum aestivum*) cropping system under Mid-Hill conditions of N-W Himalayas. Indian J. Weed Sci., 31 (3&4): 204- 209.

Pandey, A.K., N.N. Singh, and H.S. Gupta (2002). Maize based cropping systems in the hills of India. p 142-147. *In* N.P. Rajbhandari, J.K. Ransom, K. Adhikari, and A.F.E.

Palmer (eds) Sustainable Maize Production Systems for Nepal. Proceedings of a Maize Symposium, 3-5 Dec 2001, Kathmandu, Nepal. CIMMYT and Nepal Agricultural Research Council, Nepal.

Pandey, S.P. (2000). Current soil fertility management recommendations, constraints and opportunities for maize based cropping systems in the hills of Nepal. In: Tripathi, B.P., Rajbhandari, N.P., Ransom, J.K. (Eds.), Improved Soil Fertility Management for Sustainable Maize Production – Proceedings of a Working Group Meeting of the Hill Maize Research Project. NARC and CIMMYT, Kathmandu, pp. 55-60.

Panwar, A.S. (2008). Effect of nutrient management in maize (*Zea mays*) - mustard (*Brassica compestris var toria*) cropping system in mid-hills altitude. Indian J. Agric. Sci., 78 (1): 27-31.

Pathak, S.K., B.Singh and S.N.Singh (2002).Effect of integrated nutrient management on growth, yield and economics in maize-wheat cropping system. Indian J. Agron., 47 (3): 325-332.

Pathak, S.K., S.B. Singh, R.N. Jha and R.P.Sharma (2005). Effect of nutrient management on nutrient uptake and changes in soil fertility in maize – wheat cropping system. Indian J. Agron., 50(2): 89 – 93.

Pradeep, H.M. and N.S. Hebsur (2008). Studies on Sugarcane Based Industrial Wastes on Maize Chickpea Cropping Sequence in Vertisols of North Karnataka. Karnataka J. Agric. Sci., 21(4) : 611.

Prathyusha, C., S. Hemalatha, V. Praveen Rao and G. Jayasree (2012). Effect of nitrogen on yield attributes and yield of specialty corn under pongamia + maize agri - silvi system. J. Res. ANGRAU, 40(4): 124 -127.

Prem singh, A.K. Singh, J.P. Singh, Mahander Singh and P.P. Mishra (2001). N substitution through FYM in various cropping systems under irrigated conditions. Indian J. Agron., 30 (4): 159 - 161.

Qiao, Y., S. Miao, X. Han, M. You, X. Zhu and W.R. Horwath (2014). The effect of fertilizer practices on N balance and global warming potential of maize–soybean–wheat rotations in Northeastern China. Field Crops Res., 159: 34-39.

Raina, S.K. and R.C. Jaggi (2008). Effect of sulphur in presence and absence of FYM on onion under onion –maize cropping sequence. Indian J. Agric. Sci., 78(8): 659 -662.

Rajkumara, S., Y. R. Aladakatti, P. L. Patil and A.B. Khot (2009). Nitrogen and phosphorus substitution through organics in maize-chickpea sequence under irrigated conditions Karnataka J. Agric. Sci., 22 (1): 15-17.

Rajput, S.S., M.S. Shaktawat and S.K. Intodia (2009). Residual effect of Udaipur rock phosphate sources and PYM on productivity and nutrient uptake by succeeding maize after wheat. India J. Agric. Sci., 77(3): 145 – 149.

Ram, S., K. Prasad and G.C. Sharma (2000). Production potential and economics of maize and pigeon pea based cropping systems. Indian J. Agron., 45 (2): 236-241.

Ramah, K. (2008). Study on drip irrigation in maize (Zea mays L) based cropping system. Ph.D., Thesis, TNAU, Coimbatore.

Ramah, K., P. Santhi and K. Ponnuswamy (2010). Economic viability of drip fertigation in maize (*Zea mays* L.) based cropping system. Madras Agric. J., 97 (1-3): 12-16.

Ramah, K., P.Santhi and K.Ponnuswamy (2009). Economic feasibility of drip fertigation in maize – sunflower – beet root cropping system. International J. Tropical Agric., 27 (1-2): 131–136.

Ramesh, P., N.R. Panwar, A.B. Singh and S. Ramana (2008). Effect of organic manures on productivity, nutrient uptake and soil fertility of maize (*Zea mays* L.) – linseed (*Linum usitatissimum*) cropping system. *Indian. J. Agric. Sci.,* 78(4): 351-354.

Reddy, A.K. and G. Bheemaiah (1991), Effect of cropping system and soil types on yield of maize (*Zea mays* L.). *Indian. J. Agron*., 36(2): 139-142.

Reddy, B.G and M.S. Reddy (1999). Effect of integrated nutrient management on soil available micro-nutrient in maize-soybean cropping system. J. Res. ANGRAU, 27(3): 24-28.

Reddy, B.G. and M.S. Reddy (1999). Available macronutrient status in soil as influenced by integrated nutrient management in maize-soybean cropping system. *Journal of Research ANGRAU*, 27 (4): 55-62.

Reddy, M.D., K. Aril Kumar and K.F. Rao (1996). Production efficiency of rainy season maize based cropping systems in the command areas of A.P. Indian. J. Agric. Sci., 66(5): 276 - 279.

Sahoo, S.C. and M.M. Panda (1999). Driect and residual effect of fertilizer and manures on maize – mustard cropping system. Crop Res., 17 (1): 118 – 120.

Saikia, T.P. and J. Pandey (1999). Weed shift in maize – chickpea cropping system. Indian J. Agron., 44 (2): 246-249.

Saikia, T.P., J.Pandey and G.Kulshrestha (2000). Investigation on residue of atrazine and fluchloralin in maize – chickpea and maize – Indian mustard croppig sequence. Indian J. Agron., 45: 653 – 657.

Sarkar, A.J.K., S. Lal, and G. Dev (1998). Balanced fertilizer use and liming to sustain high yield in corn-wheat rotation on arid soil. *Better Crop Inter.,* 12: 17-19.

Sathish, A., A.Y.Hugar, N.Kusagur and H.Chandrappa (2011). Effect of INM on soil fertility status and productivity of rice – maize sequence under permanent plot experiment. Indian J. Agric. Res., 45 (4): 320 – 325.

Sathish, H., Kusagur, Chandrappa and Sannathimmappa (2010). Effect of INM practices in rice – maize cropping system. Crop Res., 41 (1-3): 1-6.

Scopel, E., F. Antonie, Guerra, E. Chavez and C.Marc (2005). Impact of direct sowing mulch based cropping systems on soil carbon, soil erosion and maize yield. Agron. Sustain. Dev., 25(4): 425 – 432.

Seshadri Reddy, S., B. Shivaraj, V.C. Reddy and G. Ananda (2005). Direct effect of fertilizers and residual effect of organic manures on yield and economics of maize (*Zea mays* L.) in groundnut – maize cropping system. Crop Res., 30(1): 1-5.

Shakya, P. R., B.N. Adhikari, R. P. Upreti and S. Thapa (2002). Synchronization of Nitrogen Requirements in Maize/Millet Relay Cropping System in Mid-Hill Condition of Nepal. Proceedings of the 8th Asian Regional Maize Workshop, Bangkok, Thailand: August 5-8, 2002. P: 273 - 281.

Shanmugasundaram, R. and P. Savithri (2006). Effect of levels and frequency of $ZnSO_4$ application on its content and uptake by maize and sunflower in their sequential cropping system. Agric. Sci. Digest, 26(1): 31-34.

Sharma, A. R. and U.K. Behera (2009). Recycling of legume residues for nitrogen economy and higher productivity in maize-wheat cropping system. *Nutr. Cycl. Agroecosyst*, 83: 197-210.

Sharma, A.R., R. Singh and S.K. Dhyani (2005). Conservation tillage and mulching for optimizing productivity in maize – wheat cropping system in the outer western Himalayan region. Indian J. Soil Conservation, 33 (1): 35 – 43.

Sharma, M.P. and J.P. Gupta (1998). Effect of organic materials on grain yield and soil properties in maize – wheat cropping systems. Indian J. Agric. Sci., 68: 715 - 717.

Sharma, P., V. Abrol, C.R. Maruthi sankar and B. Singh (2009). Influence of tillage practices and mulching options on productivity, economics and soil physical properties of maize–wheat system. Indian J. Agric. Sci., 79(11): 865 – 870.

Sharma, R.D and J.R. Singh (1998). Direct and residual effects of liquid APP as a Phosphorus source in wheat – maize cropping system in an entisol of western UP. Indian J. Agron., 43(2), 223 – 230.

Shiray. Y.S., R.P. Singh and B. G. Shiv kumar (2002). Effect of nitrogen on yield attributes, yield and quality of maize in different cropping systems. Indian J. Agric. Sci., 72: 161 – 163.

Shivay, Y.S. and R.P. Singh (2000). Growth, yield attributes, yield and nitrogen uptake of maize (*Zea mays* L.) as influenced by cropping systems and nitrogen levels. *Ann. Agric. Res.,* 21(4): 494 - 498.

Singh, B.B. and H.A. Adjeigbe (2002). Improving Cowpea Cereals-Based Cropping Systems in the Dry Savannas of West Africa. In: Challenges and Opportunities for Enhancing Sustainable Cowpea Production, Fatokun, A., S.A.Tarawali, B.B. Singh, P.M. Kormawa and M.Tamo (Eds.). IITA, Ibadan, Nigeria, pp: 276 -284.

Singh, D., D.S. Rana, R.N. Pandey and K.Kumar (1999). Response of fertilizers in maize–wheat – cowpea cropping system. Indian J. Agron., 44 (2): 242-245.

Singh, K. and U.C. Sharma (1989). Nitrogen need of maize - potato cropping sequence in acid soils of Megalaya. Indian J. Agric. Sic., 59: 157-161.

Singh, M.K., S.K. Pal, R. Thakur and U.N. Verma (1995). Integrated nutrient energy management in maize - wheat cropping system. J. Farming systems Res. Dev., 1 (1&2): 31 - 40.

Singh, S. and A.K.Sarkar (2001). Balanced use of major nutrients for sustaining higher production of maize – wheat cropping systems in acidic soils of Jharkhand. Indian J. Agron., 46 (4): 605-610.

Singhal, S.K., V.K. Sharma and R.D. Singh (2012). Effect of inorganic and biofertilizers (VAM and PSB) on yield of maize and wheat cropping sequence and soild fertility. Indian J. Agric. Res., 46(2): 167 – 172.

Srinivas , A. and M. Srinivasa raju (1998). Rainy season maize – based cropping systems in Andhra Pradesh. Indian J. Agron., 43(4): 576 – 579.

Sunil kumar, V. And L.K. Idnani (2012). Crop water use, nutrient content and uptake as influenced by irrigation and nitrogen management in potato – maize cropping system. Indian J. Agric. Sci., 82 (12): 1032 – 1038.

Surendra Singh and A.K. Sarkar (2001). Balanced use of major nutrients for sustaining higher productivity of maize *(Zea mays)* - wheat cropping system in acidic soils of Jharkhand. Indian J. Agron., 46: 605-610.

Suri, V.K. and T.S.Verma (1999). Targetted yield concept for efficient and economic fertilizer use in maize – wheat cropping system and build up of native fertility in a Typic Hapludalf. J. Indian Soc. Soil Sci., 47(1): 67-72.

Sutaliya, K. and R.N. Singh (2005). Effect of planting time, fertility level and PSB on growth, yield and yield attributes of winter maize under rice - maize cropping system. Indian J. Agron., 50(3): 173 -175.

Tejeswara Rao, K. (2008). Studies on Organic Farming in Maize - Sunflower - Greengram Cropping System. M.Sc (Ag) Thesis, Acharya N.G. Ranga Agricultural University, Rajendranagar, Hyderabad.

Tiwari, R.C., P.K. Sharma and S.K. Khandelwal (2004). Effect of green manuring through Sesbania cannabina and S. rostrata and N application through urea to maize – wheat cropping system. Indian J. Agron., 49 (1): 15 – 21.

Tobias, E. H., Y. Shanchao, R. Schulz, X.Chen, F. Zhang and T.Müller (2014). Nitrogen dynamics, apparent mineralization and balance calculations in a maize – wheat double cropping system of the North China Plain.Field Crops Res., 160: 22–30.

Totawat, K.L., L.L. Somani, R. Singh and G. Singh (2001). INM in maize - wheat cropping sequence on typic haplustalfs of Western India. Annals of Arid zone, 40: 439 - 444.

Tripathi, B.P. (2002). Soil fertility management in upland maize-based cropping system in the hills of Nepal. Proceedings of the 8th Asian Regional Maize Workshop, Bangkok, Thailand: August 5-8, 2002. P: 282 - 295.

Veer Singh, Ajaya Srivastava, R K Singh and U S Savita (2011). Effect of tillage practices and residue management on soil quality and crop yield under maize (Zea mays) - based cropping system in Mollisol. Indian J. Agric. Sci., 81 (11): 1019–25.

Verma, A., V. Nepalia and P.C. Kanthaliya (2006). Effect of integrated nutrient supply on growth, yield and nutrient uptake by maize – wheat cropping system. Indian J. Agron., 51 (1): 3 – 6.

Verma, C.P., K.L. Kedar Prasad and R.N.Verma (2005). Production potential and economics of maize – wheat cropping system under different soil conditioners and fertility levels. Crop Res., 29(1): 23-27.

Verma, C.P., Kedar Prasad, H.V. Singh and R.N. Verma (2003). Effect of soil conditioners and fertilizers on yield and economics of maize (*Zea mays* L.) in maize-wheat sequence. *Crop Res*., 25(3): 449-453.

Virendra kumar and I.P.S. Ahlawat (2004). Carry over effect of biofertilizers and nitrogen applied to wheat and direct applied N in maize in wheat – maize cropping system. Indian J. Agron., 49 (4): 233 -236.

Virendra Kumar and I.P.S. Ahlawat (2006). Effect of biofertilizers and nitrogen on wheat (*Triticum aestivum*) and their after effects on succeeding maize (*Zea mays* L.) in wheat-maize cropping system. Indian J. Agric. Sci., 76(8): 465-468.

Walia, S.S. and D.S. Kler (2010). Effect of organic and inorganic sources of nutrition on growth, macro and micro nutrient uptake in maize under maize – wheat sequence. Indian J. Ecol., 37(1): 27-29.

Yadav, M.P., J. Rai, S.P. Kushwaha and G.K. Singh (2005). Production potential and economic analysis of various cropping systems for Central plains zone of Uttar Pradesh. Indian J. Agron., 50 (2): 83 -85.

# 16

# Farm Implements for Maize Cultivation

### Seedbed Preparation

After the harvest of preceding crop, the field should be ploughed deep with a mould board plough or disk plough. The ploughed field should be left exposed to the sun. As the time for sowing approaches the field is irrigated and thereafter tilled 2 -3 times using a disk harrow. A cultivator with wooden plank will make the surface smoother.

## Low draft chisel plough

The implement could be used for deep tillage upto a depth of 40 cm for opening hard soil pan. This is tractor mounted implement (35 to 45 hp). Overall dimensions are 450 x 940 x 1250 mm and weight is 42 kg. The capacity of the implement is 1.4 ha /day. Cost of the implement is Rs. 7,500 whereas cost of operation Rs. 700/ ha.

### Ridger or ridge maker

Ridger is used to form ridges required for maize sowing in well tilled soils. The ridger can also be used for earthing up, making field furrows and channels. A ridger, commonly called as a ridger plough, consists of a single share and double mould board attach to steel or wooden beam. Normally, there is a provision to vary width of mould board. A wider mould board will results in flatter ridges as compared to narrower mould board. The ridgers are available in bullock drawn and tractor drawn units. The bullock drawn units have a single furrow bottom and tractor drawn units have 2 - 5 furrow bottoms. Depending on its design and adjustments, a ridger can form flat or conical types of ridges. Flat ridges are used in areas having less water. So that water remains on the flat portion for a longer duration and is available to the crop. Conical ridges are used in areas having excess of water to allow its quick drainage.

## Ridge packer

The ridges need to be compacted before planting for stability. This operation is done by ridge packer. The equipment is made of M. S. sheet formed in trapezoidal shape. It has three packing sections which pack the ridges and make it compact to make moisture available for longer duration. The three point linkage is provided to attach the implement with tractor. This is very simple in construction and can be made at local workshop. A small peg is provided on each packer to make continuous groove on the ridge top for dibbling seeds. The seed are sown manually and covered with soil.

## Dibbling/ planting machine

**Rotary dibbler:** The dibbler is operated by pushing it manually into the soil in the direction of travel. The jaws enter into soil and automatic dropping of seed takes place. Well - prepared seedbed is required for its operation. This machine is developed at CIA, Bhopal 1981 - 83. It saves 57 % labour, operating time and 58 % on cost operation compared to conventional method of hand dibbling behind country plough. It places the maize seed at 2.0 cm depth with operating speed 2.6 km/ hour having field capacity 0.042 ha/ hour.

**Tractor mounted seed fertilizer drill - cum-planter**: It is a planting attachment to the commercially available tractor operated seed-cum-fertilizer drills suitable for sowing maize. It is developed by PAU, Ludhiana. It saves 60 % labour operating time and 10 - 20 % cost of operation compared to conventional method of sowing behind the country plough. It also results in 5 to 10 % increase in yield compared to conventional method of sowing. The seed rate for maize is 25 kg/ ha. It has field capacity 0.30 ha/ hour at speed of 2.5 - 3.0 km/ hour. The efficiency is 60 %. The fuel consumption is 3.0 liters/hour.

**Tractor mounted ridge planter for winter maize:** It is a tractor - operated implement suitable for planting crops like maize on one side of ridges. In this, a planting unit is mounted on the commercially available ridger. The furrow opener is moved to sow seeds only on left side of a ridge. This implement was developed at PAU, Ludhiana during 1984-85. It saves 70 % on cost of operation compared to conventional method of sowing behind the country plough. It also results in 10 % increase in yield compared to conventional method of making ridges by tractor operated ridger and planting manually. The seed rate for maize is 25 kg/ ha. It has field capacity 0.20 -0.25 ha/ hour at speed of 2.0 -2.5 km/ha. The efficiency is 60 - 70 %. The fuel consumption is 3.01itres/hour.

**Tractor drawn cultivator mounted seed planter**

This planter can be mounted on tractor drawn cultivator which is commercially available (35 hp tractor). This is useful for line sowing of crops like groundnut, sorghum, maize, pulses and paddy. Suitable for sowing larger area in less time before the moisture is depleted in the dry tracts. The overall dimension is 2350 x 1125 x 1150 mm and weight is 425 kg. The capacity of the implement is 4 ha /day. The cost of the implement is Rs. 12,000 (Planter attachment to the cultivator) whereas cost of operation Rs.420/ ha.

**Broad bed former-cum-seed planter**

For forming broad beds and line sowing simultaneously, the crops like groundnut, sorghum, maize and pulses. It is suitable for low rainfall regions. This is tractor mounted implement (35 hp tractor). The overall dimension is 2240 x 1980 x 1050 mm and weight is 375 kg. The capacity of the implement is 3 ha /day. The cost of the implement is Rs. 25,000 whereas cost of operation is Rs. 520/ ha.

**Basin lister cum seed planter**

For forming 1250 x 30 x 15 cm sized basins along with the line sowing of the crops like groundnut, sorghum, maize and pulses. It is suitable for low rainfall regions. This is tractor mounted implement (35 hp tractor). The overall dimension is 2350 x 2150 x 1050 mm and weight is 525 kg. The capacity of the implement is 3 ha /day. The cost of the implement is Rs. 35,000 whereas cost of operation is Rs. 550/ ha.

**Power tiller operated cup seeder**

It is useful to sow seeds in rows with power tillers for crops like maize, Bengal gram and pulses. This is power tiller mounted implement (8-10 hp tractor). The overall dimension is 1880 x 1200 x 800 mm and weight is 110 kg. The capacity of the implement is 1.6 ha /day. The cost of the implement is Rs. 10,000 whereas cost of operation is Rs. 200/ ha.

### Weeding machine/ equipment

After sowing of seed, the weeds start emerging and compete with the crop. The weeders are necessary equipments to control the weeds. The weeders can be manual, animal drawn or tractor drawn. To cover large area in short time, tractor drawn weeding equipment is preferred. The tractor drawn weeder frame is consisting of two angle iron of 6 mm thickness 5 x 5 x 332 cm size of one rear shaft. The shaft is provided with number of holes to adjust the working distance between two rows. The shovel 50 x 50 cm shaped is welded with M. S. flat, which is 75 cm in length and 2.5 x 5 cm size. This shaft is given a shape of L size and the hovel is attached to shaft. Total number of shovels is five for maize, which work 75 cm from each other. It initially covers five rows but on successive turn it covers only four rows. One shovel is overlapping to previous one.

### Long handled weeder

It is used for weeding in row crops for removing shallow rooted weeds. It has been designed ergonomically for easy operation. Useful in dry land and garden land crops at a soil moisture content of 8 to 10 per cent. This is manually operated implement. Power requirement is one labour. The overall dimension is 1100 x 650 x 1050 mm and weight is 3 kg. The capacity of the implement is 0.05 ha /day. The cost of the implement is Rs. 450 whereas cost of operation is Rs. 1600/ ha.

## Harvesting

Maize is used either as grain or cobs. A different machine is required for each type of harvest. In areas where maize is grown as a cash crop, grain is harvesting shelling capabilities. In traditional farming systems the crop is often harvested by hand and cobs are stored in traditional structures. Quite often the crop is left standing in the field long after the cobs have matured, so that the cobs may lose moisture and store more safely after harvest. Manual harvesting of maize does not require any specific tool. It simply involves removing the cob from the standing stalk. The work time averages 20 to 30 days per hectare. To improve their drying, it is often recommended to remove the husks from the cobs. Maize husking is usually a manual task carried out by groups of women and children.

The first mechanical harvester to detach ears of maize from the standing stalks, the corn snapper was built in North America in the middle 19$^{th}$ century. This was followed by the development of corn pickers which incorporated a mechanism for removing the husks from the harvested ears. The first animal drown maize pickers were replaced by tractor drawn units (1 to 2 rows) and then tractor mounted units (1 row). Finally came the development of self propelled units capable of harvesting from upto 4 rows. A specific feature in maize harvesters is the header, which leaves the stalks standing as it removes the ears. The rates of work can vary from 2 hours/ha with a 3 row self propelled harvester to 5 hours/ha with a tractor mounted single row unit. Harvest losses range from 3 to 5 % but they may be upto 10 to 15 % under adverse conditions. Depending on the situation, a single row harvester can be employed effectively on upto 20 hectares or more. But the use of a multirow machine demands several tens of hectares to be economically effective. Specially designed for harvesting maize as grain, the corn seller was initially a corn husker in which the husking mechanism was replaced by a threshing one (usually of the axil type). Corn sellers are self propelled machines of the 3 to 6 row type with capacities of 1 to 2 hours per hectare.

Threshing traditional maize shelling is carried out as a manual operation. Maize kernels are separated from the cob by pressing on the grains with the thumbs. According to the operators ability the work rate is about 10 kg per hour. Outputs upto 23 kg hour can be achieved with hand held tools (wooden or slotted metal cylinders). Another threshing method, sometimes applied in tropical countries, involves putting cobs in bags and beating them with sticks. Outputs achieved prove attractive but bags deteriorate rapidly. Now a days many small maize shellers equipped with a rotating cylinder of the peg or bar type, are available in market. Their output ranges between 500 and 2000 kg per hours and they may be driven from a tractor power requirements vary from 5 to 15 hp according to equipment involve.

Corn pickers are used in areas where ear corn is needed for livestock feed. The corn picker may be classified according to construction as self propelled having its own engine, transmission and drive wheels; mounted- component oarts are mounter on the tractor framework. The picker is operated by the tractor PTO shaft. Trailed PTO driven pickers with its weights carried primarily on its own carriage wheels.

The most important function of a corn picker is the removal of the ear from the stalk. This action is accomplished by moving two inclined roughened rotation iron rolls down the standing row of the corn. The ear is picked off while the rest of the stalk is pulled between the rolls. Different makes of pickers do not differ widely in design with respect to the gathering mechanisms and the snapping rolls. The snapping rolls are the sole harvesting mechanisms in some corn harvester, which are referred to as corn snappers. Pickers differ widely in husking mechanism design. The husking mechanism on various corn pickers may be classified as follows:

1. **Continuous husking bed:** Husking rolls are merely extensions of the snapping rolls.
2. **Separate husking bed:** Husking mechanism is placed apart from the snapping rolls.

Two-row corn pickers may have two separate husking beds of either the continuous or separate type or they may have one combined husking bed, necessarily of the separate type. All pickers use a fan blast to perform a cleaning function. The most elaborate pickers have a shelled corn saver that retains the corn shelled by the husking rolls while discarding the dirt, silk and husks.

**Manually operated maize sheller**

This sheller is made from short length of 62.5 mm diameter M.S. pipe 0616 SWG. Its fins are made from 18 gauge M.S. sheet and bent at $70^0$. These fins are tapered ends of the fins provide two variable openings 39 mm at one side and 26.5 mm on the other side. This arrangement helps to handle different size cobs. The sheller held in the left hand and the maize cob is inserted by right hand, then these are rotated in opposite directions. The fins remove the kernels from the cob. About 15 -23 kg of kernels can be shelled in an hour with shelling efficiency of more than 99 %.

**CIAE Multi crop thresher**

It consists of spike tooth cylinder aspirator type blower and sieve shaker. Two top covers, three concaves, three sieves, variable cylinder speeds (7-21 m/s) are provided for threshing of different crops. It is suitable for threshing wheat, maize, sorghum, rice, gram, pigeon pea, soybean, mustard, sunflower, safflower and linseed crops. It saves 26 to 39 % labour operating time and 22 % on cost of operation compared to threshing by single crop thresher. This thresher operates at 8.4 m/ s cylinder speed for threshing of maize crop with the output capacity 1635 kg/ hour and power consumption of the machine is 3.3 kw. This thresher has threshing and cleaning efficiency 99.90 and 99.3 % respectively for maize crop.

**Dehusking and shelling:** The dehusking and shelling of maize are labour and time intensive operations. The farmer after harvesting the crop stacks it, along with the ears on the border of the field and postpone dehusiking and shelling even for three to four months. This postponement exposes the crop to a variety of risk resulting from unfavorable weather. If there are untimely rains after harvesting the crop, a number of fungi develop in the grains and the toxins produced are dangerous to human health. The development of power operated machines for dehusking was initiated long back. The machine is capable of dehusking four to five quintals of ears per hour. After that the manually operated two machines were developed, with the capacity of 1 q/ hour and 1.5 q/hour respectively.

**Dehusker-cum -sheller:** The output of the machine varies from 60 to 65 kg of shelled grain per hour. It needs three persons, two for operation and one for feeding. The power transmission involved a set of spur gears. It was costly affair to procure and maintain the gears. The machine has been further modified to perform dehusking and shelling by using a single cylinder. So that the use of spur gear is eliminated. The output of this machine is 60-70 kg of shelled grain per hour and it needs only two persons for operating the machine, as the friction is reduced to a great extent. The shelling efficiency is 85 -90 %.

### Husker sheller for maize

It dehusks and shells maize cobs. Power requirement is 7.5 hp electric motor / diesel engine. The overall dimension is 3000 x 710 x 1450 mm and weight is 3 kg. The capacity of the implement is 100 quintals of cob/day. The cost of the implement is Rs. 75000 whereas cost of operation is Rs. 30/ hour.

### Corn combine

Maize combine differ little from small grain combines. The essential difference is the construction of the row crop head that replaces the cutter bar head. The ear snapping mechanism for a combine corn header differs from that for conventional corn pickers. Stripper plates keep the ear from being pulled down into rolls and causing kernel shelling and loss. The snapping rolls can be very aggressive and will work well in the higher stalk moistures encountered by corn combines. Some of the newer models of corn pickers can be equipped with this type of snapping rolls. Losses in maize combine harvester are as follows:

i. **Loose ear loss:** Ear separated from the stalk and lying on the ground out of reach of the gathering mechanism.

ii. **Picker ear loss:** Ears left on the ground that were attached to the stalk but escaped the snapping rolls while being picked up by the gathering points.

iii. **Shelled corn loss:** Shelled corn lying on the ground as a result of shelling occurs at the snapping rolls.

iv. **Threshing loss:** The kernels lost because they were not shelled from the cob.

v. **Separating loss:** Kernels were not separated from the husks, silks, etc.

## References


Abou El-Nour, E.A.A. (2002). Growth and nutrient contents - response of maize to foliar nutrition with micronutrients under irrigation with Saline water. J. of Biol. Sci., 2: 92-97.

Acharya, C.L. and P.D. Sharma (1994). Tillage and mulch effects on soil physical environment, root growth, nutrient uptake and yield of maize and wheat on an alfisol in north west India. Soil and Tillage Research, 32 (4): 291 - 302.

Al-Kaisi, M.M. and X.Yin (2004). Stepwise time response of corn yield and economic return to no tillage. *Soil and Tillage Research,* 78: 91–101.

Caires, E.F., F.J. Garbuio, S.G. Chuka, Barth and J.C.L. Correa (2008). Effect of soil acidity amelioration by surface liming on growth and yield of no-till corn, soybean, and wheat. *European J, Agron.,* 28: 57-64.

Carrillo, A.R., A.R. Limongi, L.G. Auilla and P.L.M. Cespedes (1995). Validation of technologies in maize on varieties, population densities and fertilizer treatment under no-tillage conditions. *Memorias de la III Reunion Latinoamericana y XVI Reunion de la Zon Andina de Investigadores on Maize. Cochabamba, Santa Cruz, Bolivia.*

Dogra, P.B.P. Joshi and N.K. Sharma (2002). Economics analysis of tillage practices for maize cultivation in the Himalayan Humid sub – Tropics. Indian J. Soil Cons., 32(2): 172 – 178.

Dogra, P., B.P. Joshi and N.K. Sharma (2002). Economics analysis of tillage practices for maize cultivation in the Himalayan Humid sub-Tropics. Indian J. Agric. Sci., 75(7): 414 – 416.

Drury, C.F, C.S. Tan, W.P. Renolds, T.W. Welacky, S.E. Weaver, A.S. Hamill and T.J. Vyn (2003). Impacts of zone tillage and red clover on corn performance and soil physical quality. *Soil Science Society of America Journal* 67: 867–77.

Fischer, RA, F. Santiveri and IR. Vidal (2002). Rotation, tillage and crop residue management for wheat and maize in the sub-humid tropical highlands: II. Maize and system performance. *Fields Crop Research* 79 (2-3): 123–37.

Gajri, P.R., V.K. Arora and M.R. Chaudhary (1994). Maize growth responses to deep tillage straw mulching and farmyard manure in a coarse textured Soils of NW India. Soil Use Manage., 10: 15-20.

Halvorson, A.D., A.R. Mosier, C.A. Reule and W.C. Bausch (2006). Nitrogen and tillage effects on irrigated continuous corn yields. Agron. J., 98: 63-71.

Jat, M.L., A. Srivastava, S.K. Sharma, R.K. Gupta, P.H. Zaidi, H.K.Rai and G. Srinivasan (2005). Evaluation of maize - wheat cropping system under double-no-till practice in Indo-Gangetic Plains of India. Japan Association for International Collaboration of Agriculture and Forestry, 28 (1): 25 - 42.

Johnson, W.G., M.S. Defelice and C.S. Holman (1997). Application timing affects weed control with metolachlor plus atrazine in no-till corn (*Zea mays* L.). Weed Tech., 11(2): 207-211.

Kamble, H.G., N.P.S. Sirohi, R.R. Singh and S.M.Senger (2004). Improved implements for maize cultivation, Intensive Agriculture, 42 (7-8): 3-10.

Karunatilake, U., E.H.M. Van and R.R. Schindelbeck (2000). Soil and maize response to plow and no-tillage after alfalfa-to-maize conversion on a clay loam soil in New York. *Soil and Tillage Research,* 55: 31–42.

McGahen, J.H. (1973). The performance of maize hybrid over varying plant populations as affected by cultural tillage treatments of chemically weeded fields. Dissertation Abstracts International, 34: 1341.

Mishra, U., A.N. Ussiri David and R. Lal (2010). Tillage effects on soil organic caron storage and dynamics in corn belt of Ohio, USA. Soil and Tillage Research, 107 (2): 88 - 96.

Monneveux, P., E. Quillerou, C. Sanchez and C.J. Lopez (2005). Effect of zero tillage and residues conservation on continuous maize cropping in a subtropical environment (Mexico). *Plant and Soil,* 279 (1-2) 95–105.

Mukundam, B., S. Srividya and V. Raja (2011). Influence of rice zero tillage – maize system on productivity and soil fertility status. Andhra Agric. J., 58 (2): 133 – 139.

Mynavathi, V.S., N.K. Prabhakaran and C. Chinnusamy (2008). Evaluation of mechanical weeders in irrigated maize. *Indian J. Weed Sci.,* 40 (3&4): 210-213.

Pramanik, S.C. (1999). In-situ conservation of residual soil moisture through tillage and mulch for maize (Zea mays) in tropical BayIslands. Indian J. Agric. Sci., 69 (4): 254-257.

Rana (2010). Mechanical weeding in maize. Crop Res., 41 (1-3): 21-23.

Romeo, V. Labios, Jocelyn, D. Labios, Leonardo, L. Tamisin, Manuel, Q. Esguerra, Romulo, C. Cambaya, Jupiter, S. Tividad and J.O. Manalo (2002). Tillage practices in corn Production. Proceedings of the 8th Asian Regional Maize Workshop, Bangkok, Thailand: August 5-8, 2002. P: 296 – 306.

Saha, D. Chakraborty, A.R. Sharm,a, R.K. Tomar, S. Bhadraray, V. Sen, U.K. Behera, J.J. Purakayastha, R.N. Garg and N. Kalra (2010). Effect of tillage and residue management on soil physical properties and crop productivity in maize – Indian mustard system. Indian J. Agric. Sci., 80 (8): 679 – 685.

Sarma, C.K. and R.C. Gautam (2010). Weed growth, yield and nutrient uptake in maize (*Zea mays* L.) as influenced by tillage, seed rate and weed control methods. *Indian J. Agron.,* 55(4): 299 - 303.

Sharma, C.K. and R.C. Gautam (2006). Effect of tillage, seed rate and weed control methods on weeds and maize. Indian J. Weed Sci., 38 (1&2): 58 - 61.

Sharma, P., V.Abrol and G.R. Maurthi Shankar (2009). Effect if tillage and mulching on the crop productivity and soil properities in maize – wheat rotation. Res. Crops, 10 (3): 536 - 541.

Singh, R., A.R. Sharma and U.K. Behera (2007). Tillage and crop establishment practices for improving productivity of maize under different weed control methods. Indian J. Agric. Sci., 77(11): 731 – 737.

Srividya, S., K. Chandrasekhar and R. Veeraraghavaiah (2011). Effect of tillage and herbicide use on weed management in maize. Andhra Agric. J., 58 (2): 123 - 126.

Tony, J.V. and J.D. Janovick (2001). Potassium placement and tillage system effects on com response following long - term no- till. Agron. J., 93: 487-495.

Veer Singh, A. Srivantava, R.K. Singh and U.S. Savita (2011). Effect of tillage practices and residue management on soil quality and crop yield under maize based cropping system in mollisol. Indian J. Agric. Sci., 81(1): 1019 – 1025.

Veer Singh, Shri Ram, Amit Bhatnagar and Uma Shnkar Savita (2011). Effect of tillage methods on soil properities and productivity of quality protein Maize- wheat system. Indian J. Agron., 56 (2): 83-87.

# 17

# Rainfed Maize Cultivation

In areas having semi-arid climate provided the soils are deep with heavy texture (alluvial soils of Uttar Pradesh and Bihar, black soils of Malwa plateau, Bundelkhand and Udaipur regions and red and black soils of northern Description Telangana region). Maize is also grown in some of the drought prone areas such as Dungarpur district in Rajasthan as the runoff from the uncultivated lands contributes substantially for improving the moisture regime in the maize fields lying on the lower side of the watershed. With the development of new technology, the dryland maize has shown promise in certain non – traditional environments. It has already become a popular crop in the semi-arid red soils of Bangalore region. Similarly high yield levels have been consistently obtained in humid areas such as Bhubaneswar and Ranchi.

## Soil and Moisture Conservation Practices

Suitable land treatments are very essential for maize crop in order to achieve efficient use of applied inputs. The choice of these measures depends largely upon soil and rainfall characteristics. In the sub-mountane region of Punjab and Jammu and Kashmir, the practice of land levelling has been found to be very suitable for conservation of soil and moisture resources. In fact, this measure is a pre-requisite to the use of inputs like fertilizer, hybrid seed, etc. in Dehra Dun valley the system of zing terracing has been found to be suitable. In this system maize is taken in the upper 75 per cent area of the plot which is kept slopy to provide quick drainage and paddy is taken in the remaining 25 per cent area which is levelled to accommodate the runoff from the upper portion of the plot.

For the uplands of Bhubaneswar and Ranchi region, tati terrace has been found to be suitable. It is however, essential to provide suitable waste weirs in each terrace to avoid water logging problem. In the black soils of Malwa plateau, surface drainage is very crucial. A combination of community as well as individual land treatments (i.e. diversion draub, field drain, wasteweir, etc.) have been found to be suitable. Control of gullies through gabion or loose boulder drop structures is essential to prevent its further encroachment in the cultivated lands. For Telangana, Bundelkhand and Udaipur regions, the system of graded bunding or conservation drain is suitable for safe disposal of runoff as well as for conservation of soil resource. In red soils of Bangalore region, the systems of graded bunding or graded border strip (with a provision of grassed waterway on boundary lines) have been found to be suitable.

## Cropping Systems

Depending upon the rainfall distribution, maize can be grown either in an intercropping system or in a double cropping system (Table 1). The areas which receive assured winter rains (Hoshiarpur, Rakh Dhiansar, Dehra Dun, Varanasi and Udaipur) maize crop is followed by a second crop under dryland conditions. In black soil at Jhansi and Indore which receive low rainfall during post – rainy season, intercropping system is found to be more assured. Such a system is found to be suitable also in red soils at Bangalore, Medak, Bhubaneswar and Ranchi due to undependable rains during post-rainy season. At Bangalore, however, a double cropping system of fodder maize – finger millet has also been found to be promising due to favourable rains during summer months.

**Table 1. Suitable cropping systems with maize at different locations**

| Location | Cropping system | |
|---|---|---|
| | Rainy | Post rainy |
| **A. For double cropping** | | |
| Hoshiarpur | Maize | Wheat / chickpea |
| Rakh Dhiansar | Maize | Wheat / Barley |
| Dehra Dun | Maize | Wheat / chickpea |
| Varanasi | Maize | Chickpea/ Mustard |
| Udaipur | Maize | Chickpea / Erica |
| **B. For Intercropping** | | |
| Indore | Maize + soybean | |
| Jhansi | Maize + pigeon pea | |
| Bhubaneswar | Maize + pigeon pea | |

| | |
|---|---|
| Ranchi | Maize + pigeon pea |
| Medak | Maize + pigeon pea |
| Bangalore | Maize + pigeon pea |

*The choice of crop depends upon the soil moisture in upper 15 cm at the sowing time.

At Rakh Dhiansar (Jammu), productivity and net return were higher in maize – wheat sequence followed by green fodder – toria + gobhi sarson. Maize – toria –wheat and green gram – toria – wheat achieved higher land utilization efficiency under dryland condition (Jamwal, 2001).

**Table 2. Performance of maize based crop sequences under dryland condition**

| Cropping sequence | Grain yield (kg/ha) | | Land use efficiency (%) | Production efficiency (kg/ha/day) | Net return (Rs/ha) | B:C ratio |
|---|---|---|---|---|---|---|
| | Kharif Maize | Rabi crops | | | | |
| Maize - wheat | 2520 (5051) | 2700 | 73.9 | 18.70 | 14175 | 1.16 |
| Green manuring - Toria - wheat | 624 (2728) | 1592 | 60.3 | 12.35 | 3234 | 0.34 |
| Green gram - Toria - wheat | 366 (Gg) 208 (T) (3740) | 2491 | 82.0 | 16.87 | 4875 | 0.47 |
| Green manuring – wheat | (3053) | 3257 | 49.3 | 10.17 | 3345 | 0.31 |
| Maize - Toria - gram | 1873 (2726) | 160 (T) 214(G) | 71.2 | 10.48 | 2650 | 0.28 |
| Maize – Toria + Gobhi sarson | 1833 (2393) | 155 (T) 527 (Gs) | 71.0 | 12.66 | 3227 | 0.28 |
| Maize - Toria - wheat | 1973 (4256) | 250 (T) 1908 | 82.0 | 14.18 | 3415 | 0.31 |
| Green fodder – Toria + Gobhi sarson | 29410 (4894) | 281 (T) 583 (Gs) | 65.7 | 20.39 | 17908 | 2.06 |
| CD 5 % | 888 | - | - | - | - | - |

Figure parentheses are wheat equivalent yield. Gg: Green gram, T: Toria, G: gram, Gs: Gobhi sarson.

## Tillage operation

At least 2 – 3 shallow tillage operations during summer season are found to be essential in all the maize growing areas. In situations like Dehra Dun and Bangalore, deep ploughing operation has also given a significant yield advantage through weed control and efficient moisture conservation. Similarly, bed and furrow or ridge furrow systems have been found to be suitable in semi-arid regions of Udaipur, Jhansi as well as in sub-humid regions of Bhubaneswar, Ranchi and Varanasi.

At least two intercultural operations are found to be essential for maize crop to control weeds and conserve soil moisture. In the sub-montane regions of Hoshiarpur the existing practice in which a plough furrow is made in the standing crop of maize as a part of intercultural operation is also found to be beneficial for increasing the yield through better moisture conservation as well as safe disposal of excess runoff (Table 3).

**Table 3. Effect of inter –terrace land treatment on yield of maize at Hoshiarpur**

| Land treatment | Grain yield (q / ha) | | | Average |
|---|---|---|---|---|
| | 1975 | 1976 | 1978 | |
| Control | 16.0 | 17.1 | 12.0 | 18.4 |
| Haloding* | 36.5 | 23.9 | 20.5 | 27.0 |

*A plough furrow is made in between the crop rows at about 4-week stage.

Similarly, in semi-arid region of Bangalore, the existing practice of tied-ridging (which is achieved by manual spading or through bullock driven ridger) is very efficient or trapping of subsequent rains. In the red soils of Telengana, ridging operation as a part of top dressing (with the help of country plough) has been found to be useful for moisture conservation. At Hoshiarpur, the practice of organic mulching has also been found to be useful for increasing the yield of maize crop as well as for achieving better plant stand of the crop of wheat in sequence.

### Choice of genotype

The relative advantage of hybrid over the local maize has been varying from place to place. Based upon the trials carried out during the last decade, the various locations can broadly be divided into two categories. Areas where hybrids gave distinct advantage: The improved hybrids have been giving a distinctly superior performance over the local variety at Dehradun, Bangalore and Medak (Table 4).

**Table 4. Comparison of hybrid and local variety of maize at selected locations**

| Locations | Name of the hybrid | Grain yield (q / ha) | |
|---|---|---|---|
| | | Local | Hybrid |
| Dehradun | Ganga 2 | 17.8 | 34.9 |
| Bangalore | Deccan | 31.8 | 56.6 |
| Medak | Deccan 101 | 18.8 | 27.9 |
| | **Average** | **22.8** | **39.8** |

Hybrid Ganga 2 and improved composite A 51 – 54 have shown good performance in non-traditional areas of Ranchi and Bhubaneswar respectively. The data clearly bringout that hybrid should be extensively popularised in the maize growing areas of Dehradun valley, Bangalore and Telengana. Similarly improved composites / hybrids should be introduced in the non-traditional areas of Bhubaneswar and Chotanagpur regions.

Areas where hybrids gave only marginal advantage: At a number of locations, the existing hybrids and composite have shown just 10 per cent yield advantage over the local variety (Table 5). These hybrids did not also fit properly into the existing double cropping system due to relatively longer duration (about 100 – 110 days) as compared to the local variety which usually matured in about 85 – 90 days. Therefore at the above locations improved management practices should be popularised with the existing local varieties until suitable hybrids / composites are evolved.

**Table 5. Comparison of hybrid / composite with local variety of maize at selected locations**

| Location | Name of the hybrid | Grain yield (q / ha) | |
|---|---|---|---|
| | | Local | Hybrid |
| Hoshiarpur | JML – 22* | 29.0 | 30.2* |
| Varanasi | Ganga – 2 | 18.1 | 20.6 |
| Udaipur | Deccan | 8.6 | 10.9 |
| Rakh Dhiansar | Ganga – 2 | 9.3 | 10.1 |
| Jhansi | Diara* | 9.7 | 11.9* |
| Indore | Ganga – 5 | 22.2 | 22.6 |
| | Average | 16.2 | 17.7 |

*Composite was evaluated instead of hybrid

## Seeding and Spacing

Yield of maize crop remains the same over a wide range of plant populations provided a uniformity in plant spacing is maintained. On an average 6 – 8 plants per $m^2$ have been found to be optimum for good yields.

The overall plant population (seed rate) depends to a large extent on the choice of cropping system. For a pure crop of maize about 20 kg per ha is required for achieving optimum plant stand. The quantity of seed may be reduced to 15 g per ha in case maize is intercropped with pigeonpea in 2:1 row ratio. This seed rate may be further reduced to about 10 kg per ha if maize is intercropped with soybean in 1:3 row ratio.

A high flexibility in crop geometry has been found in case of maize. The row width could be kept anywhere between 30 to 60 cm depending upon the seeding devices. However, a minimum spacing of 15 cm must be maintained within the rows; otherwise, percentage of barren plants will increase. On an average, a crop geometry of 45 x 20 cm is found to be optimum for most of the areas.

## Weed Management

Timely weed control is another essential operation for maize. Two weeding operations at about two weeks and four weeks, stages must be carried out to minimize the competition effect. In situations where labour availability is a major constraint, it is advisable to use the herbicide atrazine at the rate of 1.0 kg a. i. per hectare as a pre-emergence application.

**Table 6. Effect of integrated weed management on weeds and yield of rainfed maize**

| Treatments | Weed density (No. $m^2$) | WCE (%) | Grain yield (q/ha) | Stover yield (q/ha) |
|---|---|---|---|---|
| Alachlor 2 kg a.i/ha | 281 | 45.1 | 31.3 | 52.3 |
| Alachlor 2 kg a.i/ha + HW | 226 | 55.9 | 33.4 | 55.4 |
| Atrazin 1 kg/ha | 358 | 30.1 | 22.6 | 40.8 |
| Atrazin 1 kg/ha + HW | 306 | 40.2 | 25.0 | 43.2 |
| Pendimethaline 1 kg /ha | 322 | 37.1 | 23.0 | 40.8 |
| Pendimethaline 1 kg /ha + HW | 273 | 46.7 | 25.3 | 42.4 |
| Black gram as smother crop | 290 | 43.4 | 24.8 | 43.5 |
| Dain cha as smother crop | 377 | 26.4 | 25.1 | 42.6 |

| Treatments | Weed density (No. $m^2$) | WCE (%) | Grain yield (g/ha) | Stover yield (a/ha) |
|---|---|---|---|---|
| HW on 20 & 40 DAS | 201 | 60.7 | 33.6 | 55.4 |
| Earthing up | 309 | 39.6 | 31.4 | 52.0 |
| Weed free | 0 | 100 | 37.0 | 60.9 |
| Weedy check | 512 | 0 | 11.0 | 20.4 |
| CD | - | - | 2.4 | 6.4 |

At Kumargan, (Uttar Pradesh), two hand weeding at 20 and 40 DAS proved effective in controlling weeds (WCE 60.7 %) and higher yield of maize followed by pre-emergence application of alachlor @ 2.0 kg a.i /ha + HW on 30 DAS (Alok et al., 2012).

## Fertility Management

Keeping in view the soil, genotype and rainfall pattern the doses of fertiliser recommended for maize crop under dryland conditions as various locations are presented in Table 7. One third nitrogen with full dose of $P_2O_5$ and $K_2O$ should be applied at sowing time. Half of the remaining nitrogen should be applied at about four weeks after sowing. The remaining nitrogen can be avoided if season is dry. Wherever needed, the entire quantity of zinc and lime should be applied as a basal dose.

**Table 7. Recommended fertiliser doses for dryland maize at different locations**

| Location | Genotype | *Fertiliser dose (kg/ha) | | | Remarks |
|---|---|---|---|---|---|
| | | N | $P_2O_5$ | $K_2O$ | |
| Dehradun | Hybrid | 80 | 40 | 20 | |
| Bangalore | Hybrid | 75 | 50 | 25 | Also zinc sulphate (10 kg / ha) |
| Medak | Hybrid | 60 | 30 | 0 | |
| Ranchi | Hybrid | 80 | 40 | 20 | Also lime (3 tonnes / ha) once in 5 year |
| Bhubaneswar | Composite | 80 | 40 | 20 | Also lime (3 tonnes / ha) once in 5 year |
| Hoshiarpur | Local | 40<br>80 | 20<br>40 | 10<br>20 | Loamy sand Sandy loam to clay loam |
| Rakh Dhisnsar | Local | 40 | 20 | 0 | |
| Indore | Local | 40 | 20 | 0 | |
| Udaipur | Local | 40 | 20 | 0 | |
| Varanasi | Local | 40 | 20 | 0 | |

*Delete $P_2O_5$ and $K_2O$ if FYM is applied more than 10 tonnes / ha

A balanced application of nitrogen, phosphorus and potash is very important. Such a practice has resulted in more than 25 per cent increase in yield without any additional cost one fertilizer input (Table 8).

**Table 8. Effect of balanced use of fertilisers on maize (ORP, Bangalore 1977)**

| Fertiliser use | Grain yield (q/ha) |
|---|---|
| Farmers method | 28.5 |
| Balanced use of N: $P_2O_5$ : $K_2O$ (3:2:1) | 37.8 |

*The total cost of fertiliser input was kept equal in both the treatments.

Response to nitrogen has been varying depending upon soil type, genotype and rainfall pattern. At Hoshiarpur, 80 kg nitrogen per ha has been found suitable in sandy loam to clay loam soils whereas only 40 kg nitrogen per ha was sufficient in loamy sand soil (Table 9). In these soils, $P_2O_5$ and $K_2O$ should be applied only as per the soil test. Application of P and K is not necessary in case adequate farm yard manure (more than 12 tonnes per ha) has been applied in the field.

**Table 9. Response of maize to nitrogen in two different soil types**

| Nitrogen level (kg/ha) | Grain yield (q/ha) | |
|---|---|---|
| | Loamy sand | Sandy loam |
| 0 | 12.6 | 20.7 |
| 40 | 21.7 | 28.9 |
| 80 | 20.0 | 34.1 |
| 120 | 19.7 | 31.2 |

In red soils at Bangalore, the application of 10 kg zinc sulphate per ha is also found to be beneficial for maize in addition to the application of N, P and K (Table 10).

**Table 10. Response to zinc application at Bangalore**

| Treatment | Grain yield (q/ha) | | | | Average |
|---|---|---|---|---|---|
| | 1974 | 1975 | 1976 | 1977 | |
| Control | 34.0 | 64.7 | 24.8 | 60.2 | 45.9 |
| Zinc* | 42.0 | 63.1 | 33.0 | 74.3 | 53.1 |

*Zinc was applied @ 10 kg/ha. N, $P_2O_5$ and $K_2O$ were applied in equal quantity to both the treatments.

Similarly, application of lime at the rate of 3 tonnes per ha (once in five years) was essential for high yields of maize in acidic soils of Bhubaneswar (Table 11).

**Table 11. Effect of lime application on maize at Bhubaneswar**

| Treatments | Grain yield (q/ha) | | Average |
|---|---|---|---|
| | 1975 | 1976 | |
| Control | 13.8 | 8.0 | 10.9 |
| Lime* | 27.5 | 27.8 | 27 |

*Lime was applied as paper mill sludge to provide an equivalent quantity of 2.5 tonnes of $CaCO_3$ /ha

Placement of fertiliser at sowing as well as at top dressing is very essential. Maize crop is usually sown behind country plough. Therefore, at sowing time the fertiliser placement can be achieved through tube attached to plough. Similarly, at the time of top dressing the fertiliser should be preferably placed through the tube method rather than the usual practice of broadcasting / spreading it on the surface without any soil cover. In Hoshiarpur region, top dressing of urea can be done even by broadcasting method provided it is carried out just before the haloding operation.

## Moisture Conservation

**Table 12. Effect of moisture conservation treatments on maize yield**

| Treatment | Soil loss (kg/ha) | Runoff (%) | Grain yield (kg/ha) | Fodder yield (kg/ha) | Harvest indesx (%) |
|---|---|---|---|---|---|
| Sowing along slope | 2580 | 55.6 | 1303 | 4400 | 23.3 |
| Sowing across slope | 1860 | 17.3 | 1380 | 4243 | 24.7 |
| Redges and furrow sowing | 900 | 7.9 | 1307 | 4743 | 22.4 |
| Flat sowing & ridging after 30 DAS | 970 | 8.6 | 1477 | 5903 | 21.6 |
| Opening of ditches across slope (10 m interval ) | 1366 | 11.7 | 1233 | 4697 | 21.9 |
| Compartmental bunds (5 m) | 1370 | 11.9 | 1190 | 3780 | 23.5 |
| LSDCP <0.05 | 147 | 1.2 | 10.7 | 6.5 | 0.24 |

At Arjia (Rajasthan), flat sowing and ridging at 30 DAS was superior with significantly higher grain yield of 1477 kg/ha, fodder yield of 5903 kg/ha, net return of Rs. 4093 /ha and B:C ratio of 1.56 (Jat et al., 2012).

## Intercropping

In rainfed satiation, the sowing of maize is generally done with the onset of monsoon rains. Most suitable temperature for germination is 21° C and for growth is 32° C (Jain, 1973). Extremely high temperature, moisture stress and low humidity during flowering damage the flowering and the foliage, desiccation of pollen and interfere with the pollination resulting in poor grain formation. Thus rate of development of maize from planting to anthesis is a function of temperture rather than photosynthesis (Brouwer et al., 1970). Negative relationship of HTU with yield was reported in baby corn (Thavaprakash et al., 2007). Greater reduction in yield of long season maize varieties under delayed sowing situations was attributed to decrease in season length might have an effect by of reduction in their potential yield (Johannes, 1996). Datta and Bandyopadhyay (2006) reported that 12.5 to 33.9 % yield reduction in grrundnut by intercropping with maize was due to the depressing effect of maize on groundnut. Minimum loss with maize might be due to its elastic response to change in plant population, less light competition effect, staggering of peak demands for growth factors and early maturing of maize (Padhi and Panigrahi, 2006).

**Table 13. Grain yield of maize in African babul based Agroforestry**

| Agroforestry system | Annual return (Rs/ha) | Tree population / ha | Estimated volume of tree ($m^2$/tree) | Yield (t/ha) | |
|---|---|---|---|---|---|
| | | | | Grain | Stover |
| Sole maize cropping | - | - | - | 1.67 | 3.78 |
| Intercropping of maize with African babul at 3.0 x 3.0 m | 1369 | 1111 | 0.06 | 2.46 | 4.37 |
| Intercropping of maize with African babul at 4.2 x 3.0 m | 1654 | 794 | 0.10 | 2.63 | 4.46 |
| Intercropping of maize with African babul at 5.4 x 3.0 m | 1800 | 617 | 0.14 | 3.09 | 4.72 |
| CD 5% | - | - | - | 0.25 | 0.41 |
| N levels (kg/ha) | | | | | |
| 40 | - | - | - | 1.98 | 3.79 |
| 80 | - | - | - | 2.54 | 4.40 |
| 120 | | | | 3.03 | 4.81 |
| CD 5 % | - | - | - | 0.10 | 0.11 |
| P levels (kg/ha) | | | | | |
| 30 | - | - | - | 2.39 | 4.20 |
| 60 | - | - | - | 2.64 | 4.42 |
| CD 5 % | - | - | - | 0.08 | 0.09 |

(Subrahmaniayam and Bhee Maiah,2003)

At Rajendra Nagar (Andhra Pradesh), grain and stover yields higher under intercropping irrespective of tree spacing than sole maize cropping. Under intercropping situations, tree spaced at wider spacing (5.4 x 3.0 m) showed significantly more grain yield than the other tree spacings. Application of higher doses of nitrogen (120 kg/ha) and phosphorus (60 kg/ha) recorded higher grain and stover yields of maize. Maximum monetary return (Rs.18330 /ha) was obtained with maize when intercropped with African babul at 5.4 x 3.0 m under higher dose of N (120 kg/ha) and P (60 kg/ha) applied (Subrahmaniayam and Bhee Maiah, 2003).

**Table 14. Productivity and profitability of maize based vegetable cropping systems under rainfed condition**

| Inter cropping | Maize green cob yield (kg/ha) | Vegetable yield (kg/ha) | Maize equivalent yield (kg/ha) | Net return (Rs/ha) | B:C ratio |
|---|---|---|---|---|---|
| Maize + cowpea (Bodi) | 7421 | 3506 | 24248 | 28991 | 1.92 |
| Maize + runner bean | 7588 | 1696 | 14373 | 6182 | 1.21 |
| Maize + cucumber | 6749 | 5078 | 27062 | 37232 | 2.22 |
| Maize + ridge gourd | 3800 | 7236 | 32745 | 50740 | 2.63 |
| Maize + bitter gourd | 7083 | 3061 | 21775 | 25415 | 1.88 |
| Maize + snake gourd | 8764 | 4286 | 22480 | 26777 | 1.91 |
| Sole maize | 10326 | - | 10326 | 6139 | 1.31 |
| CD 5 % | 1997 | - | 5976 | - | - |

(Behera et al., 2010)

At Phulbani (Orissa), vegetables viz., trailing cowpea, runner bean, cucumber, ridge gourd, bitter gourd and Snake gourd were planted in basins spaced at 120 cm in both the sides. Sole maize was planted at 60 x 30 cm. The systems composed 100 % population of vegetables and 75 % population of maize. High value viny vegetables can be cultivated using maize as live stake and gave additional yield of vegetables from field of maize through intercropping (Behera et al., 2010). At Wadura (Jammu and Kashmir), grain yield of sole maize was higher than the intercropping but the maize equivalent yield was higher (4.87 t/ha) in maize + French bean (2:2) intercropping system with higher net return (Rs. 16462/ha) and B:C ratio (1.96) (Singh et al., 2008).

## Pest management

The available genotypes are highly tolerant to pests and diseases. However, stem-borer infestation may be noticed in some instances. In early stages, the damage by this pest is localized in randomly distributed patches in field. Therefore, efforts should be made to use the pesticides only upto about one meter area around the affected patches. It is also advisable to use the granular formulation of pesticide. Otherwise the spraying operation would destroy the natural parasite of pests like aphids and hence the problem of aphids is likely to get accentuated.

## Relative contribution of production factors

Trials at research stations in Telengana region (ICRISAT, 1976) have consistently brought out that the grain yield levels of maize can be increased from 4.5 to 34.7 q per ha provided a complete package of practices is adopted (Table 15). Among the various practices, improved variety alone contributed only 1.8 q of grains per ha whereas fertiliser alone contributed 14.3 q per ha. A combination of seed and fertiliser resulted in an increase of 17.7 q per ha. Efficiency of these inputs, were further increased when management practice like land treatment and placement of fertiliser were properly carried.

**Table 15. Relative contribution of production factors on yield of rainfed maize**

| *Production factors | | | Grain yield (a / ha) | |
|---|---|---|---|---|
| variety | Fertiliser | Soil & Water management | Maize | Pigeon pea |
| T | T | T | 4.5 | 3.2 |
| Imp | T | T | 6.3 | 5.0 |
| T | Imp | T | 19.0 | 4.5 |
| Imp | Imp | T | 22.2 | 5.4 |
| Imp | Imp | Imp | 34.7 | 6.4 |

Traditional (T): Local variety, 10 tonnes FYM / ha; Flat system of sowing, local seeding implements. Improved (Imp.): SB 23 variety; 80 kg N + 35 kg $P_2O_5$ per ha; Broadbed and furrow system; Improved equipment for seeding and placement of fertiliser (ICRISAT, 1976 – 77).

Similarly, at Hoshiarpur yield levels can be increased by about 75 % through the adoption of improved management essentially included in additional input of 40 kg N per ha and provided spacing under improved management does not affect the yield potential. It is required mainly for convenience in inter culture / top dressing operations.

## References

Alok, M., Nulpur, M., B. Singh and A.K. Singh (2012). 1NW in maize under rainfed conditions. Indian J. Dryland Agric. Res. Dev., 27(1) : 70-73 .

Ameta, G.S. and Mahender Singh (2005). Studies on use of thiourea spray for yield advantage in rainfed maize. International J. Tropical Agric., 23 (1-4): 307 - 310.

Anilkumar and K.S. Thakur (2004). Effect of integrated nutrient management on promising composite maize (*Zea mays*) varieties under rainfed mid-hill conditions of Himachal Pradesh. *Indian Journal of Agricultural Sciences*, 74 (1): 40-42.

Arora, S. and M.S. Hadda (2003). Soil moisture conservation and nutrient management practices in maize-wheat cropping system in rain-fed North-western tract of India. Indian Journal of Dryland Agriculture Research and Development, 18: 70-74.

Avasthe, R. (2011). Response of rainfed maize to nitrogen management in mid hill acidic soils of Sikkim. Indian J. Agron., 56 (3): 232 - 236.

Balyan, J.K., B.S.Kumpawat, M.L.Jat and L.K.Jain (2006). Integrated nutrient management for higher productivity and energy use in rainfed maize in southern Rajasthan. Curr. Agric., 30 (1-2): 83-88.

Behera, B., S.K. Mohanty, A. Mishra, C.R.Subudhi and P.K. Mishra (2010). Performance of high value viny vegetables with maize as live stake under rainfed conditions of Eastern Ghats Orissa. Indian J. Dryland Agric. Res. Dev., 25 (2): 170 – 172.

Behra, B., G.S. Singh, P.C. Pradhan and P.C. Senapati (1998 b). Weed management in runner bean + maize intercropping under rainfed condition. Indian J. Agric. Sci., 68 (10): 697 - 698.

Bhopal Singh, C.M.Singh, M. Bhargava and R.D. Sood (1991). Integrated weed control in rainfed maize. Himachal J. Agric. Res., 17(1): 1-2.

Biradar, S.A. and S.M. Hiremath (2008). In situ green manuring of intercropped legumes on the performance of maize - chickpea / safflower cropping system under rainfed condition. Karnataka J. Agric. Sci., 21 (4): 607.

Dasaraddi, S.V. (1998), Effect of in situ incorporation of legumes intercropped with maize in maize-safflower relay cropping system under rainfed condition. *M. Sc. (Agri.) Thesis,* Univ. Agric. Sci., Dharwad, Karnataka, India

Datta, D. and P. B and yopadhyay (2006). Production potential of intercropping of groundnut with pigeonpea and maize under varios proportion in rainfed Alfisols of west Bengal. Indian J. Agron., 51 (2): 103 – 106.

Deshmukh, S.C., O.P. Girothia and O.P. Verma, (2002). Integrated nutrient management for sustainable rainfed maize production in vetisols. *Indian Journal of Dryland Agricultural Research and Development*, 17 (2): 144-147.

Dilshad, M.D., M.I. Lone, Ghulam Jilani, M. Azim Malik, M. Yousaf, Rizwan Khalid and Fakhra Shamim (2010). Integrated Plant Nutrient Management on Maize under Rainfed Condition. Pakistan Journal of Nutrition, 9 (9): 896-901,

Dwivedi, S.K. and G.K. Shrivastava (2011). Planting geometry and weed management for maize + black gram intercropping system under rainfed vertisols. Indian J. Agron., 56 (3): 202-208.

Faroda, A.S., N.L. Joshi, R. Singh and A. Saxena (2007). Yield and economicsof intercropping maize under rainfed conditions. Indian J. Agric. Sci., 76 (9): 554 – 556.

Gaur, B.L. (1991). Effect of time of FYM application and N rates on the yield of rainfed maize. Indian J. Agron., 36 (1): 107 – 108.

Gaur, B.L. (2002). Performance of improved technologies of rainfed maize in on-farm trials of watershed area. Indian Journal of Dryland Agriculture Research and Development, 17: 74-75.

Gaur, B.L. (1991). Effect of time of FYM application and N rates on the yield of rainfed maize. *Indian Journal of Agronomy*, 36 (1): 107-108.

Ghuman, B.S. and H.S. Sur (2001). Tillage and residue management effects on soil properties and yield of rainfed maize and wheat in subhumid subtropical climate. Soil Tillage Res., 58: 1 – 10.

Gupta, R. and M.P. Sharma (2010). Effect of inorganics and organics on soil characteristics and crop yield under rainfed maize - Gobi sarson sequence. Indian J. Dryland Agric. Res. Dev., 25 (2): 142 - 153.

Hadda, M.S. and S. Arora (2006). Soil and nutrient management practices for sustaining crop yields under maize-wheat cropping sequence in sub-mountain Punjab, India. Soil Environ., 25(1): 1-5.

Hammer (2003). Managing the risk of growing dryland maize in the northern region. pp 112-119. In: Birch, C. J. and Wilson, S. R. (Eds.) Versatile Maize, Golden Opportunities, Proceedings, 5th Australian Maize Conference, 18-20 February 2003, Toowoomba, Australia, Maize Association of Australia, Darlington Point, NSW.

Haque, M., Sharma, R.P. Prasad and Shambhu (2008). Weed control in maize based intercropping system under rainfed condition. *ISWS Biennial Conference on Weed Management in Modern Agriculture,* February 27-28, 2008, Pusa, Bihar. pp.118.

Heisey, P.W. and G.O. Edmeades (1999). Maize production in drought-stressed environments: technical options and research resource allocation. World Maize Facts and Trends 1997/1998.

Hulihalli, U.K. and S. P. Dinesh kumar (2011). Effect of INM practices on growth and yield of maize under rainfed condition. Andhra Agric. J., 58(3): 262 - 266.

Jaipaly, S.K. Samdal, V.K. Suri and A.K. Dhiman (2008). Validation of fertilizer adjustment equation based on yield target concept and complimentary use of organic and biofertilizers along with inorganic fertilizers in rainfed maize – wheat system in wet temperate zone of Himachal Pradesh, Indian J. Agric. Sci., 78 (6): 490 – 494.

Jamwal, J.S. (2001). Productivity and economics of different maize based crop sequences under dryland condition. Indian J. Agron., 46 (4): 601-604.

Jamwal, J.S. (2005). Productivity and economics of maize – wheat cropping system under integrated nutrient supply system in rainfed areas of Jammu. Indian J. Agron., 50 (2): 110 – 112.

Jamwal, J.S. (2006). Effect of integrated nutrient management in maize on succeeding winter crops under rainfed conditions. Indian J. Agron., 51 (1): 14 – 16.

Jat, M.L., G.R.M. Sankar, K.S. Reddy, S.K. Sharma, M. Kumar, P.K. Mishra, P. Singh, J.K. Balyan and L.K. Jam (2012). Efficient moisture conservation practices for maximizing maize productivity profitability, EVE and resource conservation in a semi-arid Incetisol. Indian J. Soil Cons., 40 (3): 218 – 224.

Jat, M.L., P. Singh, J.K. Balyan, A. Gupta and L.K. Jam (2006). Energy of different in situ moisture conservation practices in maize. Indian J. Agric. Sci., 75 (7): 414 – 416.

Kandasamy, O.S and C.N. Chandrasekhar (1998). Competitive efficacy of chemical and mechanical methods of weed management in rainfed maize (*Zea mays*). *Indian Journal of Weed Science,* 30 (3/4): 201-203.

Khanday, B.A., R.C. Thakur, and M.K. Kushu(1993). Nitrogen, FYM and zinc interaction on maize. *Haryana Journal of Agronomy*, 9 (1): 79-81.

Khandey, B.A. and R.C. Thakur (1990). Response of rainfed maize to fertilization. Indian J. Agric. Sci., 60: 631 – 633.

Kumar, A. and K.S. Thakur (2004). Effect of INM on promising composite maize varieties under rainfed mid hill conditions of Himachal Pradesh. Indian J. Agric. Sci., 74: 40 - 42.

Kumar, A. and K.S. Thakur (2004a). Effect of INM on Promising composite maize varieties under rainfed mid hin conditions of Himachal Pradesh. Indian J. Agric. Sci., 78(1): 27 – 31.

Kumar, S., C.R. Rawat and N.P. Melkania (2005). Forage production potential and economics of maize (Zea mays) and cowpea (Vigna unguiculata) intercropping under rainfed conditions. Indian J. Agron., 50 (3): 184 - 186.

Lidhoo, C.K. (2007). Appropriate weeding tools for dryland maize of Jammu region. *J. Res. SKUAST*, 6: 29-33.

Lingaraju, B. S., S. B. Marer and S. S. Chandrashekar (2008). Studies on Intercropping of Maize and Pigeonpea Under Rainfed Conditions in Northern Transitional Zone of Karnataka. Karnataka J. Agric. Sci., 21 (1): 1-3.

Mankotia, B.S., R.C. Thakur, Jagdish Kumar and Sanjay Kumar (2006). Use of Lantana and Eupatorium as green manure in rainfed maize – wheat system in North –west Himalayas. Indian J. Weed Sci., 38 (3&4): 263 - 266.

Manoj Kumar and M. Singh, (2003). Effect of nitrogen and phosphorus levels on yield and nutrient uptake in maize under rainfed condition of Nagaland. Crop Res., 25(1): 46-49.

Marer, S.B., B.S. Lingaraju and G.B. Shashidhara (2007). Productivity and economics of maize and pigeonpea intercropping under rainfed condition in northern transitional zone of karnataka. Karnataka J. Agric. Sci., 20: 1-3.

Mishra, A.S. and S. Elamathi (2009). Response of maize (*Zea mays* L.) + legume intercropping system to different weed control practices. Madras Agric. J., 96 (7-12): 322-324.

Mishra, C.M. (1993). Performance of maize (*Zea mays* L.) varieties to fertility levels under rainfed condition of Madhya Pradesh. *Indian. J. Agron.,* 38: 483-485.

Moolchand, S., S. Prabhukumar, C.V. Sairam and Arunkumar (2009). Evaluation of different weed management practices in rainfed maize on farmers' fields. Pakistan J. Weed Sci. Res., 15:183-189.

Narayanasamy, M.R., V.Veerabadran, C.Jayanthi and C.Chinnusamy (1994). Plant density and nutrient management for rainfed maize in red soils. Madras Agric. J., 81: 248-251.

Narwood, C. A. (2001). Dryland corn in western Kansas; effect of hybrid maturity, planting date and plant population. Agron. J., 93 (3): 540-574.

Njoroge, K., B. Wafula and J.K. Ransom, (1996). Characterization of drought stress in the major maize production zones of Kenya. Proc. Sym. Developing Drought and Low N tolerant maize- March, 25-29, 1996 ,CIMMYT, El Batan, Mexico. pp.121-124.

Obi, M.E. and P.O. Ebo (1995). Effect of organic and inorganic amendments on soil physical properties and maize production in a severally degraded sandy soil. *Bio. Tech.,* 51(2-3): 117-123.

Paliwal, A.K., R.N.Singh, R.K.Mishra, R.K. Dwivedi and S.K. Chaudhary (1999). Effect of nitrogen and zinc application on the growth, yield and nutrient uptake in rainfed maize (*Zea mays* L.). Advances Plant Sci., 12: 223-226.

Pandey, I.B., V. Bharati and S.S. Mishra (2003). Effect of maize (*Zea mays)*-based intercropping systems on maize yield and associated weeds under rainfed condition. Indian J. Agron., 48 (1): 30-33.

Paradkar, V. K. and R.K. Sharma (1993). Effect of Nitrogen fertilizer on maize (*Zea mays* L.) varieties under rainfed conditions. *Indian J. Agron., 38*(2): 303-304.

Paradkar, V.K., R.K. Sharma, O.P. Rathor and V.K. Rastogi (1993). Performance of fodder maize (*Zea mays* L.)+grain maize intercropping system under rainfed conditions. Indian J. Agron., 38(3): 455-457.

Parmar, D.K. and Sharma Vinod (2001). Nitrogen requirement of single hybrid maize (*Zea mays* L.) under rainfed condition. *Ind. J. Agric. Sci.,* 71(4): 252-254.

Prasad, K. and A. Rafey (1996). Weed control in maize + soybean intercropping system under rainfed condition. Madras Agric. J., 83: 443 - 445.

Prasanna kumar, A.S. Halepyati , B.T. Pujari and B.K. Desai (2007). Effect of integrated nutrient management on productivity, nutrient uptake and economics of maize (Zea mays L.) under rainfed condition. Karnataka J. Agric. Sci., 20 (3): 462-465.

Rajashekhara Rao, B.K., K.L. Sahrawat, S.P. Wani and G. Pardhasaradhy (2010). Integrated Nutrient Management to Enhance On-Farm Productivity of Rainfed Maize in India. International Journal of Soil Science, 5: 216-225.

Ramesh, G. and T. Nadanassababady (2005). Impact of herbicides on weed and soil ecosystem of rainfed maize (*Zea mays* L.). *Indian J. Agric. Res*., 39(1): 31-36.

Rameshwar and C.M. Singh (1997). Economy in fertilizer use in maize – wheat sequence under rainfed condition in Himachal Pradesh. Indian J. Agric. Sci., 67 (12): 564 – 567.

Rana, K.S., R.K. Shivaran and Ashok Kumar (2006). Effect of moisture conservation practices on productivity and water use in maize based intercropping systems under rainfed condition. Indian J. Agron., 51(1):24-26.

Rana, K.S. and R.S.Choudhary (2006). Productivity, nitrogen uptake and water use in maize (*Zea mays* L.) - mungbean intercropping system as affected by nitrogen levels under rainfed conditions. Crop Res., 32(3): 306-308.

Rana, K.S. and R.K. Shivran (2003). Growth and yield of maize (*Zea mays* L.) as influenced by cropping system and moisture conservation practices under rainfed conditions. *Ann. Agric. Res. New Series*, 24(2): 350-353.

Reddy, U.S. (2003). Integrated weed management in rainfed maize. M. Sc (Agri) Thesis. ANGR Agricultural Univerity, Hyderabad, Telengana, India.

Rout, D. and M.R. Satapathy (1996). Chemical control in rainfed maize (*zea mays*). *Indian J. Argon.,* 39: 166–167.

Rout, D. and M.R. Satapathy (1996). Chemical weed control in rainfed maize (*Zea mays*). *Indian J. Agron.,* 41(1): 51-53.

Sachan, S.S. And U.S. Gangawar (1996). Effect of row spacing, moisture conservation practices and levels of nitrogen on growth and yield of maize (*Zea mays* L.) under rainfed conditions. Indian J. Soil Conservation, 24(2): 125-127.

Saha, G.P. and V.C. Srivastava (1992), Relative efficiency of chemical and cultural were control in rainfed maize. *Indian J. Agron.,* 37(4): 818-819.

Saini, J.P. (2000). Efficacy of atrazine as post-emergence herbicide for weed control in maize (*Zea mays*) under rainfed conditions. Back ←Abstract ↓Descriptors ↓Top ↑Indian Journal of Agricultural Sci., 70 (11): 801-803.

Sakthivel, N., A. Balasubramanian, S. Radhamani and P. Subbian (2003). Effect of in situ moisture conservation practices and intercropping system on yield of rainfed maize in western zone of Tamil Nadu. Madras Agric., J. 90 (7-9): 411-415.

Sakthivel, N., S. Radhamani, A. Balasubramanian and P. Subbian (2003a). Comparative performance of different maize based intercropping systems and planting patterns under rainfed situation. Madras Agric. J., 90 (7-9): 416-420.

Sangakkara, U.R., P.S.R.D. Bandaranayake, J.N. Gajanayake and P. Stamp (2004). Plantpopulations and yield of rainfed maize grown in wet and dry seasons of the tropics. Maydica, 49: 83-88.

Sari, V.K. and U. Puri (1997). Effect of phosphorus application with and without FYM rainfed maize – wheat – maize sequence. Indian J. Agric. Sci., 67(1): 13 -15.

Sarita Jha, V.K. Suri and T.S. Verma (1997). Response of rainfed maize to integrated nutrient management based on targeted yield concept in *a Typic Haplustalf. J. Indian Soc. Soil Sci.,* 45(3): 603- 605.

Seerat, A.Y., G.N. Dhanapal, M.A. Shankar, N. Krishnamurthy, M.R. Umesh and H.M. Manjumatha (2012). Effect of INM on growth and yield of rainfed maize in Alfisol. Mysore J. Agric. Sci., 46 (4): 913 -916.

Senapati, S.C. and S.D. Sharma (2007). Effect of buffer strips on soil erosion and yield of maize in rainfed slopy uplands in north-western plateau zone of Orissa. Indian J. Soil Cons., 35 (1): 47 - 49.

Sharma, A.R., A.S. Toor and H. Sur (2000). Effect of interculture operations and scheduling of Atrazine application on weed control and productivity of rainfed maize (*Zea mays* L.) in Shiwalik foot hills of Punjab. *Indian J. Agril. Sci.,* 70(1): 757-761.

Sharma, C.K. (2007). Growth and development studies in maize as affected by integrated weed management under rainfed condition. Indian J. Agron., 52 (4): 321 - 324.

Sharma, N.N., S.R. Paul and D. Sharma, (2000). Response of maize (*Zea mays* L.) to nitrogen and phosphorus under rainfed conditions of hill zone of Assam. *Ind. J. Agron.*, 45(1): 128-131.

Sharma, A.R., Singh Ratan, S.K. Dhyani and R.K. Dube (2010). Tillage and legume mulching effects on moisture conservation and productivity of rainfed maize (*Zea mays*)–wheat (*Triticum aestivum*) cropping system. Indian J. Agron., 55 (4): 244-247.

Sheoran, P, V. Sardana, S. Singh and B. Bhushan (2010). Bioeconomic evaluation of rainfed maize based intercripping systems with black gram under different spatial arranagements. Indian J. Agric. Sci., 80 (3): 244 – 247.

Shivran, R.K. and K.S. Rana (2003). Growth and yield of maize was influenced by cropping system and moisture conservation practices under rainfed conditions. Annals Agric, Res., 24 (2): 350 – 353.

Singh, M., S. Prabhukumar, C.V. Sairam and A. Kumar (2009). Evaluation of different weed management practices in rainfed maize on farmers' fields. Pakistan J. Weed Sci. Res., 15 (2-3): 183-189.

Singh, U., A.A. Saad and S.R. Singh (2008). Production potential, biological feasibility and economic viability of maize based intercropping systems under rainfed conditions of Kashmir valley. Indian J. Agric. Sci., 78(12): 1023 – 1027.

Singh, V.K. and R.P. Bajpai (1991). Intercropping in maize (Zea mays) under rainfed condition. *Indian Journal of Agronomy,* 36: 398-399.

Singh, V.P. (2000). Planting geometry in maize (Zea mays) and black gram (Phaseolus mungo) intercropping system under rainfed low hill valley of Kumaon. Indian J. Agron., 45(2): 274-278.

Singh, V.P. and V.K. Singh (2001). Productivity potential and economics of maize and soybean intercropping under rainfed low hill or valley situation of Uttaranchal. Indian J. Agron., 46: 27 - 31.

Solanki, N.S. (2010). Radiation use and rain water use efficiency in maize based intercropping system under rainfed condition. In: Agrometeorological services for farmers (Ed. Vyas pandey). Anand Agricultural University, Anand, P: 59 - 66.

Spaner, D., D.E. Mather and R.A. Brathwaite (1996). Yield and quality attributes of rainfed green immature corn in Trinidad. Hort.., 6(2): 131-134.

Subramaniam, M.V.R. and G. Bhee maiah (2003). Yield economics of rainfed maize in African babul (*Fadherbia albida*) based agro forestry system under different fertility levels. Indian J. Agric. Sci., 73(3): 138 – 141.

Sujatha, M.G. and B.S. Lingaraju (2007). Integrated Nitrogen Management in Maize Under Rainfed Condition. Karnataka J. Agric. Sci., 20 (4): 658 - 898.

Sujatha, M.G., B.S. Lingaraju, Y.B. Palled and K.V. Ashalatha (2008). Importance of integrated nutrient management practices in maize under rainfed condition. Karnataka J. Agric. Sci., 21(3): 334-338.

Suri, V.K. and U.K. Puri (1997). Effect of phosphorous application with and without FYM on rainfed maize – wheat – maize sequence. Indian J. Agric. Sci., 67 (1): 13 – 15.

Terarwal, J.P., B.R am and D.S. Mena (2011). Effect of INM on productivity, profitability nutrient uptake and soil fertility in rainfed maize. Indian J. Agron., 56(4): 373 – 376.

Tetarwal, J.P., Ram and D.S. Meena (2011). Bioeconomic feasibility of soybean intercropped with urdbean and maize under various row proportions in rainfed situations of south – eastern Rajastha. J. Oilseeds Res., 28 (2): 115 – 119.

Thakur, D.R. and Y. Sharma (1996). Integrated weed management in rainfed maize (*Zea mays* L.). Indian Weed Sci., 28 (3&4): 207-208.

Thakur, R.C., R.D.Sood, M.Bhargava, B.Singh and K.K. Mahajan (1992). Balanced fertilization for increasing food grains production under rainfed conditions of Himachal Pradesh. Fert. News, 37: 29 - 34.

Thiagalingam, K., N. Gould and P. Watson (1991). Effect of tillage on rainfed maize and soybean yield and the nitrogen fertilizer requirements for maize. *Soil and Tillage Research,* 19: 47–54.

Torres, V.R., J.H. Davila, A.B. Mendoza, F.R. Godina and R.K. Maiti (2004), Importance of agronomic characteristics in the grain yield of maize under irrigated and rainfed conditions. Crop Res., 27(2 & 3): 169-176.

Uwah, D.F., M.G. Solomon and F.I. Nya (2005). Response of rainfed popcorn (*Zea mays* L.*everta*) to nitrogen and inter-row spacings in south eastern rainforest of Nigeria. *J. Applied Chemistry Agric. Res., 9*: 21-29.

Vadivel, N., P. Subbian and A. Velayutham (2001). Effect of integrated nitrogen management practices on growth and yield of rainfed winter maize. Indian J. Agron., 46(2): 250 – 254.

Venkata Lakshmi, K. and A. Balasusbramanian (2009). Supplemental irrigation, green manuring and nitrogen levels on growth components, dry-matter accumulation and physiological parameters in dryland maize. Madras Agric. J., 96 (7-12): 370-373.

Verma, T.S. (1991). Influence of methods of farmyard manure application on maize under rainfed conditions. *Crop Res.,* 4 (1): 161-164.

Vishalu, Nagaraju, Nanjappa, Kalyanamurthy, Devakumar and Kalaraju (2009). Effect of plant density and fertilizer levels on yield and economics of hybrid maize under rainfed condition. Crop Res., 37 (1-3): 49 - 51.

# 18

# Baby Corn Cultivation

Baby corn is a popular vegetable in USA, Europe and Southeast Asia. The demand for baby corn is rapidly increasing in urban areas in India. Baby corn is not a separate type of corn like sweet corn or popcorn and any corn type can be used as baby corn. It is delicacy, which can be profitably used in prolific types of corn i.e. those types , which bear two or more ears per plant. The shank with unpollinated silk is baby corn. The economic product is harvest just after the silks emerge (1-2 cm long). Baby corn has immense potential as a salad and as cooked vegetable. It is used as an ingredient in ChopSuey (Chinese dish), soups, deep fried baby corn with meat, rice and other vegetables. Large number of dishes may be prepared from baby corn. Baby corn is highly nutritive. The nutritive value of baby corn is given in Table 1. Since only immature cob is harvested as the economic produce, the crop meant as baby corn can be harvested within 50-55days. Thus in the areas adjoining cities or other urban areas (peri-urban agriculture) multiple crop of baby corn can be raised which would fetch greater income to the farmers. Baby corn can be effectively used as both a nutritious vegetable and as an export crop to earn valuable foreign exchange. After harvest the still young plants may be used as fodder for cattle.

**Table 1. Nutritive value of baby corn (per 100 g of edible portion)**

| Nutrient | Content |
|---|---|
| Moisture (g/ 100 g) | 7.37 |
| Crude protein (g/ 100 g) | 10.04 |
| Crude fat (g/ 100 g) | 4.43 |
| Crude fibre (g/ 100 g) | 2.40 |
| Ash (g/ 100 g) | 1.34 |
| Total carbohydrates (g/ 100 g) | 81.97 |
| Energy (K.cal/100 g) | 375.67 |
| Total soluble sugars (g/ 100 g) | 0.14 |
| Calcium (mg/100 g) | 17.76 |
| Phosphorus (mg/ 100 g) | 197.89 |
| Iron (mg/ 100 g) | 2.73 |
| Thiamine | 0.50 |
| Riboflavin | 0.08 |
| Ascorbic Acid | 11.00 |

Baby corn is a young finger like unfertilized cob of maize with two to three centimeter emerged silk preferably harvested within 1-3 days of silk emergence depending upon the growing season. It can be eaten raw or as salads, chutney, vegetables, pickles, kheer, Chinese preparations, etc. The desirable size of baby corn is 6 to 11 cm in length and 1.0 to 1.5 cm in diameter with regular row arrangement. The most preferred colour by the consumers / exporters is generally creamish to very light yellow. It has great potential for export. Thailand is the major baby corn producing and exporting country in the world while India has emerged as one of the potential baby corn producing countries because of the low cost of production as compared to many other countries. In late 1970's people in USA and western countries started to consume raw cobs called baby corn. Change in foot habit from non vegetarian to vegetarian aggravated the consumption of vegetables especially baby corn. Baby corn production, being a recent development has proved as an enormously successful venture in countries like Thailand and Taiwan.

**Table 2. Baby corn producing, exporting and importing countries**

| Particulars | Countries |
|---|---|
| Major baby corn producing countries | Thailand, Taiwan, Guatemala, South Africa, Zambia, Zimbabwe, Peoples Republic of China, Nepal, Japan, Sri Lanka and India |
| Major baby corn exporting countries | Thailand, Taiwan, Guatemala, South. Africa, Zambia, Zimbabwe and Peoples Republic of China |
| Major baby corn importing countries | USA, Japan, Hong Kong, Singapore, Australia, Malaysia, Canada, Saudi Arabia, New Zealand, European countries and India |

Fresh baby corn (sweet corn) is highly perishable in hot weather. For every 5°C increase in pulp temperature sugar breakdown to starch doubles, meaning that baby corn can completely lose flavour in a short period of time. Hence, growers are suggested that cultivation of baby corn be taken in areas near to main market or having proximity to tourist places. It can be cultivated in other places also provided there are assured transport facilities for lifting and carrying the same to the consumption markets.

Dass *et al.* (2004) reported that the cultivation of baby corn is getting popular due to its deliciousness, high nutritive value and great potential for export. Comparatively baby corn cultivation is more remunerative than the grain maize in a given period of time. In the presence of assured market and infrastructure for processing, at least double income can be generated as compared to common grain crop. Further, baby corn cultivation is a solution to problems emerging from continuous rice-wheat system like lowering of water table, *Phalaris minor* (weed) and degradation of soil health. Single cross early maturing hybrids, prolific in nature and having medium height are the preferred genotypes. While three to four baby corn crops can be taken up in North India, round the year production is possible in South India.

**Selection of suitable varieties**

No variety has been exclusively bred for baby corn purpose in India. However, early and prolific varieties like Him 123, VL 42, and Him 129. Early composite and hybrid variety Prakash, PEHM 1 and PEHM 2 can successfully be cultivated as baby corn (Table 3). At the time of selection of cultivar preference is to be given for short stature and prolific cultivars. Hybrids are preferred over open pollinated varieties because hybrids are more uniform in flowering. Thus they may require only 4-5 times plucking. Contrary to this nonuniform flowering of varieties leads to prolonged harvesting. Short stature materials can be well accommodated in higher plant densities.

**Table 3. List of cultivars recommended for cultivation for baby corn purpose and their growing season**

| Variety | Hybrid | Recommended season | Recommended region | Time of harvest | Yield (cobs/ha) |
|---|---|---|---|---|---|
| Him 123, VL42, HIM 129, VL Makka 16, Chandan Safed, Makka-2, Chandan Makka-1, Chandan Makka-3,Navin, Diara-3, D-765, Farm Sumeri, Kanchan, MCU-508, VL Makka-41, Kiran(J 660), VL Makka-88, Pusa Comp-1, surya, Gujarat Makki-1, Madhuri (sweet corn), Panch Ganga, Megha,Gurat Makai 2, Birsa Makai-1,C-14, JM.8, Shakti-1, Amar (D-941), Aravali Makka-1, Gujarat Makkai-4, Narmada Moti, Priya sweet | Ganga 9, Him 128, Deccan-1 DHM-107, Pusa Early Hybrid Makka-1, Pusa Early Hybrid Makka-2, Jh-3459, Pusa Early Hybrid Makka-3, Vivek Hybrid-9, Parkash | Kharif and Rabi | Across the Country in Kharif & Peninsular India in Rabi | 45-50 Days in kharif 50-55 Days in Rabi | 1,50,000 |

Short duration, prolific, single cross hybrid with medium height should be selected. HM-4, Prakash, etc are suitable hybrids for baby corn. To earn better returns in shortest period from baby corn crop, early maturing single cross hybrids are suggested with silking period of 45-50 days in kharif, 75- 80 days in spring and 120-130 days in winter in North Indian states. Baby corn is cultivated with high plant populations. In general, with high plant density, lodging has been observed. Therefore, hybrid with medium plant height and loding resistance is preferred. Cultivars producing more than one ear per plant are desirable for higher yield of baby corn. As baby corn is cultivated in higher plant density, fertilizer responsive cultivars are more suitable for the purpose. Varieties possessing no barrenness trait are desirable for higher yield. For getting better fodder quality, genotypes grown for baby corn cultivation must possess the stay green traits. After the picking of baby corn, left over plant material is used as green fodder. To accommodate more plant per unit area, erect leaves is a very good trait. This allows better interception of light and hence, enhances net photosynthesis. The length of dehusked ear preferably not exceeding 10 cm with 1.0 -1.5 cm diameter. Husked to dehusked ratio should be 5-6: 1. Harvesting period should be within 10- 12 days in kharif / spring and 20 days in winter crop. Each plant should give 3-4 pickings of baby corn. Each baby corn should maintain desirable size and colour. Traits contributing to higher baby corn yield are as follows:

**HM 4**

The pedigree is HKI 1105 x HKI 323. It is released by CCS HAU, Karnal during 2005. It is recommended for Haryana, Delhi, Punjab and Eastern Uttar Pradesh. It is medium maturity baby corn, orange, flint, medium tall, tolerance to frost and cold. The average yield is 6000- 6800 kg/ha.

**VL Baby corn 1**

The parentage is [{(VL 16 x Murulia) x VL 16} x VL 16] and is released by VPKAS, Almora during 2005. It can be grown across the country. It is extra early maturing, yellow, prolific with average yield of 3800 - 4000 kg/ha with baby corn yield as 1200 - 1300 kg/ha.

**VL 42**

It is a double cross hybrid and produced baby corn yield of 1428 kg/ha with an average of more than 1 - 2 ears per plant. It is recommended for cultivation at the national level. It is extra early, orange, flint and resistance against stem borer as well as leaf blight disease. It is earliest in producing baby corn (45 days) than other varieties. Fresh weight per green cob at harvest is 35 - 40 g. Fresh weight per baby corn is 4.5 - 5.5 g. Diameter of fresh baby corn is 3.5 - 4.0 cm and length of fresh baby corn is 6.5 - 7.0 cm. Green fodder yield is 20.0 - 25.0 t/ha.

**COBC 1**

The parentage is composite involving UMI 836 and UMI 836-1-2. Duration is 55 - 65 days (fresh cobs harvest). It is recommended for all maize growing areas. It is suitable for both rainfed/irrigated conditions. It yields 6.7 tonnes of tender cobs and 32.3 tonnes of green fodder / ha for irrigated and 3 tonnes of tender cobs per ha for rainfed condition. Special features are 2 to 3 cobs / plant, possessing delightful sweet taste and flavour. It is suitable for inland and export market. Green fodder has good forage value. Sheath colour is green. Node colour is green. Mid rib is white colour. Cob size is small. Grain colour is yellow. Nature of kernels is dent.

**Sonal**

This is baby corn hybrid and suitable for cultivation in all seasons. It is ready to harvest within 55-56 days. It is resistant to pest and diseases. Thus minimum attacks occur during the cultivation. The grains are rich in phosphorus and calcium and less in cholesterol content. It finds its application in food processing, salad, soup and in many other recipes. It is developed by Super Agri Seeds Private Limited, Secunderabad, Andhra Pradesh.

**Table 4. Verities on yield and economics of baby corn**

| Variety | Barrenness (%) | Baby corn yield (kg/ha) | Green fodder yield (t/ha) | Net return (Rs/ha) | B:C ratio |
|---|---|---|---|---|---|
| VL makka 16 | 8.8 | 1103 | 24.2 | 18322 | 2.11 |
| VL makka 41 | 9.1 | 1031 | 19.1 | 15082 | 1.91 |
| VL makka 42 | 8.4 | 1428 | 19.8 | 24980 | 2.48 |
| VL makka 88 | 8.9 | 1046 | 23.3 | 16717 | 2.00 |
| VL Amber (pop corn) | 11.6 | 846 | 26.1 | 12751 | 1.78 |
| Local (Dhyari) | 11.2 | 814 | 23.6 | 11376 | 1.78 |
| CD 5 % | - | 74 | 2.0 | 2172 | 0.09 |

(Pandeyetal.,2002)

At Almora (Uttranchal), significantly higher baby corn yield (1428 kg/ha), net return (Rs. 24980/ha) and B:C ratio (2.48) were recorded in VL makka 42. Maize composite VL makka 16 was next best variety for baby corn yield and monetary returns (Pandey et al., 2002).

**Soil**

The soil requirement, land preparation and crop management practices for baby corn are the same as that for sweet corn and popcorn. For baby corn two systems are used.

## Sowing

One system uses standard populations of about 58000 plants / ha, where the top ear is left on the plant for grain corn or sweet corn and subsequent ears harvested for baby corn. The second system uses high plant populations at a spacing of 45 cm x 20 cm with 2 plants per hill, having a population density of 175000 plants/ha, where all ears are harvested for baby corn. The standard plant populations produce yields of about 46.5 q unhusked ears (4.65 q husked ears) / ha, while the high populations produce yields of about '93-106 q unhusked ears (9.3-10.60 q of husked ears) / ha. A seed rate of 20-25 kg/ha is recommended. This would thereby realize more number of cobs and consequently more returns to the farmers.

It can be sown round the year in southern India. In northern India, it can be sown from February to November. It can be grown through transplanting in furrows in December-January in northern India. For this purpose nursery should be raised in November. Generally, August to November planting yields best quality baby corn. Sowing should be done on southern side of the ridges with 60 cm spacing with 15 - 20 cm plant spacing depending upon plant type (erect/spreading). Optimum seed rate is 22-25 kg/ha. The baby corn crop can be raised continuously for several months if staggered sowing of 10 days interval is adapted. The first sowing can be done even under protected structures (poly houses) and its availability can be made even before the normal season.

**Table 5. Plant density on yield and economics of baby corn**

| Plant densities | Barrenness (%) | Baby corn yield (kg/ha) | Green fodder yield (t/ha) | Net return (Rs/ha) | B:C ratio |
|---|---|---|---|---|---|
| 50 x 18 cm (111000 plants) | 6.9 | 900 | 20.3 | 12384 | 1.76 |
| 50 x 15 cm (133000 plants) | 9.0 | 1108 | 23.4 | 17592 | 2.06 |
| 50 x 12 cm (166000 plants) | 13.1 | 1148 | 24.5 | 19165 | 2.14 |
| CD 5 % | - | 110 | 2.7 | 3800 | 0.17 |

(Pandey et al.,2002)

At Almora (Uttranchal), plant barrenness was markedly increased from 6.o to 13.1 % with increase in plant density from 111000 to 166000 plants/ha. Baby corn and green fodder yields, net return and B:C ratio were increased with plant density (Pandey et al., 2002).

**Table 6. Plant spacing on baby Corn**

| Plant Spacing | Cob yield (q/ha) | Baby corn yield (q/ha) | Fodder yield (q/ha) | Net return (Rs/ha) | B : C ratio |
|---|---|---|---|---|---|
| 60 x 20 cm | 98.5 | 16.3 | 365.0 | 3806 | 2.37 |
| 40 x 20 cm | 100.7 | 20.8 | 419.3 | 50843 | 3.00 |
| 60 x 10 cm | 107.7 | 18.7 | 475.0 | 47089 | 2.60 |
| 40 x 10 cm | 102.5 | 19.1 | 501.0 | 48670 | 2.52 |
| CD 5% | 4.5 | 2.0 | 15.4 | - | - |

(Thakur et al.,1998)

At Bajaura (H.P), plant spacing of 40 x 20 cm gave significant higher baby corn (17.4 q/ha), cob with husk (86.89q/ha) yields. However maximum green fodder yield and N uptake by plants were recorded in 40 x 10 cm planting (Thakur et al., 1998).

**Table 7. Effect of land configurations on baby corn**

| Land configuration | Barrenness (%) | Baby corn yield (kg/ha) | Fodder yield (t/ha) | Net return (Rs/ha) | B : C ratio |
|---|---|---|---|---|---|
| Flat bed sowing | 9.33 | 817.9 | 23.87 | 32940 | 2.23 |
| Flat bed sowing + earthing up 30 DAS | 8.35 | 865.8 | 27.42 | 34830 | 2.21 |
| Broad bed and furrow system | 7.08 | 997.5 | 30.83 | 42150 | 2.44 |
| Paired row sowing on flat bed | 8.83 | 976.7 | 28.77 | 41250 | 2.78 |
| Paired row sowing on broad bed and furrow system | 616 | 1069.2 | 32.58 | 45120 | 2.64 |
| CD 5 % | 1.15 | 32.15 | 2.21 | 1542 | 0.21 |

(Panwas and Munda,2006)

At Umiam (Meghalaya), maximum net return (Rs. 45120) was realized with broad bed and furrow system method while benefit cost ratio was higher with paired row sowing on flat bed (Panwas and Munda, 2006).

Thakur *et al.* (1998) reported maximum net return of Rs. 50,843 per ha and net return per rupee invested (Rs. 3.00) when crop spaced at 40 cm x 40 cm. The net profit and the benefit: cost ratios were higher at a spacing of 40 cm x 20 cm than other spacing of 40 cm x 20 cm and 40 cm x 15 cm (Sahoo and Panda, 1999). Pandey *et al.* (2002) concluded that the plant densities 1,33,000 and 1,66,000 plants/ ha recorded significantly more net returns and benefit : cost ratio than 1,11,000 plants / ha.

### Seed treatment

Seeds should be treated with fungicides and insecticides before sowing to protect it from seed and soil borne diseases and some insect-pests. Bavistin + Captan in 1:1 ratio @ 2 g/kg seed is done for Turcicum leaf blight (TLB), Banded leaf and sheath blight (BLSB), Maydis leaf blight (MLB). Apron 35 SD @ 4 g/kg seed is recommended for Brown stripe downy mildew (BSDM). Captan 2.5 g/kg seed can be done for Pythium Stalkrot. Seed treatment with Fipronil @ 4 ml/kg is recommended for termite and shoot fly.

### Nutrient management

Nutrient application should be based on soil test value. Generally 150-180:60:60:25 Kg/ha N, P, K and $ZnS0_4$ with 8-10 t/ha FYM should be applied. Full dose of phosphorus, potash and zinc and 10 % N should be applied as basal dose. The remaining dose of nitrogen should be applied in four splits viz., 20 % N at 4leaf stage, 30 % N at 8leaf stage, 25 % N before detasseling and 15 % N after detasseling to avoid losses and to meet the requirement throughout the crop cycle.

Nitrogen is applied @ 150-200 kg/ha in three equal splits as basal dose at the time of sowing, at knee high stages (20-25 DAS) and at pre-tasseling stages (40-45 DAS). On the other hand, phosphorus and potassium are applied as basal dost at the usual rate of 60 kg/ ha and 40 kg/ha, respectively.

**Table 8. Rate and time of application of nitrogen on yield and economics of baby corn**

| Treatments | Barrenness (%) | Baby corn yield (kg/ha) | Green fodder yield (q/ha) | Net return (Rs/ha) |
|---|---|---|---|---|
| **Nitrogen dose (kg/ha)** | | | | |
| 60 | 13.5 | 1290 | 211 | 26592 |
| 90 | 10.2 | 1458 | 222 | 31174 |
| 120 | 7.9 | 1573 | 227 | 33864 |
| CD 5 % | - | 108 | NS | 5623 |
| **Nitrogen scheduling** | | | | |
| All at sowing | 11.3 | 1382 | 222 | 29057 |
| All at knee high | 10.3 | 1492 | 200 | 31265 |

| | | | | |
|---|---|---|---|---|
| ½ at sowing + ½ at knee high | 10.1 | 1493 | 228 | 32202 |
| ¾ at sowing + ¼ at knee high | 10.7 | 1410 | 227 | 30053 |
| ¼ at sowing + ¾ at knee high | 10.2 | 1419 | 223 | 30135 |
| CD 5 % | - | NS | 21.6 | NS |

(Pandey et al.,2000)

At Almora (Uttaranchal), application of 120 kg N/ha gave significantly higher yield and net return compared with 60 and 90 kg N/ha respectively. Higher improvement in yield attributes, baby corn yield and net return were recorded with application nitrogen in two equal splits at sowing and at knee high stages (Pandey et al., 2000).

**Table 9. Integrated nutrient management on baby corn**

| INM practices | Plant height (cm) | DMP (g/plant) | Cob yield (kg/ha) | Green fodder yield (kg/ha) | Net return (Rs/ha) | B:C ratio |
|---|---|---|---|---|---|---|
| 100 % RDF (120:40:30 kg NPK/ha) | 177 | 54.78 | 1422 | 19433 | 12145 | 2.03 |
| 125% RDF | 207 | 60.83 | 1841 | 21500 | 16814 | 2.36 |
| 150% RDF | 227 | 64.68 | 2048 | 22667 | 18910 | 2.47 |
| Vermicompost 3 t/ha | 165 | 45.57 | 1167 | 17833 | 65913 | 1.46 |
| Vermicompost 3.75 t/ha | 189 | 51.65 | 1575 | 19933 | 10518 | 1.69 |
| Vermicompost 4.5 t/ha | 225 | 58.51 | 1989 | 22333 | 14673 | 1.90 |
| 50 % RDF + Vermicompost 1.5 t/ha | 187 | 62.18 | 1492 | 20400 | 12187 | 1.94 |
| 62.5 % RDF +Vermicompost 1.875 t/ha | 203 | 67.08 | 1893 | 22833 | 16574 | 2.20 |
| 75 % RDF +Vermicompost 2.25 t/ha | 233 | 74.34 | 2258 | 24839 | 20384 | 2.40 |
| 50 % RDF + Vermicompost 1.5 t/+ Biofetilizers | 192 | 68.60 | 1913 | 23500 | 17898 | 2.38 |
| 62.5 % RDF + Vermicompost 1.875 t/ha +Biofetilizers | 206 | 74.14 | 2284 | 26167 | 22101 | 2.60 |
| 75 % RDF + Vermicompost 2.25 t/ha + Biofetilizers | 239 | 80.08 | 2641 | 30133 | 26815 | 2.83 |
| CD 5 % | 27 | 5.12 | 258 | 1921 | 2864 | 0.21 |

(Dadarwal et al., 2009)

At Udaipur (Rajasthan), maximum plant height, dry matter production (DMP), cob yield, green fodder yield and monetary return were obtained by applying 75 % RDF + 2.25 t/ha vermicompost + biofertilizers (seed treatment of Azotobacter and Phosphobacteria @ 600 g/ha each) followed by 62.5 % RDF + 1.875 t/ha vermicompost + biofertilizers (Dadarwal et al., 2009).

**Table 10. Yield, of protein, vitamin C and carbohydrate content of baby corn as influenced by fertilizer levels**

| Fertilizer levels | Yield (kg/ha) | | Green fodder yield (t/ha) | Protein (%) | Ascorbic acid (mg/100g) | Carbohydrate (%) |
|---|---|---|---|---|---|---|
| | Cob with husk | Cob without husk | | | | |
| 150:75:40 $N:P_2O_5:K_2O$ kg /ha | 1907 | 672 | 51.50 | 10.81 | 11.83 | 43.85 |
| 200:100:50 $N:P_2:K_2O$ kg /ha | 1769 | 532 | 56.50 | 11.63 | 12.26 | 47.59 |
| 250:150:60 $N:P_2O_5:K_2O$ kg/ ha | 1937 | 610 | 58.50 | 12.68 | 11.93 | 52.01 |
| CD 5 % | 140 | 51 | 1.56 | 0.09 | 0.19 | 1.26 |

(Gulabrao, 2010)

At Dharwad (Karnataka), 250:125:60 N: $P_2O_5:K_2O$ kg/ha recorded significantly higher cob yield with husk (1937 kg /ha), contents of quality, growth and organoleptic parameters *viz.,* protein (12.68 %), total carbohydrate (52.01 %), non reducing sugar (0.324 %), reducing sugar (0.075%) and total sugar (0.416 %) compared to 200:100:50 N: $P_2O_5:K_2O$ kg/ha (Gulabrao, 2010).

### Weed management

Broad leaf weeds and most of the grasses can be controlled by pre-emergence spray of Atrazine @ 1.0-1.5 kg/ ha in 500-600 litre of water. While spraying, the person who is doing spray should move backward so that the atrazine film on the soil surface may not be disturbed. Preferably, three nozzle booms may be used for proper ground coverage and saving of time. One to two hoeings are recommended for aeration and uprooting of the remaining weeds. While doing hoeing, the person should move backward to avoid compaction of soil and to facilitate better aeration.

## Water management

First irrigation should be given very carefully. Water should not overflow on the ridges. The irrigation should be applied in furrows upto 2/3rd height of the ridges. Irrigation should be given as and when required by the crop depending upon the rains and moisture holding capacity of the soil. Young seedlings, knee high stage, silking and picking are the most sensitive stages for water stress for crops and irrigation should be ensured at these stages. Light and frequent irrigations are desirable for crop. During winter (mid December to mid February) soil should be kept wet to avoid frost injury.

At Madurai (Tamil Nadu), the applied irrigation water was 410 mm for drip irrigation of 100 % PE and 352 mm and 294 mm for drip irrigation of 80 % PE and 60 % PE respectively. The rainfall recorded during the cropping season was 138.2 mm and the effective rainfall (ER) was higher in drip irrigation of 60 % PE to the tune of 51.0 mm during growth period compared to drip irrigation regimes of 80 % PE and 100 % PE where in the effective rainfall were 31.6 mm and 19.2 mm, respectively. Total water used was higher for drip irrigation of 100 % PE (429.2 mm) compared to drip irrigation of 80 % PE (383.6 mm), 60 % PE (345.0mm) respectively. The WUE was lower in higher drip irrigation regime whereas it was reverse with limited drip irrigation of 80 % PE in which case the applied water might have been utilized more effectively (Rani and Sukumari, 2005).

**Table 11. Effect of drip irrigation regimes and fertilizer application methods on water use efficiency of baby corn**

| Treatments | Number of irrigations | Water applied (mm) | Effective rainfall (mm) | Total water used (mm) | Green cob yield (kg /ha) | WUE (kg /ha/ mm) |
|---|---|---|---|---|---|---|
| Drip irrigation of 60 % pan evaporation + Soil application of 100 % RDF of N and K | 19 | 294.24 | 51.0 | 345.2 | 4065 | 11.8 |
| Drip irrigation of 60 % pan evaporation + fertigation of 50 % RDF of N and K | 19 | 294.24 | 51.0 | 345.2 | 5885 | 17.8 |
| Drip irrigation of 60 % pan evaporation + fertigation of 100 % RDF of N and K | 19 | 294.24 | 51.0 | 345.2 | 4967 | 14.4 |

| Treatments | Number of irrigations | Water applied (mm) | Effective rainfall (mm) | Total water used (mm) | Green cob yield (kg /ha) | WUE (kg /ha/ mm) |
|---|---|---|---|---|---|---|
| Drip irrigation of 80 % pan evaporation + Soil application of 100 % RDF of N and K | 19 | 352.32 | 31.6 | 383.9 | 5386 | 14.0 |
| Drip irrigation of 80 % pan evaporation + fertigation of 50 % RDF of N and K | 19 | 352.32 | 31.6 | 383.9 | 7849 | 20.4 |
| Drip irrigation of 80 % pan evaporation + fertigation of 100 % RDF of N and K | 19 | 352.32 | 31.6 | 383.9 | 5500 | 14.3 |
| Drip irrigation of 100 % pan evaporation + Soil application of 100 % RDF of N and K | 19 | 410.40 | 19.2 | 429.6 | 4395 | 10.2 |
| Drip irrigation of 100 % pan evaporation + fertigation of 50 % RDF of N and K | 19 | 410.40 | 19.2 | 429.6 | 6515 | 15.2 |
| Drip irrigation of 100 % pan evaporation + fertigation of 100 % RDF of N and K | 19 | 410.40 | 19.2 | 429.6 | 5400 | 12.6 |

(Rani and Sukumari, 2005). Entire 'P' was applied as basal. Fetigation of N and K was given at 10, 20, 30 and 40 DAS in four splits. The irrigation was scheduled once in 3 days.

**Table 12. Husked baby corn and green fodder yield as influenced by irrigation schedules**

| Irrigation schedules | Husked baby corn yield (q /ha) | Green fodder yield (t /ha) | Total water used (mm) | Water use efficiency (kg/ ha- mm) | Net returns (Rs/ ha) | B:C ratio |
|---|---|---|---|---|---|---|
| Drip at 100% PE | 85.53 | 65.14 | 443 | 19.31 | 91101 | 4.12 |
| Drip at 75% PE | 78.66 | 51.04 | 346 | 22.70 | 77164 | 3.56 |
| Drip at 50% PE | 72.36 | 38.01 | 250 | 28.97 | 64164 | 2.93 |
| Surface irrigation | 78.92 | 51.12 | 597 | 13.23 | 81278 | 4.54 |
| CD 5% | 5.23 | 4.54 | - | 1.58 | 6454 | 0.30 |

(Anil kumar, 2005)

At Dharwad (Karnataka), drip at 100 % PE recorded significantly higher number of leaves, dry matter production, husked baby corn length, girth, number of babies / plant, husked baby corn yield (85.53 q/ha), green fodder yield (65.14 t/ha), gross returns (Rs. 112967/ha) and net returns (Rs. 91101/ha) over drip at 75 % PE, 50 % PE and surface irrigation. The quality parameters, crude protein content was higher in drip irrigation scheduled at 100 % PE, but crude fibre and reducing sugars were significantly higher in drip at 50 % PE over other irrigation schedules (Anil kumar, 2005).

## Intercropping

Baby corn being highly remunerative crop they can successfully be cultivated in periurban agriculture. However, instead of cultivating baby corn as sole crop it may be intercropped with other highly remunerative crops like marigold, tuberose, gladiolus, radish spices, pea etc. This provides additional income to the farmers from unit area and makes agriculture more sustainable.

At Lam (Andhra Pradesh), growth and yield components of baby corn such as plant height, dry matter production, number of cobs, cob yield were improved due to the application of biofertilizers along with inorganic inputs. Among the cropping systems, baby corn + vegetable cowpea recorded higher growth and yield components. This was followed by baby corn + dolichos bean (Reddy et al., 2009).

**Table 13. Yield and nutrient uptake of baby corn as influenced by intercropping systems**

| Intercropping systems | Yield (kg /ha) | | | Nutrient uptake (kg /ha) | | |
|---|---|---|---|---|---|---|
| | Green cob | Baby corn equivalent | Fodder yield | N | P | K |
| Baby corn alone | 7527 | 7677 | 21500 | 180.6 | 21.5 | 335.3 |
| Baby corn + Radish | 7437 | 9310 | 21200 | 181.3 | 21.1 | 334.6 |
| Baby corn + Coriander | 7493 | 9269 | 21300 | 181.4 | 21.3 | 336.4 |
| CD 5 % | NS | 115 | NS | NS | NS | NS |

(Thavaprakaash et al., 2005)

At Coimbatore (Tamil Nadu), sole crop situation resulted higher weed population and weed dry matter than intercropped baby corn. The results were reverse for weed smothering efficiency. Babu corn equivalent was higher (9310 kg/ha) with baby corn + radish intercropping system than baby corn alone. Intercropping systems failed to influence the green cob yield (Thavaprakaash et al., 2005).

## Intercultivation

**Table 14. Effect of growth regulators on baby corn**

| | Reducing sugar (%) | Crude protein (%) | Ascorbic acid (mg/100) | Cob length (cm) | | Days to silk | Days to first harvest | Yield / plant (g) |
|---|---|---|---|---|---|---|---|---|
| | | | | First | Second | | | |
| GA3 200 ppm | 2.43 | 12.34 | 12.48 | 107 | 98 | 52.3 | 54.0 | 72.0 |
| 400 ppm | 2.57 | 13.12 | 13.00 | 114 | 102 | 52.3 | 54.3 | 71.0 |
| 60 ppm | 2.41 | 12.54 | 12.48 | 116 | 104 | 53.3 | 55.3 | 71.0 |
| Benzyl adenine 10 ppm | 2.38 | 12.25 | 12.48 | 100 | 87 | 51.0 | 53.0 | 72.0 |
| 20 | 2.42 | 12.54 | 12.48 | 101 | 89 | 49.3 | 51.3 | 71.7 |
| 30 | 2.37 | 12.25 | 12.48 | 103 | 91 | 49.3 | 51.3 | 71.0 |
| IAA 10 ppm | 2.34 | 12.25 | 12.48 | 105 | 92 | 49.0 | 51.3 | 72.0 |
| 20 ppm | 2.38 | 12.54 | 12.48 | 104 | 93 | 49.6 | 52.0 | 71.0 |
| 30 ppm | 2.35 | 12.25 | 12.48 | 106 | 94 | 50.3 | 52.7 | 71.0 |
| Cycocel 500 ppm | 2.42 | 12.54 | 12.48 | 94 | 82 | 49.0 | 51.0 | 72.0 |
| 1000 ppm | 2.53 | 12.83 | 12.74 | 91 | 77 | 48.3 | 50.3 | 72.7 |
| 1500 ppm | 2.42 | 12.54 | 12.48 | 87 | 74 | 48.0 | 50.0 | 73.0 |

| | Reducing sugar (%) | Crude protein (%) | Ascorbic acid (mg/100) | Cob length (cm) | | Days to silk | Days to first harvest | Yield / plant (g) |
|---|---|---|---|---|---|---|---|---|
| | | | | First | Second | | | |
| Paclobutrazol 200 ppm | 2.34 | 12.25 | 12.48 | 87 | 74 | 47.0 | 49.0 | 73.2 |
| 400 ppm | 2.37 | 12.54 | 12.48 | 85 | 74 | 46.3 | 48.3 | 73.7 |
| 600 ppm | 2.35 | 12.25 | 12.48 | 83 | 68 | 46.0 | 48.0 | 74.0 |
| Control | 2.28 | 11.66 | 12.22 | 95 | 83 | 49.6 | 51.6 | 69.0 |
| CD 5% | 0.05 | 0.56 | 0.33 | 1.72 | 1.7 | 1.3 | 1.3 | 1.5 |

(Vani et al.,2004)

At Hyderabad (Andhra Pradesh), GA3 @ 40ppm (sprayed twice with tassel emergence at 5 days interval) followed by cycocel @ 1000ppm were found to be best in improving growth, yield and quality of baby corn (Vani et al., 2004).

**Table 15. Effect of plant growth regulators on quality parameters and yield of baby corn**

| Growth regulators | Total sugar (%) | Starch (%) | Protein (%) | Crude protein (%) | Vitamin A(mg /100 g) | Vitamin C (mg /100 g) | Cob yield (kg/ha) |
|---|---|---|---|---|---|---|---|
| Control (water spray) | 1.36 | 8.11 | 5.10 | 12.81 | 0.145 | 13.44 | 7291 |
| Mepiquot chloride (200 ppm) | 1.62 | 7.95 | 5.44 | 13.62 | 0.143 | 13.35 | 8003 |
| NAA (40 ppm) | 1.59 | 7.92 | 5.38 | 13.58 | 0.142 | 13.39 | 7872 |
| Putrescine (50 ppm) | 1.48 | 8.06 | 5.23 | 13.21 | 0.143 | 13.41 | 7569 |
| CD (P=0.05) | 0.08 | NS | 0.10 | 0.29 | NS | NS | 231 |

(Muthukumar et al., 2005)

At Coimbatore (Tamil Nadu), growth regulators viz. Mepiquat chloride 200 ppm and NAA 40 ppm improved quality as well as cob yield of baby corn (Muthukumar et al., 2005).

**Table 16. Influence of topping practices on yield of baby corn**

| Topping practices | Yield (kg /ha) | | Length of corn (cm) | | Diameter of corn (cm) | | Weight of corn (g) | |
|---|---|---|---|---|---|---|---|---|
| | Green cob | Green fodder | I harvest | V harvest | I harvest | V harvest | I harvest | V harvest |
| Detasseling alone | 6370 | 35000 | 9.0 | 8.5 | 1.5 | 1.6 | 9.4 | 8.5 |
| Topping beyond 9th internode | 7018 | 34200 | 9.9 | 9.3 | 1.7 | 1.6 | 10.4 | 9.3 |
| Topping beyond 10th internode | 7183 | 34400 | 10.5 | 9.5 | 1.7 | 1.6 | 10.5 | 9.5 |
| Topping beyond 11th internode | 6520 | 34800 | 9.2 | 8.6 | 1.5 | 1.5 | 9.7 | 8.6 |
| CD 5 % | 416 | NS | 0.4 | 0.4 | 0.06 | 0.05 | 0.6 | 0.4 |

(Rathika et al., 2008)

At Coimbatore (Tamil Nadu), topping beyond 10th internode favourably influenced the yield parameters *viz.,* length, diameter of cobs and corns and green cob yield of baby corn (Rathika et al., 2008).

## Protection from serious insect pests

Stem borer (*Chilo partellus*), pink borer (*Sesamia inferens*) and sorghum shoot fly (*Atherigona spp*) are serious problems in kharif, rabi and spring seasons respectively. First spray is given with 500 g Carboryl or 625 ml Endosulfan (35EC) application of insecticide in 500 litre water in the central whorl of plant. Second spray is sprayed with 750 g Carboryl or 900 ml Endosulfan (35EC) in 700-800 litre water, if required. For control of stalk borers Endosulfan 35 EC @ 2ml/litre of water is sprayed on 10-14 days old plants.

## Detasseling

To maintain the quality of baby corn, detasseling is an essential operation. It is done by removing the tassel of the plant as soon as it emerges from the flag leaf. It should be practiced row-wise. While detasseling, leaf should not be removed which will otherwise affect net photosynthesis and ultimately reduce average baby corn yield. The removal of 1 to 3 leaves along with tassel reduces 5-15 % yield of baby corn. The removed tassel should not be thrown in the field as it is nutrient rich and should be fed to the cattle.

Isolation from other types of corn does not affect the baby corn crop as the ears are harvested before pollination. However, adjacent sweet corn plantings could be affected by pollen from baby corn varieties of unlike genetic back ground. The baby corn crop planting can also be affected if some ears are left to be harvested as sweet corn.

### Harvesting

Picking should be done daily in kharif and on alternate days in winter season within 1-3 days of silk emergence from the leaf sheath depending upon the variety. Harvesting should be done when baby corn silk comes out 2.0-3.0 cm from the top of ears preferably in the morning or evening, when the baby corn moisture is higher and ambient temperature is low. In single cross hybrid plant, 3-4 pickings may be obtained.

The ears are harvested (45-50 days after emergence) when the silks are 1-2 cm long, i.e., within 1-2 days after silk emergence. Feed corn varieties are harvested at silking, while supersweet varieties may be harvested up to the time silks are about 5 cm long but still fresh. Ears quickly become too long and tough. Suitable time for harvesting of ears may be determined by sampling for size. Harvesting is usually done in the morning when the moisture is high and the temperatures are low. The picking of baby corn is to be done once in three days and generally 7-8 pickings are required depending on genotypes used.

### Yield

It depends on potential of genotypes and climatic conditions. In a good crop, on an average 15500 – 19400 kg/ha husked baby corn or 1100 – 1900 kg/ha dehusked baby corn can be harvested. Green fodder yield is about 25 – 40 t/ha that gives additional income to the growers. A number of by-products are produced in the cultivation of baby corn such as tassels, silks, husks, green plant material etc. after harvest. All these by-products are highly nutritive and can be fed to the cattle which increase milk yield.

## Post Harvest Management

The effort should be made to peel the baby corn on the same day and stored in a cool and dry place. It should be carried out in shady places having good ventilation and air circulation. Dehusked baby corn should be put in containers like plastic baskets, bags, sacks and ensure that they are not heaped. Baby corn should be transported to the processing unit at the earliest.

The husk is to be carefully removed so as not to break or damage the ear. Ears intended for processing must be carefully hand husked and de-silked. Subsequent to the removal of the ear husks, the shanks are cleared of the silks. Then the shanks are graded based on their size and colour and packed in polythene sheets before marketing.

In many cases baby corn for vegetable use is marketed without dehusking of the cobs. This reduces labour involved in processing but fetches less market price. Optimum size for market and cannery industries is 4.5-10 cm long and 7-17 mm diameter of dehusked cobs. Yellow coloured cobs with regular row arrangement fetch better market price. Harvested baby corn may be stored for 3-4 days at 10 $^{0}$C without much effect on its quality. For long term storage and distant transport, baby corn is canned in brine solution (3 %), sugar (2 %) and citric acid (0.3 %) solution and stored under refrigerated conditions. Baby corn may also be stored in vinegar. Baby corn pickle is also gaining popularity in Indian market and it already has an established international market, particularly in Europe.

## Quality traits of baby corn

Baby corn should be uniform in colour, shape and size. Colour of baby corn should be creamish or light yellow. Ears should be straight. Ears should not have slitting marks. Ears should be fresh, free from defects like rotting, over dryness, over ripening and water soaking. Trimming of baby corn should be neat and leave no scars on ears. Ovule/row arrangement should be regular and straight.

## Grading

Baby corn can be sorted and graded by machine or manually and different grades of baby corn are mentioned in table 17.

**Table 17. Different grades of baby corn**

| Grade | Length (cm) | Diameter (cm) |
|---|---|---|
| Short* | 4-7 | 1.0-1.2 |
| Medium* | 7-11 | 1.2-1.4 |
| Long** | 11-13 | 1.4-1.5 |

* International Marketing **Local Marketing

## Packaging

The packing varies from unit to unit. It can be in tin, glasses and polybags. For longer time of preservation, glass packing is the best. Glass packing has 52 % baby corn and 48 % brine solution. Some firm is using tin packing. Farmers are generally using polybags for packing.

## Processing

Baby corn can be processed to improve its shelf life. Main processing methods which can be used to improve the shelf life are canning, dehydration and freezing.

## Canning

Canning is the most commonly used method for processing of baby corn. It can be canned in brine solution and can be stored for months together and transported to far off places. The ears are usually canned at processing factories. Flow diagram is Peeled baby corn > Cleaning > Boiling > Soaking > Grading > Containing > Brine solution > Exhaust > Lid covering > Cooling > Quality inspection.

## Dehydration

Dehydration can be used to increase shelf life of baby corn for longer period. Baby corn can be cut into ½ cm round pieces and dried in oven (air oven/ vaccum oven) or can be solar dried. Dried baby corn can be packed in polythene pack / vacuum pack/ tetrapack and can be stored well for longer period. Dehydrated boby corn can be rehydrated by soaking in water and can be used in preparation of food products. Products developed using dried baby corn has been found to be acceptable organoleptically like those prepared from fresh baby corn.

## Freezing

It can be frozen and stored for long period like other frozen vegetables. Frozen baby corn can be used effectively for preparation of food products. Preparation like soups and vegetables prepared from frozen baby corn are as acceptable as preparation made from fresh baby corn. Frozen baby corn can be directly used in various preparations.

## Preservation

After filling baby corn in container, brine and water are added in cans in the proportion of 2:98 (Brine 2 % and water 98 %), alternatively, a solution of 3 % brine, 2 % sugar, 0.4 % citric acid and balance water can also be used.

**Economic returns**

Farmers can take 3-4 crops in a year depending upon the climatic conditions. The cost of cultivation is Rs. 15,420/ha. The gross return is Rs. 60,000/ha and net return is Rs. 40,000/ha.

**Marketing**

Assured marketing is pre-requisite for baby corn cultivation. So far, in India, there is no systematic market for baby corn. It can be marketed as fresh husked/dehusked young cobs, canned product and pickles. Since, baby corn candy and pickle are in great demand in USA and some European countries, India could be an important country for export of baby corn in future. M/s.Pachranga private company, Panipat, Haryana is already exporting baby corn pickles.

**Seed production**

Seed production of varieties identified for baby corn production has been taken on priority in isolations identified by the monitoring team for the purpose.

## References

Almeida, I. P., C. De, P.S. Le-Silva, M. Z. D Negreiros and Z. Barbosa (2005). Baby corn, green ear and grain yield of corn cultivars. Botucatu, Brazil: Sociedade de Olericultura do Brasil, UNESP - FCA. Horticultura-Brasileira, 23 (4): 960-964

Anil kumar .S.M. (2005). Response of baby corn to planting methods and irrigation schedules under drip. M.Sc (Agronomy) Thesis. Department of Agronomy, College of agriculture, University of Agricultural Sciences, Dharwad, Karnataka, India.

Anita Bindhani, K.C. Barik, L.M. Garnayak and P.K. Mahapatra (2007). Nitrogen management in baby corn *(Zea mays* L.*). Indian J. Agron.,* 52(2): 135-138.

Aravinth, V., G. Kuppuswamy and M. Gangapathy (2011). Growth and yield of baby corn as influenced by intercropping planting geometry and nutrient management. Indian J. Agric. Sci., 81 (9): 875 - 877.

Arun Kumar, K. (2009). Integrated nitrogen management for baby corn. M.Sc (Ag) Thesis, Acharya N.G. Ranga Agricultural University, Rajendranagar, Hyderabad, Telengana, India.

Arun Kumar, K., G. Karuna Sagar, V. Chandrika and P.M. Reddy (2009). Influence of integrated nitrogen management on Yield, nitrogen uptake, soil fertility status and Economics of baby corn. Indian J. Agric. Res., 43 (3): 227-229.

Asha Kwatra and Salil Sehgal (2007). Value added products of Maize (quality protein maize and Baby corn). In: Nation. conf. Double Maize Prod. IFFCO Foundation, ICAR, DMR, DAC and IFFCL, New Delhi.

Ashoka, P., Mudalagiriyappa, B.T. Pujari, P. S. Hugar and B. K. Desai (2008). Effect of Micronutrients with or without organic manures on yield of baby corn - chickpea sequence. Karnataka J. Agric. Sci., 21(4): 485 - 487.

Banik, P. and R.C. Sharma (2009). Yield and resource utilization efficiency in baby corn-legume-intercropping system in the eastern plateau of India. J. Sustainable Agric., 33: 379-395.

Bar-Zur, A. and H. Saadi (1990). Yield and plant characteristics of babycorn for commercial production. J. Hort. Sci., pp. 65-97.

Baskheti, D.C, M.Z.K.Warsi and S.S. Verma (2006). Baby corn: a profitable crop for farmers. Indian Farmers' Digest, 39 (3): 29 -34.

Bindhani, A., K.C. Barik, L.M. Garnayak and P.K.Mahapatra (2008). Productivity and nitrogen use efficient of baby corn at different levels and timing of nitrogen application under rainfed conditions. Indian J. Agric. Sci., 78(7): 629 – 631.

Chikkappa, G.K., J.S. Bhat and Sain Dass (2014). Specialty corn for enhancing farm profitability in peri-urban agriculture. Indian Farming, 64 (4): 55 – 62.

Dadarwal, R. S., N. K. Jain and D. Singh (2009). Integrated nutrient management in baby corn (*Zea mays*). Indian J. Agric. Sci., 79 (12): 1023-1025.

Das, S., G. Ghosh and T. Arbrachan (2005). Baby corn for higher profit and nutrition. Indian Farmers' Digest, 38 (4): 38 -43.

Das, S., G. Ghosh, M.D. Kaleem, and V. Bahadur (2008). Effect of different levels of nitrogen and crop geometry on the growth, yield and quality of baby corn (*Zea mays L.*) CV. 'Golden Baby'. ISHS Acta Horticulturae 809: International Symposium on the Socio-Economic Impact of Modern Vegetable Production Technology in Tropical Asia.

Dass, S., P. Dharam, P.Arora, K.S. Dhanju, J.C.Mehla, V.R. Singh, G. Yadav and Sarvjeet (2004). Baby corn in crop diversification, National seminar on diversification of agriculture through horticultural crops, CCS, HAU, Regional Research Station, Uchani, Karnal, pp. 21-23.

Faiguebaum, H. and C. Olivares (1995). Evaluation of effects of three spacings on baby corn cv. Sweet Boy. Ciencia-e-Investigation-agriraria, 22(1&2):15-19.

Gulabrao, S.K. (2010). Response of baby corn (*Zea mays* L.) to plant density and fertilizer levels. M.Sc (Agronomy) Thesis. University of Agricultural Sciences, Dharwad, Karnataka.

Kalloo, G. and S. Kumar (1999). Baby corn production: promising technology for South East Asian Region. SAIC News Lett., 9(2): 6- 8.

Kasikranan, S., H. Jones and A. Suksri (2001). Growth, yield, qualities and appropriate sizes of eight baby corn cultivars (*Zea mays* L.) for industrial uses grown on Oxic Paleustults Soil, Northeast Thailand. Pakistan J. Biological Sci., 4(1):32-36.

Kotch, R. S., J. H. Murphy, M. D. Orzolek, and P. A. Ferretti (1995). Factors affecting the production of baby corn. J. Vegetable Crop Prod., 1(1): 19-28.

Luikham, E., J.K. Rajan, K. Rajendran and P.S.M. Anal (2003). Effect of organic and inorganic nitrogen on growth and yield of baby corn *(Zea mays* L). Agricultural Science Digest, 23(2): 119-121.

Mahadevaiah, K., G.K., Sagar and V. Sumathi (2010). Effect of herbicides on weed dynamics and yield of baby corn in southern agro-climatic zone of Andhra Pradesh. Indian J. Weed Sci., 42 (3&4): 238-240.

Mahajan, G., R. Sharda, A. Kumar and K.G. Singh (2007). Effect of plastic mulch on economizing irrigation water and weed control in baby corn sown by different methods. African J. Agric. Res., 2(1):19-26.

Mathema, B.B. (1994). Evaluation of baby corn varieties at three plant population densities at Jayashree Khola, Gaindakot, Nawalparasi. Agro Enterprise Centre, Kathmandu, Nepal.

Mathema, B.B. (1996). Baby corn production manual. Agro-Enterprise Centre, Kathmandu, Nepal. July 15, 1996.

Miles, C. and O. Shaffner (1999). Babycorn research report. Washington State University Cooperative Extension, Lewis County. Pp. 8.

Muthukumar, V.B., K. Velayudham and N. Thavaprakaash (2005). Growth and yield of baby Corn (*Zea mays* L.) as influenced by plant growth regulators and different time of nitrogen application. Res. J. Agric. and Biological Sci., 1(4): 303-307.

Muthukumar, V.B., K. Velayudham and N. Thavaprakash (2005a). Effect of plant growth regulators and time of nitrogen application on quality and green cob yield of baby corn (Zea mays L). Madras Agric. J., 92 (7-9): 545-548.

Muthukumar, V.B., K.Velayudham and N. Thavaprakaash (2007). Plant growth regulators and split application of nitrogen improves the quality parameters and green cob yield of baby corn (*Zea mays* L.). J. Agron., 6**:** 208-211.

Najeeb, S., A.G. Rather, F.A. Sheikh, M. A. Ahanger and N.A. Teli (2011). Baby corn (*Zea mays* L.): A means of crop diversification under temperate conditions of Kashmir. Maize Genetics Cooperation Newsletter, 85: 1- 5.

Niranjan kumar, B., Shiv dhar and Ashok kumar (2012). Effect of seed rate nutrient sources and weed control methods on yield and economics of baby corn. Indian J. Agron., 57 (1): 96 – 99.

Pandey, A.K., V. Prakash, V.P. Mani and R.D. Singh (2000). Effect of rate of nitrogen and time of application on yield and economics of baby corn (*Zea mays*L.). Indian J. Agron., 45(2): 338-343.

Pandey, A.K., G.H. Ramesh and S.H. Singh (1998). Effect of genotypes on economics of baby corn. Indian J. Agron., 56(12): 167-172.

Pandey, A.K., P. Ved and Gupta 2002). Effect of integrated weed management practices on yield and economics of babycorn (*Zea mays*). Indian J. Agric. Sci., 72(4): 206-209.

Pandey, A.K., V.P. Mani, R.D. Singh and V.S. Chauhan (1998). Technologies for baby corn production in hills of north western Himalayas. Indian Farming, 47 (6): 10 -12.

Pandey, A.K., V.P.Mani, Ved Prakash, R.D. Singh and H.S. Gupta (2002). Effect of varieties and plant densities on yield, yield attributes and economics of baby corn (*Zea mays*). Indian J. Agron., 47 (2): 221-226.

Pandey, A.K., Ved Prakasha, V.P. Mani and R.D. Singh (2000). Effect of rate of nitrogen and time of application on yield and economics of baby corn (*Zea mays* L.). Indian J. Agron., 45(2): 338-343.

Panwar, A.S. and G.C Munda (2006). Response of baby corn to N and land configuration in mid hills of Meghalaya. Indian J Agric. Sci., 75(5): 293 – 296.

Pradeep Kumar, Yogesh Kumar and Akhilesh Sarraf (2004). Baby corn – A potential crop, Intensive Agric., P: 3-9.

Prasanna, B.M., N.N. Singh and R. Sarkar (1995). Baby corn. Indian Farming, 45(3): 5-7.

Rajanna, A.E. (2003). Effect of irrigation and fertility levels on growth, yield and quality of baby corn. *M.Sc. Thesis, UAS, Bangalore.*

Rajendran, K. and S.D. Sunder Singh (1994). Effect of irrigation regimes and N levels on yield water requirement, water use efficiency and quality of baby corn (*Zea mays*). *Agric. Sci. Digest,* 19: 159- 161.

Ramachandrappa, B.K., H.V. Nanjappa and H.K. Shivakumar (2004). Yield and quality of baby corn (Zea mays L.) as influenced by spacing and fertilization levels Acta Agronomica Hungarica, 52(3): 237–243.

Ramachandrappa, B.K., H.V. Nanjappa, M.N. Thimme gowda and T.M. Soumya (2004). Production management of profitable baby corn cultivation. Indian Farming (August, 2004): 3-7.

Ramachandrappa, B.K., H.V. Nanjappa and T.M. Soumya (2007). Sensory parameters, nutrient content, yield and yield attributes of baby corn varieties as influenced by stages of harvest. Mysore J. Agric. Sci., 41(1): 1-7.

Rameshwar Kumar, Y.L. Sharma, Punam, G.D. Sharma and Atul (2011). Intercropping maize (Zea mays) for grain and baby corn with wild marigold (Tagetes minuta) under organic conditions in mid-hills of HP Himalaya. Himachal J. Agric. Res., 37 (1): 91 -94.

Rani, S. and P. Sukumari (2005). Water use efficiency of baby corn as influenced by drip irrigation regimes and fertilizer application methods. Agricultural college and Research Institute, Tamil Nadu Agricultural University, Madurai, Tamil Nadu, India.

Ranjan, J. K., A.A. Sofi, Biswajit Das, Pragya, Hare Krishna, Brijpal Bisen and H. C. Joshi (2007). Baby Corn: a new vegetable for food basket Vatika from seed and Plant people, Spring -2007 (January-March). Indo American Hybrid Seeds (India) Pvt Ltd, Bangalore, India. P: 25-27.

Ranjan, J. K., N. Ahmed, B. Das, Pragya Ranjan and B. K. Mishra (2013). Green technology for production of baby corn (Zea mays L) under North-West Himalayan conditions. International J. Chem.Tech Res., 5 (2): 880-885.

Rathika, S., K. Velayudham, P. Muthukrishanan and N. Thavaprakash (2008). Effect of crop geometry and topping practices on the productivity of baby corn (*Zea mays* L.) based intercropping systems. Madras Agric. J., 95(7-12): 380-385.

Reddy V. Bali, Madhavi **G.** Bindu, Reddy V. Chenga, Reddy K. Gurava, Reddy M. Chandra Sekhar (2009). Intercropping of baby corn (*Zea mays* L.) with legumes and cover crops. Agric. Sci. Digest, 29 (4): 260- 263.

Saha, S., G.K. Appireddy, S. Kundu and H.S. Gupta (2007). Comparative efficiency of three organic manures at varying rates of its application to baby corn. Archives of Agron. and Soil Sci., 53(5): 507-517.

Sahoo, S.C. and M.M. Panda (1999). Effect of level of nitrogen and plant population on yield of baby corn (*Zea mays*). Indian J. Agric. Sci., 69 (2): 157-158.

Sahoo, S.C. and M.M. Panda (1999). Determination of optimum planting geometry for baby corn. Indian J. Agric. Sci., 69 (9): 664-665.

Sain Dass, Asha Kwatra, Dharam Pal, V.K. Yadav, Sujay Rakshit, R.P.,Singh, Santosh Huda, J.C. Mehla, N.P., Singh, S.B.Gupta, P. Kumar, Neelam Narang, K.S. Dhanju, Rajpal Singh, P.H. Zaidi and K.P. Singh (2007). Baby Corn in India, Directorate of Maize Research Publishing Co, Pusa campus, New Delhi, pp. 1-2.

Sarkar, S., A. Sarkar and B.K.Mandal (2011). Response of baby corn + rapeseed intercropping system under different moisture regions and mulch. Indian J.Agric. Res., 45(4): 314 – 319.

Shivakumar, H. K., B. K. Ramachandrappa, H. V. Nanjappa and Mudalagiriyappa (2011). Effect of phenophase based irrigation schedules on growth, yield and quality of baby corn (*Zea mays* L.). Agric. Sci., 2 (3): 267-272.

Silva, P.S., P.I. Silva, A.K. Sousa, K.M. Gurgel and I.P. Filho (2006). Green ear yield and grain yield of maize after harvest of the first ear as baby corn. Hortic. Bras., 24(2):151- 155.

Subhendu, M., M. Subimal and N. Subhadeep (2004). Effect of integrated weed management on yield components, yield and economics of baby corn (*Zea mays*). Ann. Agric. Res., 25(2): 242-244.

Sukanya, T.S., H.V. Najnappa and B.K. Ramchandrappa (2000). Effect of spacings on the growth, development and yield of baby corn varieties. Karnataka J. Agric. Sci., 12 (1-4): 10-14.

Sukanya, T.S. (1999). Effect of genotypes and spacing levels on growth and yield of baby corn (Zea mays L). M.Sc (Agri.) Thesis, Univ. Agric. Sci., Bangalore, , Karnataka, India.

Sundar Singh, S.D. (2001a). Effect of irrigation regimes and nitrogen levels on growth, yield and quality of baby corn. Madras Agric. J., 88 (7-9): 530 - 532.

Sunder Singh, S. D. (2001b). Effect of irrigation regimes and nitrogen levels on growth yield and quality of baby corn. Madras Agric. J., 88 (7-9): 367-370.

Thakur, D. R. and V. Sharma (2000). Effect of planting geometry on baby corn yield in hybrid and composite cultivars of maize (*Zea mays*). Indian J. Agric. Sci., 70 (4): 246-247.

Thakur, D. R., Om Prakash, P. C. Kharwara and S. K. Bhalla (1997). Effect of nitrogen and plant spacing on growth, yield and economics of baby corn (*Zea mays*). Indian J. Agric. Sci., 42 (3): 479-483.

Thakur, D. R., Om Prakash, P. C. Kharwara and S. K. Bhalla (1998). Effect of nitrogen and plant spacing on yield, nitrogen uptake and economics in baby corn (*Zea mays* L.). Indian J. Agron., 43 (4): 668-671.

Thakur, D.R., P.C. Kharwara and Om Prakash (1995). Effect of nitrogen and plant spacing on growth, development and yield of baby corn (*Zea mays* L.). Himachal Pradesh J. Agric. Res., 21 (1 & 2): 5 - 10.

Thakur, D.R., Vinod Sharma and V. Sharma (2000). Effect of planting geometry on baby corn yield in hybrid and composite cultivars of maize. J. Agric. Sci., 70 (4): 246-247.

Thavaprakaash, N. and K.Velayudham (2007). Effect of crop geometry, intercropping systems and INM practices on cob yield and nutrient uptake of baby corn. Asian J. Agric. Res., 1 (1): 10 - 16.

Thavaprakaash, N., K. Velayudham and V.B. Muthukumar (2005). Effect of crop geometry, intercropping system and nutrient management on cob yield and nutrient uptake of baby corn (Zea Mays L.). Madras Agric. J., 92 (10-12): 646 - 652.

Thavaprakaash, N., K.Velayudham and V.B. Muthukumar (2005). Study of crop geometry, intercropping systems and nutrient management practices on weed density and yield in baby corn based intercropping systems. Madras Agric. J., 92 (7-9): 407-414.

Thavaprakaash, N., K.Velayudham and V.B. Muthukumar (2005b). Effect of crop geometry, intercropping systems and integrated nutrient management practices on the productivity of baby corn (*Zea mays*L.) based intercropping systems. Res. J.Agrl. Biol. Sci., 1 (4): 295-302.

Thimmegouda, M.N., H.V. Nanjappa and B. K. Ramachandrappa (2007). Influence of tillage and moisture regimes with soil solarisation on weed dynamics and yield of baby corn - groundnut crop sequence. Indian J. Weed Sci., 39: 13 - 16.

Trongpanich, K. (1992). A comparison of quality and preference of some baby corn verities for canning. Kasetsart J., 26(1): 44-49.

Upadhyay, K.P., M.D. Sharma, S.M. Shakya, G. Ortiz-Ferrara, T.P. Tiwari and R. C. Sharma (2010). Performance and profitability study of baby corn and Tomato intercropping. Pakistan J. Agric. Sci., 47 (3): 183 – 193.

Vani, V.S., V. Shankaraiah, Y.N.Reddy and J.Dilip babu (2004). Effect of pre-harvest spray of different growth regulators on growth, yield and quality of baby corn. Indian J. Agric. Sci., 74 (5): 262 – 264.

Venkatesh, S., J. C.Sekhar and Sujay Rakshit (2003). Baby corn, Speciality corn technical series III, Directorate of Maize Research, Publishing Co., Pusa, New Delhi, pp. 1-4.

Verma, D. K. and Sarma, B.K., (2001). Studies on genetic parameters in baby corn in the mid hills of Meghalaya. Indian J. Hill Farm., 14(2): 123-125.

Verma, S.S., S.N. Mishra and S.K. Bhatt (1998). A preliminary observation on yield of maize cultivars for baby corn. Agric. Sci., Dig. (Karnal), 18(4): 225-227.

Ye, F.H., Y.S. Liu, Y.G. Yang; L.F.F. and Y.S. Jin (2000). Study on planting dates of vegetable corn Suynuo -1 for all the year around. Acta Agriculture Zhejiangensis, 12 (2): 78 – 83.

Zarapkar, D.S. (2006). Effect of spacing and yield on growth and yield of baby corn. M. Sc.(Agri.) Thesis. Dr. Balasaheb Sawant Konkan Krishi Vidyapeeth, Dapoli, Maharashtra, India.

# 19

# Sweet Corn Cultivation

Sweet corn is one of the most popular vegetables in the USA, Canada, Australia, Europe and other developed countries of the world. It is a very delicious and rich source of energy, vitamin C and A. It is eaten as raw, boiled or steamed green cobs/ grain. It is also used in preparation of soup, salad and other recipes. It is becoming popular in urban areas of country. Therefore, its cultivation is remunerative for peri urban farmers. Besides green cobs the green fodder is also available to the farmers for their cattle. Generally sweet corn is early in maturity. It is harvested in 70-75 days during kharif season. Green cobs are harvested after 18-20 days of pollination during kharif but the duration may vary season to season. At the harvest time the moisture is generally 70 % in the grain and sugar content varies from 11 to more than 20 %. It is becoming popular in India and other Asian countries. Sweet corn differs from other corns (field maize, pop corn and ornamental) because the kernels have a high sugar content in the milk on early dough stage. It is consumed in the immature stage of the crop. The kernels of sweet corn taste much sweeter than normal corn, especially at 25-30 %. The quality and level of sweet corn depends on the type of gene involved for sweetness.

Sweet corn is generally dull yellow and white but dull yellow colour is preferred. Its picking should be done in the morning or evening time. Green cobs should be immediately transported to the cold storage in refrigerated trucks to avoid the conversion of sugar to starch. It loses flavour if kept in high temperature after picking. Sweet corn with high sugar content should not be planted when temperature is below 16°C.

## Varieties

Development of improved varieties (Cultivars) is subject to the fresh market and industry priority. Processors and fresh market buyers have specific characteristics they see as important. Before planting growers should ensure that the cultivars they have chosen meets market requirement. The sweet corn lines/varieties should posses the following attributes:

i) High yielding, ii) Good kernel set, iii) Sweetness depending on gene type and iv) Resistance to major diseases and insects. The important breeding target under temperate conditions are high yield, early maturity owing to short growing season, drought tolerance, resistance to TLB and important the sweetness depending on gene type. Hybrid: HSC1 for J& K and HP. Composite: Madhuri, Win orange, Priya.

### Su locus

A recessive gene (Su) at this locus interferes with the conversion of sugars to starch. The sugar content of homozygous recessive genotype (Su Su) is twice that of normal maize at the green leaf stage.

### Se locus (Sugar enhancer)

The recessive gene at this locus enhances the effect of the Su gene. Homozygous recessive Se Se, Su Su genotypes have higher sugar content than Su Su genotype and maintain their quality for large period.

### Sh-2 locus (Shrunkek)

Homozygous recessive Sh2 Sh2 genotypes have higher sugar content than Su Su types. There is no conversion of sucrose to starch and the level of water soluble polysaccharides is low. Shrunken2 varieties maintain acceptable quality for a longer period during harvest compared to sugar types and a watery rather than a creamy texture.

Worldwide a number of sweet corn varieties are grown both yellow and white grain type either as fresh market cobs or for processing unit. All these varieties are not suitable under temperate conditions because of their long duration or have their feasibilities to tropical and sub - tropical world.

The notably mention of the varieties of such as composites PS 78, Madhuri, Priya and Almora have been tested evaluated and analyzed for various traits conditioning their grouping into sweet corn category.

The composite PS 78 (Mishri Makai) is the native well adopted sweet corn under temperate conditions of Kashmir. The variety is early maturing with super sweetness nature. The main bottle neck of the variety is low yielding and proneness to TLB.

The other three varieties were collected from Almora and evaluated for various precise characteristics which are prerequisite for such kind of specialty corn. The comparison of the varieties for various characters is given in Table 1.

**Table 1. Relative comparison of various traits for the sweet corn varieties evaluated under temperate conditions of Kashmir**

| Quality traits | Varieties | | |
|---|---|---|---|
| | **PS-78** | **Madhuri** | **Almora** |
| Days to 50 % tasseling | 60 | 75 | 78 |
| Days to 50 % silking | 63 | 78 | 82 |
| Days to maturity | 120 | 135 | 136 |
| Plant height (cm) | 180 | 220 | 225 |
| Cob height (cm) | 95 | 115 | 115 |
| Colour of the cob | Yellow | Yellow | Yellow |
| Taste | Super sweet | Super sweet | Sweet |
| No. of cobs/ plant | 2 | 2 | 1-2 |
| Fodder quality | Medium | Very good | Good |
| Cob yield (fresh) (q/ha) | 90 | 130 | 120 |
| Cob yield at 15 % moisture level (q/ha) | 30 | 50 | 45 |
| Fodder yield (q/ha) | 70 | 145 | 140 |
| Harvest index (%) | 0.3 | 0.25 | 0.23 |
| B:C ratio | 2.3 | 2.8 | 2.7 |

The breeding programme is in progress for development of new varieties having the combinations of desired traits such as high yield, earliness, tolerance to various biotic and abiotic stresses together with super sweetness using the basic material/germplasm of above mentioned populations.

## Seed Treatment

At Pune (Maharashtra), seed treatment, soil application at 20 DAS and slurry application of Gluconacetobacter diazotrophicus produced significantly higher yield and monetary return than other methods of application of G diazotropics (Patil et al., 2012).

**Table 2. Yield and economics of sweet corn as influenced by G. diazotrophics**

| Treatments | Fresh cob yield (kg/ha) | Green fodder yield (kg/ha) | Net return (Rs./ha) | B:C ratio |
|---|---|---|---|---|
| G diazotrophics seed treatment @ 25 g/kg of seed | 8025 | 14913 | 48942 | 2.21 |
| Soil appl. @ 6.5kg/ha in 25 kg FYM on 30 DAS | 8189 | 15156 | 50357 | 2.24 |
| Slurry appl. @ 6.25 kg/ha in 150 litre water 30 DAS | 7781 | 14655 | 460750 | 2.14 |
| G diazotrophics seed treatment @ 25 g/kg of seed + Soil appl. @ 6.5kg/ha in 25 kg FYM on 30 DAS | 9778 | 17260 | 67488 | 2.66 |
| G diazotrophics seed treatment @ 25 g/kg of seed + Slurry appl. @ 6.25 kg/ha in 150 litre water on 30 DAS | 8688 | 16412 | 56182 | 2.36 |
| G diazotrophics seed treatment @ 25 g/kg of seed + Soil appl. @ 6.5 kg/ha in 25 kg FYM on 20 DAS + Slurry appl. @ 6.25 kg/ha in 150 litre water on 30 DAS | 12281 | 19235 | 93287 | 3.27 |
| No G. diazotrophics | 5009 | 10605 | 16210 | 1.40 |
| CD 5% | 345 | 1557 | 3825 | 0.09 |

(Patil *et al.*, 2012)

## Crop Planting

Sweet and super sweet Sh2 cultivars are most exacting in their cultural requirements. Soil at planting needs to be friendly tilled, trash free with good soil moisture and to initiate germination. The minimum soil temperature needed at planting ranges between 14-16$^0$C. Sowing should be made at a depth of 25-40 mm.

**Table 3. Effect of plant population on sweet corn (Madhuri)**

| Plant population | Barrenness (%) | Kernel recovery (%) | Green cob (t/ha) | Fodder (t/ha) | Fresh Kernel yield (t/ha) | Net profit (Rs/ha) | B:C ratio |
|---|---|---|---|---|---|---|---|
| 60 x 15 cm | 18.7 | 41.6 | 9.26 | 34.84 | 3.92 | 24964 | 2.59 |
| 60 x 20 cm | 12.8 | 43.5 | 11.80 | 29.58 | 5.46 | 35293 | 3.30 |
| 60 x 25 cm | 11.7 | 46.4 | 10.51 | 24.47 | 5.03 | 29743 | 2.96 |
| 60 x 30 cm | 11.8 | 48.3 | 9.54 | 20.63 | 4.71 | 25973 | 2.77 |
| CD 5 % | 11.7 | NS | 0.49 | 2.24 | 0.46 | 2079 | 0.13 |

At Jashipur (Orissa), plant spacing of 60 x 20 cm was found optimum to get higher return and yield from sweet corn (Sahoo and Mahapatra, 2007).

## Fertilizer

Basal application of fertilizer is at sowing time, 40 mm to one side and 60 mm below seed. Apply 60 kg N, 50 kg $P_2O_5$, 30 kg $K_2O$ and 11 kg $ZnSO_4$/ha. Side dress (21 days after emergence) 1/3rd of $N_2$ (20 kg N) and 1/3$^{rd}$ (20 kg N ) at flower initiation stage.

**Table 4. Effect of varied levels of NPK on yield and yield parameters of sweet corn**

| Treatments | Cobs / plant | Grains /cob | Cob length (cm) | Cob girth (cm) | Fresh cob yield (t/ha) |
|---|---|---|---|---|---|
| RDF for grain maize (150:75:37.5 kg NPK/ ha) | 2.20 | 576 | 16.80 | 16.36 | 13.70 |
| 75% RDN + 100% RDP + 100% RDK | 1.93 | 535 | 15.26 | 14.36 | 13.54 |
| 50% RDN + 100% RDP + 100% RDK | 1.66 | 438 | 13.36 | 11.63 | 10.38 |
| 100% RDN + 75% RDP + 100% RDK | 1.80 | 541 | 16.16 | 15.80 | 12.20 |
| 75% RDN + 75% RDP + 100% RDK | 1.73 | 508 | 13.93 | 13.16 | 11.50 |
| 50% RDN + 75% RDP + 100% RDK | 1.60 | 422 | 12.46 | 11.36 | 9.61 |
| 100% RDN + 100% RDP + 75% RDK | 2.00 | 570 | 16.20 | 16.36 | 13.30 |

| Treatments | Cobs / plant | Grains /cob | Cob length (cm) | Cob girth (cm) | Fresh cob yield (t/ha) |
|---|---|---|---|---|---|
| 75% RDN + 100% RDP + 75% RDK | 1.80 | 520 | 13.86 | 13.36 | 11.94 |
| 50% RDN + 100% RDP + 75% RDK | 1.63 | 432 | 12.83 | 11.76 | 10.18 |
| 100% RDN + 75% RDP + 75% RDK | 1.96 | 532 | 15.46 | 15.40 | 12.32 |
| 75% RDN + 75% RDP + 75% RDK | 1.70 | 483 | 13.60 | 13.46 | 11.30 |
| 50% RDN + 75% RDP + 75% RDK | 1.53 | 420 | 11.73 | 10.66 | 9.48 |
| 100 % RDN + 100 % RDP + 125 % RDK | 2.26 | 583 | 17.50 | 17.13 | 13.72 |
| 75% RDN + 100% RDP + 125% RDK | 1.96 | 544 | 15.46 | 14.86 | 12.22 |
| 50% RDN + 100% RDP + 125% RDK | 1.66 | 445 | 13.13 | 12.60 | 10.68 |
| 100% RDN + 75% RDP + 125% RDK | 1.96 | 541 | 15.50 | 16.03 | 12.58 |
| 75% RDN + 75% RDP + 125% RDK | 1.73 | 510 | 16.70 | 13.63 | 11.58 |
| 50% RDN + 75% RDP + 125% RDK | 1.66 | 435 | 12.80 | 12.56 | 9.90 |
| CD 5 % | 0.34 | 57 | 3.18 | 3.62 | 1.61 |

(Arun kumar et al., 2007). RDF - Recommended dose of fertilizer (150:75:37.5 kg NPK/ha). RDN - Recommended dose of nitrogen. RDP - Recommended dose of phosphorus. RDK - Recommended dose of potassium.

At Dharwad, (Karnataka), application of P and K at recommended levels of grain maize (75 kg P /ha and 37.5 kg K /ha) along with 75 % of recommended level of N of grain maize (112.5 kg N / ha) was found to be necessary to increase the quality parameters *viz.,* non-reducing sugar, total sugar and protein content. Reduction of N below 75 % and P and K below 100 % of recommended dose of grain maize not only accounted for reduction in total sugar and protein content but also in sweetness of grains (Arun kumar et al., 2007b).

## Irrigation

Irrigation water is usually applied by furrow. As a general guide 25-50 mm per week is used to grow a crop with some moisture usually provided by rainfall. Sweet corn is more sensitive to moisture stress than field maize Under temperate conditions of Kashmir it is generally grown on marginal land as rainfed crop but if necessary life saving irrigations are provided to the crop at critical growth stages such as at flowering initiation stage.

At Rajendra nagar (Andhra Pradesh), scheduling of drip irrigation at 120 % Epan recorded higher plant height (251 cm), LAI (5.27), cob length (23.1 cm), kernels/ row (45.9), green cob (12.83 t/ ha) and green fodder (30.00 t/ ha) yields which was comparable with drip irrigation at 100 % Epan. Fertigation of N at 240 kg/ ha was on par with 200 kg N/ ha in terms of yield attributes and yield of sweet corn (Basava et al., 2012).

**Table 5. Effect of drip irrigation and N fertigation on yield attributes and yields of sweet corn**

| Treatments | Cob weight (g) | Kernels/ row | Cobs/ plant | Yield (t/ ha) | |
|---|---|---|---|---|---|
| | | | | Green cob | Green fodder |
| **Irrigation levels** | | | | | |
| Drip at 60 % EPan | 307 | 38.2 | 1.1 | 8.71 | 19.13 |
| Drip at 80 % Epan | 341 | 39.6 | 1.3 | 10.24 | 23.47 |
| Drip at 100 % Epan | 363 | 41.6 | 1.4 | 12.23 | 28.93 |
| Drip at 120 % Epan | 374 | 41.9 | 1.4 | 12.83 | 30.00 |
| Surface irrigation at 1.0 IW/ CPE | 272 | 33.8 | 1.1 | 7.72 | 17.29 |
| CD (5 %) | 17 | 1.1 | 0.03 | 0.82 | 1.48 |
| **Nitrogen levels (kg/ ha)** | | | | | |
| 120 | 302 | 36.5 | 1.1 | 8.42 | 19.93 |
| 160 | 325 | 38.5 | 1.2 | 9.32 | 22.68 |
| 200 | 346 | 40.4 | 1.3 | 11.66 | 25.72 |
| 240 | 352 | 40.8 | 1.3 | 11.99 | 26.73 |
| CD (5 %) | 15 | 0.9 | 0.03 | 0.73 | 1.32 |

## Weed Control

Normally weeds are easily controlled in sweet corn crops by using herbicides and cultivations. Poorly managed weed populations can harbour pests and diseases that can restrict plant growth and cob quality. Rotation of crops and thorough land preparation will minimize weed problem. Inter-row cultivation is usually necessary once or twice with 30 days of planting, to break soil crust and later can be combined with fertilizer side dressing and hilling up. Cultivation needs to be shallow as sweet corn roots are near the surface and must not be damaged. Atrizine @1kg/ha has been proved through agronomic trials to control all pre-emergent broad leaved weeds.

## Intercropping

At Dapoli (Maharashtra), dry pod yield, haulm yield of groundnut, green cob yield and green biomass yield of sweet corn were higher in case of sole planting and application of 125 % RDF (recommended dose of fertilizers). In case of intercropping situation groundnut + sweet corn in ratio of 3:1 and supplemented with 125 % RDF recorded higher dry pod yield while higher green cob yield was recorded at 1:1 row proportion. The maximum value of LER was recorded due to intercropping at 3:1 ratio and supplied with 100 % RDF. The net returns and B:C ratio were higher due to sole planting of sweet corn grown with 125 % RDF followed by groundnut + sweet corn in ratio of 1:1 and supplemented with 125 % RDF (Bhagat et al., 2006).

**Table 6. Groundnut (TG-26) dry pod and haulm yield, sweet corn (Sumadhur) green cob and stover yield and economics of intercropping system**

| Treatments | Yield (q/ha) of groundnut | | Yield (q/ha) of sweet corn | | Net returns (Rs/ha) | B:C ratio |
|---|---|---|---|---|---|---|
| | Dry pod | Haulm | Green cob | Stover | | |
| Groundnut 75 % RDF | 28.22 | 36.09 | - | - | 2019 | 1.05 |
| Groundnut 100 % RDF | 28.74 | 37.38 | - | - | 2304 | 1.05 |
| Groundnut 125 % RDF | 31.04 | 38.93 | - | - | 4812 | 1.11 |
| Sweet corn 75 % RDF | - | - | 127.25 | 83.83 | 61875 | 1.92 |
| Sweet corn 100 % RDF | - | - | 135.32 | 89.88 | 65878 | 1.96 |
| Sweet corn 125 % RDF | - | - | 150.44 | 101.29 | 86634 | 2.17 |
| G + S (1:1), 75 % RDF | 16.40 | 20.28 | 66.75 | 44.51 | 23867 | 1.44 |

| Treatments | Yield (q/ha) of groundnut | | Yield (q/ha) of sweet corn | | Net returns (Rs/ha) | B:C ratio |
|---|---|---|---|---|---|---|
| | Dry pod | Haulm | Green cob | Stover | | |
| G + S (1:1), 100 % RDF | 17.29 | 22.14 | 69.75 | 47.84 | 25394 | 1.46 |
| G + S (1:1), 125 % RDF | 18.69 | 23.89 | 77.60 | 52.59 | 35198 | 1.61 |
| G + S (2:1), 75 % RDF | 19.11 | 25.07 | 53.24 | 35.99 | 17562 | 1.34 |
| G + S (2:1), 100 % RDF | 20.05 | 26.02 | 54.67 | 36.60 | 18765 | 1.36 |
| G + S (2:1), 125 % RDF | 21.95 | 28.22 | 57.58 | 38.39 | 27609 | 1.50 |
| G + S (3:1), 75 % RDF | 25.40 | 32.12 | 42.68 | 28.45 | 14493 | 1.29 |
| G + S (3:1), 100 % RDF | 26.28 | 33.63 | 43.61 | 29.51 | 16690 | 1.33 |
| G + S (3:1) 125 % RDF | 26.34 | 34.96 | 44.83 | 31.01 | 20141 | 1.39 |
| CD 5 % | 2.15 | 2.03 | 9.75 | 4.23 | - | - |

Recommended fertilizer dose for groundnut (G) and sweet corn (S) was 25:50:50 and 150:50:50 kg NPK/ha respectively.

**Pest control**

The conditions that need to be met are pest is present a crop is susceptible to attack and environmental conditions favour pest increase. The integrated pest management strategy can be effectively used for the management of pests most economically with a minimum use of pesticide and reduce development of resistance. Under Indian conditions, cut worms and Turcicum leaf blight are the most severe pests.

**Harvesting**

Corn for the fresh market is generally harvested by hand or machine at higher moisture content of 70-80 % and when the kernels at the top of the cob are 75 % fill. Picking may take place over several days as the cobs seldom ripen evenly.

## Visual assessment of cobs

The number of days is temperature dependent. The varieties known to be ready by 95 days take 120-125 days when grown under temperate Kashmir conditions. Randomly assess cobs in a field when silks are browned off and dry, mature cobs usually have the largest diameter. It usually takes 4-5 weeks from onset of silking to harvest under temperate conditions. However, careful choice and local trialing of cultivars is important as yield and cob quality can vary between cultivars.

## Consignment and transport

Sweet corn lose their crispness and sweetness a few days after harvest. Hence before consigning fresh market sweet corn, it is necessary to be sure of buyer's specifications for product, packing, handling and cool chain management. Fresh sweet corn is highly perishable in hot weather. For every 5 °C increase in pulp temperature, sugar breakdown to starch doubles, meaning that sweet corn can completely lose flavour in a short period of time. Within a few days at ambient temperatures, husks also dry out very quickly, silks dry out and darken and cobs sweat and caramelize, resulting in significant quality deterioration. The recommended pulp temperature throughout the cool chain (farm to consumer) is 0-2 $^0$C.

## Export market

The super sweet cultivations have increased sweet corn's popularity in many countries such as USA, Australia and Japan where fresh packs and long life vacuum packs are popular. Bicolour and super sweet cultivars are preferred. However, buyer preferences can vary and research of market and product specification is important. Temperate conditions more particularly Kashmir climate is most popular for quality crop production. Hence, there is high potential for export of the crop.

## Processing

Sweet corn is processed into the different product like frozen cobetts (protein of whole cobs), whole cobs and whole kernels. Beside canning operations in the form of whole kernels and cream styles viz-a-viz soups, mixed vegetable packs, corn jacks. Kernel products are the most popular processed lines with frozen whole cob and cobett products exported to improve their market share. The most important to mention here is the cob quality available for processing. Hence, breeding for variety that is highly acceptable from processors point of view should be initiated. Under Kashmir conditions where there is a growing problem of unemployment, the processing unit for sweet corn has a potential to engage skilled and semiskilled people.

**Sweet corn kernel structure, chemical composition, and sensory qualities**

Sweet corn cobs constituting raw material for processing must be characterized by the highest quality of kernels. Kernel quality is defined not just by the chemical and sensory properties, but also by the mechanical parameters of kernels. This appears to fully justify joint consideration of all those properties. Sweet corn is probably a mutant of fodder corn (Orłowski, 2000). Significant differences between the two are related more to the genetics than to the structure of the kernels. As emphasized by Salunkhe and Kadam (1998), the

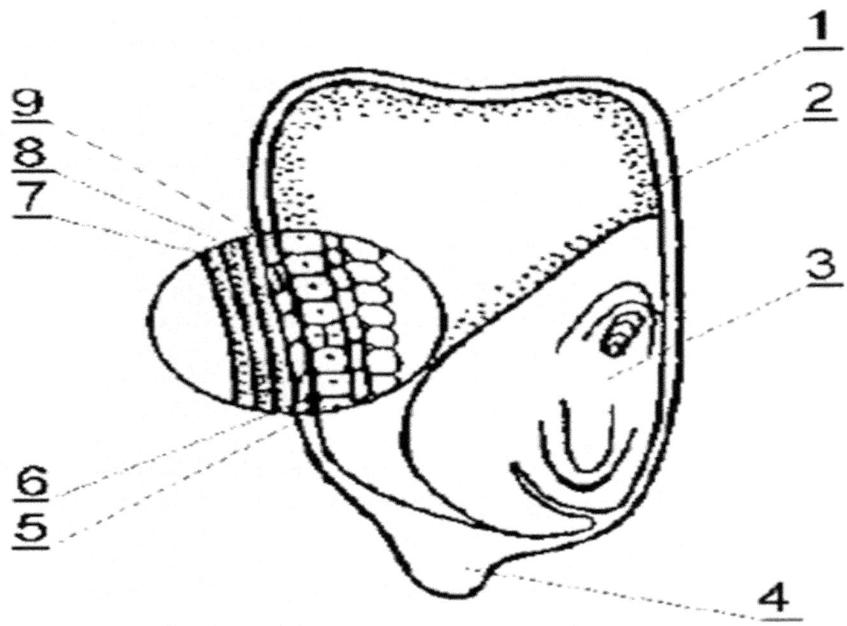


**Fig. 1. Schematic of longitudinal cross section of a sweet corn kernel: 1 – pericarp, 2 – endosperm, 3 – germ, 4 – pedicel, 5 – aleurone layer, 6 – tube cells, 7 – epicarp, 8 – mesocarp, 9 – cross cells (Salunkhe and Kadam, 1998).**

Sweet corn kernel is built of the pedicel, the pericarp, the germ, and the parenchyma (Fig. 1). The pedicel is a hard and fibrous remnant of the tissue that joins the kernel to the cob core (Szpaar and Dregiew, 1999). During kernel shearing, a part of the pedicel remains on the kernel, which has a negative effect on its nutritional value. Kernel shape is described as flattened and wedge-like, with the tip much broader than the base end by which the kernel is attached to the cob core. The kernel is deep set on an abbreviated shoot (rachis) forming the cob, and is covered with a thin pericarp. The pericarp is a component of the seed coat tissue and forms the outer layer of the kernel. The thickness of the layer determines the kernel skin level of tenderness. This feature is important in the estimation of kernel quality for processing. As compared to other varieties, sweet corn is characterized by lower thickness of the epicarp, on average 25-30 μm (Ito and Brewbaker, 1991). The epicarp is composed of a single layer of pericarp, in the cavities of which single cells of the mesocarp are located. It also includes one or two layers of cross-cells and one or more tube cells adherent to the pericarp. The germ, located obliquely at the base of the kernel, is large and constitutes about 15 % of the volume or 11.5-14 % of the weight of the kernel. Germ size, however, is cultivar-related and may constitute 5 % of the mass of the whole kernel (Puangnak, 1998). In turn, the parenchyma is the largest element of the kernel. It is in the parenchyma that the sugars, starch and water-soluble polysaccharides are accumulated in.

In the phase of consumption ripeness, the consistency of kernels is soft, delicate, creamy, and the taste is sweet and fragrant (Reyes and Varseveld, 1982). The kernel seed coat has colours from pale yellow to orange, often with a violet tint and shiny. In the genotype of sweet corn, except gene *su* (sugary) gene determining the sweet taste and tenderness of the kernel, also other genes have been discovered, responsible for enhancing the sweetness and the taste and utility qualities – the gene *se* (sugary enhancement) and the gene *sh2* (shrunken 2). Genetic combinations of the genotypes *su* and *sh2* helped in the creation of very sweet cultivars (Simonne *et al.*, 1999). In the phase of full ripeness, sweet corn has wrinkled kernels, almost completely filled with vitreous parenchyma, mostly white or yellow in colour. The reserve substance of the parenchyma is composed of amylodexstrins which are responsible for the sweet taste. The kernel has a bulbous shape, oval, wedge-shaped or angular, a smooth or wrinkled surface, and white, yellow, read or brownish colouring. It is enclosed in a cover formed from fused pericarp and seed coat, beneath which there is a layer of aleurone cells, the parenchyma and the germ. From the moment of pollination till harvest the cob of sweet corn undergoes numerous physical and chemical transformations which largely affect the taste and the quality of kernels. The taste is especially strongly affected by the transformations of sugars (Wong, 1994). With respect to the content of sugars, three types of sweet corn cultivars are distinguished: normally sweet cultivars, type *su* (sugary), with sugar content of 4-6 %, cultivars with increased sugar content, type *se* (sugary enhancement) – 6-8 % and very sweet cultivars, type *sh2* (shrunken 2) – 8-12 % (Warzecha, 2003). Apart from the content of sugars, fresh mass of sweet corn kernels contains 2.1-4.5% of proteins, 3-20 % of starch, 1.1-2.7 % of fats, 0.9–1.9 % of cellulose, 9–12 mg of vitamin C, small amounts of vitamins A, B1, B2, PP and mineral components such as: sodium, potassium, magnesium, calcium, phosphorus, iron, selenium, copper, nickel and chromium (Felczyński *et. al.*, 1999; Lee and McCoon, 1981). The chemical composition of the kernels is related to the weather conditions, ripeness, and method of storage (Salunkhe and Kadam, 1998). The nutritional value of sweet corn kernels is related to the content of water (72.7 %) and to the total content of solid parts (27.3 %). Solid parts include hydrocarbons (81 %), proteins (13 %), lipids (3.5 %), and others (2.5 %) (Hardenburg and Watada, 1986). Starch is the dominant hydrocarbon component. Sweet corn has higher nutritional value in the phase of milk ripeness. With progressing phase of ripeness, in the transition to the phase of wax ripeness the content of sugars decreases, accompanied by an increase in the content of starch (Suk and Sang, 1999). In 100 g of kernels there is about 3.03 g of saccharine, 0.34 g of glucose and 0.31 g of fructose. The content of saccharine increases and that of reducing sugars decreases as the kernels reach the optimum ripeness. The content of proteins in the kernels decreases from the surface towards the centre of the kernel. The content of proteins, free amino acids, water-soluble and insoluble hydrocarbons, increases up to the phase of wax ripeness and then gradually decreases (Azanza *et. al.*, 1996). The amounts of the particular components in various cultivars and in various phases of ripeness variable. In comparison to other cereals, sweet corn is relatively rich in oil. Approximately 90 % of the oil is accumulated in the germ. Sweet corn is more tasty than other corn species to its high content of water soluble polysaccharides. That component imparts to the kernels their tender and creamy character.

The most important parameters that affect the sensory quality of the kernels include sweetness, texture, and taste (Wong and Swiader, 1995). Sweetness depends on the content of sugars, while texture depends on a number of factors, such as tenderness of the seed coat, moisture, content of water-soluble polysaccharides. Taste, in turn, is frequently associated with the content of DMS (dimethyl sulphide). Decrease in kernel quality related to loss of taste and aroma after the harvest is a problem for the processing industry. The loss of taste in fresh or frozen state of the kernels is caused by enzyme activity (Collins and Biles, 1996). Fresh kernels are characterized by faint aroma or its total absence. Wade (1981) states that cut kernels have three characteristic types of aroma. Two of these are similar to the aroma of fresh green vegetables, and the third is a methol-type aroma. With progressing ripeness of sweet corn, the level of DMS in the kernels decreases, which is a serious problem for the processing industry due to the related considerable loss of taste of processed sweet corn products.

The consumption quality of fresh sweet corn largely depends on the content of sugars and water-soluble hydrocarbons in the kernels (Evensen and Boyer, 1986). The quality of sweet corn kernels can be determined in many ways. The basic discriminant of kernels for the processing industry is kernel hardness and taste. One of the most important factors determining the quality of kernels for processing is the use of cultivars characterized by uniform ripening. The choice of cultivar affects not only the yield of kernels cut off cobs, but also the taste quality of the kernels. Other quality factors include the colour, sweetness, and tenderness of the kernel cover. With ripening, the cover becomes harder and harder. The quality of sweet corn is correlated to the content of sugars. The transformation of sugars into starch is also related to decreasing moisture content of the kernels.

## Characterization of Sweet Corn Varieties

The choice of variety is one of the more important factors to determine whether sweet corn production is a success. Population (fixed) varieties have been largely replaced in cultivation by hybrid (heterotic) varieties, created by man in the first half of the 20th century as new forms of the crop plant. Hybrid varieties spread less, produce bigger and evenly ripening cobs, and are higher yielding compared to the popular varieties. They are characterized by high sensory qualities and are suitable for direct consumption and for the processing industry alike. They meet the requirements of the fruit and vegetable processing industry in terms of having very delicate kernel skin and kernels easy to separate from the cobs in whose mass the kernels constitute 30-40 %. Primary differences among the hybrid varieties include the duration of their vegetation period, content of sugars, and suitability for various uses. The productive value of the varieties is determined primarily by their yield capacity and the earliness of their ripening.

Producers involved in sweet corn growing for direct consumption are interested in obtaining a large number of well kernelled cobs. Those producing sweet corn for industrial processing expect a high yield of material suitable for processing. In terms of the length of the vegetation period the following groups of corn varieties are distinguished:

» early varieties (70–80 days),

» medium early varieties (85–90 days),

» late varieties (95–110 days).

Also important is the division of varieties with respect to the content of sugars in kernels at the phase of harvest ripeness. With reference to their genetic features, they are classified as:

» normally sweet, with the gene *„su-1" (sugary)*,

» with increased content of sugars, with the gene „se" or „se+" *(sugary enhanced)*,

» very sweet, with the gene *„sh-2"(shrunken 2)*.

»

Variety selection is an important consideration in sweet corn production and includes factors such as sweetness, days to maturity, seed color, size, yield potential and tolerance to pests. The Department of Agricultural Extension Service can provide a list of varieties recommended for each region.

**Table 6. Sweet corn genotypes**

| Genotype | Sweetness | Conversion of sugars to starch | Isolate form |
|---|---|---|---|
| Normal sugary (su) | Moderately sweet | Rapid | (sh2) varieties |
| Sugary enhanced (se), (se+) | Sweeter than (su) less sweet than (sh2) | Not as rapid as (su) | (sh2) varieties |
| Super sweet or shrunken (sh2) | Very sweet | Very slow | (su), (se), (sh +) varieties |

(http://attra.ncat.org/attra-pub/PDF/sweetcorn.pdf)

Modern sweet corn varieties are classified as: "normal sugary" (su); "sugary enhanced" (se) and (se+); and "shrunken" (sh2), also called "super sweet." These differ in flavour and tenderness and in the rate at which starches are converted to sugar. In general, (se) lines yield the best, followed by (sh2), and finally (su). Cross-pollination of sweet corn with other kinds of corn or with some other sweet corn genotypes can result in starchy-tasting kernels.

Generally, a minimal isolation distance of 75 m between those varieties or types is recommended whereas 210 m, however, is preferred for more complete isolation. The normally sweet varieties contain 4-6 % of sugars in fresh kernel mass, those with higher sugar content from 6 to 8 %, and the very sweet varieties, sometimes also called super sweet or extra sweet, from 8 to 12 %. Another criterion of division or classification of cultivars is the colour of their kernels which can be yellow (the larger group of varieties), white, yellow-white *(bicolor)* and red (Wong, 1994).

## Uses of Sweet Corn

Sweet corn kernels can be consumed both as fresh produce and in processed forms. In practice, most frequently three basic directions of their utilization are distinguished:

- direct consumption - cobs harvested at milk ripeness of seed, for fresh produce market.
- vegetable processing industry - cobs harvested at late-milt ripens of seed, for pickles and frozen foods.

» industrial processing - cobs harvested at full ripeness of seed, for flour, meal, etc.

## Sweet Corn Processing Technology

### Harvest of sweet corn cobs

Sweet corn for processing is harvested at a relatively immature stage as compared to field corn. Processing of corn is used to increase its shelf life but as a consequence, a significant loss of nutrients may occur via heat degradation or leaching (Scott and Eldridge, 2005). It is picked at different stages of maturity depending on the way it is to be processed. The corn for freezing is harvested at about the same stage as that for fresh market, while the corn for whole kernel pack and cream-style is harvested at a slightly later stage of maturity. For whole kernel canning and freezing, optimum kernel moisture ranges from 70 to 76 %. For cream-style canning corn, optimum kernel moisture is about 66 %. It correlated very closely with the moisture percentage and with postharvest grade evaluation (Olson, 2000). An accurate determination of the sweet corn maturity for harvest can ensure the best possible crop yield and quality. The optimum moisture for harvesting shrunken sweet corn for freezing and canning is no less than 76 % and no more than 79 %. This compares to the range for standard sweet corn of 70 -72 %. Because the shrunken sweet corn loses only about ¼ % of moisture per 24-hour period at the 76 % level as compared to 1 % /24-hour period for standard sweet corn, the harvest window for shrunken corn harvest for a processing line and results in fewer bypassed fields due to planting, mistiming or weather delay (Marshall and Tracy, 2003). However, according to Warzecha (2003), it is easier to mechanise the harvest of standard sweet corn than that of shrunken sweet corn.

The standard sweet corn compares to shrunken varieties mature to longer. Sweet corn has a very short period of optimum harvest maturity, and its quality changes rapidly close to and following the peak. Ears harvested immature will have a small diameter, a poor cob fill, and kernels that are watery and lack sweetness. At optimum harvest maturity, the kernels are plump, sweet, milky, tender, and nearly of maximum sizes. After optimum harvest maturity has been reached, the eating quality of sweet corn begins to decrease rapidly, while the husk appearance changes very little. Over mature corn is rather starchy than sweet, tough, and the kernels are often dented (Motes *et al.*, 2007). However, according to Kumari *et al.* (2007), the unfavourable correlation coefficients between the sugar content and grain weight suggested that it is difficult to obtain high-yielding sweet corn hybrids of good quality. Results of sweet corn ear size measured at different harvest date are presented in Table 7.

**Table 7. Kernels moisture content, yield ears, ear length, ear diameter, number of kernels per row, number of kernel rows and bulk density**

| Particular | Harvest date | | | | LSD (5 %) |
|---|---|---|---|---|---|
| | 1st | 2nd | 3rd | 4th | |
| Moisture content (%) | 77.41a (0.95) | 75.62ba (0.88) | 72.31c (1.05) | 69.83d (1.09) | 2.05 |
| Yield (t/ha) | 18.64a (1.15) | 17.98ba (1.21) | 16.31c (0.98) | 15.88dc (1.11) | 1.20 |
| Length (cm) | 22.21 (2.09) | | | | - |
| Max. diameter (cm) | 4.94 (0.98) | | | | - |
| Number of kernels per row (pcs) | 28.05 (1.57) | | | | - |
| Number of kernel rows (pcs) | 14.72 (1.54) | | | | - |

Standard deviation in parenthesis

The 100 husked ears measured at a first harvest date are length 22.21 ± 2.09 cm and maximum diameter 4.94 ± 0.98 cm. The yield decreased from 18.64 to 15.88 t/ha, when the moisture content of kernels decreased from 77.41 to 69.83 %. The similar decreasing of moisture content with increasing harvest maturity has been reported by Wong, (1994). At all harvest date (moisture contents), an increase of deformation with an increase in applied forces was observed (Table).

**Table 8. Means values of compression, shear and puncture force**

| Particular | Harvest date | | | | LSD (5 %) |
|---|---|---|---|---|---|
| | 1st | 2nd | 3rd | 4th | |
| Compression force (N) | 29.48a (1.78) | 35.54ba (1.32) | 42.71c (1.54) | 49.56dc (1.23) | 7.11 |
| Shear force (N) | 8.21a (0.28) | 10.41ab (0.32) | 12.34cb (0.41) | 15.21dc (0.37) | 3.21 |
| Puncture force (N) | 9.11a (0.18) | 12.65ba (0.19) | 15.28cb (0.17) | 17.23dc (0.21) | 4.56 |

Standard deviation in parenthesis

The hull rupture is marked by an audible "click", and a sudden decrease of the force occurs. The point marked by the abrupt force decrease is often called the bio-yield point and the loading was stopped once this point was reached. The measured parameters were the rupture force, when the kernel hull undergoes failure during compression, shear and puncture, the deformation up to the rupture point.

## Kernels removing from the cobs

### Cutting method for removal of kernels from raw sweet corn cobs

The food processing industry currently applies machines utilizing rotary cutter heads for severing corn kernels (Kessler and Harry, 1998)

The separating sweet corn kernels from the cob for consumption purposes is performed with the use of special cutters. To obtain kernels of high quality, the working elements of such machines should be carefully adjusted so that all the kernels are cut off as close to the cob core as possible, but without cutting off the cob husks whose presence among the kernels acquired worsens the quality of the product. The kernels should be cut off at 2/3 of their length with the kernel germs remaining on the cob core, as predominant in corn production were varieties with long cob husks. New hybrid varieties of corn are free of that defect, and to increase the amount of material acquired cutter knifes are set to the maximum length of kernels detached. Care must be taken that the cut through the kernel be smooth, without tearing the seed cover, and set so that no thick cob husks are among the kernels detached. This requires frequent sharpening of the cutter knives. Adjustment of the working elements of the kernel cutter should be made suitably to changes in dimensions and properties of the material processed. Corn cobs should oriented with their narrower end towards the cutter head, as in that position the cutter knives adapt better to correct detachment of kernels. During kernel cutter operation, it is necessary to systematically check all the moving elements of the machine, and to clean and lubricate as required. This has an immense effect on the quality of kernels detached, as with the cutter head dirty the positions of cutter knives take longer to adjust to the changing cob diameter, and they may result in increased amount of incorrectly cut off kernels.

The mechanical removal of sweet corn kernels from the cob by cutting leads to waste, loss of nutritional value, and loss of yield (Hanna *et al.,* 1988). It is not possible to obtain all the kernels of corn. The cutting operation severs the upper part of the kernel from the lower part, which remains on the cob. Thus, the hull of the kernel is broken open and a part of each kernel is wasted (approximately 20% remains on the cob), including much or the entire germ of the kernel. During subsequent wet-processing, including washing and blanching of the severed kernels, some of the corn meat is leached from its pouch and lost (Kunicki, 2003). However, the proportion of cut-off kernels is strictly related to the moisture product processing characteristics. The successful detachment of sweet corn kernels as intact kernels promises large advantages in terms of yield and effluent over cut kernels during processing, but it does require low strength kernel attachment to the cob. Niedziółka and Szymanek (2004) showed that blanching prior to the cutting process, instead of blanching after kernel cutting, resulted in a 12% increase in yield of cut kernels and quality. However, Trongpanich *et. al.* (2002) did not show a statistically significant difference, while Michalsky (1986) reported that by delaying the time of harvest until the moisture content decreased, it is possible to increase the detached part of the kernel.

Szymanek (2011) showed that the cut kernel yield increased with the increase in cutter head angular speed. The change of speed from 167.5 to 293.2 rad· $s^{-1}$ increased approximately 51 % for the Jubilee variety, 29 % for the Boston variety, and 54 % for the Spirit variety. The differences recorded between the varieties were statistically significant. An increase of cutter head angular speed from 167.5 up to 293.2 rad· $s^{-1}$ resulted in a statistically significant decrease of inferior kernel share. Highest reduction in inferior kernels in the investigated speed range was obtained for the Spirit variety, at about 64 %, whereas lowest was for Boston, at about 44 %. An increase in the amount of kernels cut, along with a change of a cutter head angular speed, also decreased sugar content loss. The most substantial decline in sugar losses within the studied range of cutter head speed was observed for the Jubilee variety, with about 48 %, while lower for Spirit and Boston was about 33 %. A change in a cutter head angular speed generally resulted in a statistically significant effect on the differences in sugar content. Increasing cutter head angular speed in the range of 167.5 to 293.2 rad· $s^{-1}$ induced a decrease in kernel mass losses subject to the plant variety. The changes were in the range from 51 to 28 %. Spirit variety had higher reduction in sugar losses.

The losses occurring in the production process have negative effects not only in economical terms, but also pose considerable environmental pollution problems. The loss of kernels could, however, be reduced by means of the respective choice of sweet corn varieties. Robertson *et al.* (1980) think that solutions to this problem are being developed by food engineers and plant breeders by modifying the mechanical separation method and by modifying the raw product processing characteristics. However, claims that reduction of the kernel moisture could be achieved by delaying the cob harvesting dates, which, in turn, reduces loss during the kernel cutting process. In addition, blanching the cobs prior to attempting the cutting process (Stewart *et al.,* 1997) and selecting the optimum operation parameters of the kernel cutter devices also reduces loss (Niedziółka and Szymanek, 2006). The method proposed by Robertson and Farkas (1982), whereby the cobs are halved along their length and the kernel is subsequently detached on reciprocally moving conveyor belts can practically reduce kernel loss to almost zero. However, this method is less efficient and consumes too much energy and has not so far been applied on industrial scale. In recent years, the growing concern about applying freezing methods, including cryo-liquids, in the food processing industry has also been reflected by the growth in manufacturing of frozen products, including sweet corn (frozen cobs or kernel). In highly technologically advanced countries, about 30 % (20 % in Poland) of the production output of sweet corn is designated for freezing and about 40 % for canned food (60 % in Poland) (Waligóra, 2006). Freezing in boiling liquids means very high convective heat-transfer coefficients, considerable temperatures and consequently very short freezing times. Unlike in classic (slow) freezing, the cryo-liquid freezing method results in a high quality of the product processed. Kernel frozen in temperatures of –4$^0$C fit long-term storage requirements without any significant change in taste and technological properties (Matheus *et al.,* 2004). The growing demand for frozen products brings about a need to develop technologies that can provide storage of fresh vegetables and fruit, while their physical properties undergo no essential changes.

## Shelling method for removal of kernels from frozen sweet corn cobs

In the shelling method for removal of kernels from frozen sweet corn cob the corn cobs prior to shelling are subjecting to blanching and freezing by spraying in liquid nitrogen. The mean values of kernel separation rate of cobs for 2, 4, 6, 8 and 10 minutes freezing time were: 55.1 %; 79.9 %, 96.7 %, 95.8 % and 97.0 % respectively.

In the range from 2 to 6 minutes freezing time we can observe the significant increases of kernels separation rate and then from 6 to 10 minutes the not significant changes of kernels separation rate. This might be due to lack of uniformity in kernels freezing which results in different kernels hardness and not complete kernels removal. The advantage of this method is the reduction of waste and increased yield of corn. During the analyses of kernels separation (Szymanek, 2011), 86.3 % of intact kernels removed after 2 and 4 minutes of freezing contained adhering cob related tissue, and 13.7 % of kernels were without such tissue. However, for intact kernel freezing for 6, 8 and 10 minutes, the relations were opposite and amounted to, respectively, 8.9 and 91.1 %. The optimal freezing time influences not only the quantity but also the quality.

Robertson *et al.* (1980), found that intact kernels have more adhering cob related tissues which is perceived as a defect of intact kernels. In the range from 2 to 10 minutes were observed the decrease of kernels damage from 7.3 to 3.9 %. The share of damaged kernels showed the same tendency as changes of kernels separation rate. The proportions of damaged kernels for 6, 8 and 10 minutes freezing time although different in value are not significantly different. Decreasing of kernels damages together with lengthen of freezing time might be due to increasing of hardness of whole kernels which results in kernels being more resistant to mechanical action of shelling unit. The mechanism of damages formation can be compared to threshing dried cob. Nguyen (1986) reported that when shelling dried cob, only the linkage between kernel and corn cob is broken. When threshing fresh cob, we have to break down two linkages: kernel - corn cob and kernel - kernel. This causes a considerable amount of broken kernels. Similar situation might occur when freezing kernels.

The delay of the corn cobs harvest date affected the sweet corn quality. The moisture content, sugars level and ear weight decreased. The bulk density (intact and cut kernels), compression, shear and puncture force, starch content, processing recovery, and cut corn yield increased. The first harvest date proved to be more advantageous due to a higher sweet corn quality for processing than the following harvest date. The analysed key attributes of the sweet corn quality for processing, that is moisture and sugars contents had the highest values at this time. When the moisture content decreased from 77.41 % to 69.83 % with delayed harvest date, a decline was observed in the sweet corn quality (increase of force in compression, shear and puncture tests, rise of the starch level). However, the following harvest date had an advantage in a higher processing recovery and a higher cut corn yield.

## References

Abdullah, O. (2008). Effect of nitrogen on fresh ear yield and kernel protein content of sweet corn (*Zea mays* L. *saccharata*) under upper Mesopotamia region of Turkey. Indian J. Agric. Sci., 78(1): 50-52.

Abdullah, O., Mehmet Simsek and A. Gulgun Oktem (2003). Deficit irrigation effects on sweet corn (Zea mays saccharata Sturt) with drip irrigation system in a semi-arid region I. Water-yield relationship. Agric. Water Mangt., 61: 63-74.

Abu-Awwad-Am (1994). Irrigated sweet corn crop production functions and efficient water use. J. Agron., 172 (3): 161-170.

Arun kumar, M.A., S.K. Gali and N.S. Hebsur (2007). Effect of different levels of NPK on growth and yield parameters of sweet corn. Karnataka J. Agric. Sci., 20 (1): 41 – 43.

Arun Kumar, M.A., S.K. Gali and R.V. Patil (2007a). Effect of levels of NPK on quality of sweet corn grown on vertisols. Karnataka J. Agric. Sci., 20(1): 44 - 46.

Arvadiya, L.K., V.C. Raj, T.U. Patel and M.K. Arvadia (2012). Influence of plant population and weed management on weed flora and productivity of sweet corn. Indian J. Agron., 57 (2): 162 – 167.

Arvadiya, L.K.., V.C.Raj, T.V. Patel, M.K. Arvadia and A.M. Patel (2012a). Effect of plant population and weed management practices on productivity of sweet corn. Indian J. Weed Sci., 44(3): 168 – 171.

Ashok Kumar, D.S. Rajgopal and Lalit Kumar (2008). Effect of vermicompost, poultry manure and azotobacter inoculation on growth, yield and nutrient uptake of sweet corn. Indian J. Agron., 34(4): 342-347.

Ashok, K.R. (2009). Production potential and nitrogen use efficiency of sweet corn (Zea mays) as influenced by different planting densities and nitrogen levels. Indian J. Agric. Sci., 79(5): 351 - 355.

Azanza, F., B.P. Klein and J.A. Juvik (1996). Sensory characterization of sweet corn lines differing in physical and chemical composition. J. Food Sci., 61: 253–257.

Azanza, F., J.A. Juvik and B.P. Klein (1994). Relationships between sensory quality attributes and kernel chemical composition of fresh-frozen sweet corn. J. Food Quality, 17: 150–172.

Basava, S., K.B. Suneetha Devi, Y. Sivalakshmi and P.S. Babu (2012). Response of sweet corn hybrid to drip + fertigation. J. Res. ANGRAU, 40(4): 101 -103.

Bhagat, S.B., S.A. Chavan, M.V. Zagade and A.V. Dahiphale (2006).Intercropping groundnut and sweet corn at different fertility levels and row proportions. *Indian J. Crop Science*, **1**(1-2): 151 - 153.

Bhatt, S.P., M. Yakadri and Y. Sivalakshmi (2010). Response of sweet corn hybrid to varying plant densities and N levels. In: Symposium on Resource Conservation Technologies to Agriculture in the context of climate change. Bapatla, A.P. 28 – 30, July 2010. PP: 25.

Chavan, S.D. (2009). Effect of organic manures and inorganic fertilizers on quality and yield of sweet corn (*Zea mays* L. *saccharata*). Indian J. Agron., 23(5): 23-28.

Chen, Y.Q., Q.L. Zhang and H.J. Chen (1993). Effect of applying NPK on yield and quality of sweet corn. J. South China Agric. Univ., 14(1): 33-35.

Cherr, C.M., L. Avila, J.M.S. Scholberg and R. Mcsorley (2005). Effects of green manure use on sweet corn root length density under reduced tillage conditions. Rene. Agric. Food Sys., 21(3): 165-173.

Chougule, S.D. (2003). Effect of different plant geometry on sweet corn growth. Maharashtra J. Agric. Sci., 34(12): 122-125.

Collins, J.K. and C.L. Biles (1996). Flavour qualities of frozen sweet corn are effected by genotype and blanching. J. Sci. Food and Agric., 72(4): 425-429.

Coşkun, M.B, Ý. Yalç and C. Őzarslan (2006): Physical properties of sweet corn seed. J. Food Engineering, 74: 523–528.

Dalvi, P.N., Bhondave, T.S., Jawale, S.M., Shaikh, A.A. and Dalvi, N.D., 2009, Effect of sources of organic manures in integrated nutrient management on yield and quality of sweet corn. *J. Maharashtra Agric. Univ.,* 34(2): 222-223.

Efthimiadou, A., D. Bilalis, A. Karkanis, B. Froud-Williams and H. Eleftherochorinos, 2009. Effects of cultural system (organic and conventional) on growth, photosynthesis and yield components of sweet corn (Zea mays *L.)* under semi-arid environment. Notilae Botanicae Horti Agrobotanici Cluj-Napoca, 37(2): 104-111.

Elvin Román-Paoli and David Sotomayor-Ramírez (2004). Soil conditioner efficacy on Lajas Valley sweet corn production. J. Agric. Univ. P.R., 88(3-4): 97-108.

Evensen, K.B. and C.D. Boyer (1986). Carbohydrate and sensory quality of fresh and stored sweet corn. *J. Amer. Soc. Hort. Sci.,* 111(5): 734-738.

Falivene, S. (1995). Improving the international competitiveness of the processing sweet corn industry in New South Wales. Final Report. HRDC project.

Fanngfupong, S., J. Duangpatara, G. Siriwitayawan and S. Choavhong (1994). Influences of sweet corn seed vigour on yield and optimum sowing rate. Kasetsart J., 27(4): 401-411.

GaGawade, D.G. (1998). Response of sweet corn (*Zea mays* L. *saccharata*) to nutrient management. *M.Sc. (Agri.) Thesis,* Dr. Balasaheb Sawant Konkan Krishi Vidyapeeth, Dapoli, Dist. Ratnagiri (Maharashtra).

Gregory E. Welbaum, Jonathan M. Frantz, Malkanthi K. Gunatilaka and Zhengxing Shen (2001). A comparison of the growth, establishment and maturity of direct-seeded and transplanted *sh2* sweet corn. Hort. Sci., 36(4):687–690.

Gzazia, J.D., Tittonell, P.A., Germinara, D. and Chiesa, A., 2003, Phosphorus and nitrogen fertilization in sweet corn (*Zea mays* L. *saccharata*). *Spanish J. Agric. Res.*, 1(2): 103-107.

Haque, M. M., A. Hamid and N. I. Bhuiyan (2001). Nutirent uptake and productivity as affected by Nitrogen and Potassium application levels in maize/sweet potato intercropping system; *Korean Journal of Crop Sciences, 48* (1): 1-5.

Hazariaka, D.S., Gupta, A.S. and Pawar, S.N., 2006, Role of Panchagavya in sweet corn as a bio-pesticide. *Ind. J. Agron.,* 36(1-3): 234-236.

Jhonston, P. (1995). Super sweet management. In: Proceedings of Sweet Corn IPM, NSW Agriculture, Cowra, NSW.

Jose, B.K., A.J. Peaason and T.M. Spiers (1991). Crop uptake of phosphorus from soil amendments with chicken manure and green waste compost in sweet corn. Proceedings of Annual Conference, Agronomy Society of New Zealand, 28: 31- 34.

Kar, P.P., K.C. Barik, P.K. Mahapatra, L.M. Garnayak, B.S. Rath, D.K .Bastia and C.M. Khanda (2006). Effect of planting geometry and nitrogen on yield, economics and nitrogen uptake of sweet corn *(Zea mays*). Indian J. Agron., 51(1): 43-45.

Khadtare, S.V., M.V. Patel, D.D. Mokashi and J.D. Jadhav (2006). Influence of vermicompost on quality parameters and soil fertility status of sweet corn (*Zea mays* L. *saccharata*). *J. Soils Crops,* 16(2): 384-389.

Kulvadee, T. and T. Chowladda (1997). Effect of harvesting period on field and quality of Cannes whole kernel sweet corn. Food, 27: 248–254.

Kumari, J., R.N. Gadag and G.K. Jha (2007). Genetic analysis and correlation in sweet corn (*Zea mays*) for quality traits, field emergence and grain yield. Indian J. Agric. Sci., 77: 613–615.

Kunjir, S.S. (2004). Effect of planting geometry, nitrogen levels and micronutrients on the performance of sweet corn (*Zea mays* L. *saccharata*) under lateritic soils. *M. Sc. (Agri.) hesis,* Dr. Balasaheb Sawant Konkan Krishi Vidyapeeth, Dapoli, Dist. Ratnagiri (Maharashtra).

Kunjir, S.S., S.A. Chavan, S.B. Bhagat and N.B. Zende (2007). Effect of planting geometry, nitrogen levels and micronutrients on the growth and yield of sweet corn. Crop Prot. Prod., 2(3): 25-27.

Lambert, K., I. Turgut and I. B. Sorenson (1994). Effect of plant population and NPK doses on fresh ear yield and yield components of sweet corn grown under Bursa condition. Turkish J. Agric. Fores., 24(3): 341-347.

Liebradt, G.D., D.L. Flor and J.S. Hapse (1991). Effect of vermicompost on nutrient uptake and quality parameters of sweet corn (*Zea mays* L. *saccharata*). *Indian. J. Agron.,* 23(10): 236-239.

Liu-Peng, Hu-Chang Hao, Dong-Shu Ting and Wang-Kong Jun (2003). The comparison of sugar components in the developing grains of sweet corn and normal corn. Agric. Sci. (China), 2: 258–264.

Nu, P., A. Chiesa and G.R. Cline (1993). Effect of levels of nitrogen, phosphorus and potassium observed that sweet corn (*Zea mays* L. *saccharata*). *Hort. Tech.*, 12(1): 118-125.

Olsen, J.K., G.W. Blight and D. Gillespie (1990a). Comparison of yield, cob characteristics and sensory quality of six supersweet corn cultivars grown in a subtropical environment. Aust. J. Exp. Agric., 30 (3): 387–393.

Olsen, J.K., J.E. Giles and R.A. Jordan (1990b). Post-harvest carbohydrate changes and sensory quality of three sweet corn cultivars. Sci. Hortic., 44: 179–189.

Orosz, F., S. Jakab, T. Losak and K. Slezak (2009). Effects of fertilizer application to sweet corn (*Zea mays* L.) grown on sandy soil. Journal of Environmental Biology, 30 (6): 933 - 938.

Parera, C.A. and D.J. Cantliffe (1994). Presowing seed treatments to enhance supersweet sweet corn seed and seedling quality. Hort. Sci., 29: 277– 278.

Patil, M.M, N.D. Dalvi, P.U. Raundal and H.L.Ghadge (2012). Effect of fertilizer levels and G. diazotropics on growth, yield, quality and economics of sweet corn. J.Agric. Res. Technol., 37(3): 364 – 369.

Pierce, L. (1989). Sweet corn in vegetables, characterises, production and marketing. John Wiley and sons, Toronto, P: 289-393.

Pradeep, K.R., K. Yogesh and A. Saraf (2005). Cultivation of sweet corn (*Zea mays* L. *saccharata*). Indian Farming, 54: 10-12.

Prasad, B.K., D.N. Singh and S.N. Singh (2003). Effect of long-term use of fertilizer, lime and manures on growth and yield of sweet corn. *J. Indian. Soc. Soil Sci.,* 34: 271-274.

Puangnak, W. (1998). Effect of hybrid, maturity and kernel structure on lipid content, composition and aroma development in sweet corn. M.S. thesis, University of Maryland, College Park.

Rai, P.K., Y. Kumar and A. Sarraf (2005). Cultivation of sweet corn. Indian Farming, 54 (10): 10 -12.

Raja, V. (2001). Effect of nitrogen and plant population on yield and quality of sweet corn. Indian J. Agron., 46(2): 246 - 249.

Raju, M.S., K.A. Singh and A. Srinivas (1998). Cultivation tips on sweet corn, Madhuri. Indian Farming, 47 (2): 19 - 20.

Reyes, F.G. and G.W. Varseveld (1982). Sugar composition and flavor quality of high sugar (shrunken) and normal sweet corn. J. Food Sci., 47: 753-755.

Robertson, G.H., D.G.Guadagni and M.E. Lazar (1980). Flavor and texture of preserved intact sweet corn: Comparison with cut sweet corn and storage tests. J. Food Sci., 45: 221–223.

Roger, F. and R.D. Dennis (1993). Developing an effective Southern pea and sweet corn inter crop system. Hort. Techn., 3(2): 178 – 184.

Ruan, R.R, P.L. Chen and S. Almaer (1999). Nondestructive analysis of sweet corn maturity using NMR. Hort. Sci., 34: 319–321.

Sahoo, C. and P.K. Mahapatra (2004). Response of sweet corn to nitrogen levels and plant population. Indian J. Agric. Sic., 74: 337 - 338.

Sahoo, C. and P.K. Mahapatra (2007). Response of sweet corn to plant pop and fertility levels during Rabi season IJAS, 77(11). 779 - 781.

Sahoo, S. C. and P. K. Mahapatra (2005). Response of sweet corn to fertility levels under on farm situation. Indian J. Agric. Sci., 75(9): 603 - 604.

Sahoo, S.C. and P.K. Mahapatra (2007). Yield and economics of sweet corn (Zea mays ) as affected by plant population and fertility levels. Indian J. Agron., 52(3) : 239 - 242.

Shah, D.A. and H.R. Dillard (2006). Yield loss in sweet corn caused by *Puccinia sorghi*: a meta-analysis. *Plant Disease,* 90: 1413-1418.

Simonne, E., A. Simonne and R. Boozer (1999). Yield, ear characteristics and consumer acceptance of selected white sweet corn varieties in the southeastern United States. Hort. Technology, 9(2): 289-293.

Singh, N., H.V. Nanjappa and B.K. Ramachandrappa (2007). Influence of nutrient levels and critical period of crop weed competition on growth, yield and weed control in sweet corn (*Zea mays* L. *saccharata*). *Mysore J. Agric. Sci.*, 41: 191-198.

Spandana Bhatt, P. (2012). Response of sweet corn hybrid to varying plant densities and nitrogen levels. African J. Agric. Res., 7(46): 6158 - 6166.

Stone, P.J., D.R. Wilson, P.D. Jamieson and R.N. Gillespie (2001). Water deficit effects on sweet corn. Part II. Canopy development. Aust. J. Agric. Res., 52, 115–126.

Suk, S.L. and H.Y. Sang (1999). Sugars, soluble solids and flavor of sweet, super sweet and waxy corns during grain filling. Korean J. Crop Sci., 44(3): 267-272.

Sunitha, N (2007). Agro-Techniques for higher yield and quality of sweet corn *(Zea mays L.)* in the southern Agro-Climatic zone of Andhra Pradesh. Ph.D Thesis, Acharya N.G. Ranga Agricultural University, Rajendranagar, Hyderabad - 500 030.

Sunitha, N., P.M. Reddy and M. Sadhineni (2010). Effect of cultural manipulation and performance of sweet corn. Indian J. Weed Sci., 42(3 & 4): 184-188.

Suthar, M., D. Singh and V. Nepalia (2012). Green fodder and cob yield of sweet corn varieties at varying fertility levels. Forage Res., 38(2): 115 – 118.

Szymanek, M. (2009). Influence of sweet corn harvest date on kernels quality. Res. Agric. Eng., *55* (1): 10–17.

Szymanek, M. (2011). Studies on shelling of frozen sweet corn. J.Food Proc. Eng., 34(3): 716–727.

Treat, C.L. and W.F. Tracy (1993). Contributions of dent corn germplasm to stalk and root quality in sweet corn. J. American Soc. Hort. Sci., 118: 885 - 889.

Trongpanich, K., S. Stonsoavapak, D. Hengsawadi and N.A. Lowitoon (2002). A comparative study on pretreatment processes of canned whole kernel sweet corn. Kasetsart J. Nat. Sci., 36 (1): 63–68.

Turgut, I. (2000). Effects of plant populations and nitrogen doses on fresh ear yield and yield components of sweet corn grown under Bursa condition. Turkish J. Agric. Forest, 24 (3): 341-347.

Ummed Singh, A.A. Saad, T. Ram, Lekh Chand, S.A. Mir and F.A. Aga (2012). Productivity, economics and nitrogen use efficiency of sweet corn as influenced by planting geometry and nitrogen fertilization. Indian J. Agron., 57 (1): 43 - 48.

Uwah, D.F., A.E. Eneji and U.J. Eshiet (2011). Organic and mineral fertilizers effects on the performance of sweet maize (*Zea mays* (L.) saccharata Strut.) in south eastern rainforest zone of Nigeria. International Journal of Agriculture Sciences, 3(1): 54-61.

Van wychen, L.R., R.G. Harvey, M.J. Vangassel, T.L. Rabdey and D.J. Bach (1999). Efficiency and crop response to gulfosinate based weed management in PAI transformed sweet corn. Weed Tech., 13:104-111.

Venkatesh, S., R. Sanjay and J.C. Shekhar (2003). Sweet corn. Directorate of Maize Research, New Delhi. P: 1-3.

Venkateshwarlu, S. and P. Raghuram Reddy (2010). Response of sweet corn and popcorn to row spacings and fertility levels. Indian J. Dryland Agric. Res. Dev., 25 (2): 110 – 111.

Vijay, J., B.P.Tuse, S.M. Jawale, A.A. Shaikh and N.D. Dalavi (2009). Effect of fertilizer levels and dates of sowing on growth and yield of sweet corn J. Maharashtra Agric. Univ., 34(1): 108 – 109.

Viswanatha, G.B., B.K. Ramachandrappa and H.V. Nanjappa (2002). Soil - plant water status and yield of sweet corn as influenced by drip irrigation and planting methods. Agric. Water Managt., 55: 85 - 91.

Wagh, D.S. (2002). Effect of spacing and integrated nutrient management on growth and yield of sweet corn (*Zea mays* L. *saccharata*). *M.Sc. Thesis* Mahatma Phule Krishi Vidyapeeth, Rahuri, Pune Dist., Maharashtra.

Waligóra, H. (2006). Harvest and utilization of sweet corn. Kukurydza, 2(28): 26-27.

Waters, L.J., R.L. Burrows, M.A. Bennett and J. Schoenecker (1990). Seed moisture and transplant management techniques influence sweet corn stand establishment, growth, development and yield. J. Amer. Soc. Hort. Sci., 115: 888 – 892.

Wong, A.D. (1994). A study of kernel composition affecting the quality of shrunken 2 sweet corn. Dissertation Abstracts International, 55(1): 10.

Wong, A.D. and J.M. Swiader (1995). Nitrogen and sulfur fertilization influences aromatic flavor components in shrunken2 sweet corn kernels. J. Am. Soc. Hort. Sci., 120 (5): 771 - 777.

Wong, A.D., J.A. Juvik, D.C. Breeden and J.M. Swiader (1994). Shrunken sweet corn yield and the chemical components of quality. J. American Soc. Hort. Sci., 119: 747–755.

Xu, Hui Lain and H.L. Xu (2000). Effect of a microbial inoculants and organic fertilizers in the growth, photosynthetic and yield of sweet corn (*Zea mays* L. *saccharata*). *J. Crop Prod.,* 3(9): 183 - 214.

Yodpet, C. (1979). Studies on sweet corn as potential young cob corn (*Zea mays* L.) *Ph.D. Thesis,* Philippines, pp. 276 - 282.

Yu Hang, D.P. Yousif and H.C. Ali (1993). Effect of planting date and fertilizer on growth, yield and other organic characters of sweet corn. Masapatonia J. Agric., 21(3): 171-187.

# 20

# Popcorn Cultivation

Popcorn is the most primitive of the surviving races of maize. This corn type is characterized by a very hard, corneous endosperm containing only a small portion of soft starch. Popcorns are essentially small-kernelled flint types. Some of the more recently developed popcorns have thick pericarps (seed coats), while some primitive semi-popcorn, such as the Argentine popcorns and have thin pericarps. It is used primarily for human consumption as freshly popped corn or as the basis of popcorn confections. Besides their use as a popular snack, ground popcorn, as flour or grits, can be used in the preparation of many traditional dishes. Isolated planting is not necessary in case of popcorn, since there are no major xenia effects on popping expansion and many types of popcorn are cross-sterile with field corn. Although conditions for growing popcorn are the same as for dent corn, special harvesting, drying and storage practices are necessary to maintain popping quality.

Popcorn plant type has some distinctive characters compared to normal corn. The plant type is lanky. The tassel is highly branched and the branches are droopy. The ear placement is higher up compared to normal corn. Often the prolificacy is more in fertile soils with good management. It is very common to find two ears per plant. In many popcorn types there is a tendency of tillering and the brace roots are also fewer compared to normal corn. In popcorn, several grain colours are grown viz., white, yellow, red, etc. However, yellow types are more common. The kernels may be either pointed (rice like) or round (pearl like). The rice type popcorn kernels are common in white grain types and are typically peaked i.e. long and pointed at the tip. Pearl type of popcorn is more common than the rice type.

It has smooth and round kernels and is common in yellow grain type. Based on the type of flalces produced on popping, popcorn is further classified as butterfly type (flat flakes) and mushroom type (compact globose flakes). The popcorn kernel's general composition is carbohydrate (principally starch), protein, fat and water. Water is stored in a small circle of soft starch in each kernel. As the kernel is heated, the water heats, builds up pressure and takes up any available room. The harder surface surrounding the starch resists the vapour pressure for as long as it can. When the outer surface gives way, the water vapour further expands, causing popcorn to explode. The soft starch pops out, the kernel turns inside out, steam inside the kernel is released and the corn pops.

## Cultivation

The land requirements and cultivation aspects of popcorn are similar to that of sweet corn.

## Variety

There are three types of popcorn kernels: white, small yellow and large yellow besides red kernel types. Generally white popcorn has a rice shaped kernel and yellow popcorn kernels are pearl shaped. The three kernel types meet different needs within the industry. Growers and processors should consider this when selecting a variety. Industry standards have not been developed for kernel size determination. A common measure is based on the number of kernels in 10 g. The large one has 52-67 kernels, medium 58-75, and small 76-105. In India there are principally three released popcorn varieties (Table 1). However, all of them are medium in size and yellow pearl kernel type. These varieties can be grown across the country during kharif season, while in Rabi they can predominantly be grown in peninsular India. Besides this another major consideration in popcorn variety selection is maturity because maximum popping potential of a variety can be achieved only if it reaches full maturity. If planting is delayed, consider an earlier maturing variety.

**Table 1. Popular popcorn varieties released in India**

| Varieties | Recommended | | Days to harvest | Yield |
|---|---|---|---|---|
| | Seasons | Areas | | |
| Amber popcorn, VL Almora popcorn and pearl popcorn | Kharif and Rabi | Across the country in kharif and peninsular India in rabi | 80-90 days in kharif and 95-100 days in rabi | 25 q/ha in kharif & 30 q/ha in rabi |

## Soil

Any soil type suitable for normal corn should produce a good popcorn crop. Soil pH should be in the range of 6.0 to 6.8. Popcorn seeds germinate more slowly than dent corn and the seedlings grow at slower rate. Thus, medium to coarse textured soils, which warm slightly faster than fine-textured soils, should improve germination, emergence and seedling establishment. The popcorn root system is less extensive than that of normal corn. Therefore high clay soils and poorly drained soils will weaken the roots, reduce yield and increase lodging.

## Land preparation

Land preparation for popcorn is similar to that for field corn. The most common method is to use two or three discings, although some heavy soil types will produce better tilth if ploughed. Discing or field cultivating is then used for seedbed preparation. If needed, a spike-tooth harrow or other finishing tools can be used with the last pass over the field to smooth the seedbed.

## Fertilizer management

Fertility requirements for popcorn are similar to normal corn. Nitrogen is to be applied at a rate of 100-110 kg/ha. Nitrogen requirements are based on expected yield. In general the rate is 85 % of the requirement for an expected yield of normal corn in the same field. Any normal corn fertilizer programme adapted to popcorn production should take into consideration popcorn's relatively poor standing ability. However, high nitrogen rates can cause lodging, especially when less than the recommended rate of potash is used. As popcorn seedlings grow slower than normal corn, a starter dose is more important than subsequent split doses. Other nutrients are supplied according to need as shown by soil test, similar to normal corn requirements. The fertilizer doses recommended are 80: 60: 40 kg/ha each of N, P and K. N is applied in three equal splits at sowing, 25 and 45 days after emergence. Zinc Sulphate is to be applied 50 kg/ha once in three seasons.

Increasing rates of N from 40 to 100 kg per ha increased the yield from 12.8 to 14.1 t/ha. Higher grain and stover yields of popcorn were obtained with 100 kg N/ha (Roy and Singh, 1986).

At Calabar (Nigeria), lime at 500 kg/ ha gave higher popcorn yields with a yield increase of 18.03 % above control. Popcorn applied with 80 kg N /ha increased the crop productivity compared to 0, 40 and 120 kg N/ha (Effa et al., 2012).

**Table 2. Effect of nitrogen and lime levels on growth parameters, yield attributes and yield of popcorn varieties**

| Treatment | Plant height (cm) | Leaf area index | Days to 50 % tasselling | Total dry matter (g) | Weight of ears/ plant (g) | Kernels / ear | Grain yield (t/ ha) | Agronomic Use efficiency of N |
|---|---|---|---|---|---|---|---|---|
| **N (kg / ha)** | | | | | | | | |
| 0 | 182.81d | 2.18d | 51.39a | 120.35d | 59.19d | 250.05d | 1.42d | 0.00c |
| 40 | 217.30c | 3.10c | 48.72b | 159.82c | 70.63c | 362.35c | 1.84c | 12.85a |
| 80 | 257.01a | 4.21a | 46.11c | 213.90a | 75.58a | 451.07a | 2.51a | 14.29a |
| 120 | 233.92b | 3.95b | 44.61d | 199.89b | 74.06b | 391.97b | 2.18b | 6.86b |
| **Lime (kg/ ha)** | | | | | | | | |
| 0 | 208.39c | 3.10a | 47.21a | 154.86c | 66.09c | 328.90c | 1.83b | 6.40b |
| 500 | 231.28a | 3.49a | 47.08a | 171.08b | 73.96a | 392.74a | 2.16a | 12.14a |
| 1000 | 228.61b | 3.49a | 48.88a | 194.54a | 69.55b | 369.99b | 1.98b | 6.96b |
| **Variety** | | | | | | | | |
| Yellow Composite | 225.88a | 3.40a | 48.06a | 179.85a | 70.52a | 367.25a | 2.03a | 9.33a |
| Ashland | 219.64a | 3.32b | 47.36b | 167.13b | 69.21a | 360.49a | 1.95a | 7.67a |

(Effa et al., 2012).

Application of 180 kg N /ha was found optimum for popcorn grown as a second crop in Hatay ecological conditions in Eastern Mediterranean region of Turkey (Huseyin and Omer, 2010).

**Table 3. Effects of N doses on grain yield and yield related characteristics of pop corn**

| Nitrogen Doses (kg / ha) | Tasseling period (day) | Plant height (cm) | Stem diameter (mm) | Ear height (cm) | Ear weight (g) | Grain yield (kg/ha) |
|---|---|---|---|---|---|---|
| 0 | 49.67A | 165.1C | 17.45B | 89.42B | 64.00C | 2953C |
| 120 | 46.29BC | 174.5B | 18.73A | 97.50A | 70.63B | 3753B |
| 180 | 46.88B | 179.2AB | 18.50A | 100.9A | 71.71B | 4185A |
| 240 | 45.63C | 185.0A | 18.53A | 103.4A | 76.21A | 4225A |

(Huseyin and Omer, 2010).

### Planting

Kharif popcorn can be planted from the last week of May to the first fortnight of July. A soil temperature of 10-13 $^0$C is necessary. Planting later than July 15$^{th}$ often results in a reduction of yield and maximum popping expansion. Early planting permits the crop to mature and reach the desired harvesting moisture of 15-18 % by early to mid-October. Rabi sowing preferably is completed in between October end and middle of November. So that the crop can establish itself before the onset of winter.

### Seed rate

Seed rates for popcorn is higher than for normal corn because of its smaller plant size and lower yielding ability. Generally, consider 25 % increase above the recommended normal corn plant densities for most soils and cultural practices. Row widths of 75-95 cm are common. With irrigation facility the population should be from 50,000 to 67,000 plants / ha. A slightly lower population is used for dry land areas. Thus recommended seed rate for common popcorn varieties in India is 12-14 kg/ha of land. The spacing recommended for planting is 60 cm x 20 cm or 45 cm x 30 cm.

In popcorn, grain yield increased with increases in nitrogen fertilizer levels and plant densities. Higher grain yield was obtained at 100 kg N/ ha with 80,000 plant /ha (Roy and Singh, 1986). Higher pop corn grain yield was obtained at narrow row spacing (40 cm) than 50 and 60 cm inter-row spacings with the higher nitrogen fertilizer level (90 kg N/ha) applied as compared to 0, 30, 60 kg N/ha (Thakur and Malhotra, 1991). Ülger (1998) reported higher popcorn yield as 5850 kg /ha obtained at 250 kg N/ ha with 15 cm intra row and 70 cm inter-row spacing. Özkan (2007) demonstrated that higher grain yield was obtained at 200 kg N/ ha with 4530 kg /ha. Gökmen et al. (2001) indicated that for maximum grain yield of popcorn, 100-150 kg N/ ha should be applied and the plant density should be 7.0 plants m$^{-2}$. Sezer and Yanbeyi (1997) demonstrated that ear characteristics were negatively affected by increases in plant densities, although plant height, ear height and grain yield increased with increases in plant densities.

**Table 4. Effects of plant densities on grain yield and yield-related characteristics of pop corn**

| Plant densities (plants /ha) | Tasseling period (day) | Plant height(cm) | Stem diameter (mm) | Ear height (cm) | Ear weight (g) | Grain yield (kg/ha) |
|---|---|---|---|---|---|---|
| 60,000 | 47.29 | 172.2B | 18.55A | 94.46B | 72.21 AB | 3581C |
| 74,000 | 47.04 | 176.4A | 18.67A | 99.04A | 73.38 A | 3877AB |
| 88,000 | 47.21 | 176.7A | 18.37AB | 98.25A | 70.21 B | 3973A |
| 102,000 | 46.92 | 178.6A | 17.63B | 100.5A | 66.75 C | 3685BC |

(Huseyin and Omer, 2010).

Higher popcorn yield as 5850 kg/ ha was obtained with 15 cm intra-row and 70 cm inter-row spacing (Ülger, 1998).

**Table 5. Effect of plant population on popcorn yield, yield components and popping expansion ratio**

| Population | Yield (kg/ha) | Kernel weight (mg) | Kernels /cob | Ears/ ha | Popping expansion ratio |
|---|---|---|---|---|---|
| 37000 | 2390 | 143 | 372 | 47400 | 32 |
| 49000 | 2510 | 141 | 377 | 50700 | 32 |
| 62000 | 2570 | 137 | 374 | 53400 | 32 |
| LSD(0.05) | 80 | 3 | NS | 1700 | NS |

(Daniel and Marr, 2005)

At Manhattan (USA), for popcorn grown on the clay pan soils of the eastern Great Plains, plant population was optimum at near 49,000 plants/ha. Popping expansion volume appears unaffected by plant population (Daniel and Marr, 2005).

## Irrigation

Irrigation of popcorn is the same as that for field corns. Some growers like to delay early irrigation to stimulate better rooting of the developing plants. However, the value of this practice has not been determined. Irrigation during pollen shed is not recommended. Irrigation should continue until the black-layer has formed in the developing kernel.

**Table 6. Effect of irrigation scheme on 4-yr average (1995 to 1998) popcorn yield components and popping expansion ratio**

| Irrigation scheme | Yield (kg/ha) | Kernel weight (mg) | Kernels / cob | Ears/ha | Popping expansion ratio |
|---|---|---|---|---|---|
| None | 2475 | 140 | 350 | 50200 | 32 |
| R1–2.5 cm | 2750 | 143 | 393 | 51200 | 33 |
| R1–5 cm | 2575 | 140 | 379 | 51200 | 31 |
| R3–2.5 cm | 2425 | 138 | 378 | 49200 | 31 |
| R3–5 cm | 2350 | 140 | 348 | 49700 | 31 |
| R1/R3–2.5 cm at each | 2775 | 141 | 398 | 51900 | 32 |
| LSD (0.10) | NS | NS | 32 | NS | NS |

(Daniel and Marr, 2005)

At Manhattan (USA), supplemental irrigation of popcorn at selected reproductive growth stages can affect popcorn yield. However, the response is sporadic. With limited irrigation at R1 (silking stage) or R3 (milk stage) yield components are little or not affected by irrigation (Daniel and Marr, 2005).

## Intercultural operations

Intercultivation with tractor or bullock drawn cultivator is usually undertaken once at knee high stage and if weed situation requires, it is followed by one or two cultivations. Mechanical cultivation will depend upon herbicide effectiveness. Most effective herbicide for weed control is application of Atrazine or Simazine @ 1 kg ai /ha. The predominant application method is with planting or before emergence. This gives weed protection up to 30-35 days after sowing.

## Intercropping

Popcorn being a highly remunerative crop can successfully be cultivated in periurban agriculture. However, instead of cultivating popcorn as sole crop it may be intercropped with other highly remunerative crops like cabbage, cauliflower, radish, spinach, coriander, potato, marigold, garlic, tuberose, gladiolus, spices, pea, etc. This provides additional income to the farmers from unit area and makes agriculture more sustainable.

## Pest management

Most of the insects that attack field corn can also attack popcorn. Insect control during the season is important, especially for Stalk borers (*Chilo partellus* Swinhoe) in kharif season and *Sesamia inferens* Walker in rabi season. Spray of Endosulphan @ 2 ml/1itre within 21 days after sowing is recommended. Popcorn is more susceptible than field corn to diseases especially stalk rots and ear rots and blue eye disease caused by three types of molds *Penicillium viridicatum, P. cyclopium* and *Aspergillus glaucus*. These pathogens not only affect the crop in the field but may affect stored grains leading to low popping volume, low germination and sometimes a low level of toxicity. This disease is controlled by proper aeration during storage. However, application of Bavistin @ 1 g a.i per litre takes care of most of the foliar diseases.

## Harvesting

Popcorns mature in 85-90 days in kharif and 95-100 days in Rabi. Since kernels are the economic product, harvest should necessarily be done on the attainment of physiological maturity, which is evidenced by the formation of a black polyphenol layer at the base of the kernels. Either ear or shelled popcorn may be harvested. Moisture at harvesting for ear corn should be 18-20 %, for shelled corn 16-18 %. Harvesting equipment should be adjusted accurately and carefully so that kernel damage is prevented. Damaged kernels do not pop properly and will cause a discount on the grain market. Depending on the proximity of the processing plant, the grain may be delivered directly at harvest or stored on the farm until delivery is requested. If on-farm storage is required, storage facilities must be clean and good aeration must be provided. Aeration is essential for extended storage. Generally, the delivery of the grain is taken at 14-15 % moisture, so the storage and aeration facilities must be able to remove some grain moisture. Relative humidity of 70 % will maintain the grain at the proper moisture. Kernels that are too dry will not pop properly. Dirt, cob remnants and other foreign material at harvest must be cleaned before storage. Proper moisture level is the reason to store popcorn properly. For day-to-day use popcorns are to be stored in airtight container and in cool cupboard. Popcorn should not be stored in refrigerator. Air inside a refrigerator contains very little moisture and can cause the popcorn to dry out.

## Marketing

Popcorn is mainly produced under contract farming, which specifies the varieties and the acreages to be planted. The grower usually buys the seed through the contracting company. Popcorn seed grown without sales contract is difficult to market and is risky for farmers who do not have appropriate storage facility with good aeration.

### Popping quality

The quality of popcorn is expressed as popped volume, shape of the popped kernels, tenderness and flavour. Only volume can be measured easily. The popped volume is perhaps most important because the commercial buyers buy on weight basis but sell the popped popcorn by volume. The conventional oil popping methods are still most common. However, air popping and microwave popping are becoming increasingly important in the retail market. Popping expansion of the commercially occupied popcorn varieties in India is just 20:1. Problems associated with low expansion volume, the high number of unpopped kernels and scorching of popped kernels make conventional popping superior to microwave popping. Popcorn variety, the moisture content of the kernels, storage conditions, type of package and microwave oven wattage influence the quality of microwave popping.

### Value addition

Popcorn while being popped can be coated with salt or sugar or cheese. Sometimes coating with chocolate is also done. This enhances the market value and consumer acceptance. The manufacturers prefer butterfly type of popping due to increased volume since popcorn after popping is sold on a volume basis. However, consumers prefer mushroom types.

### Nutritional value of popcorn

Popcorn is a good source of carbohydrate, energy and fibre. For individuals of normal weight, carbohydrate is the best source of body fuel. Fibre is not used for energy but is an important health factor, which is insufficient in present day urban diets. There are two kinds of fibre - soluble plays a role in regulating hunger, cholesterol and blood sugar and insoluble is important in gastrointestinal health. Popcorn has no artificial colour or flavour additives and is surprisingly low in calories. One cup of popped popcorn has 31 calories if eaten plain or seasoned with herbs, 133 calories if drizzled with a tablespoon of butter, margarine or oil, 34 calories if sprayed with butter-flavoured oil, 35 calories if sprinkled with butter substitute. Most Kernel Season's flavours have less than one calorie per serving.

Many medical groups feel popcorn has a nutritive value. Popcorn is a wholesome, fun food, which aids digestion by providing necessary roughage. Health and medical associations regard popcorn as an excellent mealtime complement sugar-free, fat-free and low in calories. The National Cancer Institute (NCI), USA suggests fibre in the diet to reduce the risk of some forms of cancer. The American Dental Association includes popcorn in its list of recommended sugar-free snacks. The American Diabetes Association and the

American Dietetic Association permit popcorn as a bread exchange on weight control diets. The Feingold Diet for hyperactive children permits popcorn because it contains no artificial additives. Nutrition experts agree that popcorn is an all around "good" food healthy, economical and tasty. Popcorn is a whole grain and its nutritive value is retained inside the hull until it is popped. The US Department of Agriculture has found the following nutritive values in one pint of popped popcorn. The figures may vary because of the actual weight or size of the kernels.

| | |
|---|---|
| Weight | half to 1 ounce |
| Calories | 50 to 110 calories (unbuttered) |
| Carbohydrates | 10 to 22 g |
| Protein | 2 to 4 g |
| Fat | 1 to 2 g |
| Calcium | 2 mg |
| Iron | 0.4 to 0.6 mg |
| Niacin | 0.2 to 0.6 mg |

Popcorn is a very good dieting aid. Eaten immediately before a meal, it will take the edge of an appetite. The cellulose of the popcorn hull is an indigestible carbohydrate. This cellulose is excellent dietary roughage and compares to bran flakes. Digestible carbohydrates in the popcorn provide energy and are supposed to help the body to metabolize fat.When popcorn is being used as part of a dieting programme, the butter may be skipped and just light salt with the popped kernels may provide a good taste.

**Uses of popcorn**

Principally popcorn is consumed as snacks with salt and selected spices. For domestic use automatic popcorn poppers, microwave popcorn poppers and air-poppers are available. However, popcorns can be popped with microwave oven as well. Popcorn should be stored on pantry shelf, never in the refrigerator or freezer. The popcorn should be placed in the center of the microwave oven. Microwave should be set on high setting (full power). It is always better to follow the instructions on the microwave popping bag for cooking time for specific microwave oven. Some general cooking time recommendations:

| Microwave wattage | Cooking time |
|---|---|
| 1000 Watt Oven | 1:45-2:00 minutes |
| 700 Watt Oven | 2:45-3:30 minutes |
| 650 Watt Oven | 3:00-4:00 minutes |
| 500 Watt Oven | 4:30-6:00 minutes |

## References

Adeleye, E.O. and L.S. Ayeni (2010). Effects of soil preparation methods and spacing on the growth and yield of popcorn on an alfisol in south western Nigeria. Researcher, 2(10): 1-6.

Allred Coyle, T.A., R.B. Toma, W. Reiboldt and M. Thakur (2000). Effects of moisture content, hybrid variety, kernel size, and micro wave wattage on the expansion volume of microwave popcorn. Int. J. Food Sci. Nutr., 51: 389–394.

Amusa, N.A., J.E. Iken and F.O. Fayenuwo (2005). The incidence of field diseases and vertebrate pests of popcorn (*Zea mays everta*) varieties cultivated in forest agro-ecologies of Nigeria. *World J. Agric. Sci., 1* (2): 173-177.

Ashok kumar (2009). Influence of varying plant population and nitrogen levels on growth, yield, economics and nitrogen use efficiency of pop corn. Crop Res., 37: 19 - 23.

Babak Peykarestan, S. M.R. Seify, A. M. Alijani and M. Najafi (2012). Zinc dose and sowing date effects on popcorn (KSC 604 P.C.) yield components. International J. Agric. Crop Sci., 4 (1-4): 930-938,

Babak Peykarestan, S.M.R. Seify, A.M.M. Sanavi and H. Omidi (2012). Nitrogen fertilizer levels and sowing date effects on popcorn (KSC 604 P.C.) yield components. International J. Agric. Res. Rev., 2 (3): 218-226.

Babak Peykarestan, S.M.R. Seify, M. S. Fadaei and M. Abdoli (2012). Potassium dose and sowing date effects on popcorn (KSC 604 P.C.) yield components. International J. Agric. Res. Rev., 2 (4): 425-432.

Choudary, J. and D. Singh (2006). Effect of plant population and fertilizer levels on yield and economic viability of pop corn. Agron. Digest, 6 & 7: 14 - 17.

D'Croz-Mason, N. and R. Waldren (1990). Popcorn production Guide G78–426. Univ. of Nebraska Coop. Ext. Serv., Lincoln.

Dhuyvetter, K.C. and D.W. Henson (1994). Center pivot irrigated popcorn. Misc. Folder 996 (revised). Kansas State Univ. Coop. Ext. Serv. Manhattan.

Dofing, S.M., M.A.T. Compton and J.S. Buck (1990). Genotype x popping method interaction for expansion volume in popcorn. Crop Sci., 30: 62–65.

Effa, E.B., D.F. Uwah, G.A. Iwo, E.E. Obok and G.O. Ukoha (2012).Yield Performance of Popcorn (*Zea mays* L. *everta*) under Lime and Nitrogen Fertilization on an Acid Soil. J. Agric. Sci., 4(10): 12-19.

Eze, S.C. and I.U. Obi (2008). Response of UNN popcorn variety to N. P. K fertilizer and lime. In Proceedings of 42nd Annual Conference. Agricultural Society of Nigeria (ASN), Ebony State. 19-23, October, 2008, pp. 101-105.

Gana, A.K., J.A. Adigun, K.O. Adejonwo, B.W. Ndahi and L.D. Busari (2007). Evaluation of pre-emergence herbicides for popcorn (*Zea mays* L.) production in the northern guinea savanna of Nigeria. African Scientist, 9 (1): 9-12.

Gokmen, S., O. Sencar and M.A. Sakin (1999). Response of popcorn (*Zea mays* L. everta) to nitrogen rates and plant densities. *Turkey J. Agric.*, 25 (4): 15-23.

Huseyin, G. and K.K. Omer (2010). Nitrogen dose and plant density effects on pop corn grain yield. African J. Biotechnology, 9(2-5): 3828-3832.

Mahajan, V. and H.S. Gupta (1998). Pop corn for north eastern hills. Indian Farming, 47 (6): 19.

Özkan, A. (2007). Effects of different nitrogen levels on grain yields, agricultural properties and some quality parameters of two popcorn (*Zea mays everta* Sturt.) cultivars under Cukurova conditions. C.Ü. Naturel Sci. Institute. Ph.D Thesis. p. 111.

Roy, R. K. and K. S. P. Singh (1986). Response of popcorn (*Zea mays*) to plant population and nitrogen. Indian J. Agron., 31(1): 88-92.

Sabri, G.K., S. Ozer and A. S. Mehmet (2001). Response of popcorn (Zea mays everta) to nitrogen rates and plant densities. Turk J. Agric For., 25: 15-23.

Sade, B., F. Kucukmumcu and H. Gayretli (1996). The determination of grain yield, some morphological traits of pop corn populations (*Zea mays L. everta*) in Konya ecological conditions. J. Selçuk Univ. Agric. Faculty, 9(11): 130-143.

Sezer, I, and S. Yanbeyi (1997). Plant density and nitrogen fertilizer effect on grain yield, yield components and some plant characters of pop corn in Çaramba plain. Turkey 2. Field Corps Congress, 22-25 September 1997, Samsun. pp. 128-133.

Sheshagiri (1998). Agronomic investigations on pop-corn. PhD. thesis, University of Agricultural Sciences, Bangalore, India.

Singh, D.P., D.S. Prasad, U.K. Prasad and N.N. Sharma (1982). Response of popcorn to nitrogen and plant population. Indian J. Agron., 27 (2): 189-190.

Tekkanat, A. and S. Soylu (2005). Determination of important quality characters and grain yield in popcorn cultivars. J. Selçuk Univ. Agric. Faculty, 19(37): 51-60.

Thakur, D.R. and V.V. Malhotra (1991). Response of Popcorn (Zea mays everta) to Row Spacing and Nitrogen. Indian J. Agric. Sci., 61(8): 586-587.

Tran,T.T., G.S. Srzednicki and R.H. Driscoll (1999). Effects of aeration on the quality of popcorn Agric. Eng. Int.: The CIGR J. Sci. Res. Dev. Vol. 1, Dec. 1999.

Ülger, A.C. (1998). The effects of nitrogen doses and intra row spacingon grain yield and some agronomical characters of popcorn (*Zea mays eveta* Sturt). J. Agric. Fac. Ç.Ü., 13(1): 155-164.

Umesha, A. A. G. Bandi, S. B. Yogananda and Kiran Kumar (2001). Performance of popcorn varieties on their yield under varying levels of plant densities. Crop Res., 22(3): 350-353.

Yogananda, S.B., B. Shivaraj, A. Gowda and N. Krishnamurthy (1999). Effect of plant population, phosphorus and potassium levels on growth and yield of popcorn (*Zea mays* L.). Mysore J. Agric. Sci. 33: 21–25.

Ziegler, K.E. and B. Ashman (1994). Popcorn in Specialty Corns. Hallauer, Ed., A.R. Department of Agronomy, Iowa State University, CRC Press,

Ziegler, K.E., P.B. Ashman, G.M. White and D.S. Wysong (1988). Popcorn production and marketing Natl. Corn Handb. NCH-5. Purdue Univ. Coop. Ext. Serv., West Lafayette, In

Ziegler, K.E., R.B. Ashman, G.M. White and D.B. Wysong (1984). Popcorn Production and Marketing, Cooperative Extension Service, Purdue University, West Lafayette, IN, A publication of the National Corn Handbook Project NCH-5.

Ziegler, K.E., W.D, Guthrie and D.C. Foley (1987). Registration of BSP1C1 and BSPW1C1 popcorn (Maize) germplasam. Crop Sci., 27: 1318-1319.

21

# High Oil Corn Cultivation

Most of the normal maize lines have 3-4 % oil content. In general, lines with more than 6 % oil are considered high oil lines. About 95 % of the total oil is in the germ. When the oil percent increases the starch decreases. The wet milling industries are still in advantage with high oil content corn. In USA the high oil corn is cultivated on contractual basis and remunerative price is paid to the farmers. In India, its cultivation is not economical because it is not sold on premium basis. Generally in normal maize crop, 15-20 % population of high oil hybrids is used as pollen parent and there is detasseling of the normal corn plant. Due to xenia effect there is an increase of oil in normal maize and its cultivation is done in isolation. The corn oil has low content of saturated fatty acid and is considered to be one of the best quality cooking oil. In India, more than 60000 tonnes of corn oil is made available for various uses.

Normal corn has usually around 3.5 % oil, principally in the embryo. High-oil corn (HOC) has nearly twice the oil concentration (greater than 6 %) and substantially higher protein than conventional corn. This makes high-oil corn grains amenable to higher oil extraction as a byproduct of corn refining. High-oil corn also serves as an excellent feed ingredient for livestock and poultry producers. The industry discovered the potentiality of extraction of oil from the germ. The first commercial production of corn oil took place in 1889. Importance of corn oil as an ingredient in margarine, salad oils, cooking oils and a large number of food products is well known. Less known is the use of high-oil corn to enhance the feeding rations of animals.

High-oil corn differs from normal corn with respect to kernel anatomy and composition. Normal corn has 82 % endosperm and 12 % germ. The germ (including embryo and scutellum) in HOC is proportionately larger than in conventional corn. Normal corn kernel has 8 % protein in endosperm and 16 % protein in the germ. The germ in HOC has nearly 1 % higher total protein and it also contains higher levels of essential amino acids like lysine, threonine and methionine (Table 1). It is estimated that 80 to 84 % of the total oil is located in the germ, 12 % in the aleurone and 5 % in the endosperm. If the tiny piece of germ is cut out and squeezed on a piece of paper, the corn oil may be observed.

**Table 1. Essential nutrition values (% dry matter basis) for commercial and high-oil corn germ**

| Nutrient | Commercial corn grain | High-oil corn grain |
|---|---|---|
| Crude fat (%) | 4.00 | 7.25 |
| Protein (%) | 8.86 | 9.75 |
| Energy (Kcal/lb) | 1777 | 1858 |
| Lycine (%) | 0.26 | 0.30 |
| Methionine (%) | 0.19 | 0.23 |
| Total starch (%) | 71 | 66 |

The first high-oil corn varieties were selected in 1896 at the University of Illinois. Over the years genetic selection has steadily improved agronomic performance and yield of corn hybrids with higher oil content. Corn kernel oil concentration ranges between 1.2 and 21.3 % on weight basis. Commercial hybrids grown in the US Corn Belt typically contain 4-5 % oil. These selections have narrowed the yield gap between high-oil corn and normal corn hybrids, though in general high-oil corn hybrids yield 5-10 % lower than normal corn. The corn germ consists of scutellum and embryo. The embryo contains primary axis, plumule, coleoptile, mesocotyl, adventitious and primary roots and the coleorhiza. The scutellum in normal corn makes up about 10-12 % of total kernel weight but it contains 83-85 % of total kernel oil. Pericarp tip cap and endosperm contain 10-13 % of total oil. The scutellum consists of epithelial parenchyma cells and vascular tissues. Oil synthesis takes place in developing scutellum tissue deposits, oil bodies or spherosomes in the parenchyma cells of the scutellum. Immediately after pollination kernel development starts very fast with higher rate of endosperm development as compared to germ. Within 15 days after pollination (DAP) oil can be detected in developing embryos and it continues to increase till 45 DAP and then remains constant to physiological maturity or black layer stage. High-oil inbreds show a high rate of oil accumulation from 15-45 DAP as compared to normal inbreds.

## Extraction of corn oil

Corn oil extraction is not an isolated event in any corn-based refinery. Thus the process of oil extraction goes along with other processes like production of starch, corn sweeteners, etc. Basic step for all these processes is inspection and cleaning. The shipments are inspected for contamination and cleaned thrice to remove cob parts, dust, chaff and foreign materials before steeping, the first processing step begins. In this step of steeping, clean corn grains are soaked at 50°C water for 30-40 hours in big steel tanks called steep tank. During this process the grains swell and soften. After steeping, the corn is coarsely ground to break the germ loose from other components. The ground corn, in water slurry, flows to the germ separators. In this process of germ separation, cyclone separators spin the low density corn germ out of the slurry. The germs are pumped into a screen and washed separately to remove stray amount of starch in the mixture. Oil is recovered from corn germ by expelling, solvent extraction or a combination of expelling and solvent extraction. Wet milled germ is preferably expelled from an oil content of 50-60 % down to 20-25 % and finally extracted with hexane to a residual oil content of 1-2 % in the spent corn germ flakes. The oil is then refined and filtered into finished corn oil. The germ residue is saved and used as a component of animal feeds. The starch component of the grain is further processed to give a number of products.

Crude corn oil contains 95 % triglycerides besides containing minor compounds like free fatty acids, waxes, phospholipids, pigments, and odourous compounds. Before making the corn oil acceptable to the consumers these components are removed from the crude oil through a refining process. This process involves several steps: formation of sodium soaps of the free fatty acids, removal of the emulsion containing the soaps and phospholipids by centrifugation, removal of waxes by chilling, removal of pigments by contact with bleaching clays, removal of odours by high-vacuum distillation at 225 to 260°C. The fatty acid fraction is recovered by heating the emulsion in the presence of sulphuric acid and is sold as an ingredient for use in beef and poultry rations.

## Quality of corn oil

The oil refining process gives corn oil an excellent frying quality and resistance to smoking or discolouration. Corn oil is valued for its high energy value. Energy value of corn oil is 884 kcal as against 86 kcal in case of equal amount of corn meal. Quality of cooking oil is usually associated with proportion of unsaturated to saturated fatty acids. Higher proportion of unsaturated fatty acids in oil represents better quality of oil. Standard method to measure quality of corn oil is ratio of unsaturated linoleic acid to other major saturated fatty acids and iodine number. Refined corn oil contains 95 % triglycerides. Corn Belt corn typically has 60 % linoleic acid (18:2), 20 % oleic acid (18:1), 18 % palmitic acid (16:0), 2 % steric acid (18:0), small amounts of linoleic acid (18:3) and trace amounts of other fatty acids. Comparison of nutrient content of edible vegetable oils shows that corn oil has highest proportion of unsaturated fatty acids (58.7 %) and significantly low proportion of saturated fatty acids (12.7 %) (Table 2). Proportion of monosaturated fatty acids is very low in corn oil as compared to other edible oils. Ratio of unsaturated linoleic acid to other major saturated fatty acids for normal corn is about 1.67, 2.31 for sunflower, 1.36 for soybean and 0.12 for palm oil, which contains large amounts of saturated fatty acids. Iodine number, a measure of degree of unsaturation of corn oil, has a negative correlation with total oil concentration.

The negative association between oil percentage and iodine number results from a decrease in linoleic and an increase in oleic fatty acids. It has been found that two factors affect fatty acid composition of corn oil. These are genotype and environment. Changes in the inbreds of corn hybrids influence the fatty acid composition of commercial corn hybrids. Oil quality and quantity are found to vary over years and locations. The fatty acid composition of corn oil is affected by position of kernels on the ear. Palmitic and linoleic acid content of the oil increases for kernels from the base to the tip of the ear. On the other hand, oleic fatty acid content of the kernels decreases from base to tip. Sampling of kernels in the central portion of the ear is recommended for samples to be analyzed for fatty acid composition. Presence of significant amount of phytosterols (968 mg/100 g) in corn oil improves its quality significantly by reducing the amount of cholesterol absorption from food. Corn oil also provides a good source of vitamin E (21.11 mg ATE/100 g) and tocopherol (14.3 mg/100 g). These are responsible for increased shelf life of corn oil. Four isomers of tocopherols are found in corn, viz. α-, β-, γ-, and δ-. Out of these α-tocopherol is most common.

**Table 2. Nutritive value (per 100 g edible portion) of various vegetable oils**

| Constituents | Corn oil | Peanut oil | Olive oil | Canola oil |
|---|---|---|---|---|
| Energy (kcal) | 884 | 884 | 884 | 884 |
| **Minerals** | | | | |
| Iron (mg) | - | 0.03 | 0.38 | - |
| Zic (mg) | - | 0.01 | 0.06 | - |

| | | | | |
|---|---|---|---|---|
| Phosphorus (mg) | - | - | 1.0 | - |
| **Vitamins** | | | | |
| Vitamin E (mg ATE) | 21.11 | 12.92 | 12.40 | 20.95 |
| Tocopherol, alpha (mg) | 14.30 | 11.60 | 11.90 | - |
| **Lipids** | | | | |
| Fatty acids, total saturated (g) | 12.70 | 16.9 | 13.50 | 7.1 |
| 16:0 | 10.00 | 9.5 | - | 4.0 |
| 18:0 | 1.80 | 2.2 | 11.0 | 1.8 |
| 20:0 | 0.2 | 1.4 | 2.2 | 0.7 |
| Fatty acid total monosaturated (g) 18:0 undifferentiated | 24.20 | 46.2 | 73.7 | 58.9 |
| Fatty acid total polyunaturated (g) | 58.7 | 32.0 | 8.4 | 29.6 |
| 18:2 undifferentiated | 58.0 | 32.0 | 7.9 | 20.3 |
| 18:3 undifferentiated | 0.7 | - | 0.5 | 9.3 |
| Phtyosterols (mg) | 968 | 207 | 221 | 0 |

**Uses of high oil corn**

Most commonly corn oil is used as salad oil, cooking oil and in margarine. Main reason for higher acceptance of corn oil over other vegetable oils like soybean oil, cotton seed oil etc. is due to its flavour and stability during storage and cooking, without addition of synthetic antioxidants. Corn oil contains on an average 14.3 % tocopherols. These act as natural antioxidants by protecting double chemical bonds from oxidation. Tocopherols along with low levels of linoleic acid (18:3) are responsible for stability of corn oil flavour. Its bland flavour and high smoke point are mainly responsible for its popularity. Refined corn oil has very high smoking point at below 232°C as compared to other cooking oils (Table 3). Another reason for popularity of corn oil is its high content of unsaturated fatty acids, which reduces blood cholesterol levels.

**Table 3. Smoking points of corn oil and its use**

| Group | Cooking method | Oil type | Specific smoking point |
|---|---|---|---|
| Below 160° C | Low heating baking, light sauté, pressure cooking | Unrefined corn oil | Below 160° C |
| Below 190° C | Baking sauté, Stirry, wok cooking | Refined corn oil | Below 232° C |

Commercial corn oil can significantly reduce the amount of cholesterol absorbed from food. Phytosterol in corn oil is responsible for lower cholesterol absorption in the body from cholesterol containing food materials and it increases the percentage of high density lipids (HDL) in blood. Large numbers of cosmetic products like moisturizer, hair conditioner and stain remover has been released in the market which is principally based on corn oil. Crude corn oil may also be used as surfactant. Corn oil has been used to increase the efficiencies of fungal bio-control agents. Besides this, corn oil may also be used for various purposes like removal of rust from iron skillet, prevention of car doors from freezing, glue and stain remover, cleaning of home appliances, etc. However, these uses are not very common and regular in nature.

Principally corn oil is used for food and cosmetic purposes. However, high-oil corn can also be used to enhance the feeding rations of animals. Corn with higher levels of oil and lesser proportion of carbohydrate are obviously higher in energy because the calorie content of oil is approximately 2.5 % as great as that of carbohydrates. When oil level is higher (more than 6 %) in feed rations the animals usually show a significant feed response over normal rations. Lambs on high-oil, high-protein rations gained 7 % more weight, retained more nitrogen and required 6 % less feed per pound of grain than the normal diet. High-oil corn diets fed to cows during last 30 days of gestation resulted in significantly greater body weight gains and higher fat levels in the colostrum than normal diets. In laying hens, diet of 17% protein with high-oil corn leads to better egg to feed ratio than normal ration, and egg production and yield increased with high protein high-oil diets. High-oil diet is also superior in weight gain to feed ratio. Skin and plasma pigmentation was also improved in chicks fed with high-oil diets. Lactating dairy cattle showed 12 % greater intake of dry matter for high-oil diet. On an average, advantage of 7 % or more with high-oil corn rations in chicken, swine, dairy cattle and sheep over normal rations. In feed rations, high-oil corn plays an important role because of the following reasons:

1. High-oil corn has nutritional advantage for feeding because of greater energy per unit of feed.

2. Protein concentration in high-oil hybrids is higher and protein quality is enhanced because of the larger scutellum size compared with normal corn. Increase in amino acid, lysine, increased from 0.244 % in normal corn to 0.297 % in high-oil corns.

3. High-oil corn hybrids usually have higher proportion of yellow pigments (carotenoids, xanthophylls, etc.), which are beneficial in poultry rations.

4. The high-oil trait in produce is a consistent trait and can easily be monitored by using nuclear magnetic resonance spectroscopy (NMR), near infrared reflectance (NIR) or NIR Transmission (NIRT) procedures. Thus oil concentration of corn lots during transport from producer to user can be monitored effectively.

**Points for high-oil corn vs normal corn**

Breeding for higher oil content is inherently self-defeating because synthesis of oil is an energy wasting process compared to starch synthesis (an input of approximately two calories is required to create oil containing one calorie). High-oil corns experience yield loss per hectare as compared to low-oil corn. Because corn is regularly bought and sold in the market on a weight basis, few would be willing to produce and sell high-oil types if they produce less tonnage unless a premium is paid. Therefore, efforts are being made to enhance oil content without compromise to yield potentiality of the genotype. The average grain yield of the high-oil hybrids reduced by 5 % compared to normal hybrids. Grain yield of high-oil hybrids usually decreases as oil level increases greater than 8 %. Grain yield decreases at 5-10 %, (some cases may be as high as 20 %) can be expected with HOC hybrids. Variation in total oil content has not been found to be associated with total protein content but was positively correlated with percent germ protein and relative concentration of tryptophan in the kernel.

Association of increased oil content with reduced ear length, smaller ear diameter, lighter kernel weight, reduced plant height and ear height and early flowering has been reported. The association of reduction in grain yields with increased oil concentration may relate to the mutual exclusiveness of high-oil concentration and high grain yield. The genotype of the sporophyte greatly determines the amount of oil produced by high-oil hybrids so that these hybrids may have a more efficient energy trapping system. Oil content and calorie yield per plant are not always associated in hybrids.

**Xenia effects**

The term xenia refers to immediate influence of pollen grains on the non-maternal tissues or germ and endosperm of seed plants. Xenia effects exist in corn for total oil content. Female parent had the largest influence on oil percentage of the kernel, and pollen source (xenia) had a consistent, relatively small effect on the same. The xenia effect observed possibly resulted from an increase in grain size and increased oil concentration in the germ as high-oil content of the pollen source increases the proportion of germ in the kernel. The same relationship holds true for the percentage of oil in the germ. The xenia effects in high-oil corns can cause problems in growing performance trials with hybrids ranging in oil levels from 4-10 %. Under open pollinated conditions, the oil levels of the lower oil hybrids will be increased and the oil levels of the higher oil hybrids will be reduced because of xenia effects.

**Environmental effect on oil content**

Locations, date of planting and crop management have been reported to influence oil content in high-oil genotypes partially. Higher doses of N, P and K influence grain yield and in turn may influence oil production per hectare. Though environmental factors affecting oil values of high-oil corn hybrids reported effect of genotype on oil values seems to be greater. Caloric yield increases may be small. For example a hybrid with 8 % oil will produce approximately 3 % more calories per acre than a hybrid with 4.5 % oil where both have the same yield. Marketing opportunities and prices must be known before planting. There is no established market for high-oil corn. So price fluctuation must be anticipated before planting high-oil corn to the field. It is always recommended to sow high-oil corn with a buy-back agreement with the industries.

**Production technology**

Production technology of high-oil corn is not different from normal corn. However, taking the advantage of xenia effect, oil content in normal hybrids may be enhanced by adopting top cross method of producing high-oil corn. This method involves blending two types of materials: pollinators (donor of high-oil trait) and normal commercial hybrids (recipient). On an average 7-8 % of the plants are of the high-oil corn pollinator, while the remaining 92-93 % of the plants are adopted, high performing commercial hybrids. Pollen from the pollinator transfers high-oil trait to the kernels of the detasseled commercial hybrids. Thus, the xenia effect is exploited to induce high-oil trait in adopted commercial hybrids. If tassels of donor and recipient hybrids are distinctly different (pink and green) then 3-8 % seeds of donor may be mixed with the recipient hybrids and planted together. At the time of flowering, before anthesis recipient hybrids may be detasseled and allowed xenia effect to work. Otherwise, donor lines are to be sown in separate rows in 1:5 to 1:10 ratio (depending on the pollen produced by the donor lines) and recipient hybrid lines may be detasselled at appropriate time. Additionally high-oil donors may also be used in hybrid combinations, in which one parent will be used as high-oil line. High-oil corn can be grown successfully in kharif, rabi and zaid seasons without any difficulty provided assured irrigation facility is available.

**Land selection**

High-oil corn like normal maize can be successfully grown in most parts of the country from plains to hilly regions up to an elevation of 2700 m. It can successfully be grown in well-drained soils with pH of 5.5-7.0. However, it may be grown in all types of soils from sandy to heavy clay. Saline and alkaline soils should be avoided since the crop suffers adversely right after germination. Places where high-oil corns are to be grown must have facility for 5-6 irrigations. Moisture stress, particularly, at the time of anthesis adversely affect the yield and quality of the produce.

**Land preparation**

High-oil corn requires thorough disc ploughing followed by land leveling. Thereafter ridges and furrows are to be laid out with an inter row distance of 60-75 cm. It is recommended to incorporate FYM @ 5-6 t/ha during last ploughing.

**Sowing**

Date of sowing has to be suitably adjusted by making best use of natural precipitation. In irrigated areas, it is most desirable to complete the sowing 10-15 days before onset of rain. The best planting time is June-July in kharif (rainy) and September-October in rabi (winter) season. An early sown crop will require 1-2 irrigation before the regular monsoon sets in. During rabi season, optimum sowing time is middle of October to first week of November.

**Seed rate**

It is preferred to use fresh seeds every year. Recommended plant population for optimum yield is 55000-60000 plants/ha with row spacing of 60-75 cm and intra row spacing of 15-30 cm. This will require 10-11 kg seeds/ha.

**Table 4. Effect of plant population on yield, oil and protein of high-oil corn**

| Plant population (Nos/ha) | Yield (t/ha) | Oil (%) | Protein (%) |
|---|---|---|---|
| 62500 | 9.84 | 7.9 | 9.4 |
| 7500 | 10.16 | 7.9 | 9.3 |
| 87500 | 10.03 | 7.9 | 9.3 |
| 100000 | 9.97 | 7.9 | 9.3 |
| CD 5 % | NS | NS | NS |

(Adee et al., 2007)

Increasing the plant population did not increase yield but increased seed cost, thus net value decreased as plant population increased, at least above 75,000 plants/ha (Table 4).

Based on the net values calculated, the optimum plant population for the high-oil corn hybrids we used is between 62500 and 75000/ha (Adee et al., 2007).

### Seed treatment

Seed treatment with Imidachloprid 70WS @ 5 g/kg seed takes care of insect pests up to 30 DAS and fungicide treatment helps to prevent the attack of damping-off fungi.

### Method of sowing

Two seeds per hill are dibbled manually or mechanically one third from the top on the side of the ridge. Sowing on ridges serves the twin purpose of water conservation and protection from water logging in the initial stages to which high-oil corn is highly susceptible. Planting depth should be 3-4 cm for all. Plants are thinned to one plant per hill 10-12 days after emergence.

**Table 5. Effect of planting date on corn yield, oil, protein and starch contents**

| Planting date* | Yield (bushels /acre) | Oil content (%) | Protein content (%) | Starch content (%) |
|---|---|---|---|---|
| April 14 | 161.8 A | 6.6 A | 7.7 A | 57.6 A |
| May 10 | 155.4 A | 6.6 A | 8.0 A | 57.6 A |
| June 3 | 124.1 B | 6.2 B | 8.4 B | 57.4 A |

(Joe and Dale, 2002). Significant differences (5 %) noted by different letters in same column

At Nashua (USA), earlier planting had a positive effect on yield and oil content, with negative effect on protein concentration (Joe and Dale, 2002).

### Nutrient management

Fertilizer must be applied according to soil test results (Table 6). A general recommendation would be to apply a total of 110-130 kg of nitrogen, 60-80 kg of phosphate ($P_2O_5$) and 55-65 Kg potash ($K_2O_5$) per ha. For early plantings, when cold conditions are often encountered (16 °C), use a "starter" fertilizer (e.g., 85 kg of 18-46-0 / ha) for improving seedling vigour, stand establishment and early plant growth is recommended. Fertilizer should be applied in 1 or 2 bands approximately 7-8 cm to the side and 5-7 cm below the seed.

**Table 6. Fertilizer requirements of high oil corn**

| Fertilizer timing | Nitrogen (kg/ha) | Phosphorus (kg/ha) | Potassium (kg/ha) |
|---|---|---|---|
| Total recommended | 110-130 | 60-80 | 55-65 |
| Broadcast 3 weeks before planting | 40-50 | 60-80 | 55-65 |
| Top dress 2 weeks after emergence | 25-40 | 0 | 0 |
| Band placement (at flowering) | 25 | 0 | 0 |

Nitrogen rate significantly affected yield, grain oil and protein content and net value. Grain oil content decreased as N rate increased. This decrease was less than 0.3 % (3.2 %) from the lowest N rate (90 kg/ha) to the higher N rate (225 kg/ha). Oil content did not vary by more than 0.5 % due to N rates. Protein content increased as N rate increased, from 8.9 % at 90 kg/ha to 9.6 % at 225 kg/ha (Adee et al., 2007).

**Table 7. Effect of N rate on yield, oil and protein of high-oil corn**

| N rate (kg/ha) | Yield (kg/ha) | Oil (%) | Protein (%) |
|---|---|---|---|
| 90 | 3.683 | 8.1 | 8.9 |
| 135 | 4.013 | 7.9 | 9.3 |
| 180 | 4.165 | 7.8 | 9.5 |
| 225 | 4.165 | 7.8 | 9.6 |

(Adee et al., 2007)

**Weed management**

A field that does not contain noxious weeds must be selected. Crop rotation is important for limiting the build up of troublesome weeds. Shallow cultivation should be used in concert with chemicals for weed control. The crop should remain weed-free during the early stages of plant growth; otherwise, yields might be substantially reduced. Efficient weed control is achieved for 30-35 days through a spray of the herbicide Atrazine @ 1 Kg ai/ha 1-2 days after the initial irrigation.

**Water management**

Depending on the soil type the number of irrigations varies from 4-5 irrigations in heavy soils and 7-8 irrigations in light soils. Irrigate at least 2.5-4.0 cm a week in order to obtain high yield. The most critical time period to have adequate moisture is during tasselling and silking.

### Intercultivation

Top dressing of urea is done 25 days after emergence after running a cultivator in between the rows that will help weed control and better root aeration apart from soil water conservation. Subsequently, a ridger is run to prevent plant lodging and minimize volatilization losses of urea. Top dressing with urea is followed by irrigation.

**Table 8. Grain yield, and oil and protein contents (%) of a high oil TC Blend with different defoliation treatments from V3/4 to R3**

| Defoliation treatment | | Yield (t/ha) | Oil (%) | Protein (%) |
|---|---|---|---|---|
| Stage | Percent | | | |
| No defoliation | - | 13.08 | 7.6 | 9.0 |
| V3/4 | 100 | 12.55 | 7.3 | 9.1 |
| V6/7 | 100 | 8.25 | 7.1 | 9.3 |
| V9/10 | 50 | 11.88 | 7.3 | 8.7 |
| | 100 | 9.71 | 7.3 | 8.6 |
| V12/13 | 50 | 11.43 | 7.6 | 8.7 |
| | 100 | 5.52 | 7.0 | 8.8 |
| V15/16 | 50 | 10.43 | 7.4 | 8.8 |
| | 100 | 1.33 | 4.6 | 9.8 |
| VT/R1 | 25 | 11.95 | 7.5 | 8.8 |
| | 50 | 9.78 | 7.3 | 8.9 |
| | 100 | 0.71 | 4.7 | 15.1 |
| R3 | 25 | 12.33 | 7.4 | 8.7 |
| | 50 | 10.48 | 7.3 | 8.7 |
| | 100 | 3.93 | 4.8 | 14.3 |
| LSD (0.10) | - | 2.10 | 0.6 | 1.4 |

(Thomison and Nafziger, 2003)

At Illinois, grain yields decreased as leaf removal increased in severity and as time of defoliation neared tassel emergence. The greatest yield reductions caused by defoliation occurred at or near flowering, with complete defoliation at flowering reducing yields by 95 % or more. Grain oil content was usually less affected by defoliation than grain yield and it typically required defoliation sufficient to reduce yield by more than 50 % and sometimes as high as 70 % in order to decrease oil content. Grain oil content was reduced as much as 30 to 40 % with complete defoliation during late vegetative, flowering, and early grain fill whereas protein content was increased as much as 50 % in high-oil corn (Thomison and Nafziger, 2003).

**Table 9. Grain yield, and oil and protein content of a high-oil (HO) TC Blend and its check hybrid with different defoliation treatments levels**

| Corn type | Defoliation | | Grain yield | Oil | Protein |
|---|---|---|---|---|---|
| | Stage | Percent | (t/ha) | (%) | (%) |
| HO TC Blend | No defoliation | - | 8.83 | 7.3 | 7.4 |
| (SK2650-2) | VT | 50 | 5.33 | 6.8 | 7.4 |
| | | 100 | 0.0 | NA | NA |
| | R3 | 50 | 5.98 | 6.4 | 7.8 |
| | | 100 | 3.49 | 5.1 | 11.9 |
| | R5 | 50 | 6.87 | 6.6 | 7.7 |
| | | 100 | 5.03 | 5.8 | 8.6 |
| Check hybrid | No defoliation | - | 8.58 | 3.8 | 8.0 |
| (Pfister 2650) | VT | 50 | 4.89 | 3.4 | 8.5 |
| | | 100 | 0.0 | NA | NA |
| | R3 | 50 | 5.61 | 3.2 | 8.9 |
| | | 100 | 3.63 | 2.5 | 12.3 |
| | R5 | 50 | 6.55 | 3.3 | 8.3 |
| | | 100 | 5.28 | 3.0 | 9.5 |
| LSD (0.05) | - | - | 0.64 | 0.8 | 1.1 |

(Thomison and Nafziger, 2003). NA: Not available.

At Illinois, SK 2650-2 variety recorded higher grain yields, oil percent and protein content than Pfister 2650 hybrid (Thomison and Nafziger, 2003).

At Ohio (USA), corn plants were 100 % defoliated at V5 and 50 and 100 % defoliated at V13. Defoliation similarly affected grain yield and oil concentration of HOC TC Blends and NOC hybrids. Complete defoliation at V13, averaged across years and hybrids, was the most damaging defoliation treatment, reducing grain yields 29 %. Defoliation did not significantly alter grain oil concentration of HOC grain parents (74.5 g/ kg) or NOC (34.5 g /kg). Defoliation caused similar responses in ear weights and oil concentrations in both components (the pollinator and the grain parent) of the HOC TC Blends (Mangen et al., 2005).

**Table 10. Yield, ear yield components and grain oil concentration of a high-oil corn (HOC) TC Blend and its normal-oil corn (NOC)**

| Hybrid | Defoliation treatment | Yield (mg/ha) | Kernels/ ear | 300-kernel weight | Grain oil content (g/kg) |
|---|---|---|---|---|---|
| NOC conventional hybrid | 100 % at V5 | 10.43 | 637 | 72.6 | 33.1 |
| | 50 % at V13 | 11.33 | 659 | 72.5 | 32.1 |
| | 100 % at V13 | 9.38 | 621 | 65.1 | 30.6 |
| | UTC | 13.66 | 690 | 76.4 | 32.5 |
| HOC TC Blend | 100 % at V5 | 9.99 | 621 | 69.9 | 74.1 |
| | 50 % at V13 | 10.73 | 668 | 64.3 | 73.8 |
| | 100 % at V13 | 8.17 | 621 | 61.0 | 73.3 |
| | UTC | 11.53 | 694 | 68.5 | 75.0 |
| LSD (0.05) | Hybrid (H) | NS | NS | NS | 40.0 |
| | Defoliation (D) | 1.35 | 39 | 6.0 | NS |

**Pest and disease management**

A preventive spray of Endosulfan 35 EC @ 2 ml/litre of water is given to 10-14 days old plants to take care of stalk borers *Chilo partellus* in kharif and *Sesamia inferens* in rabi. Disease problems tend to be sporadic. Troublesome diseases include leaf blight and post flowering stalk rots. Use crop rotation, and avoid sequential planting in adjacent fields to minimize disease. Cultivars varieties tolerant or resistant to disease. A spray with Bavistin @ 1 g/litre of water takes care of most of the foliar diseases.

**Corn oil in Indian**

In view of the availability of a variety of oilseed crops for edible purpose, in India, corn oil is not very popular for domestic uses. In USA corn oil constitutes 4.3 % of vegetable oil production and about 7.2 % of domestic oil utilization. Corn oil does not constitute a part of conventional food recipe in our country. This is the principal reason for less popularity of corn oil as cooking oil. Besides, high price of corn oil is a limiting factor in popularization of corn oil in our country. Moreover public awareness is also lacking in our country regarding the health benefit associated with corn oil. Yearly in India nearly 2.6 thousand quintals of corn oil is produced in principal corn-based factories (Table 5). Average cumulative extraction capacity of these factories is 2.41 thousand tonnes per day. On an average these factories can extract 3.0-3.5 % oil from corn, which otherwise contain at least 5-6 % oil. Thus there is immense scope to increase the production of corn oil only by improving the efficiency of oil extraction from corn germs. With increased health consciousness and expansion of cosmetic industries in India demand for corn oil is bound to increase in near future. In addition, high-oil corn may also find its due place in feed rations of the country.

Under NATP programme, two high-oil populations, HOP-1 and HOP-2 have been developed with 7.0 % oil content. These two populations are in the process of registration. When these populations are sown as male in a ratio of 6: 1 with promising hybrids and hybrids are detassled at flowering, it is possible to achieve higher oil percentage without compromising with yield parameters. A proper market of high-oil corn is to be created so that the farmers receive good premium of their produce. Majority of the maize-based factories are located in states of Andhra Pradesh, Gujarat, Maharashtra, Madhya Pradesh, Haryana and Punjab (Table 11). Main maize producing states like Bihar and Uttar Pradesh do not have any such factory. This is a serious lacuna, which has to be addressed by adopting appropriate policy decision by the government.

**Table 11. Some Indian corn based factories**

| Name of factories | Location |
|---|---|
| M/s. Bharati starch | Jamunanagar, Haryana |
| M/s. Anil starch products Ltd. | Ahmedabad, Gujarat |
| M/s. Gujarat Ambiya proteins Ltd | Ahmedabad, Gujarat |
| M/s. Ridhi Sidhi products Ltd | Ahmedabad, Gujarat |
| M/s. Sahayadri starch | Sangli, Maharashtra |

| | |
|---|---|
| M/s. Universal starch Chem Allied Ltd. | Dondaicha, Maharashtra |
| M/s. Raja Ram products | Mandsaur, Madhya Pradesh and Mohali, Punjab |
| M/s. Laxmi starch | Indore, Madhya Pradesh |
| M/s. Sukhjit starch Industries Ltd. | Phagwara, Punjab |
| M/s. Sukhjit starch Industries Ltd. | Nizamabad, Andhra Pradesh |

High-oil corn (HOC) typically contains from 60 to 80 g /kg oil on a dry weight basis. Some livestock producers prefer HOC as a feed source because the grain contains approximately 4.5 % more energy per kilogram than conventional corn grain and can replace more expensive dietary sources of fat and protein (Alexander, 1988). HOC improves feed efficiency and increases rate of gain over conventional corn (Lambert, 2001). The TopCross grain production system is used to produce corn with 65 g/ kg of oil on a dry matter basis (U.S. Grains Council, 2002). Corn seed is planted as a physical mixture (TC Blend) containing 91% high-yielding, cytoplasmic male sterile (CMS) hybrid (grain parent) seed and 9 % pollinator seed (Edge, 1997).

## References


Adee, E.A., E.D. Nafziger and L.E. Paul (2007). Optimizing plant population and nitrogen rate for high-oil corn. Online. Crop Management doi:10.1094/CM-2007-0831-01-RS.

Edge, M. (1997). Seed management issues for TopCross high oil corn. p. 49–55. *In* J.E. Cortes (ed.) Proc. Annu. Seed Technol. Conf., 19th, Ames, IA. 18 Feb. 1997. Seed Sci. Cent., Iowa State Univ., Ames.

Genter, C.F., J.F. Eheart and W.N. Linkous (1956). Effects of location, fertilizer, and rate of planting on the oil and protein contents of corn grain. Agron. J. 48: 63-67.

Huang Showen, Sun Guifang, Jinjiyun, Yuoyubao and Heping (2004). Effect of nitrogen, phosphorus and potassium application on grain yield and quality of grain and quality of high oil and starch maize. Res. Rep., 5: 6-7.

Joe, M. and E.F. Dale (2002). Planting Date Effects on Yield and Grain Composition of High Oil Corn. Iowa State University, Northeast Research and Demonstration Farm ISRF01-13. P: 1-2.

Joe, M., E.F. Dale and P. Ken (2013). Planting Date Effects on Yield and Grain Composition of High Oil Corn. Iowa State University, Northeast Research and Demonstration Farm, ISRF01-13.

Lambert, R.J. (2001). High-oil hybrids. Pages 131-154 in: Specialty Corns. A. R. Hallauer, ed. CRC Press, Boca Raton, FL.

Mangen, T.F., P.R. Thomison and S.D. Strachan (2005). Early season defoliation effects on topcross high-oil corn production. Agron. J., 97: 823–831.

Strachan, S.D. and S.L. Kaplan (2001). Responses of high-oil corn to rootworm beetles during pollination. Agron. J., 93:1043–1048.

Thomison, P.R. and E. D. Nafziger (2003). Defoliation affects grain yield, protein and oil of Top Cross high-oil corn. Online. Crop Management doi: 10.1094/CM-2003-1027-01-RS.

Thomison, P.R., A.B. Geyer, B.L. Bishop, J.R. Young and E. Lentz (2004). Nitrogen fertility effects on grain yield, protein, and oil of corn hybrids with enhanced grain quality traits. Online. Crop Management doi:10.1094/CM-2004-1124-02-RS.

Thomison, P.R., A.B. Geyer, L.D. Lotz, H.J. Seigrist and T.L. Dobbels (2002). TopCross high oil corn production: agronomic performance. Agron. J., 94: 290-299.

Thomison, P.R., A.B. Geyer, L.D. Lotz, H.J. Seigrist and T.L. Dobbels (2003). TopCross high oil corn production: select grain quality attributes. Agron. J., 95: 147-154.

Wang Y. (2007). Effect of sowing date and harvest time on high oil corn quality. Annual Report. Jilin College of Science and Technology of Agriculture, Jilin. USA.

# 22

# Integrated Pest Management

Maize is a host crop for a complex of pests that may cause economic injury throughout the growing season. Despite the number of pests that feed on maize, the field generally exhibits minimal pest problems because the normal activity of most pest populations is too low to inflict significant injury and economic losses. Management of pest populations on maize requires a combination of preventive and timely responsive actions based on the risks associated with various cultural practices and pest activity observations collected by periodic field inspections.

Maize is attacked by over 250 species of insect and pests (Mathur, 1991). Out of these four tissue borers viz., maize stalk borer, pink stem borer, shoot fly and Asiatic corn borer are regular and serious pests for maize. In general, insect pests are more damaging in tropical environment than in the temperature environments because of the favourable weather conditions (James, 2003). Stalk borers are the principal pests in all the top 25 maize growing countries of the world. Potential losses due to insect pests on global basis are estimated to be in the order of 14 - 18 % (Oerka, 2002). In India, *Sesamia inferens* (Walker) was predominant and distributed in Andra Pradesh, Karnataka, Tamil Nadu, Maharastra, Uttar Pradesh, Assam, Bihar, Delhi, Punjab, Orissa, Madhya Pradesh, Haryana, West Bengal and Andaman Islands. Shoot flies *Atherigona soccata* and *A. naqui* are serious on spring sown maize in India and also becoming problem in Karnataka (Joshi *et al.,* 2005). The leaf feeder *Mythimna separata* (Walker) in Bihar and Uttar Pradesh, *Spodoptera* spp. in Andhra Pradesh and Tamil Nadu and *Colemania sphenerioides* in Rajasthan are assuming seriousness in maize. Sucking insects, *Perigrinus maidis, Rhopalosipum maidis* and *Chilotoba acuta* in Peninsular India and Haryana during rainy season, termites *Odentotermes* in rainfed areas, *Helocoverpa armigera* at milking stage in Madhya Pradesh and coastal Andhra Pradesh are the new problems in the maize cultivation, which needs immediate attention.

**African maize stem borer**

*Busseola fusca* (Fuller) causes damage by destruction of the growing point in the whorl (dead heart), loss of photosynthetic leaf area due to foliar feeding, lodging due to burrowing in the stem and extensive damage to young kernels due to feeding of larvae from the second and third generations. Damage is critical when the growing points of young plants are completely damaged when there is extensive lodging due to stem boring and when second and third generation larvae feed directly on the cob. Yield loss as high as 40 % has been attributed to stem borers. Adult moths generally have dark brown, patterned forewings and white to grey-brown hind wings. In wet, cold conditions adult moths develop darker colouring. Adult moths are nocturnal and have a wing span of 25 to 40 mm. Females are generally slightly larger than males. Females usually mate on the night of emergence and will lay on the subsequent 3 to 4 nights. Females lay eggs in batches of 100 to 200. Each female can lay up to 1000 eggs in a life time. Adults have an area of dispersal of up to 1.6 km and occasionally further. Eggs are light yellow in colour, spherical and approximately 1 mm in diameter. Eggs are laid in a single column on the stem under leaf sheaths. Eggs hatch in about 10 days. Young larvae are generally dark brown, purple or black in colour. They migrate to the whorl to feed. There are usually 6 larval instars although 8 are possible in unfavorable conditions. Larvae mature in about 35 days and pupate in the stem. Prior to pupating, larvae cut a small hold in the stem which enables the adult moth to emerge. Pupae are generally 25 mm in length and shiny yellow brown to dark brown in colour. Males are usually smaller than females. Pupae are found in galleries within the stem. The pupae stage lasts from 2 to 3 weeks. In favorable conditions, the life cycle can take 7 or 8 weeks. In maize crops, female moths prefer to lay on plants less than two months old and on the youngest unfolded leaf.

Exposing crop residue to direct sunlight or using crop residue for livestock feed or compost can reduce the incidence of diapausing larvae. Pheromone and light traps can be used to trap adult moths. Modifying sowing periods to avoid periods of heavy infestation can reduce crop damage. Intercropping with legumes or cassava can reduce the incidence of stem borer infestation on maize. There are a number of natural enemies of *Busseola fusca* including various Hymenoptera parasitoids of the larvae, pupae, and eggs. Napier grass and Sudan grass are used as trap crops. They are important fodder crops and produce a gummy substance that traps moths and prevents over 80 % of larvae from reaching adult hood.

- Collection and destruction of the stubbles which are left in the field or heaped in one corner of the field since they act as a source of infestation as the larvae hibernate in them.
- Removal of dead hearts.
- Release of *Trichogramma chilonis* 8 cards / ha at 12 and 22 DAS. Alternatively spray profenophos 50 EC @ 1.5 ml/litre of water.
- Intercropping of maize with suitable varieties of cowpea is an ecofriendly option for reducing the incidence.
- Larvae are naturally controlled by larval parasitoid *Cotesia* sp.
- Need based application of carbofuran 3 G into whorls of infested plants at the rate of a pinch of granule per plant.

**Pink borer**

*Sesamia inferens* (walker), Noctuidae, Lepidoptera infestation is serious in peninsular region. In Northern part of India, it is prevalent in Rabi season. The larvae attack almost all parts of the plant such as leaf, stem, tassel and ear. The larvae have migrating tendency and may attack a number of plants. The larvae feed under the leaf sheath and remaining there in the early stage of growth. Later, they make in road into the central shoot causing the death of central leaf. Due to larval feeding, the grown up plants show many slit like oval elongated holes on the leaf blades. They also fore tunnel inside the stem and exit holes at the surface. The tunnels are generally filled with excreta. The decaying shoots in grown up plants cause cob to rot. Thus it causes complete loss of grain. Larvae also feed on immature cobs and tassels. The moths lay bead like eggs in 2 - 3 longitudinal rows on the sheath of bottom leaves of young maize plant. Eggs are cream coloured when laid. It turn steel gray before hatching. The incubation period is about a week. The full grown larvae are 25 - 30 mm long and purple from dorsal side. The larval period is from 3 to 4 weeks. The pupae are robust dark brown colour having a powdery appearance. The pupation occurs in the stem or in between the stem and leaf sheath. The pest has 4 - 5 generation a year. The dead hearts should be removed and destroyed with the larvae. Profenophos 50 EC or Phophamidon 40 EC either of them at 0.05 % would effectively check the borer.

- Deep summer ploughing.

- » Removal of harvested stubbles and infested plants.
- » Use resistant varieties.
- » Use trap crops like sorghum in 3 – 4 rows.
- » Optimum plant density.
- » Growing of intercrops like cowpea, soybean and red gram.
- » Release of *Trichogramma chilonis* @ 8 cards (1,50,000 parasitized eggs)/ha at 12 and 22 days after germination on weekly interval.
- » Need based application of a pinch of carbofuran 3 G in whorl of infested plants is recommended. If cannot be applied late in the season because of residues in the grains.

**Armyworm**

It is also called as African armyworm (Okalambo) and Rice armyworm. The Latin names are *Pseudaletia unipuncta* (Haworth) and *Spodoptera exempta* (Walker). Entire leaves and young plants can be consumed during severe infestation. 'Window pane' damage to leaves caused by newly emerged larvae, which feed by scraping the leaf surface without perforating it. Ragged leaf appearance from the leaf edges inwards, also due to feeding. Under severe infestation the entire foliar tissue might be lost, with only midribs remaining. Lower leaves are consumed first. Large number of caterpillars migrating across fields and roads. The 6$^{th}$ instar larvae are responsible for over 80 % of feeding.

Adult moths have a wingspan of 40 mm. Forewings are generally sand, grey or brown in colour with a distinctive white dot in the centre. Hind wings are pale brown with a slightly darker area on the posterior edge. Adults feed on flower nectar or decaying fruit. Adults are active at night and therefore rarely seen. Females lay eggs in rows or clusters of 50 to 150. Female may lay up to 2000 eggs in a lifetime. Adult life span can reach 17 days. Eggs are laid on leaf sheaths, leaves or around the base of the plant. Leaf blades are often rolled around the egg masses. Eggs are white, bead like and smaller than a pin head and hatch in 3 to 4 weeks. Larvae pass through 6 instars which can take up to 4 weeks. First instar larvae range from 2 to 4 mm in length and move in a looping motion. Mature larvae range from 25 to 35 mm in length. Larvae are smooth-bodied and can be dark green, brown or black with alternating light and dark stripes running along the body. Larvae feed nocturnally and hide under crop debris during the day time. Larvae burrow 2 to 3 cm below the soil surface to pupate. Pupae are reddish-brown in colour and range from 15 to 20 mm in length.

It is predominantly nocturnal and therefore is often not detected until the crop is damaged. Moth traps can be used to detect presence of adult moths. Fields should be regularly inspected for presence of eggs, frass, and young larvae. Larvae are predominantly nocturnal. Inspections should include the soil around plants and under crop debris where larvae hide during the day. High bird presence in the field may reflect heavy armyworm infestation. Ploughing a deep ditch with steep edges or filled with water or insecticide can form a physical barrier to migrating larvae. Insecticides can be applied to the crop at early larval stages if infestation levels are high. However, application of insecticide is ineffective once larvae are mature as most crop damage has been completed. Various pyrethroids, carbamates and organophosphates are approved for its control. The egg parasite *Telenomus minimus* and species of braconid wasps such as *Apanteles flaeviceps, A. marginiventris,* and *A. militaris* have been used to successfully manage its infestation. *Bt* (*Bacillus thuringiensis*) toxins and products that mimic the molting hormone ecdysone have been used to manage larvae populations. Dusting of 5 % malathion or spraying of profenophos 50 EC @ 0.05 % at fortnightly interval checks the insect.

**Leaf aphid**

Plants affected by *Rhopalosiphum maidis* (Fitch) may become stunted, show a conspicuous yellowish mottling and turn reddish as they mature. Young plants that have been infected seldom produce ears. Yellowish mottling will turn to red with reduced ear growth. The role of this insect as a vector of the sugarcane mosaic virus, maize dwarf mosaic virus and maize leaf-fleck virus makes it a pest of considerable economic importance. Piercing of the leaves and sucking of plant fluids by the insect causes some yellowish mottling, but this damage is seldom of economic importance. Sugary droplets excreted by the aphids favour the development of black moulds and make the plants sticky. These insects usually attack maize plants at the end of the midwhorl stage. Their colonies may completely cover emerging tassels and the surrounding leaves, preventing pollen release. In severe outbreaks the ear shoot is also infested and seed set may be affected. The small, greenish blue adults females do not lay eggs but give birth to nymphs. In crowded colonies winged forms are produced that eventually migrate to other plants. Skins that have been shed give the colonies a whitish appearance. The nymphs and adults of aphid infest the maize crop from August to October. Use only recommended doses of fertilizers. Release of *Crysoperla carnea* @ 50000/ha twice at weekly interval when infestation is noticed. Spray imidacloprid 17.8 SL @ 0.30 ml/lit. or malathion 50 EC @ 2.5 litre / ha. or profenophos 50 EC @ 1.5ml/ lit. or phophamidon 40 EC at 0.05 % would be effective against aphids. Maize ecosystem is endowed with good number of natural enemies of aphids, prominent among them are coccinellids, syrphids and chrysopids. Because of copious production of pollens these predatory insects thrive well and often control the aphid population and do not allow the aphid to cause much economic loss.

**Sugarcane leaf hopper**

The pest *Pyrilla perpusilla* (walker) is prevalent in Uttar Pradesh, Punjab, Karnataka, Tamil Nadu and Delhi. The eggs are laid in clusters of 60 - 90. Both nymphs and the adults sit on the lower surface of the leaves. In severe outbreaks, these may be seen spread over all parts of the plants. The insects suck the leaf sap and the infested leaves turn yellow, brown and wilted. The honey dew they secrete invites fungus, which reduces the photosynthetic activity of the plant. They cause general debilitation in the plants thus reducing the yield potential. In maize ecosystem, Pyrilla population is controlled at egg stage by *Tetrasichus pyrillae* parasite. The nymphs and adults are controlled by an ectoparasite *Epiricania melanoleuca.* The adult is black and white, which lay 700 - 1200 eggs in a single egg mass near cocoon from which it emerges. The neonates stick to the abdomen of the nymphs and adults and suck their haemolymph. The adults and nymphs are debilitated and rendered sterile.

**Corn stunt leafhoppers**

*Dalbulus maidis* and *Dalbulus spp.* feed and suck the plant juices. Transmission of virus and spiroplasmas, including corn stunt spiroplasma (*Sprioplasma kunkelii*) resulting in corn stunt. Sooty mould accumulation is seen where leafhoppers deposit sticky honey dew while feeding. Symptom is stunting of plant, premature wilting of leaves, chlorotic stripes on young leaves, leaf purpling, excessive tillering and cobbing and formation of barren cobs. Continous, overlapping maize cropping favour this pest incidence. Late planting and dry climatic conditions favor insect development. The eggs are less than 1 mm in length and 0.2 mm wide. A few days after oviposit the eggs develop a tuft of diagnostic microfilaments from the proximal end of the egg, which can be seen with a 10 x hand lens. At 7-10 days the eggs became easier to see as they become white and have red eyespots. Eggs are typically deposited on mid-veins of leaves of young maize plants generally in the whorl. Several eggs will occasionally be oviposited in a row.

Egg stage lasts 9-12 days in summer but several weeks in cooler weather. There are five translucent, yellow nymph instars. Development time for nymphal instars can take from 2.5 days at 27°C to 33 days at 10°C. The first instars are pale yellow with black eyes. They are less than 1mm long. From the second instar through to the fifth instar, they have a pair of irregular, black spots on the interior margin of the last two abdominal tergites. The last instar is a little less than 4 mm long. The nymphs feed at the base of the leaves in the whorl or between the leaf bases on the lower stem. The adults are light yellow with uneven, black spots on the abdomen. The spots are larger and more numerous on leafhoppers in high altitudes and cool climates. Two large spots are located on the head over the ocelli, which are greater in diameter than the ocelli. At times there may be several other dark spots on the head. Adults range in length from 3.7- 4.3 mm. Adults life span ranges from 45 to almost 180 days, depending on temperature. Lower temperatures favor longevity. During a lifetime adult females can lay over 600 eggs. Number of eggs deposited at one time range from 3 to 14. Adults secrete honeydew during feeding which is occasionally fed on by ants. Total life cycle time is approximately 25-30 days at optimum temperature of 25°C.

If the corn stunt leafhopper transmits diseases to the maize plant, particularly maize planted late, yield loss could be severe. In irrigated fields with continuous maize cropping, severe corn leafhopper infestation can occur with subsequent pathogen transmission causing significant yield loss. It multiply rapidly reaching their peak by February - March when the maize plants are the tasselling stage. Cold and cloudy weather increases the infestation which prefers well fertilized plants with broad leaves. Monitoring for leafhopper incidence during the first two months of crop growth is particularly important. Avoiding continuous maize monoculture and managing volunteer maize plants during the off-season will help reduce leafhopper and corn stunt incidence. Removing plants (rouging) infected with corn stunt early in the season can reduce disease spread. Local maize varieties are often more resistant to corn stunt than introduced varieties. Yellow maize varieties are often more resistant. Several *Hymenoptera* species are natural enemies of leafhopper nymphs, adults and eggs. A somewhat expensive control method for sweet corn crops is the use of reflective mulch. This repels the adult corn leafhopper. Insecticides can reduce the incidence of leafhoppers. Seed treatments with systemic insecticides have been shown to reduce corn leafhopper populations, but incidence of corn stunt disease has only been negligibly reduced. Reinfestations occur rapidly, particularly when other fields of corn in the area are being harvested, and leafhoppers are searching for other corn plants.

**Cutworm**

Stems of seedlings cut at ground level or slightly below the surface *by Agrotis ipsilon (Hufnagel), Agrotis spp., Peridroma saucia (Hubner), Chorizagrotis auxiliaries (*Grote) and other species. Holes in leaves and leaf margins removed, caused by feeding. Cavities in the base of the stem due to feeding, resulting in plants prone to wilting and lodging. Brownish-black larvae is found in the soil around damaged plants. Pearly white round eggs on plant stems or on ground around seedlings with worm frass (faeces). Planting maize in rotation with other host crops such as soybean and rye encourage build up of cutworms. Dark grey, black or brown coloured moth with markings on the front wings, wingspan of 40 to 50 mm, length 20 to 30 mm. Females are darker than males and begin egg laying 5 to 11 days after emergence. Adults are nocturnal and lay eggs on vegetation, on moist ground around plants or in cracks in the ground. During the day time, adults hide in vegetation. Eggs are laid singularly or in clusters (up to 2000) which is pearly white, round, diameter of 0.5 mm. Eggs hatch during the 3 to 11 day stage, depending on temperature. Larvae are grassy in appearance and can range in colour from grey to brown or black depending on species. Larvae generally have two spotted yellow stripes running down the back. There are 6 larval instar stages. Larvae can be 6 mm when hatched and up to 50 mm when mature. They can overwinter and continue feeding for up to 20 weeks, depending on available food. The pupal stages are shiny, brown, smooth and about 20 mm in length. Cutworms can overwinter as pupae, which are found under the soil surface. Upto four generations may occur during the growing season depending on species and latitude. Cutworms are polyphagous with a very wide host range, including various vegetable and cereal crops such as asparagus, bean, beet, cabbage, castor bean, cotton, grape, lettuce, peanut, pepper, potato, radish, spinach, squash, strawberry, tobacco, rice, rye and tomato. Damage is most severe when a large number of cutworm larvae are present at seedling emergence. Each larva can destroy up to 4 plants. Yield loss as high as 47 % has been recorded due to cutworm damage. Ploughing fields 3 to 6 weeks prior to planting will reduce cutworm infestation, by removing alternate hosts and disturbing eggs and overwintering pupae and larvae in the soil. Cutworms have many natural enemies including parasitic wasps, grasshoppers, viruses and fungal pathogens. Under severe infestation, insecticides may need to be applied. Insecticide should be applied when damage is at the early stages when 3 to 10 % of the crop is damaged. Incidence of cutworms is best assessed by using a flashlight at night to monitor cutworm presence and damage.

- » Digging near damaged seedlings and destruction of larvae
- » *Bacillus thuringiensis* a widely available caterpillar bio-pesticide is very effective in the control of cut worm.

### *Diabrotica* beetles and rootworms

Root clipping and tunneling by the worms of *Diabrotica balteata (*LeConte)*, D. virgifera, D. longicornis, D. speciosa* and *Diabrotica* spp. may results in tilted, curved stems or lodged plants. Lodging may begin around the midwhorl stage and continue to maturity. In seedlings the beetles eat portions of the leaf margins (*D. balteata*), leave irregular holes in the tender leaves and occasionally cut the plant at the base. Leaf feeding on older plants seldom has economic significance. At flowering, however, the beetles feed on green silks (*D. balteata*) and pollen leading at times to poor seed set. *D. virgifera* and *D. longicornis* are reported as vectors of maize chlorotic mottle virus (MCMV) and Stewarts's wilt bacteria. In the soil around infested plants and in their root systems, one can find the white or creamish-coloured, thing, soft, small worms. These have three pairs of legs, brown heads and vary in length at maturity from 3 to 10 mm. After three molts the larvae develop into soft, white pupae inside cells in the soil, from which the beetles emerge. These vary in size (6 to 9 mm) and colour pattern according to species. They lay most of their eggs at soil depths of up to 10 cm. The number of generations in a year (ranging from one to four) varies with the species and latitude.

**European maize borer**

It is also known as Corn moth *Ostrinia nubilalis (Hubner)*. Larvae feed on tassels, whorl and foliar tissue, including midribs. Older larvae burrow into the stalk, ears and kernels. Severe infestation of stalks can result in broken tassels and lodging. Infestation of ears leads to reduced yields and loss of commercial value. Boring damage on the ears can also make them vulnerable to fungal infection. The signs are horizontal rows of perforations in young leaves, stunted plants, lodging and patchy plant stand, excessive frass (faeces) in the collar of the plant. Female moths are pale yellow to light brown in colour with a wing ranging from 25 to 35 mm. Male moths are usually slightly darker and smaller with a wing of 20 to 25 mm. Adult moths are nocturnal and remain motionless in border vegetation during the day. Adult longevity ranges from 18 to 24 days. Females begin oviposition 4 days after emergence and continue oviposition for up to 14 days. On average, females lay 20 to 50 eggs a day. Females can mate multiple times resulting in increased fecundity. Minimum temperature required for flight is 13 to 15°C. Eggs are oval in shape (1mm in length and 0.75mm in width) and flattened. Eggs are initially creamy white in colour but darken with age. Eggs are deposited in masses of 5 to 30 eggs on the underside of leaves where they overlap with fish scales. Eggs hatch in 4 to 9 days. Larvae have 6 instars. Body length during the first instar is approximately 1.5 mm, expanding to approximately 20 mm in the final instar. Larvae are generally light brown, greyish or pinkish in colour. Larvae in the final instar stage can overwinter in tunnels in the maize stem. Pupae are generally light brown in colour and measures 13 to 17 mm in length and 2 to 4 mm in width. There can be 1 to 4 generations per growing. Boring of the stem can lead to serious crop lodging, while boring in the ear can lead to direct yield loss, cob rots and loss of commercial value. Boring of the ear can also make the ear vulnerable to fungal infection and ear rots that produce mycotoxins. The yield loss as high as 6 % are incurred per borer per plant. Light and pheromone traps can be used to detect and confirm the identity of adult moths. Host resistance is available in many maize hybrids. Hybrids have been developed with strong stalks that resist lodging after being bored by the European maize borer. Parasitoid wasps such as *Trichogramma* species that attack the eggs have been used to control this pest. Use of *Bacillus thuringiensis* products can suppress pest infestations. Under sever infestation application of chemicals, particularly in granular form in the whorl, can provide effective means of pest management. Chemicals can also be applied as a liquid spray to coincide with egg hatch.

» Hand picking and destruction of larvae.

» Installation of pheromone traps @ 10 / ha for monitoring purpose.

» Release of *Trichogramma chilonis* 5cc / ha

» Natural enemies present in maize ecosystem are Trichogramma, Braconids, Tachinids and NPV.

**Fall armyworm**

*Spodoptera frugiperda* (J.E.Smith) causes foliar, whorl, ear and tassel damage. Stems cut at the base. The signs are 'window pane' feeding damage caused by newly emerged larvae, which feed by scraping the leaf surface without perforating it. Burrowing damage on the ear through the husk. Cold, wet springs reduce effectiveness of natural predators thus favoring the fall armyworm. Late planted and late maturing maize is more vulnerable to damage. Adult moths are 20 to 25 mm long with a wingspan of 30 to 40mm. Forewings are shaded grey to brown, often mottled with a conspicuous white spot on the extreme tip. Hind wings are silvery white with a narrow dark border. Adults are nocturnal and active during warm, humid evenings. Females lay eggs in clusters of 50 to 2000 eggs in a lifetime. The adult lifespan is estimated to be 10 days. Eggs are white, pinkish or light green in colour and spherical in shape. Eggs are usually laid on the underside of leaves. Larvae emerge 3 to 5 days and migrate to the whorl. There are six larval instar stages. In the 2nd and 3rd instar stages larvae are often cannibalistic, resulting in only one larva in the whorl. Mature larvae are 30 to 40mm in length and vary in colour from light tan to green to black. Larvae are characterized by several subdorsal and lateral stripes running along the body. Dark, elevated spots bearing spines occur dorsally along the body. They have four large spots on the upper surface of the last segment. Larvae mature in 14 to 21 days. Pupation 2 to 8 cm below the soil surface. Cocoons are generally oval and 20 to 30mm in length. Pupae are reddish brown and measure 13 to 17 mm in length. Pupation usually takes 9 to 13 days. In optimum conditions the entire lifecycle can be completed in 30 days. Maize crops can normally support two generations. Optimum temperature for larval development is 28° C, although the egg stage and pupal stage require slightly lower temperatures.

The adult moth is similar to *Spodoptera exempta* (Walker) and *S. littoralis* (Boisduval), but can be distinguished by its distinct, dark veins on the hind wing and pale orbicular stigma present on the male forewing. Larvae are characterized by the distinct white Y-shaped mark on the top of the head. Unlike armyworm larvae, fall armyworm larvae feed during both the day and night. Signs of feeding and crop damage are similar to other armyworm species (*Spodoptera*), earworms and maize borer damage. Individual larvae need to be inspected to confirm identity. Presence of host species (preferably maize) and favourable climatic conditions dictate the prevalence and severity of fall armyworm. Grassy fields attract adult females for oviposition. The fall armyworm has a wide range of hosts including maize, rice, sorghum, sugarcane, cotton, alfalfa, peanuts, tobacco, and soybean, in addition to various wild grasses. However, gramineous plants are preferred. Late planted maize and advanced growth stages are more vulnerable to its damage. Under severe infestation yield loss has been documented ranging from 25 to 50 %.

Pheromone traps can be used to determine incidence of adult moths and disrupt mating during the whorl stages. Plant early to avoid periods of heavy infestation later in the season. Plant early maturing varieties. Rotate maize with a non-host. Reduced tillage methods often result in an increase of natural predators and parasitoids. Parasitic wasps, natural predators and pathogens help to control the population of this pest. The egg parasitoid *Telenomus remus* is frequently introduced to effectively control egg mass of this insect. Insecticide application should be considered when 25 % of plants show signs of feeding damage. Insecticides recommended include pyrethroids, carbamates and organophosphates. Certain varieties of *Bt* maize provide adequate control of this pest. Certain maize varieties with a thick epidermis are also resistant to this insect attack.

» Ensure thorough ploughing to expose pupae to natural enemies and avian predators

» Collection and destruction of egg masses and different stages of larvae

» Apply neem cake @ 250 kg/ha to reduce the adult emergence from pupae

» Use light trap @ one/ha during night hours to monitor the adult moth activity in and around maize fields

» Avoid growing maize after maize crop

» Need based spray of chloropyriphos 20 EC @ 2 ml/litre or thiodicarb 75 WP @ 2 gm/litre or flubendiamide 480 SC @ 0.30 ml/litre or emamectin benzoate 5 SG@ 0.40g / litre or spinosad 45 SC @ 0.30 ml/litre or quinalphos 25 EC 1 litre /ha should be given.

### Stem borer

*Chilo partellus* (Swinhoe), Pyralidae, Lepidoptera is widely distributed in maize agroeco system. The pest is prevalent during kharif. It infests the plant as all stages. The major loss is caused at the early stage of the plant. The leaves after hatching, feed on leaves where they form small irregular holes and later bore into stem to make tunnel. The caterpillar cut the growing point of the maize plant resulting in the drying up of the control shoot and subsequently formation of dead heart. The pest egg active from March to October and has 6 - 7 overlapping generations. During winter, it under goes hibernation in larval stage in the stubbles of stalk. In peninsular India, the winter is not severe and therefore the pest remains active round the year. The moths lay eggs in clustered on the under surface of the leaves. The eggs are naked, flat and cream in colour. Each cluster has 10 - 50 eggs arranged in 2 - 3 rows. There is no overlapping of eggs. The average fecundity of a female moth is 250 - 300 eggs during its life span. The longevity of adults varies from 2 - 10 days. The incubation period at 27°C is 4 - 5 days and at low temperature it is increasing upto 8 days. The larval period varies from 14 - 28 days. After passing through 6 instars, the full grown larva makes exit hole in the stem and pupate inside it. The moth emerges from the stem through the exit hole. Destruct stubbles, straw, weed and alternate host to control this pest. Use higher seed rate and pull out and destroy early stages affected plants. Release eggs of *Trichogramma chilonis* @ 50000/ha in 4 - 5 times at weekly interval starting with complete germination of crop. Spray fenvelerate 20 EC or cypermethrin 10 EC @ 500 ml/ha. The pest can be checked if chlorpyriphos 20 EC @ 2 ml/litre or profenophos 50 EC @ 1 litre/ha emulsion is sprayed 15 days after sowing.

### Flea beetle

*Chaetocnema pulicaria* (Corn) beetles scratch between the leaf veins, causing lesions. The worms feed on the roots, causing lodging. The signs are white, thin lesions on the upper leaf surface and lodging. The small (1 mm - 2 mm) shiny, black beetles deposit their eggs in soil, where the tiny root worms develop into white, delicate pupae. The adults can fly and typically have swollen hind legs that are well adapted for jumping.

### Billbugs

Lesser billbug *(Nicentrites testaceipes* Casey*)* and large billbug *(Geraeus senilis)* feed in the whorls, scratching sections of the leaves. The grubs feed on the roots, weakening them. The damage signs are white flecks on leaves and lodging. Forage crops preceding maize favours development these bugs. From midwhorl stage to near tasseling, one can observe in the whorl abundant, grayish, lesser billbugs 3mm to 4mm long and large billbugs 5mm to 7 mm long. When disturbed these drop into the whorl or onto the ground and remain motionless for some time. These insects pass through egg (white and bean-shaped), larva (white, legless, humpbacked grubs with brown heads), pupa (white and soft), and adult (referred to as weevil, snout beetle or billbug).

**Neotropical corn borer**

*Diatraea lineolata* (Walker) feeds on the whorl and destruct the growing point of young plants. Tunneling of the stems, causing interference with nutrient and water translocation. Boring in the stems leading to plant lodging. Boring in the ears, which can increase prevalence of ear rots. The common signs are dead heart, perforations and feeding damage on emerging leaves and crop lodging. Reduced tillage methods, which enable the larvae to overwinter in the diapause stage. Mild winters, which favour survival of overwintering larvae. Cream-colored moths with straw-colored forewings and white hindwings. They have a wingspan of 20 – 42 mm. Females are larger than males. Fresh eggs are yellow, but develop a red band just before hatching. Eggs are deposited in a scale-like, overlapping manner (similar to fish scales) on the upper and lower leaf surface with white with prominent brown/black spots on each segment. Larvae have a yellow/brown prothoracic shield and are 20 to 25 mm long when mature. Larvae in diapause are pale-coloured. Pupae are dark brown with two rounded processes on their heads. Diapause may occur during a dry period when food quality declines. In these cases, mature larvae will diapause in the lower part of the plant stem and pupate with the first rain. Two to three generations can occur per growing season. It causes an average of 25% economic crop loss to borers in susceptible regions. Moth traps can be used to monitor the incidence of adults, while fields should be regularly inspected for presence of eggs and larvae. Chemical or biological control should commence when 25 % of plants have eggs. Management of crop residue infested with overwintering larvae will reduce larvae numbers in the subsequent season. Early planting can help avoid periods of heavy pest infestation later in the season. Planting early maturing varieties also reduces the number of pest generations per crop. Crop rotation with non-hosts can reduce pest incidence. Chemical can be applied either as a spray or in granular form in the whorl. Application of pesticide to the soil at planting, which will protect the plants for one month. There are several natural parasitoids including various *Trichogramma* wasp species that parasitize eggs. The larval parasite *Cotesia flavipes* has also been used to provide control.

**Oriental corn borer**

*Ostrinia furnacalis* (Guenee) feeds on leaves and in the whorl. Boring in the stems, tassels, ears and midribs which can lead to lodging, breaking of tassels and dropping of ears. The adults, brownish or straw-coloured moths, deposit their oval, light-yellow, overlapping, scale-like eggs in clusters of variable sizes on the upper side of leaves or on the husk. Underside of leaves scarified close to empty egg clusters. Perforations and 'shot-gun' damage on leaves emerging from the whorl. Large amounts of larval excrement in the leaf axis. Boring tunnels within the stems, tassel and ears. The factors such as mild cold/winter periods, reduced tillage enables the larvae to overwinter, continuous cultivation of hosts and lack of natural predators favour this insect development. Adults are brownish or straw-coloured with a wingspan of 30 mm. Males are slightly darker than females and have a tapering abdomen. In a lifetime a female can deposit up to 1500 eggs. Adults can live for up to 11 days and are prolific fliers, frequently covering up to 1.5 km in a lifetime. Eggs are light yellow or cream in colour, turning black as the time for hatching approaches. Eggs hatch in 3 to 5 days. There are generally 6 larval instars. Young larvae are pink or a yellowish-gray in colour. Mature larvae are white with wart-like black spots on each body segment. Mature larvae are up to 50 mm long. Larval stages usually last between 17 and 30 days. Pupae are located wherever the last feeding occurred and can be found in tunnels in the stem, in the ears, or between the sheath and shank. Pupation usually lasts for 6 to 9 days. In favorable conditions this species has a 30-day lifecycle. During the growing season, it completes 1 to 3 generations. Yield losses ranging from 20 to 80 % have been reported due to this borer. Moth traps (light and pheromone) can be used to estimate incidence of the pest. Leaves should be regularly monitored for presence of egg masses and feeding larvae. Control of this species should be implemented if more than 3 egg masses are observed per 100 plants. Management of maize stems and crop residue following harvest can reduce the number of larvae that overwinter. Various natural enemies including various predators, parasitoids and pathogens play an important role in controlling populations of this borer. In particular, wasps of the *Trichogramma* species have been introduced to successfully control this borer infestation. Earwigs (*Euborellia annulata*), flower bug (*Orius tantillus*) and *Xanthopimpla stemmator* are also important parasitoids of this borer. Use of *Bacillus thuringiensis* (Bt) products can suppress pest infestations. For successful treatment *Bt* should be applied in the whorl when larvae are at the first instars. Chemicals can be used to effectively control severe pest infestations. Insecticide -treated seed should also be used to protect against stem borer damage. Maize varieties with varying levels of resistance to this borer are available.

**South western maize borer**

*Diatraea grandiosella* (Fabricius) feeds in the whorl causing dead heart. Tunneling of the stems and ears causing interference with nutrient and water translocation. Boring in the ears, which can increase prevalence of ear rots. Girdling of the stem near the soil surface, causing lodging. The important signs are pinholes and perforations in young leaves emerging from the whorl, tunneling in the stems, dead heart and ears and lodging. Reduced tillage methods will allow larvae to overwinter in infested crop stems. Mild temperatures and low rainfall during the off season will aid the survival of larvae in diapause. Adult moths are white to straw coloured and are about 20 mm in length. Moths have a large mouth part, nocturnal, hiding in vegetation during day time. Females oviposit between 100 and 400 eggs in a 5 day lifespan. Fresh eggs are cream in colour and oval in shape (3mm in length) and are deposited on the upper leaf surface in clusters of 2 to 5. Second generation eggs are commonly deposited near the ear. As eggs mature they develop three parallel, reddish-orange rows and hatch in 3 to 5 days. When larvae first emerge they are reddish in colour, but change to dull white with prominent brown and black spots on each body segment. Immature larvae feed in the whorl while mature larvae bore into the stem. Larvae can be up to 40mm in length. Pupae are dark brown in colour and about 25 mm in length and emerge as adult in 7 to 10 days. There are two to three generations in a year. The economic threshold for control is when larvae are observed on 35 % of plants. Early planting can help avoid periods of heavy pest infestation, particularly the second and third generations. Planting early maturing varieties also reduces the number of pest generations per crop. Harvesting early can reduce yield loss due to lodging. Management of infected stalks left in the ground can reduce the number of larvae that overwinter to subsequent seasons. Chemicals can be applied either as a spray or in granular form in the whorl. The most effective biological control agent is the late season wasp *Trichogramma minutum. Apanteles diatraea* is another important control agent. Hybrids with resistance to this borer have been developed and should be sown in areas where the pest is a major problem. *Bacillus thuringiensis* (*Bt*) maize hybrids have also been shown to provide effect control of this borer.

**Wireworms**

Families Elateridate (*Melanotus, Agroties,* and *Dalopius* are among the important genera) and Tenebrionidae (*Eleodes* spp.) bore and feed plant roots lead crop lodging and injury to base of stem. The common signs are patchy seedling emergence, wilting and tillering of seedlings and lodging of older plants. Adult beetles are brownish to black with a hard, elongated, flat shell and are 5 to 20 mm in length. Adults are active fliers and females burrow into the soil to oviposit. Adult lifespan can last 10 to 12 months. Adults make an audible clicking noise when flipping over after they have been turned upside down. Eggs are tiny, round and pearly white. Several hundred eggs can be deposited by adult females in moist soil in grassy areas. Eggs are laid singly and widely scattered amongst the soil particles. This stage can last from few days to a few weeks. Larvae are known as wireworms. Initially wireworms are white, soft and 10 mm long. Mature larvae are about 40 mm long, shiny, smooth, sluggish, deep yellow or brownish and hard but flexible. Larvae have three pairs of short, underdeveloped legs and obvious segmentation. Larval development can take several months to several years depending on the species. Pupae are white, soft and located in weak cells in the soil. Diapause of larvae or adult beetles in the ground occurs during cold periods. Larvae migrate vertically and horizontally in the soil according to soil moisture. They are sensitive to heat. Inspection of soil around damaged seedlings will reveal wireworms. In dry weather, the larvae may be quite deep in the soil. The threshold for chemical control is more than 10 larvae present per square meter of soil or more than 3 present per meter of row-length. Solar-bait stations can be used to monitor wireworms. Two stations per acre are sufficient. Preventative insecticides should be considered if there is an average of one wireworm found per bait station. Cultural control is important for wireworms. Flooding or drying of the soil following harvest will kill many wireworms. Natural enemies such as parasitic nematodes, fungi and birds, should be encouraged. Crop rotation with alfalfa will help reduce infestation levels. Chemical control can be effective if mixing chemicals with the top soil, soil fumigants or seed treatments.

**Shoot fly**

It causes severe losses in spring sown maize in Northern plains of India. Among these *Atherigona soccata* (Randani) and *A. naqvii* (stoyskal) are serious pests in Northern India. *A soccata, A. bidens, A. falcate, A. orientalis and A. punctata* occur occasionally in the Southern region. *A. soccata* lays eggs on under surface of the lower leaves and stalk while *A. nagvii* predominantly lays in the cracks and crevices in the soil around and leaves of young seedlings. The maggots damage from 3 leaf stage upto 25 days after sowing. The newly emerged maggots crawl along the leaf surface, then they enter the leaf whorl, feed on the plant tissues and move down. Browning of the central axis which gradually dries upto form dead heart, manifests the infestation. These symptoms usually after 5 - 7 days of egg laying. The infestation cause development of tillers and the plant shows stunted growth. Older plants do not show dead hearts but the damaged leaves get interwoven with the central leaf and show scorching and distorted symptoms. The plants usually show poor growth and bear small size ear with few grains. The shootfly infestation varies from 69 to 97 % in spring crop. Use higher seed rate and pull out and destroy affected plants along with maggot / pupa. Seed treatment with imidacloprid 200 SL @ 6 ml / kg of seed reduces the incidence significantly. Apply quinalphos 5 G @ 20 kg/ha or spray 1.5 litres of quinalphos 25 EC/ ha. phorate 10 G @ 0.5 kg a.i/ha in furrows is effective against this pest which needs a close watch since rabi maize is grown with irrigation which favours the spread of the pest.

### Earworm

*Helicoverpa zea* (Boddie) *p*rincipally cause damage to the maize ear. It causes direct injury to kernels. The common symptoms are presence of small spherical white eggs in silks or husks and kernel damage. It is distributed through out the world. Upon careful inspection of fresh maize silks or young husks, one may find the small spherical white eggs, which have a shiny appearance and are laid by robust, brownish moths. Mature larvae vary widely in colour, from brown to green. Because the larvae are cannibalistic, seldom does more than one remain in the ear. Full-grown larvae (4 to 5 cm long) drop to the ground and pupate in earthen cells a few centimeters below the soil surface. Although they may occasionally feed in the leaf whorl or on the tender tassels, the larvae principally damage the maize ear. They begin feeding soon after emergence, concentrating on the silk channel. In addition to causing direct injury to kernels, they provide a portal for infection by ear rot fungi. After pollination, dried tassel with the eggs must be removed and destroyed. The varieties with tight husk offer resistance to the larvae making inroad into the cob often the damage to the grain is not much yet the infestation marks reduce the market value of the green cobs. Apply any one of the following 20$^{th}$ day after panicle emergence.

- » Malathion 5 D @25 kg/ha
- » Phosalone 4 D @25 kg/ha
- » Neem seed kernel extract 5%
- » Azadirachtin 1%

### Ear maggot

*Euxesta stigmatias* Loew (Diptera: Ulidiidae) is the most damaging to maize. Other species include *E. major, E. sororcula* and *E. mazorca.* Green-bodied flies with four dark bands on the wings. Typically they have a blue metallic thorax and reddish-brown eyes. Adult flies are 0.5 cm long. The banded wings slide together with quick strokes as they move on the foliage of the maize plant. Adults feed on pollen, nectar, plant sap and glandular exudates. Mating generally occurs at dusk and dawn. Eggs are usually deposited in clumps of 10 to 40. Duration of the egg stage is typically between 2 and 4 days. Maggots are cylindrical and elongated with rounded posterior that tapers to a pointed head with a pair of black mouth hooks. Mature larvae are typically 1cm long when fully grown. There are three larval instars. Maggots flip vigorously when disturbed. Larval development lasts 10 to 21 days. Larvae pupate in dark brown cocoons. Pupation can occur in the soil or silk channels. Pupation takes 7 to 11 days. General life cycle is completed in 24 – 32 days. Larvae will feed in the whorl and bore into the growing point causing dead heart. Multiple tillers may form which do not bear ears. The ovipositing adult easily enters ears with previous damage from birds or insects. The larvae will eat the silks and this will affect pollination, reducing yield. The insides of kernels may be eaten, leaving only the seed coat. Monitoring is most effective in the early morning or early evening. Insecticide (organophosphates and pyrethroids) application should occur in the cool of early morning or late afternoon. Some maize cultivars are moderately resistant to this pest due to high levels of maysin in the corn silk. Predators like earwigs, mites and minute pirate bugs control eggs.

**Spider mites**

*Tetranychus* spp., *Paratetranychus* spp. and *Olygonychus* spp. can damage maize from the seedling stage to maturity. Spider mite injury is inflicted through piercing and sucking on the foliar tissue. As the colonies of mites increase in size, they cause lower leaves of the maize plant to become dry. The mites then migrate to the upper leaves. The presence of small, faint, yellow blotches on the lower leaves. On the underside of damaged leaves, one can observe tiny green to reddish brown mites protected by a delicate web secreted by the adults (*Paratetranychus* spp.) whose eggs have a pearl-like appearance. Mites (which are more closely related to spiders than insects) go through a larval and two nymphal instars and multiply very quickly in hot, dry weather. It is distributed throughout world.

» Conservation of natural enemies such as predatory mites and anthocorid bugs

» Spraying of dicofol (Kelthane 18.5 EC) 0.0185 % or propargite 57 EC @ 2ml/litre or dusting with sulphur or spray application with wettable sulphur powder.

**Sugarcane borer**

*Diatraea saccharalis* (Fabricius) damages 'pinhole' by feeding and perforations on leaves unfolding from the whorl. Damage to the growing point of young maize plants causing unfolding leaves to wilt, die and turn white, symptoms commonly referred to as 'dead heart'. Damage to the growing point also results in stunted growth or plant death. Severe boring of the stems weakens plants leading to crop lodging and interference with water and nutrient translocation. Boring in the ears, resulting in kernel damage and increased vulnerability to ear rots. Adult moths are yellowish or straw coloured with several narrow brown lines running along the length of the forewing. Hind wings in females are white but slightly darker in males. Wingspan of females ranges from 25 to 40 mm and in males from 18 to 30 mm. Adults are nocturnal and remain hidden in vegetation during the daytime. The adult stage usually lasts between 3 to 8 days. Females deposit eggs in the evenings for up to 4 days. Females can deposit up to 700 eggs in a lifetime. Eggs are deposited in clusters of up to 50 on both the upper and lower surfaces of leaves. Eggs are oval, scale-like and deposited in an overlapping fashion similar to roof tiles or fish scales. Eggs are white when deposited, turning reddish or black close to the time of hatching. Eggs hatch in 4 to 9 days. Larvae tend to hatch simultaneously and migrate to the whorl to feed. There are usually 6 instar stages. Young larvae are white in colour with black heads and 1 to 2 mm long. Later stage larvae burrow into the stems, ears, and tassels and are creamy in color with prominent dark brown spots bearing stout hairs on each body section. Overwintering larvae lose their brown spots and are predominantly creamy in colour. Mature larvae are up to 30 mm long. Larval development can take between 25 and 35 days. Pupation occurs wherever larvae of the final instar happen to be feeding. Pupae are brown in colour and 16 to 20 mm long. Pupation generally takes 7 to 9 days. In warm conditions, the life cycle is 25 days, although with cool conditions the life cycle can last up to 200 days. There may be 4 to 5 generations are possible per growing season depending on prevailing weather conditions. Control measures should be implemented when egg clusters are observed on 25 % of plants. Management of infested crop residue following harvest will reduce the amount of overwintering larvae. In irrigated fields, flooding following harvest can kill overwintering larvae in crop debris. Rotation with a non-graminaceous species can reduce pest infestation. Several natural predators and parasitoids including *Trichogramma* wasp species help maintain pest populations. Insecticides can be used to provide effective control of sugar cane borer under heavy infestation. Granular application of insecticides in the whorl can limit damage due to this borer. Cultivate varieties show good levels of resistance to this borer.

**White grubs**

*Phyllophaga spp. and Cyclocephala spp. (annual white grubs)* causes root damage, resulting in wilting of seedlings, poor plant growth, lodging or seedling death. Adult beetles are pale yellow to dark brown, with a light brown abdomen. Legs are dark brown. Adult antennae are 10-segmented. Adults are 1.5 to 2 cm long and 1.2 to 2.3 cm wide. Females tend to be wider than males. Beetles emerge from pupae in the soil in response to the onset of the rainy season or soil disturbance. Adults typically emerge at dusk at temperatures between 27 and 30 °C and are active night fliers. Adults commonly feed in trees, but return to grasslands and cultivated fields to lay eggs following mating. Female adults can continue ovipositing for over 100 days. Up to 60 eggs can be oviposited at a time. Freshly laid eggs are oval, white, about 3 mm long and 2 mm wide. After three to four days, eggs become spherical and smooth and are about 2 mm in diameter. Mature eggs are brownish-black with a diameter of about 4 mm. Eggs are deposited singly in weedy fields or grasslands several centimeters (5-15 cm) below the soil surface. Duration required for eggs to hatch varies considerably from 6 days to over 50 days in some cases. Larvae are creamy white and C-shaped. They measure 2 to 3 mm when young, but grow to nearly 4cm when mature. Mature larvae appear slightly swollen and semi-transparent. Larvae have prominent brown heads, mandibles and five segmented antennae. Larvae have three pairs of spiny, well developed legs and a shiny tip to the abdomen. Larvae undergo three instar stages. Pupae are soft and white to yellowish in early stages, gradually becoming brown. Pupae generally occur in cells in the soil and range from 1.5 to 3 cm in length and width is 1.5cm. Pupal development takes 30 to 40 days. Most species of white grub complete their lifecycle in one year. Crop rotation with deep-rooted legumes such as alfalfa, sweet clover or sunflower. Promotion and conservation of natural predators and pathogens. Reducing or pruning trees that attract adults bordering the crop. Use of light and pheromone traps to trap adults prior to egg laying. Heavy nitrogen fertilization adversely affects first instar larvae. Trees that harbor adults can be sprayed with insecticide or shaken down to collect and destroy the beetles. Spores of the pathogens *Bacillus popilliae* and *B. lentimorbus* can be used to inoculate the soil. Effective parasitic wasps include *Tiphia* species, *Myzinum* species and *Pelecinus polyturator*, as well as the fly *Pyrgota undata.* Green muscardine fungi, *Metarhizium anisopliae* and nematode, *Steinernema glaseri* have also been effective against white grubs. Granular soil treatment with fensulfothion, quinalphos, phorate, carbofuran or isofenphos during planting has been effective in suppressing if the pest is severe. Malathion 5 D @ 25 kg/ha should be dusted in furrows, racked and mixed with soil.

### Hairy caterpillar

The insect's scientific name is *Dicrisia oblique (*Walk). The infestation of hairy caterpillar is active during the month of June to September. Young hairy larvae are gregarious and feed upon green matter of maize leaves. Grown up caterpillars eat away. The larva leaves behind only the veins. Fresh silk is also sometimes damaged resulting in barren partially filled grains. Collect and destroy the egg masses and caterpillar of early instars remaining congregated on the under surface of leaves. For advanced stage, dispersed larvae, spray chlorpyriphos @ 1.0 litre or DDVP 500 ml /ha. Apply methyl parathion 2 % or malathion 5 D or quinalphos 1.5 % dust @ 25 kg / ha on the periphery of infested field to check migration of the larvae.

### Termites

*Odontotermies obesus* (Rambur), *Macrotermes spp.*, *Allodontermes spp.* and *Microtermes besi* (Holmgr) is prevalent in Bihar, Madhya Pradesh, Punjab and Uttar Pradesh. It is considered as minor pest, but in heavily infested areas the crop gets serious damaged. Termites remain quite active in sandy and sandy loam soils during dry season. They attack maize after germination and also at all other stages of the crop growth. The early infestation is evident by the presence of dead plants. Such plants are easily pulled out along with chewed roots. The grown up plants show complete wilting, root loose strength and usually the plants fall over the ground. Earthen galleries appear of the outer surface of the infested stem and the inner hollow space of termites attack the main root system. Stem is also filled with the soil. The incidence of termites is sporadic. The longer a field has been cultivated, the greater will be the yield losses caused by these insects. It is common in Africa and India. Avoid using undecomposed or raw cow dung in the field. Remove all stubbles of previous crops. Irrigate the field at frequent intervals. Mixing of quinalphos 1.5 % dust @ 25 kg / ha or Chlorpyriphos 1.5 % dust @ 20 kg/ha in the soil before sowing is effective.

» Clean cultivation delays termite attack.

» Care should be taken to avoid partially decomposed manures.

» If the termite incidence is in patches, spot application of fipronil granules or chlorpyriphos in the root zone.

» Irrigation of field also reduce termite for a short period.

**Climbing cutworm**

With the introduction of the winter maize in Bihar, the insect *Rhyacia herculea* (Corti and Draudt), Noctuidae, Lepidoptera) has become a serious pest in this area. The caterpillars attack the 20 - 30 days old plants. They feed voraciously on the apical portion and margin of the leaves, consuming all but the mid-rib. Feeding takes place during night and larvae remain hidden inside the leaf whorl - during the day. The faecal pellets in the whorls mark the characteristics of larval presence. The losses cause by the pest varies from 12 to 34 %.

**Phadka grass hopper**

*Heiroglyphus nigrorepletces* (Bol.) is a serious pest of kharif maize in arid region. The grass hoppers lay the eggs in the loose soil on the bunds. The nymphs feed on the grass. As the hoppers grow, they invade the maize crop and more amount of biomass is consumed per hopper. In the later stage of their growth they voraciously feed the leaves and produce lots of faecal matter in the whorl thus rendering the crop unfit even for fodder. The crop gives withered look. Based on the leaf area consumed by the hopper during its development, it caused 18 % crop loss. Most species can fly long distance, but those having short wings disperse only by jumping. Adults deposit their eggs in the soil of grassy areas or pastures.

Nymphs go through five to seven instars, depending on the species. This insect geographically distributed in arid and semi arid regions throughout world. Provide hot weather condition with mould board plough to expose the eggs to sun and natural enemies. Newly hatched nymphs first feed gregariously on the grass where the eggs are laid. Therefore, their control is possible by dusting with quinalphos 1.5 % dust @ 25 kg / ha or Chlorpyriphos 1.5 % dust @ 20 kg/ha or methyl parathion @ 20 kg / ha for well grown nymphs and adults.

**Surface grasshopper**

Young maize seedlings in October are attacked by the surface grasshopper (*Chrotogonus sp*) which cuts the leaves at the ground level, causing gaps in the field. Trimming sides of field bunds and exposing egg masses by summer ploughing for birds to pick them. Dusting malathion 5 D @ 25 kg/ha or quinalphos 1.5 % dust @ 25 kg / ha early in the morning once or twice at an interval of a fortnight would control the pest.

**Grey weevil**

Adults of grey weevil (*Myllocerus* sp) which eat away the leaf margins can be checked with 5 % malathion dust @ 25 kg/ha. Mix any of the quinalphos 5 G 15 kg/ha or phorate 10 G 20 kg/ha with sand to make up a total quantity of 50 kg and apply in the leaf whorls on the $20^{th}$ day of sowing. If granular insecticides are not used, spray quinalphos 25 EC 1 lit/ha or chlorpyriphos 20 EC @ 1 litre/ha or profenophos 50 EC @ 1.5ml/litre on the $20^{th}$ day of sowing (500 litres of spray fluid/ha).

**Seed corn maggot**

*Hylemya platura, Hylemya* spp, *Delia platura* destroys seed, resulting in poor germination and stands. Fly with dark grey body, black legs, 5 – 7 mm. Similar to house fly, but smaller. Eggs hatch in 1 to 9 days. White/yellow/cream-coloured maggot, pointed head, round tail, 5 –7 mm, segmented (12). Development time lasts between two and three weeks. Damage is characterized by the presence of maggot tunnels inside the seed or loss of the entire seed content. In addition, the cotyledons and first leaves of the seedling may be deformed and weak. The eggs are deposited in crevices or surface soil close to seeds. There are normally 4 to 5 generations per each year. If stand loss is severe, replanting may be necessary. Insecticide treatments are with thiamethoxam or clothianidin as seed treatment. Soil-applied carbofuran 3 G could be broadcast before planting. It requires incorporation at 5 cm depth.

## Birds and rodents

Various birds and rats will attack maize. This leads to patchy seedling emergence. These also damage developing ear or stored grain. Birds and rodents usually eat seeds or plants, especially early in the season and thus reduce plant density. They can also damage maize late in the season by eating grain off the mature ear. Seed planted too shallow or soil not firmed over the seed. Poor husk cover, so the grain on the ear was easily attacked by birds. Lodging allowing rodents to reach ears. Maize planted out of season or a very early variety, so it was the only food supply for birds and rodents in this area. The common management practices are change the planting method, apply a bird repellent to the seed, poison rodents in burrows near the field, plant a variety with improved seed cover, double the maize (bend the stalks below the ear after the R5 stage) to limit bird damage, plant varieties of similar maturity for the region and employ a bird scarer.

## Stored grain insect pests

The post harvest problems began once the crop has attained physiological maturity and drying in the field upto 100 % losses have been reported in maize due to *Sitophilus oryzae* L. (Singh *et al.,* 1975). The losses assessed in stored maize due to *Rhizopertha dominica* (Fabricius) are from 33.1 to 52.9 % (Pandey and Singh, 1975) and by *Sitotroga cereallella* (Olivier) from 70.2 to 90.0 %. Phenolic compounds located within the aleurone layer or pericarp of kernel appears to involve in increased kernel hardness associated with thickness of pericarp (Senetal., 1994).

### a. Angoumois grain moth

Young larvae of *Sitotroga cerealella* penetrate and feed inside maize grain. This insect may also infest the crop in the field prior to harvest. Observation of moths flying around infested grain stores. Their presence is especially evident if ears or shelled grain are disturbed. It spreads throughout the world. Small, yellowish or straw-coloured moths, which are 9 mm long and have a fringe along the posterior margin of their wings, can be observed flying around infested grain stores. The moths tend to lay clusters of scale-like eggs, which turn from white to red near hatching, between two close surfaces (between the kernel and reduced glumes at the base of the kernel). The young larvae are tiny and white. In their last larval instar, the larvae prepare a round exit hole for the moth, leaving the outer seed wall (pericarp) only partially cut as a flap over the hole.

## b. Grain borers

*Rhyzopertha dominica (*lesser grain borer) and *Prostephanus truncatus* (Horn) (larger grain borer) affect both stored grain and maize ears in the field. It occurs primarily in the tropics and subtropics. The small, nearly cylindrical beetles are reddish brown to dark brown. Their head is turned down and appears to be covered by a hood (*Prostephanus truncatus*). The larger grain borer is 3 to 5 mm long and cylindrical in shape. The lesser grain borer is 2 to 3 mm long and slim. Both are capable of flying. The beetles lay their eggs on the grain or in the floury dust. The tiny, whitish grubs, feed on the dust or crawl into and feed on slightly damaged kernels. Pupation takes place inside the kernels and emerging adults then cut their way out of the kernels. These insects infest both stored grain and maturing maize ears in the field. In a very short time, the adults produce large quantities of floury dust as they bore into and feed on the grains. Damaged grains can readily be identified since they are usually covered by a film of this dust.

## c. Grain weevils

*Sitophilus zeamais* Motschulsky *(maize grain weevil), Sitophilus oryzae (rice weevil) and Sitophilus granarius* (Linnaeus) *(granary weevil)* infest stored grains. The larvae develop into pupae inside the kernel. Removal of husks in the field will reveal weevils and the irregular punctures they make in kernels during feeding or egg laying (*Sitophilus zeamais*). The rice weevil, smallest of the three species, rarely flies but is otherwise similar to the maize weevil, which is a strong flier. The granary weevil also resembles the maize weevil, but cannot fly and lacks its four reddish spots on the wing covers. In kernels dislodged from the cob, one can readily observe thread-like galleries made by the legless, stout, whitish grubs as they feed inside the kernels.

## d. Indian meal moth

*Plodia interpunctella* (Hubner) is the pest of stored maize does not infest maize in the field. Moths about 10 mm long, with three bands (a narrow reddish brown one and a wider one of the same colour separated by a whitish band), can be observed flying around infested grain stores. The moths lay their eggs on the surface of the maize grain. Upon inspecting the surface of stored grain, one can find dense silken webbing produced by light-yellowish larvae, which eventually acquire a greenish or pinkish tint. Full-grown larvae leave the inside of the grain and pupate in white, silken cocoons on the outside of the infested grain mass.

## References

Aboul Eid, H. Z. and A.I. Ghorab (1981).'The occurrence of *Heterodera zeae* in maize fields in Egypt. *Egyptian Journal of phytopathology*, 13: 51 - 61.

Akob, C.A. and F.K. Ewete (2007). The efficacy of ashes of four locally used plant materials against maize stem borer in Cameroon. *International Journal of Insect Science*, 27 (1): 21 - 26.

Ali, Y. (2007). Study on population dynamics of insect pests of maize. M.Sc. (Hons.) Agric. Thesis, Dept. of Entomology, NWFP Agric. Univ. Peshawar.

Anonymous (2001). Integrated pest management package for maize. Directorate for Plant Protection, Quarantine and storage, Faridabad – 121 001. P: 15.

Beres, P. and F. Lisowicz (2005). Effectiveness of *Trichogramma* species in maize protection against European corn borer (*Ostrinia nubilalis*) in ecological farms. *Progress in Plant Protection,* 45: 47 - 51.

Chabi-Olaye, A., C. Nolte, F. Schulthess and C. Borgemeister (2005). Abundance, dispersion and parasitism of the stem borer, *Busseola fusca* (Lepidoptera: Noctuidae) in maize in the humid forest zone of southern Cameroon. *Bulletin of Entomological Research* 95: 169 - 177.

Cook, S.M., Z.R. Khan and J.A. Pickett (2007). The use of push-pull strategies in integrated pest management. *Annual Review of Entomology*, 52: 375 - 400.

Cugala, D., F. Schulthess, C.P.O. Ogol and Omwega (2006). Assessment of the impact of natural enemies on stem borer infestation and yield loss in maize using selected insecticides in Mozambique. *Annual Society of Entomology* (France), 42 (3-4): 503-510.

Daves, C.A., W.P. Williams and F.M. Davis (2007). Impact of plant resistance on southwestern corn borer (Lepidoptera: Crambidae) biology and plant damage. *Journal of Economic Entomology,* 100: 969 - 975.

Davidson, R.H and W.F. Lyon (1987). *Insect Pests of Farm, Garden and Orchard.* Hoboken, NJ: John Wiley & Sons, New York.

Flynn, J.L., T.E. Reagan and E.O. Ogunwolu (1984). Establishment and damage of the sugarcane borer (Lepidoptera: Pyralidae) in corn as influenced by plant development. *Journal of Economic Entomology,* 77: 691 - 697.

Gesell, S. and D. Calvin (2002). Wireworms as pests of field crops. Extension Fact Sheets.

Pennsylvania State University College of Agricultural Sciences. http://www.ento.psu.edu/extension/factsheets/wireworms.htm (7 January 2007).

Gethi, M., C. Mutinda and A. Diallo (2001). Stem Borers in Maize: A Natural Stress and Progress Towards Host Plant Resistance. Seventh Eastern and Southern Africa Regional Maize Conference, 11-15 February 2001. Pp. 45 - 48.

Godfrey, L.D., S.D. Wright, C.G. Summers and C.A. Frate (2006). Corn: seed corn maggot. *UC IPM Pest Management Guidelines.* UC ANR Publication 3443. University of California, California.

Harris, K.M. and K.F. Nwanze (1992). Busseola fusca (Fuller), the African maize stalk borer: a handbook of information. International Crops Research Institute for the Semi-Arid Tropics, Patancheru, India.

Heinrichs, E.A., J.E. Foster and M.E. Rice (2000). Maize Insect Pests in North America. *Radcliffe's IPM World Textbook.* University of Minnesota, Minnesota.

Henrik, S. and P. Peeter (1997). Reduction of stemborer damage by intercropping maize with cowpea. Agric. Ecosyst. Environ., 62: 13 - 19.

Holt, J., W. Mushobozi, R.K. Day, J.D. Knight, M. Kimani, J. Njuki and R. Musebe (2006). A simple Bayesian network to interpret the accuracy of armyworm outbreak forecasts. *Annals of Applied Biology,* 148: 141 - 146.

Hussain, M., S.R/ Chughtai, H.I Javed, H.N. Malik and MH. Munawwar (2006). Performance of locally constituted quality protein maize hybrids: A fortune for malnourished people and feed industry in Pakistan. *Asian Journal of Plant Science*, 5(2):385 - 389.

Irfan, U. K., M. Nawaz, F. Said, K. Sohail and Subhanullah (2015). Integrated pest management of maize stem borer, *Chilo partellus* (Swinhoe) in maize crop and its impact on yield. *Journal of Entomology and Zoology Studies*, 3(5): 470 - 472.

King, A.B. and J.L. Saunders (1984). The Invertebrate Pests of Annual Food Crops in Central America. Overseas Development Admin., London.

Kuhlman, D.E. and R.G. Dedert (1987). Insect Pests of Maize in Zambia. Urbana – Champaign, IL: University of Illinois, Illinois.

Kullik, S.A., M.K. Sears, D.G.R. McLeod, L.L. Gualtieri and A.W. Schaafsma (2005). Phenology and field biology of black cutworm (Lepidoptera: Noctuidae) in Ontario no-till corn. *Journal of Economic Entomology*, 98: 1594 - 1602.

Li, Y., W.T. Wilsey, C.R. Weeden and A.M. Shelton (2001). Cutworm (*Peridroma saucia*) – Life Cycle. *Pests in the Northeastern United States.* Cornell University, Cornell, USA.

Maredia, K.M. and J.A. Mihm (1991). Sugarcane borer (Lepidoptera: Pyralidae) damage to maize at four plant growth stages. *Environmental Entomology*, 20: 1019 - 1023.

Maurya, R.P. (2009). Integrated insect pest management in maize. *Indian farmers' Digest*, 42 (4): 36 - 37.

Nafus, D.M. and I.H. Schreiner (1991). Review of biology and control of the Asian corn borer, *Ostrinia furnacalis*. *Tropical pest management*, 37: 41-56.

Nuessly, G., K. Pernezny, P. Stansly, R. Sprenkel and R. Lentini (2006). Corn Leafhopper: *Dalbulus maidis.* University of Florida, Florida, USA.

Ortega A. (1987). Insect pests of Maize: A Guide for Field Identification. CIMMYT, Mexico.

Panwar, V.P. (1995). Agricultural insect pests of crops and their control. Kalyani Publishers, New Delhi, India.

Power, A.G. (1990). Leafhopper response to genetically diverse maize stands. Entomol. Exp. Applied, 49: 213 - 219.

Pradyumn kumar, J.C. Sekhar, P. Lakshmi Soujanya and Jaswinder Kaur (2014). Management of insect pests of maize. *Indian Farming*, 64 (4): 30 – 36.

Prke, E.C. (2002). Crop losses due to insect pests in major crops. In: Crop protection compendium. CAB International, Walling ford, U.K.

Rice, M.E. (1993). Insect Pests of Corn. Iowa State University, Iowa, USA.

Saeed, M.Q., U.R. Anis, J. Noora, Habibullah and A. Farhad (2006). Evaluation of different insecticides against maize stem borer, *Chilo partellus* in Peshawar Valley. *Sarhad Journal of Agriculture*, 22(1): 117 - 120.

Samantha, M., C. Zeyaur, R. Khan and A.R. John (2007). The Use of Push-Pull Strategies in Integrated Pest Management. A *Review of Entomology*, 52: 375 - 400.

Sidebottom, J., J. Owen, D. Hundley and J. Neal (1998). White Grubs. Scouting Fraser Fir in North Carolina. Publication AG 573. North Carolina Cooperative Extension Service, North Carolina State University, USA.

Singh, K., N.S. Agarwal and G.K. Girish (1975). Studies on damage of *sitophiles oryzae* and its effect on the germination of seeds of various Hyvs of wheat & maize. Indian J. Ent., 37(2): 146 - 149.

Sloderbeck, P.E., R.A. Higgins and L.L. Buschman (1996). Southwestern Corn Borer. Entomology Corn Insects. Kansas State University Cooperative Extension Service, USA.

Sparks, A.N. (1979). A review of the biology of the fall armyworm. *The Florida Entomologist*, 62: 82 - 87.

Thakur, S.S. and J. Kumar (1999). Estimation of avoidable losses caused by various insect pests in maize. *Journal of Insect Science* (*Ludhiana*), 12: 154 - 155.

Umali, M.M. (2005). Management of the Asian corn borer. Seed quest News Release No. 11, P: 444.

Upadhyaya, V.K., B.N. Mishra and U.S. Yadav (1991). Insect pests of rabi maize: preventing their proliferation. *Indian Farming*, (7): 27 - 28.

Wakman, W., N. Nonci, M. Yasin, M.S. Pabbage and Surtikanti (2002). Biological control of Asian corn borer using *Trichogramma evanescens* and *Beauveria bassiana. In:* Proc. of 8th Asian Regional Maize Workshop. Bangkok, Thailand. 5-8 August 2002. Pp. 413 - 426.

Wyckuys, K.A.G. and R.J. O'Neil (2006). Population dynamics of *Spodoptera frugiperda* Smith (Lepidoptera: Noctuidae) and associated arthropod natural enemies in Honduran subsistence maize. *Crop Protection,* 25: 1180 - 1190.

# 23

# Integrated Disease Management

An important constraint for maize production in the tropics is the ubiquitous incidence of diseases at both pre- and post - flowering stages. The importance and magnitude of damage depends on the interaction between the host plant and the respective pathogen under a given set of environments. Worldwide, the losses due to maize disease are estimated about 9.4 %, in Asia 12 % and in Africa 14 % (James, 2003). On Country basis losses are estimated to 13.2 % in India (Payak and Sharma, 1995). Considering the tropical environment, which allows for 2 to 3 plantings of maize per year, losses may be higher than those reported elsewhere. Major maize diseases include leaf blights, stalk rots, ear rots and kernel rots. Although some diseases can be controlled by a single practice, such as planting a resistant hybrid, most diseases require a combination of practices to ensure that economic damage is kept to a minimum. Once a disease has been identified, its management depends on understanding its causes, the factors that favour disease development, which plant parts are affected and when the disease organisms are spread.

## i. Foliar diseases

### a. Banded leaf and sheath blight: *Rhizoctonia solani*

It is caused by the Basidiomycete fungi *Rhizoctonia solani* (Teliomorph: *Thanatephorus cucumeris).* The disease develops on leaves and sheaths and can spread to the ears. Characteristic symptoms include concentric bands and rings on infected leaves and sheaths that are discoloured, brown, tan or grey in colour. Typically, disease develops on the first and second leaf sheath above the ground and eventually spreads to the ear causing ear rot. Ear rot is characterized by light brown, cottony mycelium on the ear and the presence of small, round, black sclerotia (compact mass of hyphae that can survive in unfavorable conditions). Ears dry prematurely and caking of the ear sheaths is common. *R. solani* does not produce spores and is generally identified by characteristics of the mycelium and sclerotia. Mycelium is colourless when young, but assumes a light brown colour as it matures. Under microscopic examination hyphae are multinucleate, hyaline, septate and branch at right angles. Hyphae are typically constricted and septate at the point of branching. Septae contain a doughnut-shaped pore that enables nuclei, mitochondira and septae to migrate between cells. Hyphae are 4 to 15μm wide.

Humid conditions and irrigated fields are highly favorable for the disease. Additionally, high crop densities impact disease severity. In addition to maize, it is also pathogenic to rice, wheat, sorghum, bean and soybean. *R. solani* survives in the soil and on infected crop debris as sclerotia or mycelium. Sclerotia are known to survive for several years in the soil. The fungi spread by water (flooding), irrigation, movement of contaminated soil and plant debris. At the onset of the growing season, in response to favorable humidity and temperatures (15 to 35°C), fungal growth is attracted to freshly planted host crops by chemical stimulants released by growing plant cells. The fungi overwinter as sclerotia or in infected crop debris. In India, yield losses as high as 40% have been recorded. Application of fungicide is economically viable when susceptible varieties are grown and climatic conditions favor disease severity. Treatment of the soil with fungicide prior to planting can reduce survival of the pathogen and reduce disease severity. Several micro-organisms are known to parasitize *Rhizoctonia* species, including several fungi (*Trichoderma, Gliocladium* and *Laetisaria* species), bacteria (*Pseudomonas* and *Bacillus*) and nematodes (*Aphelenchus avenae*). Addition of these organisms to planting material or soils infested with *Rhizoctonia* may reduce disease pressure. Composting of hardwood on *Rhizoctonia*-infested soil has been found to reduce disease severity, apparently by promoting the growth of *Trichoderma* and other antagonistic micro- organisms. Fields should be well drained prior to planting. Seeds should be planted on raised beds to avoid water contact and promote faster development.

» The pathogen has wide host range and hence suitable crop rotations with non-host crops should be followed.

» The infected plants should be carefully disposed off prior to sclerotia formation.

» Use hybrids / composites with moderate level of resistant to BLB viz., PMH 4, PMH 5, PEHM 1.

» Stripping of basal 2 -3 leaves along with their sheath considerably lowers incidence without affecting grain yield.

» Treat the seed with peat-based formulation of bioagent *Pesudomonas fluorescens* 16 g / kg of seed or soil application of 7 g/litre of water.

» Spray tolcophos methyl (Rhizolex 50 WP) 10 g/ 10 litre of water or Validamycin (Sheethmar) 2.7 ml/litre of water at first appearance of disease.

**b. Common rust: *Puccinia sorghi***

The causal organism is *Puccinia sorghi.* In India the disease is prevalent in Jammu and Kashmir, Himachal Pradesh, Uttar Pradesh, Rajasthan, Maharastra, Andra Pradesh, Tamil Nadu, Karnataka, Bihar, West Bengal, Madhya Pradesh, Haryana and Sikkim. Circular to elongate golden brown or cinnamon brown, powdery, erumpent pustules appear on both leaf surfaces. As the crop matures brownish black pustules containing dark thick walled two celled teliospores develop. In severe cases infection spreads to sheaths and other plant parts. The yield loss is ranging from 11.2 to 33.6 % in Diara region in Bihar due to this disease. It is found in most subtropical, temperate and highland environments with high humidity. Yield losses in excess of 50 % have been recorded under severe disease pressure. Urediospores of common rust are reddish- brown, spherical to ellipsoid, measuring 21-30 x 24-33μm. Each urediospore is binucleate. Teliospores are dark brown to black in colour, smooth, oblong (14-25 x 28- 46μm), two celled and attached to pale brown pedicels that are up to 80 μm long.

The life cycle of *P. sorghi* involves two hosts (maize/teosinte and *Oxalis* species) and five spore stages (teliospores, basidiospores, spermatia, aeciospores and urediniospores). In tropical or subtropical regions urediospores can overwinter and serve as the primary source of inoculum in subsequent seasons. Urediospores are disseminated by wind over vast distances (hundreds of kilometers) and frequently spread from tropical/subtropical regions to temperate regions in spring and summer when maize is cultivated.

The sexual stage of the life cycle occurs predominantly in tropical and subtropical regions. Teliospores are unable to overwinter in most temperate regions. Urediospores germinate on leaves at temperatures ranging from 4 to 30°C, although the optimum temperature range is between 15 and 17°C. Generally, urediospores germinate within 6 hours. Infection structures penetrate the host via stomata on leaf surfaces, stems and ear sheaths. Chlorotic spots as a result of infection can arise in as little as 24 hours in favorable climatic conditions. Optimal temperatures for pustule formation are between 15 and 20° C and in favorable weather conditions the disease cycle can be as short as 5 days. Damage is most serious in early infected plants and when disease spreads to leaves above the ear, which contribute most to grain filling. Late planted maize is particularly vulnerable. Under severe infections, sugars may be diverted from the stalks for grain filling, leading to crop lodging. Sweet corn varieties are particularly susceptible to common rust compared to dent maize varieties.

Plant hybrids like Deccan, Ganga-5, Deccan Hybrid Makka-103 and DHM - 1 which are resistant to this disease to minimize the disease intensity. Spray Mancozeb 2.5g/lit can be taken (0.4 %) as soon as first symptoms are observed and it can be repeated at 10 days interval till flowering. Scouting for rust should be conducted 2 weeks prior to tasseling. If wet weather is not expected, rust is unlikely to increase in severity. Cultivate early maturing varieties that limit secondary cycles of disease and can avoid periods of heavy disease pressure later in the season. Moderate temperature (16 – 25 °C) and high relative humidity favour disease development and spread.

### c. Curvularia leaf spot: *Curvularia lunata*

Curvularia leaf spot is caused by the fungi *Curvularia lunata* (wakker) var. *aeria, C. pallescens* and *C. masculans.* The disease appears in the form of flecking which later develops into larger lesions. The lesions in general are round to oval, separate or coalescent, 1-6 mm in diameter. The centre of each lesion is straw coloured to light brown, which is surrounded by a dark brown margin.

The disease is prevalent in hot, humid maize areas and can damage the crop significantly. It is one of the most improtant diseases of maize in tropical maize growing areas. It may attack any parts of the plant at any stage of growth. The optimum temperature fore growth of pathogen is 28 - 32°C. Cpper oxychloride, copper oxychloride + Zineb and benmyl were more effective against the disease. The disease control can be acheived by two application of 0.5 % difolatan as preventive measure and just after appearance of disease. Seed treatment can be given with Thiram or Captan @ 2g/kg reduce seed infection. Two spraying of Captafol @ 2 g/lit of water is recommended to control the disease.

### d. Gray leaf spot: *Cercospora zeae-maydis*

It is caused by fungal species in the genera *Cercospora.* Yield losses in excess of 70 % have been caused by GLS. Initially, *C. zeae-maydis* was considered the sole causal agent of GLS. However, recent molecular investigations show that GLS is caused by two genetically distinct but morphologically indistinguishable *Cercospora* species. Initially they were considered 'sibling species' (*C. zeae-maydis* Group I and *C. zeae-maydis* Group II). In 2006, *C. zeae-maydis* Group II was classed as a separate species, *Cercospora zeina*, while *C. zeae-maydis* Group I retained the name *Cercospora zeae-maydis*. GLS can be easily identified by characteristic rectangular, tan-coloured lesions that are contained within leaf margins, as the fungi is not able to penetrate sclerenchyma tissue in the leaf veins. As lesions mature, they assume a graying cast due to sporulation of the fungi. Lesions are typically 2-4 × 10-60 mm in size and usually develop on the lower leaves, gradually spreading upwards on the plant during the season. Under severe disease pressure, entire leaves can be blighted and lesions can develop on cob sheaths.

Conidia are hyaline, obclavate, slightly curved and contain 3 to 10 septa. Conidia are 5-10 x 30-100μm in size and are borne on conidiophores that emerge in fascicles from the leaf stomata. Conidiophores are 3 to 8 septate, olivaceous to brown and 4-8 x 70-130 μm in dimension. It occurs throughout maize growing regions of the world where the growing season is characterized by high humidity and temperatures between 22 and 30 °C. Overcast, cloudy days favor disease severity. It is endemic in regions where maize is cultivated intensively. High levels of nitrogen fertilisation have been associated with increased disease severity. As the pathogen is host specific and only able to survive between seasons in infected crop debris, the disease tends to be more prevalent in regions where reduced tillage techniques have been adopted. The fungi are host specific and overwinter as mycelium and stromata in infected maize plant debris left on the soil surface. At the onset of the following growing season, fungi in infected crop debris begin sporulating. Conidia are then wind and rain splash disseminated to freshly-planted maize fields. Conidia infect leaves through the stomata and colonize leaf tissue, giving rise to characteristic symptoms. Secondary cycles of disease are initiated by conidia produced within lesions. Under severe infection, leaves may be completely blighted and killed up to 30 days prior to crop maturity. Cultivation of resistant varieties is the practical means of disease management. Fungicide such as triazole and benzimidazole have been shown to provide most effective control of GLS when applied at lesions first observed and 30 days after the first application depending on prevailing climatic conditions. Management of infected crop residue following harvest can reduce inoculum levels in subsequent seasons. In some regions early planting is recommended to avoid periods of high disease pressure.

### e. Macrospora leaf stripe: *Stenocarpella macrospora*

It is caused by the pathogen *Stenocarpella macrospora,* syn. *Diplodia macrospora.* This disease has not been reported to cause economic damage, but causes some damage in commercial maize plantings. *Stenocarpella macrospora* is mostly an ear-rotting agent, but under appropriate climatic conditions can cause foliar damage. Symptoms consist of necrotic lesions along the veins. These lesions resemble spotting produced by bacteria or by *Exserohilum turcicum* (leaf blight). When held against the light, *S. macrospora* lesions exhibit a distinct narrow yellow margin not present in lesions caused by other pathogens. Under humid conditions, the black, spore-producing pycnidia formed in the necrotic lesions release spores in fine black threads and the lesion splits. It is found in hot and humid areas.

**f. Maydis leaf blight: *Bipolaris maydis***

It is caused by *Bipolaris maydis* (Nisikado). Young lesions are small and diamond shaped. As they mature, they elongate. Lesions may coalesce, producing a complete "burning" of large areas of the leaves. They vary in size and shape among inbreds and hybrids with different genetic background. Race 'O' produced tan, elongated (2-6 x 3-22 mm) lesion between the veins with limited margins, with buff to brown borders, usually attacks only leaves.

The mature lesions are upto 2.5 cm and RH 100 % for atleast 6 hours on several consecutive nights. Regions with a warm (20 to 32°C) and damp growing season are most at risk from Maydis leaf blight. Long, dry, sunny periods during the growing season are unfavorable for the disease. The pathogens can over winter in infected maize tissue on the soil surface or seed. It is most serious in warm and wet temperate and tropical areas, where yield losses close to 70 % have been reported due to the disease. Conidia are 15-20 x 70-160 μm in dimension. Conidia are olivaceous brown, spindle-shaped, and taper to round ends. Conidia are 5 to 11 septate and are characterized by bipolar germination.

The asexual stage (conidial stage) of the fungi is the most important source of inoculum, both during the season and at the start of the season. The sexual stage is characterized by production of ascospores (typically 4 to 8) in cylindrical, hyaline asci. Ascospores are 6-7 x 130-340 μm in size, dark and have 5 to 9 septa. At the onset of the subsequent growing season, in response to favorable temperatures and humidity, mycelia within the crop debris begin sporulating (producing conidia). Conidia are then wind and rain splash disseminated to freshly planted maize in the vicinity. Conidia germinate and infect plants through stomata, giving rise to characteristic lesions within which conidia are produced, which leads to secondary disease cycles. In favorable environmental conditions, the disease cycle can be completed in as little as 72 hours. Seedlings grown from kernels infected with the pathogen may die four weeks following planting. Increased application of nitrogen fertilizer and increased crop density are associated with increased disease severity. Fungicides like mancozeb, thiram and carbexlin have been reported to be effective. The pathogen could be eradicated from seed with steam air mixture is 17 minutes at 54 - 57°C to panicle. Spray with Mancozeb @ 2 g/ lit of water. Cultivate resistant varieties likes Deccan, VL 42, Prabhat, KH-5901, PRO-324, PRO-339,

ICI-701, F- 7013, F-7012, PEMH 1, PEMH 2, PEMH 3, Paras, Sartaj and Deccan 109. It is prevalent in hot (20 – 30 °C) humid, maize growing areas.

### g. Phaeosphaeria leaf spot: *Phaeosphaeria maydis*

The disease can reduce grain production in susceptible cultivars by upto 60 %. The disease manifests as lesions (0.3 - 2.0 cm) that are scattered over the leaf and are initially small, pale green or chlorotic, becoming bleached or dried with dark brown margins and round or elongate to oblong. Spots may coalesce and become irregularly shaped. The casual organism is ascomycete fungi *Phaeosphaeria maydis* (P. Henn.). Perithecia and pycnidia are spherical or globose, dark brown to black with a round ostiole from which spores can escape. Perithecia and pycnidia (74-151µm in height and 67-160 µm in length) are typically immersed within lesions. Asci produced within perithecia are straight to slightly curved, hyaline, cylindrical and contain 8 ascospores. Asci are typically 7-9 x 43-70µm. Ascospores are 3 septate with slight constriction at the septa and a swollen penultimate cell. Ascospores are hyaline, usually slightly curved and measure 3-5 x 14-18µm. Conidia produced within pycnidia are hyaline, ellipsoid, round and measure 2.4-5 x 3.2-9µm.

Humidity levels above 70 % and night time temperatures above 14°C strengthen disease severity. It is considered a potential threat to maize production in regions where high humidity and low night-time temperatures are prevalent during the growing season. The control of the disease was attained with mancozeb @ 2.4 kg/ha at 7 days interval. As the sowing dates were delayed, there was an increase in disease severity and a decrease in grain yield. Cultivate of resistant varieties is the most cost effective and practical means of disease management.

### h. Polysora rust: *Puccinia polysora*

Unlike common rust, Polysora rust is most severe in warm and humid growing conditions. Yield losses in excess of 45% have been recorded due to Polysora rust. In recent years, the incidence of rust has taken a heavy toll in majority of the cultivators grown in southern districts of Karnataka namely Mysore, Mandya, Hassan parts of Coorg, Kolar, Shimoga and Chitradurga districts and in the coastal areas of Arunachal pradesh. The pathogen is *Puccinia polysora.* Urediospores of Polysora rust are clearly ovoid (20-29 x 29-40µm) and golden in colour. Urediospores of *P. polysora* are also characterised by fewer spines and teliospores of *P. polysora* are not known to germinate. *P. polysora* is not known to have an alternative host on which the sexual stage of the life cycle is completed. The optimal temperature for urediospore germination and infection is from 24 to 28°C. Periods of extended dew or high humidity favor Polysora rust. Urediospores are the primary source of inoculum both at the start and during the growing season. Continuous maize cultivation and presence of infected crop debris at the onset of subsequent seasons will result in large levels of initial disease inoculum. Urediospores are wind dispersed.

Symptoms of Polysora rust are very similar to common rust (*P. sorghi*). However, the two can be distinguished, as pustules of Polysora rust occur predominantly on the upper leaf surface, whereas pustules of common rust occur abundantly on both leaf surfaces. Pustules (uredinia) are orangish in colour, circular (0.2 to 2 mm long), raised and most prominent on the leaf, but also occur on the stems and sheaths. When the pustules rupture, urediospores are released, resulting in the secondary cycles of the disease, giving the leaves a rusty appearance. Towards plant maturity, pustules become dark brown to black and release teliospores. Pustules that release teliospores are known as telia. Under severe disease pressure, leaves can turn chlorotic and senesce prematurely. Rupturing of pustules results in release of powdery brown urediospores. Under favorable disease conditions, Polysora rust can spread rapidly. Spores remain viable even when disseminated vast distances in excess of hundreds of kilometres. This disease is known to infect maize, plume grasses, gama grasses and teosinthe. Resistant varieties offer the most efficient and cost-effective means of disease management. Fungicides should be applied while there is a limited amount of secondary inoculum. Cultivate early maturing varieties which limit secondary cycles of the disease. Planting early can avoid periods of heavy disease pressure later in the season. Management of infected crop residue will reduce disease inoculum levels at the start of the subsequent season.

» Maintain good sanitation in the field and follow appropriate tillage.

» Destroy crop debris so as to reduce early infection.

» Grow resistant /tolerant hybrids / composites viz., HSC 1, Buland, HHM 1, HM 11, HQPM 1 and Nityashree.

» Spray Mancozeb 2 – 4 g/litre of water at first appearance of pustules. If disease is severed, three spray at 15 days intervals are recommended.

**i. Downy mildews**

Downy mildews are important maize diseases in many tropical regions of the world. They are particularly destructive in many regions of tropical Asia where losses in excess of 70% have been documented. Downy mildews are caused by up to ten different species of Oomycete fungi in the genera *Peronosclerospora, Scleropthora* and *Sclerospora.* Downy mildews originated in the Old World although they have since been introduced to many regions of the

New World. The different types of downy mildews diseases are *Peronosclerospora sorghi* (Sorghum downy mildew), *P. maydis* (Java downy mildew), *P. philippinensis* (Philippine downy mildew), *P. sacchari* (Sugarcane downy mildew), *Scleropthora rayssiae* var. *zeae* (Brown stripe downy mildew), *Sclerospora graminicola* (Graminicola downy mildew or green ear) and *Sclerophthora macrospora* (crazy top). White stripe symptoms caused by downy mildew are not limited by veins and can therefore be distinguished from signs of iron deficiency. Cool, wet and humid conditions are optimal for disease development. In favorable conditions, disease cycles are rapid, leading to severe infection and spread of disease. Perennation of infected crop debris serves as a source of inoculum in subsequent seasons. Cultivation of alternate hosts in rotation or simultaneously will build pathogen pressure. Moist soils favor oospore germination and therefore damp soil as a result of irrigation or reduced tillage techniques will encourage disease development.

» Maintain good sanitation in the field and follow appropriate tillage operations and crop rotations.

» Planting before rainy season minimizes the occurrence of the disease.

» Rogue out the infected plants and destroy them.

» Ensure destruction of infected crop debris, good weed control, reduced crop density, crop rotation with non-host crops and low seed moisture (less than 10 %) at planting time.

» Avoid maize – sorghum crop rotation.

» In case of RDM, destroy alternate host *Heteropogon meloncocarpus*, to break its life cycle.

» Use resistant / tolerant hybrids/ composites viz., PAU 352, PMH 2, PMH 4, PMH 5, Pratap Makka 3, Pratap Makka 5, Gujarat Makka 4, Gujarat Makka 6, Prakash, Aravali Makka 1 against BSDM, DMH 2, NAC 6002, NAC 6004, COH 3, COH (M) 4.

» COH(M) 5, HM 11, Aravali Makka 1 against SDM.

» Treat seed with metalaxyl (Apron 35 SD) 2.5 g /kg of seed.

» Spray systemic fungicides such as metalaxyl 2.0 – 2.5 g/ litre of water at first appearance of disease symptoms.

**Sorghum downy mildew:** The causal organism is *Perenosclerospora sorghi*. Early maize crops escape infection because by the time conidia are produced, they develop resistance on the collateral host susceptibility occurs only upto about 15 days of age. In susceptible seedling plants, less than 4 weeks after seedling infection becomes systemic in all growth, subsequent to downward growth of mycelium and colonization of shoot apex (growing point). All *Peronosclerospora* spp. induce both local and systemic infection. Malformation of tassels in infected plants. Chlorosis, white stripes, stunting with downy fungal growth on both leaf surfaces are the characteristic symptoms. Cultivate resistant varieties such as CoH (M) 5, Comp. A-9, Indimyt 345, EH-43861, KH-526, AH-36.

**Brown stripe Downy mildew:** The causal organism is *Sclerophthora rayssiae*. Disease symptoms have been observed only on leaves. They are vein limited. Wilting generally starts from the top leaves; Leaves become dull green, eventually loose colour and become dry. Chlorotic stripes, 3-7 mm wide will develop and they further extend in parallel fashion and may in severe cases cover the entire leaf lamina. Severe infection also incites blotching. The stripes in advanced stage become necrotic with purple or reddish colour and present a burnt appearance. Grow resistant varieties -Prabhat, Kohinoor, ICI-703, PAC-9401, PMZ-2 and SEEDTEC-2331.

**Crazy top downy mildew:** The causal organism is *Sclerophthora macrospora*. Disease symptoms first appear as rolling and twisting of upper leaves before mallformation of tassel. The important symptom of the disease is the partial or complete malformation of the tassel which continues until the tassel resembles a mass of narrow, twisted, leafy structures. Affected plants show excessive tillering, stunted with chlorotic stripes on leaves.

**Sugarcane Downy mildew**: The causal organism is *Peronosclerospora sacchari*. Downy growth is seen on the both leaf surfaces. Plants may be distorted with small, poorly filled ears with misshapen tassels. Symptom is characterized by local lesions and systemic infection. Initial leasions are small, round, chlorotic spots on the leaves. Systemic symptoms appear as pale yellow to white stripes or streaks at the base of the 3rd to 6th oldest leaves. Several streaks may develop on leaf and may extend on the entire leaf.

The eradication of collateral and wild hosts near maize field and rouging infected maize plants has been recommended. Destruction of plant debris is done by deep ploughing and other methods. Seed treatment with Metalaxyl at 6.0 g/kg and foliar spray of Mancozeb 2.5 g/litre or Metalaxyl MZ at 2g/litre or Metalaxyl + Mancozeb @ 2.5 g/litre of water is recommended. Use of resistant varieties like DMR 1, DMR 5 and Ganga 11 for sowing.

### j. Spot blotch: *Cochliobolus sativus*

It is also known as *Helminthosporium* leaf blight, foliar blight. The pathogen is *Cochliobolus sativus* (Ito & Kuribayashi) Drechsler ex Dastur (Anamorph: *Bipolaris sorokiniania* (Sacc.) Shoemaker). Ideal conditions for spot blotch development on the leaves are relative humidity of near 100 % with temperatures of 20 to 30°C and long periods (>12 to 18 hours) of leaf wetness caused by rainfall, irrigation or dew. It can be detected as early as three weeks after emergence. If the climate is suitable, the disease develops in severe epidemics, particularly when the crop is around heading. Since it is seed transmitted, lower leaves appear more severely damaged. The disease progresses from the lower to upper parts of the plants. Lesions on the leaves start as dark brown lesions of a few mm that extend as elongated dark brown spots greater than 1 cm. A yellowing due to toxin production is sometimes observed extending from the lesion. As the disease progresses the spots join together, forming large blotches that cover the leaves and eventually kill it. Depending on environmental conditions, the pathogen causes either spot blotch or common root rot. Symptoms caused by *C. sativus* can be found on sub-crown internodes (common root rot), stems, leaves and spikes (spot blotch). Seedling blight and death may also occur, in particular in wet and warm environments.

### k. Turcicum leaf blight: *Exserohilum turcicum*

It is a major constraint to maize production in many maize growing regions worldwide. Yield losses as high as 70 % have been attributed to Turcicum leaf blight. The pathogen is *Exserohilum turcicum* (syn. *Helminthosporium turcicum*) and teliomorph stage is *Setosphaeria turcica.* The conidia are produced profusely within mature lesions. On average, conidia dimensions are 20 × 105 µm and are olive grey in colour. Conidia are 3 - 8 septate, spindle-shaped and have a protruding hilum. There are seven physiological races of *E. turcicum* that are pathogenic to maize. These races can be distinguished based on their pathogenecity to different maize genotypes with different resistant genes. The pathogen overwinters on infected crop debris left on the soil surface and is also more prevalent where increased levels of nitrogen fertiliser are applied. Conidia are 3 - 8 septate, spindle-shaped and have a protruding hilum.

Early symptoms are oval, water-soaked spots on leaves. Mature symptoms are characteristic cigar shaped lesions that are 3 to 15 cm long. Lesions are elliptical and tan in colour, developing distinct dark areas as they mature that are associated with fungal sporulation. Lesions typically first appear on lower leaves, spreading to upper leaves and the ear sheaths as the crop matures. Under severe infection, lesions may coalesce, blighting the entire leaf.

Spores (conidia) are then disseminated by wind and rain splash to freshly planted maize. Conidia can be carried vast distances in the wind. They germinate in temperatures ranging from 17 to 27°C and during periods of extended leaf wetness (6 to 18 hours), infecting host tissue. Secondary cycles of disease occur where conidia produced in disease lesions are disseminated within the crop and to other fields by rain splash and wind. In addition to maize, *it* is known to infect sorghum, Johnson grass, gama grass, teosinthe and Sudan grass. Host resistance is the most efficient and cost- effective means of disease resistance. Rotating maize with non-host crops can reduce disease pressure. For control of this disease field sanitation and long crop rotation are recommended. The disease can be managed by fungicides - copper, moneb, mancozeb and zineb are all effective.

### l. Seedling blights: *Pythium ultimum, Fusarium verticillioides*

These are closely associated with seed rots of maize and are not widespread but are endemically present in localized areas every year. Seed may be rotted and may die prior to or shortly after germination. Seedlings that do emerge may have rotted roots and stems, resulting in less vigorous plants that may also die. Seedling blights are usually more prevalent in the seeds produced in the drought conditions or when seed damaged by an early frost before harvest is planted in cool soil or wherever misplacement of fertilizer or misuse of pesticides are involved. A number of pathogens have been reported to be associated with the seeds are *Pythium* sp, *Fusarium sp, Helminthosporium maydis, Penicillim sp, Aspergillus spp, Alternaria spp, Nigrospora sp, Trichoderma spp, Diplodia spp* and *Stenocarpella maydis* are also associated with hybrid seeds. Seed treatment with Fernasand (Thiram + Lindane) resulted in complete control of all the seed borne fungi.

- It is prevalent in sub-temperate areas where the soil temperature (less than 13°C) during planting.
- Plant good quality, healthy and injury free seed to ensure good germination.
- Ensure proper seed bed preparation, planting of seed in warm fairly moist soil (more than 15.8 °C).
- Ensure correct placement of fertilizers, herbicides and other pesticides.
- Treat the seeds with thiram / captan @ 2.5 g/kg of seed.

### m. Brown spot: *Physoderma maydis*

It is generally of minor importance although serious localized outbreaks have been reported from tropical areas. It caused loss upto 24.5 to 27 %. It is caused by *Physoderma maydis* (miyabe). Symptoms develop of leaf blades and consist of small chlorotic spots, arranged as alternate bands of diseases and healthy tissues. Spots on the mid-ribs are circular and dark brown while lesions on the laminae continue as chlorotic spots. Nodes and internodes also show brown lesions. In severe infections, these may coalesce and induce stalk rotting and lodging.

Deep ploughing of infested debris as it resulted in little or no disease. Soil amendments with rice bran and rice straw were most effective in controlling the disease due to increased microbial activity in the soil.

» Maintain good sanitation in the field and follow recommended tillage practices.
» Plough down crop debris.
» Follow crop rotation with non-host crops.
» Use resistant / tolerant hybrids / composites recommended for the region.
» Treat the seed with fungicide benomy (Benlate) or carbendazim (Bavistin) or oxycarboxin (Plantavax) 2.5 g/kg of seed.
» Spray copper fungicides @ 3 g /litre of water in the whorls of maize plant twice a week for 4 weeks before silking.

**n. Bacterial leaf stripe: *Pseudomonas andropogonis***

Primary lesions develop first on the lower leaves. These are amber to olive coloured. There are initially long, narrow with parallel sides. Later these become brown and necrotic under favourable conditions spread upwards ultimately showing chlorotic striping of the upper leaves. It is caused by *Pseudomonas* and ropogonis. The bacterium grows upto a temperature of 41°C. Prolonged warm and wet weather results in severe disease incidence.

**o. Bacterial leaf blight**

The water-soaked lesions developing from the whorl later burn brown of grey or white. Sometimes these lesions are characterised by red borders. After maturity or tasselling new by borne lesions were not seen. Lesions coalesce during wet weather forming large necrotic areas. The infected leaves shred after maturity. Plant tops turn brown and are stunted. Outside the stalk turn brown to black and inside brown and emanates foul odour. *Pseudomonas avenae* (Manns) causes this disease. The disease is controlled by 100 ppm streptomycin spraying.

### p. Tropical rust

*Physopella zea* (mains) causes this disease. Pulstules are round to oval in shape. Pustules colour during uredial stage is white to pale yellow in the centre. During telial stage, black colour develops around the pustules but with conspicuous ligh colour in the centre, pustules occur in groups mostly 0.3 to 1.0 μm long on upper leaf surface. Later develops purplish/ circular or oblong blotches with creamy centres, 0.6 cm in diameter. The dark chocolate brown telia usually scattered in a group around the uredia. Cultivation as resistant verities is one most effective and economic way of managing three rust diseases. The composite such as NAC 6004, NAC 6002 and hybrid Deccan 105 are resistant cultivars. Rust was effectively controlled by weekly spraying of Zineb (0.2 %) + surfactant F (40Z + 100 gals). Three spraying of Dithane M-45 at an interval of 15 days beginning from 31st January was found to be beneficial economical.

### q. Anthracnose

Anthracnose caused by the fungus *Collectotircum humgraminicolo* (Ces.) is a disease of world-wide importance in maize. The disease affects all parts of the plant and can be formed at any time during the growing season. In the seedling stage, symptoms typically consist of oval lesions often with concentric zones of enlargements. Leaf bright of juvenile plants may be severe and can even result in seedling death.

### r. Late wilt: *Cephalosporium maydis*

The causal organisms are *Cephalosporium maydis* and *Cephalosporium acremonium*. The disease kills the plant prematurely after flowering. Infected plants do not show symptoms until they reach to tasseling. Wilting generally starts from the top leaves. Leaves become dull green, eventually loose colour and become dry. In advanced stages the stalk loses its healthy green colour. Lower portions become dry, shrunken with or without wrinkling, hardens and turn purple to dark brown which in more prominent on lower internodes. When split open diseased stalks, show brown vascular bundles starting from underground portion of the roots. Diseased plants produce only ears with undeveloped shrunken kernels. In severe cases affected plants remain abortive causing 100 per cent loss. *Cephalosporium maydis* is primarily soil borne and may infect young maize plants more readily than other plants through roots or mesocotyl. In case of *C. acremonium* only vascular burdles get blackened. Use of resistant varieties like Ganga Safed 2. Fallow crop sanitation and crop rotations. Avoid water stress at flowering stage. Treat the seed with Thiram or Captan @3g/kg seed.

## ii. Stalk rots and smuts

### a. Charcoal stalk rot: *Macrophomina phaseolina*

It is caused by the fungi *Macrophomina phaseolina* (syn. *Macrophomina phaseoli, Botryodiplodia phaseoli* and *Sclerotium bataticola*). Drought and high temperature encourage disease severity. The characteristic symptoms of the disease become apparent as the plants approach maturity. The disease generally appears early after flowering. Plants affected by *M. phaseolina* show evidence of pre-mature ripening. The out sides of lower internodes become straw coloured. The pith becomes badly disintegrated. The infected stalks may split longitudinally into a mass of fibres. A distinguishing character of the disease is the presence of the small black sclerotia in the pith of the affected stalks. Roots are also invaded and show black sclerotia in the disorganised tissue.

*M. phaseolina* overwinters as sclerotia in the soil and can remain viable for several years. In dry and hot conditions fungi infect the roots of maize plants and colonize the lower stalk, eventually giving rise to characteristic symptoms. It is a soil borne disease. Sclerotia typically measure 0.05 to 0.22 mm in diameter. Isolates of *M. phaseolina* that infect maize do not form conidia. Soil temperatures of 30-42°C increase disease severity. High crop densities and poor soil nutrient status also favor disease severity. *M. phaseolina* infects an extremely wide range of hosts, including sorghum, soybean, cucurbits and various weed species. Plant varieties with a good stalk strength. Regular irrigations particularly during flowering time should be provided. Use resistant varieties like DHM 103, Ganga Safed - 2 and avoid sowing of susceptible varieties like DHM 105. Seed treatment with Carbendazim or Thiram 3 g/kg seed is effective. Field sanitation, crop rotation should be followed. Application of potassium has been shown to delay senescence of pith cells and thicken their walls. Harvest when the crop is mature (30 % moisture content) as delayed harvesting can result in extensive crop loss due to lodging.

### b. False head smut: *Ustilaginoidea virens*

False head smut (*Ustilaginoidea virens)* occurs very rarely in hot, dry or humid areas. The fungus commonly infects rice flowers more than maize. Symptoms differ from those of other smuts of maize. False head smut produces neither tassel malformation nor ear infection, as does true head smut (*Sphacelotheca reiliana*); only a few isolated male florets in the tassel show dark-green masses of spores (sori). False head smut also differs from common smut (*Ustilago maydis*) in that no galls are produced.

### c. Gibberella stalk rot: *Gibberella zeae*

Infected plants typically wilt, leaves turn dull grayish-green and the lower stalk turns from dark green to straw-coloured. The internal pith of the lower stem disintegrates and goes soft. When split open, the stalks exhibit a reddish discolouration. Often black perithecia or mycelium growth can be observed at lower stalk nodes where the plant is infected.

Conidia of the asexual stage of *Gibberella zeae* (Anamorph: *Fusarium graminearum*) are hyaline, slightly curved, tapering at both ends and 3 to 5 septate. Conidia measure 4-6 × 10-30 μm. The perfect stage is produced within infected lesions, usually late in the season. Pertithecia of *G. zeae* are blackish and contain 8 ascospores. Ascospores are hyaline, up to 3 septate, slightly curved, tapered at both ends and measure 3-5 × 10-30 μm. The fungus frequently overwinters as perithecia and ascospores only mature at the onset of the subsequent season when they are the primary source of disease inoculum. Occasionally chlamydospores are formed in lesions. It infects a range of other cereals including wheat, barley, oat and rye. It is also known to infect species of ***Lycopersicon, Pisum, Trifolium*** and ***Solanum***. Ascospores produced in perithecia are wind and rain splash disseminated to freshly planted maize. Cultivation of resistant and tolerant varieties is the most cost effective and practical means of disease management. Avoid planting above recommended crop densities. Rotate with non-hosts, such as soybeans. Where possible, harvest early to avoid yield loss due to lodging. Manage infected stems and roots during the offseason to prevent perennation of inoculum. Control insects, such as corn borers, that may serve as vectors or cause injury to plants and make them susceptible to infection. Ensure seed is fungicide treated.

### d. Fusarium stalk rot: *Fusarium verticillioides*

Infected plants typically wilt, leaves turn dull grayish-green and the lower stalk turns from dark green to straw-coloured. The internal pith of the lower stem disintegrates and goes soft. When split open, the stalks exhibit a reddish discoloration. Often black perithecia or mycelium growth can be observed at lower stalk nodes where the plant is infected. The perfect stage of *Fusarium verticillioides* (Teleopmorph: *Gibberella fujikuroi*) is rarely. During the imperfect (asexual) stage, both macroconidia and microconidia are produced. Macroconidia are hyaline, slightly curved and tapered at both ends. They are 3 to 5 septate and measure 2-5 × 15-60 μm. Microconidia are produced prolifically and are borne in chains. They are single-celled and measure 2-3 x 5-12 μm. *F. verticillioides* is also known to be seed borne and hence is more prevalent where infected grain is used as seed.

Increased crop density and nitrogen fertilization rates favor disease severity. Crops under stress (e.g. due to foliar diseases) are more prone to stalk rots. It infects a range of cultivated crops including sorghum, sugarcane, wheat, cotton, banana, pineapple and tomato. Mycelium in infected crop debris on the soil surface produces macroconidia and microconidia, which are wind and rain splash disseminated to freshly planted maize. The disease management is similar as that of Gibberella stalk rot disease.

» Cultural practices like deep ploughing, sanitation and removal of crop debris should be done.

» Follow crop rotation with non-host crops.

» Use resistant hybrids / composites viz., PEMH 1, PEMH 2, PEMH 3, PEMH 4, PEMH 5, HM 11, HQPM 4, DHM 111, DHM 117, Pratap Kanchan 2, Pratap Makka 3 and Pratap Makka 5, Priya sweet corn.

» Maintain recommended plant population.

» Avoid water stress at flowering time which reduces disease incidence.

» Avoid high level of nitrogen and low level of potassium application. However, potassium level up to 80 kg/ha helps in minimizing the disease.

» Manage the attack of borers in maize as their injury predisposes the talk rot.

» Don not use the seed produced from infected areas.

» Use *Trichoderma* formulation in furrows after mixing with moist FYM incubated for 10 days @ 2.50 kg/ 250 kg FYM per hectare.

### e. Head smut: *Sphacelotheca reiliana*

It is caused by *Sphacelotheca reiliana.* The pathogen affects the tassels and the ears, causing characteristic symptoms that are easily recognized. Head smut can cause extensive economic damage in areas where maize is cultivated frequently and the growing season is characterized by dry and hot conditions. Incidence of head smut as high as 80 % has been documented. The pathogen is *Sphacelotheca reiliana* (syn. *Sorosporium reilianum, Ustilago reiliana*). Characteristic symptoms of head smut are visible once the maize plant has reached the tasseling and reproductive growth stages. Tassels and ears are typically malformed, overgrown and replaced by smut sori. Smut sori are usually covered by a thin tissue, which ruptures to revela masses of dry, black spores (teliospores) that are wind disseminated.

A tangled mass of vascular threads is often observed in ruptured sori. Sori can infect a few spikelets in the tassel or can infect the entire tassel leading to abnormal growth and structures. Infected ears are round, lacking in silks and the ear is replaced by masses of teliospores that leave the vascular bundles exposed and shredded. Teliospores of *Sphacelotheca reilians* are dark coloured, ranging from rusty brown to black. Teliospores are 9-14 μm in diameter, circular, spiny and are produced abundantly in sori. Teliospores germinate from basidia and sporidia that are single- celled, hyaline and 7 to 15 μm in diameter.

Dry and cool weather at the start of the season favors the disease as maize seedling growth is slow. Later in the season dry and warm conditions (21-30°C) favor disease severity. Soil moisture content of 15-25 % w/w is optimal for seedling infection. Head smut has been recorded as being more severe under nitrogen deficiency. It is known to infect maize, sorghum, Sudan grass, Johnson grass and teosinte. Teliospores are produced abundantly in the sori of smutted maize. They are wind disseminated once the sori rupture and overwinter in the soil until temperature and soil moisture levels are optimal for germination, able to survive in the soil for over three years. Teliospores germinate to form four-celled basidia. Each basidium germinates into several haploid basidiospores which bud in a similar fashion to yeast cells, forming haploid sporidia of opposing mating types (+ and -). Mating types are generally produced in equal numbers. Sporidia of opposing mating types fuse to produce dikaryotic hyphae that are infectious to maize seedlings. Infective mycelium invades the meristemic tissue of seedlings and undifferentiated floral tissue, giving rise to characteristic symptoms once the plant reaches the reproductive growth stages. Head smut is predominantly a soil borne disease. Seed borne inoculum has been documented but is not considered important. Resistant varieties are available and should be cultivated in regions where head smut is a problem. As teliospores can survive for several years in the soil, crop rotation is of limited use. Planting maize during seasons that are unfavourable to teliospore germination will reduce infection. Ensuring that the soil is well-fertilized can reduce disease severity. Removing and destroying sori before they rupture can limit the spread of inoculum. Seed treatment with systemic fungicides has been used to effectively control head smut.

### f. Pythium stalk rot: *Pythium aphanidermatum*

This pathogen *Pythium aphanidermatum* (Edson) causes severe stalk rot in maize under favourable conditions and it is widely distributed in India *viz*., Sikkim, Himachal Pradesh, West Bengal, Punjab, Haryana, Delhi, Uttar Pradesh and Bihar) with a wide host range. This disease is usually confined to a single internode just above the soil line. Stalk of infected plants are not completely broken off. The infected area becomes brown, water scaked and soft. The stalk may be twisted distorted and collapsed plants are most vulnerable when they are 6 - 7-week-old. Roots, brass roots and nodes are not affected by pathogen. The infection leads to sudden and rapid toppling of plants at pre- tasseling stage.

The internodes and leaf sheath have been found to be favourable sites of infection. Hyphae are non-septate and hyaline. Sporangia appear inflated and measure 4-20 x 50- 1,000μm. Sporangia can be branched or unbranched and can form complexes. Zoospores are kidney shaped, laterally biciliate and measure about 7.5 x 12 μm. Oogonium are spherical (22-27 μm in diameter) and terminal while antheridia are often intercallary. Oospores are thick walled and 17 to 19 μm in diameter. Typically, oospores will be observed from diseased stalk tissue. It is known to infect a wide range of cultivated crops including cereals and grasses, cucurbits, horticultural crops and cotton. It survives in the soil as mycelium or oospores. Oospores can survive in the soil for several years. They produce sporangia which release motile zoospores. Zoopores swim towards plant or root tissue in response to exudates which serve as nutrient or chemotrophic stimulants. Zoospores infect the plant directly at or just below the soil surface where free water is present, leading to the development of characteristic lesions. Mycelium within the infected host tissue gives rise to oogonium and antheridium, the sexual stage of the life cycle. The antheridium fertilizes the oognium producing oospore, which is able to overwinter in crop debris or in the soil. Temperature affects both the growth of pathogen and disease development profoundly. The pathogen causes maximum disease under high humidity and high temperature conditions. The optimum temperature to produce maximum disease intensity is 30 - 35°C. The pathogen grows well in acidic soils having pH values between 3 and 5. Stalk rots were more prevalent in fields where maize had been grown annually for 10 years. The disease could be decreased by spraying with Ridomil MZ 400. The disease can be controlled by seed treatment with metalaxyl. Soil drenching with captan 75 % @ 12 g/100 litres of water, when the crop is 5 to 7-week-old. Some resistant materials have been identified including CM 110, CM 104, CM 400, CM 600, MP 428, MO 18W, MP 420, TX 61 M, MP 305, L 153, T 202, GA 209, B 37 and GE 440. Regulate planting time between 10 and 20$^{th}$ July in North India. Provide good field drainage. Plant population should not be exceeded 50000 plants/ha. Remove previous crop debris/straw. Use resistant varieties and composites for sowing Various biological control agents such as *Gliocladium catenulatum* and *Trichoderma* species are also being investigated for control.

- Regular planting time between July 10$^{th}$ and 20$^{th}$ in North India.
- Plant population should not to exceed 50000 plants/ha in endemic areas.
- Manage good field drainage to avoid water logging that helps in zoospore disposal.
- Remove previous crop debris.
- Drench the base of the plants when crop is 5 – 7 weeks old with captan 2.5 g/litre of water.

- » Enrich the soil with biocontrol agents impregnated composts for good development of antagonists against soil borne pathogen.
- » Soil solarization, fumigation and soil drenches with BCAs and fungicides are helpful in reducing the soil inoculum.
- » Use tolerant hybrids / composites recommended for the region.

### g. Stenocarpella stalk rot: *Stenocarpella maydis*

It is common in most maize growing regions globally and yield loss exceeds 50 % in conditions that favor disease severity. *S. maydis* has been shown to cause up to 37 % loss of germination when infected seed is cultivated. Infection of the ears by this pathogen can result in production of mycotoxins that are toxic to humans and livestock. The pathogen is *Stenocarpella maydis* (syn. *Diplodia maydis*) prevalent in cool, humid temperate regions whereas *Stenocarpella macrospora* (syn. *Diplodia macrospora*) prevalent in warm, humid regions. Symptoms of this disease apparent after silking. It causes browning of the pith of the basal internodes of the maize plant. Oval, elongate lesions (up to 10 cm long) with pale cream-brown centers and dark brown borders often form on infected stalk tissue. Leaves wilt suddenly and become grayish-green. The pith in infected stalk tissue becomes discoloured and weakened and the crop stand is characterized by lodging. This stalk rot forms abundant, dark-coloured pycnidia on infected stalk internodes late in the season. Pycnidia are flask-shaped (spherical with circular protruding papillate ostiole) and multicellular. Pycnidia are 150-450 µm in diameter while the ostiole is typically 30-40 µm in diameter. Pycnidia contain pale brown, straight to slightly curved conidia with rounded ends. Conidia of *S. macrospora* are 0-3 septate, measuring 7-12 x 44-82 µm. Conidia of *S. maydis* are 1-2 septate and measure 5-8 x 15-34 µm. Occasionally pycnidia contain colourless, narrow spores that measure 1-2 x 25-35 µm. A sexual stage for these fungi has not been recorded. This disease is favored by dry weather early in the growing season followed by frequent rainfall after silking. Strong winds, rainfall and insects aid the dissemination of conidia. As the causal fungi overwinter in infected crop debris, *Stenocarpella* stalk rot is more prevalent in regions where reduced tillage is employed. As with other stalk rots, low potassium content in the soil encourages disease severity. Cultivate resistant varieties. Flint varieties are generally more resistant than dent varieties. Seeds treated with fungicide are effective in controlling early infection of the seedlings. Treatment of *S. maydis* infected seed with antagonistic *Streptomyces* species has been shown to effectively reduce disease severity. Balanced soil fertility with adequate potassium will reduce incidence of *Stenocarpella* stalk rot. Harvesting early will reduce losses due to lodging following crop maturity. Crop rotation is effective. Lower plant density results in reduced disease severity.

### h. Erwinia stalk rot: *Erwinia dissolvens*

It is one of the most important diseases of maize in tropical countries. In India four bacterial pathogens have been reported to cause stalk rot of maize namely *Erwinia dissolvens* (Rosen), *Erwinia carotovora, E. chrysanthemi* PV. Zeae (sabet) and *Pseudomonas syringae* pv. lapsa (Ark). Stalkrot caused by *P. syringae pv. lapsa* has been reported from Uttar Pradesh and Rajasthan. The basal internodes develop soft rot and give a water-soaked appearance. A mild sweet fermenting odour accompanies such rotting. Leaves some time show signs of wilting or water loss and affected plants within a few days of infection lodge or topple down. Ears and shank may also show rot. They fail to develop further and the ears hang down simply from the plant.

The development of bacterial soft rot like pythium stalk rot is favoured by high temperature (29 - 35°C) and poor air circulation. Hot and humid conditions favour the disease incidence. The duration of bright sunshine was found to be directly correlated with the disease incidence. In India the diseases were reported in Tarai areas of Uttar Pradesh and Bihar, Punjab, Delhi, West Bengal, Rajasthan, Andhra Pradesh, Jammu and Kashmir and Uttaranchal. The disease is reported to cause yield loss up to 85 %. Application of bleaching powder @ 10 kg/ha, the first at pre-flowering stage and the second 10 days later considerably checked the stalk rot incidence. Avoid water logging, field should have proper drainage. Planting of the crop on ridges than flat soil. Avoid use of sewage water for irrigated. Use of disease resistance varieties, i.e. Hybrids Ganga Safed-2, DHM 103, show significantly less disease incidence than other hybrids.

- Avoid water logging in the field.
- Plant the crop on ridges rather than on flat soil.
- Do not use sewage water for irrigation.
- Use tolerant hybrids / composites viz., PAU 352, Pratap Kanchan 2, Pratap Makka 3, Pratap Makka 5, PEMH 3.
- Apply bleaching powder (33 % chlorine) 10 kg/ha at first notice of the disease along with irrigation.

## iii. Ear rots

### a. Aspergillus ear rot: *Aspergillus niger*

This is caused by several *Aspergillus* species. Although crop damage due to Aspergillus ear rot is minimal in most years, the disease is of importance as a number of Aspergillus species (*Aspergillus niger, A. glaucus, A. ochraceus, A. toxicarius, A. Oryzae, A. flavus* and *A. parasiticus*) produce potent mycotoxins (aflatoxins) that are toxic to birds and mammals. This is characterized by the production of powdery masses of spores on and between kernels. Spores produced by *A. niger* are black in colour, while those produced by other *Aspergillus* species are typically yellow-green and turn dark green to brown as colonies mature. Typically, only a few kernels (usually injured kernels) of the ear are infected and infection usually occurs close to the ear tip. Occasionally, injured kernels may not show signs of infection but will be dull and discoloured. Hyphae are hyaline and septate. Colourless conidiophores originate from the hyphae and terminate in a vesicle. Vesicles are globose. Flask shaped phialides cover the surface of the vesicle radially or only at the upper surface (columnar). Phialides may be attached directly to the vesicle or via a supporting cell (metula). Conidia form in radial chains on the phialides. Conidia are globose, smooth to slightly rough, and 2-7µm in diameter. It may also form sclerotia which are 400-700 µm in diameter, globose and dark red to black in colour. Drought and high temperatures increase disease severity. Aflatoxins are toxic, carcinogenic compounds produced by *A. flavus* and *A. parasiticus.* Aflatoxins are toxic to birds and mammals and therefore maize grains affected by Aspergillus ear rot may not be suitable for human and livestock consumption. Livestock fed with aflatoxin-contaminated feed may lose weight and suffer from liver damage, hemorrhages and increased susceptibility to opportunistic diseases. Meat from livestock fed with aflatoxin-contaminated feed is not contaminated, as aflatoxins are metabolized and excreted. However, aflatoxins can accumulate in milk from cows fed with aflatoxin-contaminated feed. The Unites States Food and Drug Administration have set a threshold level of 0.5 parts per billion (ppb) for aflatoxin M1 in milk. Cultivate resistant varieties. Treatment of stored, infected maize with ammonia is known to reduce aflatoxin contamination. Reduce crop stress by irrigation and appropriate fertilization. Manage insect pests which damage the ears. Storage of maize at low moisture levels will reduce storage rots.

**b. Common smut: *Ustilago maydis***

It is caused by the basidiomycete fungi *Ustilago maydis.* Symptoms are easily identified by the characteristic galls that form on most aboveground parts of the maize plant, including ears, stalks, tassels and leaves. Galls are initially greenish/grey-white with a shiny periderm. As galls mature, they change to a dark grey colour and contain masses of dark grey, black spores (teliospores). As galls mature, the periderm dries and disintegrates, releasing teliospores. Galls that form at the ear tip are most distinctive and can reach 15 cm in diameter. Galls often form on the ear tip. Galls on the leaf typically form on the midrib and distort the leaf, but rarely exceed 1.5 cm in diameter. Young plants are most vulnerable as infection of the meristemic tissues can result in stunting or plant death.

Teliospores are generally spherical, 8-11 µm in diameter, dark brown to black and spiny. Teliospores are diploid and germinate to give rise to four or more fusiform, haploid sporidia. Occasionally, sporidia can be diploid. Plants can be infected by hyphae that emerge directly from the teliospore as well as diploid sporidia, which arise from the fusion of compatible haploid sporidia. This is known to infect maize and Teosinte.

Teliospores are released from ruptured galls and overwinter in the soil, where they remain viable for several years. Under favorable environmental conditions, teliospores germinate to form haploid sporidia. Sporidia of compatible mating groups then fuse to form diploid mycelia that can infect maize plants. Economic loss as high as 20 % has been attributed to the disease when susceptible varieties have been grown under conditions favoring disease severity. Varieties that are resistant to this smut are widely available and offer the most cost-effective means of disease management. Avoiding mechanical damage to plants will reduce plant injury, which is the primary means of infection by the fungi. Controlling insect damage will also limit plant injury. Remove galls before they rupture will limit the spread of teliospores. A well-balanced fertilizer regime will reduce disease severity. Application of phosphorous reduces disease incidence.

### c. Gibberella ear rot: *Gibberella zeae*

It is also known as red rot. *Gibberella zeae* (Anamorph: *Fusarium graminearum*) produce mycotoxins that are harmful to humans and livestock. Ear infection is initially characterized by the presence of white mycelium on the ear tips that gradually moves towards the base of the ear. The mycelium turns a distinctive reddish-pink colour in infected kernels. Early infection can result in entire ears being colonized by reddish- pink mycelium, extensive kernel rotting and husks adhering tightly to ears. Bluish- black perithecia may also form on the husks.

Damage is caused by direct loss of yield due to ear rot and by the production of mycotoxins that are harmful to humans and livestock. Storing grain at high moisture levels and temperatures leads to storage rot and production of mycotoxins. Damage is most critical when mycotoxins are produced at levels that are toxic and unsuitable for human and livestock consumption. Consumption of infected maize can lead to diseases known as mycotoxicoses. Mycotoxins produced by this include deoxynivalenol, zearalenone and zearalenol. The Unites States Food and Drug Administration recommends that deoxynivalenol levels do not exceed 1 ppm for human consumption. Cultivate resistant varieties. Varieties with tight husks appear to be more vulnerable to this ear rot. Application of fungicide early in the season can limit ear infection. Management of insect pests (ear borers) will reduce infection of the ear through injury sites caused by insect feeding. Store grains at low moisture (below 15 %). Manage of infected crop debris.

### d. Fusarium ear rot: *Fusarium verticillioides*

It is caused by *Fusarium verticillioides* (Teleopmorph: *Gibberella fujikuroi*), *F. proliferatum* (teleomorph: *G. fujikuroi* mating population D) and *F. subglutinans* (teleomorph: *G. fujikuroi* mating population E). The causal pathogens produce mycotoxins that are harmful to humans and livestock. This pathogen also causes stalk and seedling blights of maize.

It is characterized by cottony mycelium growth that typically occurs on a few kernels or is limited to certain parts of the ear. Mycelium is generally white, pale pink or pale lavender. Infected kernels typically display white streaking (also known as 'starburst' symptoms) on the pericarp and often germinate on the cob. Typically, infection occurs close to ear tips and is commonly associated with damage and injury caused by ear borers. Under severe infestation, the entire ear appears withered and is characterized by mycelium growth between kernels. Storing grain at high moisture and temperatures levels leads to ear rot and production of mycotoxins. *Fusarium verticillioides* and *F. proliferatum* produce mycotoxins known as fumonisins.

» Maintain recommended plant population and follow recommended crop production practices including balanced use of fertilizers and tillage practices.

» Ensure early planting and timely harvesting of the crop.'

» Timely irrigation reduces drought stress.

» Adopt practices to minimize the insect and mechanized damage at harvesting. The damaged grains should be removed and destroyed.

» Use tolerant / resistant hybrids / composites for cultivation.

» For storage, dry and store the grains properly at 13 % or less grain moisture.

» Keep storage facilities clean.

### e. Penicillium ear rot: *Penicillium oxalicum*

It is commonly caused by *Penicillium oxalicum, P. chrysogenum, P. glaucum, P. cyclopium* and *P. funiculosum.* It is characterized by a distinct light blue-green powdery mold that grows between kernels and on the ear surface. Infected kernels are typically bleached and streaked. Infection of stored grain results in blue-green discolouration of the embryo. It is septate, hyaline hyphae up to 5μm in diameter. Conidia are borne on long conidiophores that typically branch in a 'broomlike' manner. Conidia are single- celled, round (resembling glass beads), up to 5 µm in diameter, usually green in colour and form in chains at the tips of conidiophores. It occurs on ears that have been damaged mechanically or by insect injury or infested with insect borers. The disease can also arise where maize is stored at high moisture levels, particularly if it was observed prior to harvest. Humidity above 80 % after grain fill leads to increased disease severity. Varieties with poor sheath coverage (exposed kernels) are more prone to infection. Conidia are predominantly wind and rain splash disseminated and infect ears through wounds. The ear is colonized and secondary sporulation occurs, giving rise to the characteristic green-blue colour of the mold. The fungi overwinter in the soil and on alternate hosts. Damage is caused by rotting of kernels leading to direct yield loss. Rotting of stored grain can lead to post-harvest losses. Contamination of grain by ear rots can also result in lower market prices. Some species are known to produce the mycotoxin ochratoxin, which is considered carcinogenic and acutely toxic to mammalian kidneys. Grow resistant varieties. Minimize mechanical damage to ears. Shelled maize should be stored below 13 % moisture while unshelled ears should be stored at below 18 % moisture. Store clean maize by eliminating infected ears to prevent storage rots. Control insect pests to reduce damage to ears.

### f. Stenocarpella ear rot: *Stenocarpella maydis*

It is caused by *Stenocarpella* species and commonly found in hot, humid maize growing regions. It can produce various mycotoxins that are harmful to birds. The pathogen is *Stenocarpella maydis* and *Stenocarpella macrospora*. Early infection of ears (just after flowering) results in development of husks that are bleached (straw coloured) and dry although the maize plant remains green and healthy. Ears are also shrunken, bleached and chaffed, with gray-white cottony mycelium growth between and over the kernels. Infected kernels are typically dull grey to brown. Symptoms usually first appear at the base of the ear, before the entire ear is colonized by the fungi.

Infected ears may be totally rotted and light in weight. Late in the season, abundant, black, miniature pycnidia form on kernel and ear tissues. Pycnidia are flask-shaped (spherical with circular, protruding papillate ostiole) and multicellular. Pycnidia are 150-450 µm in diameter, while the ostiole is typically 30-40 µm in diameter. Pycnidia contain pale brown, straight to slightly curved conidia with rounded ends. Conidia of *S. macrospora* are 0-3 septate measuring 7-12 x 44-82µm. Conidia of *S. maydis* are 1-2 septate and measure 5-8 x 15-34 µm. Occasionally, pycnidia contain colourless, narrow spores that measure 1-2 x 25-35µm. A sexual stage for these fungi has not been recorded. Wet weather immediately following silking increases disease severity. It can emerge as a problem post-harvest if grain is stored at high moisture content (above 20%). The fungi overwinter as pycnidia in infected maize stalk debris. During wet weather, conidia are produced and are rain splash disseminated to silks. The fungi then grow down the silks and infect the ear. Generally, ears are most vulnerable to infection in the three-week period following silking. *S. maydis* produces the mycotoxin diplodiatoxin and *S. macrospora* produces the mycotoxin diplodial, which are both harmful to birds. Livestock may refuse grain that is severely affected by Stenocarpella ear rot. Cultivate of resistant varieties. Crop rotation is effective at reducing disease severity. Destruction of infected crop debris (stalks, ears) following harvest will reduce the amount of overwintering disease inoculum. Store grains at below 15 % moisture content.

### g. Mycotoxin

Maize grain is a good substrate for mold infection and production of potentially dangerous mycotozn harmful to both human and animals. Fungi are associated with grain moldness but most common are *Aspergillus flavour, A parasiticus, Fusarium graminerarum, F. moniliforme, pencilliume sp.* and *Diplodia maydis*. These fungi produce different types of mycotozins, consumption of which can even lead to death in acute cases. Grain storage practices can be altered to decreases the likelihood of post- harvest mycotoxin development. Cultural practices that tend to expose plants to greater drought stress will lead to higher levels of aflatoxins. The fungal flora responsible for mycotoxin production is widely distributed in nature and is favoured by high temperature. Temperatures ranging from 26.7 to 37.8°C and relative humidity of 85 % (18 to 20 % moisture in the grain) are optimum for fungal growth and toxin production growth of these fungi does not occur below 12 to 13 % moisture in the grain. In order to minimize the level of mycotoxin in maize, the following practices should be followed. Use recommendation of crop production practices. Do early planting. Give timely irrigation to reduce drought stress. Minimize insect damage. Harvest is to be done early. Avoid kernal damage during harvest. Dry and store corn properly. Regular and critical inspection grain of fungi and the treated grains can be used only for lilvestock and poultry feed. Dispose off infected grains after screenings and do not feed infected grains to animals.

## iv. Viruses and mollicutes

### a. Corn stunt: *Spiroplasma kunkelii*

The pathogen is *Spiroplasma kunkelii.* Corn stunt is transmitted primarily by the cicadellid leafhoppers *Dalbulus maidis* (corn leafhopper) and by *D. elimatus* (Mexican corn leafhopper). Other less important *Dalbulus, Graminella, Exitianus, Stirellus* and *Euscelidius* leafhopper species may also transmit *S. kunkelii.* Symptoms include chlorosis of leaf margins, followed by reddening or purpling of older leaves. Chlorotic stripes often form at the base of younger leaves, which also turn purple at the tips. Chlorotic stripes may also enlarge to cover the entire leaf. At low temperature, chlorosis and purpling may be restricted to the base of upper leaves. Foliar symptoms appear close to tasseling. Infected plants are stunted due to shortening of internodes and numerous, barren ear shoots develop. Often, plants tiller and root excessively and tassels are sterile. Plants can be completely barren or suffer from poor seed set and reduced ear size. Under severe infection, plants can die prematurely. It is motile prokaryote that lacks true cell walls. Morphology ranges from helical filaments that measure 0.1 - 0.15 × 3-10 μm to spherical cells that are 0.3-0.8μm in diameter. *S. kunkelii* is visible under phase contrast microscopy of plant juice or abdominal smears from vectors. *S. kunkelii* is a facultative anaerobe and is phloem-limited. It is highly resistant to penicillin *in vitro*. *S. kunkelii* can be readily cultured on several media where optimal growth occurs at 30-32°C and with a pH of 7.2-7.4. The disease has also been recorded from elevations more than 2000 m above sea level. Identity of *S. kunkelii* can be confirmed using ELISA tests. Additionally, stomata of epidermal strips of infected leaves often have rounded guard cells, while subsidiary cells often fuse into adjoining epidermal cells. Infected leafhoppers are reported to survive in greater numbers than uninfected ones. Once acquired by the leafhopper following feeding on infected maize plants, *S. kunkelii* undergoes an incubation period of 2 to 3 weeks in the vector. Cultivate resistant varieties to leafhoppers will reduce the risk of infection.

Management of insect vectors during the season will reduce rates of transmission. Destruction of volunteer plants during the off-season by removing infected plants or discing will reduce potential sources of inoculum. Planting early will avoid disease severity.

### *b. Maize dwarf mosaic virus*

It is caused by various strains of the Sugarcane mosaic virus *Potyvirus*. These viruses are transmitted predominantly by several genera of aphids, but can also be transmitted mechanically and through seed. Symptoms vary widely depending on host genotype, time of infection and viral strain. Generally, infected plants develop distinct chlorotic mosaics, mottles or streaks on green tissues (typically observed on young leaves). Infected plants are characterized by stunting and shortening of upper internodes. Ear development can be arrested, leading to incomplete grain filling and direct yield loss. Plants infected early may produce nubbins or can be totally barren. It is single stranded RNA potyviruses in *Potyviridae*. They are flexuous and rod-shaped, measuring 12 x 750 nm. Antiserum kits are commercially available to confirm virus identity. Maize dwarf mosaic and sugarcane mosaic virus occur where aphid vectors are prevalent and where alternate hosts are cultivated. Use of insecticides, biological control agents and cultural practices (management of aphid refuges) that reduce aphid populations. Excessive use of nitrogen fertilizer favours aphid reproduction. Excess application of nitrogen above recommended levels should be avoided.

Removal of volunteer plants and management of alternate hosts, such as Johnson grass and graminaceous weed species, will reduce sources of inoculum. Use of certified, virus- free seed will limit seed-borne transmission of this virus.

*c. **Maize streak virus***

It is transmitted by species of *Cicadulina* leafhoppers. Disease symptoms first become apparent about a week following infection. Small (0.1 to 2 mm in diameter), chlorotic, circular spots arise on the basal sections of young leaves that emerge after infection. Spots contrast sharply with surrounding healthy green leaf tissue. Numbers of spots increase with plant growth and can coalesce.

In susceptible varieties, spots enlarge parallel to leaf veins, forming distinctive elongated, chlorotic streaks distributed evenly over the leaf surface. Infection with by this virus commonly results in crop stunting and barren plants. This virus can be detected using ELISA test kits. It is single stranded and non-enveloped. Under the electron microscope, characteristic geminate isometric particles (doublets) are visible. It has quasi-icosahedral particles that measure 18 × 30 nm. Development period of *Cicadulina* species ranges from 22 to 45 days, depending on prevailing climatic conditions. Temperatures between 20 - 35°C are considered optimum for development. It is known to infect wheat, oat, sugarcane, millet, rice, barley, rye and sorghum. Vectors acquire the virus following feeding on an infected plant. A latent period of 6 to 12 hours is required before the vector can transmit the virus to another host plant. The virus does not propagate in the vector, although vectors transmit the virus for life even after molting. The virus overwinters in a wide range of graminaceous grasses and cereal hosts. Cultivate resistant varieties. Control of vector populations through insecticide.

## References

Adejumo, T.O., U. Hettwer and P. Karlovsky (2007). Occurrence of *Fusarium* species and trichothecenes in Nigerian maize. *International Journal of Food Microbiology,* 116: 350-357.

Adenle, V.O. and K.F. Cardwell (2000). Seed transmission of *Peronosclerospora sorghi*, causal agent of maize downy mildew in Nigeria. *Plant Pathology*, 49:628-635.

Agrios, G.N. (1988). *Plant Pathology.* Third Edition. Academic Press, Inc., San Diego.

Ahmad, Y., A. Hameed and M.Asim (1996). Effect of soil solarization on corn stalk rot. *Plant and Soil*, 179: 17-24.

Al-Sohaily, I.A., C.J. Mankin and G. Semeniuk (1963). Physiologic specialization of *Sphacelotheca reiliana* to sorghum and corn. *Phytopathology*, 53: 723- 726.

Beckman, P.M. and G.A. Payne. 1983. Cultural techniques and conditions influencing growth and sporulation of *Cercospora zeae-maydis* and lesion development in corn. *Phytopathology,* 73: 286-289.

Boude, M.R. (1982). Epidemiology of downy mildew disease of maize, sorghum and pearl millet. *Tropical Pest Management*, 28: 220-23.

Bradbury, J.F. (1991). *Spiroplasma kunkelii.* IMI Descriptions of Fungi and Bacteria (No. 105) Sheet 1047. Surrey, UK: CABI Bioscience.

Brenna, W and J.E.F. Figueiredo. 2005. Biological control of *Stenocarpella maydis* in maize seed with antagonistic *Streptomyces* sp. isolates. *Journal of Phytopathology.* 153: 623-626.

Byrens, K.J., J.K. Pataky and D.G. White (1989). Relationships between yield of three maize hybrids and severity of southern leaf blight caused by race of *Bipolaris maydis. Plant Disease,* 73: 834-840.

CAB (1970), Distribution maps of plant diseases. Edition 5. Map 237. Wallingford, UK: CAB International

CAB (1974), *Sugarcane mosaic virus*. Distribution Maps of Plant Diseases. Edition 4 (October), Map 229. Wallingford, UK: CAB International.

CAB (1977), *Gibberella fujikuroi.* Distribution Maps of Plant Diseases. Edition 5 (October), Map 102. Wallingford, UK: CAB International.

CABI (1978), Distribution maps of Plant Diseases. Edition 4, Map 279. Wallingford, UK: CAB International.

CABI (1985). *Macrophomina phaseolina.* Distribution maps of Plant Diseases. Edition 1 (August), Map 566. Wallingford, UK: CAB International.

CABI (1986). *Ustilago zeae.* Distribution Maps of Plant Diseases. Edition 6 (October), Map 93. Wallingford, UK: CAB International.

CABI (1997). *Maize streak gemini virus*. Distribution maps of plant disease. Edition 1 (December), Map 739. Wallingford, UK: CAB International.

CABI (1998). *Gibberella zeae.* Distribution Maps of Plant Diseases. Edition 1 (October), Map 763. Wallingford, UK: CAB International.

CABI (2002). CAB International. 2002. Crop Protection Compendium. Wallingford, UK: CAB International.

CABI (2003). Distribution Maps of Plant Diseases. Edition 6. Map 346. Wallingford, UK: CAB International

Carpane, P., I.G. Laguna, E. Virla, S. Paradell, L. Murua and M.P. Gimenez-Pecci (2006). Experimental transmission of corn stunt spiroplasma present in different regions of Argentina. *Maydica,* 51:461-8.

Carson, M.L. (1999). "Helminthosporium" Leaf Spots and Blights. In Donald G. White (ed), *Compendium of Corn Diseases.* St. Paul, Minnesota: The American Phytopathology Society. Pp. 15-16.

Carson, M.L. (2005). Yield loss potential of *Phaeospharia* leaf spot of maize caused by *Phaeosphaeria maydis* in the United States. *Plant Disease,* 89: 986-988.

Carson, M.L. and M.M. Goodman (2006). Pathogenecity, aggressiveness and virulence of three species of *Cercospora* associated with gray leaf spot of maize. *Maydica,* 51: 89-92.

Casa, R.T., L. Zambolim and E.M. Reis (1998). Transmission and control of *Diplodia* in maize seeds. *Fitopatologia Brasileira,* 23: 436-441.

Casela, C.R. (1998). The Phaeosphaeria leaf spot. In: C. Casela, B. Renfro and F. Krattiger (eds.), *Diagnosing Maize Diseases in Latin America.* ISAAA Briefs No. 9. Ithaca and Brasilia: ISAAA /EMBRAPA. Pp. 15-17.

Chawla, G. and A. N. Srivastava (1995). Response of certain maize cultivars to *Heterodera sorghi. Indian J. Nematology,* 25: 28-29.

Chen, T.A. and C.H. Liao (1975). Corn stunt spiroplasma: Isolation, cultivation and proof of pathogenecity. *Science,* 188:1015-1017.

Chitwood, B. G. and M.C. Toung (1960). *Meloidogyne* from Taiwan and New Delhi. Phytopathology, 50: 631-632.

CIMMYT (2004). *Maize Disease: A guide for field identification.* Mexico, D.F. Claflin, L.E. (1999). Mollicutes. In Donald G. White (ed), *Compendium of Corn Diseases.* St. Paul, Minnesota: The American Phytopathology Society. Pp. 8-9.

Contreras, M.R. (1978). Soluble solids, pith senescence, and stalk rot development in tropical maize population. Dissertation Abstracts International, 38: 3980.

Crous, P.W., J.Z. Groenewald, M. Groenewald, P. Caldwell, U. Braun and T.C. Harrington (2006). Species of *Cercospora* associated with gray leaf spot of maize. *Studies in Mycology,* 55: 189-197.

Cullen, D., R.W. Caldwell and E.B. Smalley (1983). Susceptibility of maize to *Gibberella zeae* ear rot: relationship to host genotype, pathogen virulence, and zearalenone contamination. *Plant Disease,* 67: 89-91.

De Nazareno, N.R.X., P.E. Lipps and L.V. Madden (1993). Effect of levels of corn residue on epidemiology of gray leaf spot in Ohio. *Plant Disease,* 77: 67-70.

Deadman, M., H. Al-Hasani and A. Al-Sadi (2006). Solarization and biofumigation reduce *Pythium aphanidermatum* induced damping-off and enhance vegetative growth of greenhouse cucumber in Oman. *Journal of Plant Pathology,* 88: 335-357.

Do Amaral, A.L., F.K. Dal Soglio, M.L. De Carli and J.F.B. Neto (2005). Pathogenic fungi causing symptoms similar to *Phaeosphaeria* leaf spot of maize in Brazil. *Plant Disease,* 89: 44-49.

Dodd, J.LO. (1980). The role of plant stresses in development of corn stalk rots. *Plant Disease,* 64: 533-537.

Donald, G.W. (1999). Compendium of corn diseases. St. Paul, Minnesota: The American Phytopathology Society. Pp. 47-49.

Durojaiye, J.A., R. Ikotun and J.M. Fajemisin (1987). The survival of propagules of *Macrophomina phaseoli* and *Diplodia maydis* in Nigerian soils. *Journal of Basic Microbiology,* 27: 67-73.

Fininsa, C. (1996). Effect of bean and maize intercropping on bean common bacterial blight and rust diseases. *Int. J. Pest Manage.*, 42: 51-54.

Flett, B.C., N.W. McLaren and F.C. Wehner (2001). Incidence of *Stenocarpella maydis* ear rot of corn under crop rotation systems. *Plant Disease,* 85: 92-94.

Frederiksen, R.A. and B.L. Renfro (1977). Global status of maize downy mildew. *Annual Review of Phytopathology,* 15: 249-275.

Hernandez, J.R., M. Yasem de Romero, C.G. Diaz and J.C. Ramallo (2002). First report of *Puccinia polysora* on corn in Argentina. *Plant Disease*, 86:187.

Huber, D.M. and R. D. Watson (1974). Nitrogen form and plant disease. *Annual Review of Phytopathology,* 12:139-165.

Jeger, M.J., E. Gilijamse., C.H. Bock and H.D. Frinking (1998). The epidemiology, variability and control of the downy mildews of pearl millet and sorghum, with particular reference to Africa. *Plant Pathology,* 47: 544-69.

Karlen, D. L., D. C. Arny, and L. M. Walsh (1973). Incidence of chocolate spot (Pseudomonas syringae), northern corn leaf blight (Helminthosporium turcicum) and lodging of corn as infl uenced by soil fertility. *Communications in Soil Science and Plant Analysis*, 4: 359 - 368.

Kozakiewicz, Z. (1992). *Penicillium oxalicum*. IMI Descriptions of Fungi and Bacteria, 1992 (no. 111) Sheet 1107. Surrey, UK: CABI Bioscience.

Kumar, R. and I.S. Singh (2002). Inheritance of resistance to banded leaf and sheath blight (*Rhizoctonia solani* f.sp. *sasakii*) of maize (*Zea mays* L.). In: G. Srinivasan, P.H. Zaidi, B.M. Prasanna, F. Gonzalez, and K. Lesnick (eds.), Proceedings of the 8th Asian Regional Maize Workshop on "New Technologies for the New Millennium", Bangkok, Thailand. August 5-8, Mexico, D.F.: CIMMYT. Pp. 356-65.

Kyetere, D.T., R. Ming, M.D. McMullen, R.C. Pratt, J. Brewbaker and T. Musket (1999). Genetic analysis of tolerance to maize streak virus in maize. *Genome*, 42: 20 - 26.

Latterell, F.M. and A.E. Rossi (1983). Gray leaf spot of corn: A disease on the move. *Plant Disease Reporter,* 67: 842-847.

Lipps, P.E. and D. Mills (2002). Northern corn leaf blight. Extension Fact Sheet AC-20-02. Columbus: Ohio State University, USA.

Lipps, P.E. and D.R. Mills (2007). *Diplodia* ear rot of corn. Ohio State University Extension Fact Sheet AC-0046-01.

Louie, R. (1999). Diseases Caused by Viruses. In: Donald G. White (ed), *Compendium of Corn Diseases.* St. Paul, Minnesota: The American Phytopathology Society. Pp. 49-55.

Martinez, C., C. Roux, A. Jauneau, and R. Dargent (2002). The biological cycle of *Sporisorium reilianum* f. sp. *zeae*: an overview using microscopy. *Mycologia,* 94: 505 - 514.

Meena Shekhar and Sangit (2005). Integrated management system of micotoxin in maize. Farmers Forum, 5(10): 22 - 24.

Meena Shekhar, Sangit Kumar, Shrvan Kumar and K.S.Hooda. (2014). Disease management in maize. Indian Farming, 64 (4): 22 – 28.

Michaelson, M.E. (1957). Factors affecting development of stalk rot of corn caused by *Diplodia zeae* and *Gibberella zeae*. *Phytopathology*, 47: 499 - 503.

Miller, S.A., P.E. Lipps and R.C. Randall (2007). Common Smut of Corn. Ohio State University Service. Extension Factsheet HYG-3119-96.

Naidoo, G., A.M. Forbes, C. Paul, D.G. White and T.R. Rocheford (2002). Resistance of Aspergillus Ear Rot and Aflatoxin Accumulation in Maize F1 Hybrids. Crop Science, 42: 360 - 364.

Ngiwe, C. (1974). Effect of *Diplodia zeae* and *B-Phomopsis* on the germination of seeds of maize (*Zea mays*). *Plant Disease Reporter*, 58: 414 - 415.

Nyvall, R.F. (1999). Field crop diseases. Iowa State University Press, Ames. P: 247 - 305.

Olatinwo, R., K. Cardwell, A. Menkir, M. Deadman and A. Julian (2004). Inheritance of Resistance to Stenocarpella (Earle) Ear Rot of Maize in the Mid-Altitude Zone of Nigeria. *European Journal of Plant Pathology,* 105: 535-543.

Olatinwo, R.O., K.F. Cardwell, M.L. Deadman and A.M. Julian (1999). Epidemiology of *Stenocarpella macrospore* (Earle) Sutton on maize in the midaltitude zone of Nigeria. *Journal of Phytopathology,* 146: 347 - 352.

Onions, A.H.S. (1966). Aspergillus flavus. In: IMI Descriptions of Fungi and Bacteria, (No.10) Sheet 91, Surrey, UK.

Oteifa, B. A. and D. M.Elgindi (1982). Relative susceptibility of certain commercially important cultivars to existing biotypes of *Meloidogyne incognita* and *Meloidogyne javanica,* Nile delta, Egypt'. Proceedings of the Third Research and Planning Conference on Root-knot Nematodes, *Meloidogyne* spp. Region VII. Coimbra, Portugal, P: 157-159.

Pal, D. and S.A.K.M. Kaiser (2001). Effect of agronomic practices on Maydis leaf blight

disease of maize. *Journal of Mycopathological Research,* 39: 77-82.

Pant, S.K., Pramodkumar and V.S. Chauhan (2001). Effect of turcicum leaf blight on photosynthesis in maize. *Indian Phytopathology*, 54: 251-252.

Pataky, J.K. (1999). Common rust. In: Donald G. White (ed.), *Compendium of Corn Diseases.* St. Paul, MN: The American Phytopathology Society. Pp. 35-38.

Payak, M.M. and B.L.Renfro (1968). Combating maize disease. *Indian Farmer Digest*, 1: 53-58.

Payak, M.M. and R.C. Sharma (1983). Disease rating scales in maize in India. In: Techniques of Scoring for Resistance to Important Diseases of Maize. Indian Agricultural Research Institute, New Delhi, pp. 1-4.

Payak, M.M. and R.C. Sharma (1995). Maize disease and approaches to their management in India. *Trop. Pest Management*, 3(1): 302 - 303.

Pemezny, K. and T. Kucharek (2003). Rust diseases of several legumes and corn in Florida. Belle Glade, FL: IFAS Extension, University of Florida.

Punja, Z.K. and R. Yip (2003). Biological control of damping-off and root rot caused by *Pythium aphanidermatum* on greenhouse cucumbers. *Canadian Journal of Plant Pathology,* 25: 411-17.

Saxena, S.C. (2002). Bio-intensive integrated disease management of banded leaf and sheath blight of maize. In: Proc. of 8th Asian Regional Maize Workshop: New Technologies for the New Millennium, Bangkok, Thailand: August 5-8, 2002. Mexico, D.F.: CIMMYT. Pp. 380-390.

Shaner, G.E. and D.H. Scott (1998). Stalk rots of corn. Purdue University Cooperative Extension Service. Extension Sheet BP-59.

Sharma, R.C., S.K. Vasal, F. Gonzalez, B.K. Batsa and N.N. Singh (2002). Redressal of banded leaf and sheath blight of maize through breeding, chemical and biocontrol agents. In: Proc. of 8th Asian Regional Maize Workshop: New Technologies for the New Millennium, Bangkok, Thailand: August 5-8, 2002. Mexico, D.F.: CIMMYT. Pp. 391-97.

Shukla, D.D., C.W. Ward and A.A. Brunt (1994). *The Potyviridae.* Wallingford, UK: CAB International.

Singh, B., S.C. Saxena and A. Joshi (2005). Integrated management of banded leaf and sheath blight of maize. *Indian Farmers' Digest*, 38 (10): 9 -10.

Singh, B., S.C.Saxena, K.Singh and K.Vishunavar (2005). Seed borne disease of maize and their management. *Indian Farmers' Digest*, 38 (11): 26 -27.

Singh, N.B., P.P. Singh and K.P.P. Nair (1986). Effect of bacterial activity and ptoductivity of associated maize. *Expt. Agric.*, 22: 339 – 344.

Smith, D.R. and B.L. Renfro. (1999). Downy Mildews in *Compendium of Corn Diseases*. Edited by Donald G. White. St. Paul: The American Phytopathology Society. Pp 25-32.

Sneh, B., L. Burpee and A. Ogoshi (1991). *Identification of Rhizoctonia species.* St. Paul, MN: APS Press.

Tenuta, A. 2006. Identifying Corn Ear Molds. Ontario Ministry of Agriculture, Food and Rural Affairs.

Thakur, R.P. and K. Mathur (2002). Downy mildews of India. *Crop Protection*, 21: 333-345.

Trenbath, B.R. (1993). Intercropping for the management of pests and diseaes. *Field Crops Res.*, 34: 381-405.

Tsai, J.H. (1987). Bionomics of *Dalbulus maidis* (DeLong and Wolcott). A vector of mollicutes and virus (Homoptera: *Cicadellidae*). In K. Maramorosch and S.P. Raychaudhuri (eds.), *Mycoplasma diseases of crops: basic and applied aspects.* New York: Springer-Verlag. Pp: 209-221.

Tsai, J.H. and J.W. Miller (1995). Corn stunt spiroplasma. Plant Pathology Circular No. 373. Florida Agriculture and Consumer Services, Division of Plant Industry.

Virla, E.G., C.G. Diaz, P. Carpane, I.G. Laguna, J. Ramallo, L. Geronimo Gomez and M.P. Gimenez-Pecci (2004). Preliminary evaluation about corn yield losses caused by 'Corn stunt spiroplasm' (CSS) in Tucuman,Argentina. *Boletin de Sanidad Vegetal – Plagas,* 30: 403-413.

Wallin, J.R. and D.V. Loonan (1977). Temperature and humidity associated with sporulation of *Helminthosporium maydis* race T. *Phytophathology,* 67: 1370 - 1372.

Wang, J., M. Levy and L.D. Dunkle (1998). Sibling species of *Cercospora* associated with gray leaf spot of maize. *Phytopathology,* 88: 1269 - 1275.

Ward, J.M.J., E.L. Stromberg, D.C. Nowell and F.W. Nutter, Jr. (1999). Gray leaf spot: A disease of global importance in maize production. *Plant Disease,* 83: 884-895.

Ward, J.M.J., M.D. Laing and F.H.J. Rijkenberg (1997). Frequency and timing of fungicide applications for the control of gray leaf spot of maize. *Plant Disease*, 81: 41-48.

24

# Integrated Nematode Management

Several biotic and abiotic factors contribute in limiting the yield potential of maize crop qualitatively as well as quantitatively. Among them are disease, insect pests and nematodes. More than 100 species of plant parasitic nematodes have been found associated with this crop in different parts of the world which account for considerable losses. Maize cyst nematode *(Heterodera zeae),* lesion nematodes *(Pratylenchus spp.),* root knot nematode (*Meloidogyne sp.*), sorghum cyst nematode (*Heterodera sorghi*) and stunt nematode (*Tylenchorhynchas vulgaris*) cause unprofitable growth of the crop, thus resulting in considerable loss in yield (Srivastava and Sethi, 1986a). Among them, plant parasitic nematodes are responsible to cause 10.2 % losses on maize (Sasser and Freckman, 1987). Plant parasitic nematodes viz., cyst nematodes, lesion nematodes, root knot nematodes, stunt nematode and spiral nematode (Helicotylenchus spp.) has been found to be associated with maize (Patel et al., 2000). These nematodes apart from causing losses by themselves interact with other disease causing agents and adversely affect the quality and quantity of maize production. However, the maize cyst nematode, is considered as one of the most important nematode pests of maize in India and abroad (Singh and Rathore, 2001).

## a. Maize cyst nematode

Maize cyst nematode *(Heterodera zeae)* is considered as the important nematode pest of maize in India (Srivastava *et al,* 1995). The nematode, first reported from Udaipur (Rajasthan) by Koshy *et al* (1970) on maize is now known to be widely spread in maize growing areas of northern, central and western parts of India, especially Jammu and Kashmir, Himachal

Pradesh, Punjab, Haryana, Delhi, Rajasthan, Uttar Pradesh, Bihar, Gujarat, Madhya Pradesh and Maharashtra. Darekar *et al,* 1981; Sharma *et al,* 1984; Bajaj and Bhatti, 1984; Makadia *et al,* 1988). Its presence in high altitude areas in the States of Jammu and Kashmir (Srivastava and Kaushal, 1986; Khan *et al,* 1989) and Himachal Pradesh (Srivastava and Kaushal, 1991) has also been reported. Besides its widespread occurrence in India, the nematode has also been detected in Egypt, Pakistan and the USA (Aboul-Eid and Ghorab, 1981; Maqbool, 1981; Sardanelli *et al,* 1981) and in Thailand (Chinnasri *et al,* 1995). The hosts of *Heterodera zeae* are mainly confined to the Graminae family. Besides, *Zea mays* (type host), *Avena sativa, Echinochloa colona, Hordeum vulgare, Panicum* spp., *Setaria italica* and *Triticum aestivum* are also reported as hosts (Srivastava and Swaruo, 1975). Some more graminaceous plants viz., *Coix lachryma, Eleucine corocona, Oryza sativa, Secale cereale, Zea mexicana* (Sharma and Swarup, 1984) and *Vetiveria zizanioides* (Lal and Mathur, 1982) were also added to the host lists subsequently. Certain weeds and grasses such as *Achyranthes aspera, Cyperus rotundus, Dicera muricata, Eclipta prostrata, Parthenium hysterophorus, Trianthema pertulacestrum* and *Urochloa panicoidea* were found to harbour the cyst populations though in small numbers in maize fields (Parihar *et al., 1991).*

It is a multivoltine species and reproduces parthenogetically. Males are generally rare and very sporadic. Temperature plays an important role in the biology of this nematode (Srivastava, 1980). A temperature of 25°C was found most favourable for juvenile emergence (91 % emergence) from cyst (Srivastava and Sethi, 1985b). A temperature range of 25 to 30°C was more favourable for embryonic and post-embryonic development of nematode (Shahina and Maqbool, 1989) but th optimum temperature for reproduction is appeared to be 33°C (Hutzell and Krusberg, 1990). Its development is greatly influenced by soil temperature. The time taken for completion of one generation ranged between 17 and 22 days depending on maize cultivars and soil temperature (27 - 38°C). It may complete four to five generations during one maize cropping season under Indian conditions (Srivastava and Sethi, 1985a). Soil type is another important environmental factor which influences the reproduction of the nematode which prefers moderately light soil (sandy loam) for its reproduction and any addition of clay proportion to soil mixtures results in a significant decline in cyst production (Srivastava and Sethi, 1984a).

The pathogenicity of maize cyst nematode has been demonstrated on maize in India (Srivastava, 1980) and in the USA (Krusberg, 1988), respectively. Srivastava and Sethi (1984b) found that plant growth reductions are directly correlated with initial nematode population densities. Even though the above ground symptoms are not very diagnostic, the crop infected exhibited a poor, patchy and unthrifty growth with stunting and pale yellow foliage. The infected roots appear bushy and poorly developed. The diseases plants tassel earlier and bear smaller cobs with relatively fewer grains. During kharif season, it could cause a loss of about 11.0 to 21.6 % in maize cultivars Ganga - 5 and Deccan - 103, respectively, in sandy loam soils (Anonymous, 1987). Application of nitrogenous fertilizers mitigated the crop damage caused by this nematode and enhanced the plant growth and increased crop yield several fold (Anonymous, 1988 and 1989).

Penetration by second stage juveniles of this nematode into maize roots started within 12 hours of inoculation, reaching maximum by the fifth day (Srivastava and Sethi, 1985a). Even though penetration was noticed everywhere in the developing roots, the merismatic and elongation zones were more preferred sites for penetration (Kaul and Sethi, 1982a). Following root penetration by second stage juveniles, the endodermis wall became thick, the cytoplasm of pericycle and phloem cells became dense and granular and vascular bundles were pushed aside towards the endodermis resulting in the formation of three to four big and wide giant cells. The cortical cells were ruptured as third and fourth stages developed and cavities were formed in which the nematodes were lodged (Srivastava and Sethi, 1984c). Poly-nuclear condition and irregular thickening of the giant cells were also produced by this nematode (Mishra *et al.,* 1985). In infested field plots, under maize - cowpea-wheat rotation sequences, the nematode attained its high peak population in October / November, which coincided with maturity of the maize crop suggesting host specificity (Srivastava and Sethi, 1986b). Very little population of J2s was observed in the soil during December to June. In Maryland (USA), eggs and / or J2 inside cysts have been recorded to survive in the field during winter with no detectable mortality at different soil depths (Krusberg and Sardanelli, 1989). The entire contents of the cyst do not empty out even in one year. Thus, this survival strategy of the nematode, coupled with the number of generations during the crop season (four to five generations), gives an indication of the damaging potential of this nematode. Moreover, being multivoltine, it may obtain high reproduction potential.

Physiological variations have been recorded with this nematode population obtained from different geographical areas or within regions of a country. Three biotypes of were distinguished in Egypt on the basis of different reproductive potential on some maize cultivars (Kheir *et al.,* 1989). Ringer *et al.* (1987) indicated that populations from Egypt, India and the USA may differ in their abilities to reproduce on certain plants. Bajaj and Gupta (1994) also identified three host races in Haryana population of this nematode using vetiver and maize as host differentials. The three host races were multiplying on maize only (Ambala population), multiplying on vetiver only (Sonepat population) and multiplying on both maize and vetiver (Hisar population). Association of more than one population of ecto/endo-parasitic nematodes, root knot and lesion or soil fungi with this nematode have been reported under field conditions. Often the disease symptoms have been reported to be aggravated in such complexes (Kaul and Sethi, 1982). In a wilt disease complex *(Cephalosporium maydis),* the presence of this nematode has been reported to cause enhanced wilting of the plants (Singh and Siradhana, 1988). Similarly in prior inoculations of fungus, this nematode may contribute towards root damage (Anonymous, 1993). Prior establishment of any of the nematode species *Heterodera zeae, Meloidogyne incognita* and *Tylenchorhynchus vulgaris* either singly or in combination, significantly reduced invasiveness of the other species. Penetration of juveniles of both cyst and root knot nematodes was more whenever they were inoculated earlier (Kaul and Sethi, 1982a). Individual effects of various nematode species occurring in mixed inoculations were not only modified by nematode species involved or host plant or prior establishment of any one species, but also by the initial densities of nematodes involved in the interaction (Kaul and Sethi, 1982b).

**Table 1. Integrated management of maize cyst nematode on sweet corn through bio-agents and botanicals**

| Treatments | Plant growth characters | | Nematode parameters | | |
|---|---|---|---|---|---|
| | Shoot weight (g) | Root weight (g) | Cyst / 100 cc soil | Cyst/ plant | Final larvae population/ 100 cc soil |
| Paecilomyces lilacinus 2 % w/w + Karanj leaves powder 2 g/plant | 36.00 (41.18) | 25.00 (40.85) | 11.00 (52.17) | 11.50 (52.58) | 240.00 (60.00) |
| Paecilomyces lilacinus 2 % w/w + Neem leaves powder 2 g/plant | 38.50 (50.98) | 27.00 (52.11) | 8.50 (63.04) | 8.50 (64.95) | 200.00 (66.67) |
| Pochonia chlaymydosporia 2 % w/w + Neem leaves powder 2 g/plant | 37.00 (45.10) | 25.75 (45.07) | 10.50 (54.35) | 11.25 (53.61) | 210.00 (65.00) |
| Acephate 75 SP 1 % w/w + Phorate 10 G 1 kg /ha | 40.25 (57.84) | 29.00 (63.38) | 6.00 (73.91) | 6.00 (75.26) | 180.00 (70.00) |
| Untreated check | 25.50 | 17.75 | 23.00 | 24.25 | 600.00 |
| CD 5 % | 2.05 | 1.48 | 1.65 | 1.62 | 28.61 |

(Baheti et al., 2017). Initial nematode population: 600 larvae /100 cc soil.

At Udaipur (Rajasthan), *Paecilomyces lilacinus* at 2 % w/w coupled with neem leaves powder at 2 g/plant was found most effective for the management of maize cyst nematode followed by *Pochonia chlamydosporia* + neem leaves powder and *Paecilomyces lilacinus* + karanj leaves powder (Baheti et al., 2017).

Seed dressing with different nematicides both under pot and field conditions, emergence of juveniles of *Heterodera zeae* was increased while their penetration into maize roots was inhibited to a considerable extent. Aldicarb used as 0.75 and 1.5 % and *Fensulfothion* @ 0.25 and 0.5 % were found to be more effective in reducing populations of this nematode (Srivastava, 1980). Cysts when exposed to five different concentrations (62.5 to 500 µg/ml) of Aldicarb, Carbofuran, Oxamyl and Phenamiphos for or an increase in the exposure time from three to 15 days resulted in decreased emergence of larvae from cysts (Kaul and Sethi, 1987a). However, Phenamiphos was proved to be the most effective nematicide in reducing juvenile emergence *in vitro*. Root invasions were least when juveniles of *Heterodera zeae* exposed to Carbofuran @ 10 µg / ml concentrations for six hours (Sethi and Kaul, 1987). Foliar spray of maize plants with Phenamipho @ 250 or 500 µg/ml three days prior to nematode inoculation was found more effective than post inoculation sprays (Kaul and Sethi, 1987b). Pre-sowing soil application of Phenamiphos resulted in higher reduction in cyst population of this nematode (Kaul and Sethi, 1988a, b). Carbosulfan (25 ST) used as seed treatment at the rate of 2 - 3 % w/w and Sebuphos (3G) applied as soil application at the rate of 2 kg a.i./ha have been found to be very effective in increasing crop yield and reducing nematode population (Srivastava and Sethi, 1989a). The carbosulfan seed treatment accompanied with Carbosulfan foliar spray resulted in additional control of the nematode (Srivastava and Sethi, 1989b; Srivastava and Jagan Lal, 1993). The potential of carbosulfan seed treatment proved to be a very effective approach in reducing the total requirement as also the cost of the chemicals against this nematode on maize.

Crop rotation is an old but effective and widely used management practice for reducing nematode populations in the soil. There is good scope for adopting crop rotation as a control for this nematode as host range of this nematode is restricted to graminaceous plants only. Since this nematode is host specific (Srivastava and Sethi, 1986b), monoculturing of maize should be avoided as it is likely that under continuous maize cropping system, nematode populations may increase considerably, ultimately resulting in significant yield loss. Two year rotation with non-host crops (pulses and oilseeds) can be fruitful as it would bring down the nematode population below economic threshold levels. Summer fallowing and keeping the fields free from weeds can also help in checking the survival and perpetuation of this nematode population in off seasons. Two to three deep ploughings during April/May at 10 - 15 day intervals would also reduce the nematode population to a considerable extent.

The use of nematode resistant and tolerant varieties is a most effective and economical management approach. Cultivars Ageti-76 and Karnal - 1 have been found to be moderately resistant against this nematode in Haryana (Kali Ram, 1989) and varieties such as Sharad

White, Gauhar Azam and Composite - 15 are moderately resistant (Shahina *et al.,* 1989). The use of oil cakes and other organic soil amendments have also been proved effective in managing populations of this nematode. Combination of mustard cake and tobacco dust at the rate of 2.5 q/ha is as efficient as Carbofuran soil treatment at the rate of 2 kg a.i./ha in terms of increased crop yield (Anonymous, 1987). A few nematophagous fungi such as *Cataneria, Verticillium* and *Gleocladium* have been isolated from the cyst of this nematode. Also, *Arthrobotrytis coincides* and *Monacrosporium salinum* have shown some potential to kill second stage juveniles of this nematode *in vitro* (Anon, 1987), but their real potential as biocontrol agents is still to be exploited and explored for the management of this nematode.

### b. Root lesion nematodes

Root lesion nematodes (*Pratylenchus spp.)* are cosmopolitan in maize fields and consistently found associated with maize crop wherever it is grown. Several species of lesion nematodes are reported to be associated with poor growth of maize. Though *Pratylenchus brachyurus, Pratylenchus zeae* and *Pratylenchus penetrans* are the most commonly encountered species in subtropical and tropical regions. *Pratylenchus zeae, Pratylenchus thornei* and *Pratylenchus delattrei* are considered important on maize in India. This nematode is the most frequently encountered species in maize field, yet the above ground disease symptoms produced by nematodes are not highly specific. Poor and stunted plant growth, reduced root and shoot weights and leaf chlorosis are usually associated with lesions nematodes damage. Nematode damage to the fibrous root system can result in extensive destruction of cortical parenchyma, severe root pruning and proliferation of lateral roots. Mechanical breakdown of cells and necrosis of stellar and cortical tissues resulting in formation of cavities are also attributed to nematode damage. The presence of small brown to black lesions on the root surface is the most important symptoms produced by lesion nematodes. Under green gram - maize - chickpea sequence, *Pratylenchus thornei* populations decreased on green gram and increased on maize and chickpea (Walia, 1982).

Temperature, soil type, moisture and tillage operations are important environmental factors which greatly affect the development and reproduction of the nematode species as well as disease development. Higher temperature (30°C) is mostly preferred by *Pratylenchus zeae* for its reproduction while *Pratylenchus brachyurus* and *Pratylenchus penetrans* favour an optimum temperature of 20°C for their penetration and development. Soil type and tillage operations have also been recorded to affect the lesion nematode population dynamics. Most *Pratylenchus* species thrive well in a wide range of soil types, but Joganathan and Sivakumar (1975 and 1976) reported higher population densities of *Pratylenchus delattrei* in black sandy clay loam soil and brown sandy loam than in any other soil type. Moisture is yet another important factor that affects the development of species of *Pratylenchus.* Generally, the damage to the maize plants increases with the decreasing moisture levels. Egunjobi (1974) found pathogenicity of *Pratylenchus brachyurus* on maize increased during the rainy season.

Assessment of exact estimates of losses due to lesion nematodes has not been made possible under field conditions due to presence of mixed populations of the nematode in the field. Also, precise evaluation of losses in maize by lesion nematode is hampered by secondary infection of nematode lesions by fungi and bacteria (Egunjobi, 1974). But significant reductions in plant growth following inoculations of *Pratylenchus zeae* and *Pratylenchus thornei* obtained under greenhouses conditions. Yield increase of 33 to 128 % has been obtained following the application of nematicides (Walters, 1979). Treatment with nematicides increased crop yield by 10 to 54 % (Bergeson, 1978; Norton *et al.,* 1978; Lordello *et al.,* 1983). Similarly, in Nigeria, *Pratylenchus brachyurus* has been reported to be responsible for 28.5 % yield reduction. The reduction in yield was correlated with a 50 % increase in nematode density (Egunjobi, 1974). The population of lesion nematodes may increase considerably under continuous maize cropping, ultimately resulting in significant yield losses (Maqbool and Hashmi, 1986; Reversat and Germani, 1985). Also, lesion nematodes have wide host range which can affect the selection of crop used to control the nematode in crop rotation. In addition, the presence of weed hosts in a field can strongly influence lesion nematode densities in maize fields (Egunjobi, 1974; Stradioto *et al.,* 1983). The involvement of other microorganisms in disease complex in crop loss of maize caused by lesion nematodes has not been studies fully. When maize seedlings were inoculated with *Pratylenchus zeae* either alone or in combination with maize mosaic virus, the effect of the pathogen was greatest when nematode and virus were inoculated simultaneously or when nematode was inoculated seven days prior to the virus (Nath *et al.,* 1978).

### c. Root knot nematode

Root knot nematode (*Meloidogyne spp.*) is another important nematode associated with maize crop in India. Chitwood and Toung (1960) were the first to report the association of *Meloidogyne africana* with maize crop from India. Krishnamurthy and Elias (1967) also recorded maize as a host of M*eloidogyne incognita.* Though associations of *Meloidogyne africana, Meloidogyne arenaria, Meloidogyne incognita, Meloidogyne javanica* and *Meloidogyne hapla* have been recorded on maize, *Meloidogyne incognita* and *Meloidogyne javanica* have been detected damaging maize in almost all the maize growing regions of the world. Although root knot nematodes are frequently encountered in maize fields, information on their biology, nature of damage and economic losses on maize is meager. However, indirect observations suggest that these species are of economic importance. Pathak and Yadav (1980) observed significant reductions in shoot and root weights of maize seedlings at $2J_2$/g soil. However, no loss of grain yield was reported. They also noticed yellowing of leaves and poor growth of plants in patches. The above ground symptoms included stunting of infected plants, leaf chlorosis and patchy growth. The root galls produced by the root knot nematodes are often small, terminal or sub-terminal. The nematode completes its life cycle in about 30 days. In *Meloidogyne incognita,* four races have been identified which reproduce well on maize with some cultivars exhibiting specificity to a specific race (Oteifa and Elgindi, 1982; Lopez, 1981). Losses in maize were severe by simultaneous inoculations of *Meloidogyne inognita* and maize-mosaic virus as compared to prior inoculations of nematode of either of the pathogens alone (Khurana *et al.,* 1970). The mosaic symptoms appeared earlier and nematode reproduction was greater when both pathogens were together than when alone (Goswami and Raychaudhari, 1978).

### d. Cyst forming nematode

*Heterodera sorghi,* a cyst forming nematode, was first reported on sorghum from Ghaziabad and Allahabad districts of Uttar Pradesh (Jain *et al.,* 1982). Subsequently, its occurrence was also reported from Haryana, Punjab, Delhi, Himachal Pradesh, Jammu and Kashmir, Andhra Pradesh and Maharashtra. Srivastava and Sethi (1987) reported maize to be the most preferred host of *Heterodera sorghi.* Although there was no field data available on exact yield losses caused by *Heterodera sorghi* in maize, experimental work indicates significant reduction in growth of maize cultivar Deccan - 103 observed at an inoculum level of $4J_2$/g soil of *Heterodera sorghi* under green house conditions (Srivastava and Chawla, 1990). The nematode is able to complete one generation ($J_2$ to $J_4$) in 24 days at 28 - 36°C on maize and in a crop season, it may complete more than four generations during *Kharif* season (Srivastava and Chawla, 1991). The nematode could also reproduce on barley, wheat, rice and pearl millet. Chawla and Srivastava (1995) reported maize cultivars (D - 851, R - 17, W - 101, Navin, VL - 88 and D – 765) are least susceptible against *Heterodera sorghi.*

### e. Other nematodes associated with maize

Certain other important nematodes have also been found associated with maize crop, but in most cases their role in limiting maize production has not been fully determined. The cereal cyst nematode (*Heterodera avenae),* stunt nematode *(Tylenchorhynchus* spp.) and lance nematode *(Hoplolaimus* spp.) are the most important species worth mentioning.

i. **Cereal cyst nematode:** Since *Heterodera avenae* first record on maize from Ajmer (Rajasthan) by Swarup *et al.,* (1964) and its subsequent report of occurrence in maize fields in Egypt (Ibrahim *et al,* 1986), few other reports have come out suggesting maize either as an efficient or inefficient host for the nematode. However, in recent years, cultivation of winter maize is becoming popular in north India. It is possible that this nematode may assume some importance in the near future in limiting winter maize production in our country.

ii. **Stunt nematode**: Among the several species of stunt nematodes associated with maize, *Tylenchorhynchus vulgaris* appears to be the most important. The species, first described from maize by Upadhyay *et al* (1972a), occurs most frequently in the sandy loam soils of Andhra Pradesh, Bihar, Delhi, Haryana, Himachal Pradesh, Punjab, Rajasthan and Uttar Pradesh (Upadhyay and Swarup, 1976). It has a wide host range and the most suitable hosts belong to family Graminae. The nematode completes its one life cycle in 25 - 27 days at 25 - 30°C (Upadhyay *et al.,* 1972a), while at 30 ± 2°C, it took 15 - 18 days (egg to egg) on maize cv. Ganga-5 (Siyanand *et al.,* 1982). A population level of 1000 or more nematode per kg soil was found to adversely affect plant growth (Upadhyay and Swarup, 1981). However, it is *to* be pathogen to all the seven maize cultivars tested (Jain, 1982). Reduction in shoot and root length, fresh and dry weight was maximum at 10,000 inoculums level. Loam, clay loam and sandy loam soils were most suitable for nematode reproduction (Upadhyay and Swarup, 1972b).

Significant reduction in plant growth of maize was obtained when *Tylenchorhynchus vulgaris* was inoculated in combination with *Pratylenchus zeae* or *Fusarium moniliforme* than when either of the organisms was inoculated alone. However, the multiplication of the nematode was not significantly affected in the presence of *Pratylenchus zeae* or *Fusarium moniliforme* (Upadhyay and Swarup, 1981). Application of Carbofuran as seed treatment were found effective for initial protection of maize in field plots predominantly inhibited with this nematode and the subsequent build up of nematodes could be prevented by granular application in soil (Singh and Bindra, 1978).

iii. **Lance nematode:** *Hoplolaimus indicus* is the most frequently encountered species associated with maize than any other species (Haider and Nath, 1992). Maize plants showing stunted and patchy growth were observed harbouring large populations of *Hoplolaimus indicus*. The nematode recorded the greatest loss in plant growth than the other nematode species studied (Mukhtar *et al.,* 1993). Significant growth reductions and stunting of maize plants have been recorded at 100 nematode per 500 g of sandy loam soil. The nematode acts as vagrant endoparasitic causing root lesions, thickening of cell wall and formation of tunnels in the cortical region. The nematode completes its one life cycle (egg to egg) in 39 - 43 days on maize cv. Ganga - 5 at 30°C ± 2°C (Siyanand *et al.,* 1982). The nematode population density was observed to be highest in October while it was lowest in June. The availability of feeder roots and temperature are important factors for population build up of this nematode (Haider and Nath, 1992).

## Interaction

Infections of *Fusarium moniliforme* on maize roots parasitized with *Hoplolaimus indicus* which aggravated root damage. In combined inoculations, the plants showed minimum top growth and maximum disease symptoms. After 60 days, the plants were prone to wilting during day time (Nath *et al.,* 1974). The severity of disease increased when nematode inoculum followed seven days after inoculation of soil with *Fusarium moniliforme* (Mukhtar *et al.,* 1993). Besides the above, a few other plant parasitic nematodes have also been found associated with maize crop but in most cases, their role / importance in limiting maize production are not fully studied. However, species of *Belonolaimus, Criconemella, Helicotylenchus, Rotylenchulus longidorus, Paratrichodorus* and *Xiphinema* are of limited importance.

## References

Baheti, B.L., M. Dodwadiya and S.S. Bhati (2017). Eco-friendly management of maize cyst nematode, Heterodera zeae on sweet corn (Zea mays L. saccharata). Journal of Entomology and Zoology Studies, 5(6): 989-993

Bajaj, H.K. and D.S. Bhatti (1984). Occurrence of *Heterodera zeae* in Haryana State, India. Haryana Agri. Univ. J. Res., 14: 551.

Bergeson, G.N. (1978). Control of the lesion nematode (*Pratylenchus* spp.) in corn with Carbofuran. Plant Disease Reporter, 62: 295 - 297.

Darekar, K.S., M.V.Kadam, D. S. Ayri and S. K. Dorge (1981). Occurrence of maize cyst nematode, *Heterodera zeae* in Maharashtra State. J. Maharashtra Agri. Univ., 6 (2): 156.

Egunjobi, O. A. (1974). Nematodes and maize growth in Nigeria I. Population dynamics of *Pratylenchus brachyurus* in about the roots of maize and its effect on maize production at Ibadan. Nematologica, 20: 91-93,

Goswami, B.K. and S.P. Ray Chaudhuri (1978). Interaction of root-knot nematodes and viruses in maize and tobacco in India. In: Proceeding of International Congress of Plant Pathology *(3 ed).* G. F. R. 16-13 August.

Hutzell, P. A. and L. R. Kursberg (1990). Temperature and the life cycle of *Heterodera zeae.* J. Nematol., 22 (3): 414-417.

Ibrahim, I.K.A., M.A. Rezk and A.A.M. Ibrahim (1986). Occurrence of the cyst nematodes *Heterodera avenae, Heterodera daverti* and *Heterodera rosii* in northern Egypt. J. Nematol., 18: 614.

Jain, K.K., C.L. Sethi, G. Swarup and A.N. Srivastava (1982). *Heterodera sorghi* sp. new cyst forming nematode parasitising sorghum. Revue Nematol., 5: 201-204.

Kaul, R.K. and C.L. Sethi (1982a). Effect of simultaneous inoculations and prior establishment of *Heterodera zeae, Meloidogyne incognita* and *Tylenchorhynchus vulgaris* on penetration of cyst and root-knot nematodes into maize roots. Indian J. Nematol., 12 (1): 79-85.

Kaul. R.K. and C.L. Sethi (1982b). Interaction amongst *Heterodera zeae, Meloidogyne incognita* and *Tylenchorhynchus vulgaris* on *Zea mays.* Indian J. Nematol., 12: 91-98.

Kaul, R.K. and C.L. Sethi (1987a). Effect of some systemic nematicide on the *in vitro* emergence of larvae from the cysts of *Heterodera zeae.* Indian J. Nematol., 17(2): 214-217.

Kaul, R.K. and C.L. Sethi (1987b). Penetration and development of *Heterodera zeae* in *Zea mays* as influenced by foliar spray of phenamiphos and oxamyl. Indian J. Nematol., 17(2): 237-240.

Kaul, R.K. and C.L. Sethi (1988a). Soil application of systemic nematicides in relation to emergence, penetration and development of *Heterodera zeae.* Indian J. Nematol., 18(1): 1-5.

Kaul, R.K. and C.L. Sethi (1988b). Effect of time of application of certain systemic nematicides on final cyst production and fecundity of *Heterodera zeae.* Indian J. Nematol., 18(1): 134-135.

Kaul, T.N., M. Kumar and Z.S. Brar (1994). A physiological analysis of growth, dry matter partitioning and grain yield of transplanted winter maize (*Zea mays* L.) in relation to N-management. *J. Res.,* Punjab Agricultural University, 31(1): 9-14.

Khair, A.M., A.A. Farahat and S.K. Abadir (1989). Studies on the corn cyst nematode (CCN) *Heterodera zeae* in Egypt. IV. Variation in development and reproduction of four different populations of some corn cultivars. Pakistan J. Nematol., 7: 69-73.

Khan, E., S. Ganguly and Y. Lalitha (1989). New Records on the occurrence of cyst nematodes in India. Int. Nematol. Network Newsl., 6(2): 19-20.

Khurana, S.M.P., B.K. Goswami and S.P. Raychaudhary (1970). Interaction of maize mosaic with root-knot nematode, *Meloidogyne incognita* in maize (*Zea mays* L.). Phytopath. Z., 69: 267-272,

Koshy, P.K. and G. Swarup (1971). Distribution of *Heterodera avenae, Heterodera zeae, Heterodera cajani* and *Anguina tritici* in India. Indian J. Nematol., 1(2): 106-111.

Koshy, P.K., G. Swarup and G.L. Sethi (1970). Heterodera zeae n. sp. (Nematoda: Heteroderidae) a cyst forming nematode on *Zea mays.* Nematologica, 16: 511-516.

Krusberg, L.R. (1988). The corn cyst nematode. *Heterodera zeae* in the United States'. In: Advances in Plant Nematology. P: 171-175.

Krusberg, L.R. and S. Sardanelli (1989). Survival of *Heterodera zeae* in soil, field and in the laboratory. J. Nematol., 21(3): 347-355.

Lal, A. and V.K. Mathur (1982).Occurrence of *Heterodera zeae* on *Vetiveria zizanioides.* Indian J. Nematol., 12(2): 405-407.

Lordello, R.R.A., F. Sawazaki, A.I.L. Lordello and J.A. Sobrinho (1983). Control of *Pratylenchus* spp. on maize with systemic namticides and castor bean cake. Sociedade brasileira de Nematologica, P: 241-250.

Makadia, B.M., D.K. Jagani and D.J. Patel (1988). Occurrence of *Heterodera zeae* on maize in Gujarat. Indian J. Nematol., 18(1): 130.

Maqbool, M. A. (1981). Occurrence of root-knot and cyst nematodes in Pakistan. Nematol. Medit., 9(2): 211-212.

Maqbool, M.A. and S. Hashmi (1986). Population trends of parasitic nematodes in different cropping sequences and the effect on yield of corn cv. AZAM. Int. Nematol. Network Newsl., 3: 34-38,

Morton, D.C., J. Tollefson, P. Huiz and S.H. Thomas (1978). Corn yield increases relative to non-fumigant chemical control of nematodes. J. Nematol., 10: 160-166.

Naganathan, T.G. and C.V. Sivakumar (1975). Host parasite relationships and influence of soil types on the lesion nematode, *Pratylenchus delattrei* Luc, 1958, on maize. Indian J. Nematol., 5: 162-169,

Naganathan, T.G. and C.V. Sivakumar (1976). Control of the lesion nematode, *Pratylenchus delattrei* Luc, 1958 on maize. Indian J. Nematol., 6: 32-38.

Nath, R.P., M.G. Haider and S.S. Prasad (1974). Combined effect of *Hoplolaimus indicus* and *Fusarium moniliforme* on maize plant. Indian J. Nematol., 4: 90 - 93.

Parihar, Aruna, B.S. Yadav and A.U. Siddiqui (1991). Reaction of some cultivated and weed plants to the corn cyst nematodes, Heterodera zeae. Indian J. Nematol., 21: 94-95.

Reversat, G. and G. Germani (1985). Recherches sur la pathopgenicdes nematodes assicies auf mais four rager an Senegal. Revue de Nematologie, 8 (1): 27 - 30.

Ringer, C.E., S. Sardanelli and L.R. Krusberg (1987). Investigations of the host range of the corn cyst nematode, *Heterodera zeae* from Maryland. Annals of Applied Nematology, 1: 97-106.

Sardanelli, S., L.R. Krusberg and A.M. Golden (1981). Corn cyst nematode, *Heterodera zeae* in the United States. Plant Disease, 65 (7): 622.

Sasser, J.N. and D.W. Freckman (1987). World Perspective on Nematology: The role of the Society. In: Vistas on Nematology: a Commemoration of the Twenty fifth Anniversary of the Society of Nematologists. J.A. Veech and D.W. Dickson, eds. Society of Nematologists, Inc. Hyattsville, M.D. P: 7-14.

Sethi, C.L. and R.K. Kaul (1987). Nematicidal exposure of larvae of *Heterodera zeae* and their development in maize roots. Indian J. Nematol., 17(2): 261-263.

Sethi, I. and O.S. Bindra (1978). Effect of Carbofuran as seed treatment and soil granular treatment alone and in combination on nematodes associated with maize'. Indian J. Nematol., 8 (1): 89-91.

Shahina, F. and M.A. Maqbool (1989). Embryonic and post-embryonic development stages of *Heterodera zeae.* on *Zea mays.* Pakistan J. Nematol., 7(1): 3-16.

Shahina, F., M. Abid and M.A. Maqbool (1989). Screening for resistance in corn cultivars against *Heterodera zeae.* Pakistan J. Nematol., 7: 75-79.

Sharma, N.K., Damanjeet and S.B. Sharma (1984). Incidence of maize cyst nematode, *Heterodera zeae* in Himachal Pradesh. Indian J. Nematol., 14 (2): 198.

Sharma, S.B. and G. Swarup (1984). Cyst forming Nematodes of India. Cosmo Publications, New Delhi.

Singh, H. and B.S. Rathore (2001). Estimation of yield losses in maize due to maize cyst nematode, Heterodera zeae. In: National Congress on Centenary of Nematology in India- Appraisal and Future Plans held at IARI, New Delhi. P: 157.

Patel, N.B., R.G. Patel, A.D. Patel, H.V. Patel, B.A. Patel and D.J. Patel (2000). Occurrence and distribution of Tylenchorhynchus vulgaris and Pratylenchus zeae in maize growing areas of Panchmahal district of Gujrat. Indian Journal of Nematology.; 30:95-96.

Singh, S.D. and B.S. Siradhana (1988). Effect of nematodes on late wilt of maize. Indian Phytopath., 41: 173-176.

Siyanand, A.R. Seshadri and D.R. Das gupta (1982). Investigation on the life-cycle of *Tylenchorhynchus vulgaris, Pratylenchus thornei* and *Hoplolaimus indicus* individually and in combined infestations in maize. Indian J. Nematol., 12: 272-276,

Srivastava, A.N. (1980). Studies on the maize cyst nematode. *Heterodera zeae.* Ph.D. Thesis submitted to P. G. School, IARI, New Delhi.

Srivastava, A.N. and C.L. Sethi (1985b). Effect of temperature on larval emergence of *Heterodera zeae*. Indian J. Nematol., 15 (2): 242-243.

Srivastava, A.N. and C.L. Sethi (1986a). Seasonal population fluctuations of *Heterodera zeae, Rotylenchulus reniformis* and *Tylenchorhynchus vulgaris* in maize-cowpea-wheat rotation. Indian J. Nematol., 16 (1): 16-18.

Srivastava, A.N. and C.L. Sethi (1984a). Effect of soil type on reproduction of *Heterodera zeae.* Indian J. Nematol., 14 (1): 72.

Srivastava, A.N. and C.L. Sethi (1984b). Relationship of initial population of *Heterodera zeae* with plant growth of maize and nematode reproduction. Indian J. Nematol., 14 (1): 110-114,

Srivastava, A.N. and C.L. Sethi (1984c). Histopathology of maize roots infested with maize cyst nematode, *Heterodera zeae.* Indian J. Nematol., 14 (2): 205-206.

Srivastava, A.N. and C.L. Sethi (1985a). On the larval penetration and biology of *Heterodera zeae* on maize. Indian J. Nematol., 15(1): 18-20.

Srivastava, A.N. and C.L. Sethi (1986b). Plant parasitic nematodes infesting maize and millets. Indian Phytopath., 39: 160-161.

Srivastava, A.N. and G. Chawla (1991). Life cycle of *Heterodera sorghi* on maize. *Indian J. Nematol.,* 21: 153-155.

Srivastava, A.N. and G. Chawla (1990). Pathogenicity and reproduction of *Heterodera zeae on Zea mays.* Indian J. Nematol., 20: 202-206.

Srivastava, A.N. and Gopal Swarup (1975). Preliminary studies on some graminaceous plants for their susceptibility to the maize cyst nematode *Heterodera zeae*. Indian Nematol., 5(2): 257-259.

Srivastava, A.N. and Jagan Lal (1993). Effect of split application of Carbofuran and Sebuphos on development of *Heterodera zeae* and growth of *Zea mays*. Indian J. Nematol., 23: 20-21.

Srivastava, A.N. and K.K. Kaushal (1986). Occurrence and distribution of *Heterodera zeae* and *heterodera sorghi* in India. In: Proceeding of Nat. Conf. Plant parasitic Nematodes, New Delhi, p. 34.

Srivastava, A.N. and K.K. Kaushal (1991). *Heterodera zeae* at high altitudes in Himachal Pradesh, India'. *Indian J. Nematol.* 21: 32-35.

# 25

# Harvesting

The maize crop is harvested when the husk has turned yellow and the grains are hard enough having not more than 20 % moisture. The appearance of the plant may be misleading, particularly in the case of high yielding hybrids and composites whose grains are dry, while the stalk and leaves may be still green. Harvest date should be determined by crop maturity, not by the calendar. Plan to harvest fields with potential lodging or harvest loss problems (such as stalk rot or deer damage) first. All field shelled corn with more than 15 % moisture must be dried for safe storage. The ideal kernel moisture level at which to harvest for dry grain storage is 25 %. Maize normally dries approximately 0.75 to 1 % per day during favourable drying weather (sunny and breezy) during the early, warmer part of the harvest season - from mid September through mid-October. By late October to early to mid-November, field dry-down rates usually drop to probably no more than 0.5 % per day. By mid-to late November the rate drops to 0.25 % per day and after that giving, drying rates are negligible. Dry-down rates can also be estimated in terms of Growing Degree Days (GDDs). Generally it takes 20 to 30 GDDs to lower grain moisture each point from 30 % down to 20 %. In September, accumulation of GDDs average 10 to 15 per day. In October, the accumulation drops to five to 10 GDDs per day. These estimates are based on generalizations, however, and some hybrids may vary considerably from this pattern of dry-down.

Monitoring harvest losses is an important part of the harvesting process. Ear corn losses from in front of the combine (pre-harvest losses) should be subtracted from the total harvest loss estimate. Keep in mind that most harvest losses occur at the gathering unit. Drought

-induced stalk lodging and insect problems reduce the yield potential of many maize fields if harvesting is delayed much beyond maturity. Ear drop damage may be high in some years as a result of extensive European corn borer damage. Losses increase by 1 to 2 % for each week of harvest delay.

Ears are removed from the standing crop. Harvested ears are dried in the sun before shelling. In the case of the late-sown crop, farmers prefer to harvest the whole plants and pile them and the ears are removed later. Maize stalks are used as a cattle-feed and fuel. In fact, no part of the maize plant, even the cobs from which the grains have been removed, is left unused. Maize grown for fodder should be harvested at the milk to early-dough stage. The earlier harvested crop is likely to yield less and have lower protein content. For silage, however, the late dough is preferred. Both power- and hand-operated low-priced maize shellers are available indigenously. These shellers are considered more efficient than hand-shelling or beating with sticks, the common practice in northern India. Farmers using hybrid maize should not save their own seed for their next crop, as the advanced-generation hybrid seeds are likely to lead to a yield reduction of 25-30 %. However, farmers can save seed from composites and the open-pollinated varieties, when grown in isolation. At least seeds from 500 to 1,000 ears of the best-yielding, normally-spaced plants resistant to prevalent diseases and pests should be bulked. Ears should be dried, shelled and treated with an insecticide and the treated seed is necessary, as the untreated seeds on ears are at times badly attacked by stored-grain pests and the germination is markedly reduced.

## Manual harvesting

In traditional farming systems, the crop is often harvested by hand and cobs are stored in traditional structures. Quite often the crop is left standing in the field long after the cobs have matured. So that, cobs may lose moisture and store more safely after harvest. Manual harvesting of maize does not require any specific tool. It simply involves removing the cob from the standing stalk. The work time averages 20 to 30 days per hectare. To improve their drying, it is often recommended to remove the husks from the cobs. Maize husking is usually a manual task carried out by groups of women and children.

## Mechanical harvesting

The first mechanical harvester to detach ears of maize from the standing stalks, the corn snapper was built in North America in the mid $19^{th}$ century. This was followed by the development of corn pickers which incorporate a mechanism for removing the husks from the harvested ears. The first animal drawn maize pickers were replaced by tractor drawn units (1 or 2 rows) and then tractor mounted units (1 row). Finally the development of self propelled units came which capable of harvesting cobs from up to 4 rows. A specific feature in maize harvesters is the header, which leaves the stalks standing as it removes the ears. The rates of work can vary from 2 hours / hectare with a 3 row self propelled harvester to 5 hours per hectare with a tractor mounted single row unit. Harvest losses range from 3 to 5 % but they may be up to 10 to 15 % under adverse conditions. Depending on the situation, a single row harvester can be employed effectively on up to 20 hectares or more. But the use of a multirow machine demands several tens of hectares to be economically effective. Specially designed for harvesting maize as grain, the corn sheller was initially a corn husker in which the husking mechanism was replaced by a threshing one (usually of the axel type). Corn shellers are self propelled machines of the 3 to 6 row type with capacities of 1 to 2 hours per hectare.

## Threshing

Traditional maize shelling is carried out as a manual operation. Maize kernels are separated from the cob by pressing on the grains with the thumbs. According to the operator's ability, the work rate is about 10 kg per hour. Outputs up to 23 kg hour can be achieved with hand held tools (wooden or slotted metal cylinders). Another threshing method, sometimes applied in tropical countries, involves putting cobs in bags and beating them with sticks. Outputs achieved prove attractive belt bags deteriorate rapidly. Nowadays many small maize shellers equipped with a rotating cylinder of the peg or bar type, are available in the market. Their output ranges between 500 and 2000 kg per hours. They may be driver from a tractor power take off or have their own engine. Power requirements vary from 5 to 15 hp according to equipment involve.

Shelled grain weights can be adjusted using a grain shrink table. Shrink represents both the moisture loss and a 0.5 % dry-matter loss encountered during dry and grain handling. To estimate the amount that a given wet weight of maize will lose during the drying process, multiply the wet weight by the shrink factor from the table. Monitor debris and cracked maize in the grain as harvesting progresses. Debris and cracked corn lower grain quality and increase the potential for spoilage of stored maize.

## Yield potential

The C4 crops have a much higher yield potential than the C3 crops. Compare maize with sunflower, maize yields of 12 to 15 t/ha are not unusual under irrigation, whereas 3 to 3.5 t/ha yields are an exception for sunflower. The C4 crops produce less complicated nutrients e.g.

sugar and starch, whereas the C3 crops produce more complex and higher quality nutrients like oil and protein. Considerable variation is observed in grain yield which depend upon the variety, the amount of the fertilizer used, the rainfall pattern, etc. Under irrigated conditions and recommended cultural practices, on an average yield 4 t/ha in the Indo-Gangetic Plains is not uncommon. In Peninsular India and at higher elevations, a mean yield of 5 -7 t/ha has frequently been obtained. Under low fertility and rainfed conditions with poor-yielding varieties, a grain yield of about 1 - 2 t/ha is obtained.

## References

Egrinya Eneji, A. (2009). Agronomy of tropical crops. Studium press, LLC, Houston, Texas, P: 28 - 41.

Hoeft, R.G., E.D. Nafziger, R.R. Johnson and S.R. Aldrich (2000). Modern corn and soybean production. MSCP Publ., Champaign, IL.

Javad Hussain, H.S. and V. Sethuraman (1980). Maize. Directorate of Agriculture, Chenpauk, Chennai, P: 92.

# 26

# Types of Corn

Maize is classified into 7 groups by based on the endosperm of kernels. They are flint, pop, waxy, and sweet. Each type of corn has unique kernel characteristics that determine its use and how it is grown. Because most specialty corn hybrids (include white, dent, waxy/ oil and pop corn) are grown under contract, it is advisable to identify a market before planting. Also, some specialty corn processors specify certain hybrids and cultural practices they want growers to use. Contracts for growing specialty hybrids usually offer a premium over the yellow dent price to compensate for the lower yield potential and the special handling required ensuring high grain quality. Most specialty corns must be grown in isolation from conventional corn fields to prevent cross pollination. High grain and silage yield potential, high feed value and availability of adapted superior hybrids account for the widespread use of yellow dents. Yellow dents have the highest content of carotene (vitamin A) of the cereal grains.

### Pop corn (*Zea mays averta* Sturt)

The grain is used for human consumption and is the basis of pop-corn confections. Its cultivation is mainly confined to the New World. Popcorns are essentially small - kernelled flint corns and is among the most primitive of the surviving races of corn. Kernels contain a hard endosperm with only a small portion of soft starch. Popcorns are generally either pearl or rice types. Pearls have smooth, rounded crowns and rices are pointed. Heating the kernel turns the moisture inside the soft starch in the center into steam that explodes the kernel inside out. The greater the expansion, the higher the quality. Popcorn is not only high in fibre, potassium and vitamins, it is also low in calories and fat. A cup of unsweetened popcorn contains between 31 and 55 calories.

### Dent corn (*Zea mays indentata* Sturt)

It is popularly known as dent corn because of dent formation on the top of the kernel having yellow or white colour. The depression or dent in the crown of the seed is the result of rapid drying and shrinkage of the soft starch. This is the most common type of maize grown in the USA.

### Flint corn (*Zea mays indurata* Sturt)

The endosperm in this type of maize kernel is soft and starchy in the center and completely enclosed by a very hard outer layer. The kernels are round on the top. The colour may be white or yellow. It was discovered by Europeans with early maturity and can germinate in cold and wet soil. It is grown in Europe, Asia, Central America and South America as well as it is principal type of grain corn grown in India.

### Flour or soft corn (*Zea mays amylacea* Sturt)

It possesses a soft endosperm. Kernels are soft and of various colours, but white and blue are most common. They are like fruit kernels in shape. It is widely grown in the USA and South Africa.

### Sweet corn (*Zea mays saccharata* Sturt)

The sugar (20 % on dry basis at green ear stage) and starch make the major component of the endosperm, which results in sweetish taste of the kernels before they attain the maturity. After maturity, the kernels become wrinkled. The cobs are picked up green (with 70 % moisture) for canning and table purposes. The crop is mainly grown in Northern half of the USA and Canada. Yellow is predominant grain colour and are rich in Vitamin A and C.

### Pod corn (*Zea mays tunicata* Sturt)

Each kernel is enclosed in pod or husk in an ear which is enclosed in husks like other types of corn. It is a primitive type of corn not grown commercially.

### Waxy corn (*Zea mays ceretina* Kulesh)

Kernels have waxy appearance. Starch is entirely composed of amylopectin in contrast to common dent corn starch with 78 % amylopectin and 22 % amylose. Waxy corns are of Chinese origin. The mutations of dent corn have resulted in evolution of waxy corn in USA. The waxy corn hybrids developed in USA with starch similar to tapioca are grown for supplying raw materials for speciality products of the wet milling starch industry for textile and paper sizing and corn oil. It is highly suitable as a food thickener. Waxy corn has also been considered as a potential animal feed. Feeding with waxy corn hybrids have occasionally shown a benefit, but they have not been consistent. Changes in feed efficiency and production have been insignificant, but usually in favour of the waxy corn hybrids. Yields of waxy corns are generally lower than those of yellow dent corn. Waxy corns must be grown in isolation from other corn types to maintain purity standards for industrial use. Keeping the harvested grain separated by 12 to 16 border rows from the rest of the field is considered adequate isolation. The 5 % contamination with regular corn pollen is considered the upper limit.

### Ear corn

Ear corn can be cribbed safely when the grain moisture is 21 per cent or less. However, with cold weather and narrow (four-foot), well-ventilated cribs, corn may be stored when grain moisture is several percentage points higher. Use table to convert earn corn yields to shelled corn equivalents. For example, four tons of ear corn at 21 per cent of grain moisture is equivalent to 8,000 divided by 77.6 of shelled corn.

### White Corn

White corn types are equal to yellow types in carbohydrate content but are deficient in vitamin A. White types are grown primarily for direct human consumption in foods such as breakfast cereals and grits. Yields of white hybrids generally are not competitive with yellow dent yields. Typical high-yielding yellow dent hybrids produce 5 to 10 % more grain yield than the best white hybrids. White and yellow downgrades either type at the market.

### High – Lysine Corn

Although normal dent corn contains about 9.4 per cent protein, the quantity of two essential amino acids, lysine and tryptophane, is below nutritional requirements for humans and nonruninant (single-stomached) animals, such as pigs and chickens. High-lysine corn corrects this deficiency and may be advantageous in swine feeding rations. Adapted verities of high-lysine corn are limited in number and have generally been lower in grain yield than normal dent verities. The softer kernels of high-lysine corn are more vulnerable to breakage at harvest, which can lead to higher incidence of kernel or ear rot. Production fields must be grown in isolation from other corn fields to ensure protein quality. Cross pollination results in normal dent corn. A separation distance of 300 feet is recommended.

### High – Oil Corn

High-oil corn (HOC) is attractive as a livestock feed because it has greater energy value than normal yellow dent corn and can replace more expensive dietary sources of fats and proteins. HOC contains 50 to 100 % more oil than normal yellow dent corn, which averages about 4 % oil on a dry weight basis. Feeding trials with HOC indicate that it has improved feed efficiency and results in increased rate of gain over conventional corn.

The Top Cross grain production system, currently the most widely used method of producing HOC in the United States, involves the planting of a blend of two types of corn. One type, referred to as the grain parent, is a cytoplasmic male sterile (produces no viable pollen) elite hybrid that comprises 90 % to 92 % of the seed in the blend. The second type, comprising 8 % to 10 % of the seed in the blend, is a high oil pollinator. The high oil pollen contains genes that cause production of kernels with larger than average embryos. Because most of the oil and several essential amino acids are in the embryo, the increased embryo size of HOC results in greater oil content and enhances grain protein quality.

## References

Egrinya Eneji, A. (2009). Agronomy of tropical crops. Studium press, LLC, Houston, Texas, P: 28 - 41.

Halluer, A.R. (1994). Specialty corns. Department of Agronomy, Iowa State University, Ames, Iowa.

Hoeft, R.G., E.D. Nafziger, R.R. Johnson, and S.R. Aldrich (2000). Modern corn and soybean production. MSCP Publ., Champaign, IL.

Javad Hussain, H.S. and V. Sethuraman (1980). Maize. Directorate of Agriculture, Chenpauk, Chennai, P: 92.

# 27

# Fodder Maize Cultivation

Maize is a domesticated form of a wild grass, first cultivated over 5,000 years ago in tropical Mexico that produces an adaptable and productive grain. It was introduced to the UK in the early part of the twentieth century. Today the bulk of maize production occurs in the US, China and Brazil. In the more northerly climates like that of the UK where there is insufficient summer warmth to ripen the crop, maize is grown largely for forage. Here it nearly all is ensiled to produce a quality, high-energy silage that complements grass silage for the winter feeding of livestock.

Forage maize is relatively easy to grow and being drought tolerant is a consistent provider of high yields with minimum input from the farmer. From a practical point of view, forage maize also spreads the silage making season (maize is ensiled in September or October) long after most grass silage making has finished. As a crop, forage maize has further advantages on the farm. However, forage maize is not completely without its difficulties being low in both protein and minerals and as a crop sometimes difficult to fit into the cropping rotation.

The production of quality fodder crop is very important for up keeping of livestock health. At present, the country faces a net deficit of 62.76 % green fodder, 23.46 % dry crop residues and 64 % feeds as a result crude protein and total digestible nutrients (TDN) deficiency are 28.44 and 24.04 % respectively and the trend could well continue in the future. Maize being one of the most important dual crops grown widely in kharif season for grain as well as for fodder in India has an edge over other cultivated fodder crops due to its high production potential ability, wider adaptability, quick growing nature, succulency, palatability, excellent fodder quality, free from toxicants and it can be safely fed to animals at any stage of crop growth.

Maize fodder can be used at any crop growth stage. Its quality is adversely affected after anthesis. To maintain the fodder quality the detasseling is advised to the farmers for better digestibility and palatability. By grazing this fodder to the milch cattle, their milk is increased. The tall, leafy and longer duration cultivars are most preferred for maize fodder cultivation. The cultivation of maize for fodder can be done round the year. Very high seed rate is used. Generally the farmers grow composite varieties or advance generation of hybrid seed which is economical to the farmers.

The harvesting window for maize is wider than for grass silage as maize holds its digestibility as forage for longer. A good crop will typically yield between 30 and 50 tonnes of fresh matter per hectare (between 9 and 15 tonnes of dry matter at 30 % moisture content). Maize produces palatable silage that livestock like but which also produces much less effluent than grass.

The crop is grown in over 0.9 million ha in different parts of the country throughout the year. It is a C4 plant having high fodder production capacity in short duration. Maize is generally grown in the areas with rainfall of 50 cm, but higher yields are achieved in 120–150 cm rainfall areas. It is susceptible to water logging. In the early stage up to 35 days after sowing, the crop is drought tolerant.

## Season

Maize can be grown throughout the year for fodder purpose in all districts under irrigated condition.

## Soil

The crop prefers deep, fertile, organic matter rich, well drained soils with pH 5.5–7.5. Maize does not prefer heavy clay low land soils. The alluvial soils of Uttar Pradesh, Bihar and Punjab are suitable for raising maize crop.

## Varieties

The composite varieties and hybrids such as African Tall, Ganga 5, J 1006, Pratap chari-6 and APFM 8 are recommended fodder cultivation.

**Table 1. Yield potential and quality characters of African tall and Ganga 5**

| Characters | African Tall | Ganga 5 |
|---|---|---|
| Green fodder yield (t/ha) | 42.00 | 34.00 |
| Crude protein (%) | 9.80 | 10.63 |
| Dry matter (%) | 17.65 | 14.27 |
| Plant height (cm) | 302.00 | 231.00 |
| Number of leaves | 13.30 | 11.50 |
| Leaf length (cm) | 81.30 | 74.80 |
| Leaf breadth (cm) | 8.15 | 8.50 |
| Stem thickness (cm) | 1.77 | 1.69 |
| Leaf-stem ratio | 0.21 | 0.17 |

**Table 2. Yield and quality parameters of fodder maize varieties**

| Varieties | Green forage yield (t/ha) | Dry matter (%) | | Crude protein (%) | | Crude fibre (%) | |
|---|---|---|---|---|---|---|---|
| | | I harvest | II harvest | I harvest | II harvest | I harvest | II harvest |
| JH-3459 | 38.12 | 21.19 | 21.13 | 8.72 | 7.43 | 24.77 | 28.67 |
| Parkash | 30.14 | 20.55 | 20.91 | 7.00 | 6.70 | 26.80 | 28.77 |
| PMH-2 | 40.14 | 21.22 | 22.47 | 8.46 | 8.31 | 26.33 | 23.13 |
| J-1006 | 46.67 | 24.28 | 24.69 | 7.44 | 5.83 | 26.13 | 23.87 |
| African tall | 30.99 | 22.24 | 22.61 | 7.14 | 5.54 | 30.20 | 28.67 |

### African Tall Composite

The variety was bred at MPKV, Kolhapur and is a composite of seven genotypes (H-611 C, H-611, H-611 (R)C3, K-III × EC-573 (R12) C3, Ukiri Comp A (F) C5 × Ukiri Comp A, (F) C3, Chitedge Comp A and Ilonga Comp) developed through modified mass selection technique. It has more dry matter and crude protein content, more number of leaves/plant, more leaf area, good grain and seed yield potential than other grain varieties. The variety is resistant to foliar diseases and stem borer. The average plant height is 260 cm and provide 60–70 t/ha green fodder and 30 q/ha grain. It is released for cultivation throughout the country.

### APFM-8

This variety was developed at ANGRAU, Hyderabad in 1997 and is suitable for cultivation in south zone of the country. This is a synthetic variety derived from Varun (V- 41) and Palampur local varietal cross advanced by mass selection. It is leafy, non lodging, orange grain variety, plant height 180– 200 cm, sturdy plants type with dark green leaves. The seed to seed maturity is 90–95 days during *kharif* and 105–110 days in winter. It provides 35 t/ha green and 7.5 t/ha dry fodder.

### J-1006

The variety was developed at PAU, Ludhiana by crossing 'Makki safed 1-DR' × 'Turpeno PB' and released in 1992 for cultivation in Punjab. It is resistant to Maydis blight, brown striped downy mildew and stem borer.

### Pratap Makka Chari 6

This variety was developed by MPUA&T, Udaipur by compositing 11 early to medium white seeded entries. The variety was released in 2008 for north-west zone of the country covering Punjab, Haryana, western Uttar Pradesh, Rajasthan and Uttarakhand. Average green forage yield is 35 t/ha. It is a medium tall variety with relatively low ear placement. Its stem is strong, medium thick and resists lodging. The ears are long, thick with a tight husk cover. The grains are medium, flint to semi dent and orange-yellow. It matures in 90–95 days. Its green fodder yield potential is 45–50 t/ha. It is released for cultivation in Punjab.

At Rajendra Nagar (Andhra Pradesh), fodder of baby corn was found to be superior with respect to crude protein while crude fibre and IVDMD were commendable. Though the fodder yield was less in baby corn, the revenue (Rs. 39750/ha) was higher from baby corn. The net return and B:C ratio were in the order of baby corn (Rs. 39750/ha; 3.54) > fodder crop (Rs. 5179/ha; 1.40) > seed crop (Rs.5446, 1.35) > green cob (Rs. 3459, 1.23) (Shanti *et al.,* 2012).

**Table 3. Green fodder, dry fodder and seed /cob yield of maize under different use pattern**

| Variety | Green fodder yield (q/ha) | Dry fodder yield (q/ha) | Seed /cob yield (q/ha) | Crude protein (%) | Crude fibre (%) |
|---|---|---|---|---|---|
| **Fodder maize** | | | | | |
| J 1006 | 182.5 | 30.6 | 0 | 7.4 | 46.12 |
| Karvir | 188.9 | 31.1 | 0 | 6.8 | 35.46 |
| Ganga safed | 172.0 | 28.7 | 0 | 8.2 | 33.75 |
| **Baby corn** | | | | | |
| Vivek 17 | 78.8 | 14.0 | 5005 | 9.0 | 32.17 |
| Vivek 11 | 86.9 | 15.5 | 5218 | 8.5 | 28.74 |
| Baby corn 1 | 78.8 | 14.0 | 4933 | 8.7 | 38.40 |

| Variety | Green fodder yield (q/ha) | Dry fodder yield (q/ha) | Seed /cob yield (q/ha) | Crude protein (%) | Crude fibre (%) |
|---|---|---|---|---|---|
| **Green cob** | | | | | |
| Harsha | 80.0 | 13.9 | 4780 | 7.4 | 39.39 |
| Ashwini | 72.5 | 11.8 | 4290 | 8.2 | 38.61 |
| Maduri | 92.8 | 15.8 | 4460 | 8.0 | 32.60 |
| **Seed crop** | | | | | |
| BH 1576 | 0 | 16.5 | 3118 | 5.3 | 33.16 |
| BH 40625 | 0 | 17.7 | 2930 | 6.0 | 35.53 |
| Amber pop corn | 0 | 15.3 | 1868 | 5.8 | 44.44 |
| Check (African tall) | 193.3 | 30.5 | 0 | 7.6 | 32.65 |
| CD 5 % | 15.0 | 5.3 | 511 | 0.5 | 5.28 |

### Field preparation

Plough the field twice with an iron plough and three or four times with country plough. Normally 4–5 harrowings are required before taking up planting with the onset of monsoon. Form ridges and furrows using a ridger at 30 cm apart and form beds of size 10 or 20 $m^2$ depending on the availability of water and slope of the land.

### Seed

About 20 kg seed/ha should be sown 5 cm deep into soil for good germination, seedling growth and vigour. Transplanting should be avoided as the plant cannot cope up with main crop stand. A population of 125000 - 130000 plants /ha at harvest is optimum for realizing higher yields.

### Sowing

As maize does not have tillering capacity to adjust to variations in plant stand, optimum seed rate for forage production is important. Row spacing and seed rate had significant effects on yield potentials of maize. The seed rate of 40 kg/ha can be adopted for fodder purpose. Treat the seeds with 3 packets (600 g) Azospirillum inoculant and 3 packets (600 g) of phosphobacteria or 6 packets of Azophos (1200 g) before sowing. The spacing recommended is 30 x 15 cm.

**Table 4. Seed rate on growth, yield and economics of fodder maize**

| Seed levels (kg/ha) | Plant height (cm) | LAI | Yield (t/ha) | | Net return (Rs/ha) | B:C ratio |
|---|---|---|---|---|---|---|
| | | | Green fodder | Dry fodder | | |
| 40 | 217 | 4.63 | 40.48 | 10.85 | 25380 | 1.33 |
| 60 | 218 | 5.91 | 43.20 | 11.50 | 26310 | 1.35 |
| 80 | 218 | 6.92 | 43.93 | 11.56 | 25890 | 1.31 |
| CD (5 %) | 1.1 | 0.38 | 0.25 | 0.22 | - | - |

(Barod et al., 2012)

Seed rate of 60 kg /ha increased LAI, green fodder and dry fodder yields as well as monetary benefit of fodder maize and on par with 80 kg/ha (Barod et al., 2012).

**Table 5. Influence of spacing on growth and seed yield of forage maize (African tall)**

| Spacings | Plant height (cm) | Stem diameter (mm) | Leaves / plant | Cobs/ plant | Cob length (cm) | Seed/ cob | 100 seed weight (g) | Seed row /cob | Seed yield (kg/ha) |
|---|---|---|---|---|---|---|---|---|---|
| 45x 30 cm | 208 | 7.09 | 15.16 | 1.02 | 23.58 | 265 | 36.09 | 10.48 | 2535 |
| 45 x 45 cm | 208 | 7.12 | 14.03 | 1.03 | 23.89 | 319 | 37.90 | 10.94 | 2893 |
| 60 x 30 cm | 206 | 7.69 | 14.62 | 1.06 | 23.78 | 323 | 39.04 | 10.91 | 3948 |
| 60 x 45 cm | 214 | 8.13 | 14.25 | 1.07 | 25.39 | 345 | 40.56 | 11.04 | 4755 |
| 75 x 30 cm | 209 | 8.22 | 14.01 | 1.09 | 26.31 | 345 | 41.04 | 11.18 | 5167 |
| 75 x 45 cm | 210 | 9.01 | 14.64 | 1.19 | 26.88 | 379 | 41.34 | 11.23 | 5327 |
| CD | 3.9 | 0.34 | 0.98 | 0.04 | 1.72 | 27 | 1.29 | 0.42 | 62 |

(Channakeshava et al., 2000)

At Bangalore (Karnataka), planting African tall maize at 75 x 45 cm spacing recorded significantly higher seed yield (5327 kg/ha) as compared to all other spacing levels (Table 5) 75 x 45 cm spacing recorded maximum net return Rs. 56974 /ha over other treatments during kharif due to higher grain yield in red sandy loam soil (Channakeshava et al., 2000). High seed rates with narrow row spacing tended to produce higher fodder yield. Higher seed rate (75 kg/ha) and narrow row spacing (15 cm) resulted in higher fodder yield (Malik et al., 2007).

**Fertilizers**

Apply and spread FYM or compost at 25 t/ha on unploughed field along with 10 packets of

Azospirillum (2000 g) and 10 packets of Phosphobacteria (2000g) inoculum or 20 packets of Azophos (4000 g) and incorporate the manure into the soil during ploughing. Apply NPK fertilizers as per soil test recommendation as for as possible. If soil testing is not done, application of 60–120 kg N, 40– 60 kg P and 40 kg K/ha is recommended for fodder crop. Early maturing varieties require less quantity than full season maturity crops. It is also advisable to apply 20 kg Zinc Sulphate/ha along with basal dose of fertilizer. One fourth of nitrogen and entire quantity of phosphorus, potassium and zinc should be applied 5–7 cm deep before sowing. The rest of the dose is applied at knee high stage and after emergence of flag leaf but before tassel emergence.

Addition of nitrogen fertilizer significantly increased plant height, stem diameter and LAI in forage maize. Phosphorus fertilization has no significant effect on the growth attributes. Application of nitrogen allowed the crop to reach 50 % tasseling earlier. Phosphorus on the other hand has no effect on days to 50 % tasseling. Nitrogen fertilization significantly increased the forage dry matter yield. In contrast, phosphorus has no effect on the dry matter yield. Nitrogen significantly increased the protein content of forage maize, on the other hand phosphorus has no effect on the crude protein content. Neither nitrogen nor phosphorus has a significant effect on the crude fiber content (Hani et al., 2006).

Green and dry fodder yields were increased with successive increment of N. Progressive increase in maize fodder yield was reported up to 200 kg N/ha (Gaurkar and Bharad, 1998; Amar Singh *et al.,* 2000). Significant increase in green forage yield of corn with increase in nitrogen rate up to 160 kg/ha (Kesin et al., 2005).

Application of 200:100:75 kg NPK/ha caused significantly higher seed yield of 4251 kg/ha of fodder maize as compared to application of either 100:50:25 kg NPK/ha or 150:75:50 kg NPK/ha (Table 6). Application of higher dose of fertilizers gave higher net return of Rs.44384/ha (Channakeshava et al., 2000).

**Table 6. Influence of fertilizer levels on forage maize (African tall)**

| Fertilizer levels (kg NPK/ha) | Plant height (cm) | Stem diameter (mm) | Leaves / plant | Cobs/ plant | Cob length (cm) | Seed/ cob | 100 seed weight (g) | Seed row / cob | Seed yield (kg/ha) |
|---|---|---|---|---|---|---|---|---|---|
| 100:50:25 | 202 | 7.47 | 14.22 | 1.03 | 23.59 | 293 | 36.12 | 10.52 | 3938 |
| 150:75:50 | 208 | 7.81 | 14.59 | 1.06 | 24.97 | 322 | 39.32 | 10.99 | 4122 |
| 200:100:75 | 215 | 8.35 | 14.29 | 1.14 | 27.36 | 373 | 42.55 | 11.39 | 4251 |
| CD | 7.6 | 0.24 | 0.98 | 0.04 | 1.21 | 19 | 1.29 | 0.30 | 37 |

(Channakeshava et al., 2000)

**Table 7. Zinc levels on fodder maize**

| Nitrogen levels (kg/ha) | Plant height (cm) | LAI | Yield (t/ha) | | Net return (Rs/ha) | B:C ratio |
|---|---|---|---|---|---|---|
| | | | Green fodder | Dry fodder | | |
| 0 | 216 | 5.62 | 40.76 | 10.93 | 25480 | 1.33 |
| 10 | 219 | 6.02 | 44.22 | 11.66 | 2642 | 1.34 |
| CD (5 %) | 0.9 | 0.31 | 0.21 | 0.19 | - | - |

(Barod et al., 2012)

Application of 10 kg $ZnSO_4$/ha recorded higher plant height, LAI, green fodder and dry fodder yields with higher monetary return over control in fodder maize (Table 10) (Barod et al., 2012).

**Water management**

Irrigate immediately after sowing and give life irrigation on the third day and thereafter once in 10 days. To ensure high and stable yield, it is desirable to give 1 or 2 irrigation at critical stages (flowering and grain filling stages). Jagdev Singh *et al,* (1989) on sandy loam soil of Hissar during summer season observed significantly higher green fodder and dry matter yield at IW/CPE ratio of 0.5 and 0.75 over 0.25 ratio. Similar results were reported by Rana and Malik (1981). Mansfield *et al,* (1990) at Belvatagi during rainy season in clay loam soil. Plant height of the fodder maize was significantly higher due to irrigation at 0.9 IW/CPE ratio (220.1 to 225.9 cm) over 0.7 and 0.5 ratio and unirrigated control. Maximum forage yield was recorded at 0.7 IW/CPE ratio (445.0 q /ha). Manhi and Shukla (1992) observed that green forage yield was more with the application of irrigation at 75 % available soil moisture over 50 % available soil moisture and no irrigation during rainy season at Jhansi. Bhup Singh *et al,* (1989) noticed on sandy loam soil at Hissar during summer that the plant height and other growth attributes number of leaves, leaf area and leaf area index increased with irrigation at 0.75 IW/CPE ratio over 0.5 ratio.

At Shambat (Sudan), neither green fodder yield nor quality of fodder maize was significantly affected by watering intervals (Table 8) (Muna et al., 2011).

**Table 8. Effects of watering, nitrogen and planting methods on green fodder yield and quality of fodder maize for two seasons**

| Watering interval | Green fodder yield (t/ha) | CP (%) | CF (%) | Ca (%) | Mg (%) | K (%) | P (%) |
|---|---|---|---|---|---|---|---|
| Watering every 10 days | 50.0 | 6.6 | 34.7 | 0.42 | 0.27 | 0.84 | 0.19 |
| Watering every 20 days | 38.6 | 7.1 | 34.6 | 0.36 | 0.31 | 0.83 | 0.20 |
| LSD 5 % | NS | NS | NS | NS | NS | NS | NS |

(Muna et al., 2011)

## Defoliation

At Bahir Dar (Ethiopia), grain yield, stover yield and yield components of maize as well as dry matter production of under sown forage legume (Vicia *villosa*) was significantly affected by the rate of maize leaf defoliation. An optimum rate of maize leaf defoliation without affected the grain yield, stover yield and dry matter yield of the defoliated leaf of maize and dry matter yield for the under sown forage crops was harvested at the rate of 25-50 % of defoliation of maize leaf. Therefore defoliation of maize leaf upto 50 % at the time of feed shortage did not have adverse effect on grain and stover yields and under sown forage legume (Halima and Chauhan, 2003).

**Table 9. Grain yield, stover yield and dry matter yield of defoliated leaf of maize and dry matter yield of under sown forage legume as affected by rate of maize leaf defoliation at various growth stages**

| Rate of leaf defoliation | Grain yield of maize (t/ha) | Stover yield of maize (t/ha) | Dry matter yield of defoliated leaf of maize (t/ha) | Dry matter yield of under sown forage legume (t/ha) |
|---|---|---|---|---|
| 0 | 6.78 | 16.01 | 0.0 | 1.28 |
| 25 | 6.59 | 14.62 | 0.47 | 1.61 |
| 50 | 5.64 | 11.67 | 1.12 | 1.55 |
| 75 | 4.99 | 9.83 | 1.92 | 1.77 |
| 100 | 3.68 | 8.41 | 2.06 | 2.00 |
| CD 1 % | 1.08 | 2.23 | 0.36 | 0.38 |
| **Growth stages** | | | | |
| Late whorl | 6.75 | 12.07 | 1.07 | 1.65 |
| Silking | 5.32 | 11.75 | 1.10 | 1.67 |
| Milking | 3.32 | 12.50 | 1.20 | 1.60 |
| CD 1 % | 1.88 | 2.86 | NS | NS |

(Halima and Chauhan, 2003)

## Weed management

Hoeing and weeding are done as and when necessary.

## Intercropping

Maize and cowpea are adaptable to wide range of environment and can provide nutritious fodder under rainfed conditions when grown in association. Intercropped of maize and bean in different planting ratio significantly affected the quantitative and qualitative characters of the forage. Higher yield of green fodder (65.7 t /ha) was obtained by sowing the crops in ratio of 100:100. Higher grain yield (9.0 t /ha) for maize was recorded from 75 + 25 % ratio, maize and cowpea. Higher grain yield for cowpea (3.9 t/ ha) was recorded from 50 + 100 % ratio, maize and cowpea, respectively. Higher crude protein (19.65 %) was produced by the cowpea sole cropping and lower from the maize sole cropping (12.11 %). Higher land equivalent ratio (2.26) was obtained by sowing the crops in ratio of 100:100 and higher crude protein was obtained by harvest time in milky stage (15.2 %) (Dahmardeh *et al.*, 2009). Fodder Maize can be intercropped with CO 5 or COFC 8 Cowpea at 3:1 ratio and harvested together to provide nutritious fodder. TNAU vermicompost at 5 t/ha + 75 % recommended dose of fertilizer for intercropping of maize and cowpea produces green fodder yield of 105 t/ha/year (3 crops/ year) which is sufficient to maintain 7 adults and 3 young cattle. The dung obtained from these animals can be used for on farm production of 19.4 tonnes of vermicompost per year.

**Table 10. Green fodder, dry matter and crude protein yield of maize + cowpea intercropping system under rainfed condition**

| Treatments | Green fodder yield (q/ha) | | Dry matter yield (q/ha) | | Crude protein yield (q/ha) | | Net return (Rs/ha) | B:C ratio | LER |
|---|---|---|---|---|---|---|---|---|---|
| | Maize | Cowpea | Maize | Cowpea | Maize | Cowpea | | | |
| Sole maize | 293.1 | 0 | 66.4 | 0 | 7.6 | 0 | 5135 | 1.78 | - |
| Sole cow pea | 0 | 179.2 | 0 | 36.5 | 0 | 6.1 | 2364 | 1.36 | - |
| Maize + cowpea (1:1) | 202.8 | 45.3 | 43.1 | 9.1 | 4.6 | 1.5 | 3580 | 1.53 | 0.94 |
| Maize + cowpea (1:2) | 188.7 | 61.7 | 40.5 | 12.4 | 4.4 | 1.7 | 3859 | 1.57 | 0.98 |
| Maize + cowpea (1:3) | 178.4 | 75.8 | 39.3 | 15.3 | 4.5 | 2.5 | 4162 | 1.62 | 1.03 |
| Maize + cowpea (1:4) | 159.1 | 84.6 | 35.9 | 17.3 | 4.0 | 2.8 | 3869 | 1.58 | 1.01 |
| Maize + cowpea (2:1) | 293.2 | 45.1 | 66.5 | 9.1 | 7.5 | 1.5 | 7113 | 2.04 | 1.25 |
| Maize + cowpea (2:2) | 280.8 | 80.1 | 66.1 | 16.9 | 7.6 | 2.7 | 8346 | 2.21 | 1.41 |
| Maize + cowpea (3:1) | 260.7 | 34.4 | 60.4 | 6.7 | 7.0 | 1.1 | 5183 | 1.75 | 1.08 |
| Maize + cowpea (3:3) | 207.5 | 56.7 | 45.6 | 11.4 | 5.1 | 1.9 | 4244 | 1.62 | 1.03 |
| Maize + cowpea (4:1) | 246.3 | 29.8 | 57.8 | 5.7 | 6.2 | 0.9 | 4416 | 1.64 | 1.02 |
| CD (5 %) | 18.5* | | 7.3* | | 0.6* | | 710 | - | - |

(Sunil kumar et al., 2005) * CD for total of both maize + cowpea

At Jhansi (UP), intercropping of maize and cowpea in the row proportion of 2:2 recorded significantly higher total green fodder (338.3 q/ha), dry matter (83.0 q/ha) and crude protein yield (10.3 q/ha), net return (Rs.8346/ha) as well as land equivalent ratio (1.41) over other intercropping systems (Table 10)(Sunil kumar et al., 2005).

**Table 11. Performance of forage maize based intercropping systems**

| Intercropping | Green fodder yield (t/ha) | Dry matter yield (t/ha) | Crude protein yield (t/ha) | Cost of cultivation (Rs/ha) | Net return (Rs./ha) | B:C ratio |
|---|---|---|---|---|---|---|
| Maize (sole) | 35.2 | 8.52 | 0.72 | 9310 | 8290 | 0.89 |
| Cowpea (sole) | 23.8 | 4.44 | 0.80 | 824 | 8420 | 1.02 |
| Rice bean (sole) | 22.9 | 4.17 | 0.76 | 7790 | 8240 | 1.10 |
| Cluster bean (sole) | 20.6 | 4.03 | 0.75 | 7480 | 6940 | 0.93 |
| Maize + cowpea (2:2) | 43.2 | 9.57 | 1.10 | 8780 | 16100 | 1.84 |
| Maize + rice bean (2:2) | 41.5 | 9.06 | 1.06 | 8550 | 15320 | 1.71 |
| Maize + cluster bean (2:2) | 36.2 | 8.33 | 0.87 | 8400 | 11420 | 1.36 |

(Sharma et al., 2008)

Intercropping of forage maize (African tall) with cowpea (EC 4216) or rice bean (Bidhan 1) both in 2:2 row ratio recorded significantly higher total green fodder, dry matter, crude protein, net returns and B:C ratio (Sharma et al., 2008).

## Cropping systems

At Udaipur (Rajasthan), significantly higher net return of Rs. 5807/ha was obtained by maize (fodder) – gram sequence followed by sorghum (fodder) – mustard (Rs. 4889/ha) and Pearl millet (fodder)- barley (Rs. 3387/ha) sequence (Jain and Sharma, 2005). At Uruli Kanchan (Maharashtra), legume based fodder cropping sequence of berseem in rabi season, sorghum in summer and maize in kharif season has given significantly higher net monetary return (Rs. 58562/ha/year), B:C ratio (1.93), profitability (Rs. 275.24/day/ha) and maize fodder equivalent (7329 q/ha) (Kauthale et al., 2007).

## Harvesting

Harvest the crop when the cob is in the milky stage. The best indicator of forage quality is the milkline score. The optimum milkline score is 2.5. This is a grain moisture percentage of approximately 63 to 68 % which is approximately 32 to 37 % dry matter. A milkline score between 2 and 3 is considered acceptable. At the grain fill stage, there is a visual line of separation between the starchy tip and milky base of individual grains. This line is known as the milkline and moves down the grain towards the cob over a period of 6 weeks. When the milkline is halfway down the score is said to be 2.5 (0 = top of grain before the milkline becomes visible and 5 = milkline at the base of the grain).

**Table 12. Influence of harvesting stages on fodder maize**

| Harvesting stages | Plant height (cm) | Leaves / plant | Internode / plant | Length of internode (cm) | Stem thickness (cm) | Fodder yield (q/ha) | |
|---|---|---|---|---|---|---|---|
| | | | | | | Green | Dry |
| Silking stage | 155.6 | 15.7 | 13.1 | 13.6 | 1.16 | 280.6 | 66.9 |
| Milking stage | 164.4 | 16.0 | 13.4 | 15.4 | 1.24 | 319.5 | 82.1 |
| CD 5 % | 8.4 | NS | NS | 1.0 | NS | 23.1 | 7.9 |

Tasselling stage of maize was found to be the optimum time for harvesting both the crops for green fodder than at milking stage (Kumar and Banumurthy, 2001). Katiyar and Singh (1999) reported crude protein percent varying between 3.72 and 12.82 in leaf and stem of maize crop. Higher crude protein percent in maize fodder was reported at age of 50 days (Amador and Boschin, 2000). Begum (1994) and Sharma et al. (1997) found that half of the population of maize harvested for green fodder at tasseling stage does not have detrimental effect on grain production of the left over maize crop.

## Maize silage

Corn harvested for sillage yields one-third more feed nutrients per acre than corn harvested for grain. Corn in the full dent stage produces 50 % more feed than in the milk stage and 100 % more feed than in the silking stage. Corn harvested in the milk or silking stage results in poorer quality silage because of its high moisture content. One of the most important steps in producing quality maize silage is to harvest at the proper moisture. The storage structure determines the proper moisture level at which to harvest. Desired moisture levels for different structure are as follows. Sealed airtight silos are 55 to 60 %. Bag silos are 60 to 70 %. Upright silos are 62 to 68 %. Trench silos are 65 to 70 %. Ideally, moisture levels in the silage should be monitored at harvest to prevent harvesting the crop outside the desired range. If moisture testing is not feasible, then estimate the crop moisture by the stage of crop development.

**Table 13. Nutritive value of corn silage harvested at various stages of maturity and conditions**

| Stage of maturity | Dry matter (%) | Crude protein (%) | Net energy of lactation (Kcal/kg) |
|---|---|---|---|
| Pre-silk | 10.0 | 12.4 | 281 |
| Silk | 15.0 | 11.3 | 290 |
| Milk | 21.0 | 7.0 | 304 |
| Over-ripe | 45.0 | 9.1 | 281 |
| Droughty-few ears, stunted | 25.0 | 9.9 | 281 |
| Non-pollinated-growthy mature | 27.0 | 7.6 | 318 |

At Islamabad (Pakistan), higher crude protein values were found in leaves as compared to other portions. Younger plants contained more crude protein as compared to the matured ones. Crude fibre content of various fractions of the plant ranged between 19.12 and 35.60 % with maximum values in the bottom portion of the stem. Matured plants contained more crude fibre (Azim et al., 1989).

Kernel milk line can serve as an indicator of whole plant moisture levels. As kernels start to dent, as separation between kernel starch and milk can be seen. The firm starch is deposited in the crown (outer) area of the kernel and milk occupies the basal area of the kernel. This appears as a whitish line separating the two areas. As the crop matures, this kernel milk line moves down the kernel and the whole plant moisture declines. When this line reaches the midpoint of the kernel, 90 % of the final kernel dry weight has been achieved and silage yields reach a maximum. At this point, the stover part of the plant has good digestibility and the moisture is usually in the desired range for storage in airtight silos.

A higher moisture level and a slightly earlier harvest are recommended for bunker silos. Harvest time can be predicted by monitoring the progression of the milk line. When the milk line reaches the base of the kernel, a black layer forms and the crops is physiologically mature. Silage harvest should not be delayed beyond the black layer point because the silage gets too dry, the kernels tend to harden and the digestibility of the stover declines rapidly.

The desired chopping length for maize silage is 1.56 to 1.88 cm. If silage is harvested when the crop moisture is lower than desired, consider chopping finer than normal to promote good packing and to minimize air pockets in the silage.

## Maize as fodder

Maize is an excellent crop in terms of biomass production. Since the production as well as productivity of maize is increasing in our country the availability of biomass from maize is also increasing by the same magnitude. Maize straw is used as animal fodder since the ancient times. However, the fodder quality of green maize is far excellent. Amongst the non-legume cultivated fodders, maize is the only fodder which produces better nutritional quality along with good quantity of biomass. It is commonly grown as a summer fodder in the north western regions of the country particularly Punjab, Haryana and Western Uttar Pradesh. Its quality is much better than sorghum and pearl millet since both sorghum as well as pearl millet possesses anti-quality components such as HCN and oxalate, respectively. Secondly baby corn is ready for harvest approximately 2 months after sowing. This means the baby corn as well as maize fodder is available in bulk approximately 2 months after sowing the crop.

## Maize silage

Maize can be conserved as silage. It has to have 30 % dry matter to be ensiled successfully. There is no need to dry out the plant material any more than that, so wet weather is not such a constraint as it is with making hay. This means the crop can be cut any time, depending on when it was planted.

Silage is the product from a series of processes by which cut forage of high moisture content is fermented to produce a stable feed which resists further breakdown in anaerobic storage. The objective is to retain or augment the nutrients present in the original forage and deliver silage accepted by livestock. This is usually attained through an anaerobic fermentation dominated by lactic acid bacterial. Good silage make from tropical forages has a pH less than 5.0, the percent of total nitrogen which is ammonia ($NH_3N:N$) of less than 15 %, lactic acid which is 50 % or more of the total organic acids and butyric acid content of not greater than 0.5 % of the total dry matter. When forage is put into a sealed container such as a pit covered with plastic, a drum or a plastic bag, the container is called a silo. A silo has to be completely sealed against air and the forage material must be chopped and compressed in the silo to ensure the fast development of anaerobic conditions and a rapid fall in pH. In these conditions, lactic acid bacteria, which convert some of the sugars in the plant into the pleasant tasting lactic acid, prevail over undesirable bacterial such as clostridia which produce butyric acid, which is unpalatable to livestock, and moulds, which cause rotting of the silage. Good silage has a sweet smell and cattle, goats and sheep will readily eat it. Silage can be made quite cheaply and easily, provided it is done correctly.

**Advantages of silage**

Silage acts as a fodder bank which ones made could be used round the year. There are numerous advantages of silage making. Some of these are listed below:

» It is used during the scarcity of green forages called lean periods.

» It provides round the year supply of nutritious fodder.

» Silage is as nutritious as green fodders as it preserves the nutrients in their original form and hence it is as good for animal feeding as green forages itself.

» Could help in reducing the shortage of green fodder in the country.

» The labour cost in dairy farming is significantly reduced by using only silage as fodder as 4-5 persons can easily manage a flock of 40-50 cattle heads, since maximum labour is consumed in harvesting the green forages.

» The entire crop is harvested in a single step for making silage as is the case with baby corn and sweet corn. Baby corn as well as sweet corn stalks is the best fit fodder for silage making as the entire field is harvested in one go. One time harvesting is beneficial in many ways since we can harvest the crop at the appropriate time and at the same time the field became available for the timely sowing of the next crop.

» Palatability increases as hard stems when fermented into silage become soft and better utilized by the dairy animals.

» Green forages may possess some anti-nutritional components e.g. HCN in sorghum, oxalate in pearl millet and sometimes nitrate in maize. The anti-quality components are either destroyed or lowered during silage fermentation, for example nitrates, if present, were reported to be lowered in silage as compared to the green forages. HCN is almost destroyed whereas; oxalates were also reported to be reduced to half in silage. During silage fermentation the stem of the crop became soft, which helps in easy digestion by animals.

» Seeds of the most common weeds are destroyed during silage fermentation thereby reducing the problem of dispersal of these seeds with cow dung as farm yard manure.

## Method of Silage Making

### Digging of pit

A pit is dug up at a suitable position in the farm. The location of the pit is quite important. It should be near to the cattle shed at some higher and sloppy ground so that rain water could not enter the silage and spoil it. The water table of the ground should be deep to avoid seepage. The pit should be little bit slanting so that the rain water could easily flow away to one side to avoid spoilage during rainy season.

### Size of the pit

Size of the pit depends upon the availability of fodder as well as size of herd. However, if sufficient fodder is available then the number of animals and time duration for which the silage is to be used should be considered. The size of some rectangular silo pits along with their capacity is given below:

| Length (m) | Breadth (m) | Depth (m) | Quantity of fodder in the pit (q) |
|---|---|---|---|
| 3 | 3 | 2 | 95 |
| 7 | 3 | 2 | 223 |
| 10 | 3 | 1.5 | 350-400 |

According to thumb rule for determining the capacity of the pit, if an animal need 20 kg silage daily then to feed 5 animals for a period of 90 days, the size of the pit should be 3 m x 3 m x 2 m. In other words, one cubic meter pit can accommodate about 5-6 quintals of green fodder.

### Shape of the pit

The shape of the pit is also important. It should have slanting walls with narrow base and broad opening. This type of shape is beneficial for filling the pit as this helps in maximum exclusion of air and ease in removing the silage from the pit.

## Preparation of the pit

Plaster the walls of the pit with cow dung. The farmers may also prepare a *pucca* silo pit. Cover the plastered pit with polythene. The base of the pit should not be covered by plastic sheet, rather it should be covered by straw so that the excess moisture, if present, or juice could be absorbed efficiently.

### Filling of pit

Chaff the crop into 5-8 cm pieces and start filling the pit. Spread the chaffed fodder upto a height of 1 foot and then press it. This process should be repeated after each filling. The major precaution during filling the pit is to exclude as much air as possible from the chaffed fodder. This is achieved by pressing the material by manual labour or mechanically by using a tractor. Care should be taken that material on the sides and edges are properly compressed. By doing so, keep on adding the material till the heap is around one meter above the ground level. Finally add some material in the central portion of the heap and then trample it.

### Sealing of pit

Now cover the heap with a polythene sheet. Seal the edges of the sheet by cow dung. Spread about 10-15 cm layer of straw on the sheet followed by 5-7 cm layer of earth. It should then be plastered with a layer of clay or cow dung. Any cracks in the cover, which develops subsequently, should immediately be plugged as to avoid entry of air or water into the pit. Prepared in this way, the silage is ready for feeding after 35-40 days of sealing the pit.

## Recognition of Well-Fermented Silage

### Colour

Properly prepared silage is recognized from its colour. The colour of the well-fermented silage is bright light green or dull yellow, whereas that of the poorly fermented silage is olive, blue green or dark brown.

### Smell

The smell of the well-fermented silage is like that of vinegar whereas poorly fermented silage is foul smelling. In the properly fermented silage the soluble carbohydrates are converted to lactic acid, whereas in the poorly fermented silage, butyric acid is the end product of fermentation which is primarily responsible for the bad smell. The poorly fermented silage should not be fed to the animals and should be discarded.

### Storage period

Properly prepared silage can be preserved for a long period. If properly covered silage could be stored as long as 10-12 years or so. Once opened, it should be used regularly, and should be consumed within 3-4 months.

## Feeding of silage

Open the pit from one side after removing the earth and straw. Each time, a uniform layer of silage is removed vertically (from top to bottom) depending upon the daily need. Do not open the whole pit at once. Cover the opened side immediately after removing the silage, to avoid any exposure by air/moisture. The top portion may contain moulds which should not be used for feeding. The animals may take some time (3-4 days) to adapt to the silage feeding, therefore feed 5-7 kg of silage along with some other fodder for the initial period. Once adapted, the cattle can be put on silage exclusively. The approximate quantity of silage to be fed to the animals is given below:

| Type of animals | Quantity of silage /head |
|---|---|
| Buffalo | 25-30 kg |
| Milch cows | 25-30 kg |
| Pregnant cows | 15-20 kg |
| Bulls | 20-25 kg |
| Heifers | 10-15 kg |

Silage could play a significant role in reducing the green fodder shortage of the country. Silage made from specialty corn could immensely help in expanding the dairy sector particularly in the peri-urban regions of the country.

## Maize stover

The nutritional quality of maize stover is poor. It is made up cellulose, hemicelluloses, lignin, etc. Cellulose (30-50 % of total feedstock dry matter) is a glucose polymer linked by ß–1,4 glycosidic bonds. The basic building block of this linear polymer is cellubiose, a glucose-glucose dimer (dimer: two simpler molecules-monomers- combined to form a polymer). Cellulose is almost completely digestible by ruminants as they contain the enzyme cellulase responsible for cellulose breakdown. Hemicellulose is a short, highly branched polymer of five-carbon (C5) and six-carbon (C6) sugars. Specifically, hemicellulose contains xylose and arabinose (C5 sugars) and galactose, glucose and mannose (C6 sugars). Hemicellulose is considered partially digestible. Lignin (15–25 % of total feedstock dry matter), a polyphenolic structural constituent of plants, is the largest non-carbohydrate fraction of lignocellulose.

Unlike cellulose and hemicellulose, lignin cannot be digested and is the indigestible fodder component. More the amount of lignin present in the stover, more will be the fodder considered unfit for utilization as animal fodder. Ash (3-10 % of total feedstock dry matter) is the residue remaining after ignition (dry oxidation at 575 ± 25°C) of herbaceous biomass. It is composed of minerals such as silicon, aluminum, calcium, magnesium, potassium and sodium. Other compounds present in lignocellulosic feed stocks are known as extractives. These include resins, fats and fatty acids, phenolics, phytosterols, salts, minerals, and other compounds. In general, corn stover is about 38 % cellulose, 26 % hemicellulose and 19 % lignin. Maize stover, therefore, contains low levels of crude protein content and high levels of indigestible carbohydrates such as lignin.

**Table 14. Detailed compositional analysis of corn stover (% DM)**

| Composition | Minimum | Maximum | Mean | SD |
|---|---|---|---|---|
| Cellulose | 31.3 | 41 | 37.5 | 2.8 |
| Structural glucan | 33.8 | 41 | 37.5 | 2.2 |
| Hemicellulose | 20 | 34.4 | 26.1 | 4.8 |
| Xylan | 19.8 | 25.8 | 21.7 | 2.1 |
| Arabinan | 1.7 | 6.1 | 2.7 | 1.6 |
| Galactan | 0.7 | 3 | 1.6 | 1 |
| Mannan | 0.3 | 1.8 | 0.6 | 1.1 |
| Total lignin | 15.8 | 23.1 | 18.9 | 2.6 |
| Acid soluble lignin | 1.9 | 3.6 | 2.9 | 0.9 |
| Acid insoluble lignin | 13.6 | 19.8 | 16.4 | 3.1 |
| Acid detergent lignin | 3.1 | 5 | 4.1 | 1.3 |
| Crude protein | 3.5 | 8.7 | 4.7 | 2.2 |
| Ash | 4.2 | 7.5 | 6.3 | 1.2 |

Maize stover possesses very little moisture ranging from 9-12 per cent. Lignin is structural carbohydrate which is almost indigestible to ruminants. Compared to wheat straw maize stover contain more protein, but higher lignin content. This is perhaps the reason why wheat straw is considered more popular as animal fodder compared to maize stover. Maize is considered the best fodder among non-legumes as it contains sufficiently higher quantities of protein and palatability. The quality of maize declines with maturity. However, the deterioration in quality was more severe in leaves compared to stems. The leaves which are succulent at physiological maturity become hard and develop many indigestible components with maturity. Soluble solids rapidly decrease and lignin and xylan increase shortly after grain physiological maturity.

**Table 15. Changes in composition of corn stalk and leaf with crop maturity**

| Composition | Late dent | Physiological maturity | Post physiological maturity |
|---|---|---|---|
| **Corn stalk** | | | |
| Structural glucan | 35 | 35 | 35 |
| Xylan | 16 | 22 | 23 |
| Lignin | 15 | 20 | 19 |
| Protein | 3 | 4 | 4 |
| Soluble solids | 15 | 4 | 4 |
| **Corn Leaf** | | | |
| Structural glucan | 18 | 23 | 32 |
| Xylan | 2 | 17 | 22 |
| Lignin | 4 | 13 | 16 |
| Protein | 8 | 8 | 4 |
| Soluble solids | 35 | 8 | 6 |

(Pordesimo *et al.*, 2005)

### Urea treatment of maize stover

Urea treatment of maize stover is an easy and effective method used since long for enrichment of nutritional quality of dry roughages. The method is simple and easy to conduct. The chronological events are listed as:

- » The stover is chopped to small pieces measuring 5-7 cm in size
- » The chaffed fodder is collected in a heap
- » Dissolve 3.5 kg urea in a 50-60 kg of water. It may carried out a big size *tub* easily available in dairy farms
- » Spray the urea solution uniformly over one quintal of chaffed stover. Simply, the solution may be pored over the fodder and then the fodder to be mixed thoroughly

» Store the mixed stover in a closed room or the form of a *kup.*

» Exactly 10 days after storage the treated stover is ready for feeding to the cattle.

» Before feeding, the stover is spread on ground for some time to allow the gas to escape. The stover is now ready to be used as fodder.

Urea treatment is a well-demonstrated method of improving the nutritive value of low quality stovers, by the effect of the ammonia ion on the cell wall. The ammonia ion swells and hydrolyses the cell wall carbohydrates and phenolic monomers (Chesson *et al*., 1983). The final effect is an increase of the digestibilities of dry matter (DM) and cell wall, of the nitrogen content and of DM intake (Oji *et al*., 1977).

The effect of the urea treatment on the nutritive value of roughage is the result of two processes which occur within the treated forage: (i) ureolysis which turns urea into ammonia through and enzymatic reaction that requires the presence of the urease enzyme; and (i) the effect of ammonia on the cell walls on the forage. Several factor such as urea doses, moisture and temperature affect the effectiveness of urea treatment.

**Nutrient content of urea treated maize stover**

The nutritional quality of urea treated maize stover is drastically enhanced compared to normal stover. A significant increase was observed in crude protein content. The increased microbial biomass in the treated stover may contribute significantly towards higher crude protein concentration. The neutral detergent fiber (NDF) showed a significant reduction in the urea treated maize stover. NDF is an important parameter which is directly related to dry matter intake by the animals. A fall in NDF therefore showed that animal intake of the treated stover is higher compared to normal stover.

**Table 16. Comparative chemical composition of urea treated maize stover**

| Variables | Untreated maize stover | Urea treated maize stover |
|---|---|---|
| Dry matter (%) | 91.5 | 95.6 |
| Crude protein (% dry matter) | 5.83 | 7.67 |
| NDF (% dry matter) | 86.8 | 73.9 |

(Yirga *et al*., 2011)

Thus it could be concluded that maize is an excellent crop which could effectively be utilized as a feed and fodder crop. Specialty corn is going to play an important role in the socio economic perspective of the rural folk. Baby corn and sweet corn cultivation will substantially add up to the income of the farmers as specialty corn is sold at reasonably good price in the market, where the green fodder will boost the dairy industry. The silage making is breakthrough technology which could provide a quantum boost to the dairy sector. The urea treatment of maize stover is a simple technique suitable for small and marginal farmers, whereby they can easily enhance the milk production potential of their cattle.

## References

Agath, B.M. M.G. Jethwa, V.D. Khanpara, B.B. Kaneria and R.K. Mathukia (1997). Effect of harvesting stages and levels of nitrogen and phosphorus on yield of fodder maize. GAU Res. J., 23 (1): 88 - 90.

Almodares, C.A., M. Jafarinia and M.R. Hadi (2009). The effects of nitrogen fertilizer on chemical compositions in corn and sweet sorghum. American Eurasian J. Agric. and Env. Sci., 6 (4): 441 - 446.

Amar Singh, A.K. Vyas and A.K. Singh (2000). Effect of nitrogen and zinc application on growth, yield and net return of maize (*Zea mays* L.). Ann. Agric. Res., 21(2): 295-297.

Andrew, L., Fletcher, M. Shane, R. Wilson, and R.M. John (2006). Using plastic mulch to increase maize silage production in a cool climate with a short growing season. New Zealand Inst. For Crop & Food Res. Ltd. Private Bag 4704, Christchurch 8140, New Zealand.

Argillier, O., Y. Hebert and Y. Barriere (1995). Relationship between biomass yield, grain production, lodging susceptibility and feeding value in salage maize. Maydica, 40: 125-136.

Asif Iqbal, Muhammad Ayub, Nadeem Akbar and Riaz Ahmad (2006). Growth and forage yield response of maize-legume mixed cropping to different sowing techniques. Pakistan J. Agric. Sci., 43(3-4): 126-130.

Ayub, M., M.A. Nadeem, M.S. Shara, and N. Mahmood (2002). Response of Maize (Zea mays L) fodder to different levels of Nitrogen and Phosphorus. Asian J. Plant Sci., 1 (4): 352-354.

Ayub, M., R. Mahmood, A. Tanveer and M.S. Sharar (1999). Effect of seeding density on fodder yield and quality of two maize varieties. Pakistan J. Biol. Sci., 2 (3): 664-666.

Azim, A. Z. Naseer and A. Ali (1989). Nutritional evaluation of Maize fodder at two different vegetative stage. AJAS, 2 (1): 27 – 34.

Azim, A., A.G. Khan, MA Nadeem and D. Muhammad (2000). Influence of maize and cowpea intercropping on fodder production and characteristics of silage. Asian-Australasian J. Animal Sci., 13(6): 781-784.

Begum, A. (1994). Manipulation of population for better grain and forage yield in rabi maize (*Zea mays* L.). MSc (Agriculture) Thesis, Acharya N. G. Ranga Agricultural University, Hyderabad.

Channakeshava, B. C., K. P. Rama Prasanna and B. K. Ramachandrappa (2000). Influence of spacing and fertilizer levels on seed yield and yield components in African tall fodder maize. Karnataka J.Agric. Sci., 13(2): 343-348.

Channakeshava, B.C., B.K. Ramachandrappa and K.P.R. Prasanna (2007). Influence of micronutrients and plant growth regulator on seed quality parameters in African tall fodder maize. Curr. Res., 28: 11 - 12.

Channakeshava, B.C., K.P.Ramaprasanna and B.K.Ramachandrappa (2007a). Effect of plant growth regulators and micronutrients on growth components and seed yield in African tall fodder maize (*Zea mays* L.). Agric. Sci. Digest, 27(1): 38-40.

Chaudhary, D.P., Ashwani Kumar, S.M. Sapna, P. Srivastava and R.S. Kumar (2012). Maize as fodder? an alternative approach. Directorate of Maize Research, Pusa Campus, New Delhi, Technical Bulletin 2012/04. p. 32.

Chittapur, B.M., H.B. Babalad, S.M. Hiremath, GA Kulkarni and D.R Sardeshpande (1994). Intercropping of fodder legumes in grain maize. J. Maharashtra Agric. Univ., 19(3): 454-455.

Choudhary, S.K., R.B. Thakur and S.K. Thakur (2003). Effect of relative proportion of fertility on the performance of maize + berseem intercropping system. Forage Res., 30 (1): 22-24.

Cox, W.J., S. Kalonge, D.J.R. Cherney and W.S. Red (1993). Growth, yield and quality of forage maize under different nitrogen management practices. Agron. J., 5: 341-347.

Dahmardeh, M., A. Ghanbari, B. Syasar and M. Ramroudi (2009). Effect of intercropping maize (*Zea mays* L.) with cow pea (*Vigna unguiculata* L.) on green forage yield and quality evaluation. Asian J. Plant Sci., 8: 235-239.

Dudhat, M.S., M.G. Savalia and M.V. Ramdevpatra (2004). Response of forage maize to nitrogen and phosphorus levels. Forage Res., 30 (1); 34 – 35.

Eltelib, H.A., M.A. Hamad and E.E. Ali (2006). The effect of Nitrogen and Phosphorus fertilization on growth, yield and quality of forage maize (*Zea mays* L.). J. Agron., 5(3): 515-518.

Gandhi, A.W., M.R. Wagan, I. Rajpar and N.B. Sial (2008). Response of fodder maize of foliar and soil applied Zinc. RCM report. Department of Soil Science, Faculty of crop production, singh Agriculture University, Jandojan, Pakistan.

Gangaiah, B. (2004). Effect of animal urine application on growth and yield of fodder maize. Indian J. Agric. Sci., 74 (12): 678 – 679.

Gasim, S.A. (2001). Effect of nitrogen, phosphorus and seed rate on growth, yield and quality of fodder maize (Zea mays L.). M.Sc. Thesis, Faculty of Agriculture, University of Khartoum, Khartoum, Sudan.

Gaurkar, S.G. and G.M. Bharad (1998). Effect of plant population detopping and nitrogen levels on growth and yield of maize (*Zea mays* L.). PKV Res. J., 22(1):136-37.

Girija Devi, L. (2002). Forage yield of maize as influenced by nitrogen levels and biofertilizers. Forage Res., 27 (4): 263 – 266.

Gupta, B.K., B.L. Bhardwaj and A.K. Ahuja (2004). Nutritional value of forage crops of Punjab. Punjab Agricultural University Publication, Ludhiana.

Halima, H. and S.S. Chauhan (2003). Effect of rate of maize leaf defoliation at various growth stages on grain, stover and yield components of maize and under sown forage production. Indian J. Agric. Res., 37 (2): 136 – 139.

Hani, A.E., M.A. Hamad and E.E. Ali (2006). The effect of nitrogen and phosphorus fertilization on growth, yield and quality of forage maize (*Zea mays* L.). J. Agron., 5(3): 515-518.

Hassan, S.W., F.C. Oad, S.D. Tunio, A.W. Gandahi, M.H. Siddiqui, S.M. Oad and A.W. Jagirani (2010). Impact of nitrogen levels and application methods on agronomic, physiological and nutrient uptake traits of maize fodder. Pakistan J. Bot., 42(6): 4095-4101.

Igbal, A., M. Ayoub. H. Zaman and R. Ahmed (2006). Impact of nutrient management and legumes association on agro qualitative traits of maize forage. Pak. J. Bot., 38: 1079-1084.

Jadhao, S.L., S.K. Podghan, P.G. Ingole, S.D. Malvi and N.B. Gore (1995). Effect of nitrogen levels on growth, yield and quality of two fodder maize varieties. PKV Res. J., 19: 188-189.

Jain, K.K. and S.K. Sharma (2005). Effect of kharif fodder crops on productivity and economics of succeeding rabi crops. Forage Res., 30(4): 189 – 191.

Jat, A.S., R.C. Dadheech and C.M. Parihar (2006). Direct and residual effect of herbicides, phosphorus levels and phosphate soubilizers used in wheat on the productivity and profitability of wheat – fodder maize system. Annals Agric. Res., 27 (2): 154 - 158.

Jat, R.S. and I.P.S. Ablawat (2004). Effect of vermicompost, biofertilizers and phosphorus on growth, yield and nutrient uptake by gram and their residual effect on fodder maize. Indian J. Agric. Sci., 74(7): 359 – 361.

Jayanthi, C., C. Chinnusamy, V. Veerabadran and P. Rangasamy (1994). Production potential of compatible fodder cereal-legume mixtures in the North-Western Zone of Tamil Nadu. Madras Agric. J., 81(8): 420-422.

Joshi, Y.P. and K. Kumar (2007). Response of forage maize to the levels of nitrogen and seed rates. Forage Res., 33 (3): 171 - 173.

Kauthale, V.K., P.S. Takawale and D.L. Sale (2007). Production potential and economics of different cropping sequence under irrigated conditions of western Maharashtra. Forage Res., 33(3): 164 – 167.

Koul, B.G. (1997). Effects of sowing methods, nitrogen levels and seed rate on yield and quality of fodder maize (*Zea mays* L.). M.Sc. thesis, Faculty of Agriculture, University of Khartoum, Khartoum, Sudan.

Krishna, A., S.V. Raj Khelka and R.A. Sambasiva (1998). Effect of planting pattern and nitrogen on fodder maize (*Zea mays*) inter cropped with cowpea (*Vigna unguiculata*). Indian J. Agron., 43(2): 237 - 240.

Kumar, K.R. and V.B. Bhanumurthy (2001). Effect of staggered sowing and relative proportion of cowpea on the performance of maize + cowpea intercropping system. Forage Res., 27(2): 105 -110.

Kumar, N. and N. Ramawat (2006). Production potential and economics of maize + guinea grass cropping in different proportions under sub tropical conditions. Forage Res., 32(3): 159 – 162.

Kumar, P. and N.K. Prasad (2003). Biological and economical sustainability of forage maize + cowpea intercrop. Indian J. Agric. Sci., 73(6): 341 – 342.

Macelli, L.C., V.D.P. Pereira, J.O. Machado and J.C. Barbosa (1996). Effect of intercropping of late maize with pigeon peas *(Cajanus cajan* L.) on some plant and soil properties. Clientifica – Jaboticabal, 24 (2): 283-294.

Madhusudhan Reddy, D. and V.B. Bhanumurthy (2010). Fodder, grain yield, nitrogen uptake and crude protein of forage maize as influenced by different nitrogen management practices. IJBSM, 1(2): 69-71.

Mahdi, S.S., B. Hasan and Singh lal (2012). Effect of seed rate, nitrogen and zinc levels on yield and economics of fodder maize in temperate conditions of western Himalayas Indian J. Agron., 57 (1): 86-88.

Mahdi, S.S., B. Hasan, R.A. Bhat and M.A. Aziz (2010). Yield and economics of fodder maize as influenced by nitrogen, seed rate and zinc as influenced by nitrogen, seed rate and zinc under temperature conditions. Forage Res., 36(1): 22-25.

Mahdi, S.S. B. Hasan, Singh lal and M.A. Ganie (2012). Quality of fodder maize and soil health influenced by nitrogen, seed rate and zinc in Kashmir valley. Indian J. Soil Conservation, 40 (2): 147 - 151.

Muller, L.P., P.A. Manfron, O.S. Santos, S.L.P. Mederios, V. Haut, D.N. Dourado and E.B. Fagan (2005). Growth and bromatologic composition of hydroponic maize fodder at different dates of harvest and sowing densities. Zootecnia Tropical, 23(2): 105 - 119.

Muna, E.K., E.A.A. Eltayeb and O.E. Sayda (2011). Effect of nitrogen, intercropping with Lablab bean (*Lablab purpureus*) and water stress on yield and quality of fodder maize. J. Sci. Tech., 12 (03): 55 – 66.

Nanjundappa, G., B. Shivaraj, S. Sridhara and S. Janarjuna (2000). Effect of organic and inorganic sources of nutrients alone and in combination on the growth and yield of fodder maize. Mysore J. Agric. Sci., 34: 247 - 250.

Nanjundappa, G., B. Shivaraj, Sridharasand and S. Janarjuna (2000a). Effect of organic and inorganic source of nutrients alone and in combination on the growth and yield of fodder maize. Mysore Journal of Agricultural Sciences, 34: 239-240.

Nanjundappa, G., G.R. Manare and M.K. Badiger (1994). Yield and nutrient uptake of fodder maize as influenced by nitrogen and potassium. Indian J. Agron., 39(3): 473-475.

Nyamagonda, S.S., S.S. Angadi and B. Yenagi (2002). Effect of seed proportion mixtures on yield and quality of forage maize – legume mixed cropping system. Forage Res., 28(2): 94 -97.

Omer, E.A. (1998). Effects of farmyard manure and urea fertilization on the growth and forage yield of two maize *(Zea mays L.)* cultivars. M.Sc. Thesis, Faculty of Agriculture, University of Khartoum, Khartoum, Sudan.

Patel, A.S., A.C. Sadhu, M.R. Patel and P.C. Patel (2007). Influence of nitrogen and zinc on maize. Forage Res., 34 (4): 209 - 212.

Patel, J.R. and S.Rajagopal (2001). Production potential of forage maize with legumes under intercropping system. Indian J. Agron., 46 (2): 211 - 215.

Patil, S.K., A.A. Pisal and J.S. Deale (1992). Response of fodder maize to biofertilizers. Indian J. Agron., 37: 356 - 357.

Ranzende, P.M.D.E. and MAP. Ramalho (2000). Forage production of intercropped maize and soybean cultivars. Experimental Agric., 36(1): 91-100.

Rashid, M. and M. Iqbal (2012). Effect of phosphorus fertilizer on the yield and quality of maize (*Zea mays* L) fodder on clay loam soil. J. Animal Plant Sci., 22(1): 199-203.

Reddy, C.M., K.B.S. Devi and M.A. Sultan (2004). Fodder based intercropping for higher biomass and quality fodder production. Forage Res., 29(4): 217 – 219.

Sahrawat, K.L., T.J. Rego, S.P. Wani and G. Parahasaradhi (2008). Sulfer, boron and zinc fertilization effect on grain and straw quality of maize and sorghum in semiarid and tropical region of India. J. plant Nutr., 31: 1578 - 1584.

Shanti, M., D. Nagalakshmi, R. Balaji Naik, V. Chandrika and Ch. Chiranjeevi (2012). Study of forage quality of various maize cultivars produced under different use patterns. Forage Res., 37(4): 234 – 237.

Sharma, B.K., D.R. Thakur and S.K. Bhalla (1997). Evaluation of maize fodder varieties for dual purpose (grain + fodder) under different plant spacings. Forage Res., 23(4): 221-224.

Sharma, R.P., A.K. Singh, B.K. Poddar and K.R. Raman (2008). Forage production potential and economics of maize (Zea mays) with legumes intercropping under various row proportions. Indian J. Agron., 53 (2): 121-124.

Shukla, V.K., K.L. Tiwari, S.B. Agrawal and R. Anay (2001). Efficacy of atrazine in forage maize under varying nitrogen levels. *JNKVV Res. J.,* **35**(1/2): 70.

Sunil kumar, C.R. Rawat and N.P. Melkania (2005). Forage production potential and economics of maize and cowpea intercropping under rainfed conditions. Indian J. Agron., 50 (3): 184 - 186.

Sunitha, S. and L. Sreekantan (1992). Performance of cowpea + fodder maize intercropping in summer rice fallows. Indian J. Agron., 37 (4): 795 - 798.

Surve, V.H. and M.K. Arvadia (2012). Performance of fodder sorghum (Sorghum bicolor L.), Maize (Zea mays L.) and cowpea (Vigna unguiculata (L.) Walp.) under sole and intercropping systems. International J. Agric Res Rev., 2 (1), 28-31,

Tiwana, U.S., K.P. Puri, B.L. Bhardwaj and S. Pritpal (2007). Effect of seed rates and row spacings on the fodder yield of maize under different sowing dates. Range managt. Agrofo., 28 (2B): 309 - 310.

Tripathi, S.B. and C.K. Hazra (2002). Forage productivity of winter maize under legume covers as influenced by nitrogen fertilization. Forage Res., 28(2): 55 – 58.

Tripathy, R.K., L. Pradhan and B.S. Rath (1997). Performance of maize and cowpea forage intercropping system in summer. Indian J. Agron., 42(1): 38-41.

Verma, S.S., A. Kumar and V. Singh (1999). Effect of nitrogen levels and row spacing on nitrogen content and uptake in fodder maize. Forage Res., 25(2): 139-140.

Wadsworth, G. (2002). Forage Maize Fertilizer Requirement. Potash Development Association, Brixtarw, pp. 25-29.

Yirga, H., S. Melaku and M. Urge (2011). Effect of concentrate supplementation on live weight change and carcass characteristics of Hararghe Highland sheep fed a basal diet of urea-treated maize stover. Livestock research for rural development. 23(12): 11-14.

# 28

# Seed Production in Maize

Breeders have developed several high-yielding hybrids of maize in recent years. These have made a major contribution to increase food production. Several of these hybrids are attractive to seed industry, because of higher profit margin as well as the farmers have to buy the hybrid seed each year. Use of hybrid maize has resulted in the development of new enterprises like production, processing, sale and distribution. The seed industry consists of several components like research, production, quality control and marketing. The plant breeder who develops the strain provides specific genetical and morphological characters based on which the cultivar can be identified. Production of hybrid seed consists of 3 stages.

**Breeder's seed**

It is directly controlled by the breeder. It is generally produced in limited area, either cy hand pollination or in isolation under the supervision of the breeder and monitored by a team.

**Foundation seed**

It consists of production of single crosses by sowing male and female parents in 2: 4 row ratio. Detasseling is done in female rows after removing off-types, diseased and unwanted plants, if any. Certification are maintained under the guidance of monitoring team consisting personnel's from the National seed corporation, seed Certification Agencies, breeders from agricultural universities and ICAR nominees.

### Certified seed

Male and female single crosses are sown in 2:6/ 2:8 ratio. Female plants are detasseled at appropriate time after removing all off-types, diseased and undesirables plants, if any. Certification standards are maintained as per norms under the guidance of the monitoring team. The seed obtained on female rows is called the certified seed. It is labeled as hybrid seed and sold to the farmers for commercial cultivation.

## A. Quality seed production of open-pollinated maize varieties

### Season

Maize is cultivated during June – July, September- October and January - February seasons. Sowing during November – December is suitable for seed production, since seed maturity stage will not coincide with rainfall. During early stage of this crop, water stagnation should be avoided.

### Land requirements

There are no specific land requirements for maize seed production. Well drained sandy red or black cotton soil is suitable. Selected fields should be free of volunteer maize plants and well-drained. Volunteer plants are unwanted plant growing from seeds that have fallen from the plants of previous maize crop. These plants may be of some other maize variety and cause genetic contamination. The soil should be well-aerated and suitable for maize growing.

## Isolation requirements

Maize is normally cross-pollinated by wind. Isolation of seed fields from other maize fields is, therefore, necessary to prevent foreign pollen contamination. The seed fields must be isolated at least by 400 m for foundation seed class and 200 metres for certified seed class, from fields of other varieties of maize and fields of the same variety not conforming to varietal purity requirements for certification.

### Field preparation

After presowing irrigation, prepare the land as soon as it is ploughable to a good tilth. Usually one ploughing, two to three harrowing followed by levelling, is sufficient to prepare the field to the desired tilth. Spread 12.5 tons of FYM/ ha or composted coir pith and apply ten packets of Azospirillum in the field. Prepare ridges and furrows with 60 to 75 cm spacing. Formation of ridges and furrows enables saving of irrigation water.

**Source of seed**

Obtain nucleus/ breeder's/ foundation seed from a source approved by a seed certification agency.

**Sowing**

The best time for sowing maize is two weeks prior to the onset of the monsoon. Maize is usually sown from the second week of June to mid-July. A row to row spacing of 60 to 75 cm and plant to plant spacing of 20 to 22.5 cm can be adopted. For sowing one ha of certified seed field, 10 kg of foundations seed is needed. To control seed borne pathogens of downy mildew (fungal disease) treat the seed with carbendazim or thiram at 2 g /kg seed. One day after seed treatment with fungicide, treat the seeds with 600 g of Azospirillum with rice gruel and shade dry for 15 minutes. Azosprillum treatments help in the fixation of atmospheric nitrogen in the soil and provide nourishment to the growing plants. Maize is sown in rows with the help of a maize planter or is dibbled by hand in furrows. The depth of seeding should be from 5 to 6 cm. Sow the seeds at $1/3^{rd}$ of ridge from the bottom.

**Fertilization**

Soil be analysed by a soil testing laboratory and that the final application of fertilizer be based on the results reported. Apart from FYM applied before ploughing, the total requirement of the major nutrients per hectare is 120 to 150 kg nitrogen, 50 to 60 kg phosphorus and 40 to 50 kg potash. About 40 to 50 kg nitrogen and the entire quantity of phosphorus, potash and zinc should be applied as a basal dose, before or at the time of planting. For best results, place the fertilizers to the side and below the seed level. A dose of 40 to 50 kg N/ha should be applied as a side dressing when the plants are 25 days after sowing. The last application (40 to 50 kg N/ha) should be made just before flowering. Inadequate fertilization will result in undersized seed and low yields. Undersized seed will be rejected during processing. Profits can be increased by proper fertilization.

Any nutrient deficiency in maize will adversely affect the growth of the plant. Generally due to micronutrient deficiency, the yield is reduced. Zn and Mg deficiency occurs largely in leaves. You can directly see the faded young shoot due Zn deficiency. And in between the veins of mature leaves yellow lines are seen. To overcome these deficiencies, apply 25 kg zinc sulphate per hectare. Mg deficiency will cause yellowish symptoms between edges and veins of basal leaves. The whole plant will have yellowish appearance due to Fe deficiency. To overcome the above deficiencies broadcast the micronutrient mixture @

12.5 kg/ha mixed with 40 kg sand, after sowing the seeds. Observe the plants closely for micronutrient deficiency and apply the required micronutrients.

## Irrigation

Maize is sensitive to excess water and drought conditions. Schedule irrigations properly to provide adequate moisture in the soil. Immediately after sowing, irrigate the field. Then give life irrigation on third day and irrigate depending upon the condition of the field. Generally for maize, up to 30 days, less number of irrigation is given. Afterwards, it is necessary to irrigate once in 10 days. At this stage, the female cobs will be in silky stage. During this stage, if you fail to irrigate, the silk will not emerge from the cob, leading to poor pollination and seed setting and seed production is affected. Drainage of the maize field is as important as irrigation. Keep the bunds open so that water does not stand for more than a few hours.

## Weed management

Timely weeding and intercultural operations are essential. Intercultivation should not be more than 3 to 5 cm deep so that the roots are not damaged. The last interculture operation should also include earthing up of the crop. Pre-emergence soil application of simazine (tafazine 50 %) or atrazine (atrataf 50 %) at 1.5 kg /ha in 1000 litres of water is recommended. There should be no weeds at the time of weedicide application. Afterwards do not disturb the soil for four to five weeks. Take two hand weeding on 25 and 45 days after sowing if herbicide is not applied. After weeding, apply fertilizer as top dressing and do the earthing up operation.

## Plant protection

Control measures as per the following schedule should be adopted in seed fields.

| Name of insect | Control measures |
|---|---|
| Stem borer<br>Pink borer | Apply lindane 6 G or carbofuran 3 G granules in whorls after 20 - 25 days of sowing or Spray the crop after 20, 35 and 50 days of sowing with carbaryl 50 WP at 1.5 kg/ha or Quinalphos 25 EC at 2.0 litre/ha or Endosulphan 35 EC at 1.50 litre/ha. |
| Hairy caterpillars, army worms, hoppers and maize beetles | Apply Malathion 5 D 25 kg/ha or spray endosulphan 35 EC at 1.5 litres/ha or Quinalphos 25 EC at 2 litre/ha. |

| | |
|---|---|
| Shoot fly | Apply lindance 6 G granules at 20 kg per ha or phorate 10 G granules at 15 kg/ha in the soil before sowing or spray metasystox 0.25 % solution at 600 litre/ha at an interval of 10 days. |
| Weevils and aphids | Spray Quinalphos 25 EC 1.00 litre or Carbaryl 50 WP 1.00 kg/ ha in 500 litres of spray fluid on the 20th day of sowing. |
| Grass hopper | Infested crops should be dusted with Malathion 5 D 25 kg/ha. |
| **Name of Disease** | **Control measures** |
| Seedling blight<br>Leaf blight | Use seed treated with carboxin at 2.5 g/ kg of seed. Spray zineb at 2.5 kg /ha 3 to 4 times, depending upon the intensity of disease, at an interval of 9 to 10 days. |
| Downy mildew and Brown stripe | Spray dithane M-45 @ 1.5 kg in 500 litres of water per hectare. |
| Bacterial stalk rot | Apply bleaching powder at 3.3 g in 10 litre of water att the base of the plant (800 litre per hectare) |
| Leaf spot | Spray Mancozeb or Captan 1.00 kg /ha when the disease intensity is high. |
| *Pythium stalk rot* | Make appropriate arrangement for drainage in the field. Spray 2.5 kg captan or thiram in 1000 litres of water on the stems. Use balanced fertilization. |
| *Cephalosporium* Stalk rot<br>Charcoal rot<br>Leaf rust | Rogue out the affected plants.<br>Spray dithane Z-78 at 2 to 3 kg in 1000 litres of water per hectare 3 to 4 times. |
| Head smut | Use seed treated with carboxin at 2.5 g / kg of seed. Apply Phomasan 40 mg per square metre area |
| Fusarium kernel rot<br>Cephalosporium ear rot<br>Gibberella ear rot<br>Rhizoctonia ear rot<br>Penicillium ear rot<br>Aspergillus ear rot | Sort out the affected ears. |
| Helminthosporium leaf blight | Spray zineb 2.50 kg /ha in 1000 litres of water at an interval of 7 to 10 days. |

## Roguing

Not much of rouging is required in open-pollinated varieties. Generally roguing is done three times in maize. However, roguing is carried out depending upon the necessity. These varieties have broad genetic base and are phenotypically uniform for important economic characters. Some variability, however, may still exist. Roguing for obvious off-types such as obviously tall or dwarf, etc. should be completed before pollen shedding. In addition to obvious off-types, malformed and diseased plants affected by stalk rot should be rogued from time to time as required. Roguing helps in maintaining the genetic purity of seeds. It is necessary to know the distinguishing features of the variety for effective roguing. First roguing should be done during vegetative stage, based on the height of the plant, colour of petiole and colour of leaf. During flowering stage, second roguing is done based on colour of tassel and silk. During roguing at flowering stage, the off types should be removed away from the field immediately. Otherwise it will contaminate the silk and affect the genetic purity. Finally, before harvest, based on colour of seed and cob characteristics, roguing can be done. During drying of the cobs, roguing of cob based on seed colour and seed row will maintain the genetic purity.

## Harvesting

Harvest the crop when the outer cover of the cob turns from green to white colour. Maize ears can be harvested at relatively high moisture content (30 to 35 %) if the facilities for artificial heated air drying are available. If such facilities are not available, the harvesting has to be delayed until the seed moisture content has been reduced to 15 %. An early harvest prevents losses in the field due to bird damage, stalk breakage, ear rots, etc. It is an important operation in maize to maintain genetic purity. Farm women can attend to this work efficiently. Remove very small sized cobs. The cobs which show difference in seed rows and seed colour should be identified and removed. The cobs with very less seeds also should be removed. After harvest, sort out all off-type maize ears, particularly those showing different colours and texture and the diseased ears, before placing them in bins to dry.

After drying, the ears are once again examined and any off-type or diseases ear removed before shelling. The certification standards require bin inspection of maize ears before shelling. Therefore, shelling should be undertaken after the approval of seed certification agency has been obtained. After shelling the seeds are processed at processing plants.

Seeds can be separated by mechanical means. For this operation, moisture content should be around 15 % for seed and 25 % for cob. By maintaining the moisture at this level, mechanical damage can be avoided. Seeds can also be separated from cob by beating with sticks. If the seeds are not in appropriate moisture content, occurrence of mechanical damage may occur to an extent of 48 %. Seeds with mechanical damage may invite fungal attack. After shelling, dry the seeds. Mechanical seed driers can be used for this purpose. During drying, air temperature should not go beyond 40°C. Likewise avoid drying under the hot sun and the seeds can be dried in the morning or evening hours under mild sun. Quality seeds can be obtained by size grading the seeds with 18/64" round perforated sieves. Thereby you get good quality seeds of uniform size. Average seed yield varies from 2500 to 3000 kg/ha.

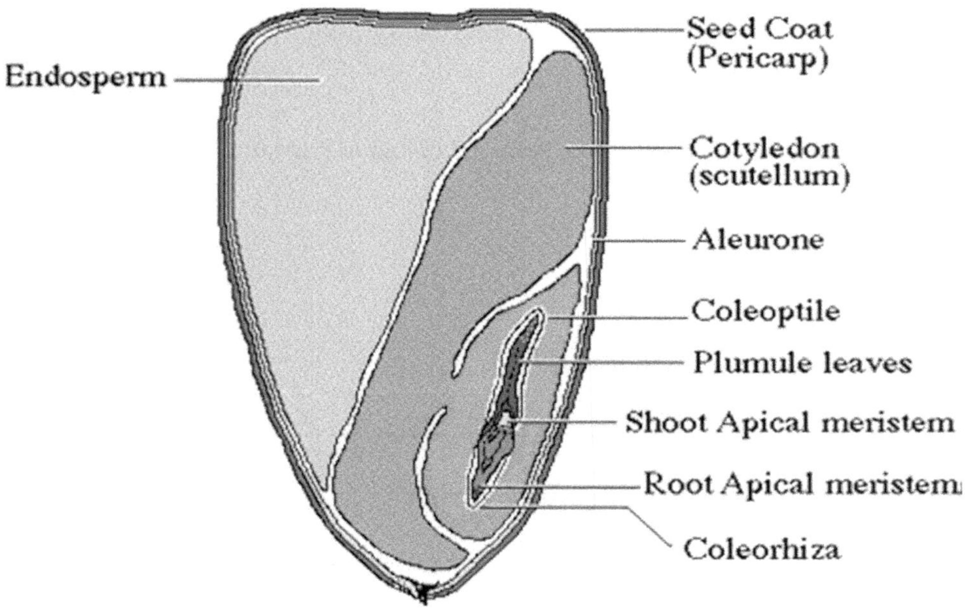


**Seed**

## Seed storage

Dry the seeds to 12 % moisture content and treat with Captan or Thiram 75 % wettable powder @ 70 g per 100 kg seeds mixed in 500 ml of water. Treated seed packed in cloth bag can be stored for more than one year. Seeds dried up to 8 % moisture, treated with above chemicals and packed in moisture vapour proof polyethylene bag can be stored for 1½ years. Instead of toxic chemicals, you can also use nontoxic chemicals like halogen mixture for seed treatment. This will reduce the environmental pollution and ensures safety to the person engaged in seed treatment. Seeds treated with halogen mixture @ 3 g /kg can also be stored well. Prepare the halogen mixture by mixing bleaching powder ($CaOCl_2$) + Calcium Carbonate ($CaCO_3$) + arappu leaf powder (*Albizzia amara*) in 5:4:1 ratio and keeping in vapour proof jar for one week. Then use it for seed treatment. Halogen mixture is nontoxic.

## Seed certification

**Table 1. Field standards (certified seed)**

| Factor | Maximum permitted |
|---|---|
| Off types plants that have shed or shedding pollen when 5 % or more than 5 % of the female flowers are in receptive stage. | 0.20 |
| Pure seed (minimum) | 98.0 % |
| Inert matter (maximum) | 2.0 % |
| Other crop seeds (maximum) | 5/kg |
| Other distinguishable varieties based on kernel colour and texture (maximum) | 5/kg |
| Weed seeds | Nil |
| Germination (minimum)-Inbreds | 80 % |
| Moisture content (maximum) | 12.0 % |
| Moisture pervious container | |
| Moisture vapour poof container | 8.0 % |

## B. Quality seed production in maize hybrids

Yield of any maize hybrid goes down in $F_2$ generation by about 20 to 40 % in comparison to its $F_1$ yield, because of the disorganisation in the set up of genes. It is therefore essential to produce hybrid seed fresh every time for supply to the cultivators. There are three major steps involved in seed production.

a. Increase of inbred seeds.

b. Production of single cross seed.

c. Production of double cross seed.

### a. Increase of inbred seed

For this, the inbreds are grown in isolation. An increase field of an inbred line must be isolated by not less than 400 m from any maize with the same kernel colour and texture, and not less than 600 m from maize with different kernel colour or texture. Isolation distance from maize of the same variety (code designation) not conforming to varietal purity requirements for certification is 400 m. Differential blooming dates are permitted for modifying isolation distances, provided 5 % or more of the plants in the seed field do not have receptive silks when more than 0.20 % of plants in the field within the prescribed isolation distances are shedding pollen. Inbred lines are relatively true breeding strains. Rigorous rouging for off-type plants before pollen shedding is therefore, essential. Start removing distinctly tall and vigorous plants from the knee-high stage. At preflowering stage, rogue out off-type plants which are easily distinguishable on the basis of plant characteristics such as leaf shape, size, plant height, etc. Continue rouging during flowering stage to remove plants differing in tassel or silk character. Every effort must be made to rogue such plants before pollen shedding. Final rouging should be done to remove stalk rot affected plants. Off type plants are rouged out regularly. Inbreds on an average produce about 225 to 450 kg seeds / ha. As there are four inbreds involved in a double cross, four isolation blocks are needed for seed increase.

### b. Production of single cross seed

Single cross seed is produced by growing two inbred lines in isolation block, in the ratio of two male rows followed by four female rows. Male inbred should be a profuse pollen producer and its flowering should coincide with that of the female inbred. The single cross seed is produced on inbred lines. It is the result of crossing between two specified inbred lines. In the crossing fields, the usual planting pattern is two rows of female inbred to one row of male inbred. Female inbred line is detasseled before pollen shedding to ensure cross-pollination form the male inbred line. The seed is saved only from rows. The seed is produced in the following manner.

The female inbred should be vigorous and high yielding. A foundation single cross production field must be isolated by not less than 400 m from any maize with kernel colour and texture the same as that of the seed parent and not less than 600 m from maize with kernel colour or texture different from that of the seed parent. Isolation distances from maize of the same variety (code designation) not conforming to varietal purity requirements for certification is 400 m. Differential blooming dates are permitted for modifying isolation distances, provided 5 % or more of the plants in the seed parent do not have receptive silks when more than 0.20 % of plants in the field, within the prescribed isolation distance, are shedding pollen. The planting ratio between male and female rows is kept at 2 : 4 to facilitate subsequent rouging and detasseling operations, it is necessary to mark male rows at both ends with pegs. Check bags carefully before putting them in the planter to avoid mixtures of inbred seeds. Mark both ends of the male rows. Avoid planting mistakes e.g., mistakes in planting ratios, etc. The seed rate required for female parent inbred is 10 kg/ha and male parent inbred is 5 kg/ha.

Rogue the fields in a manner similar to that of inbred lines. Single cross is the result of crossing two inbred lines. In order for the two parents to cross, one of them, the seed parent, has to be detasseled so that its silk will only be fertilized by the pollen from the other parent. In crossing blocks, the parent that furnishes the pollen is called the pollen parent, or the male, and the one that is detasseled is called the female parent, or seed parent. All tassels must be removed from the female rows before they have shed any pollen. Single cross seed from fields where careless detasselling has been done is bound to perform poorly. A close check, therefore, should be exercised during the flowering period.

Detasselling is started when the tassel is well out of the leaf sheath, but before the anthers have shed pollen. This, in most cases, will be one or two days after the tassels are first visible. In certain strains, however, the tassels may begin to shed pollen even before they emerge. In such cases, it is necessary to open out leaf-whorl and remove the tassel. Every seed field should be carefully detasseled at uniform frequency intervals. Detasselling have to be done every day until it is complete.

Holding the stalk with left hand, a little below the tassel, take a firm holds of the entire tassel in the right hand. Remove tassel by a steady upward pull and throw tassel on ground as you walk on. A loose or imperfect grasp of the tassel may cause a portion of it to e left on the stalk. These portions are difficult to see and will produce contaminating pollen if left on the plant. Do not break or remove leaves, as removal with reduce yields and will also result in lower quality of seed produced.

Grasp the entire tassel so that all pollen bearing parts are fully removed. Immature detasselling should be avoided. It may cause a few spikelets being left, which may emerge and shed pollen. Also the top leaves are likely to be pulled out, leading to reduction in yield or attack of disease. Do not hold tassel too low on the stalk so as to prevent pulling out of plant tops. Once detasselling starts in a field it must be repeated daily in all weather. A fixed time should be observed every day. Be particular to start detasselling from the same side every day, in the case of a large field. Mark all the male rows at both ends by driving long wooden markers in the ground, or by some other suitable means. The markers should be painted white.

Look out for suckers (tillers) on female plants and also for lodged or damaged plants in female rows, as they are likely to pass unnoticed during detasselling. Plants on the verge of shedding frequently have leaves surrounding the tassel. A gentle shaking of the plant will reveal the tassel and enable the detasseller to grasp the tassel. Instruct the detasseller to drop the tassels on the ground after removing them and not to carry them in hand, as this may involve the danger of contaminating receptive silks. Put an experienced detasseller in charge of this operation. He should walk behind the other detassellers and check that no tassels are left.

As there are two single crosses involved for making a double cross hybrid, two isolation blocks are needed for single cross seed production. Female rows should be harvested separately and stalked. The grain should be separated after shelling. Grading and processing work should be done before it is bagged and sealed. Yield of single cross seed is two thirds of the female parent yield (only 4 rows out of 6 rows are harvested for seed). The two male rows being used for consumption. If the yield of an inbred is 340 kg/ha, the yield of single cross seed would be 225 kg/ha.

## c. Production of double cross hybrid

Crosses are made between two single crosses to produce double cross seed. For this, only one isolation block is needed. Selection of male and female parents is made strictly on the same principle as is done for production of single cross seed. Moreover that single cross is treated as female parent who will produce better type of double cross seed. The ratio of male and female row is 2: 6. Increased ratio of female row is of the result that male single cross parent produces more pollen because of hybrid vigour than the inbreds. The area harvested for seed is only three fourth.

If the yield of the single cross treated as female is 3700 kg/ha, the yield of the double cross seed will be 2800 kg/ha which is more than the yield of single cross seed. For production of one hybrid, seven isolated plots (four inbreds, two single crosses and one double cross) are needed. Hence few hybrids of wide adaptability are released rather than many hybrids of narrow adaptability.

Even though the hybrid seed production methods are same as that of varieties, some additional techniques have to be followed. The hybrid seeds are produced by crossing female and male lines and harvesting the seeds from female line. Maize hybrid seed production there is no male sterility system in female plants.

**Table 2. Female and male parents of few maize hybrids**

| Hybrid | Female | Male |
|---|---|---|
| COH (M) 4 | (UMI 190 x UMI 1285) | UMI 112 |
| COH (M) 5 | Single cross hybrid of UMI 285 | UMI 61 |

## Season

Maize is cultivated during June – July, September – October and January – February seasons. Sowing during November - December is suitable for seed production, since seed maturity stage will not coincide with rainfall. During early stage of this crop water stagnation should be avoided. Rains during seed maturation phase will cause discolouration due to fungal attack thereby seed quality is reduced.

### Isolation distance

Since maize is a cross pollinated crop, it is necessary to isolate seed field from other maize fields. For this isolate the hybrid seed crop from fields of other maize varieties with same seed colour by 200 m distance. For other varieties with different seed colour give 300 m isolation distance. Isolation distance from maize of the same variety (code designation) not conforming to varietal purity requirements for certification shall be 200 m and from the fields of other hybrids having common male parent and conforming to varietal purity requirements of certification at least by 5 meters. Distances less than 200 m may be modified by planting border rows of the male parent, if the kernel colour and texture of the contaminating maize are the same as that of the seed parent (Table 3).

**Table 3. Minimum number of border rows to be planted for modifying isolation distances less than 200 m based on size of the seed field and its distances from the contaminant**

| When the field area (in hectares) growing the concerned seed parent is | | | | | | | | |
|---|---|---|---|---|---|---|---|---|
| More than : | 4 | 6 | 8 | 10 | 12 | 14 | 16 | |
| Up to : 4 | 6 | 8 | 10 | 12 | 14 | 16 | or over | |
| And the distance (in metres) of the seed parent from other maize with kernel colour and texture the same as that of the seed parent is at least. | | | | | | | | Then the minimum number of border rows required is : |
| 200.0 | 195.0 | 190.0 | 185.0 | 180.0 | 175.0 | 170.0 | 165.0 | 1 |
| 187.5 | 182.5 | 177.5 | 172.5 | 167.5 | 162.5 | 157.5 | 152.5 | 2 |
| 175.0 | 170.0 | 165.0 | 160.0 | 155.0 | 150.0 | 145.0 | 140.0 | 3 |
| 162.5 | 157.5 | 152.5 | 147.5 | 142.5 | 137.5 | 132.5 | 127.5 | 4 |
| 150.0 | 145.0 | 140.0 | 135.0 | 130.0 | 125.0 | 120.0 | 115.0 | 5 |
| 137.5 | 132.5 | 127.5 | 122.5 | 117.5 | 112.5 | 107.5 | 102.5 | 6 |

| | | | | | | | | |
|---|---|---|---|---|---|---|---|---|
| 125.0 | 120.0 | 115.0 | 110.0 | 105.0 | 100.0 | 95.0 | 90.0 | 7 |
| 112.5 | 107.5 | 102.5 | 97.5 | 92.5 | 87.5 | 82.5 | 77.5 | 8 |
| 100.0 | 95.0 | 90.0 | 85.0 | 80.0 | 75.0 | 70.0 | 65.0 | 9 |
| 87.5 | 82.5 | 77.5 | 72.5 | 67.5 | 62.5 | 57.5 | 52.5 | 10 |
| 75.0 | 70.0 | 65.0 | 60.0 | 55.0 | 50.0 | 45.0 | 40.0 | 11 |
| 62.5 | 57.5 | 52.5 | 47.5 | 42.5 | 37.5 | 32.5 | 27.5 | 12 |
| 50.0 | 45.0 | 40.0 | 35.0 | 30.0 | 25.0 | 20.0 | 15.0 | 13 |

This table applies to all sides of the seed field exposed to contamination, whether located directly opposite or diagonally. Border rows must be planted in the seed field or adjacent to it, but in no case separated by more than 5 metres from the seed field. Border rows must be planted at the same time as the rest of the seed field. Border rows should be shedding pollen when silks in the seed parent are receptive. Border rows must be planted all along and opposite to the contaminating maize. There should not be gaps in the border rows. Border rows must have been planted using the seed rate and spacing adopted for the seed crop. The area planted under border rows is taken into consideration when modifying the isolation distance.

Seed fields having diagonal exposure to contaminating fields should be planted with border rows in both directions of exposure. If two hybrid seed fields, with different pollinator parents, are within the isolation distance of one another, border rows are necessary for each of them in order to avoid contamination of the respective seed parent. Natural barriers such as tall thick trees, buildings, etc. between the seed and contaminating fields, cannot take the places of border rows. Border rows must be planted with seed used for planting male rows in the seed field. Seed saved from male rows of the previous production of the same cross cannot be used for planting of border rows or for planting within the isolation distance. The isolation distance continues to be 300 m if the kernel colour or texture of the contaminating maize is different from that of the seed parent or if the contaminating field is planted with sweet or pop corn. In this case, modification of isolation distance by planting border rows will not be permitted. Differential blooming dates are permitted for modifying isolation distances, provided 5 % or more of the plants in the seed parent do not have receptive silks when more than 0.50 % of plants in the field, within the prescribed isolation distance, are shedding pollen.

**Land requirement**

For maize hybrid seed production, well drained sandy red or black cotton soil is suitable. The selected fields should be free from volunteer maize plants.

### Field preparation

Though no fixed rules can be laid down to govern tillage of the land, successful maize cultivation requires good seed bed. Different types of soils require different types of tillage. So main objective of preparation of seed bed should be to obtain good tilth. The surface soil should be in a condition that seed can be placed in contact with the moisture and soil. While preparing the seed bed, the soil should be left for some time in such a condition that air will circulate freely. Improved tillage implements are available for preparation of seed bed. Ploughing should be done when the soil conditions permit such operation. Plough the field 5 to 6 times to get fine tilth. Spread 12.5 tons of FYM or composted coir pith and 10 packets of Azospirilluim in the field. Prepare ridges and furrows with 60 cm spacing. Ridges and furrows enable the saving of irrigation water.

## Seed rate

The seed rate varies according to the size of the seed, the object of raising the crop and keeping quality of grain. For one hectare, 5 kg female and 2 kg male line seeds are needed. Treat the seeds with Azospirillum biofertilizer as described for varieties.

### Seed treatment

To control seed borne pathogens of downy mildew (fungal disease) treat the seed with carbendazim or thiram at 2 g/ kg seed. One day after seed treatment with fungicide, treat the seeds with 600 g of Azospirillum with rice gruel and shade dry for 15 minutes. Azosprillum treatments help in the fixation of atmospheric nitrogen in the soil and provide nourishment to the growing plants.

### Sowing

Ridges and furrows at 60 - 75 cm spacing are to be formed. 2 seeds per hill on the sides of ridges adopting a spacing of 25 - 30 cm between the plants are sown. During sowing, a planting ratio of 4 female lines and 2 male lines (4:2) should be followed. Four female line rows should alternate with two male rows. Sow three border rows all around the seeds field with male line seeds. The seeds of hybrid maize should be sown about 5 cm deep. The greater depth will affect the germination and the growth of the crop. If the planting ratio is 6: 2, the seed rate female parent is 12 to 14 kg/ha and male parent is 4 to 5 kg/ ha.

### Irrigation

Immediately after sowing, irrigate the field. Then given life irrigation on third day and provide subsequent irrigations depending upon the condition of the field. Generally for maize, upto 30 days, less irrigation is given. Then it is necessary to irrigate once in 10 days. Smooth silk will emerge out from the cob for fertilization. During this stage, you have to irrigate so that, the silk will come out smoothly from the cob and seed setting will occur.

**Fertilizer application**

Apply 25 tonnes of FYM per ha. Then apply inorganic fertilizers at 100 kg N, 100 kg P and 50 kg K per ha as basal dose. On 20$^{th}$ and 40$^{th}$ day after sowing, apply 50 kg N as top dressing. Fertilizers should be mixed together and should be applied in a separate furrow which should be about 8 cm away from the furrow in which seed is to be sown. It should never be applied in the same furrow in which seed is to be sown, otherwise germination will be badly affected.

Any nutrient deficiency in maize will adversely affect the growth of the plant. Generally due to micronutrient deficiency, the yield is reduced. Zn and Mg deficiency occurs largely in leaves. To overcome the above deficiencies you have to broadcast the micronutrient mixture @ 12.5 kg/ ha mixed with 40 kg sand, after sowing the seeds.

**Weed management**

Take two hand weeding on 25 DAS and 45 days after sowing. After weeding, apply fertilizers as top dressing and do the earthing up operation. Herbicides can also be applied. Spraying of Atrazin 500 g mixed in 1000 lit water and life irrigation on third day control the weeds. One to two hand weedings, with Khurpi will be needed along the rows and 3 - 4 hoeings with cultivator in between the rows. When plants are young, the depth of hoeing should be between 5 - 8 cm but when plants become 60 - 90 cm high, the depth should be reduced to 2 - 5 cm to avoid root damage.

**Earthing up**

Earthing up the plants is the most important operation in the cultivation of maize. Earthing up means placing of earth near the roots of plants. The objective of earthing up is to provide anchorage of the lower whorls of adventitious roots above the soil which then begin to function as absorbing roots. It prevents the plants from lodging. It also helps to drain away water through channels made in this way in rainy season as maize cannot with stand accumulation of water. The operations is done when the plant attain a height of 60 - 120 cm. It can be done with the help of desi plough. Furrows thus should be connected with the surface drains, so that rain water may flow out easily.

## Roguing

Rogue the off type plants as described for varieties. Do the rouging operation separately in female and male lines. Remove the male plants in female lines and female plants in male lines, plants of other varieties and off-type and diseased plants. Plant or seed deviating significantly from the characteristics of a variety as described by the breeder in any observed respect. During roguing at flowering stage, the off types should be removed away from the field immediately. Otherwise it will contaminate the silk and affect the genetic purity.

## Plant protection

Generally insect and disease attack is very less in maize. These practices are similar to those of open-pollinated varieties.

## Detasselling

Male inflorescence of maize plant is called tassel. Removal of tassels from female plants immediately after emergence is called detasselling. Detasselling should be continued for 14 days. It is done to enable crossing of female parent with male. Detasselling is done by holding the stalk with left hand, a little below the tassel, taking a firm hold of the entire with the right hand and removing tassel by a steady upward pull. Entire tassel should be removed without leaving any portion within the leaf. The removed tassels should be buried at the same place itself. Remove the tassels perfectly without damage to the leaves. Removed tassels should not be left on the field they may also shed the pollen causing contamination.

## Harvest

Judge the optimum harvesting stage as that of varieties and harvest the male and female parental lines separately. First, the male lines should be harvested fully, after the removal of cobs of male lines from the field, the female lines should be harvested. The seeds harvested from the female line is the hybrid seed. All other practices such cob sorting, cob shelling and seed grading are similar to those of open-pollinated varieties. By adopting the above techniques, 1000 - 2500 kg / ha of hybrid seed can be harvested depending upon the area of production.

## Seed certification

Seed certification guarantees the quality of seed as it ensures that the certified seed has the genetic, physical, physiological and seed health qualities. Genetic purity means that the seed gives rise to a plant which conforms to the varietal characteristics of the variety. The physical purity means that the seed is free from stones, broken seeds, straw bits and leaf bits etc. Physiological quality is measured by germination and seed health envisages freedom from pest and diseases.

Seed certification is being done in many stages. It starts from verifying whether seeds were obtained from authenticated source, verification of isolation distance and inspection during plant growth, flowering, harvesting, processing and bagging. Also seed samples are drawn from the seed lot and sent to seed testing lab to test whether the seeds are possessing required physical purity and germination. Then certification tag is issued. Colour of the tag is blue for certified seeds. Only those seeds harvested from fields having prescribed field standards and possessing required seed standards are certified by the Certification Agency. Seeds thus certified are offered for sales.

**Table 4. Field standards (Certified seed)**

| **Factor** | **Maximum permitted** |
|---|---|
| Off types (maximum) | **0.5 %** |
| Shedding tassel (maximum) | **1.0 %** |

**Table 5. Seed standards (Certified seed)**

| **Factor** | **Maximum permitted** |
|---|---|
| Physical purity (minimum) | 98.0 % |
| Inert matter (maximum) | 2.0 % |
| Other crop seeds (maximum) | 10.0/kg |
| Other distinguishable varieties (maximum) | 10.0/kg (Number) |
| Weed seeds based on kernel colour and texture | Nil |
| Germination (minimum) | 90.0 |
| **Moisture (maximum)**<br>For moisture vapour pervious | 12.0 % |
| For moisture vapour poof containers (maximum) | 8.0 % |

## d. Seed production technology of single cross hybrid (HQPM 1)

### Isolation distance

If possible, the hybrid seed production should be taken either in the area where no other maize variety is planted nearby the seed production plot or at least 400-500 m distance is required between two maize genotypes to maintain the genetic purity. The parents are female 193-1 and male 163.

### Sowing time

Sowing can be taken up during first week of July in Kharif and first week of November in Rabi in northern India. Seed rate recommended is 15 kg/ ha for female parent whereas it is 10 kg/ha male. Female: Male ratio is 3:1. Row spacing (East-West direction) is 60 cm and plant spacing is 20 cm.

## Nutrient management

FYM @15 t/ha, N 200 kg, $P_2O_5$ 80 kg, $K_2O$ 80 kg and $ZnSO_4$ 25 kg/ ha are to be applied for this crop. Total FYM, full doses of phosphorus, potash and zinc and 10 % N should be applied as basal. The remaining dose of nitrogen should be applied in four splits viz., 20 % N at 4 leaf stage, 30 % N at 8 leaf stage, 30 % N at flowering stage and 10 % N at grain filling to meet the requirement throughout the crop cycle.

### Water management

Irrigation should be given as and when required by the crop depending upon the rains and moisture holding capacity of the soil. Light and frequent irrigations are desirable young seedlings, knee high stage, flowering and grain filling and 10 days after grain filling are the most sensitive stages for water stress for inbreds and irrigation should ensured at these stages. Water should not overflow on the ridges. The irrigation should be applied in furrows up to 2/3rd height of the ridges.

## Weed management

Pre-emergence application of atrazine @ of 1.0-1.5 kg a.i /ha in 600 litre water is effective for controlling weeds. While spraying, the person who is doing spray should move backward so that the atrazine film on the soil surface may not be disturbed. One day prior to earthing up, third split of nitrogen should be applied followed by hoeing. On the next day earthing up operation should be completed with movement of the person in the backward direction. This operation should be done before tasseling stage to save the crop from lodging.

### Removal of off-type plants and thinning

At early stage i.e. after 12-15 days of sowing, off-type plants and excess plants should be removed and proper plant to plant distance of 20-25 cm should be maintained to provide an equal opportunity to each plant to grow. At knee high and flowering stages (before anthesis), dissimilar plants should be removed from the male and female lines to maintain the genetic purity of seed. Dissimilar tassel bearing male plant should also be removed.

## Detasseling

Detasseling in female should be done before anthesis. It should be practiced row-wise. One person should follow to monitor the each row to check that no part of the tassel is left inside. The process of detasseling should continue for 8-10 days. While detasseling, leaf should not be removed which will otherwise reduce the photosynthesis. The removal of 1 to 3 leaves along with tassel reduces 5-15 % yield. The removed tassel should not be thrown in the field but fed to the cattle as it is nutritive fodder.

## Insect pest management

Stem borer is a serious problem in maize. It can be controlled by 1-2 spray of Carboryl or Endosulfan after 10 days and 20 days of germination. Application should be done in the central whorl of plant. 500 g Carboryl or 625 ml Endosulfan 35 EC in 500 liter water is given during first spray while, 750 g Carboryl or 900 ml Endosulfan 35 EC in 700-800 liter water during second spray, if required.

## Harvesting

Male parent should be harvested first then the female and should be kept separately. Optimum moisture content in grain at harvesting should be around 20 %. The harvested cobs should spread evenly instead of making heap.

## Stages of crop inspection

At the time of sowing, it is taken up to monitor isolation distance, planting ratio of male: female, proper sowing time, seed treatment. During pre-flowering/vegetative stage, it is done to verify the rouging and removal of off type plants. During flowering stage, it is carried out to check disease and pest infestation. During post-flowering and pre-harvest stage, inspection is done to remove the late and diseased plants. Harvesting time, it helps to see the proper time of harvesting.

## Post harvest management

The drying of the cobs should not be done either on the kuccha or pucca flour, rather it should be dried on tarpoline sheets to avoid seed injury and during night the cobs should be kept covered. To maintain the purity, dissimilar, diseased and pest infested cobs should be removed before shelling. The female cobs should be dried up to 13-14 % moisture content before shelling.

Shelling of female parent should be done earlier than male to avoid mechanical mixture. Shelling can be done manually or by power operated maize sheller. All under size, broken, damaged seeds should be removed for maintaining the quality of seed. Seed drying should be done till the moisture content of the seed is reduced to 8 % and it should be kept in aerated jute bags. Seed should be stored at cool and dry place preferably in cold storage. Poor storage conditions will lead to loss of vigour and poor germination. Marketing should be done with specifications and standards.

**Seed priming**

Seed is a living entity and is subjected to various environmental stresses which affect the quality. Delays in germination and low seed viability are the serious problems limiting the production of maize. Highly vigorous seeds germinate rapidly, uniformly and are able to withstand environmental adversity after sowing. However, the use of maize seeds of low physiological quality is a common practice under tropical and subtropical production conditions, leading to inadequate plant population in the field. Seed priming is one of the most important developments to help rapid and uniform germination and emergence of seeds and to increase seed tolerance to adverse environmental conditions (Heydecker *et al.,* 1975; Harris *et al.,* 1999). Seed priming has presented promising and even surprising results, for many seeds including the cereal seeds (Bradford, 1986a). The few studies on maize are not overemphasized and are encouraging, but more information is required before its use as a routine practice in seed technology. Harris *et al.,* (2001) reported that maize genotypes responded positively to priming, where priming was effective and represented increases ranging from 17 to 76 %. Primed and dried seeds normally have a more rapid and uniform germination when subsequently re-hydrated, especially under adverse environmental conditions (Bradford, 1986a).

At Dharwad (Karnataka), significantly minimum electrical conductivity was recorded in seeds primed with $KNO_3$ @ 0.2 % (0.69 $dSm^{-1}$). While, unprimed seeds (control) (0.75 $dSm^{-1}$) recorded significantly more electrical conductivity (Table 6). It might be due to removal of deleterious hormones, seed leachates at the time of seed invigouration. Maize seeds may be treated with $KNO_3$ (0.2 %) and hydroprimed seeds + thiram (0.25 %) for better seed quality and stand establishment in the field (Hanegave et al., 2011). Among seed priming practices, higher seed yield was obtained in bio-priming (5835) and combination of all priming methods (5885 kg/ha) as compared to farmers' practice (unprimed seeds) (Priya and Patil, 2008).

**Table 6. Influence of seed priming on seed quality parameters of maize**

| Treatment | Germination (%) | Speed of germination | Root length (cm) | Shoot length (cm) | Seedling dry weight (g) | Seedling vigour index | Electrical conductivity ($dSm^{-1}$) |
|---|---|---|---|---|---|---|---|
| Control (unprimed seeds) | 95.30 (77.52)* | 31.33 | 21.00 | 18.28 | 2.17 | 3743 | 0.75 |
| Hydro-priming | 97.33 (80.68) | 33.52 | 21.77 | 18.78 | 2.40 | 3947 | 0.71 |
| Hydro-priming +Thiram (0.25 %) | 98.10 (82.12) | 34.33 | 22.33 | 19.00 | 2.53 | 4055 | 0.70 |
| $KNO_3$ (0.2 %) | 98.67 (83.51) | 35.33 | 23.00 | 19.50 | 2.60 | 4193 | 0.69 |
| $CaCl_2.2H_2O$ (2 %) | 98.03 (81.98) | 34.00 | 22.07 | 18.60 | 2.47 | 3987 | 0.70 |
| KCI (100 ppm) | 96.67 (79.55) | 32.33 | 21.07 | 18.27 | 2.29 | 3802 | 0.72 |
| $KH_2PO_4$ (1.0 %) | 97.67 (81.30) | 34.33 | 21.67 | 18.73 | 2.39 | 3945 | 0.71 |
| CD 5 % | 1.53 | 0.75 | 0.95 | 0.48 | 0.21 | 89.19 | 0.019 |

* Figures in parenthesis indicate arcsine values

## Seed Production of Open-Pollinated Maize Varieties (OPVs)

### Reproduction in maize

The maize plant has separate male and female flowers. The male flowers, or tassels, are located at the top of the plant, whereas the female flowers (ears and silks) develop about half way up the stalk. The location of the tassel at the top of a relatively tall plant and its separation from the female flower promotes cross-pollination between plants. Commonly, pollen from a tassel will be blown by the wind to one or more plants nearby. Pollen is very small, and a single tassel may produce up to 25 million pollen grains. Shortly after pollen contacts the female silks, it germinates and grows down through the silk, eventually fertilizing the young ovule. The fertilized ovule develops into the embryonic plant within the developing seed.

### Method to produce OPV seed

When producing seed, a farmer usually wants to maintain the characteristics of a variety. Cross pollination between different maize varieties must therefore be prevented. Isolating the seed production field from other maize fields helps to achieve this. If two different varieties are grown next to each other, cross-pollination will occur between the two varieties, and the crop grown from such seed will have a mixture of the characteristics of those two varieties.

### Isolating a seed production field

Isolation of a seed crop can be done in four ways:

1. By space. Maintain a distance of at least 300 m between the seed crop and any other maize field that has a different variety planted.
2. By time. Sow your seed crop a month earlier or later than neighbouring maize fields.
3. By certified seed barriers. A barrier of genetically pure seed of the same variety may be planted within the isolation distance of the seed production field.
4. By natural barriers. Seed production plots can be established on land isolated by natural or artificial forests.

Distance and time isolation are the most commonly used approaches. The goal is to have no other maize variety shedding pollen nearby when the seed production field is flowering. Wind may carry pollen further than 300 m. Thus, if there are constant strong winds in one particular direction, the distance to the next maize field should be at least 400 m.

### Selection of suitable field

Consult with your neighbours about when and where they will plant their maize, so you can sow yours in a field that is properly isolated. Apart from considering isolation, select your best field for maize seed production and manage it well, because the value of good seed is higher than the value of grain. Choose a field where no maize has been grown during the previous year to reduce the possibility that last year's maize crop may germinate and cross-pollinate your seed maize. Farming communities may want to produce the seed of one variety for the entire community in a single field. This will help in finding a well isolated field with good soil. Also, the village can make a collective effort to look after the maize crop well.

### Managing seed production field

Prepare the soil in your field at least two weeks prior to planting. If there are any grains from previous maize crops left in the soil, they may germinate in these two weeks and you should remove them when you sow your maize seed crop. If possible, sow your maize seed crop early. Choose a plant density that is recommended for your area. A common seeding rate for maize is 20 kg/ha. Planting in straight rows will facilitate crop management and seed production operations. Apply fertilizer and remove weeds in time.

### Eliminating undesirable plants

Carefully examine your maize seed crop as it grows. You may find plants that look very different or flower much earlier or later than most other plants in the field. These plants are called off-types and they should be removed-a process known as "roguing"-before pollen shedding starts. Most farmers do not like to remove maize plants from their field, but rouging is critical to maintain varietal purity.

### Harvest and drying

During harvest and drying, be careful that the seed of your maize crop does not get mixed with seed or grain from other maize varieties. Keep only the best and healthiest ears and kernels for seed, and use the rest of the harvest as grain. Your best seed comes from healthy undamaged ears that are typical for the variety. Thus, discard off-types, rotten and damaged ears, and ears where the kernels have started to germinate or are affected by insects. Place the harvested ears on a clean and dry surface, such as concrete or plastic, and dry them well in the sun. To make sure that all kernels get exposed to the sun, spread the ears in a flat layer and turn them several times. Maize seed stores best at <12 % moisture content.

## Shelling, cleaning and storing seed

When the maize seed is dry, it can be stored on the cob or shelled. Be careful not to damage the seed when shelling. Your best seed typically comes from the middle part of the ear. After shelling, clean the seed, removing dirt and other inert matter. Remove seeds that are small, look diseased, have started to germinate, or are damaged by insects. Treat the seed with an insecticide-fungicide combination and store it in a dry and cool place. When treating the seed, follow the guidelines on the seed treatment package carefully. One good way to store seed is to place it in a jute bag, close the bag and drop it into a plastic bag, close the plastic bag and place this in another jute bag, finally closing the outer jute bag. Seed bags should be stored in a seed warehouse on wooden pallets in cool conditions away from fertilizer and chemicals.

## Quality control

The production and distribution of quality maize seed requires diligent efforts both during field production and post-harvest handling. Field inspections are commonly conducted at different crop development stages to ensure quality.

## Field inspections

A planting inspection is commonly conducted to determine that the maize seed planted is genetically pure, of known origin, and is an appropriate variety for the area. If a mechanical planter is being used, it should be checked to ensure that it is clean and free of maize seed of other types and is properly calibrated to achieve recommended seeding densities. The field should be inspected to verify that it is properly isolated and free of volunteer plants.

A second field inspection may be made during the vegetative growth phase. Isolation should be checked, along with the presence of disease, insect pests, or weed infestations. At this stage, off-type and diseased plants may be rogued. The most important field inspections are made just prior to and during flowering. At this time the maize seed field is most susceptible to genetic contamination from wind-blown pollen coming from off-type plants within the field or other maize varieties in surrounding fields. Therefore, it is essential during the pre-flowering inspection to confirm that the maize seed field has been properly rogued and is sufficiently isolated from other maize fields. Plants that are off-type or diseased, along with harmful weeds, must be removed at this time.

A pre-harvest or harvest inspection may be conducted as the crop reaches maturity and the seed has lost a significant portion of its moisture content. Off-type plants, such as those that are still green when most the other plants are dry, may be removed at this stage. At harvest, ears with different grain colour or texture from the produced variety should be removed.

### Quality control tests

Various standard tests for moisture content, germination, and physical purity can be conducted to evaluate the quality of the seed. The most common evaluation described here is the germination test, designed to determine the seed's capacity to germinate and produce normal plants when sown under appropriate conditions.

### Germination test

Germination tests may be conducted by building an open wooden box, 1 m long, 50 cm wide, and 10 cm deep. Fill the box with insect-free, loose soil. Divide it in half and plant 100 seeds in rows separated by 10 cm. Since the objective is to see how many seeds germinate, sow them one by one in a thin line about 2 cm deep. The box should be watered thoroughly and kept in a safe place away from birds and other animals. Another alternative would be to conduct the germination test in a well-prepared bed near the homestead or in a protected garden. Another option is to evaluate maize seed germination in paper towels or newspapers. With this method, 50 maize seeds are spread out in lines on moistened paper and then covered by another wetted paper. The paper is rolled and tied with a string or elastic band. The paper rolls can be placed in plastic bags or other containers and kept moist for seven days. With each of these methods, a count for normal developing seedlings is made between 7 and 10 days after sowing. A minimum of 80 % germination is the suggested standard; anything below this means the seed is not of acceptable quality.

## References

Anonymous (1999). Development, maintenance, and seed multiplication of open-pollinated maize varieties, 2nd edition. Mexico, D.F.: CIMMYT.

Agrawal, R.L. (2002). Seed technology. Oxford and IBH publishing Co. Pvt. Ltd., New Delhi. P: 130 - 144.

Anil, S.H., Ravi Hunje, H.L. Nadaf, N.K. Biradarpatil and D.S. Uppar (2011). Effect of seed priming on seed quality of maize (Zea mays L.). Karnataka J. Agric. Sci., 24 (2): 237-238.

Beck, D.L. (2004). Hybrid corn seed production. In: C.W. Smith, J. Betran, and E.C.A. Runge (eds.), Corn: Origin, History, Technology, and Production. Hoboken, New Jersey: John Wiley & Sons, Inc.

Cordova, H.S., J.L. Queme and P. Rosado (1999). Small-scale production of maize seed by farmers in Guatemala. Second edition. Mexico, D.F.: CIMMYT and El Programa Regional de Maiz (PRM) para Centroamérica y El Caribe.

Hanegave, A.S., Ravi hunje, H.L. Nadaf, N.K. Biradar Patil and D.S. Uppar (2011). Effect of seed priming on seed quality of maize (Zea mays L). Karnataka J. Agric. Sci., 24 (2): 237-238.

Harris, D., A. Joshi, P.A. Khan, P. Gothkar and P.S. Sodhi (1999). On-farm seed priming in semi-arid agriculture: development and evaluation in maize, rice and chickpea in India using participatory methods. Exp. Agric., 35: 15-29.

Harris, D., A.K. Pathan, P. Gothkar, A. Joshi, W. Chivasa and P. Nyamudeza (2001). On farm seed priming: Using participatory methods to revive and refine a key technology. Agric. Syc., 69: 151-164.

Harris, D., A. Rashid, P.A. Hollington, L. Jasi and C. Riches (2002). Prospects of improving maize yields with 'on-farm' seed priming. Pp 180-185 in: N.P. Rajbhandari, J.K. Ransom, K. Adikhari and A.F.E. Palmer (eds) 'Sustainable Maize Production Systems for Nepal': Proceedings of a Maize Symposium held, December 3-5, 2001, Kathmandu, Nepal. Kathmandu: NARC and CIMMYT.

Harris, D., A. Rashid, S. Ali and P.A. Hollington (2002a). 'On-Farm' Seed Priming with Maize in Pakistan. Proceedings of the 8th Asian Regional Maize Workshop, Bangkok, Thailand: August 5-8, 2002. P: 316 -322.

Khot, R.B. and M.K. Umrani (1992). Seed yield and quality parameters of African tall maize as influenced by spacing and level of nitrogen. Indian J. Agron., 37: 103-104.

Khot, R.B. and N. Umrani (1992a). Seed yield of maize as influenced by plant spacings and levels of nitrogen. Indian J. Agron., 37 (1): 183 - 184.

Khot, R.B., N.K.Umrani, J.S. Desale and P.S. Pol (1993). Effects of plant spacings and nitrogen levels on maize seed production. J. Maharashtra Agric. Univ., 18: 155-156.

Kumar, R., H.Govind and V.K. Sachan (2007). Techniques of quality maize seed production in north western Himalayan region. Indian Farmers' Digest, 40 (10): 16 -18.

Maruthi, K. (2011). Standardization of screen size and storability studies in sweet corn. Department of seed science and Technology, UAS, GKVK, Bangalore.

Morris, M.L., R.P. Singh, and S. Pal (1998). India's maize seed industry in transition: Changing roles for the public and private sectors. Food Policy, 23(1): 55-71.

Pandey, S., J. Kumar and B.S. Parmar (2005). Polymer seed coating on soybean and maize for seed quality enhancement and better emergence. Ann. Agric. Res., 26 (4): 509 – 513.

Priya, P. and V.C. Patil (2008). Effect of seed priming on field performance of maize based cropping systems under rainfed condition. Karnataka J. Agric. Sci., 21(4): 658 - 659.

Saayman, A.E.J., Ventr and H.A. Van De (1997). Influence of weed competition on subsequent germination and seed vigour of *Zea mays*. *Seed Sci. Tech.,* 25(1): 59- 65.

Sushil, P., J. Kumar and B.S. Parmar (2005). Polymer seed coating on soybean and maize for seed quality enhancement and better emergence. Ann. Agric. Res., 26 (4): 509 - 513.

# 29

# Genetically Modified Maize

Genetically modified (GM) maize is one of the 11 GM crops grown commercially in United States and Canada. About 85 % of the US maize crop was genetically modified in 2009. It is also grown commercially in Brazil, Argentina, South Africa, Canada, the Philippines and Spain. On a smaller scale, it is cultivated in the Czech Republic, Portugal, Egypt and Honduras. Transgenic maize has been deliberately genetically modified to have agronomically desirable traits. Traits that have been engineered into maize include resistance to herbicides and resistance to insect pests, the latter being achieved by incorporation of a gene that codes for the *Bacillus thuringiensis* (Bt) toxin. In 2009, transgenic maize was grown commercially in 11 countries, including the USA (where 85 % of the maize crop was genetically modified), Brazil (36 % GM), Argentina (83 % GM), South Africa (57 % GM), Canada (84 % GM), the Philippines (19 % GM) and Spain (20% GM).

### Herbicide resistant maize

Maize varieties resistant to glyphosate herbicides have been produced. There are also maize hybrids with tolerance to imidazoline herbicides marketed by Pioneer Hi-Breed under the trademark "Clearfield" but in these, the herbicide-tolerance trait was bred without the use of genetic engineering. Consequently, the regulatory framework governing the approval, use, trade and consumption of transgenic crops does not apply for imidazoline-tolerant maize. Herbicide-resistant GM maize is grown in the United States. A variation of herbicide-resistant GM maize was approved for import into the European Union in 2004, but such imports remain highly controversial.

## Bt maize

It is a variant of maize, genetically altered to express the bacterial Bt toxin, which is poisonous to insect pests. In the case of maize, the pest is the European corn borer. Expressing the toxin was achieved by inserting a gene from the microorganism *Bacillus thuringiensis* into the maize genome. This gene codes for a toxin that causes the formation of pores in the Lepidoptera larval digestive tract. These pores allow naturally occurring enteric bacteria, such as *E. coli* and *Enterobacter*, to enter the hemocoel, where they multiply and cause sepsis (Broderick et al., 2006). This is contrary to the common misconception that Bt toxin kills the larvae by starvation. In 2001, Bt 176 varieties were voluntarily withdrawn from the list of approved varieties by the United States Environmental Protection Agency (EPA) when it was found to have little or no Bt expression in the ears and was not found to be effective against second generation maize borers.

## Bt maize on nontarget insects

In May 1999, a laboratory trial indicates the pollen of genetically modified Bt corn presented a threat to monarch caterpillars. Critics claimed the popular media were wrong to report monarch butterflies were threatened because this experiment did not duplicate natural conditions under which monarch caterpillars may come in contact with maize pollen. In 2001, six comprehensive studies showed that Bt maize pollen does not pose a risk to monarch populations for the following reasons:

The density of Bt maize pollen that overlays milkweed leaves in the environment rarely comes close to the levels needed to harm monarch butterflies. Both laboratory and field studies confirmed this. There is limited overlap between the period Bt maize sheds pollen and when caterpillars are present. Only a portion of the monarch caterpillar population feeds on milkweeds in and near maize fields (Sears *et al.*, 2001).

Monarch populations in the USA during 1999 increased by 30 %, despite Bt corn accounting for 30 % of all corn grown in the USA that year. The beneficial effects of Bt maize on Monarch populations can be attributed to reduced pesticide use (Trewavas and Leaver, 2001). Numerous scientific studies continue to investigate the potential effects of Bt maize on a variety of nontarget invertebrates. A synthesis of data from many such field studies found the measured effect depends on the standard of comparison. The overall abundance of nontarget invertebrates in Cry1Ab variety Bt maize fields is significantly higher compared to non-GM maize fields treated with insecticides, but significantly lower compared to insecticide-free non-GM maize fields. Abundance in fields of another variety,

Cry3Bb maize, is not significantly different compared to non-GM maize fields either with or without insecticide (Marvier *et al.,* 2007).

**Preventing Bt resistance in pests**

By law, farmers in the United States who plant Bt maize must plant non-Bt maize nearby. These unmodified fields are to provide a location to harbor pests. The theory behind these refuges is to slow the evolution of the pests' resistance to the Bt pesticide. Doing so enables an area of the landscape where wild type pests will not be immediately killed. It is anticipated resistance to Bt will evolve in the form of a recessive allele in the pest. Because of this, a pest that gains resistance will have an incredibly higher fitness than the wild type pest in the Bt maize fields. If the resistant pest is feeding in the non-Bt maize nearby, the resistance is neutral and offers no advantage to the pest over any nonresistant pest. Ensuring there are at least some breeding pests nearby that are not resistant increases the chance the resistant pests will choose to mate with nonresistant ones. Since the gene is recessive, all offspring will be heterozygous and the offspring from that mating will not be resistant to Bt and therefore no longer a threat. Using this method, scientists and farmers hope to keep the number of resistant genes very low and use genetic drift to ensure any resistance that does emerge does not spread. Although mandated by law, compliance data from the EPA for 2008 showed, however, 25 % of Bt maize growers were not in compliance. The data showed noncompliance climbed to 5.29 million hectares or almost 15 % of all Bt maize grown, suggesting in some areas ample acreage does not exist to support pests without resistance to mate with any resistant pests that survived the Bt maize. The European corn borer, one of the primary insects Bt is meant to target, has been shown to be capable of developing resistance to the Bt protein.

**Cross pollination**

The non-Bt pesticide status of the refuges is being compromised by wind-borne pollen drifting into the non-Bt maize fields. Maize harvested from the supposed Bt-free zones has shown traces of Bt toxin. The levels found in the non-Bt maize decreases with distance from the Bt-maize fields indicating the pollen is wind-borne rather than another method of transfer. The concentrations in the refuge fields were found to be low-to-moderate. Possible solutions to the cross pollination problem are to plant a wider refuge field or plant varieties of maize that bloom at different times than the Bt fields do (Chilcutt and Tabashnik, 2004).

**In sweet corn for human consumption**

"Attribute" is the brand name for a line of Bt sweet corn. Seed is available only to large professional farmers who sign a stewardship agreement. The farmer must agree to not repackage or resell Attribute seed. Growers also must grow the maize exactly as directed. Herbicide-resistant sweet corn has not yet been released for sale.

### Safety issues

A 2009 study compared an analysis of blood and organ system data from trials with rats fed three main commercialized genetically modified types of maize which are present in food and feed in the world. Approximately 60 different biochemical parameters were classified per organ and measured in serum and urine after 5 and 14 weeks of feeding. GM maize-fed rats were compared first to their respective isogenic or parental non-GM equivalent control groups, followed by comparison to six reference groups, which had consumed various other non-GM maize varieties. According to them, for the 3 GMOs new side effects linked with GM maize consumption, which were sex- and often dose-dependent. Effects were mostly associated with the kidney and liver, the dietary detoxifying organs, although different between the 3 GMOs. Other effects were also noticed in the heart, adrenal glands, spleen and haematopoietic system. These data highlight signs of hepatorenal toxicity, possibly due to the new pesticides specific to each GM maize. In addition, unintended direct or indirect metabolic consequences of the genetic modification cannot be excluded. The French High Council of Biotechnologies Scientific Committee expressed that it presents no admissible scientific element likely to ascribe any haematological, hepatic or renal toxicity to the three re-analysed GMOs. A review by Food Standards Australia and New Zealand, the results were due to chance alone (Vendomois *et al.,* 2009).

### Star link corn controversy

Star link is a variety of Bt maize patented by a Aventis Crop Sciences (a subdivision of Aventis, acquired by Bayer AG in 2002), intended for use in animal feed. U.S. regulatory authorities permitted the commercial sale of StarLink seed with the stipulation that crops produced must not be used for human consumption. This restriction was based on the possibility that a small number of people might develop an allergic reaction to the Bt protein used in StarLink that is less rapidly digested than the version used in other Bt varieties. StarLink maize was subsequently found in food destined for consumption by humans. An episode involving Taco Bel taco shells was particularly well publicized. This led to a public relations disaster for Aventis and the biotechnology industry as a whole. Sales of StarLink seed were discontinued. The registration for Starlink varieties was voluntarily withdrawn by Aventis in October 2000. About 28 people reported apparent allergic reactions related to eating maize products that may have contained the Starlink protein. However, the US

Centers for Disease Control studied the blood of these individuals and concluded there was no evidence the reactions these people experienced were associated with hypersensitivity to the Starlink Bt protein. A subsequent review of these tests by the Federal Insecticide, Fungicide, and Rodenticide Act Scientific Advisory Panel points out that while "the negative results decrease the probability that the Cry9C protein is the cause of allergic symptoms in the individuals examined ... in the absence of a positive control and questions regarding the sensitivity and specificity of the assay, it is not possible to assign a negative predictive value to this". The southern portion of the U.S. maize belt planted the greatest amount of StarLink maize. It is this portion of the U.S. where maize borer damage creates the greatest economic loss to farmers. The US maize supply has been monitored for the presence of the Starlink Bt proteins since 2001. No positive samples have been found since 2004, showing that, most probably, it was possible to withdraw this GM crop without leaving traces in the environment once it has been used in the field.

## Virus resistant maize

In 2007, researchers from South Africa announced the production of transgenic maize resistant maize streak disease (MSD) caused by maize streak virus (MSV). This variety of maize is still in the research-and-development phase.

## Isolation requirements

Managing pollen drift is an important consideration in the production of specially maize and non-GMO maize as IP grain crops. Maize is a cross – pollinating crop in which most pollination results from pollen dispersed by wind and gravity. Although most of a maize field's pollen is deposited within a short distance of the field, with a 15 mph wind pollen may travel as far as one-half mile in a couple of minutes. Pollen from maize containing transgenes or genetically modified organisms (GMOs), such as Bt maize, may contaminate (by cross pollination) nearby non-GMO maize. The European Union guidelines require that foods, including grains, containing more than 0.9 % biotech material (GMSs) be labeled as genetically engineered. Producers of IP non-GMO maize, such as organic farmers, need to minimize pollen contamination by GMO maize if they are to obtain premiums. Growers can fallow several planting practices to minimize GMO pollen contamination, including use of isolation and border rows, planting dates, and / or hybrid maturity.

## Biotech interventions in maize

Both pest resistance and herbicide tolerant genes/events have also been stacked. As of now several events of the two traits have been approved in different countries and are being extensively cultivated. In early 2009, there were nine different event of GM maize in the varieties cultivated globally and three additional maize events have been authorized in at least one country worldwide, but not yet commercialized anywhere (Table 1) (Stein and Rodriguez, 2009).

**Table 1. Commercial GM maize and GM maize in the commercial pipeline worldwide**

| Developer | Product | Event name/ genes | Trait name | Unique identifier |
|---|---|---|---|---|
| **Commercialized maize events** | | | | |
| Monsanto | Yield Gard Corn Borer | MON810 | Insect resistance (to lepidopterans) | MON-ØØ81Ø-6 |
| Monsanto | Roundup Ready Corn 2 | NK603 | Herbicide tolerance (to glyphosate) | MON-ØØ6Ø3-6 |
| Monsanto | YieldGardR ootworm | MON863 | Insect resistance (to coleopterans) | MON-ØØ863-5 |
| Monsanto | YieldGard VT | MON88017 | Insect resistance (to coleopterans) | MON-88Ø17-3 |
| Dow Agro Sciences and Pioneer Hi-Bred | Herculex I | 1507 | Insect resistance (to lepidopterans) | DAS-Ø15Ø7-1 |
| Dow AgroSciences and Pioneer Hi-Bred | Herculex RW | 59122 | Insect resistance (to coleopterans) | DAS-59122-7 |
| Syngenta | Agrisure CB | Bt11 | Insect resistance (to lepidopterans) | SYN-BTØ11-1 |
| Syngenta | Agrisure GT | GA21 | Herbicide tolerance (to glyphosate) | MON-ØØØ21-9 |
| Syngenta | Agrisure RW | MIR604 | Insect resistance (to coleopterans) | SYN-IR6Ø4-5 |
| **Maize events authorized in at least one country but not yet commercialized anywhere** | | | | |
| Monsanto | YieldGard VTPRO | MON89034 | Insect resistance (to lepidopterans) | MON-89Ø34-3 |
| Monsanto | High lysine | LY038 | Crop composition (high lysine content) | REN-ØØØ38-3 |
| Syngenta | n/a | 3272 | Crop composition (amylase content) | SYN-E3272-5 |

In addition, there are five more GM maize events that have entered the regulatory system in at least one country but that are not yet authorized anywhere in the world, namely Syngenta's new lepidopteran resistant maize, Pioneer's Optimum GAT maize and three GM maize events from China (Table 2).

**Table 2. GM maize in the regulatory pipeline worldwide**

| Developer | Product | Event name/ genes | Trait name | Unique identifier |
|---|---|---|---|---|
| Syngenta | Agrisure Viptera | MIR162 | Insect resistance (to lepidopterans) | SYN-IR162-4 |
| Pioneer Hi-Bred | Optimum | GAT 98140 | Herbicide tolerance (to ALS inhibitors and glyphosate) | DP-Ø9814Ø-6 |
| n/a (China) | n/a | Cry1A | Insect resistance | n/a |
| n/a (China) | n/a | n/a | Crop composition (high lysine content) | n/a |
| n/a (China) | n/a | n/a | Crop composition (phytase enzyme) | n/a |

In addition to the above, seven new events containing traits regarding crop composition and drought tolerance are at advanced stages of research and development (Table 3).

**Table 3. GM maize in pipeline worldwide**

| Developer | Product | Event name/ genes | Trait |
|---|---|---|---|
| Monsanto | n/a | MON87754 | Crop composition (high oleic content) |
| Pioneer Hi-Bred | Optimum Acre Max1 | n/a | Insect resistance (to coleopterans) |
| Monsanto and BASF | n/a | MON87460 | Abiotic stress tolerance (to drought) |
| Dow AgroSciences | DHT | n/a | Herbicide tolerance |
| n/a (India) | n/a | cry1Ac + cp4epsp4 | Insect resistance |
| Syngenta | n/a | n/a | Abiotic stress tolerance (to drought) |
| BASF Plant Science | NutriDense | n/a | Crop composition (protein, amino acid and phytase content) |

On a global basis, in 2008, genetically engineered maize occupied 37.3 million hectares equivalent to 24 % of the global maize area of 157 million hectares (James, 2008). There has been a substantial increase in the deployment of stacked traits of Bt and herbicide tolerance, particularly in USA. The triple gene products in GM maize, featuring two Bt genes (one to control the European corn borer complex and the other to control root worm) and one herbicide trait continued to grow in adoption in USA in 2008. The area occupied by maize with insect resistant and herbicide tolerant genes and the two characters stacked together is as follows:

**Table 4. Global area of GM maize in 2008**

| S. No. | GM maize containing | Area (million hectares) |
|---|---|---|
| 1. | Insect resistant gene | 7.1 |
| 2. | Herbicide tolerant gene | 5.7 |
| 3. | Stacked traits | 24.5 |
| | **Total** | **37.3** |

In total maize was grown in 17 countries worldwide. Major countries growing GM maize are USA, Argentina, Canada, Brazil and South Africa. Other countries include Uruguay, Philippines, Chile, Egypt, Honduras and seven EU countries.

## References

Chilcutt, C.F., G.N. Odvody, J.C. Correa and J. Remmers (2007). Effects of *Bacillus thuringiensis* Transgenic Corn on Corn Earworm and Fall Armyworm (Lepidoptera: Noctuidae) Densities. *Journal of Economic Entomology,* 100: 327-334.

He, K., Z. Wang, D. Zhou, L. Wen, Y. Song and Z. Yao (2003). Evaluation of Transgenic Bt Corn for Resistance to the Asian Corn Borer (Lepidoptera: Pyralidae). *Journal of Economic Entomology,* 96: 935 - 940.

James, C. (2003). Global review of commercialized transgenic crops. Feature : Bt maize. ISAAA Briefs No: 29, P: 182.

Perry, J.N., L.G. Firbank, G.T. Champion, S.J. Clark, M.S. Heard, M.J. May, C. Hawes, G.R. Squire, P. Rothery, I.P. Woiwod and J.D. Pidgeon (2004). Ban on atriazine herbicides likely to reduce but not negate relative benefits of GMHT maize cropping. *Nature,* 428: 313–316.

Ravisankar, D., P. Muthukrishnan and C. Chinnusamy (2011). Evaluation of bio-efficacy, weed control and crop productivity in herbicide resistant transgenic corn hybrids (MON 89034 × NK603). Biotechnol. Bioinf. Bioeng., 1(3): 369 - 374.

# 30

# Uses of Maize Grain

Maize meal (ground dried maize) constitutes a staple food in many regions of the world. Maize has become Africa's most important staple food crop. Maize meal is made into a thick porridge in many cultures. Maize meal is also used as a replacement for wheat flour to make corn bread and other baked products. Masa (corn meal treated with lime water) is the main ingredient for tortillas, atole and many other dishes of Mexican food. Popcorn is kernels of certain varieties that explode when heated, forming fluffy pieces that are eaten as a snack. Roasted dried maize cobs with semi-hardened kernels, coated with a seasoning mixture of fried chopped spring onions with salt added to the oil, is a popular snack food in Vietnam. Cancha, which are roasted maize chulpe kernels, are a very popular snack food in Peru and also appears in traditional Peruvian ceviche. An unleavened bread called Makki di roti is popular bread eaten in the Punjab region of India and Pakistan.

Chicha and "chicha morada" (purple chicha) are drinks made usually from particular types of maize. The first one is fermented and alcoholic, the second is a soft drink commonly drunk in Peru. Corn flakes are a common breakfast cereal in North America and the UK and found in many other countries all over the world. Maize can also be prepared as hominy, in which the kernels are soaked with lye in a process called nitamalization or grits, which are coarsely ground hominy. These are commonly eaten in the southeastern USA, foods handed down from Native Americans, who called the dish sagamite.

The Brazilian dessert canjica is made by boiling maize kernels in sweetened milk.

Maize can also be harvested and consumed in the unripe state, when the kernels are fully grown but still soft. Unripe maize must usually be cooked to become palatable. This may be done by simply boiling or roasting the whole ears and eating the kernels right off the cob. Sweet corn, a genetic variety that is high in sugars and low in starch, is usually consumed in the unripe state. Such corn on the cob is a common dish in the United States, Canada, United Kingdom, some parts of South America and the Balkans, but virtually unheard of in some European countries. The cooked unripe kernels may also be shaved off the cob and served as a vegetable in side dishes, salads, garnishes, etc. Alternatively, the raw unripe kernels may also be grated off the cobs and processed into a variety of cooked dishes, such as maize puree, tamales, pamonhas, curau, cakes, ice creams, etc.

**Nutritional value of maize**

Amount of nutrients in 100 g of maize contains carbohydrates 9 g, calcium 9 mg, dietary fiber 2.7 g, fat 1.2 g, folate 46 g, iron 0.5 mg, magnesium 37 mg, niacin 1.7 mg, phosphorus 120 mg, potassium 270 mg, protein 3.2 g, sugars 3.2 g, vitamin A 10 g, vitamin B 15 mg, vitamin C 7 mg and energy 90 kcal.

**Table 1. Macro and micro nutrients content in maize grain**

| Nutrients | Quantity (/100 g) |
|---|---|
| **Macro nutrients** | |
| Moisture (g) | 14.9 |
| Protein (g) | 11.1 |
| Fat (g) | 3.6 |
| Minerals (g) | 1.5 |
| Crude fibre (g) | 2.7 |
| Carbohydrate (g) | 66.2 |
| Energy (K.cal) | 342 |
| **Vitamins** | |
| b-carotene (mg) | 90 |
| Thiamine (mg) | 0.42 |
| Riboflavin (mg) | 0.10 |
| Niacine (mg) | 1.80 |
| Folic acid (mg) | 20.0 |

| **Minerals** | |
|---|---|
| Calcium (mg) | 10.0 |
| Phosphorus (mg) | 348 |
| Iron (mg) | 2.3 |

**Table 2. Comparison of nutrient contents in maize to other staple foods (per 100 g portion)**

| Composition | Wheat | Rice | Sweet corn | Potato |
|---|---|---|---|---|
| Water (g) | 11 | 12 | 76 | 82 |
| Energy (kj) | 1506 | 1527 | 360 | 288 |
| Protein (g) | 23 | 7 | 3 | 1.7 |
| Fat (g) | 10 | 1 | 1 | 0.1 |
| Carbohydrates (g) | 52 | 79 | 19 | 16 |
| Fiber (g) | 13 | 1 | 3 | 2.4 |
| Sugars (g) | 0.1 | >0.1 | 3 | 1.2 |
| Iron (mg) | 6.3 | 0.8 | 0.5 | 0.5 |
| Manganese (mg) | 13.3 | 1.1 | 0.2 | 0.1 |
| Calcium (mg) | 39 | 28 | 2 | 9 |
| Magnesium (mg) | 239 | 25 | 37 | 21 |
| Phosphorus (mg) | 842 | 115 | 89 | 62 |
| Potassium (mg) | 892 | 115 | 270 | 407 |
| Zinc (mg) | 12.3 | 1.1 | 0.5 | 0.3 |
| Panthothenic acid (mg) | 0.1 | 1.0 | 0.7 | 0.3 |
| Vitamin b6 (mg) | 1.3 | 0.2 | 0.1 | 0.2 |
| Folate (μg) | 281 | 8 | 42 | 18 |
| Thiamin (mg) | 1.9 | 0.1 | 0.2 | 0.1 |
| Riboflavin (mg) | 0.5 | >0.1 | 0.1 | >0.1 |
| Niacin (mg) | 6.8 | 1.6 | 1.8 | 1.1 |

## Health benefits of maize

The high amount of fiber present in maize helps lower cholesterol levels and also reduces the risk of colon cancer. Maize if consumed in moderate quantities, has been seen to be beneficial for those suffering from diabetes. Being rich in folate, maize helps the generation of new cells, especially important before and during pregnancy. Those suffering from anemia have shown positive effects after consuming maize. The pantothenic acid presents in maize help the physiological functions of the body. Owing to the presence of thiamin, maize has been said to help in the metabolism of carbohydrates. Maize has been found to be helpful in treating kidney problems, including renal dysfunction. Regular consumption of maize, in moderate quantities, has been associated with better cardiovascular health. The beta-cryptoxanthin in maize makes it good for the health of the lungs and may even help prevent lung cancer. The insoluble fiber in maize makes it good for those suffering from common digestive ailments, like constipation and hemorrhoids.

**Table 3. Sweet corn yellow (raw seeds only) (Nutritional value per 100 g)**

| Nutrients | Content | Nutrients | Content |
|---|---|---|---|
| Energy | 360 KJ (86 kcal) | - Histidine | 0.089 g |
| Carbohydrates | 19.02 g | - Alanine | 0.295 g |
| Sugars | 3.22 g | - Aspartic acid | 0.244 g |
| Dietary fiber | 2.7 g | - Glutamic acid | 0.636 g |
| Fat | 1.18 g | - Glycine | 0.127 g |
| Protein | 3.22 g | - Proline | 0.292 g |
| - Tryptophan | 0.023 g | - Serine | 0.153 g |
| - Threonine | 0.129 g | Water | 75.96 g |
| - Isoleucine | 0.129 g | Vitamin A equiv. | 9 μg (1%) |
| - Leucine | 0.348 g | - Lutein and zeaxanthin | 644 μg |
| - Lysine | 0.137 g | Thiamine (Vit. $B_1$) | 0.200 mg |
| - Methionine | 0.067 g | Niacin (Vit. $B_3$) | 1.700 mg |
| - Cystine | 0.026 g | Folate (Vit. $B_9$) | 46 μg |
| - Phenylalanine | 0.150 g | Vitamin C | 6.8 mg |
| - Tyrosine | 0.123 g | Iron | 0.52 mg |
| - Valine | 0.185 g | Magnesium | 37 mg |
| - Arginine | 0.131 g | Potassium | 270 mg |

Maize is a major source of starch. Maize starch (maize flour) is a major ingredient in home cooking and in many industrialized food products. Maize is also a major source of cooking oil (corn oil) and of maize gluten. Maize starch can be hydrolyzed and enzymatically treated to produce syrups, particularly high fructose corn syrup, a sweetener and also fermented and distilled to produce grain alcohol. Grain alcohol from maize is traditionally the source of bourbon whiskey. Maize is sometimes used as the starch source for beer. Within the United States, the usage of maize for human consumption constitutes about 1/40$^{th}$ of the amount of grown in the country. In the United States and Canada, maize is mostly grown to feed for livestock, as forage, silage (made by fermentation of chopped green maize stalks) or grain. Maize meal is also a significant ingredient of some commercial animal food products, such as dog food. Maize is also used as a fish bait, called "dough balls". It is particularly popular in Europe for coarse fishing.

**Chemicals and medicines**

Starch from maize can also be made into plastics, fabrics, adhesives and many other chemical products. Stigmas from female maize flowers, known popularly as corn silk are sold as herbal supplements. The corn steep liquor, a plentiful watery byproduct of maize wet milling process, is widely used in the biochemical industry and research as a culture medium to grow many kinds of microorganisms.

**Biofuel**

"Feed maize" is being used increasingly for heating specialized corn stoves are available and use either feed maize or wood pellets to generate heat. Maize cobs are also used as a biomass fuel source. Maize is relatively cheap and home heating furnaces have been developed which use maize kernels as a fuel. They feature a large hopper that feeds the uniformly sized maize kernels (or wood pellets or cherry pits) into the fire. Maize is increasingly used as a feedstock for the production of ethanol fuel. Ethanol is mixed with gasoline in order to decrease the amount of pollutants emitted when used to fuel motor vehicles. High fuel prices in mid 2007 led to higher demand for ethanol, which in turn lead to higher prices paid to farmers for maize. This led to the 2007 harvest being one of the most profitable maize crops in modern history for farmers. Because of the relationship between fuel and maize, prices paid for the crop now tend to track the price of oil. Maize is widely used in Germany as a feedstock for biogas plants. Here the maize is harvested, shredded then placed in silage clamps from which it is fed into the biogas plants. This process makes use of the whole plant rather than simply utilizing the cob as in the production of fuel ethanol. A biomass gasification power plant in Strem near Gussing, Burgenland, Autrai began in 2005. Research is being done to make diesel out of the biogas by the Fischer Tropsch method. Increasingly ethanol is being used at low concentrations (10 % or less) as an additive in gasoline (gasohol) for motor fuels to increase the octane rating, lower pollutants and reduce petroleum use.

## Fodder

Maize makes a greater quantity of green mass than other cereal plants. So it can be used for fodder. Digestibility and palatability are higher when ensiled and fermented, rather than dried.

## Pellagra

When maize was first introduced into other farming systems than those used by traditional native-American peoples, it was generally welcomed with enthusiasm for its productivity. However, a widespread problem of malnutrition soon arose wherever maize was introduced as a staple food. This was a mystery since these types of malnutrition were not normally seen among the indigenous Americans, for whom maize was the principal staple food. It was eventually discovered that the indigenous Americans had learned to soak maize in alkali-water - made with ashes and lime (calcium oxide) by Mesoamericans and North Americans - which liberates the B-vitamin niacin, the lack of which was the underlying cause of the condition known as pellagra. This alkali process is known by its Nahuatl (Aztec)-derived name: nixtamalization. Besides the lack of niacin, pellagra was also characterized by protein deficiency, a result of the inherent lack of two key amino acids in pre-modern maize, lysine and tryptophan. Nixtamalisation was also found to increase the availability of lysine and tryptophan to some extent, but more importantly, the indigenous Americans had also learned to balance their consumption of maize with beans and other protein sources such as amaranth, meat and fish in order to acquire the complete range of amino acids for normal protein synthesis.

Maize was introduced into the diet of non-indigenous Americans without the necessary cultural knowledge acquired over thousands of years in the Americas. In the late 19th century pellagra reached epidemic proportions in parts of the southern USA, as medical researchers debated two theories for its origin. The deficiency theory (which was eventually shown to be true) said that pellagra was due to a deficiency of some nutrient and the germ theory said that pellagra was caused by a germ transmitted by stable flies. In 1914 the U.S. government officially endorsed the germ theory of pellagra, but rescinded this endorsement several years later when the evidence grew against it. By the mid-1920s the deficiency theory of pellagra was becoming scientific consensus and the theory was validated in 1932 when niacin deficiency was determined to be the cause of the illness. Once alkali processing and dietary variety were understood and applied, pellagra disappeared in the developed world. The development of high lysine maize and the promotion of a more balanced diet have also contributed to its demise. Pellagra still exists today in food-poor areas and refugee camps where people survive on donated maize.

**Utilization of maize**

In developed countries more than 60 % of the production is used in compounded feeds for poultry, pigs and ruminant animals. The by-products of dry milling include the germ and the seed-coat. The former is used as a source of edible oil of high quality. The seed-coat or pericarp is used mainly as a feed, although in recent years interest is developed in it as a source of dietary fibre. In industries, maize is wet milled to manufacture starch and by-products such as maize gluten is used as a feed ingredient. The maize germ processed to produce oil gives a by-product maize germ meal, used as an animal feedstuff. Fermentation also provides some alcoholic beverages.

Maize is consumed in many forms in different parts of the world, from maize grits, *polenta* and corn bread to popcorn and products such as maize flakes. The grain is fermented to give *ogi* in Nigeria and other countries in Africa. In Egypt maize flat bread, *aish merahra*, is widely produced. A similar product called *markouk* is eaten in Lebanon. Maize is also widely used to make beer. In Benin, malt is obtained by germinating the grain for about five days. A dough prepared from lime cooked maize is the main ingredient for many popular dishes such as *atole*, a beverage with a great variety of flavors, and *tamalitos*, made by wrapping the dough in maize husks and steam-cooking it for 20 to 30 minutes to gelatinize the starch. The dough is also used for *tamales*, a more complex preparation because of the number of ingredients it contains, in most cases with chicken or pork meat added to the gelatinized dough. It is also used to provide support for *enchiladas*, *tacos* (folded tortillas containing meat, etc.). Maize is also in preparation of *pupusas*, with fresh cheese placed between two layers of dough and baked like tortillas. When the dough is fried and flavoured, it yields foods such as chips and *chilaquiles*. Dough is allowed to ferment for two days, wrapped in banana or plantain leaves, to make *pozol* from which a number of drinks are made. Acid porridges usually made from fermented raw or heat-treated maize include pozol in Mexico and Guatemala, *ogi* in Nigeria, *uji* in Kenya and *kenkey* in Ghana. *Ogi* has a number of other names such as *akamu* or e*kogbona*, *agidi* and *eko* tutu. The nixtamal, or cooked maize without the seed-coat, is ground to coarse dough which is shaped into balls by hand, then wrapped in banana leaves to avoid drying and are allowed to ferment for two to three days, or more if necessary. Another major food made from decorticated, degermed and precooked maize, is *arepa* - maize bread without yeast, is round in shape consumed daily in Colombia and Venezuela,

Ahenkora *et al.* (1999) documented methods of preparation of some Ghanaian *maize based foods. Kenkey* was prepared by steeping the grains for 24 hrs at 25-26°C steeped grains were wet milled and kneaded with 1:2 meals: water. It was allowed for fermentation for 3 days at 25-27°C. Slurry was cooked and made into dough (*Kenkey* mass). The dough was moulded and wrapped with corn husks to get *ga kenkey*. It was boiled for two hours, sliced into pieces and dried for 24 hrs at 65°C to form *Kenkey*. To obtain fante *kenkey* the balls were wrapped in plantain leaves. For *thozafi* the maize was dry milled and mixed with water at 1:4 ratios and cooked with continuous stirring for 50 to 60 minutes to obtain a solid consistency.

## References

Ahmadi, M., W.J. Wiebold, J.E. Beuerlein, D.J. Eckert and J. Schoper (1993). Agronomic practices that affect corn kernel characteristics. Agron. J., 85: 615 - 619.

Basappa, G., J.B. Deshmanya and B.L. Patil (2007). Post- harvest losses of maize crop in karnataka - an economic analysis Karnataka J. Agric. Sci., 20 (1): 69 - 71.

Rao, P.S., C.P. Singh and R.P. Sharma (2001). Post- harvest lossess in maize in Mewar regions of Rajasthan, Indian Journal of Agricultural Marketing, 16: 51.

Usha singh (2014). Processing and value addition of maize for food and nutrition security. Indian Farming, 64 (4): 48 – 54.

# 31

# Status of Maize Processing

Maize is one of the staple foods of poor families. Traditionally, the grain is converted into flour in mills for making bread. Immature cobs are roasted and eaten all over the country. It is an important raw material for animal and poultry feed and corn flakes manufacturing units. But the quantity of maize utilized by these units is limited as the existing units are of small scale nature. They make only a few products having limited demand. Hence, an alternative large scale unit which can process a large quantity of maize to different value added products is required to be set up.

## Methods of maize processing

Maize is usually processed by two distinct processes, namely wet milling and dry milling. Dry milling produces grits, corn flour and minimum amount of corn meal. The technology has been standardized by Central Food Technological Research Institute (CFTRI), Mysore. The technical knowhow is available with CFTRI on price. Dry milling units have a crushing capacity of 10 t/day with a project cost of Rs.20 lakh. Such units can be set up anywhere depending upon the availability of raw material, power and suitable land.

### a. Dry milling

The maize kernels are screened, tempered with hot water/steam to loosen the germ and bran. Then it is degerminated to remove the germ. The husk is separated by means of aspirators. The degermed maize is dried to a moisture content of 15 - 15.5 % followed by sifting. It then is subjected to milling to produce grits, meal and flour. The germs separated is dried and passed through an expeller to produce the corn oil. The different products that result from dry milling are grits 40 %, coarse meal 20 %, germ 14 %, fine meal 10 %, flour 5 % and hominy feed10 %. CFTRI has developed a mini mill for dry milling of maize. The grits is the main product of dry milling process, which is used as porridge by boiling domestically. The processing units use grits for manufacture of products like ready-to-eat snacks (corn flakes), wall paper paste and manufacture of glucose by direct hydrolysis. The process flow of dry milling is as under:

Cleaning => Conditioning => Degerminating => Drying and Cooling => Grading and Grinding => Sifting and Classifying => Purifying => Drying => Packaging.

Dry corn millers process maize in one of three ways such as tempering degerming process, stone ground or non-degerming process and alkaline-cooked process. The main products of tempering / degerming process are flaking grits (large grits) 30 – 40 %, brewers grits (small grits) 20 %, corn meal (coarse powder) 12 %, corn flour (fine powder) 10 % and animal feed including germ 17 - 20 %.

**Flaking grits:** The large size grits obtained, as the top fraction is a premium product in the developed countries where corn flakes are popular as breakfast cereal.

**Brewer's grits:** Maize grits are used as adjuncts to barley malt in beer manufacture. Normally for 3 parts of barley malt, one part of grits is used. This ratio varies based on type of beer.

**Maize meal (Suji / Rava):** Coarse powdery fraction finer than grits is termed maize meal. The main end uses are in making snack foods, breads and bakery items. Considering the huge Indian market for these items selling maize rava should be easy.

**Corn flour:** The fine powder obtained during milling is termed corn flour. It can be made use of in preparations where other flours are used. Rural folks make rotis with corn flour. It is used in biscuit and bread production. It can be fortified with Soya meal resulting in a high protein, high-energy food.

**Bran/ germ:** The pericarp of maize kernel on powdering is termed bran. This is rich in proteins and finds ready market from feed industry. The germ contains oil. The small mills a mix of bran and germ is sold as cattle feed.

Alkaline cooked process is popular in Mexico. The flour obtained is called MASA flour and used in making Mexican snack foods like corn tortillas, corn (taco) shell and corn chips. Whole ground or stone-ground mills most often use white corn to make food products such as hominy grits and corn meal. These products are essentially whole ground corn with very little of the hull and germ removed.

### b. Wet milling

The technology is indigenous except for starch-gluten separation and starch washing unit which is imported through companies like Alfa Laval. The entire plant can be fabricated by the fabricators at Ahmedabad and few other parts of the country. There are a number of suppliers for setting up of the plant on turnkey basis.

Maize is generally processed to manufacture corn starch by wet milling method the world over. The by-products of starch manufacture like corn oil, corn steep liquor, gluten, etc. are the important value added products. The grain is unloaded from the trucks directly in receiving area or stored in silos. The material is fed to the cleaning section by a feed conveyor. The cleaning section is housed in 3 floors, where the material is screened for debris such as sand, stones and any other foreign particles. The clean material is then sent to steeping section. It is received in a tank where it is washed with hot water first and subsequently steeped in water containing sulfur dioxide @ 0.2 % for 70 hrs at 52°C. Steeping softens the kernels and also removes some solubles. Sulphur dioxide act as preservative. The steep water produced in this process is then concentrated and fortified with vitamins, minerals to produce corn steep liquor. It is then subjected to primary and secondary grinding, wherein the germ and husk are separated. The grinders are made of stainless steel with adjustable RPM with or without pneumatic settings. There are a number of manufacturers of grinding machines in India.

The pasty mix obtained after fine grinding is pumped to water filled settling troughs, known as germ separators or degerminators. It is a 3 stage process where the slurry containing soluble husk, gluten and starch are separated from germ. The lighter density rubbery germ float on the top and is skimmed off. The germ is passed to germ drier which is finally sent to oil extraction unit. The germ contains 45 % oil and the rest is crude fibre and moisture. The starch manufacturers generally prefer to sell germ rather than own oil extraction unit.

The slurry of husk, starch and gluten is ground for better recovery of starch. The fibre washing is a 6 stage process which is carried out by DSM box. The husk is separated from the soluble starch and gluten slurry by a counter current flow system. The husk is sent to either drying section or used as animal feed in wet form. The husk is mainly carbohydrate which also contains 8 % protein.

The slurry of starch and protein is passed through a centrifugal concentrator to get the concentrated slurry. This machine is also called as milk stream thickener. The starch slurry is then passed through a 12 stage hydroclone washing system, wherein the starch is washed and concentrated simultaneously. Starch slurry usually has a moisture content of 42 %. The starch slurry thus obtained is diverted to various production lines for manufacture of liquid glucose and modified starches such as dextrose, dextrose mono hydrate, etc. The wet starch is then dried by hot air by passing through a drier. The thickened slurry is passed through a high speed centrifuge to separate the heavier starch from the lighter protein (Gluten). The protein slurry is passed through a centrifuge to get concentrated slurry of gluten. The gluten contains 65 % protein and is a good source of protein for the animals and is used in animal feed preparation. The thickened gluten slurry is further concentrated to get gluten cake with 40 % solids through a rotary vacuum filter. The cake is further dried by hot air and / or sun to bring down the final moisture content to 12 %. Alfa Laval is the main company supplying this system in India for starch units.

The dry starch has a moisture content of 11-12 %. The slurry containing gluten is passed through a rotary vacuum filter in which a portion of the moisture is removed followed by drying in a hot air drier (175 $^{0}$C). The dried gluten thus obtained has a moisture content of 12 %. The main product of wet-milling of maize is starch. The gluten is mainly used for poultry feed. The average recovery of various products and co-products of maize during the wet milling are starch 60 - 62 %, gluten 8 - 9 %, germ 6 - 7 % and husk 22 - 24 %.

The starch is raw material for various ancillary industries like dextrose monohydrate, dextrins, saccharin etc. For manufacture of further derivatives of starch, ancillary units need to be attached with starch manufacturing units. The wet milling has developed into an industry that seeks optimum use and maximum value from each constituent of the maize kernel. In addition to starch and the various other products, and edible corn oil, the industry has become an important source of well-defined specialised ingredients used in feed formulation industry.

Effluent treatment plant is an essential component of a starch industry. It should be set up as per the norms of State Pollution Control Board. It has been made mandatory to set up an ETP in all starch manufacturing units.

## Critical factors in setting up a maize processing plant

The viability of a maize processing plant depends upon the availability and uninterrupted supply of raw material to the unit. On an average, a unit with a crushing capacity of 100 tonnes/ day will require about 30000 tonnes of maize per year (assuming 300 days of operation of the plant). Hence, the availability of raw material is one of the important considerations in deciding the location of maize processing unit. It should not be a problem for a unit of above capacity to procure the raw material during Kharif season. The plant will be able to procure major portion of its raw material requirement within the radius of 200 km. Yellow dent corn which is most suitable for wet milling for manufacture of starch and other by-products.

In India maize is mainly a Kharif crop. The harvesting is done in the month of September- October. As mentioned earlier, the farmers sell major portion of their produce immediately after harvesting. The remaining portion is released to the market depending upon the demand and market trends till February to March. As the traders procure maize from the farmers, the unit will be able to source the raw material from traders for another 2-3 months. Hence, there will be difficulty in procuring local maize for about 3-4 months during June to September. This can be overcome if the units procure maize in bulk during the Kharif and store in silos for the lean months. The possibility of obtaining it from adjoining states is also available. States like Bihar, Tamil Nadu, Karanataka, Andhra Pradesh and Punjab produce winter maize. The maize is generally packed in gunny bags in bulk. The most common mode of transportation is by lorries.

### Land

Land requirement of starch manufacturing unit is very high, as it requires large area to set up plant and machinery and effluent treatment plant. There should be enough land for disposal of treated waste water. A unit with crushing capacity of 100 tonnes/day should have at least 10 acres of land. However, if available at reasonable price, the unit may acquire upto 15 acre of land to meet future expansion requirements. In this model, we have assumed 10 acre of land for assessment of viability. As per the Industrial Policy Guidelines and Incentive Rules- 1999, the land is also allotted by State Government to the units on lease basis.

### Water

The water requirement for the wet milling industry is relatively large with an average use of 4 cum per tonne of crushing per day. For a wet milling unit of 100 tonnes capacity, therefore about 4 lakh litre of water/ day is required. The site where wet milling units are set up should have a good source of water, preferably a perennial river. As the unit also generate high amount of sewage water, which require to be disposed off properly. In case the water is to be sourced from ground, the water table should be high and the areas should fall in white category of unrestricted use.

**Power**

The average power requirement is about 170 - 250 units per day per tonne of maize crushing. The milling unit requires uninterrupted power supply and hence a DG set is required as standby arrangement.

**Steam**

The steam requirement is 1 ton / ton of maize crushing. The units manufacturing starch by wet milling in states like Maharashtra and Gujarat use coal for production of steam. However, coal as a source of energy for boilers is not permitted other states. Hence, oil fired boilers need to be installed. This may raise the cost of operation of the unit to some extent.

**Maize starch pilot plant**

The first step of the process is corn steeping which is carried out in the steep tank. After about 48 hours steeping time, the corn kernels are forwarded by means of a screw conveyor into a mill for coarse grinding. There the corn kernels are broken up to release the germs without damaging them. The addition of steep water during the grinding generates a so called corn mash. The mash is then carried over to a cyclone for germ separation. Afterwards the mash is fine grinded in an impact mill for complete disruption of the corn kernels. In this process step, starch is liberated from the cells. The resulting starch suspension still contains cell debris and bran components which need to be separated. This is carried out by means of centri-sieves: starch granules pass through the sieve while coarse bran components remain on it. The resulting starch milk is then applied to a fine meshed bend screen under high pressure for separation of fine fibres. Then the fibre-free starch slurry is directed to a hydro-cyclone plant for concentration and separation of dissolved proteins. This installation utilizes density differences between starch and proteins, whereas starch possesses the higher specific density. The concentrated and refined slurry still needs to be centrifuged for dewatering. In the final step the purified starch cake is dried by means of a flash dryer to obtain the end product: dry starch, a white coloured powder. List of equipments required are Steeping tanks, mills, screen bends, hydrocyclones, centrifugal machines and dryers. List of equipments required are steeping tanks, mill, screen bends, hydrocyclones, centrifugal machines and drier.

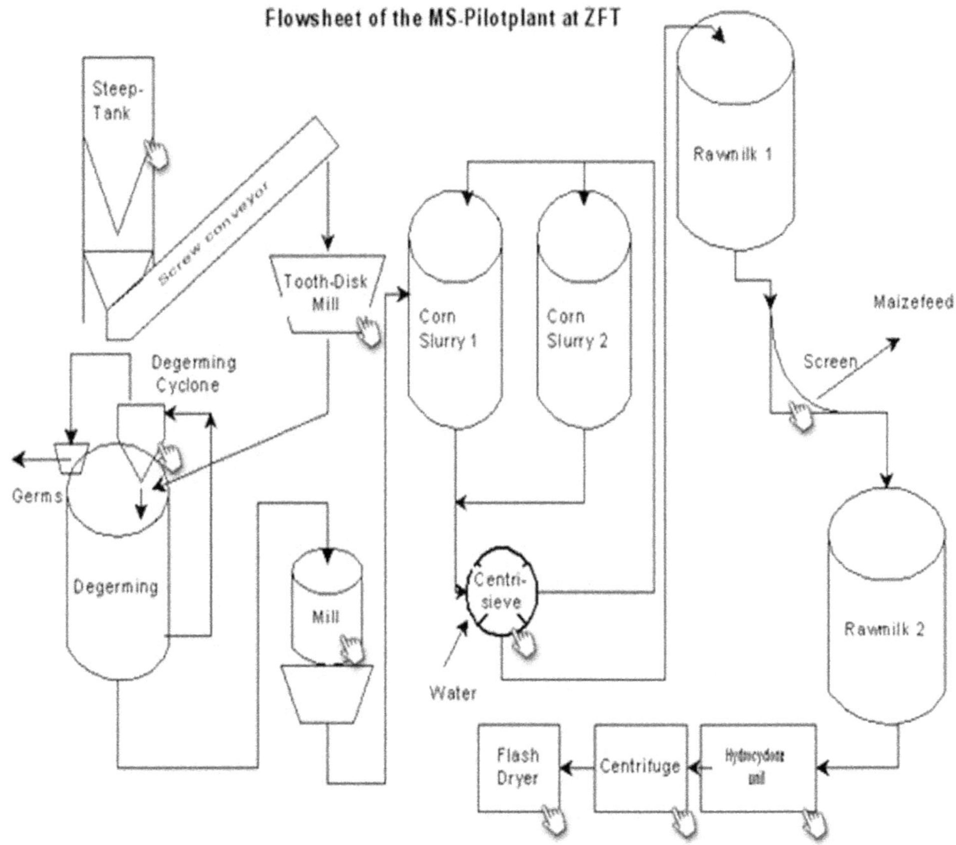


The capacity is 100 - 600 kg per charge. This is the pilot plant used for maize starch extraction. This plant is located on two floors. The process starts on the upper left side of the image and then proceeds to bottom right. The first step of the process is maize steeping which is carried out in the steep tank. After about 48 hours steeping time the maize kernels are forwarded by means of a screw conveyor into a mill for coarse grinding. There the maize kernels are broken up to release the germs without damaging them. The addition of steep water during the grinding generates a so-called maize mash. The mash is then carried over to a cyclone for germ separation. Afterwards the mash is fine-grinded in an impact mill for complete disruption of the maize kernels. In this process step, starch is liberated from the cells. The resulting starch suspension still contains cell debris and bran components which need to be separated. This is carried out by means of centri-sieves: starch granules pass through the sieve while coarse bran components remain on it. The resulting starch milk is then applied to a fine-meshed bend screen under high pressure for separation of fine fibres. Then the fibre-free starch slurry is directed to a hydro-cyclone plant for concentration and separation of dissolved proteins. This installation utilizes density differences between starch and proteins, whereas starch possesses the higher specific density. The concentrated and refined slurry still needs to be centrifuged for dewatering. In the final step the purified starch cake is dried by means of a flash drier to obtain the end product: dry starch, a white coloured powder.

## References

Rao, P.S., C.P. Singh and R.P. Sharma (2001). Post- harvest losses in maize in Mewar regions of Rajasthan, I*ndian Journal of Agricultural Marketing*, 16: 51.

Usha singh (2014). Processing and value addition of maize for food and nutrition security. Indian Farming, 64 (4): 48 – 54.

# 32

# Value Addition of Maize Grain

The starch is the main product of a maize processing unit, which is consumed in various other industries like food, pharmaceuticals, textiles, paper, restaurants, etc. The other products include gluten, germ, fibre (husk) and corn steep liquor. Gluten has great demand in animal feed industry because of its high protein content (70 %). Germ is expressed to extract germ oil which is low cholesterol containing edible oil. Fibre is used by animal feed manufacturers. It has demand in wet form itself for animal feed. Corn steep liquor is one of the substrates for culture media for manufacturing of antibiotics and other microbial production systems.

In India, Mumbai, Delhi, Ahmedabad and Kolkata are the major markets for processed maize products. Other important markets include Bhopal, Hyderabad, Chandigarh, Lucknow, Bangalore, etc. Most of the starch manufacturers of Gujarat, Maharashtra, Punjab, etc., have their marketing offices in Mumbai. Hence, Ahmedabad and Mumbai are the major trading centres for corn starch in India. Maize processors directly market their products to the consumers like pharmaceutical industries, hotels, textiles, paper industries, etc. and through traders as well. Starch and gluten have good export potential as well. India exports these products to Sri Lanka, South East Asian countries, Bangladesh and South Africa.

Nearly 25 pharmaceutical units have been set up at Baddi in Solan district and each of them consumes nearly 4-5 tonnes of starch per annum, which comes to nearly 200 tonnes of starch requirement per year. Further, Dabur India has set up a manufacturing unit of glucose from dextrose mono hydrate (DMH). The DMH is obtained from maize starch through a series of process including liquefaction, incubation, filtration, ion exchange, evaporation, crystallization and drying. Currently, this unit purchases about 20 tonnes of DMH per day from maize processing units of Gujarat and Punjab. Its annual requirement of this product is 2000 tonnes. A downstream DMH manufacturing unit can be set up easily attached to a starch manufacturing unit. Similarly, liquid glucose (LG) also has good demand in India and a LG manufacturing unit can also be set up along with a starch manufacturing unit from maize. In addition, there are a few textile manufacturing units located in the state which require low grade starch. This demand is being met from Ahmedabad and Mumbai. The procurement price of such units goes up due to the addition of transportation cost of Rs.2.0 to Rs.2.5 per kg of starch. The husk can be sold locally, but the unit may face difficulty in selling corn steep liquor as there is no unit in the state which produce antibiotics and microbial products. Corn oil can be marketed as there is good demand for it.

**a. Corn starch**

Because of the increased need for grain focused on corn-varieties suitable for the starch industry. These corn offer improved attributes such as increased starch content, easy processability and special qualities of the starch itself. There has been a positive trend during the past two decade in the wet milling industry. Corn has been the major source of starch (83 %) followed by potato (6 %), cassava (6 %), wheat (4 %) and rice (1 %). Maize is utilised mainly for ethanol production in the developed countries like USA and EU, whereas in the rest of the world, it is either used as a staple human food or manufacture of starch and its derivatives. The global production of starch from all sources was 48.5 million tonnes in the year 2000. USA with the largest starch industry contributes 51 % followed by EU (17 %) and the rest by others. During the same period 39.4 million tonnes of starch was derived from maize, whereas potato and wheat contributed 2.6 and 4.1 million tonnes respectively. The world demand for starch products is growing at an annual rate of 4 %. The demand for starch syrups is higher than dry starches in developed countries, whereas in developing countries, the situation is reverse. Due to the steady growing demand, the total world output of dry starches and syrups were estimated to reach 71 million tonnes and 37 million tonnes, respectively by 2010. EU is the major exporter of both native and modified starches, followed by US and Thailand. The largest starch consumers are US, EU, China and India. The world per capita demand for starch is 8.4 kg/annum, whereas India's per capita demand is 0.4 kg/annum. The wet milling industry in India is limited to certain pockets such as Gujarat, Maharashtra, Madhya Pradesh, Punjab, Karnataka and Chhattisgarh. There are about 17 wet milling units with a crushing capacity of about 3400 tonnes of maize/day. The state-wise number of wet milling units and the installed capacity is given in the Table 1.

**Table 1. Wet milling units in India**

| S. No. | Name of the State | No. of units | Installed capacity (tonnes maize/day) |
|---|---|---|---|
| 1 | Gujarat | 6 | 1,350 |
| 2 | Maharashtra | 5 | 1,050 |
| 3 | Madhya Pradesh | 3 | 450 |
| 4 | Karnataka | 1 | 300 |
| 5 | Punjab | 1 | 100 |
| 6 | Chhattisgarh | 1 | 150 |
| | **Total** | **17** | **3,400** |

Gujarat is the largest producer of starch, having six units with a total crushing capacity of 1350 tonnes of maize per day, followed by Maharashtra with 5 units and capacity of 1050 tonnes and Madhya Pradesh with 3 units and capacity of 450 tonnes maize. The average processing capacity of the units in India is 200 tonnes of maize / day. There are plants with as high crushing capacity as 400 tonnes/day. However, there is no plant in the country with crushing capacity of less than 100 tonnes/ day. The selection of technology is very important as regards to the viability of the unit is concerned.

### b. Corn ethanol

Ethanol produced from corn as a biomass through industrial fermentation, chemical processing and distillation. Corn is the main feedstock used for producing ethanol fuel in USA and it is mainly used as oxygenate to gasoline in the form of low level blends and to a lesser extent, as fuel for E85 flex fuel vehicles. There are two main types of corn ethanol production: dry milling and wet milling. The products of each type are utilized in different ways. In the dry milling process, the entire corn kernel is ground into flour and referred to as "meal". The meal is then slurried by adding water. Enzymes are added to the mash that converts starch to dextrose, a simple sugar. Ammonia is added to control the pH and as a nutrient for the yeast, which is added later. The mixture is processed at high temperatures to reduce the bacteria levels and transferred and cooled in fermenters. This is where the yeast is added and conversion from sugar to ethanol and carbon dioxide begins.

The entire process takes between 40 to 50 hours, during which time the mash is kept cool and agitated in order to facilitate yeast activity. After the process is complete, everything is transferred to distillation columns where the ethanol is removed from the "stillage". The ethanol is dehydrated to about 200 proofs using a molecular sieve system and a denaturant such as gasoline is added to render the product undrinkable. With this last addition, the process is complete and the product is ready to ship to gasoline retailers or terminals. The remaining tillage then undergoes a different process to produce a highly nutritious livestock feed. The carbon dioxide released from the process is also utilized to carbonate beverages and to aid in the manufacturing of dry ice.

The process of wet milling takes the corn grain and steeps it in a dilute combination of sulphuric acid and water for 24 to 48 hours in order to separate the grain into many components. The slurry mix then goes through a series of grinders to separate out the corn germ. Corn oil is a by-product of this process and is extracted and sold. The remaining components of fiber, gluten and starch are segregated out using screen, hydroclonic and centrifugal separators. The gluten protein is dried and filtered to make a corn gluten meals co-product and is highly sought after by poultry broiler operators as a feed ingredient. The steeping liquor produced is concentrated and dried with the fiber and sold as corn gluten feed to in the livestock industry. The heavy steep water is also sold as a feed ingredient and is used as an environmentally friendly alternative to salt in the winter months. The corn starch and remaining water can then be processed one of three ways: fermented into ethanol, through a similar process as dry milling, 2) dried and sold as modified corn starch or made into corn syrup.

The production of corn ethanol uses water in two ways such as irrigation and processing. There are two types of ethanol processing, wet milling and dry milling and the central difference between the two processes is how they initially treat the grain. In wet milling, the corn grain is steeped in water and then separated for processing in the first step. Dry milling, which is more common, requires a different process. According to a report by the National Renewable Energy Laboratory, Over 80 % of U.S ethanol is produced from corn by the dry grind process.

**Environmental and social issues**

Since most U.S. ethanol is produced from corn and the required electricity from many distilleries comes mainly from coal plants, there has been considerable debate about how sustainable corn-based bio-ethanol could be in replacing fossil fuels in vehicles. Controversy and concerns relate to the large amount of arable land required for crops and its impact on grain supply, direct and land use change effects as well as issues regarding its energy balance and carbon intensity considering the full life cycle of ethanol production and also issues regarding water use and pollution due to the increase expansion of ethanol production. The initial assumption that biofuels were good for the environment because they had a smaller carbon foot print is in debate because it is possible that the production of grain alcohol and therefore E85, may actually have a greater environmental impact than fossil fuel. That view says that one must consider:

- » The impact of fertilizers and carbon requiring inputs vs carbon offsetting byproducts like distillers grains.
- » The carbon footprint of the agricultural machinery runs to plant and harvest, and to spread chemicals in between.
- » The environmental impact of those chemicals themselves, including fertilizers and pesticides necessary for efficient mass production of the grains used.
- » The larger amounts of energy required to ship and process the grains and turn them into alcohol versus the more efficient process of converting oil into gasoline or diesel.
- » Even resources such as water, needed in huge amounts for grain production, can have serious environmental impact, including ground water depletion, pollution, runoff and algae blooms from waste runoff.

Current corn-based ethanol results in a 19 % reduction in greenhouse gases and is better for the environment than other gasoline additives such as MTBE. Ethanol produced today results in fewer greenhouse gas (GHG) emissions than gasoline and is fully biodegradable, unlike some fuel additives.

- » Today, on a life cycle basis, ethanol produced from corn results in about 19 % reduction in GHG emissions relative to gasoline. With improved efficiency and use of renewable energy, this reduction could be as much as 52 %.
- » In the future, ethanol produced from cellulose has the potential to cut life cycle GHG emissions by up to 86 % relative to gasoline.
- » Ethanol blended fuels currently in the market – whether E10 or E85 – meet stringent tailpipe emission standards.
- » Ethanol readily biodegrades without harm to the environment and is a safe, high performance replacement for fuel additives such as MTBE.

Ethanol from corn and cellulosc, is both much better fuels for the environment. Ethanol derived from sugar beet as used in Europe or sugar cane as grown in Brazil in industrial scale is generally seen as having a very positive $CO_2$ balance with up to 80 % reduction in well to wheel $CO_2$. Corn ethanol directly emits 52 % less greenhouse gas than gasoline. However this study does not take into account the greenhouse gasses involved in production and transportation.

### Economic impact of corn ethanol

The use of ethanol for fuel has had a damaging impact on food markets, especially in poorer countries. In the United States, ethanol is mostly made from yellow corn and as the market boomed for alternative fuel, yellow corn went up in price. Many farmers saw the potential to make more money and switched from white corn to yellow corn. White corn is the main ingredient of tortillas in Mexico and as the supply dropped, the price doubled, making the base of most Mexican foods unaffordable. Many people see this as unacceptable and want no overlap between food crops and fuel crops. The Renewable Fuels Association confirms ethanol production does in fact increase the price of corn by increasing demand. It is cited as a positive economic effect for US farmers and tax payers, but does not elaborate on the effect for other populations where field corn is part of the staple diet.

### Alternatives to corn as a feedstock

Remnants from food production such as corn stover could be used to produce ethanol instead of food corn. The use of cellulosic biomass to produce ethanol is a new trend in biofuel production. Fuels from these products are considered second generation biofuels and are considered by some to be a solution to the food verses fuel debate. The possibility of using this material has been acknowledged by the scientific community and the political community as well.

## c. Corn oil

Corn oil is extracted from the germ of corn. Its main use is in cooking, where its high smoke point makes refined corn oil valuable frying oil. It is also a key ingredient in some margarine. Corn oil is generally less expensive than most other types of vegetable oils. 25.40 kg of corn contains 3.42 kg of corn oil (2.8 % by weight). Breeders have developed high-oil varieties. However, these varieties tend to show lower field yields, so they are not universally accepted by growers. Corn oil is also a feedstock used for biodiesel. Other industrial uses for corn oil include soap, salve, paint, rust proofing for metal surfaces, inks, textiles, nitroglycerin and insecticides. It is sometimes used as a carrier for drug molecules in pharmaceutical preparations.

### Constituents of corn oil

Refined corn oil is 99 % triglyceride, with proportions of approximately 55 % polyunsaturated fatty acid, 30 % monounsaturated fatty acid and 15 % saturated fatty acid. Of the saturated fatty acids, 80 % are palmitic acid (lipid number of C16:0), 14 % stearic acid (C18:0), and 3 % arachidic acid (C20:0). Over 99 % of the monounsaturated fatty acids are oleic acid (C18:1 c). 98 % of the polyunsaturated fatty acids are theomega-6 linoleic acid (C18:2 n-6 c, c) with the 2 % remainder being the omega-3 alpha-linolenic acid (C18:3 n-3 c, c, c).

## Steps in production

**Raw material:** The average bottle of cooking oil contains vegetable oil, with no additives, preservatives or special flavourings. Corn oil is derived from the germ (embryo) of the kernel.

**Manufacturing process**: Corn oils, such are cold-pressed. This method, which entails minimal processing, produces light, flavourful oil suitable for some cooking needs. These oils undergo many steps beyond mere extraction to produce bland, clear and consistent oil.

**Cleaning and grinding:** In the case of corn, the kernel must undergo milling to separate the germ. The stripped seeds are then ground into coarse meal to provide more surface area to be pressed. Mechanized grooved rollers or hammer mills crush the material to proper consistency. The meal is then heated to facilitate the extraction of the oil. While the procedure allows more oil to be pressed out, more impurities are also pressed out with the oil, and these must be removed before the oil can be deemed edible.

**Pressing:** The heated meal is then fed continuously into a screw press, which increases the pressure progressively as the meal passes through a slotted barrel. Pressure generally increases as the oil is squeezed out from the slots in the barrel, where it can be recovered.

**Extracting additional oil with solvents:** Seeds with more oil are pressed and solvent treated. After the initial oil has been recovered from the screw press, the oil cake remaining in the press is processed by solvent extraction to attain the maximum yield. A volatile hydrocarbon (most commonly hexane) dissolves the oil out of the oil cake, which is then recovered by distilling the light solvent out. In using a special machine, flakes of meal are sent through wedge-shaped cells of a cylindrical vessel. The solvent then passes through the matter to be collected at the bottom. The solvent percolates through the matter which is periodically dumped and replaced.

**Removing solvent traces:** About 90 % of the solvent remaining in the extracted oil simply evaporates and, as it does, it is collected for reuse. The rest is retrieved with the use of a stripping column. The oil is boiled by steam and the lighter hexane floats upward. As it condenses, it too is collected.

**Refining the oil:** The oil is next refined to remove colour, odour and bitterness. Refining consists of heating the oil to between 40 and 85 °C) and mixing an alkaline substance such as sodium hydroxide or sodium carbonate with it. Soap forms from the undesired fatty acids and the alkaline additive and it is usually removed by centrifuge. The oil is further washed to remove traces of soap and then dried. Oils are also degummed at this time by treating them with water heated to between 85 and 95 °C), steam or water with acid. The gums, most of which are phosphatides, precipitate out and the dregs are removed by centrifuge. Oil that will be heated (for use in cooking) is then bleached by filtering it through fuller's earth, activated carbon or activated clays that absorb certain pigmented material from the oil. By contrast, oil that will undergo refrigeration (because it is intended for salad dressing) is winterized rapidly chilled and filtered to remove waxes. This procedure ensures that the oil will not partially solidify in the refrigerator. Finally, the oil is deodourized. In this process, steam is passed over hot oil in a vacuum between 225 and 250 °C), thus allowing the volatile taste and odour components to distill from the oil. Typically, citric acid at 0.1 percent is also added to oil after deodourization to inactivate trace metals that might promote oxidation within the oil and hence shorten its shelf-life. List of equipments required are metallic hoppers, storage bin for germ, dryer, oil expeller and oil filter

**Packaging the oil:** The completely processed oil is then measured and poured into clean containers, usually plastic bottles for domestic oils to be sold in supermarkets, glass bottles for exports or domestic oils to be sold in specialty stores or cans, plastic containers for food aid.

**By products/waste:** Since corn oil is derived from a small portion of the entire kernel, it creates corn meal and grits if it is dry milled and corn starch and corn syrup if it is wet milled.

**Negative health effects**

Some medical research suggests that excessive levels of omega-6 fatty acids, relative to omega-3 fatty acids, may increase the probability of a number of diseases and depression. Modern Western diets typically have ratios of omega-6 to omega-3 in excess of 10 to 1, some as high as 30 to 1, partly due to corn oil which has an omega-6 to omega-3 ratio of 49:1. The optimal ratio is thought to be 4 to 1 or lower. A high intake of omega-6 fatty acids may increase the likelihood that postmenopausal women will develop breast cancer. Similar effect was observed on prostate cancer. Other analysis suggested an inverse association between total polyunsaturated fatty acids and breast cancer risk.

### Information on maize oil

Among cereals and millets, oil is extracted only from maize. Maize contains 1.2 to 5.7 % edible oil. Varieties developed particularly for oil production contain as much as 14 %. Maize oil is widely used as a cooking medium and for manufacturing of hydrogenated oil. The oil has the quality of reducing cholesterol in the human blood like sunflower oil. The fat content of the oil is about 80 %.

### Physical characteristics

Physical characteristics are some of the determinants of consumer acceptability. The size, shape and appearance of the grain, influence the marketability of any food commodity. The kernels are often white or yellow in colour, although black, red and a mixture of colours is also found. There are a number of grain types, distinguished by differences in the chemical compounds deposited or stored in the kernel.

In a comparative study of the new QPM hybrids with normal maize, colour of QPM white hybrid was better than normal maize. The QPM hybrids were bigger in size and kernel hardness was improved, due to thick pericarp. QPM kernel cells are tightly packed with relatively few air spaces around the starch granule, resulting in increased hardness (Anonymous, 1988; Szaniel *et al.,* 1984).

A study was conducted by Sproule *et al.* (1988) to assess the characteristics of QPM with those of standard food grade maize. It was observed that QPM kernel had dark yellow colour with a greater amount of corneous endosperm and larger proportion of germ compared to food grade maize. Also, QPM recorded significantly greater test weight (27.90 kg/bushel) and density (1.40 g/cc) over food grade maize (27.31 kg/bushel; 1.33 g/cc respectively) (Sproule *et al.,* 1988).

Lopes and Larkins (1991) reported that similar kernel densities (1.11-1.16 and 1.09-1.12 g/ml QPM and normal maize genotypes respectively) and were nearly 10 per cent denser than typical 02 mutants (0.97-1.02 g/ml). QPM hybrids are observed to consist of hard, vitreous endosperm (Moro *et al.*, 1996).

Martinez *et al.* (1996) observed that water soluble solids were higher in QPM (81-90 g/

kg) than in normal maize (79-80 g/kg). Further, water soluble solids were positively correlated to kernel size, test weight and density. Test weight of QPM was higher than normal maize.

Gibbon *et al.* (2003) reported that altered starch structure was associated with endosperm modification in QPM. Proteomic analysis of QPM showed increased levels of granule bound starch synthase I in the soluble non-zein protein fraction. It was reported that increased extraction of this enzyme reflected a change in starch structure, which was manifested as shorter amylopectin branches and increased starch granule swelling. In mature kernels, these alterations in starch structure were associated with inter connections between starch granules that resulted in a vitreous kernel phenotype.

Gupta *et al.* (2003) reported that SDS-PAGE pattern of unalkylated zein from QPM, opaque-2 and normal maize showed similarity in polypeptide bonding despite significant quantitative differences amongst these cultivars. In normal unalkylated zein polypeptide of 20 KDa was absent but was found in unalkylated opaque-2. However, polypeptide of low molecular weight of 24 KDa of unalkylated zein and 22 KDa of alkylated zein of normal and one QPM cultivars, were found more intense compared to opaque-2 and other QPM cultivars and were responsible for endosperm vitreousness. Zein proteins were known to impart endosperm vitreousness and hardness and were not only reduced in all the QPM cultivars but also differed in their complexity and sub units as compared with opaque-2 and normal maize. Ratio of tryptophan and ratio of non-zein to zein were higher in opaque-2 and QPM as compared to normal maize.

## Nutrient composition

Cereal proteins have poor nutritional value for monogastric animal including humans. Cereals are low in essential amino acids such as lysine, tryptophan and threonine, lysine being the most limiting. Protein fraction such as albumins, globulins, zein, and glutelin also differ in cereals. Varietal differences in nutrient composition within the hybrids of maize were reported by several investigators.

Angel and Sotelo (1982) reported that protein content in the opaque-2 maize was higher (11.42 g/100 g) than normal maize (10.81 g/100 g) especially the hard endosperm opaque maize (10.81 g/100 g). There was no differences in the other nutrients such as fat (4.87, 5.32 and 5.19 g/100 g), carbohydrate (77.95, 76.44 and 75.82 g/100 g) and ash (1.64, 2.00 and 1.78 g/100 g) among the three types of maize normal, soft endosperm opaque-2 and hard endosperm opaque-2, respectively. Genetically, improved soft and hard endosperm corn varieties contained higher amounts of lysine (5.64 and 6.45 g/16 g N, respectively) and tryptophan (0.70 and 0.76 g/16 g N, respectively) as compared to normal corn (3.04 and 0.51 g/16 g N, respectively). However, the leucine content in both opaque-2 corns was lower (9.14 and 7.92 g/16 g N, respectively) than that of the normal corn varieties (18.85 g/16 g N).

The protein fractions of maize varieties were reported to increased for fraction 1-albumins, globulins, nitrogen components about 6.65 mg of protein, fraction v-true glutelin about 5.74 mg of protein in QPM compared to normal maize (Ortega *et al.*, 1986). QPM hybrid contains higher amounts of lysine by 33- 35 per cent, tryptophan by 60-66 per cent and threonine by 18-30 per cent. Leucine to isoleucine ratio was low in QPM hybrids, an advantage as it helps to liberate more tryptophan for more niacin biosynthesis thus helping to combat pellagra.

Proximate composition of QPM was similar to that of normal maize. However, QPM tended to have higher levels of crude protein (9.11 to 11.3% Vs 8.92 to 11.0 %), ether extract (4.3 to 5.12 % Vs 4.2 to 4.48 %), crude fibre (2.6 to 3.5% Vs 1.93to 2.14 %) and carbohydrates (71.37 to 75.4 % Vs 71.52 to 72.8 %), (Sproule *et al.*, 1988; Graham *et al.*, 1990; Ahenkora *et al.*, 1995; Osei *et al.*, 1999).

Bressani *et al.* (1990) compared QPM with 11 common maize types for fat, protein, total dietary fibre, Calcium, Magnesium, Sodium and Potassium. It was reported that protein content in common maize was 9.4 % and in QPM it was 10.2 %. Ether extract was more in QPM (5.5%) compared to common maize (4.7 %). Ash content varied from 1.1-1.4 % among the common maize. Calcium, magnesium, sodium and potassium contents were lower in QPM (23.7, 104.4, 15.5 and 361.9 mg/100 g respectively) compared to common maize (32.0, 128.4, 22.8 and 366.9 mg/100 g, respectively).

Martinez *et al.* (1996) reported that QPM contained higher amounts of protein, fat, dietary fibre and ash (10.56, 3.56 and 1.72 %, respectively) when compared to normal maize (10.44, 2.66 and 1.28 %, respectively). Starch content was lower in QPM than normal maize (82.6 Vs 84.2 %). However, QPM showed better lysine content (42 g/kg of protein) than normal maize (35 g/kg of protein). A range 3.7 to 4.2 g/100 g of protein was reported which was significantly higher than that of normal maize (2.6 to 3.1 g/100 g protein) (Ahenkora *et al.*, 1999).

A study was conducted by Omueti and Morton (1996) to evaluate the nutrient content of soya-maize extruded snacks named *soyabari*. It was prepared by adding pepper, onion, salt, palm oil, plantain and banana in different proportions. It was observed that protein content in soya and maize grits extrudates was higher (24.06 %) followed by extrudates containing plantain (22.75 %) and banana (20.13%). The banana containing extrudates gave higher crude fat and available carbohydrate content (14.2 and 40.3 %, respectively) than the extrudates prepared from soya-maize grits (2.9 and 38.5 %, respectively). The crude fibre content of all the extrudates were less (1.1-1.7 %) than market sample (3.5 %). Total ash content was higher in extrudates containing plantain (4 %) followed by maize-soya grits products (3.6 %) and banana extrudates (2.1 %). The gross energy of the extrudates ranged between 4 and 5 kcal/g and maize soya grits extrudates exhibited low energy value of 4 Kcal/g.

Zarkadas *et al*., (2000) observed that QPM genotypes contained high levels of lysine (3.43 to 4.21 g/100 g of protein) compared to commercial maize varieties which ranged from 2.9 to 3.1 g/100 g of protein.

Sharma *et al.* (2002) reported that moisture content ranged from 8.21 to 8.85 % in all the varieties (Paras, Prabhat, PAC 776, Bio 9637 and Prakash). Prakash and PAC 776 varieties showed the higher percentage of moisture (8.85 and 8.79 %, respectively). Varieties Paras and Prakash were rated superior for characters such as protein (10.23 and 9.80 %), fat (4.45 and 5.00 %) and tryptophan (0.72 and 0.70 % of protein). Bio 9637 and Prabhat were promising for carbohydrate (77.33 and 77.74 %), minerals (1.40 % in both varieties) and energy values (358 and 356 Kcal).

Habtamu *et al.* (2003) compared nutrient composition of QPM with local maize and four high yielding hybrids, the total protein content among the cultivars ranged from 7 to 11.8 per cent. It was 9.8 per cent for QPM and 10.1 per cent for local maize cultivar. However, the QPM maize protein contained 30 to 82 per cent more lysine, higher levels of arginine, tryptophan, histidine, threonine, cysteine and valine.

Mosha and Vicent (2004) compared maize and sorghum based weaning mixtures supplemented with rojo bean flour, ground sardines and peanut paste. Ten different types of weaning mixtures were prepared and evaluated for proximate composition. It was reported that proximate composition of maize based weaning mixture varied with the sorghum based weaning mixture, with respect to increased amounts of carbohydrate, fat, energy and decreased amounts of protein, fibre and mineral contents.

A study on effect of extrusion on essential amino acid profile and colour of whole grain flours of QPM and normal maize cultivars was conducted by Paes *and* Maga (2004). It was observed that extrusion reduced the essential amino acids isoleucine, leucine, lysine, threonine and valine when compared to their original flours. But, the contents of histidine, methionine, phenylalanine and tryptophan were not affected. QPM samples, either raw or extruded, were significantly higher in lysine, methionine and tryptophan compared to samples of normal maize.

**Functional properties**

The physicochemical properties of flour such as particle size, water and oil absorption capacities, swelling capacity and viscosity predict the functional properties of products. The grain constituents such as starch, fat and fibre influence the functional properties.

Mehta *et al.* (1989) studied on weaning gruels from maize/maize and green gram dhal and combined effect of fermentation and malting on bulk reduction and other physiochemical properties. The process of fermentation was carried out on plain maize flour and maize + green gram dhal flour slurries at various temperatures (35, 40 and 45°C), time intervals (8, 12, 16 and 20 hours) and slurry concentrations (15, 20 and 25 %). The results revealed a decrease in pH (from 6 to 3) with increase in titrable acidity (from 0.03 to 0.07) in all the maize based slurries as fermentation progressed, at all temperatures. Viscosity studies of cooked maize and maize green gram dhal slurries revealed reduction in viscosities at all time intervals and temperatures, in all the slurries. The best reduction in viscosity was observed in 15 per cent concentration slurries fermented for 8 hours at 35°C.

A study on water absorption capacity by Afoakwa (1996) reported that water absorption capacity influenced product cohesiveness. It was also reported that, proteins were mainly responsible for the water uptake and to lesser extent the starch and cellulose, at room temperature. Martinez *et al.* (1996) studied wet milling process of normal and QPM and reported that viscosity of QPM and normal maize hybrids were similar with 600-700 BU of viscosity.

Kikafunda *et al.* (1997) conducted an experiment to optimize viscosity of maize porridges. The study revealed that the viscosity of the maize porridges was significantly increased by increase in concentration (68.1 to 233.8 mPa.s), cooking time from 128.1 to 173.8 mPa.s and use of flour of fine particle size from 136.4 to 165.5 mPa.s. Increased shear speed and addition of groundnut paste or pasteurized cow's milk significantly reduced the viscosity of the porridges from 156.9 to 145 mPa.s, 214.5 to 87.5 mPa.s and 176.1 to 125.8 mPa.s.

Aminigo and Oguntunde (2000) studied on functional properties and nutritive composition of raw, blanched (pre treatment) and dry roasted maize. Nitrogen solubility and water absorption were reduced by roasting and this effect was more pronounced in blanched maize samples. Gelation capacity was reduced significantly by the heat treatments. Water and fat absorption were generally higher for samples blanched prior to roasting than for samples roasted without pre-treatment.

An investigation on functional properties *viz.*, viscosity of sorghum diastatic malt blends with maize flour was studied by Delgado and Serna (2000). It was reported that viscosity of the blends decreased as the concentration of sorghum diastatic malt increased. Addition of 6.7 % sorghum diastatic malt reduced viscosity by nearly 50 per cent.

A comparative study of the physicochemical properties of whole maize flour and dehulled-degermed maize flour was studied by Houssou and Ayernor (2002) to evaluate their functional behavior. It was observed that yield of flour was more in whole grain (92 %) compared to dehulled-degermed flour (65 %). However, dehulled-degermed flour showed higher water absorption capacity (197 %) than whole milled grain flour (154 %). Thus, the swelling capacity increased by one and a half times more in dehulled-degermed flour. The hot paste viscosity of dehulled-degermed flour was high compared to whole milled grain flour (505 and 370 BU, respectively).

Okezie *et al.* (2003) reported that water holding capacity of maize extract increased by addition of ripe bananas and ranged between 0.47 and 0.76 mg/g for 50 and 20 % levels of banana substitution, respectively. The values, however, initially increased from 0.72 to 0.76 mg/g for maize flour at 20 per cent level of substitution, it decreased progressively to 0.47 mg/g at 50 per cent level of banana substitution. It was also reported that water holding capacity increased from 2.82 ml/g for maize flour to 3.06 ml/g for 30 per cent banana substitution dropped again to 2.47 ml/g for 50 per cent banana substitution. The decrease in the water holding capacity of samples with high levels of banana substitution could be attributed to the coarse nature of the substituted samples.

An investigation on functional properties of three different flours/ meals and two blends of maize meal and soybean flour (ratios 9:1 and 8:2, maize: soybean) was conducted by Edema *et al.* (2005). The bulk density was lower in commercially sold soybean flours (0.38 g/ml) than in maize soybean blend flour (0.55 g/ml).

Wang *et al.* (2005) reported the effect of pericarp removal on properties of wet milled corn. The grains were steeped in 100 ml steeping solution (0.2 % sulphur dioxide and 0.5

% lactic acid). Pericarp was removed by sonication treatment in which the grains were filled with excess of water in a plastic bag. The bag was sealed with air squeezed out and placed in sonicator tank. The samples were sonicated for 3 minutes at room temperature and samples were blotted dried and manually pericarp was removed. It was reported that pericarp free samples with 6 hour steeping and 48 hour steeping exhibited lower peak breakdown (95 and 171 cp), final (1,333 and 1,358 cp), than 1 hour steeping (426, 1646 and 585 cp, respectively). In case of pericarp intact maize 48 hour of steeping declined the peak breakdown final and set back viscosity from 445 to 301 cp, 1637 to 1463 cp and 563 to 428 cp. The authors reported that reduction in viscosity was higher in pericarp free maize than pericarp intact maize.

### Value addition

Consumer demands for convenient, inexpensive, nutritious and ready to eat products result in value addition. Utilization in diversified ways by converting them into a variety of products such as infant foods, health foods/mixes, convenience foods and specialty foods, make a significant impact on consumer acceptability.

Singh *et al.* (2006) developed and evaluated acceptability of QPM and normal maize based cakes. The cake were prepared by incorporating maize flour at 40, 60, 80 % level of refined flour. The results revealed that all types of cakes were acceptable and with increase in the level of incorporation of maize flour the acceptability scores were reduced. Proximate analysis indicated that moisture content of cakes ranged from 24.10 to 25.80 % cakes prepared with QPM varieties had significantly higher protein than those from normal varieties of maize. Fat, ash and crude fibre content of all the cakes ranged from 10.57 to 12.28, 1.05 to 1.17 and 0.88 to 1.00 % respectively. Total carbohydrate content of cakes ranged from 71.21 to 75.03 g/100 g. The authors reported that the acceptability and nutritional composition data indicated that QPM can be successfully utilized for preparation of cakes.

### d. Corn syrup

Corn syrup is food syrup which is made from the starch of corn and composed mainly of glucose. It is used in foods to soften texture, add volume, prevent crystallization of sugar and enhance flavour. It is distinct from high fructose corn syrup, created when corn syrup undergoes enzymatic processing that produces a sweeter compound containing higher levels of fructose. The more general term glucose syrup is often used synonymously with corn syrup, since glucose syrup is most commonly made from corn starch.

## Steps in production

Glucose or dextrose syrup is produced from number 2 yellow dent corn. When wet milled, about 2.3 litres of corn are required to yield an average of 947 g of starch, to produce 1 kg of glucose or dextrose syrup. About 25 kg of corn will yield an average of 14.3 kg of starch, which in turn will yield about 15.1 kg of syrup. Thus, it takes about 2,300 litres of corn to produce a tonne of glucose syrup or 1524 kg of corn to produce one short ton. Formerly, it was produced by combining corn starch with dilute hydrochloric acid and then heating the mixture under pressure. Currently, it is mainly produced by first adding the enzyme ά-amylase to a mixture of corn starch and water. ά -amylase is secreted by various species of the bacterium Bacillus which enzyme is isolated from the liquid in which the bacteria are grown. The enzyme breaks the starch into oligosaccharides, which are then broken into glucose molecules by adding the enzyme glucoamylase, known also as «V-amylase». Glucoamylase is secreted by various species of the fungus *Aspergillus* which is isolated from the liquid in which the fungus is grown. The glucose can then be transformed into fructose by passing the glucose through a column that is loaded with the enzyme D-xylose isomerase, an enzyme that is isolated from the growth medium of any of several bacteria. The viscosity and sweetness of the syrup depends on the extent to which the hydrolysis reaction has been carried out. To distinguish different grades of syrup, they are rated according to their dextrose equivalent.

## High fructose corn syrup

High-fructose corn syrup is produced by milling corn to produce corn starch, then processing that starch to yield corn syrup, which is almost entirely glucose and then adding enzymes that change some of the glucose into fructose. The resulting syrup (after enzyme conversion) contains approximately 42 % fructose and is HFCS 42. The 42 % fructose is then purified to 90 % fructose HFCS 90. To make HFCS 55, the HFCS 90 is mixed with HFCS 42 in the appropriate ratios to form the desired HFCS 55. The enzyme process that changes the 100 % glucose corn syrup into HFCS 42 is as follows:

Corn starch is treated with alpha-amylase to produce shorter chains of sugars called oligosaccharides. Glucoamylase which is produced by *Aspergillus*, a fungus in a fermentation vat breaks the sugar chains down even further to yield the simple sugar glucose. Xylose isomerase (aka glucose isomerase) converts glucose to a mixture of about 42 % fructose and 50–52 % glucose with some other sugars mixed in.

While inexpensive alpha-amylase and glucoamylase are added directly to the slurry and used only once, the more costly xylose-isomerase is packed into columns and the sugar mixture is then passed over it, allowing it to be used repeatedly until it loses its activity. This 42–43 % fructose glucose mixture is then subjected to a liquid chromatography step, where the fructose is enriched to about 90 %. The 90 % fructose is then back-blended with 42 % fructose to achieve a 55 % fructose final product. Most manufacturers use carbon adsorption for impurity removal. Numerous filtration, ion-exchange and evaporation steps are also part of the overall process. The unit of measurement for sucrose is degrees Brix (symbol °Bx). Brix is a measurement of the mass ratio of dissolved sucrose to water in a liquid. A 25°Bx solution has 25 g of sucrose per 100 g of solution (25 % w/w). To put it another way, there are 25 g of sucrose and 75 g of water in the 100 g of solution. The Brix measurement was introduced by Antoine brix. A more universal measurement of sugars, including HFCS, is called dry solids. Dry solids are defined as the mass ratio of dry sugars to the total weight of the sugar solution. Since Brix is based on the refractive index of light against a sucrose molecule it is not accurate when measuring other sugars such as glucose, maltose and fructose. List of equipment required are thermostastic control blower, agitator, extractor, boiler, utensils, blender and tanks.

### e. Corn flakes

Crispy corn flakes are the most popular breakfast cereal known around the world. Toasted corn flakes added with milk make your breakfast complete and keep you going for whole day. Nutritious corn flakes are rich in iron and protein that help in making your memory sharp and increase your metabolism. Cereal corn flakes are added to various ready to eat snacks and Namkeens. Breakfast cereal technology has advanced greatly since its origins in the late 19$^{th}$ century. The latest innovation in the industry is the twin extruder. The two rotating screws clean as they rotate. This allows the dough to move more smoothly than in an extruder with only one screw. By using a twin-screw extruder, along with computers to precisely control temperature and pressure, cereals that usually require about 24 hours to make may be made in as little as 20 minutes.

**Table 2. Nutrients content in 100g**

| Nutrients | Content | Nutrients | Content |
|---|---|---|---|
| Energy | 326.65 Kcal | Fat of which saturates | 0.09 g |
| Protein | 7.4 g | Fiber | 1.6 g |
| Carbohydrate | 76 g | Sodium | 33.70 mg |
| Carbohydrate of which sugar | 6.5 g | Cholesterol | 0 g |
| Fat | 0.57 g | Trans Fat | 0 g |

## Manufacture of corn flake

The corn flake is one of the important value added products manufactured out of yellow and white maize. It is generally eaten as a breakfast cereal but the demand for this product is limited to hotels and big cities. It is a product of dry milling, which is manufactured by flaking of the major grain after extraction of germ. The flaked grain is either roasted for manufacture of corn flakes, breakfast cereal or fried to manufacture corn flakes served as snack foods. The raw flakes are also used for manufacture of beer. There are only a few companies like Mohan Meakin, Gaziabad and Kellogs, whose brands of corn flakes are very popular in Indian market. A brief process flow of corn flakes is given as under:

Receiving => Cleaning and Polishing => Milling => Bran => Cooking under pressure => flavouring => Agitation (lump breaking) => Drying => Sweating => Flaking => Roasting/ Frying => Grading => Packing

The corn grains after cleaning and polishing are milled to remove the germ and bran. Germ is utilised for extraction of corn oil. The bran is cooked under pressure in rotary steam cooker and mixed with flavouring material. The cooking is completed when the material turns out to a uniform translucent colour. The cooked material is carried to an agitator or lump breaker and finally dried in drier to moisture level of about 15 to 20 %. The dried material is kept in tempering tanks for few hours to permit the residual moisture to become equally distributed which is known as sweating. This is very essential to have uniform pressing for the flakes. The tempered material is next passed though a heavy duty flaking machine. The flakes are then immediately transferred to gas fired rotary ovens for roasting. While rotating, the flakes are continuously carried forward until they are dropped into conveyer. The roasted flakes are subjected to inspection, preferably on conveyers or tables. Then properly roasted flakes are graded and transferred to the packing bins immediately because flakes are hygroscopic. They are then packed in water resistant polythene packages or food grade waxed paper packages. A corn flakes unit of 300 tonnes per annum will require an investment of Rs.50 lakh.

## Steps in production

**Raw material:** The most important raw material in any breakfast cereal is grain. Most breakfast cereals contain other ingredients, such as salt, yeast, sweeteners, flavoring agents, colouring agents, vitamins, minerals and preservatives. The sweeteners used in breakfast cereals include malt (obtained from barley), white sugar, brown sugar and corn syrup. Some natural cereals are sweetened with concentrated fruit juice. A wide variety of flavours may be added to breakfast cereals, including chocolate cinnamon and fruit flavours. Other ingredients added to improve flavour include nuts, dried fruit and marsh mallows. Vitamins and minerals are often added to breakfast cereals to replace those lost during cooking. The most important of these is vitamin B-i, 90 % of which is destroyed by heat. The antioxidants BHA and BHT are the preservatives most often added to breakfast cereals to prevent them from becoming stale and rancid.

**Preparing grain:** Grain is received at the cereal factory, inspected and cleaned. It may be used in the form of whole grains or it may require further processing. Often the whole grain is crushed between large metal rollers to remove the outer layer of bran. It may then be ground more finely into flour. Whole grains or partial grains (such as corn grits) are mixed with flavouring agents, vitamins, minerals, sweeteners, salt and water in a large rotating pressure cooker. The time, temperature and speed of rotation vary with the type of grain being cooked. The cooked grain is moved to a conveyor belt which passes through a drying oven. Enough of the water remains in the cooked grain to result in a soft, solid mass which can be shaped as needed.

**Making flaked cereals:** The cooked grains are allowed to cool for several hours, stabilizing the moisture content of each grain. This process is known as tempering. The tempered grains are flattened between large metal rollers under tons of pressure. The resulting flakes are conveyed to ovens where they are tossed in a blast of very hot air to remove remaining moisture and to toast them to a desirable colour and flavour. Instead of cooked grains, flakes may also be made from extruded pellets in a similar manner.

**Adding coatings:** After shaping, the cereal may be coated with vitamins, minerals, sweeteners, flavours such as fruit juices, food colours or preservatives. Frosting is applied by spraying thick, hot syrup of sugar on the cereal in a rotating drum. As it cools the syrup dries into a white layer of frosting. List of equipment required are dehusker, flaker, tanks, dryer and cooker.

### f. Popcorn

Each kernel of popcorn contains a certain amount of moisture and oil. Unlike most other grains, the outer hull of the popcorn kernel is both strong and impervious to moisture and the starch inside consists almost entirely of a hard, dense type. As the oil and the water are heated past the boiling point, they turn the moisture in the kernel, which has a moisture-proof hull, into a superheated pressurized steam. Under these conditions, the starch inside the kernel geltainizes, softens and becomes pliable. The pressure continues to increase until the breaking point of the hull is reached (a pressure of about 135 psi and a temperature of 180 °C). The hull ruptures rapidly, causing a sudden drop in pressure inside the kernel and a corresponding rapid expansion of the steam, which expands the starch and proteins of the endosperm into airy foam. As the foam rapidly cools, the starch and protein polymers set into the familiar crispy puff. Special varieties are grown to give improved popping yield. Some wild types will pop, but the cultivated strain is *Zea mays averta*, which is a special kind of flint corn.

## Methods of popping

Although small quantities can be popped in a stovetop kettle in a home kitchen, commercial sale of freshly popped popcorn employs specially designed popcorn machines, which were invented in Chicago, Illinois by Charles Cretors in 1885. Cretors successfully introduced his invention at the Columbian Exposition in 1893. Cretors' invention introduced the first patented steam-driven popcorn machine that popped corn in oil. Previously, vendors popped corn by holding a wire basket over an open flame. At best, the result was a hot, dry, unevenly cooked confection. The Cretors' machine popped corn in a mixture of one-third clarified butter, two-thirds lard and salt. This mixture could withstand the 232 °C temperature needed to pop corn and it did without producing much smoke. A fire under a boiler created steam that drove a small engine which drove the gears, shaft and agitator that stirred the corn and powered the attention-attracting clown – the Toasty Roasty Man. A wire connected to the top of the cooking pan allowed the operator to disengage the drive mechanism, lift the cover, and dump popped corn into the storage bin beneath. Exhaust from the steam engine was piped to a hollow pan below the corn storage bin and kept freshly popped corn uniformly warm for the first time ever.

**Table 3. Nutrition in popcorn with air-popped and no additives per 100 g**

| Nutrients | Content per 100g | Nutrients | Content per 100g |
|---|---|---|---|
| Energy | 382 kcal | Protein | 12 g |
| Carbohydrates | 78 g | Thiamine (Vit. B1) | 0.2 mg |
| Dietary fibre | 15 g | Riboflavin (Vit. B2) | 0.3 mg |
| Fat | 4 g | Iron | 2.7 mg |

A very different method of popcorn-making can still is seen on the streets of some Chinese cities today. The corn is poured into a large cast-iron canister- sometimes called a 'popcorn hammer'- that is then sealed with a heavy lid and slowly turned over a curbside fire in rotisserie fashion. When a pressure gauge on the canister reaches a certain level, it is removed from the fire, a large canvas sack is put over the lid and the seal is released. With a huge boom, all of the popcorn explodes at once and is poured into the sack. Individual consumers can also buy and use specialized popping appliances that typically generate no more than a gallon of popped corn per batch. Some of these appliances also accept a small volume of oil or melted butter to assist thermal transfer from a stationary heating element, but others are "air poppers" which rapidly circulate heated air up through the interior, keeping the unpopped kernels in motion to avoid burning and blowing the popped kernels out through the chute. The majority of popcorn sold for home consumption is now packaged for use in a microwave oven in a popcorn bag.

## References

Ahenkora, K., A.S. Twumasi, W. Haag and B.D. Dzah (1995). Ghanian Kenkey from normal and quality protein maize: comparative composition and rat growth trials. Cereals Res. Communication, 23(3): 229-304.

Ahenkora, K., A.S. Twumasi, P.Y.K. Sallah and K. Obeng-Antwi (1999). Protein nutritional quality and consumer acceptability of tropical Ghanaian quality protein maize. Food and Nutrition Bulletin, 20(3): 354-359.

Aminigo, E.R. and A.O. Oguntunde (2000). Functional properties and nutritive composition of maize as affected by heat treatments. J. Food Sci. Techn., 37(1): 11-15.

Angel-Del, R.Alma and A. Stoelo (1982). Nutritive value of mixtures using chickpeas with wheat, triticale, normal and opaque-2 corns. J. Nutrition, 112: 1474- 1480.

Anonymous (1988). Quality Protein Maize. National Academic Press, Washington, DC, USA.

Anonymous (1998). Maize properties, processing and utilization. RESC Scientific series, 16: 10-11.

Ceylan, M. and E. Karababa (2004). The effects of ingredients on popcorn popping characteristics. Int. J. Food Sci. Technol., 39: 361– 370.

Collison, R. (1968). Swelling and gelation of starch. In: Starch and Its Derivatives. Ed. J.A. Radley, 4th edition, Chapman, London,

Delgado, B.M.L. and S.S.O. Serna (2000). Production and nutritional evaluation of liquefied weaning foods from malted sorghum, quality protein maize and other cereals. Cereal Chemistry, 77(5): 652-656.

Deshpande, S.S., K.C. Joshi, P.C. Bargale, K. Jha, V. Singh and S. Varghesi (2004). Development, acceptability and shelf life studies of soy fortified maize sattu. J. Food Sci. Techn., 41(6): 674-678.

Edema, M.O., L.O. Sanni and A. Sanni (2005). Evaluation of maize soybean flour blends for sour maize bread production in Nigeria.

Gibbon, B.C., X.L. Wang and B.A. Larkins (2003). Altered starch structure is associated with endosperm modification in quality protein maize. In: Proceedings of the Academy of Sciences of the Untied States of America, 100(26): 15329-15334.

Gopalan, C., R.B.V. Sastri and S.C. Balasubramanian (1999). Nutritive value of Indian foods. NIN, ICMR, Hyderabad.

Gupta, H.O., A. Sachdev, R.P. Johari and R.P. Singh (2003). Storage protein and its impact on nutritional quality in quality protein maize. Indian J. Agric. Biochemistry, 16(2): 67-71.

Houssou, P. and G.S. Ayernor (2002). Appropriate processing and food functional properties of maize flour. African J. Sci. Techn., 3(1): 126- 131.

Jayalakshmi, N. and U.Bangarusamy (2001). Maize-value added products. Kisan world, (11): 37 -38.

Kantha, S.S., N.S. Hettiarachchy and J.W. Erdman (1986). Nutrients antinutrient contents and solubility profiles of nitrogen, phytic acid and selected minerals in minged bean flour. Cereal Chemistry, 63: 9-13.

Kikafunda, J.K., A.F. Walker and S. Abeyasekera (1997). Optimising viscosity and energy density of maize porridges for child weaning in developing countries. International J. Food Sci. Nutrition, 48: 401-409.

Lee, Y.C. and M. McCoon (1981). Lipoxygenase and off –flavor development in some frozen foods. Korean J. Food Sci. Technol., 13: 53.

Lopes, M.A. and B.A. Larkins (1991). Cell biology and molecular genetics. Crop Sci., 31: 1655 - 1662.

Martinez, B.F., J.D.G. Figueroa and S.A. Larios (1996a). High lysine extruded products of quality protein maize. J. Sci. Food Agric., 71: 151-155.

Martinez, B.F., P.E. Sevilla and M. Bjarnason (1996b). Wet milling comparison of quality protein maize and normal maize. J. Sci. Food Agric., 71: 156-162.

Mehta, P., P. Patil and N. Parikh (1989). Studies on weaning gruels maize, maize and green gram dhal: Part II effect of fermentation and malting on physico-chemical changes, bulk reduction property and suitability for children belonging to maize eating population. Indian J. Nutrition and Dietetics, 26: 354-364.

Milan, C.J., D.R. Gutierrez, R.E.O. Cuevar, T.J.A. Garzon and M.C. Reyes (2004). Nixtamalized flour from quality protein maize (*Zea mays* L.) optimization of alkaline processing. Plant Food Human Nutrition, 59: 35-44.

Moro, G.L., J.E. Habben, B.R. Hamaker and B.A. Larkins (1996). Characterization of the variability in lysine content for normal and opaque 2 maize endosperm. Crop Sci., 36: 1651-1659.

Mosha, T.C.E. and M.M. Vincent (2004). Nutritional value and acceptability of home made maize/sorghum based weaning mixtures supplemented with rojo bean flour, ground sardines and peanut paste. International J. Food Sci. Nutrition, 55(4): 301-315.

Mouliswar, P., S. Kurien, V.A. Daniel, N.G. Malleshi and S. Venkatarao (1993). *In vitro* digestibility of protein and starch as energy food and its bulk reduction. J. Food Sci. Techn., 30: 36-39.

Mugula, J.K. (1992). Evaluation of the nutritive value of maize-soybean tempe as a potential weaning food in Tanzania. International J. Food Sci. Nutrition, 43: 113-119.

Naik, R.K., L.R. Kulkarni and S. Masur (2006). Development of QPM based value added traditional and RTE products. In: Proceedings of 38th Annual Meeting, 2006, Nutrition Society of India held at Kolkata.

NIIR (2002). Hand book on fruits, vegetables and food processing with canning and preservation, Asia Pacific Business Press Inc., New Delhi.

Okaka, J.E. and N.N. Potter (1979). Physico-chemical and functional properties of cowpea powders processed to reduce bean flavour. J. Sci., 44: 1235-1240.

Okezie, U., C.T. Akanbi, E.T. Otunola and I.A. Adeyemi (2003). Effect of addition of ripe bananas on some physico-chemical properties of maize extract. International J. Food Sci. Nutrition, 54(6): 437-445.

Omueti, O. and I.D. Morton (1996). Development by extrusion of soyabari snack sticks: a nutritionally improved soya maize product base on the Nigerian snack (kokoro). International J. Food Sci. Nutrition, 47: 5-13.

Osei, S.A., H.K. Dei and A.K. Tuah (1999). Evaluation of quality protein maize as a feed ingredient for layer pullet. J. Animal Food Sci., 8(2): 181-189.

Paes, M.C.D. and J.Maga (2004). Effect of extrusion on essential amino acids profile and colour of whole grain flours of quality protein maize and normal maize cultivars. Revista Brasileira e Mith e Sorgo, 3(1): 10-20.

Prasanna, B.M., S.K. Vasal, B. Kassahun and N.N. Singh (2001). Quality protein maize. Curr. Sci., 81(10): 1308-1318.

Rao, S.K., P.N. Achar and M.F. Rahman (2003). Biochemical changes induced in liver and serum of diplodiatoxin (*Stenocarpella maydis*) treated male and female rats. *Drug and Chemical Toxicology*, 26: 231-243.

Schoch, T.J. (1964). Swelling power and solubility of granular starches. In: Carbohydrate Chemistry, Ed. Whistler, R.L., Smith, R.J., Miller, B.E.N.J. and Wolform, M.I., Academic Press, New York, pp. 106-108.

Sharma, S., A.K. Saxena and V.K. Saxena (2002). Nutritional quality of evaluation of selected improved maize genotypes of Punjab. Indian J. Nutrition Dietetics, 39: 194-196.

Singh, V., A. Kawatra and S. Sehgal (2006). Development, acceptability and proximate composition of cake prepared from quality protein maize. In: Proceedings of 38$^{th}$ Annual Meeting, 2006, Nutrition Society of India held at Kolkata.

Sosulski, F.W. (1962). The centrifuge method for determining flour absorption in hard red spring wheat. Cereal Chemistry, 39: 344.

Stewart, B., P. Chevis and C.Perera (1997). Peroxidase isoforms of corn kernels and corn on the cob: Preparation and characteristics. Lebensmittel Wissenschaft und Technologie, 30: 192-201.

Szaniel, J., Sagi, F. and I. Palvolgi (1984). Hardness determination and quality prediction of maize kernels by a new instrument the monograph. Maydica, 29: 9- 20.

Wang, Y.J., S.W. Chong and W. Yang (2005). Effect of pericarp removal on properties of wet milled corn starch. Cereal Chemistry, 83(1): 25-27.

Zarkadas, C.G., R.I. Hamilton, Yu-Zi, Ran, V.K. Choi, S. Khanizadem, N.G.W. Rose and P.L. Pattison (2000). Assessment of the protein quality of 15 new northern adapted cultivars of quality protein maize using amino acid analysis. J. Agric. Food Chemistry, 48 (11): 5351 - 5361.

# 33

# Production Technology for Quality Protein Maize

Maize is a good source of carbohydrates, fats, proteins and some of the important vitamins and minerals. Several million people especially in the developing countries derive their protein and calorie requirements from maize. Maize has a low protein concentration with protein quality limited by deficiencies in lysine and tryptophan and has an excess of leucine which suppresses utilization of isoleucine leading to a poor growth and Kwashiorkor in children and pellagra in adults (Graham *et al.,* 1900). This also leads to poor net protein utilization and low biological value of traditional maize genotypes. To overcome this problem, the maize breeders have developed quality protein maize (QPM) by incorporating opaque-2 mutant gene, which is particularly responsible for enhancing lysine and tryptophan content of maize endosperm protein. Quality protein maize looks and taste like normal maize with same or higher yield potential, but it contains nearly twice the quantity of essential amino acids viz. lysine and tryptophan, which makes it rich in quality proteins.

**Table 1. Nutrient composition of normal maize vs quality protein maize**

| Amino acid | Normal maize grain | Quality protein maize |
|---|---|---|
| Gross energy (kcal/ kg) | 4342 | 4450 |
| Crude protein (%) | 8.94 | 10.32 |
| Methionin + cystine (%) | 0.37 | 0.48 |
| Lysine (%) | 0.26 | 0.40 |

| Amino acid | Normal maize grain | Quality protein maize |
|---|---|---|
| Threonine (%) | 0.31 | 0.38 |
| Tryptophan (%) | 0.07 | 0.09 |
| Arginine (%) | 0.40 | 0.66 |
| Isoleucine (%) | 0.29 | 0.32 |
| Leucine (%) | 1.09 | 0.92 |
| Valine (%) | 0.41 | 0.53 |
| Histidine (%) | 0.25 | 0.39 |
| Phenylalanine (%) | 0.44 | 0.43 |
| Sulphur amino acids (%) | 0.19 | 0.18 |
| Aromatic amino acids (%) | 0.51 | 0.50 |

(Bhatnagar *et al.*, 2010 and Panda *et al.*, 2011)

The quality protein maize grain contains in general 55 % more tryptophan, 30 % more lysine and 38 % less leucine than that of normal maize grain (Table 1). Quality protein maize has superior nutritional and biological value than normal maize (Panda *et al.*, 2011). The protein content was 15.44 and 8.94 % higher in quality protein maize (QPM) and compared to normal maize (NM). The essential amino acid contents such as lysine and tryptophan contents were higher in QPM compared to NM. Lysine and tryptophan contents were higher by 44 and 33 % in QPM than NM, respectively. Valine and leucine contents were 37.8 % higher and 17.9 % lower, respectively in QPM than NM. The ratio of leucine to isoleucine was lower in QPM than NM (2.90:1 vs 3.78:1). Further, the ratio of leucine plus isoleucine to valine was lower in QPM than NM (2.29:1 vs 3.62).

**Table 2. Comparisons on essential amino acid content in normal maize and QPM grain**

| Amino acid | Amino acid content (mg per g N) | |
|---|---|---|
| | Normal grain | QPM grain |
| Lysine | 177 | 256 |
| Isoleucine | 206 | 193 |
| Leucine | 827 | 507 |
| Sulfur amino acids | 188 | 188 |
| Aromatic amino acids | 505 | 502 |

| **Amino acid** | **Amino acid content (mg per g N)** | |
|---|---|---|
| | **Normal grain** | **QPM grain** |
| Threonine | 213 | 199 |
| Tryptophan | 35 | 78 |
| Valine | 292 | 298 |
| **Protein quality** | | |
| True protein digestibility (%) | 82-91 | 92 |
| Biological value (%) | 40-47 | 80 |
| Amount needed for equilibrium | 547 | 230 |

Maize consists of three main parts viz. hull or bran coat with high fibre content, germ rich in oil and starchy endosperm. The normal maize grain under Indian conditions on an average, contains 14.9 % moisture, 11.1 % protein, 3.6 % fat, 2.7 % fibre, 66.2 % other carbohydrates and 1.5 % minerals (National Institute of Nutrition, 2002). Maize kernel protein is made up of five different fractions. The percentage of different fractions to total nitrogen in maize kernel is albumin 7 %, globulin 5 %, non-protein nitrogen 6 %, prolamine 52 % and glutelin 25 % and the left over 5 % is residual nitrogen (Landry and Moureaux, 1982). Protein being the primary structural and functional component of every living cell is one of the most important ingredients that determine the quality of food and feed.

In normal maize grain the quality of protein is poor due to the presence of largest concentration of an alcohol soluble protein fraction 'prolamine' also known as 'Zein' in the endosperm. The amount of this alcohol soluble protein fraction zein is low in immature maize. They increase as the grain matures. Zein is very low in lysine and tryptophan content. Since this fraction contributes more than 50 % of the total protein, due to this obvious reason the protein of normal maize grain is low in lysine and tryptophan content. On the other hand, zein fractions contain very high amount of leucine and imbalanced proportion of isoleucine (Patterson *et al.,* 1980). On the whole, the ill proportion of all these four essential amino acids in maize due the high content of zein results into the poor quality of protein in normal maize. The high quality of protein in other fractions in other parts of maize kernel becomes recessive due to the dominance of zein in normal maize. The poor quality of protein in normal maize affects its biological value.

It was realized to improve the biological value of protein in maize genotypes. For this purpose, a new corn type known as QPM was developed by lowering the concentration of zein by 30 %. As a result, the concentration of two essential amino acids viz., lysine and tryptophan in grain was increased in QPM genotypes as compared to normal grain maize genotypes. The lower content of leucine in QPM further balances the ratio of leucine to isoleucine content (Table 2). The balanced proportion of all these essential amino acid in Quality Protein Maize (QPM) enhanced the biological value of protein.

The QPM research was initiated long back during 1970's, but it gained momentum during 1990's with continuous breeding efforts on development of high yielding hard endosperm modified opaque-2 maize germplasm by International Centre for Maize and Wheat Improvement (CIMMYT) and made it available for use in the breeding programme all over the world. In India, these germplasm accessions received from CIMMYT, Mexico were tested at different centers of All India Coordinated Research Project on Maize (AICRPM). Through acclimatization and repeated selection, hard endosperm modified opaque-2 maize inbred lines and their crosses have been identified for better protein quality and higher yield potential. The Directorate of Maize Research (DMR), New Delhi developed first QPM Composite variety, Shakti-1 with 0.63 % tryptophan in the year 1997, which was released and recommended for general cultivation by Indian farmers in 1998. The QPM research gained further momentum by launch of National Agricultural Technology Project (NATP) on QPM in 1998 by the Indian Council of Agricultural Research (ICAR). In this project, the QPM germplasm received from CIMMYT was acclimatized to suit the local agro-climatic conditions in India. The lines were evaluated for their productivity and deployed in combination breeding which led to release of first QPM three way cross hybrid, Shaktiman-1 by RAU, Pusa during 2001 followed by release of first QPM single cross hybrid, Shaktiman-2 during 2004 for their cultivation in Bihar state. In 2005, CCS HAU, Karnal released another QPM single cross hybrid, HQPM-1 which is the first yellow grain QPM single cross hybrid released for its cultivation across the country. Later in the series of QPM, Shaktiman-3 and Shaktiman-4 were released by RAU, Pusa during 2006 for their cultivation in Bihar. In 2007, another QPM single cross hybrid, HQPM-5 from CCSHAU, Karnal was released for its cultivation across the agro-ecologies of India. Further, two new single cross QPM hybrids i.e. HQPM-7 from CCSHAU, Karnal and Vivek QPM-9 from Vivekanand Parvatiya Krishi Anusandhan Shala (VPKAS), Almora has been identified during 2008. Vivek QPM-9 has a unique distinction of the first molecular marker assisted converted product of normal hybrid Vivek-9. The cultivation of QPM hybrids will ensure higher income to farmers as well as nutritionally superior food to the consumers.

The true protein digestibility of normal maize and quality protein maize is almost same, but the biological value of normal maize is just half as compared to that of QPM varieties

(Young *et al.*, 1971). Rather, the biological value of QPM is higher among all the food grains owing to the reason that all cereals except QPM are deficient in lysine, an essential amino acid and all pulses are deficient in methionine, the other essential amino acid (Yadavn *et al.*, 2008). The quality parameters viz., protein content, lysine and tryptophan content of QPM hybrids released for their cultivation in India are given in Table 3.

**Table 3. Quality characters of QPM hybrids released in India**

| Hybrid name | Hybrid type | Protein content in grain (%) | Tryptophan content in protein (%) |
|---|---|---|---|
| HQPM 7 | Single cross | 9.42 | 0.72 |
| Vivek QPM 9 | Single cross | 8.46 | 0.83 |
| HQPM 5 | Single cross | 9.80 | 0.76 |
| HQPM 1 | Single cross | 9.36 | 0.94 |
| Shaktiman 4 | Single cross | 9.98 | 0.93 |
| Shaktiman 3 | Single cross | 9.63 | 0.73 |
| Shaktiman 2 | Single cross | 9.30 | 1.04 |
| Shaktiman 1 | Three way cross | 9.60 | 1.01 |

In general, the management practices for QPM are almost same as normal grain maize. However, there are few specific considerations for higher productivity with good quality and better resource use efficiency.

### Suitable areas for cultivation

As normal maize, it can be grown in most parts of the country from plains to high altitudes of up to 2700 m. But, being a highly cross pollinated crop, there exists always chances of contamination under traditional maize growing areas and seasons. Therefore, for maintaining the purity and superior quality, spatial and temporal isolation should be ensured either through growing in non-traditional areas or seasons.

### Soils

It can be grown successfully in wide range of soils ranging from loamy sand to clay. However, soils with good organic matter content having high water holding capacity with neutral pH are considered good for better production. Being a sensitive crop to moisture stresses particularly excess soil moisture, it is desirable to avoid low lying fields having poor drainage. Therefore, the fields having provision of proper drainage should be selected for cultivation of maize.

## Selection of cultivars

A total of nine cultivars of which 7 single cross hybrid, 1 three ways cross hybrid and 1 composite of QPM having different maturity period and grain colour have been developed in India. These cultivars have been developed and find suitable for their cultivation under different agro-climatic conditions having compatibility under different cropping systems and preference of the local people. However, farmers are advised to grow single cross hybrids for higher productivity but for the areas like tribal and high risk prone regions where hybrids cannot reach in time due to seed constraints, the farmers can go for cultivation of composites.

### HQPM 1

The pedigree is HKI 193-1 x HKI 163. It is released by CCSHAU, Hissar during 2005. It has yellow and dent grains. It is responsive to higher doses of fertilizers, tolerant to frost/cold, resistance to maydis leaf blight, turcicum leaf blight and common rust and average yield is 6200 kg/ha. This hybrid has been identified for cultivation across the country. It falls under full season maturity group.

### HQPM 4

The pedigree is HKI-193-2 x HKI 161. It is released by CCSHAU Karnal druing 2010. It is well adopted across the country except Himalayan belt (Rabi). It is late maturing, yellow, semi-dent and average yield 6000 kg/ha.

### HQPM 5

The pedigree is HKI 163 x HKI 161. It is released by CCSHAU Karnal during 2007. It can be grown across the country. It matures in 88 – 90 days, orange, flint, responsive to higher doses of fertilizers, resistance to maydis leaf blight and *Chilo partellus*. The average yield is 6000 - 6500 kg/ha.

### HQPM 7

It is single cross quality protein maize hybrid. The pedigree is HK1-193-1 x HKI-161. It is released by CCSHAU, Hissar during 2008. It is suitable for Andhra Pradesh, Tamil Nadu, Karnataka and Maharashtra. It is late maturing (86-95 days), yellow, semi-flint and resistant to maydis leaf blight, TLB, common rust and stem borer. The average yield is 7000 - 7200 kg/ha. Nutritionally also, it has been found superior.

**Vivek QPM 9**

The pedigree is VQL 1 x VQL 2. It is released by VPKAS, Almora during 2008. It is quality protein maize version of the most successful single cross hybrid. It possess 0.83 % trytophan as compared to 0.59 % in Vivek maize Hybrid 9, 28.9 % increased in lysine, 23.3 % increase in Fe and 16.0 % increase in Zn over that in Vivek maize Hybrid 9. It is suitable for Jammu and Kashmir, Uttarakhand, Himachal Pradesh, Andhra Pradesh, Tamil Nadu, Karnataka and Maharashtra. It is extra early maturity with 85 - 90 days in hills and 80 - 85 days in plains. The grain is yellow, semi flint, flat and medium bold with average 1000 grain weight of 310 g. It performed better at low nitrogen level. It performed well across the zones. An average yield of 5843 kg/ha and 5435 kg/ha in Zone I and Zone IV respectively which was at par with Vivek maize Hybrid 9 in Zone I (5913 kg/ha) and Zone IV (5404 kg/ha). Plant is sturdy 190 - 200 cm tall in hills and 165 - 175 cm in plains. Ear placement is medium - low with good husk cover and medium broad leaf green. Days to 50 % pollen shed and silk in Zone I and Zone IV varies between 47.5 -49.9 days and 48.1 - 52.4 days respectively. Tassels are large and open, glume is purplish green, anther is light purple and silk is purple in colour. Ear is cylindrical and long with white cobe wood. It is tolerant to turcicum leaf blight and moderately tolerant to maydis leaf blight.

**Shaktiman 1**

The pedigree is (CML 142 x CML 150) x CML 186. It is released by RAU, Dholi during 2001. It can be grown in Bihar and Uttar Pradesh. It is quality protein maize hybrid with 1.01 % tryptophan in protein, flint and white grains. It is tolerant against to maydis leaf blight, turcicum leaf blight, banded leaf sheath blight and stem borer. The average yield level is more than 6000 kg/ha and matures in 86 - 95 days.

**Shaktiman 2**

The pedigree is CML 176 x CML 186 and is released by RAU, Dholi during 2004. It is suitable for Bihar state. It is resistant against maydis leaf blight, quality protein maize hybrid with 1.04 % tryptophan in protein. The average yield level is more than 6000 kg/ha and matures in 96 days.

**Shaktiman 3**

The pedigree is CML 161 x CML 163. It is released by RAU, Dholi during 2006. It is suitable for Bihar state. It is late maturing, orange-yellow, semi flint, tall, quality protein maize hybrid with 0.73 % tryptophan in protein and fairly tolerant against maydis leaf blight and LSM. The average yield level is more than 6000 kg/ha. It is resistant to lodging.

**Shaktiman 4**

The pedigree is CML 161 x CML 169. It is released by RAU, Dholi during 2006. It is suitable for Bihar state. Grain is white, semi flint, quality protein maize hybrid with 0.93 % tryptophan in protein and resistant against maydis leaf blight. The average yield level is more than 6000 kg/ha and matures in 96 days.

**Shakti 1**

It is composite of J 102, Cubal 502 and Antigua 2D02. It is released during 1971. It is suitable for Madhya Pradesh, Rajasthan and tarai belt of Uttar Pradesh. It is medium maturing, yellow and semi flint grains. The average yield is 4000 - 4500 kg/ha. It is gracious with 10.6 % protein and 3.45 % lysine. It matures about 76 - 85 days. It is moderately resistant to turcicum leaf blight, tolerant to pink borer and resistant to lodging.

**Rattan**

It is developed from the J 102 released by ANGRAU, Hyderabad during 1971. It is recommended for Punjab and Rajasthan. It is late maturing type (95 - 100 days). Grains are yellow, semi- flint and medium in size. The average yield is 4000 - 4500 kg /ha. The grains contain 10 % protein and 3.46 % lysine. It is tolerant to top shoot borer and downy mildew.

**Protina**

This is composite of (Jowatigua x Antigua car II) 02 x Doetox GCC) 02. It is released during 1971. It is suitable for cultivation in northern plains and foothills of Himalayas. It is medium maturing (100 - 105 days) and average yield is 4000 - 4500 kg/ha. The grains contain 11 % protein and 4.0 % lysine as against of 2.0 % lysine in normal maize. Plants are sturdy and tolerant to downy mildew and stalk rot diseases.

**Time of sowing**

It can be grown in all seasons viz., Kharif, post monsoon, Rabi and Spring. For higher productivity particularly in Kharif season, for the farmers having sufficient irrigation facilities, it is desirable to complete the sowing 12-15 days prior to onset of monsoon. However, in rainfed areas, the sowing time should be coincided with onset of monsoon. The optimum time of sowing in different seasons are given in Table 4.

**Table 4. Optimum sowing time in different seasons**

| Seasons | Optimum time of sowing |
|---|---|
| Kharif | Last week of June to $15^{th}$ July |
| Post monsoon | First fortnight of August |
| Rabi | Last week of October for intercropping and first fortnight of November for sole crop. |
| Spring | First week of February |

### Seed rate

The seed rate depends on seed size (genotype), season, sowing methods and the planter but, in general, 20 kg seed/ ha is optimum for higher productivity and input use efficiency. A plant population of about 70000 to 80000/ ha with a plant geometry of 60 to 70 x 20 cm is optimum depending upon the season.

**Table 5. Effect of plant density on growth parameters, yield and economics of QPM (HQPM 1)**

| Plant density (Plants /ha) | Plant height (cm) | LAI (60 DAS) | DMP (g/ plant) | Ear height (cm) | Grain (q/ha) | Cob (q/ha) | Stover (q/ha) | Net return (Rs. /ha) | B: C ratio |
|---|---|---|---|---|---|---|---|---|---|
| 66,666 | 213.6 | 4.45 | 194.30 | 72.5 | 57.67 | 84.68 | 142.0 | 31660 | 2.29 |
| 83,333 | 212.7 | 5.23 | 186.60 | 73.9 | 65.49 | 92.79 | 154.6 | 38684 | 2.56 |
| 1,00,000 | 218.2 | 5.99 | 181.86 | 79.2 | 69.90 | 99.79 | 169.1 | 42960 | 2.71 |
| CD (5 %) | 2.7 | 0.40 | 8.96 | 5.3 | 5.01 | 10.48 | 8.12 | - | - |

(Bisht et al., 2012)

At Pantnagar (Uttarakhand), leaf area index (LAI) and dry matter accumulation were significantly higher under high and low plant density, respectively. High plant density being at par with medium plant density produced maximum cob (99.76 q /ha) and grain (69.90 q/ ha) yields. High plant density gave the higher net return (Rs.42,960/ ha) (Bisht et al., 2012).

**Seed treatment**

To protect the seed from seed and soil borne diseases and insect-pests, it is always advisable to treat the seed with fungicides and insecticides before sowing (Table 6).

**Table 6. Seed treatment for major disease and pests**

| Disease / insect pest | Fungicide / pesticide with rate of application |
|---|---|
| Turcicum leaf blight, Banded leaf and sheath blight, Maydis leaf blight | Bavistin + captan in 1:1 ratio @ 2.0 g /kg of seed. |
| BSMD | Apran 35 SD @ 4.0 g /kg of seed. |
| Pythium stalk rot | Captan @ 2.5 g /kg of seed. |
| Termite and shoot fly | Imidachlorpid @ 4.0 g /kg of seed. |

**Method of sowing**

Mainly the maize is sown directly through seed but during winters where fields are not vacant in time (till November), transplanting can be done successfully by raising the nursery. However, the method of sowing depends on season, cropping system, soil conditions and time of planting. The sowing methods for situation specificity are described as follows:

**i. Raised bed planting**

In general the raised bed planting is best planting method for maize during monsoon and winter seasons both under excess moisture as well as limited irrigation availability conditions. Sowing should be done on the southern side of the east-west ridges/beds. Planting should be done at recommended spacing. Using raised bed planting method, 20-30 % irrigation water can be saved with higher productivity. Moreover, under temporary excess water logging due to heavy rains, the furrows will act as drainage channels and crop can be saved from excess soil moisture stress.

**ii. Zero-till planting**

Maize can be successfully grown without any preparatory tillage under no-till situation with less cost of cultivation, higher profitability and better resource use efficiency. Under such condition one should ensure good soil moisture at sowing and seed and fertilizers should be placed in band using zero-till seed-cum-fertilizer planter with furrow opener as per the soil texture. Large number of farmers particularly under rice-maize systems in peninsular and eastern India are practicing successful zero till planting and getting higher farm profitability.

### iii. Transplanting

Maize can also be established and grown successfully with transplanting during winter season. Under intensive cropping systems where field are not vacant in time and where chances of delayed planting exists. Therefore, for the situation where fields are vacated during December - January, it is advisable to grow nursery and transplant the seedlings in furrows and apply irrigation water for proper establishment. Using this technique, temporal isolation can also be maintained in maize growing regions for purity and good quality of QPM. For one hectare, 700 $m^2$ areas are required for nursery. The nursery should be raised during second fortnight of November. The 30-40 days old seedlings (depending on the growth) should be transplanted during December-January in furrows as the yield in furrows is always higher than flat.

### iv. Flat planting

For rainfed areas and on conserved moisture, flat planting can be done using seed-cum-ferti planters.

### v. Furrow planting

During spring season, it is always advisable to grow maize in furrows for proper growth, seed setting and higher productivity.

## Nutrient management

The application of nutrients depends mainly on soil nutrient status and cropping system. For getting desirable yields, the nutrient applications should be done in such a way that matched the soil supplying capacity and plant demand (SSNM approach) keeping in view of the preceding crop (cropping system). SSNM approach led to remarkably higher productivity over existing recommended practice across the ecologies. The maize in general responds well to organic manures. In general application of 10 t FYM/ ha 10-15 days prior to sowing supplemented with 150-180 kg N, 70-80 kg $P_2O_5$, 70-80 kg $K_2O$ and 25 kg $ZnSO_4$/ ha gives higher economic yields in QPM. Full doses of P, K and Zn should be applied as basal. Nitrogen application should be done in five splits for higher N use efficiency. 10 % N should be applied as basal, 20 % N at 4 leaf stage, 30 % N at 8 leaf stage, 30 % N at flowering stage and 10 % N at grain filling.

**Table 7. INM practices on growth and yield of quality protein maize (HQPM 1)**

| Fertilizer level | Cobs / Plant | Crop growth rate (g/m²/day) 47-75 DAS | Grain yield (t/ha) | Stover yield (t/ha) | Net return (Rs/ha) | B : C ratio |
|---|---|---|---|---|---|---|
| 75 % RDF | 1.04 | 11.55 | 3.06 | 5.51 | 23615 | 1.87 |
| 100 % RDF | 1.14 | 14.20 | 4.00 | 7.27 | 34352 | 2.66 |
| 150 % RDF | 1.17 | 17.78 | 4.30 | 7.90 | 37600 | 2.83 |
| CD 5 % | 0.06 | 2.06 | 0.24 | 0.55 | 2882 | 0.22 |
| **Source of manure** | | | | | | |
| Control | 1.04 | 12.22 | 2.94 | 5.48 | 24636 | 2.38 |
| FYM 10 t/ha | 1.15 | 14.70 | 4.13 | 7.46 | 35531 | 2.67 |
| Vermicompost 5 t/ha | 1.06 | 17.30 | 4.29 | 7.75 | 35400 | 2.31 |
| CD | 0.06 | 2.06 | 0.24 | 0.55 | 2882 | 0.22 |
| **Phosphorus sources** | | | | | | |
| DAP | 1.10 | 14.46 | 3.69 | 6.72 | 30773 | 2.37 |
| SSP | 1.14 | 15.02 | 3.88 | 7.07 | 32938 | 2.54 |
| CD 5 % | NS | NS | NS | NS | NS | NS |

(Singh and Nepalia, 2009). 100 % RDF: 90: 40: 0 kg NPK /ha.

At Udaipur (Rajasthan), application of 125 % RDF through urea, single super phosphate or DAP along with 10 t of FYM or 5 t of vermicompost /ha forms an ideal module of INM for higher productivity of quality protein maize (Singh and Nepalia, 2009).

At Dharwad (Karnataka), integration of 75 % RDF + *in situ* green manuring of sunnhemp (sown between two rows of maize) at 40 DAS, seed inoculation of biofertilizers (*Azospirillum* + PSB) and liquid manures like *Panchagavya* foliar spray @ 3 % at tasseling and silking stage and soil application of *jeevamrutha* @ 500 litres/ha at 40 and 60 DAS helps to save 25 % of RDF by producing grain yield on par with that of 100 % RDF (Ravi et al., 2012).

## Water management

The water management depends on season as about 80 % of maize is cultivated during monsoon season particularly under rainfed conditions. However, in areas with assured irrigation facilities depending upon the rains and moisture holding capacity of the soil, irrigation should be applied as and when required by the crop and first irrigation should be applied very carefully. In general, the irrigation should be applied in furrows up to 2/3$^{rd}$ height of the ridges/beds. Young seedlings, knee high stage, flowering and grain filling are the most sensitive stages. Irrigation in alternate furrows for water stress and irrigation should be ensured at these stages. In raised bed planting system, the crop can also be irrigated in alternate furrow to save more irrigation water. In rainfed areas, tied-ridges are helpful in conserving the rainwater for its availability in the root zone for longer period. For winter maize, it is advisable to keep soil wet (frequent and mild irrigation) during 15$^{th}$ December to 15 February to protect the crop from frost injury.

## Weed management

Pre-emergence application of atrazine @ of 1.0-1.5 kg a.i /ha in 600 litre water is effective way for control of weeds. While spraying, the person who is doing spray should move backward so that the atrazine film on the soil surface may not be disturbed. Preferably three boom flat fan nozzle may be used for proper ground coverage and saving time. One to two hoeing are recommended for aeration and uprooting of the remaining weeds, if any. While doing hoeing, the person should move backward to avoid compaction and better aeration. For areas where zero tillage is practiced, pre-plant application (10-15 days prior to seeding) of non-selective herbicides viz., Glyphosate @ 1.0 kg a.i./ ha water or paraquat @ 0.5 kg a.i./ ha in 600 litre water is recommended to control the weeds.

## Crop protection

Maize stalk borer (*Chilo partellus*) is a major pest during monsoon season throughout the country. Whereas *Sesemia inference* occurs during winter season particularly in peninsular India. For control of Chilo, foliar spray of 0.1 % Endosulfan 35 EC (700 ml in 250 litre water) 10 days after germination is effective. It can also be controlled by release of 8 Trichocards (*Trichogramma chilonis*) per ha at 10 days after germination. Intercropping of cowpea with maize is an eco-friendly option for reducing the incidence of *Chilo* on maize. Termite is also an important pest in many areas and for control of this fepronil granules should be applied @ 20 kg /ha followed by light irrigation. If the termite incidence is in batches, fepronil should be applied spot application @ 2-3 granules/plant. Apply Chloropyriphos @ 2.5 litre/ha mixed in sand followed by irrigation.

Several diseases occur during different seasons in various parts of the country. Turcicum leaf blight can be controlled by spraying of Zineb/Meneb @ 2.5-4.0 g/liter of water (2- 4 applications) at 8-10 days interval. Spray of Dithane Z-75 or Zineb @ 2.4 – 4.0 g/liter of water (2-4 applications) at 8-10 days interval after first appearance of symptoms of Maydis leaf blight disease. Three sprays of Dithane M-45 @ 2.0-2.5 g/liter beginning from first appearance of Polysora rust at 15 days interval. Seed treatment of peat based formulation of *Pseudomonas fluorescence* @ 16 g/kg of seed or as soil application @ 7 g/liter of water (soil drenching) or foliar spray of Sheethmar (Validamycin) @ 2.7 ml /liter water to control banded leaf and sheath blight. Use of bio-control agent *Trichoderma* formulation in furrows after mixing with FYM @ 10g/kg at 10 days prior to control post flowering stalk rot.

## Harvesting

Harvesting should be done at optimum moisture content (20 %) in grain to avoid post harvest losses due to store grain pests and diseases. The harvested cobs should be sun dried before shelling and should be shelled at 13-14 % grain moisture. Shelling can be done manually or by power operated maize sheller. For proper storage of the grain, drying should be done till the moisture content is reduced to 8-10 % and should be kept in aerated jute bags.

## Uses of QPM

In India, the maize is mainly consumed as feed (61 %) and food (25 %) and hence, QPM being superior in quality has better option in these sectors. The main uses of QPM as follows:

### a. For food and nutritional security

The QPM has got special distinction among the cereals due to presence of high amount of two essential amino acids viz., lysine and tryptophan and low content of non desirable amino acid (leucine). Therefore, the QPM can be utilized for diversified purposes in food and nutritional security as infant food, health food/mixes, convenience foods, specialty foods and emergency ration. It is also useful in fulfilling the protein requirements of different sections of society (infants, lactating mothers' convalescing patients, Kwashiorkor diseased, old persons and infirm, etc.) to prevent malnutrition. The anthropometric measurements i.e. body weight and arm circumference of children fed with QPM were remarkably higher compared to children fed with normal maize grain (DMR, 2006 and Singh, 2006).

### b. Providing nutritious feed

The poultry growth rate in India is very high and 49 % of the total maize production is consumed as feed in poultry sector in India. This sector has also further enormous growth potential (about 15 % / annum) that would raise the further demand of QPM (TAAS, 2008). Therefore, QPM with its high carbohydrates, fats, better quality proteins, some of vitamins and minerals, it is nutritious feed for poultry, livestock, swine, fish, etc. Use of QPM as poultry feed leads to early development of broilers, save energy and feed, and also the extra cost incurred on lysine and tryptophan fortification.

### c. Promoting maize based entrepreneurship

The nutritious products developed from QPM can replace fancied and highly priced industrial foods. These products can also be prepared in villages and thus could be a great source of rural entrepreneurship. Therefore, QPM based rural industries has a wider scope for employment generation and rural prosperity.

## Nutritional quality

Generally maize is poor in nutritional quality. But the breeding experiments resulting in improved quality protein maize has added inversely to its nutritional potential in alleviating both hunger and malnutrition. Several studies to evaluate the nutritional significance of QPM have been conducted in India and abroad.

A study was conducted by Graham *et al.* (1980) to evaluate the protein quality of endosperm and whole kernels meals of normal, opaque-2 and sugary-2 opaque-2. Eight convalescent malnourished children, 10-25 months of age were fed iso-nitrogenous and isocaloric diet. Each type of diet was fed for 9 days period (3 days adaptation period and 6 days of metabolic period), consecutively. Endosperm meals provided 83.2, 91.2 and 82.1 % of energy, while whole kernel meals provide 73.2, 68.6 and 67.2 % of energy in respective diets. Apparent nitrogen retention from endosperm meals was lower (15.1, 22.8 and 24.8 %) than from whole kernel meals (26.8, 30.1 and 31.6 %), both lower than from casein (39 %). It was reported that there was strong correlation between lysine absorbed and nitrogen retention, in both endosperm and whole kernel meal, but higher in whole kernel. Further, it was also indicated that to match the nitrogen retention from casein presumably equal to requirement, 50 per cent of the children needed to consume about 125-200 % of energy requirement as normal, opaque-2 or sugary-2 opaque-2 and meals which was impossible. However, for whole kernel meals the children needed to consume 85-110 % of energy of the respective diets.

Ortega *et al.* (1986) assessed *in vitro* protein digestibility of nixtamal, *masa* and tortilla of normal maize and QPM. It was reported that, QPM and its products had lower *in vitro* protein digestibility (82, 73, 80 and 68 %) compared to normal maize and its products (88, 82, 91 and 79 %) for raw grain, nixtamal, *masa* and tortilla, respectively. The differences in distribution of protein fractions were attributed for the variation in digestibility.

Sproule *et al.* (1988) evaluated energy and apparent digestibility of food grade maize and QPM. Weanling rats were fed diets of raw grains, tortillas and defatted tortilla chips. Both the diets were supplemented with 1 per cent casein-HCl to improve rat growth and reduce experimental error and formulated to contain 10 per cent protein and 5 % ether extract. The diets were iso-nitrogenous and iso-caloric. It was reported that feed intake and average daily gain in weights of rats fed QPM products were significantly higher than rats fed on food grade maize but, lower feed/gain ratio. Processing raw grains into tortilla and tortilla chips significantly increase the protein digestibility. The energy digestibility of tortilla and tortilla chips products of both types of maize was higher than the raw grains. This has been attributed to the reduction of insoluble fibres and increment of starch availability.

Graham *et al.* (1989) conducted a study to know the digestibility and utilization of QPM. For the study, common maize and QPM were used. Both varieties were wet milled into whole kernel meals. These diets supplied 6.4 % of energy as protein, 10 % as fat and the balance as carbohydrate. Six recovering malnourished male infants (7.9 to 18.5 months of age) were divided in to two groups receiving normal and QPM diets separately. All the subjects received the casein control diet during the first 9 days followed by the maize diets for 7 days each. The metabolic study revealed that rats fed QPM had lower nitrogen retention (32 %) than rats fed casein diet (41 %). Similar observations were made with normal maize also. Graham *et al.* (1990) evaluated the efficacy of QPM as the sole source of dietary protein and fat on 10 recovering malnourished children aged 14 to 29 months. The experiment was conducted for 9-11 months. The children received 90 % of their diet energy and 100 % of protein and fat from QPM. It was reported that the linear growth of children fed milk diet was slightly but not significantly greater than that of QPM fed children.

## References

Asiedu, E. A., P.Y. K. Sallah, S. Twumasi-Afriyie, K. Obeng-Antwi, K. Ahenkora and P. Adusei-Akowuah (2001). Agronomic and post-harvest characterization of three quality protein maize (QPM) hybrids developed in Ghana. Ghana J. Agric. Sci., 34: 57-62.

Bhargava, A., N. Chopra and H.O. Gupta (2003). Processing of quality protein maize (QPM) into *dalia* and processed flour and their synergistic effect on the growth of pre-school children. *IFCON*, Indian Agricultural Research Institute, New Delhi.

Bhatnagar, A., A.S. Bisht and S. Kaushik (2010). Quality protein maize: A future cereal crop for better protein nutrition. Indian farmers' Digest, 43 (11): 35 - 37.

Bisht, A.S., Amit Bhatnagar, M.S. Pal and Veer Singh (2012). Growth Dynamics, Productivity and Economics of Quality Protein Maize (*Zea mays* L.) Under Varying Plant Density and Nutrient Management Practices. Madras Agric. J., 99 (1-3): 73-76.

Bressani, R., V. Benavides, E. Acevedo and M.A. Ortiz (1990). Changes in selected nutrient contents and in protein quality of common and quality protein maize during rural tortilla preparation. Cereal Chemistry, 67(6): 515-518.

Dilip Singh and V. Nepalia (2009). Influence of integrated nutrient management on protein quality and content of maize (*Zea mays* L.). *Ind. J. Agric. Sci.,* 79(12): 1020 - 1022.

Graham, G.G., D.V. Glover, Romana, L.D. Guillermo, E. Morales and W.C. Maclean (1980). Nutritional value of normal, opaque-2 and sugary-2, opaque-2 maize hybrids for infants and children. Digestibility and Utilization. J. Nutrition, 110: 1061-1069.

Graham, G.G., J. Lembcke and E. Morales (1990). Quality protein maize as the sole source of dietary protein and fat for rapidly growing young children. Pediatrics, 85(1): 85-91.

Graham, G.G., J. Lembcke, E.B.S. Lancho and E. Morales (1989). Quality protein maize digestibility and utilization by recovering malnourished infants. Pediatrics, 83(3): 416-421.

Guerra, A., A. Guzman and Saldivar (2004). Effects of soybean fortification on protein quality of tortilla based diets produced from regular and quality protein maize. Plant Foods for Human Nutrition, 59(2): 45-50.

Gupta, H.O. (2003). Effect of pericarp on protein quality in maize. J. Food Sci. Techn., 40(1): 80-83.

Gupta, H.O. (2004). Improving the nutritional quality of maize after supplementation with processed soybean. J. Food Sci. Techn., 41(2): 167-170.

Gupta, H.O. and N.N. Singh (2005). Preparation of wheat and quality protein maize based biscuits and their storage, protein quality and sensory evaluation. J. Food Sci. Techn., 42(1): 43-46.

Habtamu, F., A. Girma, W. Asrat, S. Taffesse, G. Takele, K. Schlosser, H. Noetzold and T. Henle (2003). Assessment of protein nutritional quality and effects of traditional processes: a comparison between Ethiopian quality protein maize and five Ethiopian adapted normal maize cultivars. Nahrung, 47(4): 269-273.

Jaliya, M.M., U.F. Chiezey, B. Tanimu, A.C. Odunze, M.K. Othman, B.A. Babaji, B.M. Sani and H. Mani (2012). Effects of nitrogen and sulfur fertilizers on nitrogen content in soil, ear leaf, flag leaf and grain of QPM maize varieties at Samaru Zaria. J. Agric. Sci., 4 (5): 217 – 222.

Kaledhonkar, P.R. (2003). Evaluation of promising quality protein maize (*Zea mays* L.) cultivars in Konkan region. *M. Sc. (Agri.) Thesis,* Dr. Balasaheb Sawant Konkan Krishi Vidyapeeth, Dapoli, Dist. Ratnagiri (Maharashtra).

Kumari, P. and U. Singh (2004). Impact of supplementation of quality protein maize based Ladoos on the nutritional status of pregnant women. Indian J. Nutritional and Dietetics, 41: 528.

Ortega, F.I., Villgas, G. Evan and S.K. Vasal (1986). A comparative study of protein changes in normal and quality protein maize during tortilla making. Cereal Chemistry, 63(5): 446-451.

Paliwal, S. (2003). Quality protein maize in zero-tillage. *In* RWC-CIMMYT. Addressing resource conservation issues in rice-wheat systems of South Asia: A resource book. Rice-Wheat Consortium for the Indo-Gangetic Plains and International Maize and Wheat Improvement Center, New Delhi, India. P. 240-242.

Panda, A.K., M.V.L.N. Raju and S.V. Rama Rao (2011). Quality protein maize as poultry feed. Indian Farming, 60 (10): 23 - 24.

Paula, H.D., R.C. Santos, M.E. Silva, E.C.S. Gloria, M.L. Pedrosa, N.A.V. Almeida, A.S.V. Costa and L.C.C. Malaquias (2004). Biological evaluation of a nutritional supplement prepared with QPM maize cultivar BR-473 and other traditional food items. Brazilian Archives of Biology and Technology, 47(2): 247-251.

Ravi, N., R. Basavarajappa, C.P. Chandrashekar, S.I. Harlapur M.H. Hosamani and M.V. Manjunatha (2012). Effect of integrated nutrient management on growth and yield of quality protien maize. Karnataka J. Agric. Sci., 25 (3): 395 - 396.

Rodriguez, C.E.O., J. Milan-Carrillo, R. Mora-Escobedo, O.G. Cardenasvalenzuela and M.C. Reyes (2004). Quality protein maize (*Zea mays* L.) tempeh flour through solid state fermentation process. Lebensmittel - Wissen Chaft Und Technologie, 37(1): 59-67.

Singh, D. and V. Nepalia (2009). Influence of INM on quality protein maize productivity and soils of southern Rajasthan. Indian J. Agric. Sci., 79 (12): 1020 – 1022.

Sproule, A.M., Saldivar, S.O., Bockholt, A.J., Rooney, W. and D.A. Knabe (1988). Nutritional evaluation of tortillas and tortilla chips from quality protein maize. Cereals Foods World, 33(2): 234-236.

Twumasi-Afriyie, S., K., M. Ahenkora, Owusu-Akyaw and K. Oberg-Antwi (1994). Progress in quality protein maize improvement in Ghana. In: Proc. *Symp. 11$^{th}$ South African Maize Breeding.* Pietermarisburg, South Africa, March 15-17, 1994.

Twumasi-Afriyie, S., P. Y. K. Sallah, M. Owusu- Akyaw, K. Ahenkora, R. F. Soza, W. Haag, B. D. Dzah, B. D. Okal and A. Akuamoah Doateng (1994c) Development and utilization of quality protein maize in Ghana. *In: Proceedings of the International Symposium on Quality Protein Maize.* EMBRAPA, Sete Lagoas, Brazil, 1-3 December 1994.

Twumasl-Afrlyle, S., P. Y. K. Sallah, P. P. Frimpong Manso, K. Ahenkora, W.H. Haag, A. Agyeman and B.D. Dzah (1997) Performance of Ghanaian quality protein maize vadeties in international testing in Afiica, Asia, Central and South America. *In: West and Central Africa Regional Workshop, International Institute of Tropical Agriculture (IITA).* Coutonou, Benin. Republic, 21-25 April, 1997.

Valdez, C.A., J.M. Carillo, O.G.C. Valenzuela, R.M. Escobedo, L.A. Perez and C.R. Moreno (2005). Infant food from quality protein maize and chickpea: optimization for preparing and nutritional properties. International J. Food Sci. Nutrition, 56(4): 273-285.

Yadav, S. (2006). Growing quality protein maize: doubling productivity and profitability. Retrieved 12th May, 2009 from http://www.telesupport.org simplemachines/ index.

# 34

# Rabi Maize Cultivation

*Rabi* maize is grown on an area of 1.2 million ha with the grain production of 5.08million tonnes, with an average productivity of 4.00 t/ha (DACNET, 2012). It has emerged as an important crop in the non-traditional season and non-traditional areas. Andhra Pradesh, Bihar and Tamil nadu are the three largest maize producing states with 2.322, 1.02 and 0.47 million tons respectively closely followed by Karnataka, Maharashtra and West Bengal.

Cultivation of maize in winter season started in mid 60s in some pockets of Bihar and South India. Yield obtained during this season is invariably higher (>6 t/ha) than the *Kharif* season yield (2-2.5 t/ha.) due to long duration of growth and least infestation of pests and diseases. In Bihar, maize can be taken up in all the three seasons. In recent years, significant changes have occurred in maize production and utilization due to increasing commercial orientation of this crop and rising demand for diversified end users, especially for feed and industrial uses. A sizable number of districts (110 districts), in the states of Andhra Pradesh, Karnataka, Bihar, Maharashtra, Uttar Pradesh, Madhya Pradesh, West Bengal, Orissa, Gujarat, Chhattisgarh and Tamil Nadu have potential for growing winter maize.

## Choice of cultivars

The success and the level of profit from *Rabi* crop depend to a great extent on the choice of maize hybrid/composite to be grown. Farmers should therefore be encouraged to sow only high yielding hybrids suitable for *Rabi* season. The use of F1 hybrid seed is essential for realizing high yield. The recommended hybrids, in general, have given 60 to 80 % grain yield than the local varieties in most of the evaluation trials, with an average yield level of 6 t/ha.

**Table 1. List of hybrids (H) and composites (C) varieties for different states for rabi season**

| States | Hybrids/ composites |
|---|---|
| Delhi | **H:** PMH 3, Buland, NK 61, Pro 311, Bio 9681, Seed Tech 2324,HM11,HM8 |
| Punjab | **H:** PMH 3, PMH-1 , Buland, Sheetal, Pro 311, Bio 9681, NK 61, Pro 311, Bio 9681, Seed Tech 2324, HM 11,H M8 |
| Haryana | **H:** PMH 3, Buland, HM 5, NK 61, Pro 311, Bio 9681, Seed Tech 2324, HM 11, HM 2, HM 1, HM 8 |
| Uttar Pradesh | **H:** PMH 3, Buland, Pro Agro 4212, Pro 311, Bio 9681, NK 61, Seed Tech 2324, HM 8 |
| Rajasthan | **H:** Pro 311, Bio 9681, Seed Tech 2324, HM 8 |
| Madhya Pradesh | **H:** Pro 311, Bio 9681, Seed Tech 2324 |
| Gujarat | **H**: Pro 311, Bio 9681, Seed Tech 2324<br>**C:** G M 3, Ganga safed 2 |
| Andhra Pradesh | **H:** Kargil 900 M, Seed Tech 2324, Pro 311, Bio 9681, Pioneer 30 v 92, Prabal, 30 V 92, 900 M |
| Tamil Nadu | **H:** COHM 5, Prabal , Pro 311, Bio 9681, Seed Tech 2324, 30 V 92, 900 M Gold. |
| Maharashtra | **H:** Prabal, Pro 311, Bio 9681, Seed Tech 2324, 30 V 92, 900 M Gold. |
| Karnataka | **H:** Nithya Shree, DMH 1, DMH 2, 900 M, Bio 9681, Prabal, Pro 311, Bio 9681, Seed Tech 2324<br>**C:** NAC 6004, 30 V 92 |
| Bihar | **H:** Rajendra Hybrid 2, Rajendra Hybrid 1, Pro 311, Bio 9681, Seed Tech 2324, 30 V 92, 900 M. **C:** Hemant, Suwan and Lakshmi |
| Jharkhand | **H:** Pro 311, Bio 9681, Seed Tech 2324. **C:** Suwan |
| Odisha | **H:** Pro 311, Bio 9681, Seed Tech 2324 , PAC 705 |
| West Bengal | **H:** Pro 311, Bio 9681, Seed Tech 2324 |
| Himachal Pradesh | **H:** Pro 311, Bio 9681, Seed Tech 2324 |
| NEH Region | **H:** Pro 311, Bio 9681, Seed Tech 2324. **C:** NLD white |
| Chhattisgarh | **H:** PEHM 1, Pioneer 30 V 92 & 30 R 26, Bio 9681, Pro 4640, Pro 4643, 900 M |
| Assam | **C:** NLD white |

* H: Hybrids; *C: Composites

## Soils

Maize can be grown on a variety of soils ranging from sandy to clayey. But it performs best on well drained, aerated deep-loams and silt loams containing organic matter and nutrients. Highly saline, acidic, alkaline and water logged soils should be avoided for cultivation of maize crop.

## Date of sowing

The optimum date of sowing is important for winter maize so that the genotype grown can complete its life cycle under optimum environmental conditions. Generally, sowing should be completed before the end of October, preferably by mid-October. The temperature during second fortnight of October to mid November in most of the North India drop rather sharply, resulting in delayed germination and plant growth receives a major setback. Hence, any marked delay in sowing is likely to result in lower yield. Also, in late sown crops, there is an increased incidence of common rust, which is not a serious concern in timely sown crop. The extent of reduction in yield through delayed sowing however varies with the location (Table 2).

**Table 2. Recommended sowing time of *Rabi* maize in different states of India**

| States | Suitable periods of sowing |
|---|---|
| Bihar | 20 October- 15 November |
| Uttar Pradesh | 20 October- 15 November |
| Andhra Pradesh | 25 October- 20 November |
| Gujarat | 15 October- 15 November |
| Maharashtra | 20 October- 15 November |
| Tamil Nadu | 20 October- 15 November |
| Madhya Pradesh | 15 October- 15 November |
| Karnataka | 15 October- 15 November |
| West Bengal & N E region | 20 October- 10 November |
| Orissa | 20 October- 10 November |
| Punjab | 25 October- 15 November |
| Haryana | 25 October- 15 November |

In Punjab and Haryana, where the temperature at the time of sowing is low, it will be desirable to sow the crop on ridges. Sowing should be done on the southern side of the east-west ridge so that the optimum amount of sunshine is received and the seedbed remains warm.

**Plant density and seed rate**

A population of 90,000 plants/ha at harvest is desirable for realizing high grain yield in *Rabi.* A spacing of 60 cm between rows and 18 - 20 cm between plants would provide the desired plant population density. For this purpose, 20- 22 kg of seed would be needed to sow one hectare of land. Before sowing, seed should be soaked overnight in warm (45°C at the time of seed soaking). This treatment helps in obtaining better plant stand and healthy crop. Seeds should be sown 4-5 cm deep. In maize that plant population of 40,000, 65,000 or 90,000 plants /ha yielded 36.0 to 52.2, 41.7 to 64.9 and 42.3 to 72.7 q grains /ha, respectively (Bangarwa *et al.,*1989). Gollar and Patil (2000) studied the response of maize at different plant density with different spacing of Deccan 103 (hybrid) in clay loam soil at Dharwad (Karnataka). Higher B:C ratio (4.23) was observed at density of 1,11,111/ha (45 cm x 20 cm) over other plant densities.

At Dapoli (Maharashtra), spacing of 60 cm x 30 cm recorded significantly higher length of cob, girth of cob, number of grains per cob, weight of grains per cob and 1000 grain weight than other narrow spacings of 45 cm x 30 cm and 30 cm x 30 cm during *rabi* season (Dalvi, 1984). At Allahabad in *rabi* season with maize cv. Gangasafed-2 and Kissan sown at three plant densities *viz.,* 33,333, 66,6666 and 99,999 /ha. They recorded an average yield of 49.4 q /ha in Gangasafed-2 and 40.2 q /ha in Kissan averaged over cultivars, grain yield was higher in 99,999 plants /ha (Babu and Mitra, 1989). Gollar and Patil (2000) stated that plant population of 55555 /ha recorded higher total dry matter production as compared to 74074, 83333, 111111 plant population /ha. But in case of LAI it was higher in higher plant population. But grain yield was higher in higher plant population during *rabi* season.

**Table 3. Grain yield and economics of winter maize under different plant population**

| Plant densities | Grain yield (kg/ha) | Net return (Rs/ha) | B:C ratio |
|---|---|---|---|
| 55,555 plants/ha | 4425 b | 17249 b | 1.29c |
| 66,666 plants/ha | 5113 a | 25812 a | 1.44a |
| 83,333 plants/ha | 5148 a | 23423 a | 1.37b |
| LSD(0.05) | 159 | 2539 | 0.03 |

At Chitwan (Nepal), number of cobs/plant, cob length, cob diameter, number of grain rows/cob, number of grains/grain row and number of grains/cob, thousand grain weight were decreased under higher levels of plant population. The population level of 66,666 plants/ha gave significantly highest grain yield (5113 kg/ha), net return (Rs. 25812 /ha and benefit:

cost ratio (1.44) (Shrestha and Timsina, 2011).

### Seed treatment

Seed treatment with fungicide and insecticides is necessary to protect the crop from seed and soil borne diseases and insect- pests.

**Table 4. Recommended seed treatment**

| Disease/insect-pest | Fungicide/pesticide | Rate of application (g /kg seed) |
|---|---|---|
| Turcicum Leaf Blight, Banded Leaf and Sheath Blight, Maydis Leaf Blight | Bavistin + Captan in 1:1 ratio | 2.0 |
| BSMD | Aspran 35 SD | 4.0 |
| Pythium Stalk rot | Captan | 2.5 |
| Termite and shoot fly | Imidachlorpid | 4.0 |

## Method of Sowing

### i. Raised bed (ridge) planting

Generally the raised bed planting is considered as best planting method for maize during monsoon and winter seasons both under excess moisture as well as limited water availability/ rainfed conditions. Sowing/planting should be done on the southern side of the east-west ridges/beds. Planting should be done at proper spacing. Preferably, the raised bed planter having inclined plate, cupping or roller type seed metering systems should be used for planting that facilitates in placement of seed and fertilizers at proper place in one operation that helps in getting good crop stand, higher productivity and resource use efficiency. Using raised bed planting technology, 20-30 % irrigation water can be saved with higher productivity.

### ii. Zero-till planting

Maize can be successfully grown without any primary tillage under no-till situation with less cost of cultivation, higher farm profitability and better resource use efficiency. Under such condition one should ensure good soil moisture at sowing and seed and fertilizers should be placed in band using zero-till seed-cum-fertilizer planter with furrow opener as per the soil texture and field conditions.

### iii. Transplanting

Maize can be successfully cultivated by transplanting seedlings especially in north-west and eastern plain regions of the country. This practice is particularly suitable after the harvest of late paddy, early harvest of crops like sugarcane or as a companion crop with autumn sugarcane. Transplanting of maize may be practical in north-west and eastern plain to avoid delay. This is also suitable for the Diara and Tal areas where flood water recedes late.

For transplanting in second fortnight of January, the nursery should be sown from November 21 to 30 to obtain higher yields than the direct late sown crop. For transplanting one hectare area, 25 kg seed is to be sown in nursery of 1/10 ha as well as 7.5 kg N, 2.5 kg $P_2O_5$, 3.0 kg $K_2O$ and 1.0 kg Zinc sulphate may be applied to seed beds before sowing. Pre-emergence application of Atrazine @ 1.5 kg/ha is also recommended. The seedling may be transplanted on southern slope of east west ridges or in flat fields. It should be removed by edge tools. The uprooted nursery plant should be kept under shade before transplanting but it should be completed at the earliest. In flat beds, seedling should be planted in furrow and their roots needs to be covered with soil manually. Transplanting on ridges gives relatively higher yield than in flat beds. First irrigation should be given immediately and second irrigation after 8-10 days of planting. For planting of one hectare, 700 $m^2$ nursery area is required and the nursery should be raised during second fortnight of November. The age of seedlings for transplanting should be 30-40 days old and transplant in the month of December-January in furrows to obtain higher productivity.

## Nutrient Management

There are several factors that affect the productivity of winter maize. However, the fertilizer management is one of the most important factors that affect the growth and yield of maize. Maize is an exhaustive crop which requires all types of macro and micro nutrients in order to get better growth and exploit yield potential. Among the various nutrients, nitrogen is the principal nutrients which should be applied@ 150 kg N ha-1 in order to have better harvest. The efficiency of nitrogen utilization is better in *Rabi* than in *Kharif* season, primarily because of better water management and lower leaching losses. With better fertilizer response, it should be possible to substantially reduce the cost of production of every tonne of maize produced in *Rabi* season. The available quantity of farmyard manure should be applied before sowing, since a combination of organic and inorganic fertilizers give better results than the use of fertilizers alone.

The quantity of fertilizers to be applied depends mainly on soil fertility and the preceding field management. Therefore, for higher economic yield of maize, application of 10 t FYM/

ha, 10-15 days prior to sowing supplemented with 150-180 kg N, 70-80 kg $P_2O_5$, 70-80 kg $K_2O$ and 25 kg $ZnSO_4$ /ha is recommended.

Full doses of P, K and Zn should be applied as basal preferably drilling of fertilizers in bands along the seed using seed-cum-fertilizer drills. N application at grain filling results in better grain filling. Therefore, nitrogen should be applied in five splits for higher N use efficiency (Table 4).

**Table 4. Nitrogen application in five splits doses**

| Crop stage | Nitrogen rate (%) |
|---|---|
| Basal (at sowing) | 20 |
| V4 (four leaf stage) | 25 |
| V8 (eight leaf stage) | 30 |
| VT (tasseling stage) | 20 |
| GF (grain filling stage) | 5 |

**Table 5. Growth, yield attributes, yield and quality of winter maize as influenced by fertilizer management**

| $N_2O$ application (kg / ha) | Plant height (cm) | Days to Silking | Days to Maturity | Diameter of cob (cm) | Grain Yield (q/ ha) | Protein (%) in grain |
|---|---|---|---|---|---|---|
| 50 | 173.99 | 51.25 | 139.40 | 4.22 | 33.25 | 8.04 |
| 100 | 182.92 | 51.44 | 141.25 | 4.42 | 35.09 | 8.22 |
| 150 | 190.07 | 52.22 | 142.63 | 4.88 | 35.52 | 8.13 |
| CD 5 % | 2.79 | 0.46 | 0.91 | 0.06 | 0.84 | 0.03 |
| **Source of $N_2O$ application** | | | | | | |
| Control | 179.39 | 51.22 | 139.92 | 4.31 | 33.64 | 8.07 |
| FYM | 184.76 | 52.07 | 141.92 | 4.53 | 35.92 | 8.19 |
| Azospirillum | 182.82 | 51.63 | 141.44 | 4.45 | 35.31 | 8.13 |
| CD 5 % | 2.79 | 0.46 | 0.91 | 0.06 | 0.84 | 0.03 |

At Hamirpur (Uttar Pradesh), 150 kg of $N_2O$ /ha application significantly increased over 100 and 50 kg $N_2O$/ ha. However, $N_2O$ application through FYM was found statistically at par with $N_2O$ application through the *Azospirillum* in growth and grain yield (Verma et al., 2012).

At Ludhiana, cob length and number of grains / cob of maize were significantly higher due to application of 175 kg N / ha as compared to 125 and 225 kg N / ha. Grain yield of transplanted maize increased to the tune of 4.5 q / ha at 125 kg N as compared to 100 kg N / ha during Rabi season (Kaul *et al.,* 1994). Significantly higher grain yield and stover yield was recorded with increase in nitrogen application up to 150 kg / ha (Prasad *et al.,* 1987). The increasing levels of nitrogen from 100 to 200 kg / ha resulted in significant increase in number of rows per cob and cobs / ha, 1000 grain weight and grain yield of maize, likewise, other yield components showed similar trend (Mishra *et al.,* 1994, Shanti *et al.*, 1997 and Gawade, 1998). Plant height and leaf area index in winter maize increased with increasing nitrogen application from no nitrogen to 150 kg / ha (Prasad *et al.,* 1985). Prasad *et al.* (1987) found that higher plant height, leaf area index, dry matter accumulation, cob girth, number of cobs / plant, 1000 grain weight, grain yield and stover yield with increase in nitrogen application upto 150 kg / ha.

## Weed control

Light hoeing is to be given as and when necessary for better control of weeds. Broad- leaved weeds and most of the grasses can be conveniently controlled with the application of Atrazine or Simazine @ 0.5-1.0 kg a .i./ha in 1000 lit of water as pre-emergence spray. In zero-till maize production, pre-plant application (10-15 days prior to seeding) of non-selective herbicides viz. Glyphosate @ 1.0 kg a.i. /ha in 400-600 litre water or Paraquat @ 0.5 kg a.i. /ha in 600 litre water is recommended to control the weeds. Under heavy weed infestation, post-emergence application of Paraquat can also be done as protected spray using hoods.

In Ludhiana (Punjab), atrazine (0.25 and 0.5 kg/ha) applied 0, 1, 2, 3 or 4 weeks after transplanting gave complete control of grassy and broad-leaf weeds. Application of herbicide at 5 weeks after transplanting allowed some weed growth. Atrazine application at transplanting or upto 5 weeks after transplanting produced grain yield at par with two hand weeding. Grain yield of transplanted winter maize did not differ significantly due to doses and timing of application of atrazine (Sandhu and Bhatia, 1991).

## Water management

The rainfall during *Rabi* is rather inadequate for successful cultivation of high-yielding maize hybrids. In fact, timely availability of assured irrigation is one of the major factors determining the success of crop. Where soils are generally light, it is desirable to schedule the irrigation 70 % soil moisture availability throughout the period of crop growth and development. In heavy soils, a moisture level of 30% during the vegetative stage and 70 % during the reproductive and grain – filling period is desirable for obtaining optimum yield. Four to six irrigations are needed during the *Rabi* crop season. If six irrigations are given, they should be applied at the following at an interval of 20-25 days, one (essential) at the time of flowering, two after flowering and one at the early grain-filling stage. If only five irrigations are given, one irrigation at the vegetative stage may be avoided; and if only four irrigations are given, irrigation after the dough stage may be avoided. The irrigation should may, however be changed suitably if adequate rains are received.

At Bhubaneswar (Orissa), one irrigation at tassel initiation stage was more beneficial than that of at early vegetative stage. Up to three irrigation, water use efficiency (WUE) was increased linearly with increased number of irrigation. With four irrigations, the yield was higher, but WUE was lower than that of three irrigations, which might be due to increased water application resulted in increase crop water use without a corresponding increase of yield for the crop with four irrigations. The crop coefficients (Kc) at different stages of the crop were derived after computing actual water use using field water balance approach. The crop coefficients of 0.42–0.47, 0.90–0.97, 1.25–1.33, and 0.58–0.61 were derived at initial, development, mid and late season, respectively with three to four irrigation. Study showed that leaf area index (LAI) was significantly correlated with Kc values with the $R^2$ values of 0.93. When LAI exceeded 3.0, the Kc value was 1. Study revealed that the Kc values for the development and mid season stage were slightly higher to that obtained by the procedure proposed by FAO, which might be due to local advection (Gouranga and Nath Verma, 2005).

**Table 6. Grain yield and water use efficiency of winter maize with different levels of irrigation**

| Irrigation | Grain yield (kg/ ha) | Actual evapotrans piration (mm) | Water use efficiency (kg /ha/ mm) |
|---|---|---|---|
| One irrigation at early vegetative stage | 1526e | 289d | 5.24e |
| One irrigation at tassel initiation | 1930d | 294d | 6.64d |
| Two irrigation at tassel initiation + grain filling | 3222c | 342c | 9.42c |
| Three irrigation at early vegetative + tassel initiation + grain filling | 5016b | 381b | 13.14a |
| Four irrigation at early vegetative + tassel initiation + silking + grain-filling stages | 5262a | 436a | 12.05b |
| LSD (0.05) | 205 | 12 | 0.76 |

(Gouranga and Nath Verma, 2005)

## Intercropping

Maize is a most versatile crop for growing intercrops, because of the wide row it needs, providing higher income to the farmers. Short duration varieties of pulses (pea, rajmash and other beans), most vegetables and oilseed crop (soybean and linseed) can be successfully intercropped in maize. The yield of pure maize under intercropping is in no way lower but the intercrop is a bonus. It is also possible to intercrop wheat in maize. Short statured varieties of maize perform better under intercropping. The practice is particularly desirable under delayed sowing after late harvest of rice (Table 5).

**Table 7. Recommended crops for intercropping with maize**

| States | Recommended crops |
|---|---|
| North-western region (Punjab, Haryana, Delhi & Western Uttar Pradesh) | Pea, Rajmash, Lentil |
| North-eastern region (Bihar, Eastern Uttar Pradesh, Orissa, West Bengal & NE Region) | Pea, Rajmash, Potato, Lentil, Bakla and Onion |
| Southern region (Maharashtra, Andhra Pradesh, Karnataka & Tamil Nadu) | Fenugreek (Methi), Coriander, Sunflower and Cluster beans |
| Central region (Rajasthan, M.P & Gujarat) | Pea, Lentil, Onion, Garlic and Methi |

Singh and Sharma (1987) reported that intercropping of black gram, groundnut and mung bean in maize crop had no adverse effect on maize grain yield. The intercrops gave additional yields and increased the net returns.

## Cropping systems

Maize is rapidly emerging as a favorable option for farmers in South Asia as a component crop of rice based systems. Diversity of such changes are higher productivity and profitability of maize over winter rice or wheat, the two competing cereal crops grown in the winter season in South Asia (Ali et al., 2008, 2009). High yielding Rice - Maize systems can also accelerate the problem of secondary and micronutrient deficiencies, not only because larger amounts are removed, but also because the application of high rates of N, P, and K to achieve yield targets often stimulates the deficiency of secondary and micronutrients (Johnston et al., 2009). Pasuquin et al. (2007) demonstrated how diversification from Rice - Rice or Rice-Wheat systems to Rice - Maize systems can impact on nutrient use in Bangladesh.

## Plant Protection

### Insect pests

Maize grown in *Rabi* is known to be free from the attack of major insect pests which usually attack *Kharif* crop. However, pink stem borer is a major pest. To control this pest two sprays of Quinalphos 0.05 % 15 days and 35 days after germination followed by second application with Trichlorphon 5G or Cabofuran 3G @ 0.5 kg a.i/ha 20 days after germination followed by second application 20 days after first recommendation.

### Disease

Disease problem in *Rabi* is less as compared to *Kharif* maize. However, turcicum leaf blight and common rust occurs in moderate to high intensity. During *Rabi* season, post flowering stalk rots, particularly charcoal-rot occur predominantly in late sown crop, especially when temperature at maturity is high and the crop is subjected to soil-moisture stress. The best remedy to minimize yield loss due to these diseases is to grow resistant varieties/ hybrids.

### Crop modelling

At Ludhiana (Punjab), thermal time based phenological events (emergence and physiological maturity) were accurately predicted by model but it fails to predict photothermal based events (tasseling and silking). The predicted LAI matched well with observed during early vegetative phases and physiological maturity but an over estimation was observed during all other stages of growth. The predicted dry matter matched well with field observed values during early vegetative growth but an overestimation was observed during later phases of growth. The predicted grain yield ranged from 98.9 to 116.5 % of the actual observed yields under both the dates of planting. The results obtained with model showed that CERES model could be used to estimate final grain yield of winter maize under Punjab conditions during rabi season (Singh et al., 2005).

## References

Ali, M.Y., S.R. Waddington, D. Hudson, J. Timsina and J. Dixon (2008). Maize – rice cropping system in Bangladesh: status and research opportunities. CIMMYT-IRRI Joint publication, CIMMYT, Mexico.

Ali, M.Y., S.R. Waddington, D. Hudson, J. Timsina and J. Dixon (2009). J. Agric. Sci. Techn., 3(6): 35–53.

Bandyopadhyay P.K. and S. Mallick (1996). Irrigation requirement of Winter Maize under Shallow Water Table Condition in Damodar Valley Irrigation Command Area. J. Indian Soc. Soil Sci., 44 (4): 616 - 620.

Bandyopadhyay, P. K. and S. Mallick (2003). Effect of soil moisture levelson root distribution, water uptake and crop coefficient of winter maize in a humid tropic region. Food, Agriculture & Environment, 1 (3&4): 141-147.

Bandyopadhyay, P.K. and S. Mallick (2000). Effect of irrigation on water extraction pattern of winter maize under shallow water table conditions. Indian J. Soil Conserv., 18 (2): 138–142.

Banga, R.S., Tej singh and D.P. Singh (1994). Response of winter maize to irrigation and fertility levels under shallow water table conditions. Haryana J. Agron., 10: 177 - 181.

Baskheti, D.C., I.S.Singh, M.Z.K. Warsi and S.S.Verma (2005). Grow rabi maize for higher yield. Indian Farmers' Digest, 38 (11): 28 -30.

Bassoi, L.H. and K. Reichardt (1995). Dry matter production and nitrogen accumulation in maize grown during winter given nitrogen fertilizer by soil application and fertigation. Pesquisa - gropecuaria-Brasileria, 30 (12): 1361-1373.

Bhanu, K.R. and K. Maharishna (2006). Production technology of Rabi maize. Indian Farmers' Digest, 39 (10): 21 -23.

Bharati, V., Ravi Nandan, Vinod Kumar and I.B. Pandey (2007). Effect of irrigation levels on yield, water use efficiency and economics of winter maize (*Zea mays* L.) based intercropping system. Indian J. Agron., 52(1): 27-30.

Dixit, A. (1995). Economics of weed control methods in winter maize. *Agric. Sci. Digest,* 15(3): 143-145.

Dixit, A. and K.C. Gautam (1993). Effect of atrazine on photosynthesis and nitrogen metabolism in *rabi* maize. Indian J. Weed Sci., 26(3&4): 77-81.

Dixit, A. and K.C. Gautam (1996). Effect of atrazine on growth and yield of winter maize. Ann. Agri. Res., 17(2): 121-124.

Dixit, A. and K.C. Gautam (1996a). Studies on the effects of atrazine on weeds in winter maize. Indian J. Weed Sci., 28(3&4): 137-139.

Gautam, R. C., P. Pachauri, V. Singh and N. S.Gaur (2000), Response of winter maize (*Zea mays*) to irrigation schedule and tassel removal. Indian J. Agric. Sci., 70 (12): 859-860.

Gollar, R.G. (1996). Plant density, skipping irrigation at critical stages and staggered and simultaneous planting of intercrops in rabi maize. Ph.D thesis, University of Agricultural Sciences, Dharwad.

Gollar, R.G. and V.C. Patil (2000). Effect of plant density on growth and yield of maize genotypes during rabi season. Karnataka J. Agric. Sci., 13 (1): 1-6.

Gouranga Kar and Ashwani Kumar (2010). Energy balance and crop water stress in winter maize under phenology based irrigation scheduling. Irrig. Sci., 28(3):211–220.

Gouranga, K. and H. Nath Verma (2005). Phenology based irrigation scheduling and determination of crop coefficient of winter maize in rice fallow of eastern India. Agricultural Water Management, 75: 169–183.

Granados, G., C. Kitbamroong, C. Tavarasook, M. Chatasiri, P. Grudloyma, S. Chotichun and S. Boonpradab (1994). Winter maize in paddy fields research conducted in Thailand in 1992-1994. A progress Report. 26 p.

Howell, T.A., J.E. Steiner, A.D. Schneider, S.R. Evertt and J.A. Tolk (1997). Seasonal and maximum daily evapotranspirasion of irrigated winter wheat, sorghum and corn: southern high plains. Trans. ASAE, 40; 623 – 634.

Jadhav, B.S., A.S. Jadhav and S.B. Jadhav (1993). Effect of irrigation scheduling showing methods and mulching on *rabi* maize. J. Maharashtra Agric. Univ., 18**:** 31-57.

Johnston, A.M., H.S. Khurana, K. Majumdar and T. Satyanarayana (2009). J. Indian Soc. Soil Sic., 57(1): 1–10.

Kang, S., B. Gu, T. Du and J. Zhang (2003). Crop coefficient and ratio of transpiration to evapotranspiration of winter wheat and maize in a semi-humid region. Agric. Water Managt., 59: 239–254.

Kar, G and H.N. Verma (2005). Phenology based irrigation scheduling and determination of crop coefficient of winter maize in rice fallow of eastern India. Agric Water Manag., 75: 169–183.

Khehra, A.S., H.S. Brar, R.K. Sharma, B.S. Dhillon and V.V. Malhotra (1990). Transplanting of maize during the winter in India. Agron. J., 82: 41–47.

Kumar, A. and S.N. Singh (2002). Production potential and economics of winter maize based intercropping systems. Annals Agric. Res., 23(4): 532-534.

Kumar, K., J.N. Kaul and H.S. Gill (1992). Optimization of nitrogen fertilization on winter maize (*Zea mays*). Indian J. Agron., 37(2): 359-360.

Kumar, R.A., K. Chillar and R.C. Gautam (2006). Nutrient requirement of winter maize *(Zea mays)*-based intercropping systems. Indian J. Agric. Sci., 76 (5): 315-318.

Kumar, S. and A.S. Bangarwa (1997). Yield and yield components of winter maize (*Zea mays* L.) as influenced by plant density and nitrogen levels. Agric. Sci. Digest, 17: 181-184.

Kumar, S. and A.S. Bangarwa (1997a). Yield and yield components of winter maize (*Zea mays* L.) as influenced by plant density and nitrogen levels. Agric. Sci. Digest, 17: 354-359.

Latha, M.R., P. Savithri, R. Indirani and S. Kamaraj (2002). Residual effect of zinc-enriched organic manures on yield and zinc uptake in sunflower under maize-sunflower cropping sequence. Agric. Sci. Digest, 22 (2): 114 – 116.

Li, Q.Q. (2006). Studies on effective saving water patterns of winter wheat and summer maize. [Ph.D. Thesis.] Shandong Agricultural University, P.R. China.

Madegowda, T. (1997). Integrated nutrient management in rabi maize (*Zea mays* L.) under irrigated condition. *M.Sc. (Agri.) Thesis*, University of Agricultural Sciences, Dharwad.

Madhavi, B.N., R.S. Yadav, A.L. Rajput and M.S. Pandey (1994). Effect of plant geometry and nitrogen application on yield and quality of winter maize. Indian J. Agron., 39: 468 - 469.

Manish Kumar (1998). Growth, yield and water use efficiency of different winter maize *(Zea mays* L.) varieties as influenced by nitrogen and irrigation scheduling. *M.Sc. (Agri.) Thesis,* Indira Gandhi Agril. Univ., Raipur.

Manish Kumar, R.S. Tripathi and G.K. Shrivastava (2001). Quality and yield of winter maize as affected by varying genotypes, nitrogen levels and irrigation schedules. Madras Agric. J., 88 (10-12): 693-696.

Maurya, K.L., H.P. Sharma, H.P. Tripathi and Sher singh (2004). Effect of fertility levels on nitrogen and sulphur on growth, yield and quality of winter maize. Haryana J. Agron., 20(1): 80 - 81.

Maurya, K.L., H.P. Sharma, H.P. Tripathi and Sher singh (2005). Effect of nitrogen and sulphur application on yield attributes, yield and net return of winter maize. Haryana J. Agron., 21(2): 115-116.

Meana, O., H.R. Khafi, M.A. Shekh, A.C. Mehta and B.K. Davda (2007). Effect of vermicompost and N on content, uptake and yield of rabi maize. Crop Res., 33 (1 - 3): 53 – 54.

Mishra, B. N., R. S. Yadav, A. I. Rajput and S. M. Pandey (1994). Effect of plant geometry and nitrogen application on yield and quality of winter maize. Indian J. Agron., 39: 468-469.

Mishra, H.S., T.R. Rathore and U.S. Savita (2001). Water-use efficiency of irrigated winter maize under cool weather conditions of India. Irrig. Sci., 21: 27–33.

Misra, B.N., B. Singh and A.L. Rajput (2001). Yield, quality and economics as influenced by winter maize based intercropping system in eastern Uttar Pradesh. Indian J. Agron., 46 (3): 425-431.

Misra, B.N., R.S. Yadav, A.I. Rajput and S.M. Pandey (1994). Effect of plant geometry and nitrogen application on yield and quality of winter maize. Indian J. Agron., 39 (3): 268-269.

Nagalakshmi, K.V., V. Chandrasekhar and G. Subbaiah (2006). Weed management for efficient use of nitrogen in *rabi* maize. Andhra Agric. J., 53(1&2): 14-16.

Nagaraju (2006). Studies on optimization of agro-techniques to maximize productivity of winter maize and evaluation of DSSAT V3.5 CERES maize model. Ph.D thesis, University of Agricultural Sciences, Dharwad.

Nandal, D.S.P. and S.K. Agarwal (1991). Response of winter maize to sowing dates, irrigation and nitrogen. Indian J. Agron., 36 (2): 239-242.

Nirupma Singh, R.Ambika Rajendran, Meena Shekhar, S.L. Jat, Ramesh Kumar and R. Sai Kumar (2012). Rabi Maize Opportunities Challenges, Directorate of Maize Research, Pusa Campus, New Delhi. Technical Bulletin, No. 9: 32 p.

Omraj, M., H.R. Khafi, M.A. Shekh, C. Asha, Mehta and B.K. Davda (2007). Effect of vermicompost and nitrogen on content, uptake and yield of *rabi* maize. Crop Res., 33(2&3): 53-54.

Paradkar, V.K. and R.K. Sharma (1994). Response of improved maize (*Zea mays* L.) varieties to nitrogen during the winter season. *Indian J. Agron.*, 38(4): 650-652.

Parmeet, S., P. Singh and D. Joy (2007). Correlation and regression studies of winter maize and weed interactions. Indian J. Weed Science., 39: 21 - 23.

Pasuquin, J.M.C.A., J. Timsina, C. Witt, R.J. Buresh, A. Dobermann and J. Dixon (2007). The expansion of rice-maize systems in Bangladesh: anticipated impact on fertiliser demand. IFA Crossroads 2007. CD-ROM Proceedings. Bali, Indonesia, 18 December, 2007, IPNI organizer.

Porwal, M.K. (2000). Economics of weed control measures in winter maize (*Zea mays*). *Indian J. Agron.,* 45(2): 344-347.

Prodhan, H.S. (2001). Influence of appropriate cultivars and sowing dates on rabi maize productivity. Indian Journal of Agricultural Science, 71 (1): 55-56.

Ramulu, V., M.D. Reddy and A.M. Rao (2012). Response of rabi maize to irrigation schedules and fertigation levels. Agric. Sci. Digest, 30(2): 104 - 106.

Reddy, M.D., A. Srinivas and V. Gopalakrishnan (1999). Effect of scheduling irrigation on performance of winter maize in Alfisols. Indian J. Agric. Sci., 69: 285 – 286.

Rodriguez-del-Bosque, L.A., J.W. Smith and A.J. Martinez (1995). Winter mortality and spring emergence of corn stalk borers (Lepidoptera, Pyralidae) in subtropical Mexico. *Journal of Economic Entomology,* 88: 628 - 634.

Sandhu, B.S., T.S. Ahuja, K.L. Khera and C.B. Singh (1993). Irrigation and nitrogen requirement and performance of winter maize and raya intercrops in autumn sugarcane in northern India. Indian J. Water Management, 1: 75 – 78.

Sandhu, K.S. and R.K. Bhatia (1991). Chemical weed control in transplants winter maize. *Indian J. Weed Sci.,* 23(3&4): 53-55.

Sandhu, K.S., S.P. Mehra and H.S. Gill (1996). Studies on weed control in winter maize through herbicides. *J. Res. Punjab Agric. Univ.,* 23(4): 549-556.

Shakya, P. R. and K. Atreya (2001). Effect on organic and inorganic fertilizers on maize and millet cropping pattern. Hill Crops Workshop Proceedings of Summer and Winter Crops workshop finger millet, Buckwheat and Barley March 27-29, 2000 Lumle Kaski and September 12-14, 2000 Parwanipur, Bara, Hill Crops Research Programme (HCRP), Kabre, Dolakha, NARC, Khumaltar August, 2001.

Shinde, S.A., D.K. Shelke and G.L. Sawargaonkar (2009). Effect of irrigation schedules and integrated nutrient management on yield and nutrient uptake by *rabi* maize (*Zea mays* L.). International J. Plant Sci. (Muzaffarnagar), 4 (1): 24-26.

Shridhar, V., R.A. Singh and U.N. Singh (1992). Effect of fertility levels on winter maize (*Zea mays* L.) under different moisture regimes based on irrigation water: cumulative pan evaporation. Indian J. Agron., 36: 74-78.

Singh D.P., N.S. Rana and R.P. Singh (2000). Growth and yield of winter maize (*Zea mays*) as influenced by intercrops and nitrogen application. *Indian J. Agron*., 45: 515-519.

Singh, G. and O.P. Singh (1993). Intercropping compatibility of different crops with winter maize (*Zea nays*). *Indian J. Agron.,* 38: 519-522.

Singh, A.P. and P.C. Singh (2003). Effect of different weed control methods on growth and yield of rabi sown hybrid maize cv. Hybrid 4640. J. Living World, 10(2): 12-15.

Singh, A.K, G.R. Singh and T.R.S. Dixit (1997). Influence of plant population and moisture regimes on nutrient uptake and quality of winter maize (*Zea mays*). Indian J. Agron., 42 (1): 107-111.

Singh, A.K., A.K. Roy and D.P. Kaur (2007). Effect of irrigation and NPK on nutrient uptake pattern and qualitative parameter in winter maize and potato intercropping system. International J. Agric. Sci., 3 (1): 199 - 201.

Singh, B., R.K. Mahey and S.S. Mathauda (2005). Validation of Ceres-maize model for simulating growth and yield of winter maize (*Zea mays* L.) under Punjab conditions. J. Res., 42 (4): 417 – 423.

Singh, C.P., N.N. Sharma and U.K. Prasad (1991), Response of winter maize (*Zea mays* L.) to seeding date, seed furrow, mulching and fertilization with nitrogen, phosphorus and potassium. Indian J. Agric. Sci., 61(12): 889 - 892.

Singh, D.P., N.S. Rana and R.P. Singh (2000). Growth and yield of winter maize (*Zea mays*) as influenced by intercrops and nitrogen application. *Indian J. Agron*., 45: 515-519.

Singh, D.P., N.S. Rana and R.P. Singh (2000). Growth and yield of winter maize as influenced by intercrops and nitrogen application. Indian J. Agron., 45 (4): 676 - 680.

Singh, G. and O.P. Singh (1993). Intercropping compatibility of different crops with winter maize (*Zea nays*). Indian J. Agron., 38: 519-522.

Singh, J., A.S. Dhindwal, A.S. Malik and S.R. Poonia (1993). Effect of irrigation regime and nitrogen on winter maize under shallow water table condition. J. Water Management, 1: 22 - 24.

Singh, R.S., R. Sutaliya, R. Ghatak and S.K. Sarangi (2003), Effect of higher application of nitrogen and potassium over recommended level on growth, yield and yield attributes of late sown winter maize *(Zea mays* L.*). Crop Res.,* 26(1): 71-74.

Singh, S.N. and A. Kumar (2002). Production potential and economics of winter maize based intercropping systems, Annals Agric. Res., 23 (4): 532-534.

Sinha, S.P., Prasad S.M., Singh S.J. (2005). Nutrient utilization by winter maize (*Zea mays*) and weeds as influenced by integrated weed management. Indian J. Agron., 50 (4): 303 – 304.

Sinha, K.K., S.S. Mishra and S.J. Singh (1999). Yield and economics as influenced by winter maize based intercropping systems in North Bihar. Indian J. Agron., 44 (1): 30-35.

Sinha, R.B., R. Sakal, and S. Kumar (1995). Sulphur and phosphorus nutrition of winter maize in calcareous soils. J. Indian Soc. Soil Sci.43(3): 413-418.

Sinha, S.P., S.K. Prasad and S.J. Singh (2000). Effect of IWM on growth yield attributes and yield of winter maize. J. Application Biol., 10 (2): 158 – 162.

Sinha, S.P., S.M, Prasadand S.J. Singh (2005). Nutrient utilization by winter maize (*Zea mays*) and weeds as influenced by integrated weed management. Indian Journal of Agronomy, 50(4): 303 – 304.

Sinha, S.P., S.M. Prasad and S.J. Singh (2001). Response of winter maize (*Zea mays*) to integrated weed management. Indian J. Agron., 46(3): 485- 488.

Sinha, S.P., S.M. Prasad, S.J. Singh and K.K. Sinha (2003). Integrated weed management in winter maize (*Zea mays*) in North Bihar. *Indian J. Weed. Sci.,* 35 (3/4): 273-274.

Sridhar, V., R.A. Singh and U.N. Singh (1991). Effect of fertility level on winter maize under different moisture regimes based on irrigation water. Indian J. Agron., 36: 74-78.

Sridhar, V., R.A. Singh and U.N. Singh (1991). Effect of irrigation and fertility levels on nutrient content, uptake and recovery in *rabi* maize. Madras Agric. J., 78: 422-424.

Subbian, P., G.S. Thangamuthu and S.P. Palaniappan (1991). Long term response of winter maize to nitrogen, phosphorus and potassium. Indian J. Agron., 36 (4): 511-512.

Tank, D.A., S.K. Patel and V.P. Usadadia (2006). Nitrogen management in rabi maize. Crop Res. (Hisar), 31 (2): 323 – 324.

Timsina, J., M.L. Jat and K. Majumdar (2010). Rice-maize systems of South Asia: current status, future prospects and research priorities for nutrient management. Plant and Soil, 335: 65-82.

Tinh, H.N., P.H. Hao, L.Harrington and M. Read (1992). Winter maize in the Red River Delta of North Vietnam: Problem and Prospects. Asian Farming Systems Symposium. Colombo, Sri Lanka. November 1992.

Tripathi, A.K., A. Kumar, S. Nath and R.A. Yadav (2008) Weed dynamics, productivity and net monetary returns as influenced by winter maize based intercropping systems in central U.P. *ISWS Biennial Conference on Weed Management in Modern Agriculture,* February 27-28, 2008, Pusa, Bihar. p.120.

Vadivel, N., P. Subbian and A. Velayutham (2001). Effect of integrated nitrogen management practices on growth and yield of rainfed winter maize. Indian J. Agron., 46(2): 250 – 254.

Yang, G., A. Duan, J. Sun, F. Li, Z. Liu, H. Liu and Z. Liu (2009). Field Crop coefficient and water-use efficiency of winter wheat/spring maize strip intercropping. Crop Res., 111(1-2): 65-73.

Yang, Z., L. Zhou, Y. Lv and H. Li (2012). Long term effect of crop residual and inorganic fertilizers on yield and soil organic matter for a winter wheat – maize system in North China plain. Advanced Materials Res., 14: 356-360.

# 35

# Kharif Maize Cultivation

According to advance estimate it is cultivated in 8.7 million ha (2010-11) mainly during *Kharif* season which covers 80 % area.

The maize is cultivated throughout the year in all states of the country for various purposes including grain, fodder, green cobs, sweet corn, baby corn, pop corn in peri-urban areas. The predominant maize growing states that contributes more than 80 % of the total maize production are Andhra Pradesh (20.9 %), Karnataka (16.5 %), Rajasthan (9.9 %), Maharashtra (9.1 %), Bihar (8.9 %), Uttar Pradesh (6.1 %), Madhya Pradesh (5.7 %), Himachal Pradesh (4.4 %).

**Climate requirements**

Maize is a warm weather crop and grows right from sea level to over 3000 m altitudes. It requires adequate moisture and warmth from sowing to end of flowering. The optimum temperature for germination is 21 $^{O}$C, while for growth it is 32 $^{O}$C. Extremely high temperatures coupled with low humidity are not conducive for pollination. It cannot withstand frost at any stage of its life cycle. The crop prefers the areas receiving an annual rainfall of 50-100 cm. The crop also grows well in areas with annual rainfall of 250 to 400 cm (Assam and West Bengal hills), provided the fields are not flooded. The crop needs more than 50 % of its total water requirement in about 30-35 days after tasseling. Inadequate soil-moisture at the grain filling stage results in poor yield of shrivelled grains.

## Soils

Maize can be grown successfully in variety of soils ranging from loamy sand to clay loam. However, soils with good organic matter content having high water holding capacity with neutral pH are considered good for higher productivity. Being a sensitive crop to moisture stress particularly excess soil moisture and salinity stresses, it is desirable to avoid low lying fields having poor drainage and also the field having higher salinity. Therefore, the fields having provision of proper drainage should be selected for cultivation of maize. The alluvial soils of Uttar Pradesh, Bihar and Punjab are suitable for raising good maize crop. The soils with sandy loam to silty loam texture are best for the crop. Intermediate texture of loam to silt loam in surface horizon and little higher content of clay as silt loam to silty clay loam in subsoil is the most ideal. Soil pH of 7.5-8.5 supports good crop growth, as the crop is grown under rainfed conditions it is important that soil must have good water holding capacity, with proper drainage system to avoid water logging conditions.

## Land preparation

A deep summer ploughing after *rabi* crop followed by light ploughing by country plough and planking is required. For heavy soils, 2-3 ploughing are needed. In areas of high rainfall during *kharif*, ridge and furrows are made to use furrows for provision of drainage. Delayed early plant growth due to higher mechanical impedance of soil, poor crop establishment and low plant stands may be the primarily associated causes of low maize grain yields on conservation tillage methods (Fischer *et al.,* 2002).

**Table 1. Yield comparison of the different tillage management on maize**

| Tillage Management | Grain yield (kg/ha) |
|---|---|
| Zero | 5290a |
| Conventional | 5171a |
| Farmer's Practice | 4080b |

*Means having the same letter are not significantly different. (Romeo et al., 2002).

At Mindoro Occidental, the higher yield of yellow corn (var. C 818) was obtained using zero tillage compared to conventional and farmers' tillage practices. This resulted to higher net returns mainly due to lower production cost (Romeo et al., 2002).

### Time of sowing

In rainfed areas, the sowing time should be coincided with onset of monsoon. Maize can be sown during last week of June to first fortnight July during Kharif season. It is preferred to sow 10-15 days before start of rain which will give 15 % higher yield. In northern hilly regions, the crop is sown in May. As a irrigated *kharif* crop, it is sown a fortnight or a month before onset of monsoon to avoid water logging damage to seedlings due to heavy rainfall.

### Seed rate

To achieve higher productivity and resource-use efficiencies optimum plant stand is the key factor. During *kharif,* a plant population of 65,000-70,000/ha is recommended for grain production. The seed rate varies depending on purpose, seed size, plant type, season, sowing methods etc. For grain (normal and QPM) maize, seed rate of 20 kg /ha with plant spacing of 60 x 20 cm (83333 plants/ha) or 75 x 20 cm (66666 plants/ha). About 50-60 kg seed would be needed to sow one hectare. Seed should be grown 5 cm deep into soil for good germination, seedling growth and vigour. Transplanting should be avoided as the plant cannot cope up with main crop stand.

**Table 2. Effect of plant densities on yield attributes and yield of maize**

| Plant population (Plants/ha) | Plant height (cm) | Stem diameter (mm) | Ear length (cm) | Ear diameter (mm) | Grain weight / ear (g) | Grain yield (kg/ha) |
|---|---|---|---|---|---|---|
| 50000 | 184.6 | 21.6 | 19.2 | 47.2 | 190.1 | 7821 |
| 60000 | 186.1 | 20.5 | 18.3 | 46.9 | 172.9 | 8329 |
| 70000 | 184.0 | 19.6 | 17.8 | 46.6 | 157.8 | 8546 |
| 80000 | 185.7 | 19.3 | 17.1 | 46.2 | 150.5 | 9078 |
| 90000 | 188.4 | 18.8 | 16.8 | 46.1 | 142.1 | 9406 |
| 100000 | 190.8 | 18.6 | 16.2 | 45.7 | 136.0 | 9234 |
| CD 5 % | 3.9 | 0.6 | 0.6 | 0.5 | 9.1 | 571 |

At Hatay (Turkey), grain yield increased with increasing plant densities up to 90000/ha but slightly decreased at 100000/plants/ha (Huseyin et al., 2003). Farnham (2001) determined that corn grain yield increased from 10.1 to 10.8 t/ha as plant density increased from 59000 to 89000 plants/ha.

### Seed treatment

To protect the crop from seed and major soil borne diseases and insect-pests, seed treatment with fungicides and insecticides before sowing is advisable/ recommended (Table 3).

**Table 3. Seed treatment for major diseases and pests**

| Disease/insect-pest | Fungicide/Pesticide | Rate of application ($g\ kg^{-1}$ seed) |
|---|---|---|
| Turcicum leaf blight, Banded leaf and Sheath blight, Maydis leaf blight | Bavistin + Captan in 1:1 ratio | 2.0 |
| BSMD | Apran 35 SD | 4.0 |
| Pythium stalk rot | Captan | 2.5 |
| Termite and shoot fly | Imidachlorpid | 4.0 |

### Method of sowing

Commercially the crop is sown by drilling with seed drill. Under wet soil conditions, the tynes may get choked with soil resulting in blank rows. This method is quick and is suitable for light soils at all moisture levels and in other soils at optimum soil moisture. For maintaining optimum population and to avoid thinning, seeds are dibbled manually at required spacing. This method requires less seed rate and is best for costly varieties/hybrids. Dropping the seeds in the furrows opened by plough followed by planking is also one of the common methods of sowing. After germination, the seedlings are thinned out to maintain optimum plant stand. Hill sowing is the other method adopted for sowing. In this method, the seeds are dropped with the help of corn planter placed at appropriate place and pressed for dropping 3-4 seeds at a place. It also needs relatively less seed. The optimum depth of sowing is 7-8 cm in *kharif.* A seedling/hill is optimum for all maize types.

### Conventional till flat planting

Under heavy weed infestation where chemical/herbicidal weed management is uneconomical in no-till and also for rainfed areas where survival of crop depends on conserved soil moisture, in such situations flat planting can be done using seed-cum-fertilizer planters.

### Furrow planting

To prevent evaporative losses of water during spring season from the soil under flat as well as raised bed planting is higher and hence crop suffers due to moisture stress. Under such situation/condition, it is always advisable to grow maize in furrows for proper growth, seed setting and higher productivity.

### Varieties

The maize improvement programme got a boost with the establishment of All India Coordinated Maize Improvement programme in 1957. A large number of double cross hybrids and composites were developed and released from then onwards. However, since 1980's, single cross hybrids development is given more emphasis. Trishulata (three way hybrid) and Paras (first single cross hybrid) were thus released in 1991 and 1995, respectively. Mostly maize is grown during rainy season. Some cultivars require 60- 70 days to mature; others require 100-110 days to mature. Grain colour also varies from yellow to orange to white. Mainly flint types are preferred.

Maize composites/hybrids released by private companies for cultivation between 1997-2003 are 3058 (Y1402K), JKMH 2492, PRO 303, MMH 69, KH 951, SSF 9374, PRO 311, MMH 113, PAC 701, PAC 705, Bioseed 9681, 32A09, PRO 312, PRO 316, PAC 738, PAC 776, JH 3189, Co BC1, JH 3459, Makka 3, JKMH 175, Kohinoor hybrid, Seed Tec 740, DK 701 and PRO 339.

### Manures and fertilizers

The crop should be supplied with FYM or compost @ 10-15 t/ha before sowing to ensures good tilth and improve the water-holding capacity. The recommended dose of N varies with duration of the variety and its season of cultivation. For full season hybrids, 100-200 kg N/ha; for medium maturity, 80-100 kg N/ha; for early maturity, 60-80 kg N/ha; and for extra early maturity, 40-60 kg N/ha is required. Besides N, 40-60 kg $P_2O_5$/ha is necessary for realization of higher yields. The response to K fertilization is rare in India. However, in intensive cropping systems and low temperature regions of Himachal Pradesh and Jammu & Kashmir, K fertilization @ 30-40 kg $K_2O$/ha is necessary. In sandy soils and freshly levelled field the crop exhibits Zn deficiency. Therefore, it is advisable to apply 15-20 kg/ha zinc sulphate along with basal application of fertilizers. The deficiency of zinc in plants at later stages of growth, however, may be corrected by foliar application of zinc sulphate dissolved in water with half the quantity of lime (0.5 % $ZnSO_4$ + 0.25 % hydrated lime).

One fourth of the total quantity of nitrogen and entire quantity of phosphorus, potash and zinc (wherever required) should be applied 5-7 cm deep before sowing. The rest of N should be applied by side dressing 10-12 cm away from the base of the plant to avoid plant injury. Half of the total N should be applied at knee high stage (20-30 days after germination) and rest one fourth of the nitrogen should be applied after the emergence of flag-leaf, but before the emergence of tassels. Soon after side dressing, the fertilizer on the soil surface should be covered to minimize losses.

At Coimbatore (Tamil Nadu), yield, gross return and net return were higher under combination of 250:125:125 NPK kg /ha along with the foliar spray of PGR consortia @ 1.5 % twice at 45 DAS and 55 DAS. However, higher benefit cost ratio was recorded under combination of 200:100:100 NPK kg /ha along with the foliar spray of PGR consortia @ 1.5 % (Sekar et al., 2010).

**Table 4. Effect of fertilizer levels and foliar sprays on grain yield, stover yield and economics of hybrid maize**

| Fertilizer level (NPK kg/ ha) | Grain yield (kg/ha) | Stover yield (kg /ha) | Gross return (Rs/ ha) | Net return (Rs/ ha) | B: C ratio |
|---|---|---|---|---|---|
| 150: 75: 75 | 5419 | 9094 | 35247 | 20707 | 2.43 |
| 200:100:100 | 6348 | 9882 | 41056 | 25342 | 2.61 |
| 250: 125: 125 | 6546 | 10946 | 42562 | 25673 | 2.52 |
| CD 5 % | 208 | 601 | - | - | - |
| **Foliar spray** | | | | | |
| Control (no spray) | 5711 | 9489 | 37127 | 22203 | 2.49 |
| Salicylic acid (100 ppm) | 6114 | 9792 | 39622 | 24203 | 2.57 |
| Boric acid (0.3 %) | 6231 | 10167 | 40442 | 24753 | 2.58 |
| PGR consortia (1.5 %) | 6444 | 10549 | 41833 | 25867 | 2.61 |
| TNAU Panchagavya (3 %) | 6100 | 10014 | 39606 | 23002 | 2.38 |
| PPFM $10^6$ dilution | 6026 | 9830 | 39101 | 23417 | 2.49 |
| CD 5 % | 210 | 210 | - | - | - |

(Sekar et al., 2010)

**Water management**

Presently, the maize crop is grown in 20-30 % irrigated conditions only. Protective irrigations are necessary under periods of moisture stress, when grown during *kharif* under rainfed conditions. In *kharif*, 2-3 irrigations are required after sowing, whereas, spring crop requires 8-10 irrigations. A good crop of maize requires about 480-600 mm of water during its life cycle. *Kharif* maize sown before monsoon should be given a pre-sowing irrigation besides irrigation at 6-leaf stage. The other irrigations may be adjusted as per rainfall distribution. It must be ensured that the crop should not suffer from moisture stress at flowering and grain filling stages.

Excess water or water logging is equally or even more harmful. The crop is highly sensitive to excess water (water logging) and hence efficient drainage is equally important as irrigation. It is observed that water logging for more than 3-4 days may result in a yield reduction to the extent of 50 % or even complete crop failure, depending upon severity of water logging. To avoid water logging, sowing on ridges is desirable especially in *kharif.*

At Coimbatore (Tamil Nadu), during *kharif* among the different fertigation levels, higher maize grain yield of 7300 kg /ha was recorded under drip fertigation of 100 % RDF with 50 per cent P and K through water soluble fertilizer (WSF) followed by application of 150 % RDF through drip (7050 kg /ha). The yield increase over drip irrigation with soil application of fertilizer was 39 %. Drip irrigation helped to save water up to 43.65 % compared to surface irrigation method. Among the different intercrops tested, higher WUE of 21.0 kg /ha/ mm was observed under maize + vegetable coriander intercropping system (Muthukrishnan and Anitta Fanish, 2011).

**Table 5. Effect of drip fertigation levels on grain yield and water use efficiency of maize**

| Treatments | Grain yield (kg /ha) | Stover yield (kg /ha) | Water Use Efficiency (kg /ha/ mm) |
|---|---|---|---|
| Surface irrigation with soil application of 100 % RDF | 4500 | 8170 | 8.7 |
| Drip irrigation with soil application of 100 % RDF | 5238 | 8779 | 17.9 |
| Drip fertigation of 75 % RDF | 5530 | 9158 | 18.9 |
| Drip fertigation of 100 % RDF | 5964 | 9383 | 20.4 |
| Drip fertigation of 125 % RDF | 6815 | 10055 | 23.3 |

| Treatments | Grain yield (kg /ha) | Stover yield (kg /ha) | Water Use Efficiency (kg /ha/ mm) |
|---|---|---|---|
| Drip fertigation of 150 % RDF | 7050 | 10372 | 24.1 |
| Drip fertigation of 50 % RDF (50 % P and K as WSF) | 5750 | 9331 | 19.6 |
| drip fertigation of 75 % RDF ( 50 % P and K as WSF) | 6406 | 9740 | 21.9 |
| Drip fertigation of 100 % RDF (50 % P and K as WSF) | 7300 | 10633 | 24.9 |
| CD (P=0.05) | 235 | 326 | - |

(Muthukrishnan and Anitta Fanish, 2011)

**Weed control**

Weeds in maize result in an average of 30-70 % yield loss (Thakur, 1994). Weeds emerge with the germination of maize seeds and grow along with plants till the early growth period. This causes a severe crop weed competition to maize plants. The losses caused by weeds are more at seedling stage (3-6 weeks after sowing) than later stages. Two hoeings 15 and 30 days after sowing are necessary to control weeds. Mechanical hoeings should not be done after knee-high stage, as it may lead to leaf damage. In *Kandi* area of sub-montane Himalayan regions, ploughing between the rows of crop of 1m height is practiced. It not only helps in uprooting weeds, but also serves the earthing up purpose besides conserving rain water. The herbicide usage is common for weed control in maize. Pre-emergence application (within 2 days after sowing) of atrazine or simazine @ 1 kg/ha or alachlor @ 2 kg/ha has been found effective in weed control.

At Bangalore (Karnataka), integrated weed management practice of Alachlor @1.5 kg/ ha + hand weeding at 40 DAS recorded higher plant height, cob weight, 100 seed weight, grain yield and straw yield (Mallikarjuna et al., 2012).

**Table 6. Yield and yield attributing characters realized under different weed control practices**

| Weed control practices | Maize yield (q/ha) | Cob weight (g) | 100 seed weight (g) | Stover yield (ton/ha) | Plant height (cm) |
|---|---|---|---|---|---|
| Weedy check | 38.13 | 507.75 | 36.185 | 81.61 | 184.5 |
| Hand weeding | 54.05 | 603.20 | 37.54 | 97.18 | 195.73 |
| Alachlor @ 2 kg/ha | 54.80 | 704.80 | 40.40 | 83.80 | 190.40 |
| Alachlor @ 1.5 kg/ha + Hand weeding at 40 DAS | 63.63 | 950.50 | 39.94 | 97.42 | 202.50 |
| Mean | 56.61 | 791.88 | 39.29 | 94.04 | 199.19 |

(Mallikarjuna et al., 2012).

At Udaipur (Rajasthan), maximum values of yield of maize, black gram and maize equivalent yield were recorded when metolachlor @ 1.0 kg /ha was applied in maize based intercropping system (Shekhawat et al., 2002).

## Intercropping

Short duration varieties of pulse crop (black gram, green gram and cowpea), oilseed crop (soybean) and vegetables can successfully be grown as intercrop. A ratio of 2 rows of maize with 1 row of other desired crop can be adopted. In Bihar, mixed cropping of groundnut with maize is quite profitable. In certain areas, maize is intercropped with pigeon pea.

Chowdhury and Bose (1983) noticed the lower yield of intercropped soybean in maize than that of sole crop. The combined grain yield of the two crops in an intercropping system was more than the individual components. Though the yield of soybean when intercropped with maize was less than that of sole crop, the combined grain yield of the two crops in an intercrop was more than the individual components. Kanakeri (1991) recorded observations on maize intercropped with legumes (green gram, black gram, soybean and cowpea) in 1:1 and 1:2 row ratios at Dharwad. No significant differences in maize growth, yield parameters and yield were obtained compared to sole maize.

**Table 7. Effect of intercrops on grain yield and water use efficiency of maize**

| Intercropping systems | Yield (kg /ha) | | | | Water Use Efficiency (kg /ha/ mm) |
|---|---|---|---|---|---|
| | Grain | Stover | Intercrops | Maize grain equivalent | |
| Maize + vegetable coriander | 6522 | 9749 | 2308 | 8170 | 20.4 |
| Maize + radish | 6396 | 9592 | 4369 | 9617 | 20.1 |
| Maize + beet root | 5932 | 9333 | 4076 | 8827 | 18.6 |
| Maize + onion | 6233 | 9454 | 4452 | 9415 | 19.5 |
| CD (5 %) | 181 | 203 | - | - | - |

(Anitta Fanish and Muthukrishnan, 2011)

At Coimbatore (Tamil Nadu), among the different intercropping systems, radish intercropped with maize registered a higher maize grain equivalent yield (MEY) of 9617 kg/ ha during *kharif* seasons(Anitta Fanish and Muthukrishnan, 2011).

## Cropping systems

In irrigated areas of north India, barley, wheat, *toria*, potato, black gram, berseem, sugarcane, lucerne and oats are raised after maize.

**Table 8. System productivity, carbon sustainability index and carbon efficiency of different maize based cropping systems under different tillage practices**

| Tillage practices | System productivity (kg/ha) | | Carbon sustainability index (CSI) | | Carbon efficiency (CE) | |
|---|---|---|---|---|---|---|
| | 2008-09 | 2009-10 | 2008-09 | 2009-10 | 2008-09 | 2009-10 |
| Zero till (ZT) | 8146 | 7661 | 14.9 | 13.8 | 15.9 | 14.8 |
| Permanent beds (PB) | 8828 | 8238 | 16.4 | 14.2 | 17.4 | 15.2 |
| Conventional till (CT) | 7680 | 7322 | 13.2 | 11.7 | 14.2 | 12.7 |
| Cropping systems | | | | | | |
| Maize-Wheat-Mung bean | 8925 | 10108 | 11.7 | 11.5 | 12.7 | 12.5 |
| Maize-Chickpea-*Sesbania* | 7198 | 6974 | 16.3 | 15.7 | 17.3 | 16.7 |
| Maize-Mustard-Mungbean | 8687 | 6026 | 14.9 | 10.4 | 15.9 | 11.4 |
| Maize-Maize-*Sesbania* | 8062 | 7854 | 16.4 | 15.3 | 17.4 | 16.3 |

(Jat et al., 2010)

At New Delhi, permanent bed planting resulted in maximum system productivity during both the years of investigation compared to zero till and conventional till. The carbon-based sustainability index (CSI) and carbon efficiency (CE) was significantly higher under permanent beds compared to zero and conventional tills. Maize – wheat - mung bean resulted in maximum system productivity. This system involves a grain legume which not only contributed to increased yield, but also helps maintain soil health which in turn improves system productivity. CSI and CE declined in the order of Maize – Maize - *Sesbania* > Maize-Chickpea – *Sesbania* > Maize – Mustard – Mung bean > Maize – Wheat – Mung bean during first year and Maize – Chickpea – *Sesbania* > Maize – Maize - *Sesbania* > Maize – Wheat –Mung bean > Maize – Mustard – Mung bean during second year (Jat et al., 2010).

### Maize - Bengal gram cropping system

At Dharwad (Karnataka), application of FYM @ 7.5 t/ha recorded significantly higher maize (5899 kg/ha), bengal gram (1566 kg/ha) and maize equivalent yield (11650 kg/ha) and it was on par with vermicompost @ 2.5 t/ha. Among the fertilizer levels, application of 100 % RDF recorded significantly higher maize (5578 kg/ha), bengal gram (1495 kg/ha) and maize equivalent yield (10986 kg/ha) compared to 75 % and 50 % RDF which differ significantly each other. Significantly higher gross returns (Rs.69059 /ha) and net returns (Rs.51659 / ha) were realized with application of FYM @ 7.5 t/ha + 100 % RDF which were on par with vermicompost @ 2.5 t/ha + 100 % RDF with respect to gross returns. Significantly higher benefit cost ratio was recorded with poultry manure @ 1.0 t/ha + 100 % RDF (4.48) (Lingaraju et al., 2010).

**Table 9. Grain yield of maize, Bengal gram and maize equivalent yield as influenced by organics and fertilizer levels**

| Organics | Maize yield (kg/ha) | Bengal gram yield (kg/ha) | Maize equivalent yield (kg/ha) | Gross returns (Rs/ha) | Net returns (Rs/ha) | B:C ratio |
|---|---|---|---|---|---|---|
| FYM @ 7.5 t/ha | 5988 | 1566 | 11650 | 65201 | 48401 | 3.88 |
| Maize crop residue @ 5 t/ha | 4762 | 1321 | 9547 | 53288 | 39238 | 3.79 |
| In-situ green manuring of sunhemp | 5033 | 1336 | 9874 | 55154 | 41354 | 3.99 |
| Poultry manure @ 1.0 t/ha | 5665 | 1499 | 11091 | 61992 | 47942 | 4.41 |
| Vermicompost @ 2.5 t/ha | 5779 | 1509 | 11239 | 62822 | 43522 | 3.25 |

| Organics | Maize yield (kg/ha) | Bengal gram yield (kg/ha) | Maize equivalent yield (kg/ha) | Gross returns (Rs/ha) | Net returns (Rs/ha) | B:C ratio |
|---|---|---|---|---|---|---|
| No organic manure | 4325 | 1247 | 8844 | 49376 | 36326 | 3.78 |
| CD 5 % | 273 | 91 | 364 | 3384 | 3367 | 0.22 |
| **Fertilizer Levels** | | | | | | |
| 50% RDF | 4917 | 1326 | 9720 | 54278 | 39703 | 3.76 |
| 75% RDF | 5281 | 1418 | 10416 | 58210 | 43035 | 3.87 |
| 100% RDF | 5578 | 1495 | 10986 | 61429 | 45654 | 3.92 |
| CD 5% | 175 | 54 | 258 | 3384 | 3367 | 0.22 |

(Lingaraju et al., 2010). RDF: 100:50:25 N, P2O5 and K2O kg/ha

## Maize – Sunflower Cropping System

At Coimbatore (Tamil Nadu), application of zinc sulphate @ 37.5 kg /ha to first crop of maize followed by 12.5 kg /ha to all alternate sunflower crops increased the total yield of maize and sunflower together, showing beneficial residual effect of applied zinc to previous crop, which received cumulatively 25 % less zinc sulphate as compared to 12.5 kg /ha fo zinc sulphate applied to every crop in a sequence covering ten crops despite comparable yield were recorded among themselves. There was a build up of DTPA-Zn in the post harvest soil due to application of zinc sulphate, irrespective of levels, either to every crop or alternate crop over a period of time as against depletion of zinc observed in the treatments receiving only once and no zinc in the sequential experimentation (Shanmugasundaram and Savithri, 2005).

### Harvesting

Maize is harvested when husk is turned yellow and grains are hard enough with about 30 % moisture. There is no need to wait for stalks and leaves to dry, because they remain green in most of the hybrids and composites. The husk is removed from the cobs, and cobs are dried in sun for 7-8 days. Thereafter grains are removed either by beating the cobs by stick or with the help of maize sheller.

### Yield

Well managed crop gives 5-6 tonnes grain/ha in case of hybrids and 4.5-5.0 t/ha in case of composites under irrigated conditions. In case of rainfed crop, yield levels are about 2.0-2.5 t/ha for hybrids and 1.5-2.0 t/ha for composites.

## References

Abolhassan, M.H., S.J. Herbert and D.H. Putnam (2005). Yield response of corn to crowding stress. Agron. J., 97: 839-846.

Abou El-Nour, E.A.A. (2002). Growth and nutrient contents - Response of maize to foliar nutrition with micronutrients under irrigation with Saline water. J. Biol. Sci., 2: 92-97.

Abuzar, M.R., G.U. Sadozai, M.S. Baloch, A.A. Baloch, I.H. Shah, T. Javaid and N. Hussain (2011). Effect of plant population densities on yield of maize. J. Animal Plant Sci., 21(4): 692- 695.

Adeniyan, O.N. and S.O. Ojeniyi (2003). Comparative effectiveness of different levels of poultry manure with NPK fertilizer and residual soil fertility, nutrient uptake and yield of maize. Moor J. Agric. Res., 4 (2): 191 - 197.

Ahmad alias, M., H.A. Bukhsh, R. Ahmad, A.U. Malik, S. Hussain and M. Ishaque (2010). Agro-physiological traits of three maize hybrids as influenced by varying plant density. J. Animal Plant Sci., 20(1): 34-39.

Ali, R., S.K. Khalil, S.M. Raza and H. Khan (2003). Effect of herbicides and row spacing on maize. Pakistan J. Weed Sci. Res., 9(3-4): 171-178.

Ananthi, T., M. Mohamed Amanullah and K.S. Subramanian (2011). Influence of fertilizer levels and mycorrhiza on root colonization, root attributes and yield of hybrid maize. Madras Agric. J., 98 (1-3): 56-61.

Andrade, F.H., C. Vega, S. Uhart, A. Cirilo, M. Cantarero and O. Valntinuz (1999). Kernel number determination in maize. Crop Sci., 39: 453-459.

Ashok Kumar, Rajesh Singh, Lalit Kumar Rao and U.K. Singh (2008). Effect of integrated nitrogen management on growth and yield of maize *(Zea mays* L.) cv. PAC – 711. Madras Agric. J., 95 (7-12): 467-472.

Azadgoleh, M.A.E. and Z. Kazmi (2007). A study of the planting pattern and density effects on yield and physiological growth parameters in two corn cultivars. Ecology, Environment and Conservation, 13(3): 467-472.

Brenda, L., B. Lucas and M.E. Otegui (2006). Source sink relations and kernel weight differences in maize temperate hybrids. Field Crops Res., 95:316-326.

Channabasavanna, A.S., D.P. Biradar and S.G. Yelamali (2002). Effect of poultry manure and NPK on growth and yield of maize. Karnataka J. Agric. Sci., 15(2): 353 - 355.

Deshmukh, L.S., A.S. Jadhav, R.S. Jathure and S.K. Raskar (2009). Effect of nutrient and weed management on weed growth and productivity of kharif maize under rainfed condition. Karnataka J. Agric. Sci., 22(4): 889-891.

Deshmukh, L., R.S. Jathure and S.K. Baskar (2008). Studies on nutrient and weed management in kharif maize under rainfed condition. Indian J. Weed Sci., 37 (1&2): 87 - 89.

Farnham, D.E. (2001). Row spacing, plant density and hybrid effects on corn grain yield and moisture. Agron. J., 93 (5): 1049 – 1053.

Gaur, B.L., D.S. Rao and M.K. Kaushik (1991). Comparative efficacy of pre-and post emergence herbicides in controlling weeds in rainy-season maize (*Zea mays*). *Indian J. Agron.,* 36: 261-262.

Huseyin, G., S. Okan, K. Omer and K. Mehmet (2003). Effect of hybrid and plant density on grian yield and yield components of maize. Indian J. Agron., 48 (3): 203 -205.

Jadav, V.T. and D.K. Shelke (2006). Growth, yield and yield attributes of maize as influenced by land layouts, hybrids and fertilizer levels in kharif. Journal of Maharashtra Agricultural Universities, 31: 231-232.

Kassie, B.T., M.K. Van Ittersum, H. Hengsdijk, S. Asseng, J. Wolf and R.P. Rotter (2014). Climate-induced yield variability and yield gaps of maize (*Zea mays L.*) in the Central Rift Valley of Ethiopia. Field Crops Res., 160: 41–53.

Khot, R.B. (1990). Seed yield of maize cv. African tall maize as influenced by plant spacings and levels of nitrogen during Kharif. M.Sc (Agriculture) thesis, Mahatma Phule Krishi Vidyalaya, Rahuri, Maharashtra.

Kolage, A.K., S.H. Shinde and R.L. Bhilare (2004). Weed management in *kharif* maize. J. Maharashtra Agric. Univ., 29(1): 110-111.

Lingaraju, B.S., K.G. Parameshwarappa, U.K. Hulihalli and B. Basavaraja (2010). Effect of organics on productivity and economic feasibility in maize -Bengal gram cropping system. Indian J. Agric. Res., 44 (3): 211- 215.

Liu W., M. Tollenaar, G. Stewart and W. Deen (2004). Within row plant spacing variability does not affect corn yield. Agron J. 96: 275-280.

Luikham, E., J.K. Rajan, K. Rajendran and P.S.M. Anal (2003). Effect of organic and inorganic nitrogen on growth and yield of baby corn *(Zea mays* L). Agric. Sci. Digest, 23(2): 119-121.

Luque, S.F., A.G. Cirilo and M.E. Otegui (2006). Genetic gains in grain yield and related physiological attributes in Argentine maize hybrids. Field Crop Res., 95: 383-397.

Ma, G.S., J.Q. Xue, H.D. Lue, R.H. Zhang, S.J. Tai and J.H. Ren (2007). Effects of planting date and density on population physiological indices of summer corn (Zea mays L.) in central Shaanxi irrigation area. Chinese J. Applied Ecology, 18(6): 1247-1253.

Maddonni, G.A., M.E. Otegui and A.G. Cirilo (2001). Plant population density, row spacing and hybrid effects on maize canopy architecture and light attenuation. Field Crops Res., 71: 183 - 193.

Maddonni, G.A. and M.E. Otegui (2006). Intra-specific competition in maize: Contribution of extreme plant hierarchies to grain yield, grain yield components and kernel composition. Field Crops Res., 97: 155-166.

Maddonni, G.A., A.G. Cirilo and M.E. Otegui (2006). Row width and maize grain yield. Agron. J., 98: 1532- 1543.

Meysam, A. and A. Sarangi (2013). Deficit irrigation and nitrogen effects on maize growth in semi arid environment. World Applied Sci. J., 21 (11): 1687-1692.

Mohamed Amanullah, M., S. Sekar, S. Manoharan, P. Muthukrishnan, K.S. Subramanian and S. Vincent (2010). Influence of fertilizer levels and growth regulating substances on growth, nutrient use efficiency and yield of hybrid maize. Madras Agric. J., 97 (1-3): 68-72.

Mohamoud, A.I.S. and Sharanappa (2002). Growth and productivity of maize *(Zea mays* L.) as influenced by poultry waste composts and fertilizer levels. Mysore J. of Agric. Sci., 36 (3): 203-207.

Mohsan, S. (1999). Population dynamics and Nitrogen management effects on maize productivity. Ph.D dissertation. Dept. Agronomy, Univ. Agri., Faisalabad, Pakistan.

Monneveux, P., P.H. Zaidi, and C. Sanchez (2005). Population density and low nitrogen affects yield associated traits in tropical maize. Crop Sci., 45: 535-545.

Munamava, M.R., A.S. Goggi and L. Pollak (2006). Seed quality of maize inbred lines with different composition and genetic backgrounds. Crop Sci., 44: 542-548.

Muthukrishnan, P. and S. Anitta Fanish (2011). Influence of drip fertigation on yield, water saving and water use efficiency in maize (*Zea mays* L.) based intercropping system. Madras Agric. J., 98 (7-9): 243-247.

Naeem, M. (1998). Biological efficiency of maize as influenced by plant population and split application of nitrogen. Ph. D. Dissertation, Dept. Agronomy, Univ. Agric., Faisalabad, Pakistan.

Ojeniyi, S.O. (1993). Nutrient availability and maize yield under reduced tillage practices. Soil and Tillage Res., 26: 89 – 92.

Panda, R.K., S.K. Behera and P.S. Kashyap (2004). Effective management of irrigation water for maize under stressed conditions. Agric. Water Management, 66: 181–203.

Parasuraman, P. (2006). Studies on nutrient requirement of hybrid maize (*Zea mays* L.) under irrigated conditions. Mysore J. Agric. Sci., 40(1): 14-20.

Parasuraman, P., R. Prakash and B. Chandrasekaran (2008). Response of hybrid maize *(Zea mays* L.) to soil and foliar application of nutrients. Madras Agric. J., 95 (1-6): 200-202.

Parthipan, T. (2000). Nitrogen management strategies in hybrid maize (COH 3) using SPAD meter and predictions using CERES - Maize model. M.Sc. (Ag.) Thesis, Tamil Nadu Agric. Univ., Coimbatore.

Patel, H.R., G.G. Patel, Vyas Pandey and V.U.M. Rao (2011). Weather based decision for growing kharif maize in Middle Gujarat Agro-climatic zone of Gujarat state – a practical manual. AICRP on Agrometeorology, Department of Agricultural Meteorology, Anand Agricultural University, Anand, P: 13-14.

Patra, B.C., B.B. Mandal, B.K. Mandal and A.K. Padhi (1999). Suitability of maize (Zea mays) based intercropping system. Indian J. Agric. Sci., 69: 759 - 761.

Rajiv Sharma (2007). Weed management for Kharif maize in north India. Indian Farming, 59 (5): 23 -25.

Rajkumara, S., Y.R. Aladakatti, P.L. Patil and A.B. Khot (2009). Nitrogen and phosphorus substitution through organics in maize-chickpea sequence under irrigated conditions. Karnataka J. Agric. Sci., 22 (1): 15-17.

Ramah, K., P. Santhi and K. Ponnuswamy (2009). Irrigation scheduling and water use efficiency in maize (*Zea mays* L.) based cropping system under drip fertigation. Crop Res. (Hisar), 38 (1/3): 15-20.

Sankaran, N., S. Meena and N. Sakthivel (2005). Input management in maize. Madras Agric. J., 92 (7-9): 464-468.

Sawargaonkar, G.L., D.K. Shelke, S.A. Shinde and K. Shilpa (2008). Performance of kharif maize based legumes intercropping systems under different fertilizer doses. International J. Agric. Sci., 4(1): 152 - 155.

Sekar, S., M. Mohamed Amanullah, S. Manoharan, P. Muthukrishnan, K.S. Subramanian and S. Vincent (2010). Influence of fertilizer levels and growth substances on yield and economics of hybrid maize. Madras Agric. J., 97 (1-3): 58-61.

Shanmugasundaram, R. and P. Savithri (2005). Effect of levels and frequency of zinc sulphate application on yied and DTPA - Zn availability in maize-sunflower cropping system. Madras Agric. J., 92 (4-6): 252-257.

Shanwad, U.K., B.N. Aravindkumar, U.K. Hulihalli, Ashok Surwenshi, Mahadev Reddy and B.R. Jalageri (2010). Integrated nutrient management in Maize-Bengal gram cropping System in Northern Karnataka. Res. J. Agric. Sci., 1(3): 252-254.

Shilpashree, V.M, H.M. Chidanandappa, R. Jayaprakash and B.C. Punitha (2012). Effect of integrated nutrient management practices on Distribution of nitrogen fractions by maize crop in soil. Indian J. Fundamental and Applied Life Sci., 2 (1): 38-44.

Shivay, Y.S., R.P. Singh and C.S. Pandey (1999). Response of nitrogen on maize (*Zea mays*)–based intercropping system. Indian J. Agron., 44: 261-266.

Shrestha, J. and K.P. Timsina (2011). Agronomic evaluation and economic analysis of winter maize under different plant population and nitrogen rates in Chitwan, Nepal. Nepalese J. Agric. Sci., 9: 5-13.

Silva and Tuivavologi (1996). Effect of chicken manure and inorganic fertilizer on soil properties and the growth and yield of maize *(Zea mays* L). Growth on a hawallian axisol. J. South Pacific Agric., 3(1/2): 37-41.

Singh, A.V. and I.P.S. Ahlawat (1998). Studies on N economy in rainy season maize as affected by summer green gram. Crop Res. (Hissar), 16: 171 – 179.

Singh, R.N., N.N. Nath, S.K. Singh, T.K. Mohan and J.P. Shah (2006). Effect of agronomic management practices on growth, yield and quality of *kharif* maize (*Zea mays* L.) under excessive moisture conditions. Crop Res. (Hissar), 32 (3): 302-305.

Singh, Rameshwar and K.L. Totawat (2002). Effect of integrated use of nitrogen on performance of maize *(Zea mays* L) on haplustalfs of subhumid Southern Plains of Rajasthan. Indian J. Agric. Res., 36(2): 102 - 107.

Siva, P. (2007). Optimization of nitrogen dose and spacing in hybrid maize (*Zea mays* L.). M.Sc. (Ag.) Thesis, Tamil Nadu Agric. Univ., Coimbatore.

Sparlangue, T., F.H. Andrade, P.A. Calvino and C. Larry (2007). Why do maize hybrids respond differently to variations in plant density? Agron. J., 99(2):984-991.

Srikanth, M., M. Mohamed Amanullah, P. Muthukrishnan and K.S. Subramanian (2009). Nutrient uptake and yield of hybrid maize (*Zea mays* L.) and soil nutrient status as influenced by plant density and fertilizer levels. International J. Agric, Sci., 5: 193-196.

Srinivasulu, A., G. Karuna Sagar, V. Chandrika and D.S. Reddy (2008). Performance of maize genotypes under varied time of sowing under rainfed conditions in southern agroclimatic Zone of Andhra Pradesh. Agric. Sci. Digest, 28 (3): 233 – 234.

Stehli, A., A. Soldati and P. Stamp (1999). Vegetative performance of tropical highland maize in the field. J. Agron. Crop Sci., 183 (3): 193-198.

Surendra Singh and A.K. Sarkar (2001).Balanced use of major nutrients for sustaining higher productivity of maize *(Zea mays)* – wheat cropping system in acidic soils of Jharkhand. Indian J. Agron., 46: 605-610.

Suryavanshi, V.P., B.N. Chavan, K.T. Jadhav and P.A. Pagar (2008). Effect of spacing, nitrogen and phosphorus levels on growth, yield and economics of kharif maize. International J. Tropical Agric., 26 (3-4): 287 -292.

Thakur, D.R. (1994). Weed management in intercropping system based on maize (Zea mays) under mid hill conditions. Indian J. Agron., 39: 203-206.

Tollenaar, M. and E.A. Lee. (2002). Yield potential, yield stability and stress tolerance in maize. Field Crops Res., 75: 161-169.

Tripathi, A.K., A.N. Tewari and A. Prasad (2005). Integrated weed management in rainy season maize in Central Uttar Pradesh. Indian J. Weed Sci., 37 (3&4): 269 - 270.

Vega, C.R.C., F.H. Andrade and V.O. Sadras (2001). Reproductive partitioning and seed set efficiency in soybean, sunflower and maize. Field Crops Res., 72: 165-173.

36

# Spring Maize Cultivation

Spring maize has become popular owing to higher yields. Besides, their sowing periods are flexible and thus can be fitted in any cropping system. The spring crop is sown during late January to end of February.

## Soils

The crop is grown on a wide range of soils. However, it prefers well drained fertile soils. The soil should be medium textured with good water-holding capacity. It does not thrive well on heavy clays, especially low lands. It can be grown successfully in soils with pH ranging from 5.5 to 7.5. The alluvial soils of Uttar Pradesh, Bihar and Punjab are suitable for raising good maize crop. The soils with sandy loam to silty loam texture are best for the crop.

## Land preparation

A deep ploughing after *rabi* crop followed by light ploughing by country plough and planking is required. For heavy soils, 2-3 ploughing are needed. In spring and summer seasons, planting in furrows is preferred over ridge planting. This is done for achieving higher water use efficiency.

## Seed rate and spacing

In maize, each plant is important for realization of potential yields. The seed should be sown at a spacing of 75 cm x 18 cm or 60 cm x 22 cm. A seed rate of 20 kg/ha is required. Ramachandra Prasad *et al.* (1990) found that 40,500 plants/ ha had higher RGR (131 mg /g/ day) compared to 55,500 (126 mg/ g/ day) and 70,500 plants /ha (127 mg/ g/ day). Also AGR, NAR, CGR, LAD were higher in case of less plant population in red sandy loam. Number of cobs/ plant has significantly increased due to thinning at 22.50 cm. Number of grains /cob, grain weight /cob and 1000 grain weight also significantly increased. Thinning at 11.25 cm significantly increased the grain yield by 5.31 q /ha (13.39 %) and stover yield by 82.47 q /ha (55.50 %) over the thinning (Pandey *et al.,* 1999).

LAI was 1.7, 2.6 and 4.0 at tasselling for 1.7, 2.6 and 6.3 plants /$m^2$, respectively. Light interception at tasselling stage approximately 20, 30 and 40 % for 19, 35 and 63 plants/ $m^2$, respectively. Increase in plant population density increased the vegetative dry matter (Tetio-Kagho and Gardner, 1988; Tetio-Kagho and Gardner, 1988a). Russell and Zimmermann (1990) investigated the effect of 80 and 90 cm row spacing with plant densities of 40 to 75 m-2 in maize cv. Cortimaz 322A, Emeopa 2, 60-8801/U, EGO- 17 CP and BR-106. BR-106 recorded higher yield with intermediate densities of 63 to 65 plants/ $m^2$. Planting of maize at 60 cm row spacing recorded higher dry matter /plant (162.39 g) and higher yield (38.97 q /ha) over 30 cm and 45 cm row spacing which was due to availability of more area /plant for energy harvesting (Lucas and Remison, 1984).

## Method of sowing

For maintaining optimum population and to avoid thinning, seeds are dibbled manually at required spacing. This method requires less seed rate and is best for costly hybrids. Dropping the seeds in the furrows opened by plough followed by planking is also one of the common methods of sowing. After germination, the seedlings are thinned out to maintain optimum plant stand.

## Manures and fertilizers

The crop should be supplied with FYM or compost @ 10-15 t/ha before sowing to ensures good tilth and improve the water holding capacity. The recommended dose of N varies with duration of the variety and its season of cultivation. For full season hybrids, 100-200 kg N/ha; for medium maturity, 80-100 kg N/ha; for early maturity, 60-80 kg N/ha; and for extra early maturity, 40-60 kg N/ha is required. Besides N, 40-60 kg $P_2O_5$/ha is necessary for realization of higher yields. The response to K fertilization is rare in India. However, in intensive cropping systems and low temperature regions of Himachal Pradesh and Jammu and Kashmir, K fertilization @ 30-40 kg $K_2O$/ha is necessary. One fourth of the total quantity of nitrogen and entire quantity of phosphorus, potash and zinc (wherever required) should be applied at the time of sowing. The rest of N should be applied by side dressing 10-12 cm away from the base of the plant to avoid plant injury. Half of the total N should be applied at knee high stage (20-30 days after germination) and rest one fourth of the nitrogen should be applied after the emergence of flag leaf, but before the emergence of tassels. Soon after side dressing, the fertilizer on the soil surface should be covered to minimize losses.

In Pakistan, application of 10, 15, 20, 25 and 30 t / ha FYM and 4, 6, 8, 10 and 12 t / ha poultry manure significantly increases the percentage grain protein content (8.15 %) which was with 12 t / ha poultry manure application (Khan *et al.,* 2008). Increased leaf area index in maize with an increase in phosphorus rate from no phosphorus to 75 kg $P_2O_5$ / ha and higher grain yield (34 q/ha) was obtained with the application of 75 kg $P_2O_5$ / ha during summer season (Anjum *et al.,* 1997). The productivity of maize is largely dependent on its nutrient management particularly that of nitrogen as maize is a heavy feeder of nitrogen (Tyagi et al., 1998).

At Coimbatore (Tamil Nadu), application of 150 % of RDF once in 6 days recorded significantly higher grain yield (8,957 kg /ha) than 100 % of RDF (7,915 kg /ha) (Table 1). Irrespective of fertilizer levels, fertigation frequency scheduled once in 6 days registered higher grain yield and it was on par with other frequencies. No significant difference in grain yield was noticed among different fertigation frequencies. Absolute control recorded the lowest grain yield (4,680 kg/ ha) followed by surface irrigation with 100 % RDF. The increased percentage of grain yield over lower dose of RDF and absolute control was 13.2 and 91.3, respectively. Maize crop responded well to fertigation and application of nutrients up to 150 % of RDF and maximized the grain yield in hybrid maize (Sampathkumar and Pandian, 2010).

**Table 1. Effect of fertigation levels and frequencies on yield components and yield of maize during summer season**

| Treatments | Cob length (cm) | Cob girth (cm) | Rows/ cob | Grains/ cob | Cob weight (g) | Grain yield (kg /ha) |
|---|---|---|---|---|---|---|
| 100 % RDF once in 6 days | 17.8 | 14.5 | 14.8 | 507 | 129 | 7915 |
| 100 %RDF once in 9 days | 17.6 | 14.2 | 14.8 | 502 | 127 | 7858 |
| 100 % RDF once in 12 days | 17.3 | 14.2 | 14.6 | 495 | 124 | 7725 |
| 100 % RDF once in 15 days | 17.1 | 14.0 | 14.6 | 491 | 122 | 7684 |
| 150 % RDF once in 6 days | 18.8 | 14.8 | 15.4 | 546 | 144 | 8957 |
| 150 % RDF once in 9 days | 18.5 | 14.5 | 15.2 | 540 | 142 | 8882 |
| 150 % RDF once in 12 days | 18.5 | 14.5 | 15.2 | 532 | 139 | 8764 |

| Treatments | Cob length (cm) | Cob girth (cm) | Rows/ cob | Grains/ cob | Cob weight (g) | Grain yield (kg /ha) |
|---|---|---|---|---|---|---|
| 150 % RDF once in 15 days | 18.4 | 14.2 | 15.2 | 527 | | 8718 |
| Surface irrigation with100 % RDF | 16.3 | 14.0 | 14.6 | 456 | 105 | 6324 |
| Absolute control (no fertilizer) | 15.1 | 13.5 | 14.2 | 422 | 85 | 4680 |
| CD (0.05) | 1.1 | 0.8 | 0.5 | 19 | 9 | 480 |

(Sampathkumar and Pandian, 2010)

## Water management

Maize as *zaid* season crop is grown under irrigated conditions only. Spring crop requires 8-10 irrigations. A good crop of maize requires about 480-600 mm of water during its life cycle. A vigorously growing maize plant needs about 2-3 litres of water/day during peak growing period or on an average its consumptive use of water varies from 2.5 to 4.3 mm/day. The growth stages of the crop affect the water requirement to a greater extent. Irrigated crop of maize besides pre-sowing irrigation should be irrigated at all 5 critical stages. However, if only 3-4 irrigations are available, then irrigation at tasseling and 6-leaf stages may be avoided.

The value of cob diameter, cob length and grain yield (5.70 cm, 16.20 cm and 76.25 q /ha, respectively) was maximum in IW/CPE ratio of 1.0 at all the stages of crop of maize in red sandy loam soil of Bangalore during summer. Moisture stress was due to irrigation scheduled at IW/CPE ratio at 0.6 continuously during reproductive and grain filling stages resulted in the highest WUE of 142.24 kg /ha/ cm (Leta Tulu *et al.,* 1998).

Rajender Kumar et al, (1996) reported that number of cobs /plant, number of grains/ cob and stover yield was higher in no stress condition (3.19, 215.31 and 53.46 q/ ha, respectively) which was at par with water stress at dough stage (2.97, 214.39 and 53.84 q / ha, respectively) over stress at knee high, tasseling and silking stages in loamy soil at Karnal (Haryana). Water use was lower 400 mm and 400 mm in 1991 and 1992, respectively with water stress created at tasseling stage followed by water stress at knee high and silking stage and higher water use recorded was 475 and 469 mm in 1991 and 1992, respectively in fully irrigated plots. Higher net return of maize (Rs. 7459 /ha) was with water stress at dough (84 DAS) over knee high, tasseling, silking stages (Rs. 4038, Rs. 5919 and Rs. 3854/ ha, respectively). Silking stage was most critical for moisture, which recorded lowest returns (Rs. 3524 /ha) due to stress in loamy soil of Karnal during summer (Rajendra Kumar et al,, 1996). Higher number of cobs/ plant (1.53), cob length (18.13 cm), grain /cob (393.30), test weight (188.27 g) and grain yield (50.72 q/ ha) with scheduling of irrigation at 0.6 IW:

CPE ratio over 0.2 and 0.4 ratio in spring maize. Stover yield was also higher in 0.6 ratio (139.09 q / ha) during 1995 on sandy loam soil of Hissar (Tyagi *et al.,* 1998).

Drip system uniformity (Christiansen uniformity coefficients of 59, 80 and 97 %) had an insignificant effect on plant growth, plant nitrogen uptake, yield, yield components, nitrogen agronomic efficiency and apparent nitrogen loss. In the subhumid region of the North China Plain where irrigation only accounts for 21–23 % of the precipitation during the growing season of maize, a drip system uniformity that is lower than the values recommended by the current standards might be used to reduce the installation and operation costs of drip irrigation systems. The apparent nitrogen loss increased significantly with nitrogen application rate whereas the difference of yields between the treatments with nitrogen applied at 120 and 210 kg/ha did not reach a statistical significance but significantly superior over control (Zhen et al., 2014).

## Weed control

Two hoeing 15 and 30 days after sowing are necessary to control weeds. Mechanical hoeing should not be done after knee-high stage, as it may lead to leaf damage. In *Kandi* area of sub-montane Himalayan regions, ploughing between the rows of crop of 1m height is practiced. It not only helps in uprooting weeds, but also serves the earthing up purpose besides conserving rain water. Pre-emergence application (within 2 days after sowing) of atrazine or simazine @ 1 kg/ha or alachlor @ 2 kg/ha has been found effective in weed control.

## Intercropping

Crops like soybean, black gram, green gram and cowpea are raised mixed with maize. These legume crops are grown in space between 2 rows of maize. In Bihar, mixed cropping of groundnut with maize is quite profitable. In certain areas, maize is intercropped with pigeon pea.

At Kathalagere (Karnataka), treatments receiving both organic and inorganic fertilizers in kharif season, followed by only inorganic fertilizers during summer season has improved the soil fertility level. Higher rice grain yields were observed in kharif season in treatment receiving 25 % N through paddy straw and 75 % NPK through inorganic fertilizers with least in control. Similarly higher maize yields were observed in T6 receiving 50 % N through FYM and 50 % NPK through inorganic fertilizers in kharif followed by 75 % NPK through inorganic fertilizers with minimum yield in control. The uptake pattern also followed the yield of both the crops. Thus by following INM practices and inclusion of light irrigated crop like maize in summer season to avoid loss due to shortage of water and other constraints, yield, fertility levels and fertilizer use efficiency can be enhanced (Sathish *et al*., 2011).

**Table 2. Rice and Maize grain yield as influenced by integrated nutrient management in rice-maize cropping sequence**

| Nutrient Source | | Rice yield (kg/ha) | | Maize yield (kg/ha) | |
|---|---|---|---|---|---|
| Kharif (rice) | Summer (maize) | 1988 | 2006 | 1989 | 2008 |
| Control | Control | 2797 | 2734 | 672 | 1263 |
| 50 % NPK | 50 % NPK | 4813 | 4327 | 2046 | 3556 |
| 50 % NPK | 100 % NPK | 4356 | 4547 | 5195 | 4329 |
| 75 % NPK | 75 % NPK | 4953 | 4927 | 3898 | 4189 |
| 100 % NPK | 100 % NPK | 5449 | 6177 | 4831 | 4362 |
| 50 % NPK + 50 % N (FYM) | 100 % NPK | 5270 | 6279 | 6003 | 4943 |
| 75 % NPK + 25 % N (FYM) | 75% NPK | 5012 | 6367 | 4022 | 4040 |
| 50 % NPK + 50 % N (paddy straw) | 100 % NPK | 5514 | 6287 | 6199 | 4535 |
| 75 % NPK + 25 % N (Paddy Straw) | 75 % NPK | 5577 | 6542 | 3701 | 4710 |
| 50 % NPK + 50 % N (Glyricidia) | 100 % NPK | 5278 | 6272 | 4360 | 4223 |
| 75 % NPK + 25 % N (Glyricidia) | 75 % NPK | 5463 | 6461 | 3460 | 4038 |
| Farmers Practice (85:50:30 kg NPK/ha & FYM 2 t/ha) | Farmers practice (75:37.5:38.75 kg NPK/ha ) | 5501 | 5132 | 2959 | 3775 |
| CD 5 % | | 892 | 305 | 1454 | 926 |

(Sathish et al., 2011).

## Harvesting

Maize is harvested when husk is turned yellow and grains are hard enough with about 30 % moisture. There is no need to wait for stalks and leaves to dry, because they remain green in most of the hybrids and composites. The husk is removed from the cobs, and cobs are dried in sun for 7-8 days. Thereafter grains are removed either by beating the cobs by stick or with the help of maize sheller.

## Yield

Well managed crop gives a grain yield of 5-6 t/ha in case of hybrids and 4.5-5.0 t/ha in case of composites under irrigated conditions.

## References

Anjum, J.I., C.N. Ali, F.H. Sahi and Z.H. Khar (1997). Effect of plant population and fertilizer application on growth and yield of summer maize. J. Agric. Res. (Lahore), 60(4): 453-457.

Devender Singh, R.C. Tyagi, I.S. Hooda and O.P.S. Verma (1997). Influence of plant population, irrigation and N levels of growth of spring maize (*Zea mays* L.). Haryana J. Agron., 13(1): 54-58.

Devender, S., R.C. Tyagi, S.K. Agarwal and D. Singh (1998). Weed control methods in spring maize. Haryana Agric. Uni. J. Res., 28(1): 21-25.

Farhad, W., M.F. Saleem, M.A. Cheema and H.M. Hammad (2009). Effect of poultry manure level on the productivity of spring maize (*Zea mays* L.). J. Ani. Plant Sci., 19(3): 122-125.

Hussaini, M.A., V.B. Ogunlela, A.A. Ramalan, A.M. Falaki and A.B. Lalwal (2002). Productivity and water use in maize (*Zea mays* L.) as influenced nitrogen, phosphorus and irrigation levels. Crop Res., 23(20): 228-234.

Khan, H.Z., M.A. Malik and M.F. Saleem (2008). Effect of rate and source of organic material on the production potential of spring maize (*Zea mays* L.). Pakistan J. Agric. Sci., 45(1): 40-43.

Kumar, S., S.S. Saini and B.D. Sharma (2004). Effect of graded levels of nitrogen, phosphorus and plant geometry on the productivity of spring maize. Forage Res., 30 (1): 62 –63.

Lehrsch, A. Gary, R. E. Sojka and D. T. Westermann (2000). Nitrogen placement, rows pacing and furrow irrigation water positing effects on corn yield. Agron. J., 92: 1266-1275.

Leta Tulu, B., K. Ramachandrappa and H.V. Nanjappa (1998). Response of maize (*Zea mays*) to moisture stress at different growth stages in alfisols during summer. Mysore J. Agric. Sci., 32: 201-207.

Li, Q.Q. (2006). Studies on effective saving water patterns of winter wheat and summer maize. Ph.D. Thesis. Shandong Agricultural University, P. R. China.

Nadeem, M.Y., K.A.R. Khan and M. Jamil (1997). Interactive effect of salinity and nitrogen on growth and chemical composition of spring maize. J. Agric. Res., 35: 285-295.

Quanqi, C. Yuhai, L. Mengyu, Z. Xunbo, D. Baodi and Y. Songlie (2008). Water potential characteristics and yield of summer maize in different planting patterns. Plant Soil Environ., 54(1): 14 - 19.

Rajendra Kumar, D. R. Dahiya, N. K. Tyagi and Ashok Yadav (1996). Yield, yield attributes and economics of summer maize as influenced by water stress at critical stages. Haryana Agric. Univ. J. Res., 26: 259 - 565.

Russell, R. and F.J.P. Zimmermann (1990). Population density and maize cultivars. Embresa capixaba de pesquira Agropecudria, 65: 121.

Sampathkumar, T. and B.J. Pandian (2010). Efficiency of applied nutrients and spad values in hybrid maize under drip fertigation. Madras Agric. J., 97 (7-9): 237-241.

Sampathkumar, T. and B.J. Pandian (2010a). Effect of fertigation frequencies and levels on growth and yield of maize. Madras Agric. J., 97 (7-9): 245-248.

Sathish, A., V. Govinda Gowda, H. Chandrappa and Agaraja Kusagur (2011). Long term effect of integrated use of organic and inorganic fertilizers on productivity, soil fertility and uptake of nutrients in rice - maize cropping system. International J. Sic. Nature, 2(1): 84-88.

Singh, D., R.C. Tyagi, I.S. Hooda and O.P.S. Verma (1997). Influence of plant population, irrigation and nitrogen levels on the growth of spring maize. Haryana J. Agron., 13: 54-58.

Tahir, M., M.R. Javed, A. Tanveer, M.A. Nadeem and A. Wasaya, S.A.H. Bukhari and J.U. Rehman (2009). Effect of different herbicides on weeds, growth and yield of spring planted maize (*Zea mays* L.). Pakistan J. Life Soc. Sci., 7(2): 168-174.

Tripathi, R.S., G.K. Srivastava and S. Malaiya (2004). Effect of variety, sowing time and integrated nutrient management on growth, yield attributes and yield of summer maize (*Zea mays* L.). Ann. Agric. Res. New Series, 25(1): 155-158.

Tyagi, R.S., Devender Singh and I.S. Hooda (1998). Effect of plant population, irrigation and nitrogen on yield and its attributes of spring maize (*Zea mays* L.). Indian J. Agron., 43(4): 672-676.

Viswanatha, G.B., B.K. Ramachandrappa and H. V. Nanjappa (2000). Effect of drip irrigation and method of planting on root and shoot biomass, tasseling silking interval, yield and economics sweet corn (*Zea mays* Cv. Saccharata). Mysore J. Agric. Sci., 34: 134-141.

Zhen, W., L. Jiusheng and Y. Li (2014). Effects of drip system uniformity and nitrogen application rate on yield and nitrogen balance of spring maize in the North China Plain. Field Crops Res., 159: 10–20.

Zhiming, F., L. Dengwei and Z. Yuehong (2007). Water requirements and irrigation scheduling of spring maize using GIS and Crop Wat model in Beijing-Tianjin-Hebei region. Chinese Geographical Sci., 17 (1): 56-63.

## Annexure I. List of major starch manufacturing units in India

| S.No. | Name of the unit | Address | Telephone | Fax | E-mail & Website | Capacity (MT corn/ day) |
|---|---|---|---|---|---|---|
| **Existing Units** | | | | | | |
| 1 | Maize Products | P.O. Kathwada Maize Products, Ahmadabad, Gujarat - 382430 | +91-79-2871581 | +91-79-2873232 | sales@maizeproducts.com | 400 |
| 2 | Saahil organics | NR. Shailesh Park, B/H L.D. Engg. Hostel Polytechnic, Ahmadabad, Gujarat- 380015 | +91-76-6304572 | + 91-76-6307108 | saahil@ad1.vsnl.net.in http://www.saahil.com/ | 200 |
| 3 | Anil Products | Anil Road, Post Box No.10009, Ahmadabad, Gujarat - 380 025 | +91-79-220322 ext.252 | +91-79-2200731 | darshan@anilproducts.com | 300 |
| 4 | Riddhi Siddhi starch | Viramgaon, Ahmadabad, Gujarat | | | | 100 |
| 5 | Kashyap sweetners | Vapi, Ahmadabad, Gujarat | | | | 100 |
| 6 | Ambuja proteins Ltd | Heemat Nagar, Ahmadabad, Gujarat | | | | 250 |
| 7 | The Kisan Sahakari Starch Mfg, Soc. Ltd. | Mumbai-Agra Road, (Bladi-Nyahalod Fata), Deopur, Dhule, Maharashtra - 424005 (MS) | +91-256-270235 | +91-256-271232 | md@kisan-starch.com www.kisan-starch.com | 200 |
| 8 | Universal starch Chemical Allied Ltd. | Jay Palace, Hawai Mahal, Maharana Pratap Sinh Marg, Dadanagar, Dodaicha, Maharashtra - 425 498, India | 244235, 244236 | | bapusheb@hotmail.com www.universalstarch.com | 400 |
| 9 | Sahayadri starch | Sangli, Maharashtra | | | | 200 |
| 10 | Yaswant starch | Sirola, Maharashtra | | | | 200 |
| 11 | Tirupathi starch | Indore, Madhya Pradesh | | | | 200 |
| 12 | Rajaram Brothers | Mhow Neemuch Road, Mandsaur, Madhya Pradesh - 458 001 | (7422) 221135/36/43/440 | (7422) 223294 | | 150 |

| S.No. | Name of the unit | Address | Telephone | Fax | E-mail & Website | Capacity (MT corn/ day) |
|---|---|---|---|---|---|---|
| 13 | Kashyap sweeteners | Ratlam, Madhya Pradesh | | | | 100 |
| 14 | Riddhi Siddhi starch | Gogak, Near Balgaon, Karnataka | | | | 300 |
| 15 | Devji Agro Pvt Ltd | Kolapur, Maharashtra | | | | 50 |
| 16 | Sukhjeet starch and chemical Ltd | Sarai road, Phagwara, Punjab-144001 | 260216 | | | 100 |
| 17 | Rajaram Maize Products | Komal Niwas, Kailash Nagar, Rajnandgaon, Chhatisgarh - 491 441 | (7744) 225207, 224825, 501013, 503213, 223561 | | | 150 |
| 18 | Santosh starch products Ltd | 71, New cloth Market, Ahmadabad-380007 | 91-79-2141933 | 91-79-2160874 | | 100 |
| 19 | Jaychandra Agro Industries Pvt.Ltd | Mhatre pen building, B wing, II floor, Senapati bapat marg, Dadar (west), Mumbai-400028 | 91-22-24362210/ 24363418 | 91-22-24305969 | | 100 |
| | **Sub-total** | | | | | **3600** |
| **Proposed units** | | | | | | |
| 1 | DSQ Biotech | Origin Agro Starch, Kadhur, Tamil Nadu | Diversification from tapioca starch to maize starch | | | |
| 2 | Basant Dada Patil Makka Prakariya Karkhana Ltd. | Sangli, Maharashtra | New unit | | | |

**Annexure II. Major consultants/ Supplier of technology of wet milling of maize in India**

| S.No | Consultant/ Suppliers (Name & Address) | Telephone (Office) | Fax/e-mail | Type of service offered |
|---|---|---|---|---|
| 1 | National Research Development Corporation (A Government of India Entreprise),20-22, Zamroodhpur Community Centre, Kailash Colony Extension, New Delhi - 110 048 www.nrdcindia.com | 26419904, 26417821, 26480767, 26432627 | 26231877, 6460506, 6478010 info@nrdcindia.com | Turn key projects |
| 2 | M/s Halani & Associate, 4-A, Morlidhar Society, Odhav, Ahmadabad - 382415 | 22870474 | d_halani@yahoo.com | Consultant |
| 3 | M/s Geeli Machinery Works, 3/A, Mona Estate, Opp. Anil Starch Mills, Anil Road, Ahmadabad -380 025 | 22201602 22865409 (R) | 22203992 info@geelidewatering.com | Manu. & Exporters of Starch machineries and spares |
| 4 | M/s Shirke Structurals Pvt. Ltd. 72-76, Mundhwa, Pune - 411 036. | | | Construction of Maize silo and Tower cabin |
| 5 | M/s J.M. Engineers/ Engi Fab Industries Plot No. 143 Phase I, Behind New Post Office, GIDC, Batha, Ahmadabad - 382 445 | 25833674 | | Manu. Starch machinery |
| 6 | Alfa Laval (India) Ltd. Mumbai-Pune Road, Dapoli, Pune - 411 012 Mr. Dilip Gaikwad Mr. Nitin Gaikwad | 56382270 (Mumbai) 27127721 & 27127711 (Pune) | dilip.gaikwad@alfalaval.com nitin.gaikwad@alfalaval.com Homepage: www.alfalaval.com | Turnkey project and suppliers of centrifugal separators and hydroclone washing systems for starch units |

| S.No | Consultant/ Suppliers (Name & Address) | Telephone (Office) | Fax/e-mail | Type of service offered |
|---|---|---|---|---|
| 7 | SPECTRUM Engineering Pvt. Ltd. Makarpura Station Road, Maneja, Vadodara 390013, Gujarat | 0265-2642641/ 2642643 2644592/ 2332633 | E-Mail:- spectoms@ad1.vsnl.net.in Internet://www.vigorsoft.com/spectoms | Suppliers of Feed milling plants & equipments, Material handing & Storage Systems, etc. |
| 8 | Decon System 10, Sargam, Rambaug Col Paud Road, Pune- 411 038 | Tel 27120792 27121641 | 27121641 E Mail - deconsys@pn3.vsnl.net.in | Suppliers of bulk material handlng systems and equipments |
| 9 | Water Treatment and Effluent Treatment (I) Pvt. Ltd. B-115, 1st Floor, Sonal Palace, Phadk Road Dombili (E), Thane, Maharashtra – 421 201 | | | Manufacturer of water treatment and effluent treatment plant |

For Product Safety Concerns and Information please contact our EU
representative GPSR@taylorandfrancis.com
Taylor & Francis Verlag GmbH, Kaufingerstraße 24, 80331 München, Germany

www.ingramcontent.com/pod-product-compliance
Ingram Content Group UK Ltd.
Pitfield, Milton Keynes, MK11 3LW, UK
UKHW052054230926
475702UK00031B/2801
* 9 7 8 0 3 6 7 5 4 6 9 4 6 *